SOLUTIONS OF ACIDS

Solution	Density (grams/mL)	Concentration (moles/liter)	To make a liter of solution, use
95–98% H_2SO_4	1.84	18.1	1000 mL conc. H_2SO_4
6 M H_2SO_4	1.34	6.0	332 mL conc. H_2SO_4
3 M H_2SO_4	1.18	3.0	166 mL conc. H_2SO_4
10% H_2SO_4	1.07	1.09	60 mL conc. H_2SO_4
1 M H_2SO_4	1.06	1.0	55 mL conc. H_2SO_4
5% H_2SO_4	1.03	0.53	29 mL conc. H_2SO_4
69–71% HNO_3	1.42	15.7	1000 mL conc. HNO_3
6 M HNO_3	1.19	6.0	382 mL conc. HNO_3
3 M HNO_3	1.10	3.0	191 mL conc. HNO_3
10% HNO_3	1.05	1.67	106 mL conc. HNO_3
1 M HNO_3	1.03	1.0	64 mL conc. HNO_3
5% HNO_3	1.03	0.82	52 mL conc. HNO_3
36.6–38% HCl	1.18	12.0	1000 mL conc. HCl
6 M HCl	1.10	6.0	500 mL conc. HCl
3 M HCl	1.05	3.0	250 mL conc. HCl
10% HCl	1.05	2.87	242 mL conc. HCl
5% HCl	1.02	1.40	118 mL conc. HCl
1 M HCl	1.02	1.0	83 mL conc. HCl
99.7% CH_3COOH	1.05	17.5	1000 mL glacial CH_3COOH
6 M CH_3COOH	1.04	6.0	343 mL glacial CH_3COOH
3 M CH_3COOH	1.03	3.0	171 mL glacial CH_3COOH
10% CH_3COOH	1.01	1.69	97 mL glacial CH_3COOH
1 M CH_3COOH	1.01	1.0	57 mL glacial CH_3COOH
5% CH_3COOH	1.01	0.84	48 mL glacial CH_3COOH

SOLUTIONS OF BASES

Solution	Density (grams/mL)	Concentration (moles/liter)	To make a liter of solution, use
50% NaOH	1.53	19.1	1000 mL 50% NaOH
6 M NaOH	1.21	6.0	314 mL 50% NaOH
3 M NaOH	1.11	3.0	157 mL 50% NaOH
10% NaOH	1.11	2.77	145 mL 50% NaOH
5% NaOH	1.05	1.32	69 mL 50% NaOH
1 M NaOH	1.04	1.0	52 mL 50% NaOH
45% KOH	1.45	11.7	1000 mL 45% KOH
6 M KOH	1.26	6.0	512 mL 45% KOH
3 M KOH	1.14	3.0	256 mL 45% KOH
10% KOH	1.09	1.94	167 mL 45% KOH
1 M KOH	1.05	1.0	85 mL 45% KOH
5% KOH	1.04	0.93	79 mL 45% KOH
28–30% NH_3	0.90	15.0	1000 mL conc. NH_4OH
6 M NH_3	0.95	6.0	400 mL conc. NH_4OH
10% NH_3	0.96	5.63	375 mL conc. NH_4OH
3 M NH_3	0.98	3.0	200 mL conc. NH_4OH
5% NH_3	0.98	2.87	191 mL conc. NH_4OH
1 M NH_3	0.99	1.0	67 mL conc. NH_4OH
10% Na_2CO_3	1.10	1.04	110.3 grams anh. Na_2CO_3
5% Na_2CO_3	1.05	0.50	52.5 grams anh. Na_2CO_3
Saturated $NaHCO_3$	1.06	1.0	85 grams $NaHCO_3$
5% $NaHCO_3$	1.04	0.62	52.8 grams $NaHCO_3$

LABORATORY INVESTIGATIONS IN ORGANIC CHEMISTRY

LABORATORY INVESTIGATIONS IN

ORGANIC CHEMISTRY

DAVID C. EATON

Essex County College

McGRAW-HILL BOOK COMPANY

New York | St. Louis | San Francisco | Auckland | Bogotá | Caracas | Colorado Springs
Hamburg | Lisbon | London | Madrid | Mexico | Milan | Montreal | New Delhi
Oklahoma City | Panama | Paris | San Juan | São Paulo | Singapore
Sydney | Tokyo | Toronto

LABORATORY INVESTIGATIONS IN ORGANIC CHEMISTRY

234567890   DOCDOC   89432109

ISBN 0-07-018855-6

This book was set in Times Roman by Syntax International.
The editors were Denise Schanck and Steven Tenney;
the designer was Nicholas Krenitsky;
the production supervisor was Friederich W. Schulte.
Cover photograph by Reginald Wickham.
R. R. Donnelley & Sons Company was printer and binder.

Library of Congress Cataloging-in-Publication Data

Eaton, David C.
    Laboratory investigations in organic chemistry.

    Includes index.
    1. Chemistry, Organic—Laboratory manuals.
I. Title.
QD261.E37   1989   547   88-8328
ISBN 0-07-018855-6

Permission has been granted for the use in this publication of certain Sadtler Standard Spectra.[A]
All rights are reserved by Sadtler Research Laboratories, Division of Bio-Rad Laboratories, Inc.

Permission has been granted for the use in this publication of certain ^{1}H NMR spectra
from *The Aldrich Library of NMR Spectra*, Edition II, by Charles Pouchert, Aldrich
Chemical Company.

The author gratefully thanks both Sadtler Research Laboratories
and Dr. Pouchert of the Aldrich Chemical Company for this approval.

ABOUT THE AUTHOR

David C. Eaton is a professor of chemistry at Essex County College. He received his Ph.D. from New York University in 1969, receiving the school's Founders Day Award in 1970 for outstanding scholastic achievement. Also a graduate of Kalamazoo College, his undergraduate thesis on synthetic furocoumarins won an award in 1964 from the Michigan Academy of Sciences, Arts and Letters.

Dr. Eaton's interests include bridge and the impact the diminishing supplies of fossil fuels will eventually have upon petrochemical products and the world's energy needs. In these areas, he is working on two other books. The first one demonstrates the value of a detailed knowledge of probabilities in bridge bidding. The second one is a novel on the energy crisis of the future.

TO MY BELOVED FAMILY

CONTENTS

2. MELTING POINTS   65

3. RECRYSTALLIZATION   79

4. EXTRACTION OF ASPIRIN, β-NAPHTHOL, AND MOTH CRYSTALS   95

5. FERMENTATION OF SUCROSE TO ETHANOL 119

Enzymes as Catalysts

6. CHROMATOGRAPHY 127

7. POLYMERS AND PLASTICS 185

8. THE ESSENTIAL OILS OF PLANTS 213

33. CARBOHYDRATES

34. PROTEINS AND AMINO ACIDS

35. INFRARED SPECTROSCOPY:
Preparing the Samples and Interpreting the Spectra

36. NUCLEAR MAGNETIC RESONANCE SPECTROSCOPY: Preparing the Samples and Interpreting the Spectra 661

37. THE CHEMICAL LITERATURE 697

38. ORGANIC QUALITATIVE ANALYSIS 709

APPENDIXES 829

PREFACE

The world of organic chemistry is a world of plastics and petroleum, drugs and dyes, soaps and spices, perfumes and paints, cosmetics and carbohydrates, and flavorings and fermentation. It is a fascinating world, full of interest and wonder, that has affected all of us and altered and improved our lives. Many of the scientific discoveries and innovations in the last 50 years have been in the area of organic chemistry, providing us with a wondrous array of beneficial products that would have merely been dreams a few years earlier. All of us owe our very existence to organic reactions that occur within our bodies and in plants and animals. The various branches of organic chemistry have grown enormously and comprise many multibillion dollar industries, responsible for the livelihood and vitality of millions of people.

Organic chemistry is a challenging course, but with the proper approach it can also be very interesting, motivating, and inspiring. If the subject is made relevant to the students' everyday lives, they will find it much more satisfying, rewarding and meaningful, and will respond to it more enthusiastically. They will also be much more aware of, and will better appreciate, the role and importance of organic chemistry in the world around them.

The laboratory provides the best means for implementing this approach. Here the students can actually isolate, synthesize, or analyze the compounds found in the flavors they taste, the medicines they take, and the colors and clothes they wear. Thus, they are actively involved with these compounds, rather than just reading about them. But in order for this approach to succeed, two other requirements are also absolutely essential.

First, most of the experiments must be in areas of organic chemistry that are already somewhat familiar to the students and to which they can personally relate. Thus, some types of experiments in current organic lab manuals must be omitted, and others must be modified, but since there are so many interesting areas in organic chemistry, finding a sufficient number is not even a problem.

Second, and even more important, the experiments must be extremely well written in a stimulating, enjoyable, and easily understandable manner that fully arouses the interest and enthusiasm of the students. Each subject, whenever possible, must also be presented from both a scientific and practical point of view, so that the importance, relevance, and application of the topic is readily apparent. Only with these conditions can the full impact of the approach be realized.

From the beginning it was evident that this approach would be successful. Students in my classes were much more enthusiastic about the laboratory and the entire course than ever before. Some of them even voluntarily devoted up to five and six hours a day performing the experiments. In the synthesis of artificial flavorings and the steam distillation of essential oils from spices, for example, the students didn't want to produce or isolate just one flavoring or essential oil, but two or three. Some of them were eagerly working so far ahead of the others, that the experiments couldn't be written fast enough for them! Never before had it been like this! Furthermore, in the classroom the students were much more attentive, responded more readily and easily to questions, and appeared to show more of a desire, rather than a compulsion, to learn. I believe that other professors who use this approach can experience the same success.

It must be stressed that even though this manual utilizes an innovative approach, it does not sacrifice nor deviate from the necessary emphasis on laboratory techniques, which is so essential for a meaningful lab experience. The experiments clearly and thoroughly introduce the students to distillation and fractional distillation, melting points, extraction, filtration, recrystallization, chromatography, steam distillation, polarimetry, spectroscopy, handling gases and sodium metal, and other techniques. With this approach the students still develop the ability to comprehend and follow directions, handle equipment, make observations, and correctly interpret their results. Furthermore, the experiments still include the principal name reactions in organic chemistry, as well as a clear indication of the reaction mechanisms involved in the syntheses.

Some extra features of the manual also make it more valuable. The accompanying quotations preceding most of the experiments are an interesting and unique facet for a book of this type, and are included to enhance the basic philosophy of the approach. The background or discussion sections present the necessary theoretical information for the students to understand exactly what is occurring at every step in the procedure, so they will not blindly perform the experiments and become frustrated when they encounter difficulties. The questions at the end of each experiment were selected to test the students' knowledge of the experimental theory and the subject matter and to heighten even more their awareness of the topic's application to their daily lives. In order to answer some of these questions, the students will have to use other sources, their own common sense, or the knowledge obtained from other courses or from the completion of the experiment itself.

In addition, in many experiments the amount of chemical reagents has been greatly reduced. This reduction has three main benefits:

1. The high cost of chemical reagents and solvents is substantially reduced.

2. Everyone's exposure to toxic or otherwise hazardous chemicals is substantially reduced.

3. The time needed to complete some experiments is reduced.

These reduced amounts also require even more conscientiousness and care by the students for satisfactory results. My experience has been that students can generally meet this challenge. Thus, their overall enthusiasm for the approach in this book is not substantially lessened by these lower amounts.

This book also includes the infrared and ^{1}H NMR spectra for most of the compounds synthesized or isolated. Instructors can introduce students to infrared and ^{1}H NMR spectroscopy as early as Experiment 4. But since many instructors may wish to delay spectroscopy until the students have a firmer foundation, the techniques for preparing infrared and ^{1}H NMR samples have been placed toward the back of the book. The proper time for their inclusion is thus flexible.

The reader will also note from the Table of Contents that many topics have several individual experiments. For example, there are six experiments involving polymers and plastics, seven involving the essential oils of plants, three involving the isolation of caffeine, six involving fats and oils, and five involving photochemistry. **Not all the individual experiments under a certain topic have to be done.** Instead, the instructor should assign **only** those individual experiments that he or she finds to be more interesting, valuable, or worthwhile. In no way can all the experiments included possibly be completed in a one-year course.

The reader will also note from the Table of Contents that the fermentation of sucrose is presented separately and much earlier than the other experiments involving carbohydrates. This was done to give the students an interesting experiment on carbohydrates far sooner than the topic is normally covered. In fact, a number of experiments have deliberately been placed early because of their interesting nature. The instructor, of course, may use any other sequence in developing a lab schedule. To aid this effort, the type of reaction in laboratory syntheses is given with each experiment, and the experimental techniques utilized in each experiment are given toward the end of the book.

The author wishes to thank his wife, Blanche, and his daughter, Lisa, for their assistance, encouragement, and understanding. Great appreciation goes to the following reviewers for their valuable and encouraging comments and suggestions: Bill Bunnelle, University of Missouri; David R. Dalton, Temple University; Noring Hammond, Vassar College; Richard Olsen, Utah State University; Jan Simek, Cal Poly State University; Howard Smith, Vanderbilt University; Walter S. Trahanovsky; and James Wood, University of Nebraska. Special thanks go to Karen Misler, Denise Schanck, and Steven Tenney for their valuable roles in this project.

David C. Eaton

LABORATORY INVESTIGATIONS IN ORGANIC CHEMISTRY

STUDENT GUIDE FOR LABORATORY SUCCESS

Every possible effort has been made to make your organic chemistry laboratory experience an interesting, motivating, and rewarding one. But from past experience, many people have discovered a number of things each student can do to ensure success and get the most out of the laboratory.

ADVANCE PREPARATION

First, you should prepare yourself for the laboratory by carefully reading **beforehand** the assigned experiment and any other related sections of this book. For your first laboratory class meeting, this would include the sections Student Guide for Laboratory Success, Laboratory Safety, The Laboratory Notebook, and Laboratory Glassware.

This advance preparation ensures that you will be familiar with the material when you come to class and will not have to spend valuable laboratory time for this purpose. Furthermore, the instructor could have you begin the experiments almost immediately each laboratory session, and you will probably never learn much if you blindly try to follow the experimental procedure without first understanding it. Important observations and conclusions may be missed, and costly and time-consuming mistakes or accidents could also occur. Special sections of each experiment have been devoted to explain exactly what is occurring in the procedure and why, but this will be of little value to you unless you **read the experiment beforehand**.

BUDGETING TIME

Second, having previously read the experiment, you should use your time wisely in the laboratory. In this regard, you will frequently discover that you will be performing more than one experiment at a time. For example, the fermentation of sugar requires that the experiment be started and then left for one week. Many other experiments require that the reagents be refluxed or heated for several hours before the product is isolated. During this waiting period, you may find it advantageous to catch up on unfinished parts of a previous experiment. You may have to determine, for example, the yield and melting point of the product of the last experiment. This information usually cannot be accurately determined in the same period in which the material is isolated, because of the time it takes to "dry" the product or free it of solvent. Other related activities include looking up in a chemistry handbook the melting point or boiling point of a product, placing notes or calculations in your laboratory notebook, measuring out reagents that will be needed later, cleaning dirty glassware, and answering the questions given for each experiment. The point is that you should not stand around idle, waiting for something when there are other things that can be done. Your laboratory time is valuable and should be used effectively.

SEEKING HELP

Third, you should not hesitate to seek help, but should already have **really tried** to understand the material or to solve the problem. You will never become proficient in this course or in any other area of life if you continually rely on others to answer or figure out everything for you. Since other students may not be reliable sources, any needed assistance should be obtained from your instructor or a laboratory assistant.

Good luck to you in this laboratory experience. As stated in the preface, the world of organic chemistry is a world of plastics and petroleum, drugs and dyes, soaps and spices, perfumes and paints, cosmetics and carbohydrates, and flavorings and fermentation. It is a fascinating world, full of interest and wonder, that has affected all of us and altered and improved our lives. May the stimulation, enjoyment, and enthusiasm that created this book become the spirit you discover in organic chemistry, and may this course be one of the best and most memorable aspects of your college education.

LABORATORY SAFETY

A chemistry laboratory, especially an organic chemistry laboratory, can potentially be a dangerous place to work. There are flammable liquids, poisonous and corrosive chemicals, and fragile glassware that can all cause serious injuries. However, this danger can be minimized if proper safety precautions are known and followed. Working in an organic chemistry laboratory **can** be just as safe as working at home. Some possible hazards and ways to avoid and handle them are presented here. Other precautions will be found in the individual experiments at the appropriate points. Read these over carefully, and conscientiously and faithfully follow a safe and proper procedure. **Remember that if a serious accident occurs, the personal injury it may cause cannot be undone.** Also remember that because of the potential hazards, **you will not be allowed to work alone**. If help is not available, any minor accident could become a disaster. **Report any mishaps, no matter how minor, to your instructor and, whenever necessary, get prompt medical treatment.** No stigma should be attached to accidents that were unavoidable or nonintentional.

EYE SAFETY

The most important rule to follow here is this: **WITHOUT EXCEPTION, EYE PROTECTION MUST BE WORN IN THE LABORATORY AT ALL TIMES, REGARDLESS OF WHAT IS BEING DONE.** Your eyes are a very crucial and irreplaceable part of your body and are particularly susceptible to injury from spattering chemicals and flying glassware. Even if you are not actually performing an experiment, you could suffer eye injury by being near an accident caused by one of your neighbors. Ordinary prescription eyeglasses (not sunglasses nor contact lenses) are acceptable and may be used. However, they do not provide complete protection, and your instructor may rightfully insist that you wear goggles or safety glasses with side shields.

If there are eyewash fountains in the laboratory, know the location of the nearest one. If any chemical enters your eyes, go immediately to the fountain and flush your eyes and face with large amounts of water. Be sure to keep your eyelids open. If your neighbor gets something in his or her eyes, he or she may not be able to find the eyewash fountain, and you may need to be a guide. If the laboratory does not have an eyewash fountain, a piece of flexible hose attached to the faucet nozzle can also work effectively; it can be aimed upward directly into the face. With this alternative setup, make sure the flexible hoses are not removed from the faucet nozzles.

FIRES

Because flammable liquids are frequently used in an organic laboratory, there is continual danger of fire. This hazard, however, can be minimized and effectively handled by observing the following precautions and procedures.

1. Avoid using flames in the laboratory whenever possible. Most organic liquids can be effectively heated or distilled with a steam bath or hot plate, so a flame is unnecessary. If a flame must be used to heat a flammable liquid, make certain that the liquid is not in an open container but is protected by a condenser with all connections tight and free of strain.

2. Do not leave a burner on unnecessarily.

3. Do not smoke in the laboratory.

4. Do not use a flame without first checking to see if your neighbors on either side, across the station, and behind you are using flammable liquids. Many flammable organic liquids produce vapors that can diffuse and become ignited by a flame a considerable distance away. If a flame is needed, try to delay its use or move to a safe location, such as a fume hood or an unused area of the laboratory.

5. Related to rule 4, do not use a flammable liquid without first checking to see if there are any flames in the area. Also, never pour any flammable liquids into the center trough of a laboratory bench because the vapors may be carried near a flame further down the bench.

6. Know beforehand the location of the nearest fire extinguisher, fire blanket, and fire shower, and be certain you understand how to use them. Your instructor will explain or demonstrate the operation of these pieces of equipment. Remember that most models of extinguishers have a safety pin that must be pulled out before the extinguisher can be activated. Also, do not wildly use the extinguisher because the force of the blast from the nozzle may knock over containers of other flammable liquids and make the fire worse.

7. If you should have a fire, get away from it and don't panic. If the fire is a small one in a container, it can usually be extinguished by placing an asbestos pad, watch glass, or clipboard on top of the container. Otherwise use the fire extinguisher or allow the instructor or laboratory assistant to take care of it.

8. If your clothing should catch on fire, do not run, but walk toward the nearest fire blanket or fire shower. Running will fan the flames and make the fire worse. If one of your neighbors' clothes catches fire, assist him or her, if necessary, to the nearest fire blanket or fire shower. Do not allow the person to panic and run. You can minimize the risk of clothing fires by not wearing loose-fitting long sleeves and cuffs. Also, tie back extra long hair.

CHEMICAL POISONING

Almost all chemicals in an organic laboratory are poisonous and harmful even in small amounts. You can protect yourself from this danger by observing the following precautions.

1. Never allow any chemicals to come in contact with your skin. Liquids such as aniline, dimethyl sulfate, nitrobenzene, phenol, and phenylhydrazine may be rapidly absorbed through the skin and prove fatal. Others may enter through minor cuts and scratches with the same result. If any chemical should inadvertently be spilled on the skin, wash the area **immediately** and **thoroughly** with soap and water. Do not use any organic solvent to remove any chemical from the skin because it may actually increase the rate of absorption into the skin. You may want to wear disposable plastic gloves when using any especially toxic or corrosive chemical.

2. Avoid inhaling the fumes and vapors of chemicals and solvents as much as possible. The laboratory area should be well ventilated, and any noxious, volatile materials should be used in a fume hood. An efficient gas trap should be used for any noxious gases, such as hydrogen chloride, generated in a reaction. If you must smell any organic compound to determine its odor, hold the substance about 6 inches from your nose and use your hand to gently fan the vapors toward you. **Never** hold your nose over the container and inhale deeply.

3. Never place your fingers in your mouth, and always wash your hands when leaving the laboratory area. Without your knowing it, some chemicals may have been deposited on your hands. Keeping your hands away from your mouth ensures that these chemicals will not be ingested; washing your hands as soon as you have finished working will remove these substances.

4. Never bring food into the laboratory. The food may become contaminated with a hazardous material.

5. Never smoke in the laboratory. Besides being fire hazards, a cigarette, cigar, or pipe placed on a laboratory bench could easily pick up some hazardous chemical, which you could ingest or inhale.

6. Never taste any chemical unless authorized to do so.

7. Never mouth-pipet any chemicals. A suction bulb may not be as convenient as your mouth, but it is much safer. You could easily ingest harmful substances as well as inhale volatile organic liquids.

8. Know the location of the chart indicating the proper first-aid measures for poisoning.

9. Read the labels of the reagent bottles for any special dangers or precautions.

EXPLOSIONS

Explosions in an organic laboratory are rare; the few that do occur are usually quite minor compared to what you may see dramatized in the movies or on television. Nevertheless, you can reduce even the small chance of an explosion by observing the following precautions:

1. Never perform any unauthorized experiments. It is always best to check with your instructor first if you want to do any extra experiments. Even though you think you know what you're doing, you could easily combine the wrong reagents or handle them improperly. This could prove disastrous because you might unknowingly create a hazardous chemical reaction.

2. Never heat a flask or any apparatus that is not open to the atmosphere. Even with a condenser, the heating could easily increase the pressure inside to the point where the equipment could explode. If a flame was being used, any flammable liquid inside could also be ignited.

3. Always have an ice-water bath available for all exothermic reactions. The bath will allow you to cool the reaction to slow it down if it shows any signs of getting out of control.

4. Never add solids or boiling chips to a boiling hot liquid. The added solid could easily cause the hot liquid to shoot out of the reaction vessel and possibly burn you.

5. When heating a substance in a test tube, do not point it at yourself or anyone else. The hot liquid may "bump" and be thrown from the tube.

6. Never concentrate ether solutions to dryness using a flame or other source of intense heat. In addition to their being a fire hazard, ethers can react with oxygen to form unstable peroxides, which can be highly explosive.

You can detect peroxides in ethers and hydrocarbons by adding 1 mL of the material to 1 mL of glacial acetic acid to which has been added about 0.1 gram of sodium or potassium iodide. A yellow to brown color indicates the presence of peroxides. To be sure, run a blank determination. You can remove any peroxides in the ether by stirring or shaking the ether with a solution of 60 grams of ferrous sulfate and 6 mL of concentrated sulfuric acid in 110 mL of water.

7. Never heat, dry, or provide shock to unstable compounds such as diazonium salts, heavy-metal acetylides, diazo compounds, or polynitro compounds.

8. If you must work with a potentially explosive substance or mixture, always work with as little as possible and always work behind a safety shield of shatterproof glass.

CUTS

Most cuts in the laboratory are caused by glassware breaking under strain or excessive pressure. You can easily do this when attempting to separate "frozen" (stuck together) ground-glass joints, or when forcing a glass tube or thermometer into the hole of a cork or rubber stopper. You can separate frozen joints using the procedure in the section on laboratory glassware. And you can easily insert glass tubing and thermometers in stoppers by enlarging the hole in the stopper with a round file or by lubricating the glass with glycerine. But **always** protect your hands by wrapping the glass with several thicknesses of a cloth towel. Be careful when assembling laboratory equipment not to put any strain on any glassware.

Treat minor cuts by ordinary first-aid procedures after removing any obvious pieces of glass. Serious cuts accompanied by severe bleeding should be treated at a doctor's office, hospital, or infirmary. Apply direct pressure to the wound with a gauze pad or clean towel, and for arterial bleeding, which can be quite dangerous, you must also apply hand or thumb pressure to the appropriate pressure point. For the hands, the pressure point is where the pulse can be felt at the wrist; for the arms it is inside the upper arm, just below the armpit.

BURNS

Burns can result from heat or chemicals. For minor burns caused by flames or hot objects, immersing the affected area in cold water or ice will bring relief. Don't use salves or ointments, and don't disturb or open any blisters. Serious burns should be treated at a doctor's office, hospital, or infirmary.

For chemical burns flush the affected area with plenty of water. If large areas are burned, you may need to use the safety shower. Acid burns may then be treated with sodium bicarbonate paste, alkali burns with saturated boric acid solution, and bromine burns with 10% sodium thiosulfate. In each case a wet dressing of the recommended neutralizing reagent may be used. If the burn appears to be serious, it should be treated at a doctor's office, hospital, or infirmary.

LABORATORY APPAREL

A laboratory coat or rubber apron will protect your clothing from soiling and damage by chemical reagents and accidents. Don't wear loose-fitting long sleeves because you might overturn fragile glassware or ignite your clothing with a burner. Always wear shoes to protect your feet from spilled chemicals and pieces of broken glass. Sandals and open-toed shoes do not provide complete protection.

DISPOSAL OF CHEMICALS AND OTHER MATERIALS

Dispose of your organic liquids in a designated waste solvent can. If possible, don't pour them into a sink. If you must use a sink, flush it thoroughly with plenty of water, but even then some liquid may remain in the sink trap and vapors in the sink itself.

Chemicals that react vigorously with water, such as acid chlorides and alkali metals, may be decomposed in the hood by reacting with anhydrous alcohol.

Dispose of water-insoluble solids in a chemical waste jar. Do not throw them into a sink or wastepaper basket because they can release toxic vapors, clog the pipes, or start a fire.

Never throw solid nonchemical waste into the sink because it is unsightly and can possibly clog the drains, thus making those sinks temporarily unusable.

ADDITIONAL PRECAUTIONS

1. Try to be aware of any possible adverse effects of your laboratory operations.

2. Watch what your neighbors are doing. Being alert to any possible dangerous practices could prevent an accident and possible injury to both you and them.

3. Keep your laboratory space clean and orderly, along with the balance and chemical dispensing areas. Clean up all spills **immediately**. After taking whatever chemicals you need, replace the caps on the containers.

4. To prevent possible floods and water damage to your books and notes, attach water hoses to condensers and faucets securely, and remember that only a moderate flow of water is needed for adequate cooling.

CARCINOGENIC COMPOUNDS

Many compounds are carcinogenic (cancer-causing). The National Institute for Occupational Safety and Health (NIOSH) lists the following compounds as **acute** carcinogens. Because of their potential hazard, you will not use them in any experiments.

2-Acetylaminofluorene

4-Aminodiphenyl

Benzidine and its salts

Bis(Chloromethyl) ether

3,3′-Dichlorobenzidine and its salts

4-Dimethylaminoazobenzene

Ethyleneimine

Methyl chloromethyl ether

4,4′-Methylene *bis*(2-chloroaniline)

α-Naphthylamine

β-Naphthylamine

4-Nitrobiphenyl

N-Nitrosodimethylamine

β-Propiolactone

Vinyl chloride

THE LABORATORY NOTEBOOK

One important requirement in organic chemistry is to keep an accurate and thorough record of all experiments in a laboratory notebook. You thus have a permanent record of exactly what you did, what specific observations you made, what results you obtained, and what difficulties, if any, you encountered. This permanent record can be very valuable in pinpointing exactly where and why an experiment may have gone wrong and what can be done to correct the situation. It also allows you or someone else to repeat the experiment exactly at some future time and obtain the same results. This can be especially important in research. Even if you are not majoring in chemistry, keeping accurate and thorough records is important in many other professions. The laboratory notebook will prove beneficial in preparing you for this part of your future career. It will enable you to develop good habits and practices and the ability to work systematically and in an orderly way.

Although the exact format for the notebook may vary from experiment to experiment and instructor to instructor, the following guidelines and procedures are sound practices commonly observed in organic laboratory courses. Your instructor will inform you of any changes or other rules to be followed. Also presented is a sample of what a write-up for an organic laboratory experiment **might** look like. Although your write-up will be handwritten and perhaps not as neat as these printed pages, this example should be a valuable guide to the types of items, observations, and calculations that should be included in a typical experiment.

1. Use only a bound, hardcover book, approximately 8 × 10 inches, for the notebook. A loose-leaf or spiral notebook is not acceptable because the pages can more easily be accidently removed, torn out, and lost.

2. Number the book pages consecutively.

11

3. The notebook pages should have horizontal or horizontal and vertical lines.

4. Reserve the first few pages of the book for a table of contents, and keep it up to date.

5. Start every new experiment on a new page.

6. Make all entries in your notebook in ink, because it is more durable than pencil.

7. Never remove or erase anything from the book. To make corrections or changes, simply draw a line through the old data or conclusions and add the amended information. If you had to abort an experiment, state this and begin the new experiment on a new page.

8. Always bring the notebook to every laboratory session. How can you make entries in it if it is at home or in your car? Since your instructor may want to check it at any time, you should also keep it up to date.

9. Record all laboratory data, observations, and calculations directly into the notebook at the time they are obtained. Loose sheets of scratch paper, napkins, and paper towels are not permitted because they can easily be lost, destroyed, or thrown away.

10. To help maintain neatness, use only the right-hand pages for the actual write-up. Use the left-hand pages for calculations of weights, and so forth.

11. Include the following items for each write-up in your notebook. Many of these items should be written in the book **beforehand** as part of your laboratory preparation.

a. A descriptive title of the experiment
b. A book or journal reference with the page numbers
c. Any principal objectives of the experiment
d. The date
e. A balanced equation for the synthesis of any products
f. Any key side reactions in the synthesis
g. A purification scheme for any products synthesized or isolated from natural products
h. Relevant physical properties of the products, reactants, and solvents, such as the boiling point, melting point, density, and solubility
i. The exact amounts of all reagents used and the molecular weights of all reactants and products
j. An indication of the limiting reagent in the experiment, if any
k. Important facts such as the temperature, the exact time that reagents were added, how long the reaction was refluxed, the amount of solvent and charcoal used in recrystallizing, the point at which the experiment was interrupted at the end of the class period, and the length of the interruption

l. A calculation of the percentage yield and possible reasons why the yield may have been low

12. In the procedure section briefly describe **in your own words** what you did, what you observed, and any difficulties you encountered. **Do not merely copy the procedure from the laboratory text into your book.** Especially note any departures from the planned procedure. Remember to give sufficient detail so that you or someone else could later repeat the experiment exactly.

13. At the end of the experiment, you might want to comment on any difficulties encountered and suggest changes in the procedure to overcome or circumvent these problems.

Jan. 23, 1990

Pheromones from the Grignard Synthesis — _4-Methyl-3-heptanol_
and _4-Methyl-3-heptanone_

Reference: Experiment 18B in Eaton, page 378

Main Reactions for Synthesizing _4-Methyl-3-heptanol_

1.

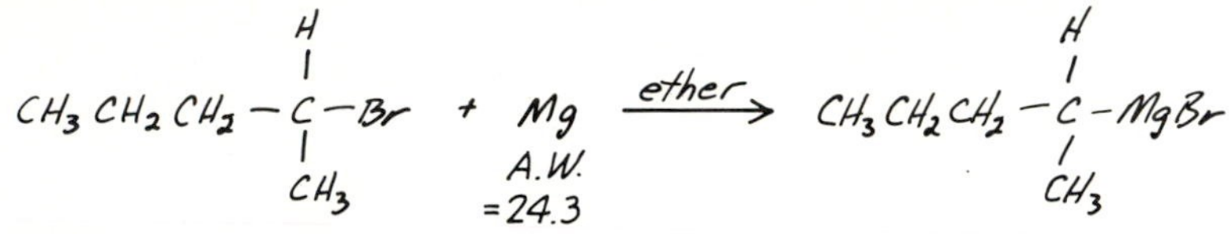

$$CH_3CH_2CH_2-\overset{\overset{\displaystyle H}{|}}{\underset{\underset{\displaystyle CH_3}{|}}{C}}-Br \; + \; Mg \quad \xrightarrow{ether} \quad CH_3CH_2CH_2-\overset{\overset{\displaystyle H}{|}}{\underset{\underset{\displaystyle CH_3}{|}}{C}}-MgBr$$

A.W. = 24.3

2-Bromopentane
Den. = 1.21 g/mL
MW = 151.05
bp = 117.4°C at 760 mm Hg

2.

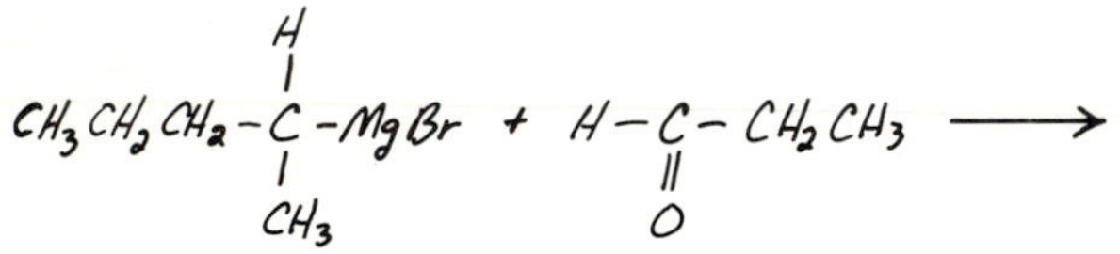

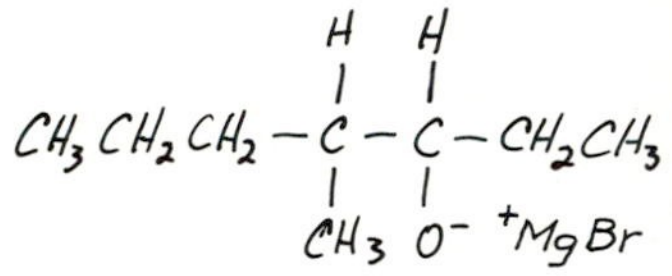

$$CH_3CH_2CH_2-\overset{\overset{\displaystyle H}{|}}{\underset{\underset{\displaystyle CH_3}{|}}{C}}-MgBr \; + \; H-\overset{\overset{\displaystyle}{\underset{\underset{\displaystyle O}{||}}{C}}}{}-CH_2CH_3 \longrightarrow$$

Propanal
Den. = 0.806 g/mL
MW = 58.1
bp = 48.8°C

$$CH_3CH_2CH_2-\overset{\overset{\displaystyle H}{|}}{\underset{\underset{\displaystyle CH_3}{|}}{C}}-\overset{\overset{\displaystyle H}{|}}{\underset{\underset{\displaystyle O^- \; {}^+MgBr}{|}}{C}}-CH_2CH_3$$

3.

$$CH_3CH_2CH_2-\overset{\overset{\displaystyle H}{|}}{\underset{\underset{\displaystyle CH_3}{|}}{C}}-\overset{\overset{\displaystyle H}{|}}{\underset{\underset{\displaystyle O^- \; {}^+MgBr}{|}}{C}}-CH_2CH_3 \quad \xrightarrow{H_3O^+} \quad CH_3CH_2CH_2-\overset{\overset{\displaystyle H}{|}}{\underset{\underset{\displaystyle CH_3}{|}}{C}}-\overset{\overset{\displaystyle H}{|}}{\underset{\underset{\displaystyle OH}{|}}{C}}-CH_2CH_3$$

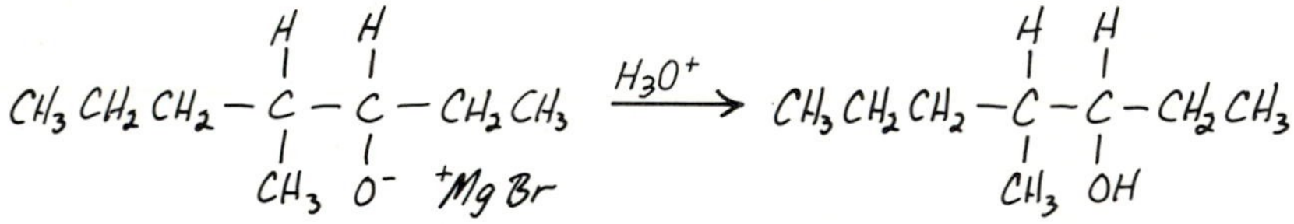

4-Methyl-3-heptanol
Den. = 0.83 g/mL
MW = 130.23
bp = 160-161°C

14

Side Reactions

$$CH_3CH_2CH_2-\underset{\underset{CH_3}{|}}{\overset{\overset{H}{|}}{C}}-MgBr \;+\; Br-\underset{\underset{CH_3}{|}}{\overset{\overset{H}{|}}{C}}-CH_2CH_2CH_3 \longrightarrow$$

$$CH_3CH_2CH_2-\underset{\underset{CH_3}{|}}{\overset{\overset{H}{|}}{C}}-\underset{\underset{CH_3}{|}}{\overset{\overset{H}{|}}{C}}-CH_2CH_2CH_3 \;+\; MgBr_2$$

Reaction cannot be prevented, but can be minimized by using a low concentration of 2-bromopentane

$$CH_3CH_2CH_2-\underset{\underset{CH_3}{|}}{\overset{\overset{H}{|}}{C}}-MgBr \;+\; H_2O \longrightarrow CH_3CH_2CH_2-\underset{\underset{CH_3}{|}}{\overset{\overset{H}{|}}{C}}-H \;+\; Mg(OH)Br$$

$$bp = 36°C$$

Reaction can be prevented by keeping the glassware, reagents, and solvent scrupulously dry.

$$CH_3CH_2CH_2-\underset{\underset{CH_3}{|}}{\overset{\overset{H}{|}}{C}}-MgBr \;+\; CH_3CH_2-\overset{\overset{O}{\|}}{C}-OH \longrightarrow$$

$$CH_3CH_2CH_2-\underset{\underset{CH_3}{|}}{\overset{\overset{H}{|}}{C}}-H \;+\; CH_3CH_2-\overset{\overset{O}{\|}}{C}-O^-\;{}^+MgBr$$

Reaction can be prevented by using freshly distilled propanal containing no propanoic acid.

$$CH_3CH_2CH_2-\underset{\underset{CH_3}{|}}{\overset{\overset{H}{|}}{C}}-MgBr \;+\; CO_2 \longrightarrow CH_3CH_2CH_2-\underset{\underset{CH_3}{|}}{\overset{\overset{H}{|}}{C}}-\overset{\overset{O}{\|}}{C}-O^-\;{}^+MgBr$$

Reaction can be partially prevented by refluxing the ether.

$$CH_3CH_2CH_2-\underset{\underset{CH_3}{|}}{\overset{\overset{H}{|}}{C}}-MgBr + O_2 \longrightarrow CH_3CH_2CH_2-\underset{\underset{CH_3}{|}}{\overset{\overset{H}{|}}{C}}-O-O^- \quad \overset{+}{M}gBr$$

Reaction can be partially prevented by refluxing the ether.

$$CH_3CH_2CH_2-\underset{\underset{CH_3}{|}}{\overset{\overset{H}{|}}{C}}-\underset{\underset{OH}{|}}{\overset{\overset{H}{|}}{C}}-CH_2CH_3 \xrightarrow[heat]{H^+} CH_3CH_2CH_2-\underset{\underset{CH_3}{|}}{C}=CH-CH_2CH_3 \quad +H_2O$$

Reaction can be minimized by neutralizing the added acid in the reaction workup.

Separation Scheme

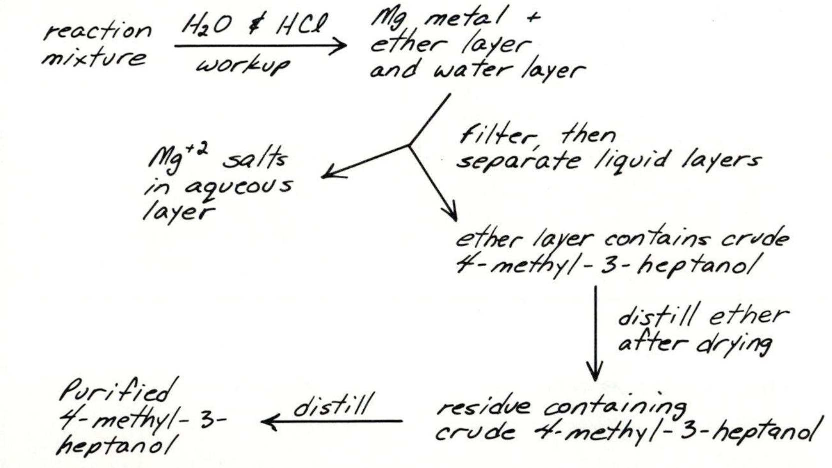

reaction mixture $\xrightarrow[\text{workup}]{H_2O \text{ \& } HCl}$ Mg metal + ether layer and water layer

filter, then separate liquid layers

Mg^{+2} salts in aqueous layer

ether layer contains crude 4-methyl-3-heptanol

distill ether after drying

Purified 4-methyl-3-heptanol $\xleftarrow{\text{distill}}$ residue containing crude 4-methyl-3-heptanol

Preparation calls for

Mg: 5.0 grams = 0.21 mole

2-Bromopentane: 19.0 mL = 0.152 mole
Propanal: 10.9 mL = 0.152 mole

16

Therefore 2-bromopentane and propanal are both limiting reagents.

Set up the apparatus shown in Figure 18-4 of Eaton, placing 5.0 grams of oven-dried Mg turnings, 75 mL of anhydrous ether, an iodine crystal to help start the reaction, and 2.0 mL of 2-bromopentane into the dried 250 mL round-bottom flask. Warmed the reaction mixture with a warm water bath, and within 3 minutes the reaction was proceeding. The ether was refluxing and the mixture was becoming a cloudy gray.

Added 17.0 mL of 2-bromopentane to the dried separatory funnel and replaced the drying tube. The reagent was then gradually added to the reaction mixture over a 30-minute period, during which time the mixture was carefully swirled periodically. As time passed, the reaction became an even cloudier gray, and the Mg metal was slowly disappearing. Several times the reaction began to get out of control, so briefly cooled it in a cold-water bath. Had to then rewarm it to continue the refluxing.

After all the 2-bromopentane was added, the reaction mixture was refluxed with occasional swirling for 20 minutes. The mixture was cooled to room temperature with a cold-water bath, and 10.9 mL of freshly distilled propanal was added to the separatory funnel. The drying tube was replaced, and the reagent was gradually added to the reaction mixture over a period of 10 minutes. During this time, the mixture was again swirled periodically. After all the propanal was added, the reaction mixture was then refluxed for 20 minutes.

Cooled the reaction mixture to 5°C and added 35 mL of distilled water through the top of the condenser. Swirled the reaction mixture and then added 15 mL of 10% HCl solution to dissolve the magnesium salts. About

1 gram of Mg metal remained unreacted at the bottom of the flask, and there were two layers of liquid present.

Gravity filtered the reaction mixture through a small wad of cotton into a 250mL separatory funnel. No Mg metal was transferred. Removed the lower layer and then washed the remaining ether layer with two 35-mL portions of 10% NaOH solution to neutralize the HCl, and once with 25mL of saturated salt solution to lower the water content of the ether. Dried the ether solution for 15 minutes over anhydrous $MgSO_4$ and decanted with filtering through a small piece of cotton in a stemless funnel.

Set up a simple distillation apparatus and distilled the ether with a steam bath until about 20 mL remained. Transferred the pale yellow remaining liquid to a 50 mL round-bottom flask and continued the distillation, using a steam bath. When the ether stopped distilling, a viscous yellow oil remained. Changed receivers and continued distilling, using a hot plate. Collected the fraction boiling between 150 and 160°C in a preweighed 25-mL Erlenmeyer flask. The yield of the colorless liquid was 7.84 grams, and the chief boiling range was 158-160°C at 741mm Hg, comparing favorably with the literature value.

$$\text{Theoretical yield} = 0.152 \text{ mole} \times 130 \text{ g/mole}$$
$$= 19.8 \text{ g}$$

$$\text{Actual yield} = \frac{7.84 \text{ g}}{19.8 \text{ g}} \times 100\% = 39.6\%$$

The low yield may be due to the side reactions, which give a lower yield of both the Grignard reagent and the final product. Not distilling the final product might prevent any dehydration of the alcohol to give the alkene. Using an inert gas atmosphere would prevent the Grignard reagent from being decomposed by carbon dioxide and oxygen.

The period is almost over, so will continue the rest of Experiment 18B next time.

Jan. 25, 1990

Experiment 18B Continued — Synthesis of 4-Methyl-3-heptanone

Main Reaction

$$CH_3CH_2CH_2-\underset{\underset{CH_3}{|}}{CH}-\underset{\underset{OH}{|}}{CH}-CH_2CH_3 + Na^+OCl^- \xrightarrow{CH_3COOH}$$

$$CH_3CH_2CH_2-\underset{\underset{CH_3}{|}}{CH}-\underset{\underset{O}{||}}{C}-CH_2CH_3 + H_2O + NaCl$$

$$MW = 128.2$$
$$bp = 156\text{-}157° \text{ at } 745\,mm\,Hg$$

Possible Side Reaction

$$CH_3CH_3CH_2-\underset{\underset{CH_3}{|}}{CH}-\underset{\underset{OH}{|}}{CH}-CH_2CH_3 \xrightarrow{CH_3COOH}$$

$$CH_3CH_2CH_2-\underset{\underset{CH_3}{|}}{C}=CH-CH_2CH_3$$

Side reaction would be aided by heat and the moderate stability of a 2° carbocation. But the weak acid catalyst might hinder this side reaction, compared with a strong acid catalyst.

<u>Preparation calls for</u>

    4-Methyl-3-heptanol: 7.0 mL = 0.040 mole

    5.25% chlorine bleach : 70 mL = 0.057 mole

    Glacial acetic acid: 10 mL = 0.18 mole

From the one-to-one mole-mole ratio between the 4-methyl-3-heptanol and the chlorine bleach, the 4-methyl-3-heptanol must therefore be the limiting reagent.

<u>Separation Scheme</u>

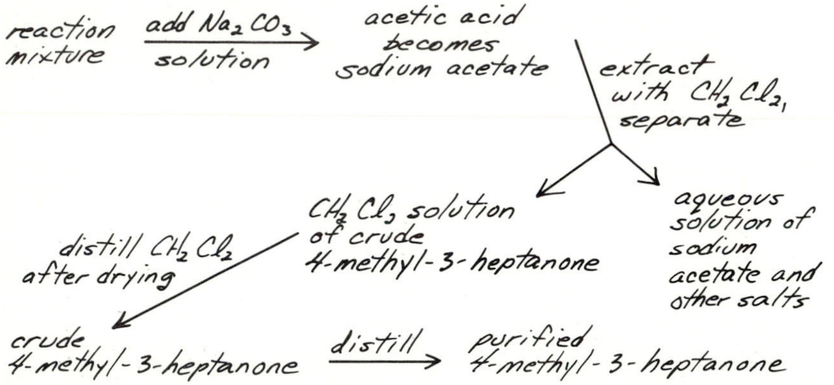

reaction mixture $\xrightarrow{\text{add Na}_2\text{CO}_3 \text{ solution}}$ acetic acid becomes sodium acetate $\xrightarrow{\substack{\text{extract} \\ \text{with CH}_2\text{Cl}_2, \\ \text{separate}}}$

CH_2Cl_2 solution of crude 4-methyl-3-heptanone

aqueous solution of sodium acetate and other salts

$\xrightarrow{\substack{\text{distill CH}_2\text{Cl}_2 \\ \text{after drying}}}$

crude 4-methyl-3-heptanone $\xrightarrow{\text{distill}}$ purified 4-methyl-3-heptanone

    Placed 7.0 mL of 4-methyl-3-heptanol and 10 mL of glacial acetic acid into a clamped 3-neck, 500-mL round-bottom flask. Placed a clamped condenser in the center neck, a 125 mL separatory funnel in one side neck, and a thermometer in the other side neck. Then filled a drying tube with moistened $NaHSO_3$, and placed the drying tube into the top of the condenser.

    Added 70 mL of 5.25% chlorine bleach to the separatory funnel. Added about 10 mL of the bleach to the 4-methyl-3-heptanol solution and swirled the reactants. The rest of the bleach was then added dropwise with periodic swirling over a 20-minute period, with the temperature kept between 47 and 51°. In the beginning, the yellow color of the chlorine disappeared as it reacted. After the bleach was added, the reaction was then continued for another 40 minutes.

The reaction mixture turned wet starch-iodide paper blue, so had to add 6 mL of saturated $NaHSO_3$ solution until no color change with the paper occurred.

Added saturated Na_2CO_3 solution slowly with stirring to the reaction mixture until no more foaming occurred. Then cooled reaction mixture, poured it into a 250-mL separatory funnel, and extracted twice with 30-mL portions of CH_2Cl_2. The CH_2Cl_2 extracts were combined and dried over anhydrous $MgSO_4$, then gravity filtered and distilled, using a hot-water bath, to yield a colorless liquid residue. This residue was then distilled using a heating mantle, to yield 2.18 grams of a colorless liquid having a chief boiling range of 154-157° at 740 mm Hg, which compares favorably with the literature value.

Theoretical yield = 0.040 mole × 128 g/mole
$$= 5.1 g$$
Percentage yield = $\dfrac{2.18g}{5.1g}$ × 100% = 43%

The low percentage yield may be due to some dehydration of the alcohol to yield the alkene, despite the lack of a strong acid present and no very high temperature. Some material was also lost to the glassware and could have been lost in extractions, due to a partial solubility of the product in the large amount of water present.

The infrared spectra of both 4-methyl-3-heptanol and 4-methyl-3-heptanone, run neat between salt plates, was practically identical to those given in the book. The only main difference was in the percentage transmission, due to the somewhat different path length through the samples. The college does not have a NMR spectrometer.

21

CALCULATION OF THEORETICAL AND PERCENTAGE YIELDS

For every chemical reaction, an equation can be written showing the conversion of the starting materials into the products. From this equation, the expected yield of a particular product can be calculated. Ideally, the amount of this product actually obtained should equal the amount calculated, but in reality this is seldom true. Competing side reactions can consume the reagents to give other, frequently undesirable products. The reaction may reach a state of equilibrium where only a portion of the starting materials have reacted. Finally, part of the desired product may be lost in the reaction workup when it is separated from all its possible contaminants.

For a chemical synthesis, it is therefore necessary to calculate the theoretical and the percentage yields. Knowing the percentage yield is important, because it indicates how efficiently the conversion of the reactants to the product has occurred. Where cost and time are important factors, the efficiency allows us to consider and compare other possible approaches in synthesizing and isolating the desired material.

To calculate the percentage yield, you must first know the theoretical yield. You can calculate it from the balanced equation and the quantities of each reactant. For example, in the synthesis of aspirin from salicylic acid and acetic anhydride, we would write

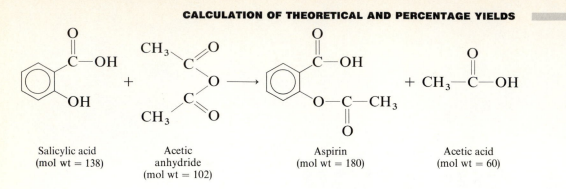

Salicylic acid
(mol wt = 138)

Acetic
anhydride
(mol wt = 102)

Aspirin
(mol wt = 180)

Acetic acid
(mol wt = 60)

This equation shows that 1 mole of salicylic acid reacts with 1 mole of acetic anhydride to yield theoretically 1 mole of aspirin. This 1:1:1 ratio of reactants and product, of course, will apply no matter what quantities are used.

You must realize that the theoretical yield of product will be based upon the **limiting reagent**, if any, in the reaction. Thus, if 0.020 mole of salicylic acid reacts with 0.025 mole of acetic anhydride, only 0.020 mole of aspirin can theoretically be produced.

Salicylic acid + acetic anhydride ⟶ Aspirin
(0.020 mole)   (0.025 mole)   (0.020 mole
                                       theoretically
                                       possible)

Remember that the limiting reagent is based upon the number of **moles, not the number of grams**. Thus, in the above example, 0.020 mole of salicylic acid would amount to 2.8 grams, which is more in terms of weight than the 2.6 grams or 0.025 mole of acetic anhydride.

With this information, we can calculate the theoretical yield for **any** reaction by using the equation

Theoretical yield = (Moles of limiting reagent)

×(mole-mole ratio of limiting reagent and product)

× (molecular weight of product)

In this example, with 0.020 mole of salicylic acid as the limiting reagent, the result would be

Theoretical yield = (0.020 mole salicylic acid)

$$\times \frac{1 \text{ mole aspirin}}{1 \text{ mole salicylic acid}} \times \frac{180 \text{ grams aspirin}}{1 \text{ mole aspirin}}$$

= 3.6 grams of aspirin

The **actual yield** is simply the weight of desired product isolated. With the actual yield and the theoretical yield, we can calculate the **percentage yield** by the equation

$$\text{Percentage yield} = \frac{\text{Actual yield}}{\text{Theoretical yield}} \times 100\%$$

In our example, if 2.7 grams of aspirin were isolated, the percentage yield would be

$$\text{Percentage yield} = \frac{2.7 \text{ grams}}{3.6 \text{ grams}} \times 100\% = 75\%$$

In some experiments, the objective is the isolation of a natural product rather than the synthesis and purification of a reaction product. In these cases, the percentage yield refers to the percentage of the substance isolated, based upon the weight of original material. For example, if 0.24 gram of caffeine were isolated from 20.0 grams of tea, the percentage yield would be

$$\text{Percentage yield} = \frac{\text{Weight of substance isolated}}{\text{Weight of original material}} \times 100\%$$

$$= \frac{0.24 \text{ gram}}{20.0 \text{ grams}} \times 100\% = 1.2\%$$

Observe that there is no calculation of the theoretical yield for this type of experiment.

LABORATORY GLASSWARE

A typical set of glassware and other equipment commonly used in most organic courses is shown in Figure LG-1. Observe that much of this glassware is the standard-taper variety. This is an improvement over the use of corks and rubber stoppers, which may cause leakage and contamination of the reagents. However, this glassware is expensive and you are responsible for it, so use it with care and respect. Mistreatment can easily make it break, which may cost you money and the time required to replace the broken items. So read this section carefully.

ASSEMBLING THE APPARATUS

Numerous illustrations are given throughout the book showing the assembly of the glassware for a particular purpose. Here are the important points to remember:

1. Carefully clamp the glassware securely to a ring stand.

2. Never apply undue pressure or strain to the glassware.

For example, when assembling the distillation apparatus in Figures 1-6 and 1-7, do not allow the condenser to temporarily hang suspended from the distilling head because it could fall off. Always clamp it before the connection is made. But do not initially clamp it so rigidly that the two pieces cannot be easily fitted together without placing a strain on the connection. The vacuum adapter in Figures 1-6 and 1-7 is also accident-prone if not clamped. Do not rely on friction to hold it in place. Also, when assembling standard-taper glassware, be sure that no solid or liquid material is present on the joint surfaces. These materials may cause the joints to leak by lessening the effectiveness of the seal. If the apparatus is heated, these materials could also cause the joints to stick or "freeze."

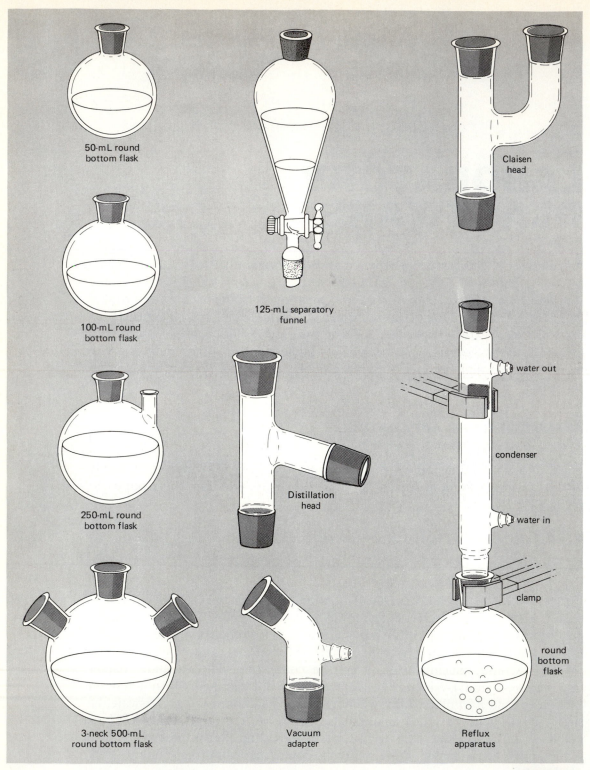

50-mL round bottom flask

100-mL round bottom flask

250-mL round bottom flask

3-neck 500-mL round bottom flask

125-mL separatory funnel

Distillation head

Vacuum adapter

Claisen head

water out

condenser

water in

clamp

round bottom flask

Reflux apparatus

FIGURE LG-1

Some commonly needed equipment in the organic laboratory.

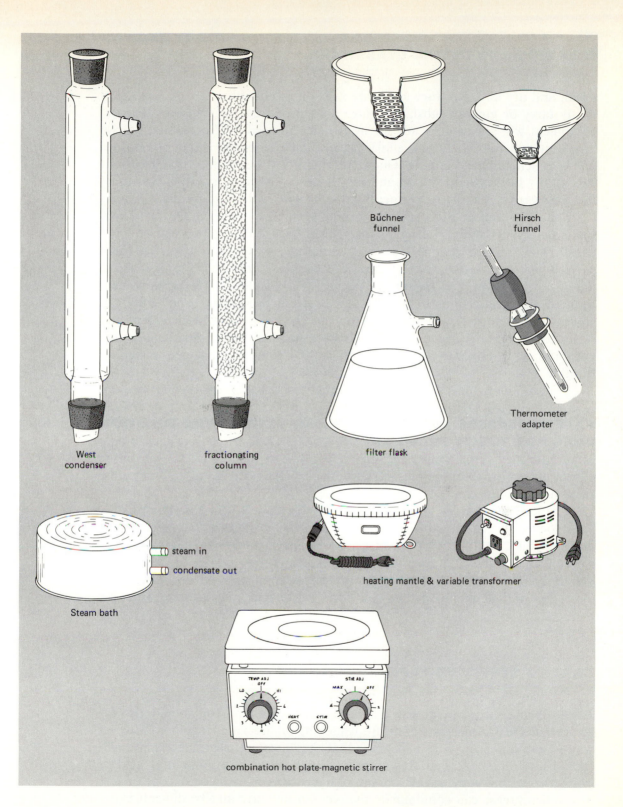

West
condenser

fractionating
column

Büchner
funnel

Hirsch
funnel

filter flask

Thermometer
adapter

steam in
condensate out

Steam bath

heating mantle & variable transformer

TEMP ADJ
OFF
LO HI
MAX OFF
STIR ADJ
HEAT
STIR

combination hot plate-magnetic stirrer

CLEANING AND DRYING GLASSWARE

Clean and dry all laboratory glassware as soon as possible after use, especially if any tarry or gummy substances have been produced. This ensures that the equipment will always be ready for use when you need it. Procrastinating can also cost you extra time and effort, because any deposited stains or residues will be much harder to remove. Glassware that is just wet with a volatile solvent can simply be allowed to drain and dry. When an aqueous solution was used, the glassware can easily be cleaned by rinsing with water and drying.

For stains and residues, several cleaning agents may be used. Ordinary synthetic detergents and cleansing powders with fine abrasives often can be quite useful. The stain or residue may also be soluble in an organic solvent such as technical- or waste-grade acetone. For monetary reasons, use only small amounts of solvent, and **never** use reagent-grade material. The wash solvent can also be effectively used several times before it becomes "spent." After the solvent has been used, you may need to wash the glassware with detergent and water to remove the residual solvent. Brown stains of manganese dioxide can generally be removed by washing with a 30% solution of aqueous sodium bisulfite.

For stubborn stains and residues that resist all other efforts, a small amount of concentrated KOH in alcohol may be used with soaking overnight. **WEAR SAFETY GLASSES AND BE VERY CAREFUL THAT THE SOLUTION DOES NOT TOUCH YOUR SKIN OR CLOTHING.** Potassium dichromate in sulfuric acid is no longer recommended, due to the hazardous impact of chromium on the environment. In addition, chromium(VI) compounds are now suspected carcinogens.

The best way to dry glassware is to simply allow it to stand overnight in your locker. Whenever possible, invert the items to permit drainage. If it is necessary to quickly dry a piece of glassware, use a suction tube connected to a water aspirator. If the item is wet with water, it may first be rinsed with a **small** quantity of **technical**-grade acetone and then dried as before. Use discretion and planning so that you don't waste time and acetone with this method.

CARE OF STANDARD-TAPER GLASSWARE

Besides being extra careful with standard-taper glassware, you should also try to prevent the ground-glass joints from sticking or "freezing" together. The best way to avoid freezing is to make sure that the joints are clean beforehand and then disassemble the apparatus immediately after use. The joints may also be lubricated with a **very small** amount of hydrocarbon-based or silicone grease. Too much grease can contaminate the reaction mixture and be difficult to remove. To properly lubricate a joint, place a **thin layer** of the grease on opposite sides of the male joint and rotate it inside the female joint so that all areas of the joints are covered. A properly lubricated joint has a clear appearance, with no visible bands or stripes. Wipe off excess grease with a paper towel.

If a ground-glass joint becomes frozen, you can sometimes loosen it by **gently** tapping the joint on the edge of the bench top. You may also try heating the joint in a hot-water or steam bath. If all this fails and the situation appears hopeless, as a last resort you may heat the joint in a flame. This **must** be done **slowly** and **carefully** because a rapid change in temperature may cause the outer part of the joint to expand rapidly and break. You want the outer part of the joint **gradually** to expand more than the inner section so that the two will separate. If necessary, your instructor will demonstrate the technique for you. He or she, in fact, may want to perform the separation for you.

E X P E R I M E N T

1

DISTILLATION

- **FRACTIONAL DISTILLATION**
- **VACUUM DISTILLATION**
- **AZEOTROPIC MIXTURES**

Double, double, toil and trouble;
Fire burn and cauldron bubble.

Shakespeare, *Macbeth,* Act IV, Scene 1

1.1 INTRODUCTION

Distillation is the principal method for purifying many organic liquids, and has long been used around the world by liquor manufacturers, the oil industry, and much of the chemical industry. Liquor manufacturers utilize it to separate the products of Scotch, rye, vodka, rum, gin, and so on, from the remains of the fermented grain, potatoes, or molasses. The oil industry has found it to be the only method of efficiently separating the components present in the millions of barrels of crude petroleum consumed each day. In this way we get the gasoline, fuel oil, kerosene, lubricating oil, petroleum jelly, paraffin wax, asphalt, and tar that we find so essential to our everyday lives. The chemical industry uses distillation to separate many of the various by-products, unconsumed reactants, and solvents present in chemical reaction mixtures.

Distillation involves converting the liquid to the gaseous state with the aid of heat, and condensing the vapor back to the liquid state by cooling. Because some components of the liquid vaporize more easily than others, the composition of the vapor will be richer in these more volatile components. Distillation takes advantage of this to individually separate the components. The details of the process are presented in the following sections.

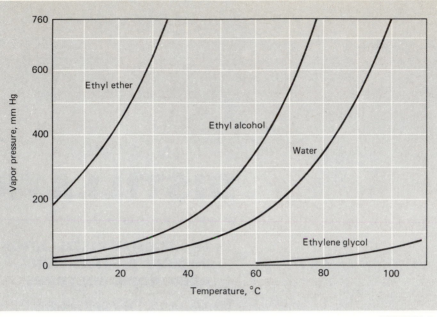

FIGURE 1-1
The variation of vapor pressure with temperature for four common liquids.

1.2 VAPOR PRESSURE

The tendency of any liquid to vaporize at any given temperature is measured by the liquid's **vapor pressure**. It is the pressure in torr or mmHg exerted by the vapor when the liquid and vapor are in equilibrium with each other.

$$\text{Liquid} \rightleftharpoons \text{vapor}$$

Vaporization occurs when liquid molecules attain enough kinetic energy to overcome the attractive forces of their neighbors. Since the amount of kinetic energy increases with temperature, heating any liquid will shift the equilibrium to the right, resulting in more molecules being vaporized and the vapor pressure increasing. This is shown for four common liquids in Figure 1-1. Notice how different the vapor pressures of these liquids are at any given temperature. This results from the large differences in the attractive forces of the various liquid molecules.

1.3 BOILING POINTS

Heating a liquid may cause the vapor pressure of the liquid to increase to the point where it exactly equals the applied pressure (usually the atmospheric pressure). When this occurs, the liquid will begin to boil.

Because of the pressure dependence of the boiling point, the atmospheric pressure or other pressure should always be recorded whenever a boiling-point temperature is taken. The variation in boiling point can even be enormous. For example, water will boil at 100.0°C at 760 torr or 1.00 atmosphere of pressure, but at only 20°C or less than room temperature if a vacuum pump is used, producing a pressure of 18 torr.

Figure 1-2 shows a nomograph for easily relating the observed boiling point of a liquid at a given reduced pressure to the observed boiling point at 760 torr. To use the nomograph, you simply determine where a straightedge intersects one of the three scales, when values for the other two scales are given. For example, assume you want to know where a compound will boil at 10 torr if its boiling point at 760 torr is 285°C. Placing the straightedge at point X on scale C and point Y on scale B yields point Z on scale A. As shown in Figure 1-2, this gives a boiling point of approximately 147°C.

Every stable organic compound will have a characteristic boiling point at 760 torr. This boiling point will be influenced by hydrogen bonding and polarity. Assuming an identical or similar molecular weight, compounds which can form

FIGURE 1-2
A temperature–pressure nomograph.

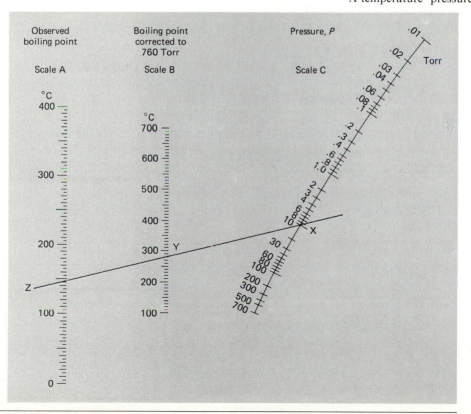

hydrogen bonds will usually boil much higher than polar compounds which cannot, and polar compounds will boil higher than nonpolar ones. Thus, ethyl alcohol, CH_3CH_2—OH, which forms hydrogen bonds, boils at 78.8°C, while methyl ether, CH_3—O—CH_3, which cannot form hydrogen bonds, boils at only -23.7°C. The polarity of methyl ether, however, makes it boil higher than propane, C_3H_8, which is nonpolar and boils at -42.1°C.

Molecular weight and molecular shape also influence the boiling point. Other factors being equal, a higher molecular weight means a higher boiling point. And long, straight-chain compounds will have higher boiling points than the corresponding branched-chain compounds. Thus straight-chain heptane boils at 98.4°C, the branched-chain 2-methylhexane boils at 90°C, while the even more branched 2,2-dimethylpentane boils at 79.2°C.

$$CH_3CH_2CH_2CH_2CH_2CH_2CH_3 \qquad CH_3{-}\overset{\overset{\displaystyle CH_3}{|}}{CH}{-}CH_2CH_2CH_2CH_3 \qquad CH_3{-}\overset{\overset{\displaystyle CH_3}{|}}{\underset{\underset{\displaystyle CH_3}{|}}{C}}{-}CH_2CH_2CH_3$$

Heptane, bp 98.4°C   2-Methylhexane, bp 90°C

2,2-Dimethylpentane, bp 79.2°C

1.4 BOILING POINTS OF MIXTURES

The boiling point of a mixture depends upon the vapor pressures of the various components. Solutes will either raise or lower the boiling point, depending upon their own vapor pressure and how they interact or bond with the main component. Generally, a less volatile solute will cause the boiling point to rise, while a more volatile one will cause the boiling point to fall. Thus, adding sugar or the less volatile acetic acid to water will cause the boiling point of the aqueous solution to rise, but adding the more volatile methyl alcohol to water will cause the boiling point to fall.

1.5 IDEAL SOLUTIONS

Many pure liquids, when mixed, form only weak intermolecular bonds with each other. Thus, each liquid exerts its own vapor pressure independent of the others. This vapor pressure can be calculated by **Raoult's law**, which states that **each volatile or easily vaporized component in a solution exerts a pressure equal to its mole fraction times the vapor pressure of the pure component**.

As an example, consider a mixture of toluene, used to make TNT, and carbon tetrachloride, sometimes used as cleaning fluid. Using Raoult's law, we have for toluene:

$$P_{tol} = P^{\circ}_{tol} \cdot X_{tol}$$

where P_{tol} is the vapor pressure produced by the toluene at a certain temperature, P_{tol}° is the vapor pressure of the pure toluene at this temperature, and X_{tol} is the mole fraction of toluene in the mixture.

In a similar manner, we have for carbon tetrachloride:

$$P_{ct} = P_{ct}^{\circ} \cdot X_{ct}$$

where P_{ct} is the vapor pressure produced by the carbon tetrachloride at a certain temperature, P_{ct}° is the vapor pressure of the pure carbon tetrachloride at this temperature, and X_{ct} is the mole fraction of carbon tetrachloride in the mixture.

Using Dalton's law of partial pressures, we have the total pressure equal to the sum of all the partial pressures. For the mixture of toluene and carbon tetrachloride, this gives us

$$P_{total} = P_{tol} + P_{ct}$$

Raoult's law and Dalton's law can be combined graphically for any mole-fraction combination and temperature. For our mixture of toluene and carbon tetrachloride, the result at 50.0°C is given in Figure 1-3. Similar diagrams may

FIGURE 1-3
A graphical application of Raoult's law.

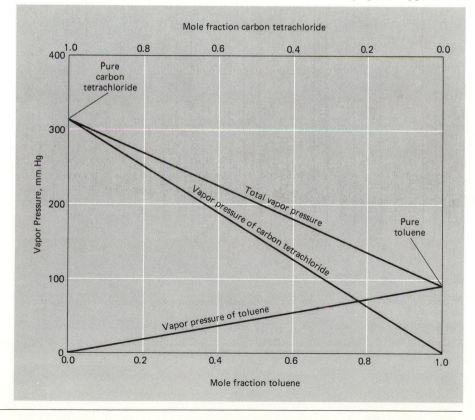

be drawn for other temperatures and ideal solutions after the vapor pressure information is obtained.

1.6 BOILING-POINT DIAGRAMS FOR IDEAL SOLUTIONS

The boiling-point diagram for any combination of toluene and carbon tetrachloride at 760 torr is shown in Figure 1-4. Other ideal solutions of two volatile components will have similar diagrams. The lower curve in Figure 1-4 shows the temperature where any combination of toluene and carbon tetrachloride will boil. By then moving horizontally at this temperature over to the upper curve, the composition of the vapor is determined.

For example, assume we have a mixture which initially has a mole percentage of 60% toluene and 40% carbon tetrachloride. When this mixture is heated, you can see from Figure 1-4 that the boiling point will be 93°C. Moving horizontally at 93°C over to the upper vapor curve, we find that it intersects at point A

FIGURE 1-4
Boiling-point diagram of toluene and carbon tetrachloride.

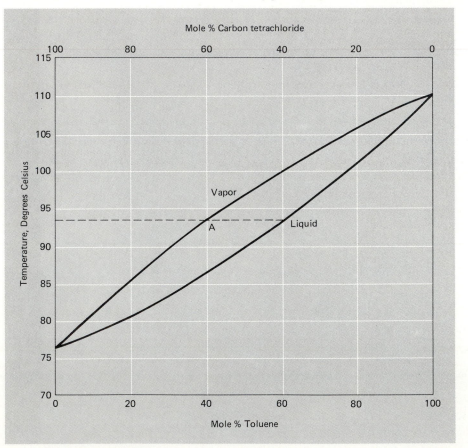

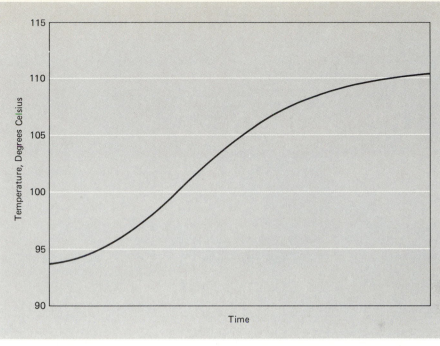

FIGURE 1-5
The boiling-point increase of a mixture of 60% toluene and
40% carbon tetrachloride in a simple distillation.

at a vapor composition of 39% toluene and 61% carbon tetrachloride. The vapor is, therefore, richer in the more volatile, lower-boiling carbon tetrachloride.

Of course, this boiling point and vapor composition will not persist during a simple distillation. As the remaining liquid continues to boil, it becomes progressively richer in the higher-boiling toluene, meaning that both the boiling point and the mole percentage of toluene in the vapor will steadily increase. The boiling-point increase is shown in Figure 1-5.

1.7 SIMPLE DISTILLATION AND FRACTIONAL DISTILLATION

Figure 1-6 shows a simple-distillation apparatus for our toluene and carbon tetrachloride mixture or for many other liquid mixtures. The liquid is distilled from the distilling flask, the vapor is liquefied in the water-cooled condenser, and the distillate is then collected in a receiver. The thermometer in the distilling head allows us to measure the temperature of the vapor just before it is liquefied and collected. From the temperature of the vapor and the volume of the distillate collected, the progress of the distillation can be monitored.

It should be evident from the last section that a simple distillation cannot completely separate toluene and carbon tetrachloride. However, a fractional distillation can.

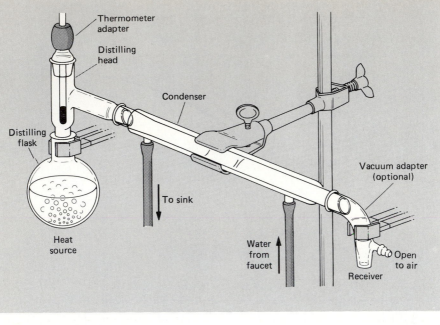

FIGURE 1-6
A simple-distillation apparatus.

A fractional-distillation apparatus is shown in Figure 1-7. Notice that it is identical to the simple-distillation apparatus except for the presence of the fractionating column between the distilling flask and the distilling head. The fractionating column can be simply a vertical tube packed with glass beads, glass helices, the commercial product Heli-Pak, or stainless steel wool sponge, or it can be one of the more elaborate columns shown in Figure 1-8.

Fractionating columns give better separations of liquid mixtures because they allow many simple distillations to occur in one operation. As the vapor travels up a column, the large surface area of the column allows the vapor to cool, condense into a liquid, and then revaporize as more heat from other vapor reaches it. This process occurs many times, and each time the vapor becomes progressively richer in the lowest-boiling component. If the column has an adequate efficiency, the vapor finally reaching the top will consist of only the lowest-boiling material. The process then continues with each less volatile substance until a complete separation is attained.

This is illustrated for our 60% toluene, 40% carbon tetrachloride mixture in the boiling-point diagram in Figure 1-9. After the first distillation, the composition of the vapor is 39% toluene and 61% carbon tetrachloride. The vapor condenses, which is indicated by the vertical drop from point A to point B in the diagram. The condensed liquid at point B then revaporizes, which is indicated by the horizontal movement from point B to point C in the diagram. As can be seen, the vapor at point C now has a new composition of 22% toluene and

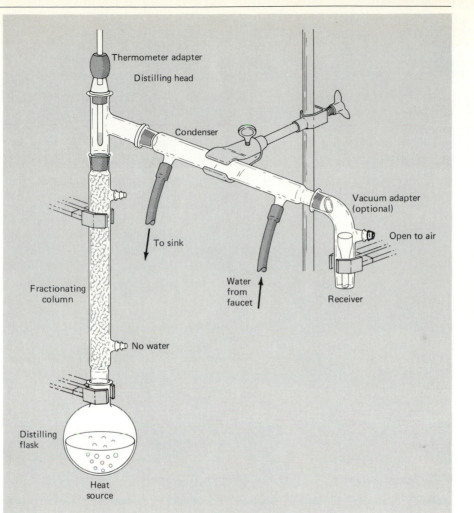

FIGURE 1-7
A fractional-distillation apparatus.

78% carbon tetrachloride, making this vapor even richer in carbon tetrachloride. This vapor then condenses, indicated by the vertical drop from point C to point D. When the liquid at point D revaporizes, we move horizontally to point E, giving us a vapor which now has a composition of 11% toluene and 89% carbon tetrachloride. Two more condensations and revaporizations bring us to point I, or a vapor having a composition of 3% toluene and 97% carbon tetrachloride. Continued condensations and revaporizations will achieve an even further separation, which, because of space limitations, cannot be adequately represented in the diagram.

Also observe from Figure 1-9 that as the liquid on the column becomes progressively richer in carbon tetrachloride, its boiling point steadily drops.

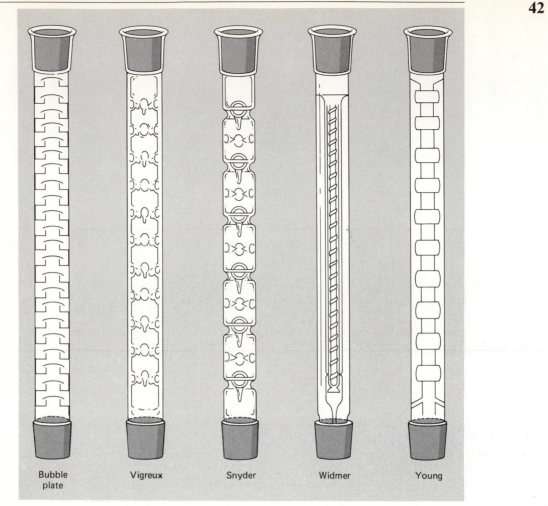

FIGURE 1-8
Some different types of fractionating columns.

Originally the liquid in the distilling flask had a boiling point of 93°C, but after the first condensation on the column, the boiling point is 87°C. Continued vaporizations and condensations lower the boiling point to 77°C, which is the boiling point of the pure carbon tetrachloride.

1.8 RATING FRACTIONATING COLUMNS

All fractionating columns may be rated according to a number of factors. The most important factor is the column's number of **theoretical plates**, meaning how many times the vapor is condensed and revaporized as it rises in the column. With proper heating, more theoretical plates mean better separation. In fact, the

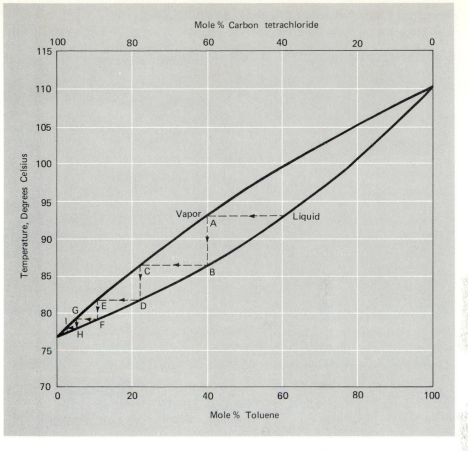

FIGURE 1-9

Boiling-point diagram of toluene and carbon tetrachloride, showing their progressive separation by fractional distillation.

number of theoretical plates needed to effectively separate components having various boiling-point differences is known and is given in Table 1-1.

Fractionating columns may also be rated in terms of their **holdup**, or the amount of liquid retained on the column. This amount of liquid will be lost in the distillation unless a high-boiling "chaser" liquid is placed in the mixture to move the held-up liquid off the column.

The term **HETP** is also important in rating fractionating columns. It means the height of the column equivalent to one theoretical plate. A low HETP is desirable to keep highly efficient columns from being inconveniently tall or having large holdups.

Table 1-2 gives the number of theoretical plates, the holdup, and the HETP for a number of fractionating columns. From the number of theoretical plates and the results of Table 1-1, the needed boiling-point differences between components for an effective separation is also given. Observe from Table 1-2 that none of these columns is able to effectively separate liquids with low boiling-

TABLE 1-1.

Theoretical Plates Needed to Effectively Separate Components Having Various Boiling-Point Differences

Boiling-point difference (°C)	Number of theoretical plates
108	1
72	2
54	3
43	4
36	5
25	8
20	10
14	15
10	20
7	30
4	50
2	100
1	200

TABLE 1-2.

Ratings of Some Fractionating Columns

Column type (20 cm long and 1 cm diameter)	Theoretical plates	Holdup (mL)	HETP (cm)	Boiling-point differences for effective separation (°C)
Metal sponge	2	4	10	72
Vigreux	2.5	1	8	60
Packed (glass beads)	3	4	7	54
Packed (metal helices)	3	4	7	54
Packed (carborundum chips)	3.3	4	6	50
Packed (glass helices)	5	4	4	36
Packed (Heli-Pak)	13	7	1.5	17

point differences. For this purpose, a spinning band column having 100 to over 200 theoretical plates is needed. It consists of a band of metal or Teflon rotating thousands of times per minute within the column.

1.9 OTHER FACTORS INFLUENCING A FRACTIONAL DISTILLATION

An efficient fractionating column can only give optimal results if the distillation is properly done. If too much heat is applied to the distilling flask, the distillation will occur too quickly and the number of condensations and revaporizations occurring on the column will be reduced. In addition, the column should be thermally insulated to prevent heat from escaping to the surroundings. Otherwise extra heat from the ascending vapor will be required.

A way to both thermally insulate a fractionating column and still observe the liquid on it is to use a glass vacuum jacket around the column. This increases the efficiency of the column by about 25 percent.

1.10 JUDGING THE EFFECTIVENESS OF A SEPARATION BY FRACTIONAL DISTILLATION

The effectiveness of the separation in a fractional distillation can frequently be determined by graphing the temperature of the vapor at the top of the column versus the volume of distillate collected. A large increase in temperature with only a slight increase in volume is indicative of an effective separation. This is shown in Figure 1-10. The lower horizontal line indicates that since the temperature is remaining constant, only the lower-boiling component is coming off the column. The sharp temperature rise following this shows an abrupt change in

FIGURE 1-10
A fractional-distillation graph, showing a complete separation of a two-component mixture.

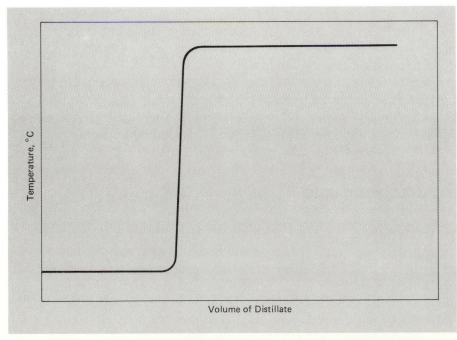

Temperature, °C

Volume of Distillate

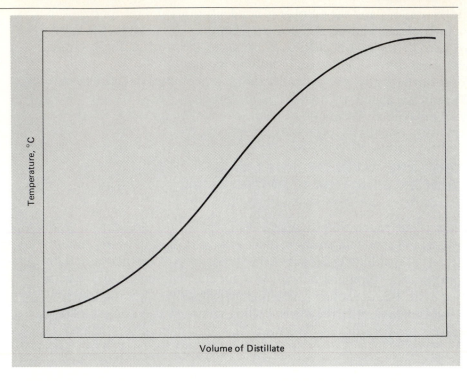

Volume of Distillate

FIGURE 1-11
A fractional-distillation graph, showing only a partial separation of a two-component mixture.

the composition of the vapor and a brief transitional period between the lower- and higher-boiling components. The upper horizontal line then shows that only the higher-boiling component is now coming off.

Contrast Figure 1-10 with Figure 1-11. In Figure 1-11 much of the graph shows only a gradual rise in the temperature. This shows that more than one component is simultaneously coming off the column, indicating that only a partial separation has occurred. In fact, the more gradual the temperature rise, the less effective is the distillation in separating the components.

1.11   AZEOTROPIC MIXTURES

Unfortunately, even highly efficient fractionating columns with many theoretical plates cannot completely separate every mixture. Some liquids, such as petroleum, will contain numerous compounds with the same or nearly the same boiling point. It is even possible for a thousand or more compounds to all have the same boiling point! In these cases, other purification methods or the independent synthesis of each desired, individual compound would have to be employed.

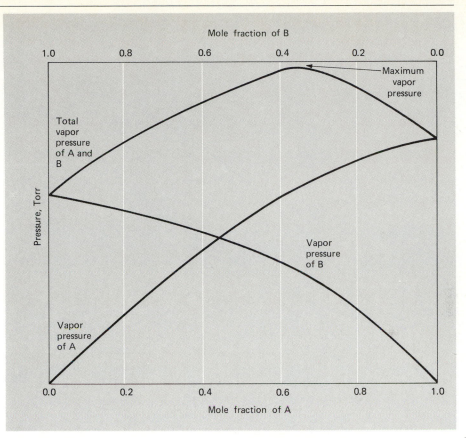

FIGURE 1-12
Vapor pressure diagram of a nonideal mixture of A and B, showing the maximum vapor pressure of the two reached at an intermediate composition.

Azeotropic mixtures are another instance where a complete separation is impossible by fractional distillation. Because of strong molecular interactions between different liquids, mixtures of these liquids will not be ideal and obey Raoult's law. Frequently, the vapor pressure of these mixtures will reach either a maximum or a minimum **not** for a pure component as expected, but for some intermediate composition. This is shown in Figure 1-12 for a maximum vapor pressure. Therefore, when this composition is reached during a distillation, the mixture will have this composition in both the liquid and the vapor. As a result, further distillation will give a condensed liquid having the same composition as before, and further separation by this method is impossible! This liquid will have a **fixed boiling point** and will act as though it were a **pure compound**!

A common example of azeotropic mixtures is ethanol and water, whose boiling-point diagram is shown in Figure 1-13. Notice how at a composition of 95.6% ethanol and 4.4% water, the liquid and vapor curves meet. Therefore,

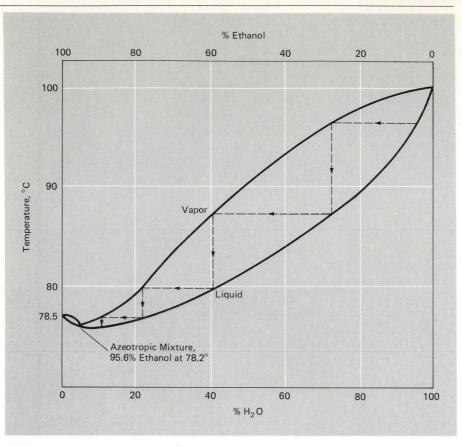

FIGURE 1-13

Boiling-point diagram for ethanol and water, showing the formation of an azeotropic mixture where the liquid and vapor curves meet.

during a fractional distillation of ethanol and water, when this composition of liquid is reached, any further distillation will be fruitless. Even with the **best** fractionating column, 100% ethanol cannot be obtained! Fortunately, it can be obtained by other methods.

Also notice from Figure 1-13 how the boiling point of the azeotropic mixture of ethanol and water is lower than that of either pure component. Ethanol and water are therefore said to form a **minimum-boiling-point azeotrope**. Other azeotropic mixtures will have their boiling point higher than that of either pure component. These mixtures are therefore called **maximum-boiling-point azeotropes**. Some examples of minimum- and maximum-boiling-point azeotropes are given in Tables 1-3 and 1-4. Minimum-boiling-point azeotropes will always result when the combined vapor pressure **exceeds** that of either pure component, such as shown in Figure 1-12. Maximum-boiling-point azeotropes,

TABLE 1-3.

Minimum-Boiling-Point Azeotropes

Component A	BP (°C)	Component B	BP (°C)	Azeotropic mixture		
				BP (°C)	Weight % A	Weight % B
Water	100.0	Acetonitrile	81.5	76.0	14.2	85.8
Water	100.0	Ethanol	78.5	78.2	4.4	95.6
Water	100.0	t-Butanol	82.5	79.9	11.8	88.2
Water	100.0	2-Propanol	82.3	80.3	12.6	87.4
Water	100.0	1-Propanol	97.3	87	28.3	71.7
Water	100.0	Dioxane	101.3	87.8	18	82
Water	100.0	Butyl acetate	126.5	90.7	27.1	72.9
Water	100.0	Pyridine	115.5	94	57	43
Water	100.0	Propanoic acid	141.4	99.1	82.2	17.8
Water	100.0	Phenol	181.8	99.5	90.8	9.2
Carbon tetrachloride	76.8	Methanol	64.7	55.7	79.4	20.6
Carbon tetrachloride	76.8	Acetone	56.2	56.1	11.5	88.5
Carbon tetrachloride	76.8	Ethanol	78.5	65.1	84.2	15.8
Chloroform	61.2	Methanol	64.7	53.4	87.4	12.6
Chloroform	61.2	Ethanol	78.5	59.4	93	7
Methanol	64.7	Acetone	56.2	55.5	12	88
Methanol	64.7	Benzene	80.1	57.5	39.1	60.9
Methanol	64.7	Toluene	110.6	63.7	72.4	27.6
Ethanol	78.5	Benzene	80.1	68.2	32.6	67.4
Ethanol	78.5	Ethyl acetate	77.1	71.8	31	69
Ethanol	78.5	Toluene	110.6	76.7	68	32
Acetic acid	118.5	Toluene	110.6	105.4	28	72

on the other hand, always result when the combined vapor pressure **is less than** that of either pure component.

One key difference in the two types of azeotropes is where the azeotropic mixture will be formed during a fractional distillation. From Figure 1-13, it should be evident that a minimum-boiling-point azeotrope will always be formed in the fractionating column and is what will first be collected in the distillate. However, analysis of a boiling-point diagram for a maximum-boiling-point azeotrope shows the exact opposite. The azeotrope is what remains in the distilling flask after all the lower-boiling-point material has been removed.

This is illustrated in Figure 1-14 for a mixture of water and formic acid. Assume we have a mixture of 50% water and 50% formic acid by weight. Distillation of this liquid does **not** move us up the liquid and vapor curves to the azeotropic mixture composition of 77.4% water and 22.6% formic acid in the

49

TABLE 1-4.

Maximum-Boiling-Point Azeotropes

Component A	BP (°C)	Component B	BP (°C)	Azeotropic mixture		
				BP (°C)	Weight % A	Weight % B
Water	100.0	Formic acid	100.8	107.2	77.4	22.6
Water	100.0	Hydrofluoric acid	19.5	111.4	64.4	35.6
Water	100.0	Ethylenediamine	116	118	77	23
Water	100.0	Nitric acid	86	120.5	32	68
Water	100.0	Perchloric acid	110	203	28.4	71.6
Water	100.0	Sulfuric acid	dec.	338	1.7	98.3
Acetic acid	118.5	Dioxane	101.3	119.5	77.0	23.0
Acetic acid	118.5	Pyridine	115.5	140	47	53
Chloroform	61.2	Acetone	56.1	64.4	78.5	21.5
Chloroform	61.2	Methyl ethyl ketone	79.6	79.9	17	83
Benzaldehyde	178.1	Phenol	181.8	185.6	49	51

FIGURE 1-14
Boiling-point diagram for water and formic acid, showing a maximum-boiling-point azeotrope.

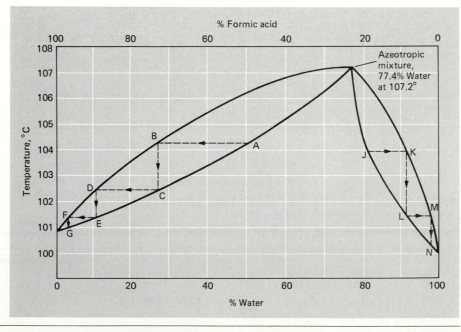

vapor. Instead we move from point A in the diagram to point G, with the vapor becoming progressively richer in formic acid. As a result, formic acid is being removed from the distilling flask, and its percentage in the distilling flask will continue to fall until the composition of the azeotropic mixture is reached. Then only the azeotropic mixture remains in the distilling flask, and it begins to distill, with any further separation by distillation being impossible.

One can also consider a mixture of 80% water and 20% formic acid by weight. Distillation of this mixture again does **not** move us up the liquid and vapor curves to the azeotropic mixture composition in the vapor. Instead we move from point J in the diagram to point N, with the vapor becoming progressively richer in water. As a result, water is being removed from the distilling flask, and its percentage in the distilling flask will continue to fall until the composition of the azeotropic mixture is again reached. The azeotropic mixture will then begin to distill, and any further separation by distillation is impossible.

Remember that not all nonideal liquid mixtures, which disobey Raoult's law, will form azeotropic mixtures. Mixtures such as methanol and water deviate significantly from Raoult's law, yet can be completely separated by fractional distillation, since the minimum and maximum vapor pressures correspond to the pure components and **not** to some intermediate composition.

1.12 TECHNIQUE OF SIMPLE AND FRACTIONAL DISTILLATION

Apparatus

For a simple distillation, the apparatus in Figure 1-6 is used, while in a fractional distillation the apparatus in Figure 1-7 is used. The distilling flask, the condenser, and the vacuum adapter, if used, should all be carefully clamped without any stress on the connecting ground-glass joints. All connecting ground-glass joints should be lightly greased with stopcock grease to ensure that they do not "freeze" together. However, do not use too much grease because it can contaminate your chemicals. When the pieces of equipment are together, the joints should look nearly transparent, with no air bubbles present. Be sure all the joints are securely together to prevent any material from escaping through leakage. No cooling water should go through the fractionating column in Figure 1-7, since then no material would distill over.

Thermometer

The thermometer in the apparatus is inserted through the rubber adapter until the bulb is located just below the side arm of the distilling head. This ensures that the bulb will be in the vapor stream and will give a correct reading.

Water lines

The water lines are attached to the condenser so that the water enters from the bottom and exits from the top. This keeps the whole outer jacket of the condenser full of water, which is important for effective cooling. The water does not have to flow through the jacket at a fast rate. In fact, a slower rate may help prevent a flood if one of the water line connections is loose and comes apart.

Boiling stones

To help prevent bumping, add several boiling stones made of unglazed porous clay plate or carborundum to the distilling flask. Bumping occurs when the liquid in the flask becomes superheated and the excess energy produces a sudden violent surge of liquid and vapor. Boiling stones help prevent bumping because they provide a source of small bubbles in their pores to promote more even boiling. If you should ever forget to add boiling stones, do not add them to a boiling hot liquid. Wait until the liquid cools down, or it may boil over on you.

Size of the distilling flask

The distilling flask generally should be about twice as large as the volume of liquid being distilled. The flask should **never** be more than two-thirds filled. The reason is that the liquid needs "room to boil," and the surface area for the boiling should be as large as possible. During a simple distillation, the liquid from an overfilled flask could also easily bump over into the condenser, possibly necessitating restarting the distillation.

If a large amount of a low-boiling liquid must be removed before distilling a small amount of a high-boiling liquid, it is best to distill most of the low-boiling liquid, then stop the distillation, and transfer the remaining undistilled liquid to a smaller distilling flask before you resume the distillation. This procedure prevents a loss of much of the high-boiling liquid at the end, due to its large volume of vapor in the large distilling flask. In addition, the large surface area of the large distilling flask may act as a condenser and make it impossible to distill the high-boiling liquid.

Heating source

A hot plate, an electric heating mantle, and an electrically heated oil bath, warm-water bath, or hot-water bath are common ways to heat the liquid in a distilling flask. The warm- or hot-water bath cannot be used to distill liquids boiling over 100°C. If the separation is easy, you can also use a steam bath to distill liquids boiling under 100°C.

Because of the fire hazard, you should seldom use a flame for a distillation unless the liquid is very high boiling and cannot be effectively heated by one of the other methods. In this case, be certain that the apparatus is tightly sealed at the ground-glass joints near the flame, to prevent a fire caused by a leak. Also keep the liquid as far away from the flame as possible.

Do not distill to dryness if an intense heating source is being used. With no liquid to carry away the heat, a distilling flask under these conditions can become very hot and may even crack. In addition, some residues left after a distillation may explode when overheated.

Rate of distillation

To save time, you can, in the beginning, rapidly heat the liquid in a distilling flask. But when the liquid begins to boil, you should lower the rate of heating so that the liquid boils gently.

Proper heating in a simple distillation should produce a uniform rate of about 1 drop per second (1–3 mL per minute)[1] of distillate being collected. For a fractional distillation, the rate will vary, but generally will be much lower than this. Too fast a rate here generally means that not enough condensations and revaporizations are occurring on the fractionating column to produce an effective separation. In addition, too fast a rate will likely flood the column with liquid, also reducing the effectiveness of the separation. If the column does become flooded, the rate of heating must be lowered to allow liquid to drain back into the distilling flask.

Too slow a distilling rate must also be avoided in a fractional distillation. Otherwise an inadequate amount of vapor may reach the thermometer, resulting in an inaccurately low temperature reading. If there is both adequate heating and effective separation, each component coming out the top of the column should give its own constant temperature. This temperature will rise for each successive component, which means the rate of heating must be progressively increased to have these components be effectively removed from the column.

Receiver

The distilled liquid may be collected in a clean, dry Erlenmeyer flask, a round-bottom flask, or some other appropriate container. To minimize vapor losses and reduce the fire hazard, use a ground-glass, round-bottom flask attached to a vacuum adapter. You can also place the round-bottom flask in an ice-water bath to further cool the distillate and lower the vapor loss.

When performing a distillation, always remember to have a number of **labeled**, clean, dry flasks available for collecting new fractions. Also, always be certain you have an open system. Otherwise the pressure inside the apparatus could build up dangerously.

1.13   VACUUM DISTILLATION

Many liquids cannot be purified by ordinary distillation at atmospheric pressure because they will decompose before their normal boiling points are reached.

---

[1] Aqueous and other strongly hydrogen-bonded liquids generally have about 20 drops per mL. Most organic liquids, however, have about 50 to 60 drops per mL.

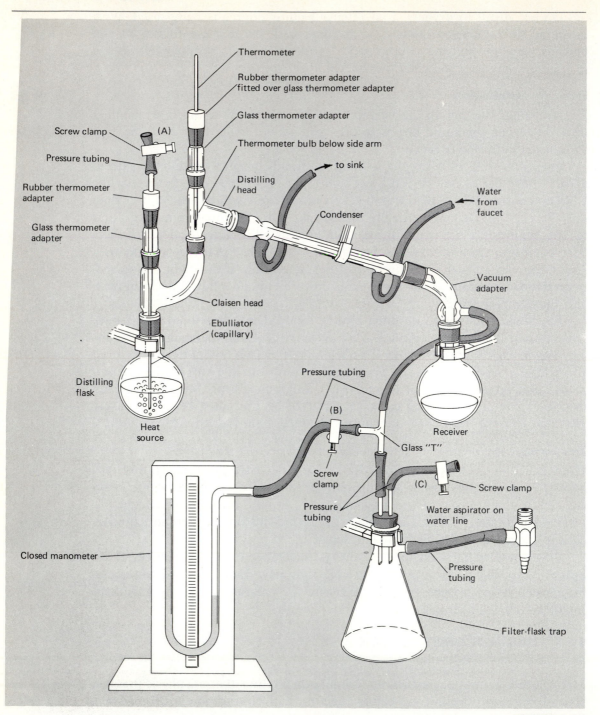

Thermometer

Rubber thermometer adapter
fitted over glass thermometer adapter

Glass thermometer adapter

Thermometer bulb below side arm

Screw clamp (A)

to sink

Pressure tubing

Distilling
head

Water
from
faucet

Rubber thermometer
adapter

Condenser

Glass thermometer
adapter

Vacuum
adapter

Claisen head

Ebulliator
(capillary)

Pressure tubing

Distilling
flask

(B)

Receiver

Heat
source

Glass "T"

Screw
clamp

(C)

Screw clamp

Pressure
tubing

Water aspirator on
water line

Closed manometer

Pressure
tubing

Filter-flask trap

FIGURE 1-15

A vacuum-distillation apparatus. The distilling flask, the condenser, the receiver, and the filter-flask trap must all be clamped.

But if these liquids are distilled at greatly reduced pressures, by a technique called **vacuum distillation**, this decomposition can frequently be eliminated. This results from the normal boiling point being greatly lowered when the external pressure is greatly lowered. As an example, glycerol boils with decomposition at 290°C at 760 torr, but boils **without** decomposition at 135°C when the pressure is lowered to 5 torr. To determine what will be the reduced boiling point at any reduced pressure, use the nomograph in Figure 1-2, as described in Section 1.3.

1.14   VACUUM-DISTILLATION APPARATUS

The apparatus for a vacuum distillation is shown in Figure 1-15. Observe from the figure that a **Claisen head** is placed between the distilling flask and the distilling head. The Claisen head prevents the collected distillate from being contaminated if bumping should send the liquid significantly upward. Severe bumping is a much more serious problem in vacuum distillation than in simple or fractional distillations.

Also notice the **ebulliator tube** in the distilling flask. The ebulliator is set near the bottom of the flask and places a certain amount of air into the system to help prevent bumping. Wooden applicator sticks and microporous boiling stones may also be used to help prevent bumping, but they are not as common. Conventional boiling stones will not work with a vacuum. If there is no ebulliator in your ground-glass kit, you can easily make one by heating a piece of glass tubing and pulling it apart about 2 to 3 cm. The glass is then scored in the thinnest portion and carefully broken there. The rubber tubing and the screw clamp above the ebulliator regulate the amount of air to be admitted.

Observe that all rubber tubing in the figure is the heavy-wall, pressure type, with the exception of the water-cooling hoses to the condenser. Ordinary tubing would easily collapse from the atmospheric pressure being so much greater than the internal pressure.

The vacuum source in the diagram may be a vacuum pump, but a water aspirator is much more practical for a class of students. Remember that a water aspirator can never give a pressure smaller than the vapor pressure of water, which is 19.8 torr at 22°C. In addition, if many students are using their aspirators simultaneously, it will be very difficult to even get a pressure this low. In this case, only a few students can work on the same water line, and it may be necessary to move to another area of the lab. If a very low pressure is needed, a vacuum pump must be used.

The pressure attained within the system is measured with a manometer. There are several types. The open-end type, shown in Figure 1-16, is filled with mercury, and the internal pressure in torr is measured using the equation

$$P_{\text{system}} = P_{\text{atmosphere}} - \Delta h_{\text{Hg}}$$

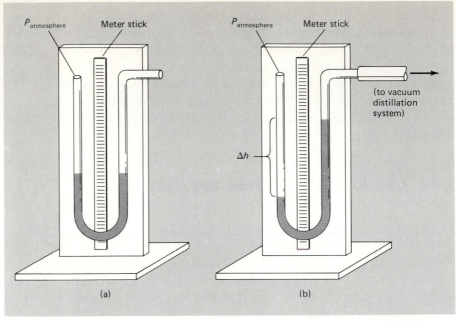

FIGURE 1-16
An open-end manometer: (*a*) open to the atmosphere at both
ends; (*b*) one end connected to a vacuum-distillation system.

The closed type, shown in Figure 1-17, is also filled with mercury, and the
internal pressure is measured using the equation

$$P_{\text{system}} = \Delta h_{\text{Hg}}$$

The closed type may also be the commercial one shown in Figure 1-18.

The trap, shown in Figure 1-15, is very important for protecting both the
manometer and the distillation apparatus from possible backups in the water
line. Notice in the figure that it also has a screw clamp and pressure-tubing
arrangement for opening the system to the atmosphere. **Always open the
screw clamp slowly**. If air is allowed to suddenly rush into the system, the
force may break the manometer, sending mercury into the room. Also when the
vacuum is started, close the clamp slowly, to prevent mercury from suddenly
being withdrawn from the manometer.

One problem with the setup in Figure 1-15 is that it is inconvenient for
changing receivers when a new fraction is coming over. The distillation must
be interrupted, the heating source removed from the apparatus, the old vacuum
released, the receiver changed, the system evacuated again, and the distillation
resumed.

To solve this problem, use an apparatus such as one of those shown in
Figure 1-19. Now all you need to do is rotate the device to collect the individual

56

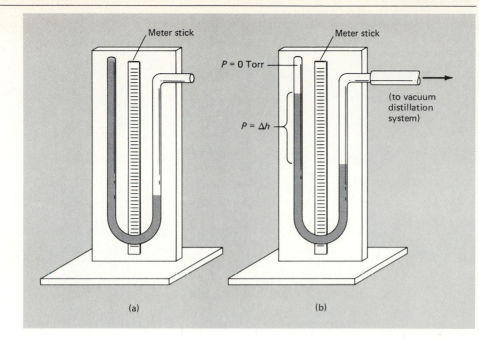

(a)   (b)

FIGURE 1-17
A closed-end manometer: (*a*) the other end open to the atmosphere; (*b*) the other end connected to a vacuum-distillation system.
FIGURE 1-18
A commercial closed-end manometer.

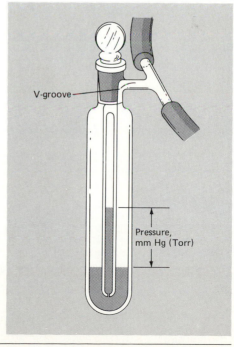

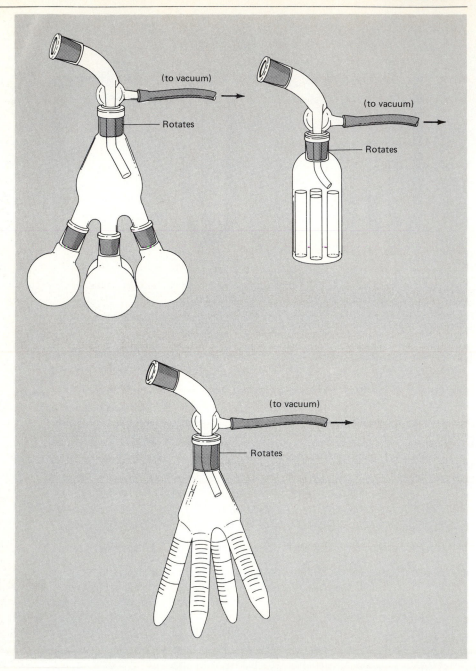

FIGURE 1-19

Three types of rotary fraction collectors for use in vacuum distillations.

fractions. From the figure, you should be able to see why these devices are frequently called "cows."

1.15 VACUUM-DISTILLATION PROCEDURE

Evacuating the apparatus

Step 1: Assemble the apparatus as shown in Figure 1-15. If more than one fraction may be collected, use one of the rotating devices shown in Figure 1-19 if available. Label all the various receiving flasks or vials and, if possible, preweigh them. Carefully check all glassware. Cracks may cause an implosion from the large pressure difference between the inside and the outside. Use only thick-walled pressure tubing, except for the water-cooling lines to the condenser. Also be certain the pressure tubing, clamps, and other key pieces of equipment are in good condition. Lightly grease the ground-glass joints, as in simple and fractional distillation. All joints should fit securely together to prevent leakage. **AS ALWAYS, WEAR YOUR SAFETY GLASSES!**

Step 2: If the liquid to be distilled contains any volatile solvent such as ether, remove it by heating the liquid on a steam bath or a warm-water bath **in the hood**. Place boiling stones in the container to help prevent bumping.

Step 3: Transfer the concentrated liquid to the distilling flask through the overhead section of the Claisen head. Replace the ebulliator tube portion of the apparatus. After the transfer, the flask should be no more than half full.

Step 4: Turn on the water to the condenser.

Step 5: Open all screw clamps at points A, B, and C.

Step 6: Fully turn on the water aspirator.

Step 7: Tighten the screw clamp at point A until the tubing is **almost** closed.

Step 8: Slowly tighten the screw clamp at point C until the tubing is completely closed. Then adjust the screw clamp at point A until a fine stream of bubbles is formed at the ebulliator tip, with the screw clamp at point C completely closed.

Step 9: Record the pressure obtained and readjust the screw clamp at point A if necessary. If the pressure is unsatisfactory, check to see if all the connections are tight. **Do not begin the distillation unless you have a good vacuum of no more than 30 to 50 torr.**

Performing the distillation

Step 10: Place the heat source into position and turn it on. When the liquid begins to boil, lower the rate of heating and adjust it to maintain a distillation rate of approximately 1 drop per second (1–3 mL per minute). Record the temperature and pressure ranges throughout the distillation. If excessive condensation occurs in the Claisen head, insulate it by wrapping in glass wool. The boiling point should remain relatively constant as long as the pressure is constant. A rapid increase in pressure is likely due to the increased use of aspirators in the area. It may also be due to the decomposition of the material being distilled. If so, a dense white fog will generally be produced in the distilling flask. Raising the temperature too rapidly is frequently the cause.

Changing receiving flasks

Step 11: Rotate the collection device of the apparatus whenever the distillation temperature indicates that a new fraction is being obtained.

Step 12: If no rotating collection device is available and a new fraction is being obtained, change the single, preweighed flask by first discontinuing the heating of the distilling flask. If necessary, lower or remove the heating source.

Step 13: **Slowly** open the screw clamp at point C. If this causes much liquid to back up into the ebulliator, slowly open the screw clamp at point A.

Step 14: After the screw clamp at point C is fully open, change the receiving flask.

Step 15: Reclose the clamp at point C and the clamp at point A, if necessary. Wait until the system has regained its former reduced pressure.

Step 16: Place the heating source back into position and continue the distillation.

Step 17: Repeat either step 11 or step 12 through step 16 as often as necessary.

Shutdown procedure

Step 18: Turn off and remove the heat source.

Step 19: **Slowly** open the clamps at points A and C.

Step 20: When the inside pressure equals the outside pressure, disconnect and turn off the water aspirator.

Step 21: Remove the receiving flask and disconnect the remaining apparatus.

1.16 THERMOMETER CALIBRATIONS AND STEM CORRECTIONS

To ensure accuracy, you should calibrate the thermometer used in a distillation. The 0°C point can be checked with a well-stirred mixture of crushed ice and distilled water, while the 100°C point can be checked with boiling distilled water.

TABLE 1-5.

Vapor Pressure of Water in Torr at Various Celsius Temperatures

Temperature	Vapor pressure	Temperature	Vapor pressure
96.0	658	98.6	723
96.2	662	98.8	728
96.4	667	99.0	733
96.6	672	99.2	739
96.8	677	99.4	744
97.0	682	99.6	749
97.2	687	99.8	755
97.4	692	100.0	760
97.6	697	100.2	765
97.8	702	100.4	771
98.0	707	100.6	776
98.2	712	100.8	782
98.4	718	101.0	788

If the pressure is different from 760 torr, the values in Table 1-5 can be used to determine the true boiling point. You can calibrate the thermometer at temperatures other than 0°C and 100°C by using the melting points of the compounds given in Experiment 2. In fact, this calibration will actually be done in Experiment 2.

The temperature on the thermometer may read **lower** than the actual boiling point because of **stem error in the thermometer**. This error occurs because the mercury in the column of the thermometer is cooler than the mercury in the bulb, and therefore has not expanded as much. To correct for this error, we use the equation

$$\text{Stem correction} = 0.000154(t - t')N$$

where

1. The stem correction is the number of degrees which must be added to the observed temperature reading.

2. 0.000154 is the coefficient of expansion for mercury in the thermometer.

3. t is the temperature read.

4. t' is half of the difference between room temperature and the observed temperature.

5. N is the number of degrees on the thermometer stem from the lower exposed level to the temperature read.

For example, assume the room temperature is 21°C, the thermometer reading is 165°C, and the lower exposed level of the thermometer is -10°C. Therefore,

$$t = 165°C$$

$$t' = \tfrac{1}{2}(165° - 21°) = 72°C$$

$$N = 165° - (-10°C) = 175°C$$

Substituting, we have

$$\text{Stem correction} = (0.000154)(165° - 72°)(175°) = 2.5°C$$

Therefore, 2.5°C would have to be added to the thermometer reading at 165°C.

Generally, a stem correction calculation is necessary only when the thermometer reading is over 100°C.

1.17   EXPERIMENTAL

A. Thermometer calibration

Check the 0°C point of your thermometer with a well-stirred mixture of crushed ice and distilled water. Then check the 100°C point by placing 10 mL of distilled water into a vertically clamped 25 × 150 mm test tube. Add a boiling stone to help prevent bumping, and boil gently with the thermometer placed in the vapor. Check the atmospheric pressure, and use Table 1-5 to determine the true boiling point of the water. Record the results in your notebook.

B. Simple distillation of methanol and water

Set up the apparatus in Figure 1-6, filling a 100-mL distilling flask with 50 mL of a 50% methanol and 50% water mixture (volume-volume). Add two to three boiling stones and follow the technique given in Section 1.12, but **do not start** distilling until the apparatus has been approved by your instructor.

Begin the distillation, heating at a rate so that about 1 drop per second (3 mL per minute) of distillate is collected in a graduated cylinder. Record your observations in your notebook, and record the temperature of the thermometer versus the volume of distillate, neatly graphing the data. At 760 torr, pure methanol boils at 64.7°C and pure water at 100.0°C. Also graph what would be the expected results if pure methanol and pure water had each been distilled separately. Stop the distillation when about 1 to 2 mL of liquid remain in the distilling flask.

C. Fractional distillation of methanol and water

Set up the apparatus shown in Figure 1-7, filling a 100-mL distilling flask with 50 mL of the previously used methanol and water mixture. If any of this mixture was lost during the first distillation, add a small amount from the reagent bottle to bring the volume up to 50 mL. If the fractionating column is not already prepared, take the wider Liebig condenser, pack it with about 1 inch of glass wool **(CAUTION: use gloves)**, then finish packing it with glass helices, glass beads, or whatever other packing material is supplied to you, followed by another layer of glass wool. Again add two to three boiling stones and follow the technique given in Section 1.12, but **do not start** until the apparatus has been approved by your instructor.

Begin the distillation, heating at a rate so that about 10 to 20 drops per minute (0.5 to 1 mL per minute) of distillate are collected in a graduated cylinder. Lower the rate if a flood of liquid is produced on the fractionating column. Record your observations in your notebook, and record the temperature of the thermometer versus the volume of distillate, neatly graphing the data. Stop the distillation when about 1 to 2 mL of liquid remain in the distilling flask or when the temperature rise has stopped.

D. Fractional distillation of either 2-propanol and water or 1-propanol and water

Remove the distilling flask and the fractionating column from your apparatus from Part C and allow them to thoroughly drain. Then, **in the hood**, slowly pour about 20 to 30 mL of acetone through your fractionating column to remove any residual water present. Allow the fractionating column to thoroughly drain and then dry. It is important for the next distillation that no liquid of any kind be present on the packing material.

Set up the fractional-distillation apparatus again, this time using 50 mL of either a 90% 2-propanol and 10% water mixture (volume-volume) or a 76% 1-propanol and 24% water mixture (volume-volume). Add boiling stones, have your instructor approve the apparatus, and again distill as in the previous fractional distillation. Record your observations in your notebook, and record the temperature of the thermometer versus the volume of distillate. Neatly graph the data. At 760 torr, pure 2-propanol boils at 82.3°C, and pure 1-propanol boils at 97.3°C. Continue the distillation until either the temperature stops rising or the results of the distillation are obvious to you.

QUESTIONS

1. From your graphs of the boiling point versus the volume of distillate, compare how effectively the mixture of methanol and water was separated with and without a fractionating column.
2. What was the result of the fractional distillation of the mixture of either 2-propanol and water or 1-propanol and water? Is there any way you can explain what happened?
3. If a liquid is found to boil at a constant temperature throughout a distillation, can you safely conclude that it is pure? Give two reasons for your answer.

4. If two miscible liquids are each found to boil at exactly the same temperature, could you safely conclude that they are identical? Why?

5. How would the observed temperature in a distillation be affected if the thermometer was located above the level of the side arm? Explain your answer.

6. How would the observed temperature be affected if a distillation was performed on the top of a mountain? Explain your answer.

7. How is the boiling point of a solution affected by a soluble nonvolatile solute such as sugar or sodium chloride? Explain your answer.

8. Why is a better separation achieved in a fractional distillation with slow heating rather than fast heating?

9. Why is it dangerous to carry out a distillation in a completely closed apparatus?

10. Explain why a packed fractionating column gives a more effective separation than an unpacked one does.

11. Why should water enter a condenser at the lowest point and leave at the highest?

12. What would be wrong with having water run through the outer jacket of a fractionating column?

13. What effect does the liquid dropping back into the distilling flask have on a fractional distillation?

14. An organic liquid begins to decompose at 120°C and has a vapor pressure at this temperature of 53 torr. Specifically, how would you distill this liquid? Why?

15. Two compounds have a boiling-point difference of 15°C. How many theoretical plates are needed to separate these substances by fractional distillation? If a glass helices fractionating column is used, how long must it be to provide this number of theoretical plates?

16. What is an azeotropic mixture? Why can't its components be separated by fractional distillation?

17. What advantage does a fractional distillation have over a simple distillation? What advantage does a vacuum distillation have over a fractional distillation?

18. Explain what could easily happen if you distilled a low-boiling, highly flammable liquid using a flame but no condenser.

19. A compound boils at 125°C at 20 torr. Approximately where would it boil at 760 torr?

20. A compound boils at 260°C at 760 torr. Approximately where would it boil at 25 torr?

21. A compound boils at 235°C at 760 torr. What pressure would be needed in a vacuum system to have this compound boil at 100°C? Could a water aspirator be used to achieve this pressure? Why?

2

MELTING POINTS

But pleasures are like poppies spread—
You seize the flower, its bloom is shed;
Or like the snow falls in the river—
A moment white—then melts forever.

Robert Burns, *Tam o' Shanter*, line 59

2.1 INTRODUCTION

When people read, hear, or think about melting or freezing, they normally associate them with temperature or heat, and the subject generally is either about food or the weather. One goes to the frozen food section of the supermarket or to the freezer section of the refrigerator. One makes a tray of ice cubes for drinks or a frozen dessert for dinner. One talks about it being freezing outside or about "melting" from the heat.

Many facts about melting and freezing are known. Melting is the process of a solid becoming a liquid by the addition of heat, while freezing is the process of a liquid becoming a solid by the removal of heat. The amount of heat added or removed during the phase change, called the **heat of fusion**, usually is in the range of 10 to 80 calories per gram. Usually this amount cannot be added or removed instantly, so melting and freezing normally occur gradually over a period of time. As a result, ice cubes cannot be made instantly when needed. The ice and snow of a blizzard remain with us for awhile, and ice cream can be brought home from the store without completely softening. For some substances, such as water, the process occurs at sharp, precisely known temperatures, whereas for others, such as glass, it takes place over a wide and frequently varying range. The addition of another soluble substance lowers the melting or

65

freezing point, which is why salt is placed on icy steps, sidewalks, and streets, and why salt is added to the ice to make homemade ice cream.

In organic chemistry, melting points are commonly used to help determine the identity of compounds. A melting point is a physical property, and if a substance is believed to be a certain compound, then its melting point must agree with the known melting point of that compound. However, literally thousands of compounds can have the same melting point. Therefore, the melting point should not be the only criterion for identification, and other evidence such as the compound's molecular formula, the way it was obtained, and its infrared and ^{1}H NMR spectra should also be used.

In organic chemistry, melting points are also useful for determining the purity of compounds. Pure organic compounds generally have sharp melting points, such as 132.5–133°C or 156–157°C. But with a soluble impurity, the same compounds melt at lower temperatures and over a wide range, such as 127–131°C or 152–155°C. The impurity which causes this change in melting point may have a melting point of its own either higher or lower than that of the major compound.

2.2 MELTING-POINT THEORY

To understand why a soluble impurity should change a compound's melting point, one must consider what is happening during melting. The solid and liquid phases of a pure compound each have individual vapor pressures, which increase as the temperature is raised. At the melting point, both the solid and liquid phases are in equilibrium, both with each other and with the vapor produced from each. This is depicted in Figure 2-1. Furthermore, at the melting point, the vapor pressure of the solid is equal to the vapor pressure of the liquid.

With a soluble impurity, this equilibrium will still be present and the vapor pressure of the solid will still be equal to the vapor pressure of the liquid. However, since the impurity lowers the vapor pressure of the liquid in solution with it, the equilibrium will attained at a lower temperature, resulting in a lower melting point.

Consider what happens when a solid mixture of 90% compound A and 10% compound B is heated. As the temperature rises, compound A begins to soften and compound B begins to dissolve in A. When the maximum amount of B has been dissolved, melting begins and the first drop of liquid appears. The composition of this liquid will be less than 90% A and more than 10% B. As more heat is applied, more of A melts, the liquid reaches a composition approaching

FIGURE 2-1

The equilibrium of the solid, liquid, and vapor phases at the melting point.

Vapor phase

Solid phase $\rightleftharpoons$ **Liquid phase**

90% A, and the melting point rises due to the increase in vapor pressure from the higher percentage of A in solution. This process continues until the solution finally becomes 90% A and 10% B. When this happens, all liquid is present and the maximum melting point of this mixture is attained.

2.3 MELTING-POINT COMPOSITION DIAGRAMS

Figure 2-2 is a melting-point composition diagram for all the possible combinations of compound C and compound D. Observe that most parts of the graph consist of two curved lines. The lower curve indicates the temperature at which melting is first **actually observed** for a certain mixture, and the upper curve indicates the temperature at which all the sample for this particular mixture has melted. Thus for a mixture of 20% C and 80% D, the first drop of **observed** liquid is formed at temperature F, and the sample has completely melted at temperature G, as indicated in the figure. Observe how the two curves are usually a considerable distance apart, reflecting the wider melting-point range for

FIGURE 2-2

The observed melting-point composition diagram for the binary mixture, C + D.

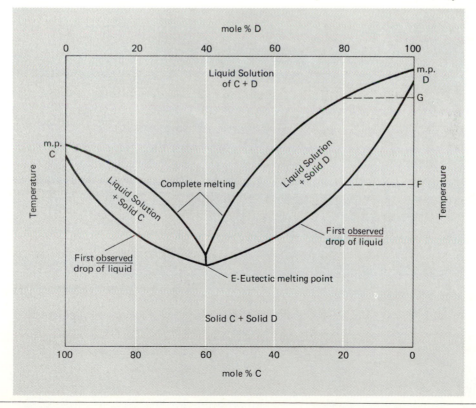

a mixture. Only at one point in between do the curves come close together with a very narrow melting-point range. This point, labeled E, represents the lowest temperature in which some combination of C and D will actually be **observed** to melt, and is called the **eutectic melting point** or the **eutectic temperature**. The composition of this mixture is called the **eutectic composition**.

The **observed** eutectic melting point occurs because there is a limit as to how much of one compound can be dissolved in the other. At this composition, the solution formed when the two begin to melt has the same composition as the original mixture. Therefore, the composition cannot change as the sample continues to melt, and the melting-point range will be very small. A mixture of this composition would thus appear to be a pure compound in terms of the melting-point range.

In the case of C and D, the eutectic melting point occurs at a composition of about 60% C and 40% D. But for mixtures of other compounds, it can occur at almost any other composition. Thus, the melting-point diagram for practically every mixture will be different, in terms of the location of the eutectic melting point and the temperatures and slopes of the curved lines at various

FIGURE 2-3

The theoretical melting-point composition diagram for the binary mixture, C + D.

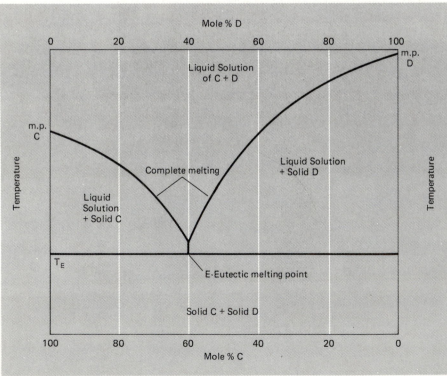

other compositions. Some binary mixtures don't even have eutectic melting points, and others have more than one. The latter case frequently occurs when the two compounds react together during heating to form a third compound.

Figure 2-2 represents the temperatures at which various combinations of compound C and compound D are **actually observed to begin melting and finish melting**. It does **not**, however, represent the theoretical melting-point composition diagram, which is shown in Figure 2-3. Observe in Figure 2-3 how the bottom curved line in Figure 2-2 has been replaced by the horizontal line T_E. All mixtures of C and D thus theoretically begin to melt at temperature T_E, which is the same as the eutectic melting-point temperature. This is typically at a lower temperature than what is indicated in Figure 2-2. The explanation for the difference is that the amount of liquid produced in the early stages of melting is very small and, as a result, is not observable. This is especially true when the percentage of one substance is quite low.

2.4   MIXED MELTING POINTS

A technique which helps prove the identity of an unknown compound is the **mixed melting point**. Suppose you have an unknown compound of melting point 132.5–133°C, which you suspect to be urea. You can determine if it is or is not urea by taking the melting point of a mixture of the unknown and urea. A melting point of 132.5–133°C for the mixture is evidence that the unknown is urea. But a melting point of, say, 127–131°C proves that the unknown is not urea, but may be cinnamic acid, mp 132.5–133°C, or one of thousands of other possible compounds having the same melting point.

The technique is not completely infallible, however. It is possible that two different compounds could both show the same melting point, even after mixing. In addition, sometimes only a modest drop in the melting point occurs, with only a modest increase in the melting-point range. To be more certain, you should take the melting point of the mixture simultaneously with the melting point of the pure unknown compound and the melting point of the pure suspected compound to note any slight deviations. It might even be necessary to vary the composition of the mixture to see if any deviations occur. Again, you should take the melting points simultaneously to better note these possible deviations.

2.5   TECHNIQUES FOR DETERMINING THE MELTING POINT

Melting-point apparatus

Melting points may be taken in several ways. In most beginning organic chemistry courses, a melting-point bath, consisting of a high-boiling, stable liquid in a glass container, is used. The high-boiling liquid can be silicone oil, dibutyl phthalate, paraffin oil, or cottonseed oil. The glass container may be a 150-mL

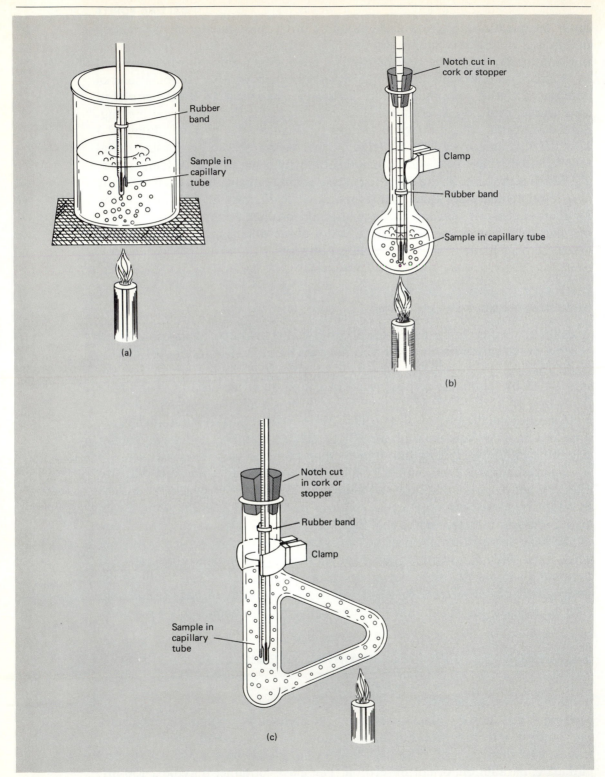

FIGURE 2-4
Various types of melting-point baths: (*a*) 150-mL beaker;
(*b*) Kjeldahl flask; (*c*) Thiele tube.

beaker, a Kjeldahl flask, or a Thiele tube. These various types of melting-point baths are shown in Figure 2-4.

All three types of melting-point baths are usually heated with a microburner or a small flame. To regulate the heating, you can vary the size of the flame and the distance from the bath. If necessary, hold the burner by its base and move it back and forth under the bath.

To maintain a uniform temperature throughout the bath when using a beaker, stir the oil with a glass rod. The Kjeldahl flask, on the other hand, must rely on convection currents. When a Thiele tube is used, only the bottom portion of the side arm is heated. Convection currents carry the heated oil up through the side arm and down the main shaft of the apparatus for uniform heating.

For all these melting-point baths, a capillary tube containing the sample is attached to a thermometer by means of a rubber band. The rubber band is made by simply cutting a small piece off the end of a piece of rubber tubing. The rubber band is then slipped up the thermometer about 3 inches. It must be kept well above the level of the hot oil, or the oil could melt the rubber and break the band. For accurate readings, the sample compound in the capillary tube is kept close to and at the level of the thermometer bulb, which is fully submerged and centered in the oil bath, as shown in Figure 2-5.

No matter which kind of bath is used, the cork or stopper holding the thermometer in place must be cut away on one side to permit the viewing of the

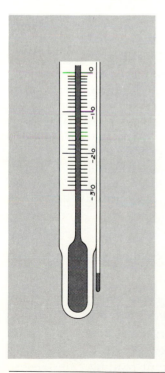

FIGURE 2-5
Closeup view of the melting-point capillary tube placed beside the thermometer, with the sample compound close to the thermometer bulb. Both the sample compound and the thermometer bulb are fully submerged and centered in the oil bath.

mercury level in this region. For a Thiele tube and a Kjeldahl flask, this opening also prevents the heating of a closed system.

As an alternative, a commercial electrical apparatus may be used for taking melting points. Four commonly available units are the Fisher-Johns hot-stage apparatus, the MEL-TEMP apparatus, the electrothermal apparatus, and the Thomas-Hoover Uni-Melt apparatus, which are all shown in Figure 2-6.

The Fisher-Johns melting-point hot stage uses no capillary tubes. Instead, small samples of the solids to be melted are pulverized and placed in different locations between two glass microscope slides. The slides are then placed in a depression on the metal block surface. For better viewing, the samples area is lit, and there is a magnifying glass mounted on the instrument above the hot stage. An electrical dial controls the temperature.

The MEL-TEMP apparatus can accommodate up to three capillary tubes. Heat is transferred to these tubes by means of an electrically heated aluminum block. The samples are illuminated through the lower port and observed through the upper port by means of a magnifying glass.

The electrothermal apparatus is like the MEL-TEMP apparatus in that the capillary tubes receive their heat from an electrically heated block. The tubes are placed in a hole in the block and viewed by means of a magnifying glass.

The Thomas-Hoover Uni-Melt apparatus can accommodate up to seven capillary tubes in a small beaker of electrically heated and stirred silicone oil. A variable transformer controls the rate of heating. A traveling periscope allows simultaneous viewing of the rising mercury column, so one need not look away from the capillary tubes to read the thermometer.

If you are using any of these electrically heated units, remember that a graph of the temperature rise versus time will usually not be linear, especially at higher temperatures. Thus, although the first 50° rise may take only 5 minutes, the next 50° rise may take 10 minutes or longer, depending upon the electrical setting.

Filling capillary tubes

Since capillary tubes are used to take most melting points, it is important to fill these tubes properly with the sample. Only thin-walled tubes, about 100 mm long and 2 mm in diameter, can be used. Any large thickness would delay the transfer of heat to the sample, resulting in the melting-point temperature reading being too high. The steps for filling the capillary tubes are as follows:

1. If the capillary tube is open at both ends, first seal one end by rotating it in the flame of the burner.

2. Place a small sample of the solid on a clean piece of paper or a clean watch glass. Then crush the sample to a powder, using a spatula. The powder will not only enter and travel through the capillary tube easier, but the larger surface-area-to-volume ratio of the powder means that heat can more quickly penetrate the entire sample, resulting in a sharper, more accurate melting-point reading.

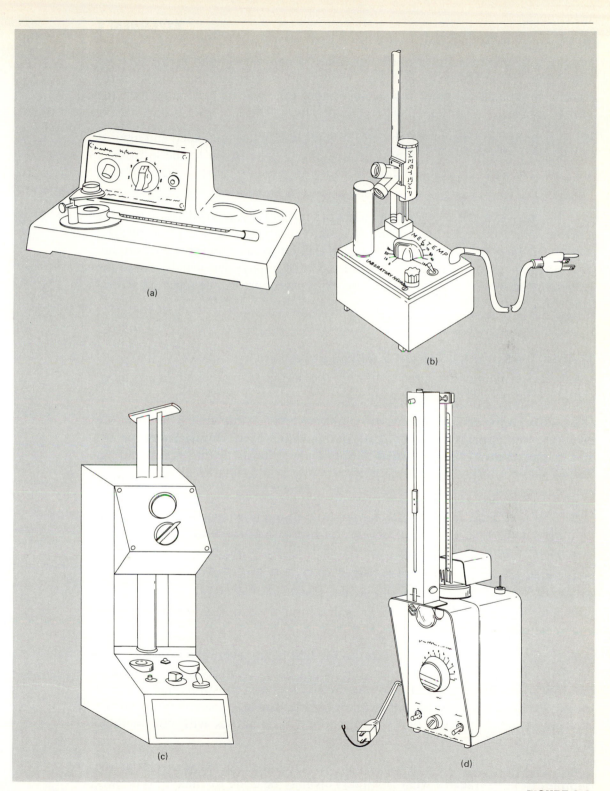

(a)

(b)

(c)

(d)

FIGURE 2-6

Various types of commercial electrical melting-point apparatus: (*a*) the Fisher-Johns hot-stage apparatus; (*b*) the MEL-TEMP apparatus; (*c*) the electrothermal apparatus; (*d*) the Thomas- Hoover Uni-Melt apparatus.

3. Scrape the powder into a mound and push the open end of the capillary tube down into the sample, until you collect a column of solid in the tube no more than one-eighth of an inch long. Too much material in the tube will produce a larger melting-point range, since more time will be needed to melt the entire sample.

4. Turn the capillary tube right side up and **gently** tap the sealed end against the bench top a number of times to move the powder down to the bottom of the tube. Do **not** use too much force because the capillary tube could break and cut you.

5. If gently tapping the capillary tube does not move the powder down to the bottom, place the tube, closed end first, into the top of a long piece of glass tubing, which is held upright on the bench top. Drop the capillary tube through the tubing, allowing it to hit the bench top. The impact will usually cause the powder to move down the capillary tube. This procedure can be repeated several times, if necessary.

Helpful hints for taking the melting point

1. Use a magnifying glass for viewing the sample being melted. Much more detail of the process will be evident.

2. If the approximate melting point of the sample is known, rapidly heat the melting-point apparatus to within 10° of this value. Then **slowly heat** the apparatus **at about 1° per minute** to obtain an accurate melting-point reading. Rapid heating during the melting point will give you results which are too high and have too large a range.

3. To save time, take the melting point of an unknown substance twice: the first time with rapid heating to obtain the approximate melting point, and the second time with slow heating, once the melting point has been approached.

4. **Never** repeat a melting point with the same capillary tube. Many organic compounds do not melt but decompose or react in other ways upon heating to give new substances. Thus, using the same capillary tube again may give you completely different results.

5. Always report **two** temperatures, the first being where the sample begins to melt, and the second being where the sample has completely melted. Remember that softening or a shrinking of the sample does **not** constitute melting. You **must** see liquid. Record **all** observations. For example, "The compound melts sharply at 143.5–144°C," or "The compound decomposes with considerable darkening at 218–223°C."

6. To save time, do a number of samples simultaneously. However, label the samples to keep track of what is what. This may be done by making one, two, or three small file scratches at the upper end of the capillary tubes. Cutting the tubes to different lengths may also be done. Be sure to record in your notebook exactly how each sample was labeled. This will prevent a possible mixup later. For easier identification, place each sample at a precise location near the thermometer bulb.

7. If doing a mixed melting-point determination, take the melting point of all the samples simultaneously for better accuracy. Again be sure to label each capillary tube.

8. Be sure the sample is completely dry. A wet sample will give you worthless results. In addition, the sample will be harder to place in the capillary tube because it will stick to the glass.

9. If doing a large number of melting points and the approximate values for the samples are known, do the lower-melting samples first, then progress to the higher-melting samples. In this way, you won't have to wait a long time for the apparatus to cool to a desired lower temperature.

10. When finished, place your apparatus upright in a safe but readily accessible section of your locker. It will be needed many times throughout the course.

2.6 CALIBRATING THE THERMOMETER

To obtain exact results in any melting-point determination, you must check the accuracy of the thermometer by taking the melting point of known standards. Temperature deviations in thermometers do exist, and it cannot be assumed that a small deviation at low temperatures will remain small at higher temperatures. This procedure of checking the accuracy is known as **calibrating the thermometer**.

At least four to six known standards of greatly differing melting points should be used for any thermometer calibration. The literature melting points of these standards and many other compounds may be found in a number of chemistry handbooks or possibly even your lecture textbook. Among the most common chemistry handbooks are the following:

1. Handbook of Chemistry and Physics, Chemical Rubber Company, Boca Raton, Florida. All compounds are listed alphabetically under the parent compound's name. Thus, ethyl acetate would be found alphabetically as the ethyl ester under its parent compound, acetic acid. This system frequently presents problems for beginning organic chemistry students, which is why they often prefer to use the other chemistry handbooks.

2. Lange's Handbook of Chemistry, McGraw-Hill, New York.

3. The Merck Index, Merck Pharmaceutical Company, Rahway, New Jersey.

4. Aldrich Catalog Handbook of Fine Chemicals, Aldrich Chemical Company, Milwaukee, Wisconsin.

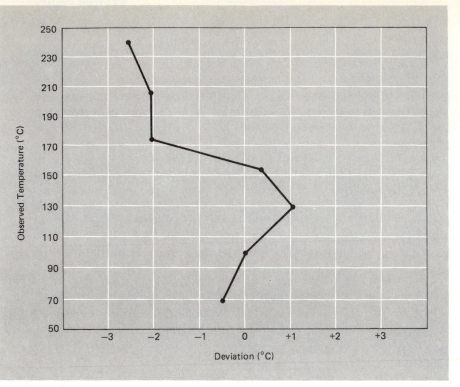

FIGURE 2-7
Thermometer calibration graph.

5. Dictionary of Organic Compounds, Oxford University Press, New York.

After the melting points of the standards have been taken and compared with the literature values, the deviations are graphed as in Figure 2-7. Once the thermometer is calibrated, give it special care, to avoid having to repeat the calibration of a replacement.

2.7 EXPERIMENTAL

A. Thermometer calibration

Look up in a chemistry handbook or other reference book the literature melting point of the following compounds; also record the name of the book and the page number. Then using these values as a guide, determine the actual melting point of the compounds on your 260° or higher-temperature thermometer, using the techniques given in the previous section. To save time, do the compounds in the order given and do two melting points together. From your data, graph a thermometer calibration curve similar to the one in Figure 2-7.

76

1. *p*-Dichlorobenzene (1,4-dichlorobenzene)

2. Naphthalene

3. Acetanilide

4. Salicylic acid (2-hydroxybenzoic acid)

5. Oxalic acid

6. *p*-Nitrobenzoic acid (4-nitrobenzoic acid)

B. Melting points of mixtures

Determine the melting point of the following substances, being sure to include both temperatures comprising the melting-point range. From your data, graph a melting-point composition diagram for urea and benzamide, similar to the one in Figure 2-2. Not enough compositions have been done for a complete diagram, but try to estimate what composition of urea and benzamide forms a eutectic mixture and what is the eutectic melting point of this mixture.

1. Pure urea

2. Pure benzamide

3. 75% urea, 25% benzamide (mole-mole percentage)

4. 50% urea, 50% benzamide (mole-mole percentage)

5. 25% urea, 75% benzamide (mole-mole percentage)

C. Melting point unknown

Determine the melting point of an unknown, which will be given to you by your instructor. The unknown may be any of the compounds below. To confirm its identity, determine simultaneously the melting point of the unknown, an actual sample of the suspected compound, and a mixture of the two. Record your results and your conclusion in your notebook.

Compound	MP (°C)	Compound	MP (°C)
Maleic anhydride	54–56	Adipic acid	152–153
Stearic acid	69–70	Benzanilide	160–161
Acetamide	81–82	*d*-Tartaric acid	172–173
1-Naphthol	95–96	4-Methylbenzoic acid	179–180
o-Toluic acid	105–106	Succinic acid	188–189
Benzoic acid	121–122	3,5-Dinitrobenzoic acid	204–205
Cinnamic acid	132–133	Anthracene	215–216
Salicylamide	140–141	4-Nitrobenzoic acid	241–242

QUESTIONS

1. In the determination of a melting point, why is it necessary to
   (a) Use a powdered rather than a crystalline sample? (Give two reasons.)
   (b) Keep the rubber band well above the level of the heating oil?
   (c) Position the capillary tube next to the thermometer bulb?
   (d) Use a new capillary tube with each new determination?
   (e) Place the sample for pulverizing on a clean surface?
   (f) Use thin-walled capillary tubes?
   (g) Use a calibrated thermometer?
   (h) Cut a notch in any cork or stopper used to hold the thermometer in place in a Thiele tube or a Kjeldahl flask? (Give two reasons.)
   (i) Use only a small amount of sample in the capillary tube?
   (j) Use a completely dry sample?
   (k) Not heat the bath too rapidly near the melting point?
2. A certain substance, melting at 188–189°C, is suspected to be 4-aminobenzoic acid. How could you be reasonably certain that the substance actually is 4-aminobenzoic acid and not another compound with the same melting point?
3. Strictly speaking, the glass of a melting-point capillary tube may be considered to be an impurity. Why then doesn't the glass lower the melting point of a compound and widen the melting-point range?
4. Using the literature melting-point values of naphthalene and oxalic acid, predict the approximate melting points of the following mixtures:
   (a) 90% naphthalene and 10% oxalic acid
   (b) 10% naphthalene and 90% oxalic acid
5. Even though a sample may be pure, why will it still have a melting-point range?
6. Which has the lower freezing point: ice or ice cream? Why?
7. Why does it take time for ice water to freeze or for ice to melt?
8. Why are a few ice cubes capable of so effectively cooling a warm glass of soda?

E X P E R I M E N T

3

RECRYSTALLIZATION

Through the hushed air the whitening shower descends.
At first thin-wavering; till at last the flakes
Fall broad and wide and fast. . . .

James Thomson, *The Seasons: Winter*

3.1   INTRODUCTION

Most products from organic reactions are initially not pure, but contain side products, unconsumed reactants, and other impurities which must be removed. If the product is a solid and the amounts of the impurities are not very large, the product can frequently be purified by a process called **recrystallization**, where a hot solution containing the product is slowly cooled, and crystals of the purified product are **slowly** and **selectively precipitated**. The procedure works because the impurities either remain dissolved in the solvent or are removed before cooling by filtration or adsorption on Norit or decolorizing carbon. Suction filtration then allows the crystals of the purified product to be collected.

The general steps for any recrystallization are as follows:

1.   Select a proper solvent.

2.   Heat the solvent to its boiling point.

3.   Dissolve the solid compound in the **minimum amount** of boiling solvent.

4.   Add decolorizing carbon if necessary.

5.   Filter the hot mixture through a heated, wide-stem or stemless funnel to remove the decolorizing carbon and any insoluble impurities. (This step may be omitted if not applicable.)

79

6. Allow the hot solution to **slowly** cool to room temperature or lower.

7. If crystals do not appear, scratch the solution against the flask with a glass rod, cool the solution further, or add a seed crystal of saved product.

8. Collect the crystals by suction filtration, using a Büchner or Hirsch funnel.

9. Rinse the crystals with a **small amount** of **cold** solvent.

10. Dry the crystals.

11. If worthwhile, obtain a second crop of crystals.

These steps will now be considered in detail.

3.2 SELECTING A PROPER SOLVENT

The selection of a proper solvent is vitally important in any recrystallization. A desirable solvent is one in which the product will be very soluble at high temperatures, but only slightly or sparingly soluble at room temperature or lower.

This large difference in solubility is crucial for success. If the product is very soluble in both the hot and cold solvents, too much of it will be lost to the solvent, resulting in a poor yield of purified material. And if the product is only somewhat soluble in the hot solvent, then a large quantity of the solvent will be required and again the amount of product lost to the solvent will be high. Figure 3-1 shows a graph of solubility versus temperature for a good solvent (line A) and two poor solvents (lines B and C). Observe the steep solubility-versus-temperature slope for the good solvent, but the moderately increasing slopes for the poor solvents.

The solvents commonly used for a recrystallization are given in Table 3-1. Note that this table now excludes carcinogenic or highly toxic solvents such as benzene, dioxane, chloroform, and carbon tetrachloride. High-boiling solvents such as dimethyl sulfoxide and N,N-dimethylformamide are also excluded because of the added difficulty in drying the purified crystals.

Although ethyl ether, methylene chloride, and petroleum ether are included in the table, they usually are not as preferred. Their low boiling points generally mean a lower temperature difference between the "hot" and cold solvents, resulting in the solute usually not having as large a difference in solubility. In addition, their lower boiling points mean a larger solvent loss to vaporization in the decolorizing and filtering steps.

The choice of the best possible solvent is often a matter of trial and error. An experienced chemist, though, is usually able to select one after only a few trials. If the product being isolated is not a new one, the chemical literature most likely will report or recommend the use of certain solvents. For new compounds, the general rule "like dissolves like" is valuable. Polar solutes would thus be more likely to dissolve in polar solvents, such as water and 95% ethanol,

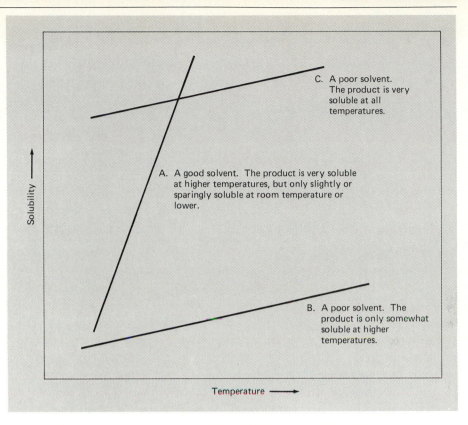

C. A poor solvent. The product is very soluble at all temperatures.

A. A good solvent. The product is very soluble at higher temperatures, but only slightly or sparingly soluble at room temperature or lower.

B. A poor solvent. The product is only somewhat soluble at higher temperatures.

Solubility

Temperature ⟶

FIGURE 3-1
Graph of solubility versus temperature for a good recrystallizing solvent and two poor recrystallizing solvents.

and less likely to dissolve in nonpolar solvents, such as cyclohexane and ligroin. For nonpolar solutes, the opposite solubility is true.

Usually a solute is polar when it is either ionic or when a large portion of the molecule is a functional group, such as —OH, —NH$_2$, —COOH, and —NHCOCH$_3$. Solubility in water and alcohol usually results from the formation of hydrogen bonds between these functional groups and the solvent. For example:

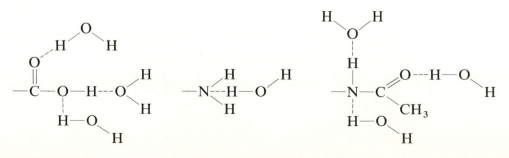

TABLE 3-1.
Common Recrystallization Solvents

Solvent	Formula	Boiling point (°C)	Freezing point (°C)	Solubility in water	Solvent polarity	Fire hazard	Inhalation toxicity
Water	H_2O	100	0	∞	Very high	None	None
95% Ethanol	CH_3CH_2OH	78	−116	∞	High	Medium	Very low
Methanol	CH_3OH	65	−98	∞	Very high	Medium	Medium
Acetone	$(CH_3)_2CO$	56	−95	∞	Intermediate	High	Medium
Tetrahydrofuran	C_4H_8O	65	−65	∞	Intermediate	High	Medium
Ethyl ether	$(C_2H_5)_2O$	35	−116	Slightly	Intermediate	Very high	Low
Ethyl acetate	$CH_3CO_2CH_2CH_3$	77	−84	Slightly	Intermediate	Medium	Medium
Methylene chloride	CH_2Cl_2	41	−95	Insoluble	Intermediate	Low	Medium
Toluene	$C_6H_5CH_3$	111	−95	Insoluble	Low	Medium	Medium
Hexane	C_6H_{14}	69	−94	Insoluble	Nonpolar	High	Low
Cyclohexane	C_6H_{12}	81	7	Insoluble	Nonpolar	High	Low
Petroleum ether	Mixture of hydrocarbons	40–60	—	Insoluble	Nonpolar	Very high	Low
Ligroin	Mixture of hydrocarbons	60–80	—	Insoluble	Nonpolar	High	Low

As the proportion of one of these functional groups in a molecule decreases and the hydrocarbon portion increases, the solute becomes less soluble in polar solvents and more soluble in less polar and nonpolar solvents. In water, for example, acetamide is soluble, but caproamide is only slightly soluble. But in toluene, acetamide is only slightly soluble, whereas caproamide is soluble.

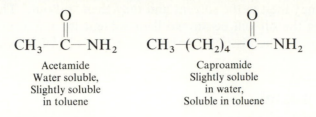

Acetamide	Caproamide
Water soluble,	Slightly soluble
Slightly soluble	in water,
in toluene	Soluble in toluene

To discover the best solvent for a new compound, place approximately 20 mg (a small spatula-tip full) of the compound in a 10×75 mm test tube. Add about 0.5 mL of the possible solvent at room temperature with stirring and note the solubility. If most or all of the compound dissolves, then the solvent is unsatisfactory, since too much of the compound is being lost to the solvent. If only a small part of the compound, however, appears to dissolve, then the solvent may be satisfactory. Place the test tube in a hot-water or steam bath and observe how much of the compound dissolves in the hot or boiling solvent. If the amount is small or moderate, the solvent is again unsatisfactory. But if most or all of the compound dissolves, a good recrystallization solvent has likely been found. If necessary, add a small amount of extra solvent to attain a complete solution, then allow the test tube to slowly cool, and compare the quantity and size of the resulting crystals with the original material. If necessary, induce crystallization by adding a crystal of the original material to "seed" the cooled solution. The inside of the test tube near the bottom may also be carefully rubbed with a glass stirring rod.

Sometimes the product being isolated will be too soluble in some solvents and not soluble enough in others. In this case, a mixture of solvents such as the ones in Table 3-2 can be useful. For example, suppose that the product is very soluble in ethanol but not very soluble in water. The product would be dissolved in the minimum amount of hot ethanol, and hot water would then be added **slowly** until crystallization or turbidity occurred. Hot ethanol would then be

TABLE 3-2.
Solvent Pairs

Methanol-water	Ether-hexane
Ethanol-water	Ether-petroleum ether
Acetone-water	Methylene chloride–methanol
Ether-methanol	Toluene-ligroin
Ether-acetone	

added **slowly** until a complete solution was again attained, and the solution would be allowed to slowly cool.

When selecting a solvent, be careful to avoid any liquid which could react chemically with the compound being recrystallized. Also avoid any liquid whose boiling point is higher than the melting point of the compound being recrystallized. Otherwise, the solid may melt in the solvent and may then "oil out." This is undesirable because often the oil will be an excellent solvent for impurities, and when the oil finally freezes, the dissolved impurities in the oil will be embedded in the crystals.

3.3 DISSOLVING THE SOLID

Once a satisfactory recrystallizing solvent has been selected, it is heated to its boiling point in an Erlenmeyer flask and is then **slowly added in small portions with stirring** to a **separate** Erlenmeyer flask containing the compound to be recrystallized. **Be sure to add only the minimum amount of the boiling solvent needed to dissolve the compound.** This ensures that the solution, when cooled, will return the maximum possible yield of product as crystals. Remember that the compound being recrystallized may dissolve only slowly in the boiling solvent, so each extra portion of solvent should not be added too quickly. It is also possible that the compound being recrystallized may contain a few insoluble impurities, and too much solvent may needlessly be added in a futile attempt to dissolve them.

Don't ever perform the reverse process of adding the compound to be recrystallized to the boiling solvent. Otherwise, it may then be impossible to tell when the solution is saturated.

3.4 DECOLORIZATION

The use of Norit or decolorizing carbon is sometimes successful in removing colored impurities from the product, but if no substantial improvement occurs, the process is not repeated. The fine carbon particles have a large, active surface, which readily attracts and adsorbs many of the resinous, polymeric, and other reactive impurities often found in organic reaction mixtures. The adsorption is not as effective at higher temperatures, but Norit is usually used in the hot liquid in order to keep the product from being recrystallized in solution.

The amount of Norit added to the hot liquid should be approximately 2–3 percent of the weight of the product being recrystallized. Any excess will only adsorb part of the product, resulting in a lower yield. Since the addition of the Norit may cause the hot mixture to boil over, it is important to cool the mixture somewhat beforehand, and then reheat to boiling after the addition. The Norit will also cause the mixture to bump, possibly throwing the hot contents dangerously out of the flask. This can be minimized by continuous stirring.

3.5 FILTERING THE HOT MIXTURE

If any insoluble impurities or decolorizing carbon is present, the hot mixture is next gravity filtered through fluted filter paper in a stemless or wide-stem, heated funnel as shown in Figure 3-2. A stemless or wide-stem funnel avoids the possible problem of clogging inside the stem from premature crystallization. A heated funnel prevents the cooling of the hot mixture during filtration, which may cause premature crystallization in the filter. You can heat the funnel by placing it at the top of an Erlenmeyer flask containing a small amount of boiling solvent. During the filtration into the flask, the hot vapors will rise up around the funnel and heat it.

A fluted filter is ordinary filter paper specially folded to provide the maximum possible surface area. This allows the filtration to occur as rapidly as possible. Fluted filter paper can be purchased commercially, or you can easily make it. One way of doing it is shown in Figure 3-3.

It is best to pour the hot mixture through the fluted filter gradually to prevent too large a buildup of solvent in the paper. The mixtures in both flasks should

FIGURE 3-2
Gravity filtration setup for recrystallization.

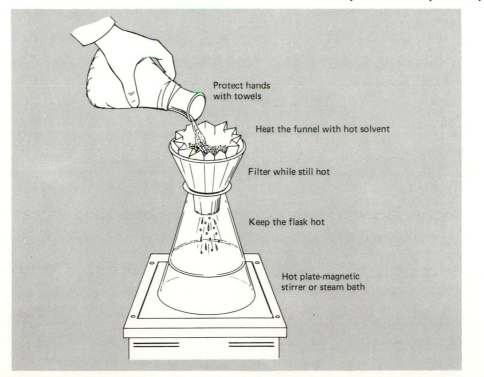

Protect hands with towels

Heat the funnel with hot solvent

Filter while still hot

Keep the flask hot

Hot plate-magnetic stirrer or steam bath

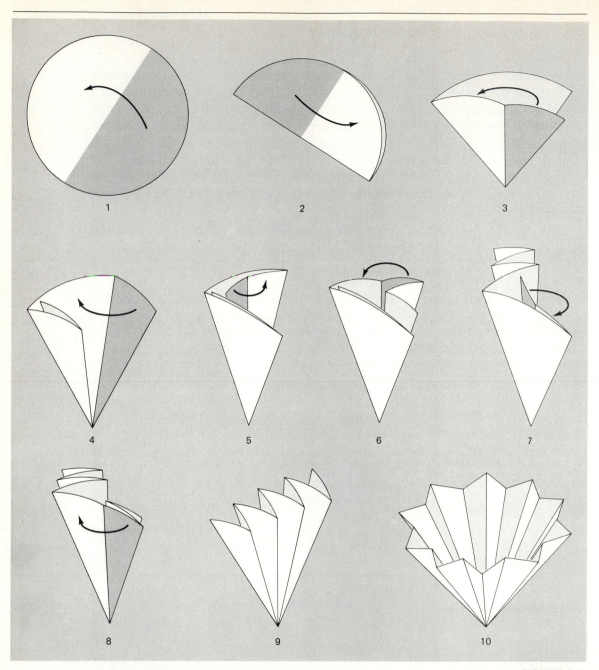

FIGURE 3-3
One way to prepare fluted filter paper.

also be kept at the boiling point to prevent premature crystal formation. If too much solvent boils away, it may be necessary to add a small quantity of hot solvent to maintain the volume.

Despite the heated funnel, crystals may still sometimes form in the filter during the filtration. If this occurs, a small amount of boiling solvent may be added to redissolve the crystals. This extra amount should be boiled off once the filtration is completed.

Remember that this hot filtration process is only necessary if the hot mixture contains insoluble impurities or decolorizing carbon. A clear and colorless hot solution with no visible solid material is ready for the next step.

3.6 COOLING AND RECRYSTALLIZATION

The Erlenmeyer flask containing the product is now stoppered and set aside, and the hot filtrate is allowed to **slowly cool without agitation**. Rapid cooling and agitation are generally undesirable because the crystal formation will occur too rapidly, resulting in impurities becoming trapped within the crystal lattice. Once room temperature is reached, the flask can be further cooled in an ice-water bath. The lower solubility of the product at this lower temperature will produce a larger yield.

If the cooled solute does not crystallize, several techniques can be employed. Further cooling in the ice-water bath may be successful. One can also scratch the solution against the inside of the flask with a glass rod. If a small amount of the original solid material was saved, a crystal of this material may be added to "seed" the cooled solution.

Once crystal formation has begun, it is important not to collect the product too early. Otherwise some material may be lost which would have separated from solution with further standing. Patience is a virtue.

3.7 COLLECTING THE CRYSTALS

When recrystallization is complete, the product is collected by suction filtration on a Büchner or Hirsch funnel. Suction filtration is usually preferred here over gravity filtration, because the difference in pressure increases the rate of liquid flow through the filter. This not only saves time, but reduces the evaporation of the solvent in the funnel, which could redeposit impurities back on the product. However, suction filtration is usually not used for filtering hot solutions. The reduced pressure in the collecting flask may frequently cause the hot filtrate to boil and the material in solution to be unwantedly deposited everywhere. In addition, clogging of the small holes in the funnel can occur, as well as the bumping and frothing of the filtrate.

The apparatus for a suction filtration is shown in Figure 3-4, and consists of the funnel in a filter flask connected to a safety trap, which in turn is connected to a water aspirator or a "house" vacuum line. Only **heavy-walled** rubber tubing can be used for the apparatus, because thin-walled tubing will collapse under vacuum from the outside pressure of the atmosphere.

If there is no "house" vacuum line, the vacuum in the apparatus is produced by the water aspirator, which consists of a side arm attached to a pipe of changing inner diameter. As the water runs through the pipe, the change in diameter changes the speed of the passing water, which, by Bernoulli's principle, creates a partial vacuum in the side arm.

A **clamped** safety trap should always be used with an aspirator. If the water pressure drops or the water is turned off while the filter flask is under vacuum, the pressure in the filter flask may be less than the pressure in the aspirator. This would cause water to flow from the aspirator into the filter flask and thus contaminate the filtrate. The safety trap stops this reverse flow. The clamp or stopcock on the trap also allows the vacuum to be quickly and easily released. **You should always do this before turning off the water, to**

FIGURE 3-4
Suction filtration apparatus.

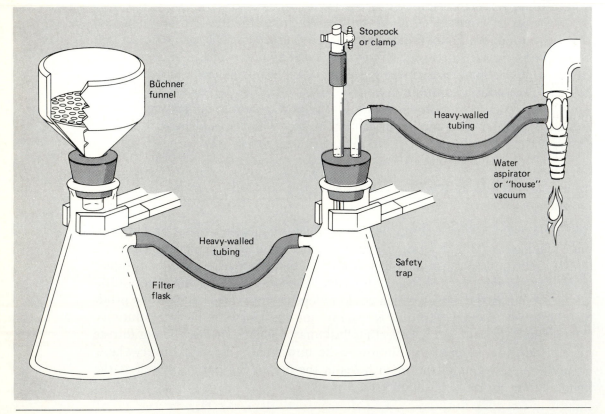

avoid the backflow of water. The released vacuum also makes it easier to disconnect the filter flask from the trap or the funnel.

The filter paper for the Büchner or Hirsch funnel should be **just** large enough to completely cover the holes while lying flat on the perforated surface. If necessary it should be trimmed so that it does not climb the sides of the funnel. Otherwise it may be difficult to create a vacuum in the filter flask and material may flow over this loose edge and not be filtered.

Suction is next applied, and a small quantity of the solvent from which the crystals will be isolated is poured over the paper. After this liquid is filtered through, the paper will cling tightly to the bottom of the funnel so that no small particles of the solid can slip under the paper and reach the filtrate below.

The crystals can now be collected on the filter paper. Frequently the crystals will lie on the bottom of the flask, and it is necessary to swirl the liquid. This will temporarily suspend these crystals in the liquid to aid in their transfer while pouring. Any crystals remaining in the flask after all the solvent has been poured may be collected by rinsing with the filtrate collected in the filter flask. Before removing the funnel from the filter flask, remember to release the vacuum by opening the clamp or stopcock on the trap.

After all the crystals are collected and no more liquid is visible in the funnel, the crystals should be carefully pressed with a clean cork under continued suction. This will help remove some additional filtrate adhering to the crystals.

3.8 WASHING THE CRYSTALS

The isolated crystals in the funnel are now washed with several **small quantities** of **fresh cold** solvent. The purpose of this washing is to remove any impurities in the solvent which are on the surfaces of the crystals. These impurities would remain on the crystals after the solvent evaporated if they weren't rinsed away.

To wash the crystals, open the stopcock or clamp on the trap to release the vacuum. Then break up the crystalline cake with a spatula, **being careful not to disturb or break the filter paper**. Next add enough fresh cold solvent to **just** cover the crystals and **carefully** stir the mixture with a spatula to ensure complete rinsing. Then reclose the stopcock or clamp on the trap, remove as much solvent as possible by suction filtration, remove any additional adhering solvent by pressing with a clean cork, and repeat the washing process at least one more time.

3.9 DRYING THE CRYSTALS

After the crystals have been washed, they must now be thoroughly dried. They should be left in the funnel until the solvent has stopped dripping; they can then be spread out on a large watch glass and left to air dry. Any large chunks which might entrap the solvent must be broken up, and the final product should either be crystalline or powdered to provide as much surface area as possible. The crystals may be considered dry when there is no further loss of weight and they do not stick to themselves or to the glass.

If necessary, you can accelerate the drying process by placing the watch glass in a drying oven. However, a number of problems are possible here.

1. The crystals may melt unless the temperature of the oven is at least 10–20° below the melting point of the crystals. Remember that the melting point of the crystals is lowered by the presence of the solvent.

2. Crystals that sublime easily should not be dried in an oven, since they could disappear by vaporization.

3. Some compounds decompose upon exposure to heat and should therefore not be oven dried.

4. If many students' samples are being dried in the same oven, the crystals might be lost due to an accident, a mixup in the samples, and so on. Always be sure to clearly label your product.

A vacuum desiccator containing a drying agent may also be used to accelerate the drying process. Although there is no danger from heat here, there are other possible problems.

1. Crystals that sublime easily should not be dried in a vacuum desiccator, since the reduced pressure makes vaporization more likely.

2. The vacuum desiccator could possibly implode. It should not be used unless it is placed in a protective cage.

3. The release of the vacuum could blow your product all over the desiccator. Instead of a watch glass, use an Erlenmeyer flask and release the vacuum slowly.

3.10 SECOND CROPS

The filtrate in the filter flask contains additional product which frequently can be worthwhile to collect. Do this by boiling off most of the solvent and allowing the concentrated filtrate to cool slowly. The crystals obtained, however, are usually not as pure as those from the first crop.

3.11 LIMITATIONS OF RECRYSTALLIZATION

As valuable and useful as recrystallization can be for effectively purifying many solid compounds, one must realize that the method cannot be indiscriminately used for all solid mixtures. In general, a recrystallization is successful when there are **only small** amounts of impurities present. With larger amounts of impurities, the method can succeed, but frequently more than one recrystallization will then be necessary. This is unfortunate because of the extra time spent and the lower and lower yield of product in each successive operation. Remember that each time part of the product is being lost to the recrystallizing solvent. In fact, if a mixture contained approximately equal amounts of a compound and an impurity with nearly the same solubility behavior, then the two could not be separated by recrystallization. In this case, some other purification method, such as column chromatography, would have to be utilized.

3.12 EXPERIMENTAL

A. Solubility tests

Determine the solubility of benzoic acid, sodium benzoate, resorcinol, anthracene, and stearic acid in each of the following solvents: water, 95% ethanol, and ligroin. The structure of each of these solutes is given below.

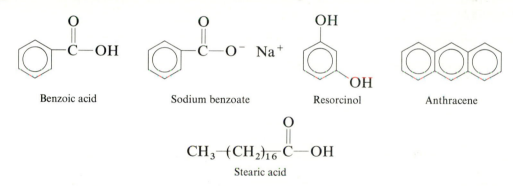

| Benzoic acid | Sodium benzoate | Resorcinol | Anthracene |

Stearic acid

To determine the solubility, place about 20 mg (a small spatula-tip full) of each finely crushed solute into a **separate, labeled** 10 × 75 mm test tube. Unless some test tubes are reused, a total of 15 will be required. Add 0.5 mL of water, 95% ethanol, or ligroin to the appropriate test tubes and stir with a glass rod, being sure the rod is clean in each case. Neatly record your observations in table form in your notebook, using the following definitions:

Soluble: Most or all of the solute dissolved.

Slightly soluble: A small portion of the solute dissolved.

Insoluble: None of the solute appeared to dissolve.

If the solute is insoluble or only slightly soluble in the solvent, heat the test tube in a steam bath or a hot-water bath (heated by a hot plate and **not** by a Bunsen burner; remember the fire hazard, especially with ligroin). Neatly record in table form in your notebook the solubility of the solute in the hot or boiling solvent, using the above definitions. If any of the solutes now dissolve, allow the hot solutions to slowly cool to room temperature and compare the quantity, size, and shape of the resulting crystals with the original material, neatly recording the results in your notebook. From your results, indicate which solvent, if any, you would consider the best one to use for recrystallizing each solute. Also indicate the reason for your decision. What possible alternative solvents or solvent combinations might you try for recrystallizing each solute? In terms of polarity, indicate why these alternatives might likely be successful.

B. Recrystallization of impure benzoic acid

Place 3.0 grams of impure benzoic acid[1] in a clean 125-mL Erlenmeyer flask and gradually add boiling hot water (heated on a hot plate) with stirring until no more benzoic acid appears to dissolve. Remember that any sawdust in your sample will not dissolve, so do **not** needlessly add excess water. No more than about 60 mL should be necessary.

Pure benzoic acid is colorless, so if your mixture has color, cool it slightly and add about 0.1 gram of decolorizing carbon. **(DO NOT ADD DECOLORIZING CARBON TO A BOILING LIQUID.)** Reheat and continuously stir the hot liquid to prevent bumping.

Filter the hot mixture through a fluted filter paper, using the technique given in Section 3.5. If the filtered solution still has color, repeat the treatment with decolorizing carbon. Stopper and set aside the Erlenmeyer flask containing the hot, colorless filtrate, and allow the hot filtrate to slowly cool to room temperature, undisturbed. When no more crystals appear to form, further cool the flask in an ice-water bath for at least 15 minutes.

Collect the crystals by suction filtration, using the procedure given in Section 3.7, and wash them with two small portions of **cold** water, using the procedure given in Section 3.8. Allow the crystals to thoroughly air dry on a watch glass and, in the meantime, do the decolorization of brown sugar portion of the experiment (Section 3.12C).

Determine the yield and the percentage recovery of the dry crystals, showing the calculation setup in your notebook. Also determine the product's melting point and literature melting point, and record the results in your notebook. When finished, submit the product to your instructor in a clean vial or bottle, clearly labeled with the following information:

1. Your name

2. The experiment number

3. The name of the product and its structure

---

[1] A mixture of 100 grams of benzoic acid, 10 grams of sodium chloride, and perhaps a variable amount of sawdust and 0.2 gram of a dye such as Congo red.

4. The melting point or boiling point actually observed

5. The actual yield in grams and the tare of the container

C. Decolorization of brown sugar

Brown sugar is partially refined sucrose, $C_{12}H_{22}O_{11}$, and can be decolorized easily with decolorizing carbon.

Dissolve 10 grams of **dark** brown sugar in 65 mL of water in a 125-mL Erlenmeyer flask. Save 5 mL of the solution for comparison purposes; then pour 15 mL of the remaining solution into **each** of four **labeled** 50-mL Erlenmeyer flasks. To two of the flasks, add 0.3 gram each of decolorizing carbon. To the remaining two, add 0.1 gram each of decolorizing carbon. Heat one of the flasks with 0.3 gram of decolorizing carbon and one of the flasks with 0.1 gram of de-colorizing carbon to near the boiling point for two minutes with stirring. Then filter each hot solution through fluted filter paper into **separate, labeled** 50-mL Erlenmeyer flasks. Also stir the contents of the two unheated flasks for the same amount of time as the two heated ones; then filter the contents through fluted filter paper into **separate, labeled** 50-mL Erlenmeyer flasks. Compare the appearance of the five solutions, and record the results in your notebook. How have the amount of decolorizing carbon and the heating affected the color?

Sugars generally are quite difficult to crystallize. Therefore, this portion of the experiment will stop here.

QUESTIONS

1. During a recrystallization, a yellow solution of a compound was treated with charcoal and the mixture was gravity filtered through fluted paper. When cooled, the crystals obtained were gray, although the compound was known to be colorless. Explain what could account for the gray color of the crystals and what you would do to obtain a colorless product.
2. Explain why washing an impure solid with a cold solvent is not as good as recrystallization for removing all the soluble impurities.
3. What criteria make a solvent satisfactory for a recrystallization?
4. Why are benzene and dimethyl sulfoxide generally not advisable to be used for a recrystallizing solvent?
5. Why is the final product from a recrystallization isolated by suction filtration rather than gravity filtration?
6. If a little decolorizing carbon does a good job in removing the impurities during a recrystallization, why shouldn't you use a lot to do an even better job?
7. Why should decolorizing carbon not be added to a solvent that is at or near its boiling point?
8. Why is it important to use a safety trap in a vacuum filtration?

9. Why is the quality of product obtained from second crops generally not as good as that from first crops?

10. Why is a fluted filter used for hot filtrations?

11. What can cause premature crystal formation, and how can the problem be solved after it has occurred?

12. In a recrystallization, why does one add the solvent to the solid rather than the solid to the solvent?

13. How can the purity of a recrystallized solid be assessed?

14. Why isn't suction filtration as preferred as gravity filtration for hot solutions?

15. What is wrong with using a narrow, long-stem funnel in the filtration of a hot solution in a recrystallization?

16. Why is it advisable not to rapidly cool a hot solution during a recrystallization?

17. Explain why the collected crystals in a recrystallization are washed with cold solvent rather than warm solvent.

EXPERIMENT 4

EXTRACTION OF ASPIRIN, β-NAPHTHOL, AND MOTH CRYSTALS

DRYING AGENTS

He had been eight years upon a project for extracting sunbeams out of cucumbers, which were to be put in vials hermetically sealed, and let out to warm the air in raw inclement summers.

Jonathan Swift, *Voyage to Laputa,* Chapter 5

4.1 INTRODUCTION

Millions of people daily prepare a pot of coffee or brew a cup of tea. In so doing, they are performing a process called **extraction**, where a solvent such as water selectively removes certain ingredients from a natural product. This process provides us with many flavorings such as vanilla, almond and orange extracts, and spearmint and peppermint oils. Certain dyes, drugs, and perfumes are also obtained from natural sources by this method.

But extraction does not have to be limited to natural products. In chemistry, it can also mean the **selective transfer** of any solute or impurity from one solvent to another. In this way, the solute or impurity may be effectively separated from the other components in a mixture. This process is called **liquid-**

95

FIGURE 4-1
A separatory funnel containing a water-insoluble organic solvent
and either water or an aqueous solution.

liquid extraction and usually occurs by shaking a water-insoluble organic solvent, such as ethyl ether or methylene chloride, with water or an aqueous solution in a separatory funnel (see Figure 4-1). The shaking temporarily mixes the two insoluble liquids and allows the selective transfer to occur. The separatory funnel is then placed upright in a ring stand, the two liquids are allowed to separate, and the lower liquid is removed through the bottom to isolate it from the other one. The liquid containing the separated substance can then be distilled or otherwise removed to allow the substance itself to be isolated.

4.2 DISTRIBUTION COEFFICIENTS

Frequently in a liquid-liquid extraction, however, the substance is not completely transferred from one solvent to another with one extraction. As a result, a number of extractions may be required for an **essentially** complete transfer. Many substances are somewhat soluble in **both** the organic and the aqueous liquids and will be distributed between the two liquids so that the concentration in each liquid is roughly proportional to the solubility in each liquid. In equation form, the result is

$$(1) \quad \frac{C_o}{C_w} \cong \frac{S_o}{S_w}$$

where C_o = concentration of the substance in the organic liquid in grams/100 mL

$\quad\quad$ C_w = concentration of the substance in the aqueous liquid in grams/100 mL

$\quad\quad$ S_o = solubility of the substance in the organic liquid in grams/100 mL

$\quad\quad$ S_w = solubility of the substance in the aqueous liquid in grams/100 mL

The ratio of C_o to C_w is known as the **distribution coefficient** or **partition coefficient** K.

$$(2)\ \ K = \frac{C_o}{C_w}$$

The coefficient will be a constant at a particular temperature for the substance and the two liquids. It will also be independent of the amount of the substance actually dissolved and the amounts of the two liquids present. Thus, it doesn't matter if the liquid originally containing the substance was saturated with it or had only a small amount of what could theoretically be dissolved. In addition, it doesn't matter if 100 mL of an aqueous solution is extracted with 25 mL, 50 mL, or any other amount of the insoluble organic liquid: K will still remain the same in all these instances.

The coefficient K can have many different values, depending upon what the substance is and exactly what the two liquids are. If $K > 100$, this means that practically all of the substance is dissolved in the organic layer. This means that just one extraction would be necessary to essentially transfer all of this substance from the aqueous layer to the organic layer. On the other hand, if $K < 0.01$, this means that practically all of the substance is dissolved in the aqueous layer. Again, just one extraction would be necessary. Usually though, K will be somewhere between 0.1 and 10, making several extractions necessary for an essentially complete transfer.

Assuming K is not very high or very low, it is always better to use several small portions of the extracting solvent rather than one large portion. The reason is that significantly more of the substance will usually be extracted for the same total volume of the extracting solvent. As an example, caffeine may be extracted from coffee or tea into water, and then separated from the other components in the coffee water or tea water by extraction with methylene chloride. In this instance, the distribution coefficient is 7.8 at 22°C. Suppose we have 200 mL of coffee water or tea water containing 2.00 grams of caffeine, and we extract it with 100 mL of methylene chloride. Using equation (2), we have the following, where x represents the number of grams of caffeine being extracted into the methylene chloride and $2.00 - x$ represents the number of grams of caffeine remaining in the water.

$$K = 7.8 = \frac{C_o}{C_w} = \frac{x \text{ grams/100 mL}}{(2.00 - x) \text{ grams/200 mL}}$$

$$7.8 = \frac{200x}{(100)(2.00 - x)}$$

$$2x = 7.8(2.00 - x)$$

$$2x = 15.6 - 7.8x$$

$$9.8x = 15.6$$

$x = 1.59$ grams or 79.5% of the caffeine extracted into the methylene chloride

We now repeat the calculation, using instead two 50-mL portions of methylene chloride.

$$7.8 = \frac{x \text{ grams/50 mL}}{(2.00 - x) \text{ grams/200 mL}}$$

$$7.8 = \frac{200x}{(50)(2.00 - x)}$$

$$4x = 7.8(2.00 - x)$$

$$4x = 15.6 - 7.8x$$

$$11.8x = 15.6$$

$x = 1.32$ grams of caffeine extracted into the first methylene chloride portion

Using the remaining 0.68 gram of caffeine in the water, we have for the second 50-mL extraction:

$$7.8 = \frac{x \text{ grams/50 mL}}{(0.68 - x) \text{ gram/200 mL}}$$

$$7.8 = \frac{200x}{(50)(0.68 - x)}$$

$$4x = 7.8(0.68 - x)$$

$$4x = 5.30 - 7.8x$$

$$11.8x = 5.30$$

$x = 0.45$ gram of caffeine extracted into the second methylene chloride portion

Adding the two amounts together, we have 1.77 grams or 88.5% of the caffeine

TABLE 4-1.

How the Number of Extractions with a Total Volume of 100 mL of Methylene Chloride Affects the Percentage of Caffeine Obtained from 200 mL of Water

Extractions		Caffeine extracted (%)	Extra caffeine extracted (%)
1	100-mL	79.5	—
2	50-mL	88.5	9.0
3	33.3-mL	91.8	3.3
4	25-mL	93.5	1.7
5	20-mL	94.5	1.0
6	16.7-mL	95.1	0.6
7	14.3-mL	95.5	0.4
8	12.5-mL	95.8	0.3
9	11.1-mL	96.1	0.3
10	10-mL	96.3	0.2

extracted into the methylene chloride. The two 50-mL extractions have thus given us 0.18 gram or 9% more caffeine, compared to the one 100-mL extraction.

Repeating the calculations with three $33\frac{1}{3}$-mL extractions, four 25-mL extractions, and so on, we obtain the results in Table 4-1. Observe that, although the total amount of caffeine extracted continues to rise, the extra amount obtained becomes progressively smaller. This observation also holds for other solutes and solvents. Therefore, after two or three extractions, one usually has to consider if the extra smaller amount obtained by further extractions is worth all the extra time and effort. Most likely, it isn't. If one were working with a very expensive or valuable substance, it would instead be more worthwhile to use multiple extractions with a larger total volume of the extracting solvent.

4.3 SELECTING AN EXTRACTING SOLVENT

Any solvent chosen for an extraction should meet the following criteria:

1. It should be insoluble or only slightly soluble with the solvent of the solution being extracted. As a result, ethyl alcohol, which is soluble with water in all proportions, would not be a good solvent for extracting an aqueous solution.

2. It should have a favorable distribution coefficient for the substance being extracted and an unfavorable distribution coefficient for any other components in the mixture. Otherwise, a selective transfer of just the substance being extracted will not occur. In addition, the solvent may not dissolve much of the substance, making a large volume and perhaps many extractions necessary for an essentially complete transfer.

3. It should be able to be easily removed from the extracted substance after the extraction. Since the removal is often by distillation, the solvent should therefore have a reasonably low boiling point. As a result, high-boiling silicone oil would not be a good choice as an extraction solvent.

4. It should be chemically inert to the extracted substance, other components in the mixture, and the solvent of the solution being extracted. There are, however, many exceptions to this rule, where a reaction between an extracting solution and the substance is desired. For example, a water-insoluble amine may be selectively extracted into water by its reaction with dilute hydrochloric acid to form a water-soluble amine salt.

$$R-NH_2 \ + HCl \longrightarrow R-NH_3^+ Cl^-$$

Water-insoluble   Water-soluble
amine   amine salt

Without this reaction, the desired selective transfer could not occur.

5. It should be reasonably safe to work with and relatively inexpensive.

A list of some common solvents used in extraction is given in Table 4-2. Benzene and chloroform have now been removed from this list because of their carcinogenic nature.

Ethyl ether is the organic solvent most frequently used in liquid-liquid extractions. It has the advantages of relative inertness, high solvent power for most organic compounds, and a low boiling point, which makes it easy to

TABLE 4-2.

Extraction Solvents

Solvent	BP	Density (g/mL)	Solubility in water (g/100 mL)	Fire hazard	Toxicity
Petroleum ether	40–60°	0.64	Low	High	Low
Ligroin	60–80°	0.68	Low	High	Low
Hexane	69°	0.66	0.02	High	Low
Ethyl ether	35°	0.71	6	High	Low
Toluene	111°	0.87	0.06	Medium	Medium
Water	100°	1.00	Infinite	None	None
Saturated aqueous NaCl	109°	1.20	Infinite	None	None
Methylene chloride	40°	1.33	2	Low	Medium

remove. Its disadvantages include its high flammability and solvent losses due to its high volatility.

Methylene chloride is also frequently used. It has the advantages of relative inertness, high solvent power for most organic compounds, low flammability, and a low boiling point, which makes it easy to remove. Its disadvantages include its higher toxicity and solvent losses due to its high volatility.

Petroleum ether, ligroin, hexane, and toluene are not used as often, primarily since they do not dissolve most organic compounds as well as ethyl ether or methylene chloride. Toluene also has a much higher boiling point, making it more difficult to remove.

4.4 EXTRACTION TECHNIQUE

The separatory funnel is the main piece of equipment used in extraction. Before the separatory funnel is filled, it is usually placed in an iron ring attached to a ring stand, as shown in Figure 4-2. Observe that the iron ring is partially

FIGURE 4-2

The separatory funnel placed in a cushioned iron ring.

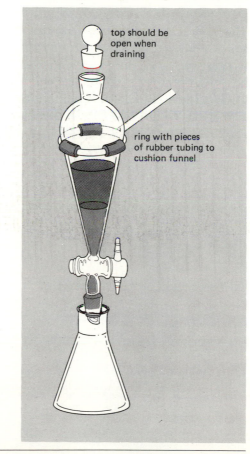

top should be
open when
draining

ring with pieces
of rubber tubing to
cushion funnel

wrapped in small pieces of rubber tubing. This cushions the separatory funnel and protects it against breakage if it should be placed in the iron ring too hard. If your iron ring doesn't already have this cushioning, cut three or four $1\frac{1}{2}$-inch pieces of thick rubber tubing and slit each open along its length. Slip them over the ring and wrap them with small lengths of wire to hold them permanently in place.

If your separatory funnel has a **glass** stopcock, remove it, clean it, and lightly grease it to avoid the leakage of liquid and the "freezing" of the joints. But avoid the middle area around the hole, since the grease can block the hole and contaminate your liquids. Reassemble the funnel, being sure the stopcock will turn without difficulty. After using the funnel, clean and grease the glass stopcock again to prevent it from "freezing" in place. **Teflon stopcocks need not be greased.**

To perform an extraction, place a separatory funnel having a capacity at least twice as large as the solution being extracted into the iron ring; **be certain the stopcock is closed**. Pour in the solution to be extracted; **be sure it is at room temperature or lower**. Then add the extracting liquid. The volume of the extracting liquid will vary, but will generally be about one-third of the volume of the solution being extracted. **Be sure the separatory funnel is filled no higher than about three-fourths of its height**. There must be sufficient room for the thorough and efficient mixing of the two liquids during the shaking process.

Stopper the top of the separatory funnel, pick it up, and invert it, **holding the stopper firmly in place**, as illustrated in Figure 4-3. **FOR SAFETY, BE SURE THE STOPPER AND THE STOPCOCK ARE ALWAYS POINTED AWAY FROM YOURSELF AND OTHER PEOPLE.** Vigorously shake the funnel for a few seconds, stop, and with the funnel still inverted, open the stopcock. This

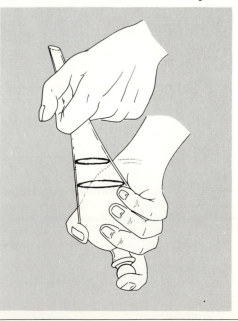

FIGURE 4-3
The proper way to shake
and vent the separatory funnel.
Be sure the funnel is not
aimed at yourself or a neighbor.

allows any built-up internal pressure from highly volatile solvents such as diethyl ether and methylene chloride to be released. If this pressure were not released, it could build up to the point where it could force out the stopper. The liquid inside would also be released, quite likely onto you, and the jolt could also cause you to drop the funnel. The pressure problem becomes even more pronounced when carbon dioxide gas is produced from the extraction of acidic compounds with sodium carbonate or sodium bicarbonate solution. Here, less vigorous shaking and more frequent venting are necessary until the reaction is largely completed.

After the internal pressure is released, close the stopcock, again vigorously shake the funnel for a few seconds, stop and again open the stopcock to release any internal pressure. Repeat this process for one to two minutes, venting the funnel less often when the pressure buildup is less intense, as noted by the diminished sound of the gas being released. After one to two minutes, the equilibrium distribution of the solute between the two liquids should be reached, and further shaking will give no further extraction.

Place the separatory funnel upright in the ring and remove the stopper. Otherwise the two liquids will not be able to properly drain. Allow the liquids to separate into two well-defined layers; then open the stopcock and collect the lower liquid in a clean beaker or flask. When the interfacing between the two layers approaches the stopcock, drain the liquid more slowly and **immediately** close the stopcock when the interfacing reaches it. Otherwise, a complete separation will not be attained. If the upper liquid no longer needs to remain in the separatory funnel, pour it off through the top into a clean beaker or flask. If the upper liquid were also removed through the stopcock, it would become contaminated by any of the lower liquid adhering to the stem of the separatory funnel.

The lower layer in an extraction will be the one having a higher density. Thus, water would be the lower layer in extractions with ethyl ether or hexane, but it would be the upper layer in extractions with methylene chloride. However, solutes of high molecular weight and concentration can sometimes raise the density of one solution enough to alter the normal arrangement. If in doubt, one way to test which layer is which is to add a few drops of the lower layer to a small test tube containing a little water. The solubility or insolubility in the water identifies the layer. It is also recommended that, unless otherwise stated, **both** layers in an extraction should be saved until the desired product has been isolated and identified. This prevents discarding the wrong layer through error or ignorance.

4.5 DRYING AGENTS

Usually after an extraction, the organic layer must be dried to remove any small amounts of water dissolved. Otherwise this water will contaminate the solute in the organic layer and may adversely affect future reactions of the solute. The water may also react with liquid solutes during distillation, inhibit the crystallization of solid solutes, and will ruin the sodium chloride cells or plates used to take infrared spectra of the solutes.

To dry an organic solution, add a small amount of a powdered drying agent such as anhydrous magnesium sulfate to the separated organic layer to form about a 2-mm layer in the bottom of the flask or beaker. The drying agent will remove the water by reacting with it to form a hydrated compound such as magnesium sulfate heptahydrate, $MgSO_4 \cdot 7H_2O$. Allow the solution to stand for about 10 to 15 minutes at room temperature with occasional swirling. If the drying agent can be easily swirled in the mixture, you have added enough. But if it clumps together or sticks to the glassware, the solution is still wet and a little more should be added. Try to avoid adding a large excess, or you will have a lower yield from the extra solvent being adsorbed onto the agent. A solution when dry will also appear perfectly clear. You may then remove the drying agent by either **carefully** decanting the solution from it or gravity filtering the solution through fluted filter paper.

Sometimes the drying procedure must be repeated to obtain a dry solution. If this is necessary, decant or gravity filter the still wet liquid into another **dry** container, add more fresh drying agent, and continue as before.

Some commonly used drying agents are listed in Table 4-3. Observe that all of them may be rated in terms of their capacity, efficiency, rate, and limitations. **Capacity** refers to how much water the drying agent will remove per unit weight. **Efficiency** means how effectively or completely the drying agent will remove all the water in solution before the equilibrium of the drying reaction is reached. **Rate** is how rapidly the drying will occur, and **limitations** refer to any compounds the drying agent cannot be used with or any special conditions necessary for success.

Observe that molecular sieves or aluminosilicates are the best all-around drying agent, having a high rate and capacity, an extremely high efficiency, and no real limitations. They can be used for practically all organic compounds.

Anhydrous magnesium sulfate and sodium sulfate are also good drying agents, and can be used for practically all organic compounds. Magnesium sulfate has a high rate and capacity, but a somewhat lower efficiency than molecular sieves. Magnesium sulfate often picks up moisture from the air during storage, which lowers its capacity, but it can be fully reactivated by mild heating. Sodium sulfate has a high capacity but is not as rapid or efficient as either molecular sieves or magnesium sulfate. Because hydrated sodium sulfate releases its water above 32°C, the drying agent must be used at room temperature or lower to be effective.

Anhydrous calcium chloride as a drying agent works moderately fast with a medium capacity and efficiency. But it reacts or forms a complex with many organic compounds, including carboxylic acids, phenols, alcohols, amines, amides, amino acids, ketones, aldehydes, and esters. Therefore, it cannot be used

TABLE 4-3.
Common Drying Agents

Drying agent	Acidity	Hydrated products	Capacity	Efficiency	Rate	Limitations	Use
Molecular sieves	Neutral	—	High	Very high	Fast	—	General
$MgSO_4$	Neutral	$MgSO_4 \cdot H_2O$ $MgSO_4 \cdot 7\,H_2O$	High	Medium to high	Fast	—	General
Na_2SO_4	Neutral	$Na_2SO_4 \cdot 7\,H_2O$ $Na_2SO_4 \cdot 10\,H_2O$	High	Low	Slow	Room temperature or lower	General
$CaCl_2$	Neutral	$CaCl_2 \cdot H_2O$ $CaCl_2 \cdot 2\,H_2O$ $CaCl_2 \cdot 6\,H_2O$	Medium	Medium	Medium	Room temperature or lower; cannot be used with most types of organic compounds	Hydrocarbons, organic halides, some ethers
$CaSO_4$	Neutral	$CaSO_4 \cdot \frac{1}{2}\,H_2O$ $CaSO_4 \cdot 2\,H_2O$	Low	High	Fast	—	General
K_2CO_3	Basic	$K_2CO_3 \cdot 1\frac{1}{2}\,H_2O$ $K_2CO_3 \cdot 2\,H_2O$	Medium	Medium	Medium	Many base-sensitive compounds	Alcohols, esters, ketones, amines, nitriles

to dry solutions of these compounds, unless their removal is also desired. Because the hexahydrate loses water above 30°C, this drying agent must also be used at room temperature or lower to be effective.

Calcium sulfate (Drierite) is often used as a drying agent in desiccators because of its high rate and efficiency. However, it has a low capacity and, as a result, is not used as often as other drying agents for drying organic solutions.

Potassium carbonate is also not used as often as other drying agents for drying organic solutions. It has only a medium rate, efficiency, and capacity, and, due to its basicity, cannot be used with many base-sensitive compounds.

4.6 SATURATED SALT SOLUTIONS

If a large amount of water is present in the organic solvent after an extraction, most of it may be removed before using a drying agent by shaking the organic solvent with a saturated solution of sodium chloride or potassium carbonate. The higher ionic strength of the salt solution makes this solution even less compatible with the organic solvent, making the water less soluble in the solvent.

Saturated salt solutions are also valuable for making water-soluble organic compounds much less water soluble and, therefore, much more easily extracted into organic solvents. The water-soluble organic compound is said to be **salted out** from the aqueous layer into the organic layer. The much higher ionic strength of the salt solution is again responsible.

Saturated salt solutions are also commonly used to remove water-soluble unwanted components from organic solvents such as ethyl ether. This process is called **washing**. The advantage over regular water is that the higher ionic strength of the salt solution dramatically reduces the solubility of the organic solvent in the aqueous phase. Therefore, part of the solvent is not lost to the aqueous phase, as it otherwise would be.

Salt solutions may even be used to isolate an organic compound when no extraction is involved. For example, soap (see Experiment 29) and other sodium salts of organic acids may be precipitated from aqueous solutions by the addition of sodium chloride. Here the **common-ion effect** is being used to greatly lower the solubility of the organic salt.

4.7 EMULSIONS

Frequently, after the separatory funnel is shaken, the liquids present will not separate completely, due to the formation of an **emulsion** or a colloidal suspension of one liquid in the other. This happens especially when any viscous, gummy or alkaline material is present in the solution.

If the volume of the emulsion is relatively small, sometimes it can be temporarily ignored, in the hope that it will disappear with later extractions. But if its volume is large, it must first be broken up or destroyed before the extraction process can continue. This sometimes requires much time and effort and, as a result, emulsions can be a real problem.

Fortunately, a number of techniques are available for dealing with emulsions.

1. If a particular solution is known beforehand to form emulsions, the extraction should be performed by gently swirling, rather than vigorously shaking, the liquids in the separatory funnel. Of course, since swirling gives much less efficient mixing than shaking, a longer swirling time of 3 to 5 minutes or more is necessary.

2. Test the tendency of the two liquids to form an emulsion by initially only swirling or gently shaking them in the separatory funnel. If emulsion formation appears, it might be a problem, so continue to only swirl or gently shake the liquids for a longer time.

3. If a large emulsion has formed, sometimes it will break up if it is merely allowed to stand undisturbed for a few minutes. If this doesn't work, **gently** swirling the liquids in the separatory funnel or **gently** stirring the emulsified layer with a stirring rod may be successful.

4. If the methods in technique 3 are unsuccessful, adding a saturated aqueous sodium chloride solution with stirring may work. The higher ionic strength of the aqueous liquid now makes it even less compatible with the organic liquid, hopefully forcing a separation. If the organic liquid has a density close to, but less than, the density of the aqueous liquid, the higher density difference from the added saturated salt solution may also bring success.

5. If the emulsion results from a small difference in the densities of the two liquids, the addition of either pentane or methylene chloride may also bring success by raising the density difference. Pentane should be added when the density of the organic liquid needs to be lowered, while methylene chloride should be added when the density of the organic liquid needs to be raised.

6. If possible, start the extraction again with a different organic liquid.

7. Centrifuge or gravity filter the emulsified material. Gravity filtration, however, may be messy and should be done only if other methods fail.

4.8 CONTINUOUS EXTRACTIONS

Sometimes the desired compound in an extraction is only slightly soluble in the extracting solvent. You can avoid using very large quantities of the extracting solvent by using a process called **continuous extraction**.

For liquid-liquid extractions, the apparatus in Figure 4-4 or Figure 4-5 is used, depending upon whether the extracting solvent has a density higher or lower than the solution being extracted. In both cases, the extracting solvent is distilled from the boiling flask, is condensed, and then travels through the solution being extracted, thus removing a small amount of the desired solute. The extracting solvent and the removed solute then go to the boiling flask, where the solute remains. The extracting solvent, however, is redistilled and thus is continuously reused until an essentially complete extraction is attained.

FIGURE 4-4
A light-solvent extractor.

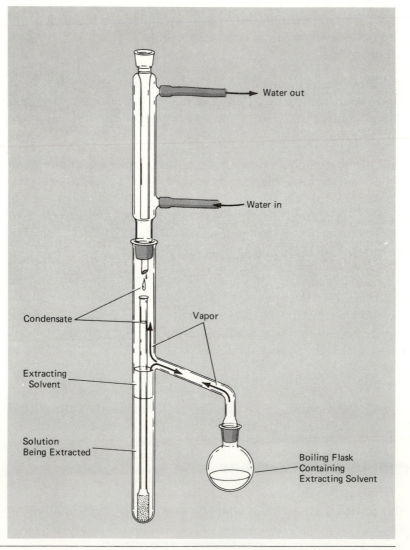

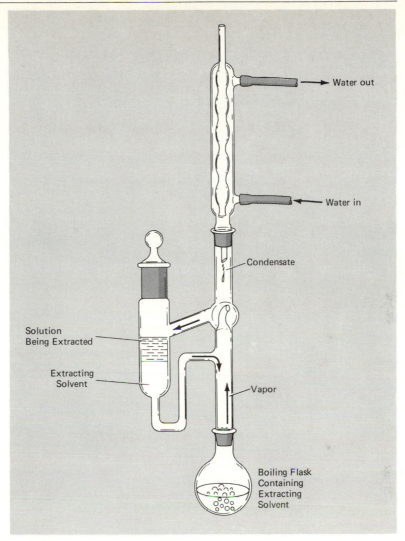

Water out

Water in

Condensate

Solution
Being Extracted

Extracting
Solvent

Vapor

Boiling Flask
Containing
Extracting
Solvent

FIGURE 4-5
A heavy-solvent extractor.

For solid-liquid extractions, the Soxhlet apparatus in Figure 4-6 is used. The solid to be extracted is placed in a **porous** thimble in the central Soxhlet chamber, and the extracting solvent is placed in the boiling flask. The extracting solvent is distilled and is then condensed into the porous thimble, where it extracts the desired solute from the solid. After the central chamber fills enough to reach the top of the siphon arm, the solution is emptied from the central chamber into the boiling flask by a siphoning process. The solute again remains in the boiling flask, but the extracting solvent is again distilled and can thus be continuously reused until an essentially complete extraction is attained.

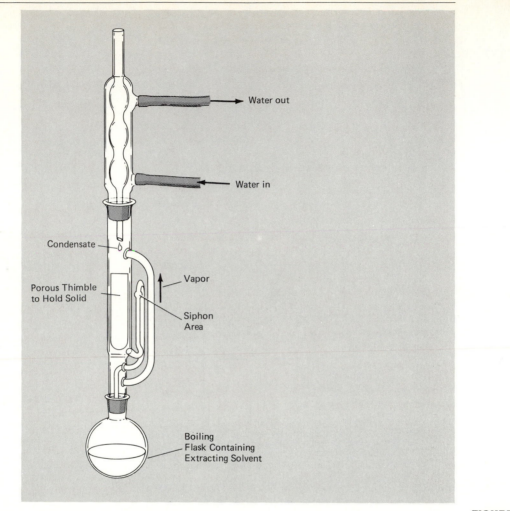

Water out

Water in

Condensate

Porous Thimble
to Hold Solid

Vapor

Siphon
Area

Boiling
Flask Containing
Extracting Solvent

FIGURE 4-6
A Soxhlet extractor.

4.9   SUMMARY OF THE LIQUID-LIQUID EXTRACTION PROCEDURE

1. Add the solution to be extracted to a separatory funnel having a capacity at least twice as large as the amount of solution. Be sure the solution is at room temperature or lower, the stopcock is closed, and a glass stopcock has been properly greased.

2. Add the extracting solvent. Its volume will vary, but will generally be about one-third of the volume of the solution being extracted. Be sure the separatory funnel is filled no higher than about three-fourths of its height.

3. With one hand on the stopper, invert the separatory funnel, as illustrated in Figure 4-3.

4. With the separatory funnel aimed away from yourself and other people, briefly shake it, then stop and open the stopcock to release any internal pressure. Look for any signs of an emulsion and if they are present, then either swirl or gently shake the funnel for 3 to 5 minutes or longer, periodically stopping and opening the stopcock to release any internal pressure. If no signs of an emulsion are present, shake the funnel more vigorously for 1 to 2 minutes, periodically stopping and opening the stopcock to release any internal pressure.

5. Place the separatory funnel back into the iron ring, remove the stopper, and separate the two liquid layers. The lower layer should be removed through the stopcock. If the upper liquid no longer needs to remain in the separatory funnel, pour it off through the top.

6. If necessary, extract the original mixture one or two more times with fresh solvent.

7. If the extracting solvent is organic, combine the extracts and, if necessary, shake with any other necessary aqueous solutions, such as saturated salt solution.

8. Dry the organic extract over anhydrous magnesium sulfate or some other suitable drying agent.

4.10 BACKGROUND

In this experiment, you will separate by extraction a mixture of three common household items: aspirin, β-naphthol (found as a part of many dyes), and naphthalene (used as moth crystals). The structures of these organic compounds are shown below.

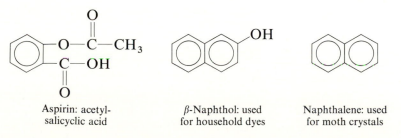

Aspirin: acetyl-
salicyclic acid

β-Naphthol: used
for household dyes

Naphthalene: used
for moth crystals

All three compounds are soluble in ether and either insoluble or only slightly soluble in water. But only aspirin is soluble in sodium bicarbonate solution, which converts the molecule to its water-soluble sodium salt by the reaction

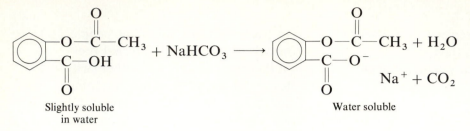

Slightly soluble
in water

Water soluble

This reaction occurs as a result of the moderate acidity of the hydrogen in the carboxyl group, COOH, and allows aspirin to be separated from the other compounds. Acidification of the sodium bicarbonate solution then regenerates the original molecule, which can be isolated by suction filtration.

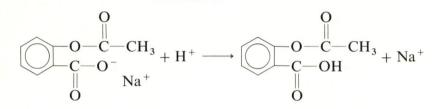

Because the ester group in aspirin can react with water under both alkaline and acidic conditions, the compound should be isolated as quickly as possible. Otherwise, as shown below, the resulting product will be salicylic acid or a mixture of aspirin and salicylic acid with a different melting point.

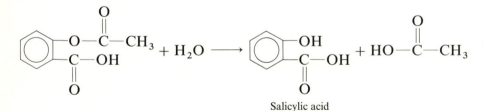

Salicylic acid

β-Naphthol can be separated from naphthalene because of its solubility in sodium hydroxide solution, which converts the molecule to its water-soluble sodium salt by the reaction

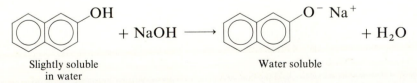

Slightly soluble
in water

Water soluble

This reaction occurs as a result of the slight acidity of the hydrogen in any hydroxyl group attached to an aromatic ring. Acidification of the sodium hydroxide solution then regenerates the original molecule, which can be collected by suction filtration.

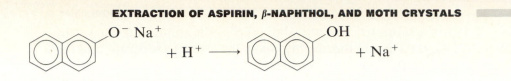

Only the naphthalene now remains in the ether solution, and it is easily recovered by first drying the solution and then evaporating the solvent.

4.11 PROCEDURE

Place 4.0 grams of the prepared mixture in a 125-mL Erlenmeyer flask, add 40 mL of ether **(CAUTION: NO FLAMES)**, dissolve by swirling, and pour the solution into a separatory funnel supported on a ring stand. Rinse out the flask with a small amount of additional ether and add this to the original solution. Add 30 mL of a 5% sodium bicarbonate solution, stopper and invert the funnel, holding it in your hands with the stopper held in place as shown in Figure 4-3, and briefly shake. Stop and, with the funnel still inverted, open the stopcock to release any interval pressure. Close the stopcock and briefly shake again. Stop and again open the stopcock. Continue this operation until no substantial internal pressure continues to develop from the carbon dioxide gas being produced.

Place the separatory funnel back on the ring stand, remove the stopper, and draw off the lower aqueous layer into a **labeled** 125-mL Erlenmeyer flask. **Gently** heat this aqueous solution in a warm-water bath to expel any ether present. (Remember that prolonged heating at unnecessary high temperatures may hydrolyze the aspirin.) Cool the solution and **cautiously** add 3 M HCl with stirring until a pH of 1–2 is indicated by Hydrion paper and a precipitate has formed. Test the pH by transferring a drop of the liquid to the Hydrion paper, using a stirring rod. Do **not** dip the Hydrion paper into the liquid because the dye in the paper will contaminate your product.

Collect the crystals by suction filtration, using a little **fresh cold** water to rinse out any residue in the flask. Wash the crystals with a small amount of **fresh cold** water and allow them to air dry.

The residual ether solution in the separatory funnel is next shaken with 30 mL of 5% sodium hydroxide solution for extraction of the weakly acidic β-naphthol. It may be necessary to add fresh ether to the separatory funnel before this step if a large quantity was lost by evaporation in the first extraction. Again be sure to periodically stop and open the stopcock to release any internal pressure.

Place the separatory funnel back on the ring stand, remove the stopper, and draw off the lower aqueous layer into a **labeled** Erlenmeyer flask. **Gently** heat this aqueous solution to expel any ether present, and then **cautiously** acidify with 3 M HCl to a pH of 1–2 with cooling and stirring. Collect the resulting crystals by suction filtration.

Add 15 mL of a saturated, aqueous sodium chloride solution to the separatory funnel, shake the mixture thoroughly, allow the layers to separate, and then draw off and discard the lower aqueous layer.

Transfer the remaining ether containing the naphthalene to a **labeled** Erlenmeyer flask, add a small amount of anhydrous calcium chloride (about one-tenth of the volume of liquid), and allow the solution to stand over the drying agent for 15 to 20 minutes with occasional swirling. Decant the liquid, filtering it through a very small wad of cotton in an ordinary funnel into a dried, **labeled** Erlenmeyer flask. Evaporate the collected liquid in a hot-water or steam bath **in the hood** to obtain your crystalline naphthalene.

Each of the products should now be recrystallized. To recrystallize aspirin, dissolve it in ether, add an equal volume of petroleum ether, and then allow the solution to stand undisturbed in an ice bath. β-Naphthol is best recrystallized from hot water, and naphthalene is best recrystallized from hot 95% ethanol. **Be sure you don't lose track of which product is which.**

Collect each purified product by suction filtration and allow it to thoroughly air dry. Determine the yield, the melting point, and the literature melting point of each. Record the results in your notebook. Then calculate the percentage recovery of each product, based upon the total amount of original starting material. Add your percentages to obtain the total percentage recovery.

At your instructor's option, determine the infrared and ^{1}H NMR spectra for each of your products. Each infrared spectrum may be taken with the compound in a KBr pellet, each ^{1}H NMR spectrum of aspirin and β-naphthol in deuterated chloroform, and the ^{1}H NMR spectrum of naphthalene in carbon tetrachloride. Compare your spectra with those in Figures 4-7 through 4-12.

When finished, submit your products, properly labeled, to your instructor.

FIGURE 4-7
The infrared spectrum of acetylsalicylic acid in a KBr pellet.
Spectrum courtesy of Sadtler Research Laboratories,
division of Bio-Rad Laboratories, Inc., copyrighted 1974.

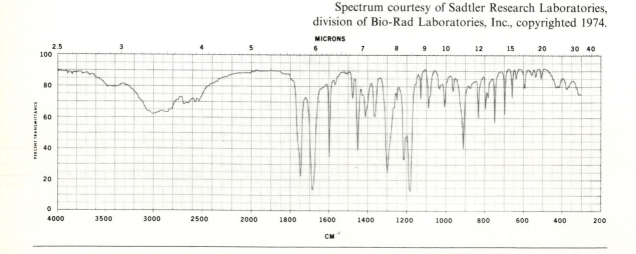

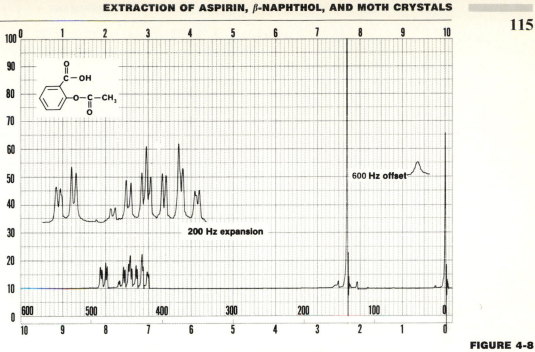

FIGURE 4-8

The ¹H NMR spectrum of acetylsalicylic acid in deuterated chloroform. Spectrum courtesy of Aldrich Chemical Company, *The Aldrich Library of NMR Spectra,* edition II, by Charles Pouchert.

FIGURE 4-9

The infrared spectrum of β-naphthol in a KBr pellet. Spectrum courtesy of Sadtler Research Laboratories, division of Bio-Rad Laboratories, Inc., copyrighted 1970.

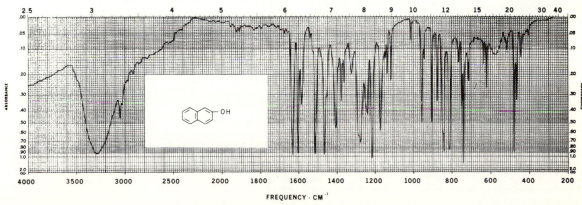

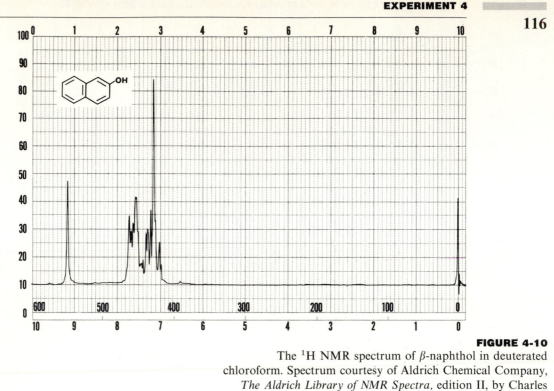

FIGURE 4-10
The ^{1}H NMR spectrum of β-naphthol in deuterated
chloroform. Spectrum courtesy of Aldrich Chemical Company,
The Aldrich Library of NMR Spectra, edition II, by Charles
Pouchert.

FIGURE 4-11
The infrared spectrum of naphthalene as a capillary cell melt.
Spectrum courtesy of Sadtler Research Laboratories, division of
Bio-Rad Laboratories, Inc., copyrighted 1966.

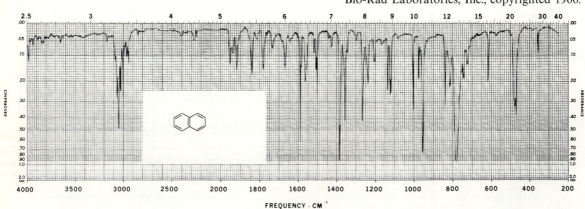

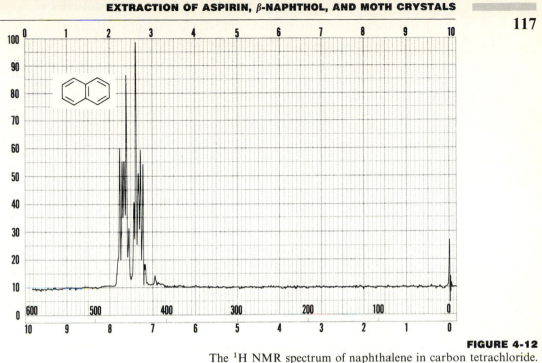

FIGURE 4-12

The ¹H NMR spectrum of naphthalene in carbon tetrachloride.
Spectrum courtesy of Aldrich Chemical Company, *The Aldrich
Library of NMR Spectra,* edition II, by Charles Pouchert.

QUESTIONS

1. Compare the basic strength of sodium bicarbonate versus sodium hydroxide.
2. Why is β-naphthol insoluble in sodium bicarbonate solution?
3. What would have happened if you had first extracted the entire mixture with sodium hydroxide solution?
4. Account for the greater amount of internal pressure in the aspirin extraction versus the β-naphthol extraction.
5. Why would there be any internal pressure at all in the β-naphthol extraction?
6. How could you be certain that the ether layer was on top and the aqueous layer was on the bottom in an extraction?
7. Why would ethyl alcohol not be a good solvent to use with water in an extraction?
8. Why would heavy mineral oil not be a good solvent in an extraction?
9. What is the purpose of using a saturated sodium chloride solution in an extraction with ether?

10. Why are anhydrous magnesium sulfate, anhydrous sodium sulfate, and anhydrous calcium chloride used in ether extractions?

11. Considering that the percentage recovery of your three compounds totals less than 100%, indicate where material most likely was lost. If necessary, how could you recover this material?

12. The distribution coefficient for solute A is 5.0 between methylene chloride and water. If 10.0 grams of A is dissolved in 100 mL of water, what weight of A would be removed by a single extraction with 100 mL of methylene chloride? By two 50-mL extractions? By three 33-mL extractions?

13. If one is performing multiple extractions of an aqueous solution, what practical advantage would a heavier-than-water organic solvent have over a lighter-than-water solvent?

14. What is an emulsion? Give several ways of breaking an emulsion. How can one try to avoid an emulsion?

15. If a compound is only slightly soluble in the extracting solvent, by what process can the compound be extracted without using an excessive amount of solvent?

FERMENTATION OF SUCROSE TO ETHANOL

ENZYMES AS CATALYSTS

Drink, drink, let the toasts start.
May young hearts never part.
Drink, drink, drink.
Let every true lover salute his sweetheart.

Dorothy Donnelly, Refrain from "The Drinking Song,"
from *The Student Prince*

INTRODUCTION

What do bread, cheese, alcoholic beverages, and a few other foods have in common? The answer, of course, is fermentation: Yeast or some mold or culture reacts with sugars and starches to produce carbon dioxide, ethyl alcohol, or other organic products. The reaction may affect only the appearance and texture of the food, as with bread, or completely change the food itself, such as with alcoholic beverages and dairy products.

The making of bread, cheese, and alcoholic beverages by fermentation are among the oldest chemical processes used by people. No one can say for certain when alcoholic beverages were first produced, but it is known that people fulfilled their desire for intoxication very early in history by making ethyl alcohol the very first synthetic organic chemical used. Bread has influenced history more than any other food and has played an important role in the rise and fall of nations. Bread riots unseated emperors in ancient Rome, and the French

119

Revolution was spurred in part when hungry French people cried out for bread in 1789 and only received the unthinking reply "Let them eat cake."

But even though fermentation had been known and used for centuries, it was not until the 1800s that the process began to be understood scientifically. In 1810, Gay-Lussac showed that the fermentation of the sugar glucose could be expressed by the equation

$$C_6H_{12}O_6 \longrightarrow 2\,CO_2 + 2\,CH_3CH_2OH$$

in which one molecule of sugar is decomposed to yield two molecules of carbon dioxide and two molecules of ethyl alcohol. It had been known that yeast, malt, or certain molds or cultures were required for fermentation, but the exact manner in which the process took place was the subject of much controversy. Finally, Louis Pasteur showed that fermentation was the result of the metabolic activity of microorganisms within the yeast, malt, or mold, and that different kinds of fermentations were caused by different microorganisms.

Pasteur was also the first to realize the true purpose of fermentation. He stated that "fermentation is the consequence of life without air." It allows organisms to survive and reproduce without using air and provides the chemical energy needed by these organisms. Although carbon dioxide and ethyl alcohol are the usual products, hydrogen gas, methane gas, carboxylic acids, other alcohols, and various other compounds can also be produced. The products formed depend upon the organism involved.

Fermentation is now known to result from enzymes produced by the organism. The enzymes are proteins, which act as highly active and efficient catalysts in the fermentation process and other organic reactions. Each enzyme is specific and catalyzes only one step in a reaction. Thus, if a complex organic reaction has 20 individual reaction steps, then 20 different enzymes are required.

The sugars and starches utilized in fermentation come from a number of sources. Much industrial or commercial ethyl alcohol comes from blackstrap molasses, a syrupy mixture of sucrose and impurities remaining after pure table sugar is crystallized from the extracted juice of sugar cane and sugar beets. After fermentation, the ethanol is removed by distillation. (Not enough industrial alcohol is produced by this process, however, and most of it is derived from petroleum.) With alcoholic beverages, the source depends on whatever beverage is being produced. Thus wine usually comes from grapes, blackberries, or other fruits, beer or ale from corn, rice, or barley, and "hard" beverages from the distillation of fermented grain, potatoes, and molasses. The characteristic flavor of the beverage is not due to ethyl alcohol but to the other substances present in or added to the source of the beverage.

When starch or sucrose is the starting material for an alcoholic beverage, a small quantity of other alcohols is also frequently obtained. These alcohols are called **fusel oil**, meaning inferior liquor. This mixture will vary, but it usually consists of propyl alcohol, isobutyl alcohol, 2-methyl-1-butanol, and

3-methyl-1-butanol. Fusel oil is responsible for the headaches or the "hangover" commonly associated with alcoholic beverages.

Ethyl alcohol is a unique substance, because it is highly taxed when used in alcoholic beverages but is not when used industrially. This tax, which amounts to over $20 per gallon, presents a special problem, which is solved by making industrial ethyl alcohol unfit to drink through the addition of **denaturants**. Denaturants are substances such as methanol, kerosene, and methyl isobutyl ketone, which are either poisonous or unpalatable and cannot be easily removed, but they do not impair the industrial utility of the alcohol. When required, pure undenatured ethyl alcohol is available for chemical purposes, but its use is strictly controlled by the federal government.

Undenatured ethyl alcohol in the home can be found not only in alcoholic beverages but also as a solvent for flavorings, pharmaceuticals, and medicinal products. This is necessary because water cannot adequately dissolve most of the components in these products, and ethanol is one of the few organic liquids that is relatively nonpoisonous, palatable without an overpowering taste, and a good solvent.

Denatured ethyl alcohol in the home can be found in perfumes, colognes, shaving lotions, and certain brands of rubbing alcohol. In industry, it is widely used as a reagent in chemical syntheses and as a solvent for products, chemical reactions, and recrystallizations. As a result, it is undoubtedly the most important alcohol and, except for petroleum and benzene, the most important organic liquid.

BACKGROUND

In this experiment, with the use of yeast, you will ferment sucrose or table sugar to obtain ethanol. Then by fractional distillation and salting out with anhydrous potassium carbonate, the ethanol will be concentrated. Since the reaction takes some time, the reagents must be mixed and then left for one full week.

Sucrose has the formula $C_{12}H_{22}O_{11}$ and consists of one molecule of glucose combined with one molecule of fructose. As the following equations show, the enzyme invertase, found in yeast, cleaves the sucrose molecule into glucose and fructose, which are then converted by zymase into ethanol and carbon dioxide. The zymase in the yeast is a group of at least 22 separate enzymes, each of which catalyzes a specific step in this complex second reaction.

Besides sucrose and yeast, the fermentation solution will contain a small amount of Pasteur's nutrient, a mixture of potassium phosphate, calcium phosphate, magnesium sulfate, and ammonium tartrate. Pasteur found that these salts enhanced yeast growth and formation. Enhancement occurs when the six carbon sugars couple with phosphoric acid to give a combination that is more easily degraded into carbon dioxide and ethanol.

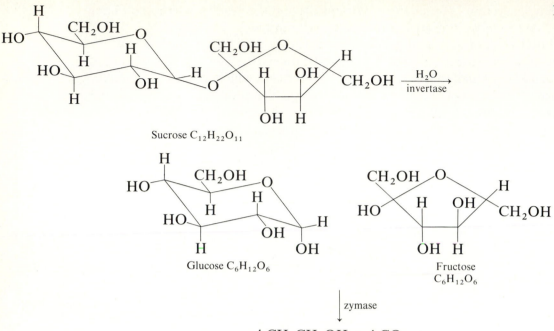

$$4 \text{ CH}_3\text{CH}_2\text{OH} + 4 \text{ CO}_2$$

In the experiment, atmospheric oxygen will be excluded from the reaction mixture by the addition of mineral oil to the carbon dioxide receiving bottle. If this were not done, the ethanol produced could be further oxidized to acetic acid or even all the way to carbon dioxide and water. Wine or apple juice changing to vinegar is an example of this oxidation reaction.

Note that the fermentation process is inhibited by the ethanol produced, and it stops when the concentration of ethanol reaches about 12 percent by volume. This is why most wines are approximately 12% alcohol. Only by fractional distillation can more concentrated ethanol solutions be obtained from fermentation, but even here the limit is 95% ethanol due to the formation of an azeotropic mixture (see page 47).

After the distillation, you may notice that the ethanol obtained does not smell like the 95% ethanol found in the lab. The reason is the other fermentation products formed in the reaction, which include acetal derivatives of acetaldehyde, propyl and isopropyl alcohol, and higher alcohols. For industrial alcohol, which must be as chemically pure as possible, this "yeasty" odor can be removed by treatment of the distilled alcohol with activated charcoal.

PROCEDURE

Place 50 grams (0.146 mole) of sucrose, 350 mL of water, 40 mL of Pasteur's nutrient, and 10 grams of fresh yeast into a 500-mL flask. Shake the mixture vigorously and fit the flask as shown in Figure 5-1 with a one-hole rubber

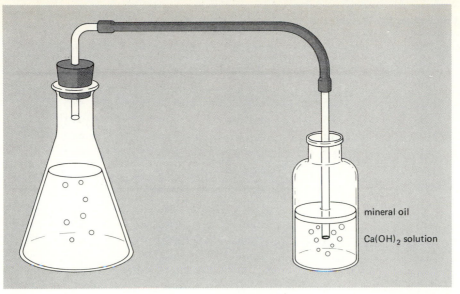

FIGURE 5-1
Fermentation apparatus.

stopper and a glass tube leading to a beaker or bottle containing limewater below a protective layer of mineral oil. Gas bubbles should shortly be seen in the limewater solution, and a precipitate should begin to form. Allow the mixture to stand at room temperature for one full week or longer if gas bubbles continue to appear.

After the fermentation is completed, carefully decant the supernatant liquid from the sediment, making sure that none of the sediment is removed. Some liquid will remain behind with the sediment. **If** the decanted liquid is not clear, it may be suction filtered through filter aid or Celite in a Büchner funnel. To prepare the filter aid layer in the funnel, place about 2 tablespoons of filter aid in a beaker with about 100 mL of water. Stir vigorously and pour the mixture into a Büchner funnel containing paper, as in an ordinary suction filtration. A thin layer of filter aid will be deposited onto the filter paper. After the water in the filter flask is discarded, you can suction filter the decanted fermentation liquid to remove any tiny yeast particles.

Transfer the clear liquid to a 500-mL round-bottom flask and set up a fractional-distillation apparatus (see page 41). Distill the mixture slowly and carefully in order to obtain the best separation possible and collect the fraction boiling between 78° and 90°. The residue remaining in the distilling flask and the fractionating column may be discarded.

Salt out the alcohol from the collected distillate by adding 10 grams of anhydrous K_2CO_3 per 50 mL of distillate. Stir well, allow the solid to settle, and then decant the liquid into a separatory funnel. Draw off and discard the lower water layer from the separatory funnel, and pipet 10.0 mL of the

TABLE 5-1.

Density and Percentage by Weight and Volume of Ethanol in Water at 20°C

Density	% by Weight	% by Volume	Density	% by Weight	% by Volume
0.989	5.0	6.2	0.856	75.0	81.3
0.982	10.0	12.4	0.843	80.0	85.5
0.975	15.0	18.5	0.831	85.0	89.5
0.969	20.0	24.5	0.818	90.0	93.3
0.962	25.0	30.4	0.815	91.0	94.0
0.954	30.0	36.2	0.813	92.0	94.7
0.945	35.0	41.8	0.810	93.0	95.4
0.935	40.0	47.3	0.807	94.0	96.1
0.925	45.0	52.7	0.804	95.0	96.8
0.914	50.0	57.8	0.801	96.0	97.5
0.903	55.0	62.8	0.798	97.0	98.1
0.891	60.0	67.7	0.795	98.0	98.8
0.879	65.0	72.4	0.792	99.0	99.4
0.868	70.0	76.9	0.789	100.0	100.0

remaining ethanol into a clean and dry Erlenmeyer flask, preweighed to the nearest 0.001 gram. Reweigh the flask, determine the density of your ethanol, and determine the percentage of ethanol by weight and volume, using the information in Table 5-1.

From the total volume of the salted-out ethanol and its density and percentage by weight, determine the actual yield of ethanol in grams and the theoretical and percentage yields, showing the calculations. Submit the ethanol in a properly labeled container to your instructor.

QUESTIONS

1. Why is it theoretically impossible for your ethanol obtained from the fractional distillation to be 100% pure? How is 100% pure ethanol obtained industrially?
2. Why does the ethanol salt out from the water when K_2CO_3 is added?
3. Write the equation for the complete combustion of ethanol.
4. Write the equation for the precipitate produced when the gas formed in the fermentation reaction is bubbled through limewater.
5. What possible commercial uses does the gas given off in the experiment have?
6. The "proof" of an alcoholic beverage is defined as double the percentage of alcohol in the drink by volume. From this definition, what percentage of alcohol is present in 86 proof whiskey? What proof should be given to wine

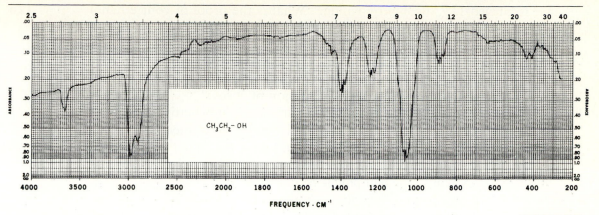

CH₃CH₂—OH

FREQUENCY · CM⁻¹

FIGURE 5-2
The Infrared spectrum of ethanol in a 10-cm gas cell.
Spectrum courtesy of Sadtler Research laboratories, division
of Bio-Rad Laboratories, Inc., copyrighted 1966.

FIGURE 5-3
The ^{1}H NMR spectrum of ethanol, neat.

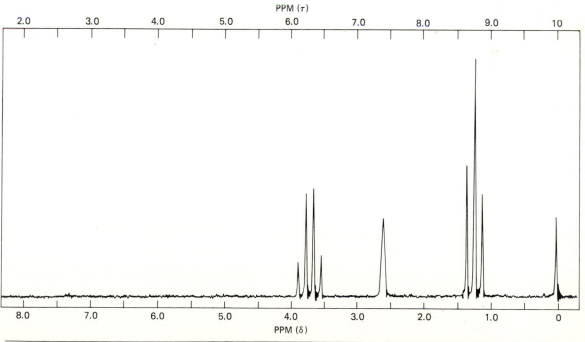

PPM (τ)

PPM (δ)

that is labeled 12% alcohol by volume? What is the proof of your ethanol handed in at the end of this experiment?

7. What is meant by the term "absolute alcohol"?
8. What happens if a fermentation reaction is exposed to the air?
9. Why is no alcohol found in bread after fermentation and baking?
10. Why is no ethanol formed in the fermentation of cheese?
11. The theoretical yield of ethanol in this experiment is substantially less than the original weight of sucrose. How can this loss be accounted for?
12. The infrared spectrum of ethanol is shown in Figure 5-2. What contaminant in your ethanol sample would damage the equipment if you took an infrared spectrum? How would the equipment be damaged?
13. The ^{1}H NMR spectrum of ethanol is shown in Figure 5-3. What contaminants in your ethanol sample would cause a deviation from this spectrum?

6

CHROMATOGRAPHY

6.1 INTRODUCTION

In 1903, a Russian botanist, Michael Tswett, extracted a mixture of pigments from green leaves and then washed the mixture with petroleum ether through a glass tube packed with powdered calcium carbonate. As the mixture passed down the chalk-filled tube, the pigments became separated, forming a number of distinctly colored zones. The name **chromatography**, meaning the graphing of colors, was given by Tswett to this separation technique.

The word "chromatography" is still used today, even though the technique is more commonly used with colorless compounds. In its various forms, the method has proven to be a tremendous breakthrough for modern chemistry, since it allows complex mixtures to be readily separated, even when only very small quantities of materials are used.

The four types of chromatography which will be considered in this book are column, paper, thin-layer, and gas-phase chromatography. Each type has in common a solid or liquid stationary phase, upon which the mixture is adsorbed, and a liquid or gaseous mobile phase, which desorbs and carries the components of the mixture along the stationary phase. Because each component of the mixture differs in how strongly it will be adsorbed or "held on" by the stationary phase, and desorbed by the mobile phase, each one will move at differing rates and will be separated as a result. Due to the wide array of stationary and mobile phases available, the method is applicable for any type of mixture which may be encountered. The basic details for each of the four types are presented in the following sections.

6.2 COLUMN CHROMATOGRAPHY

Column chromatography is the separation technique utilized by Tswett in 1903. In this method a glass tube, such as a buret, is packed as shown in Figure 6-1 with

127

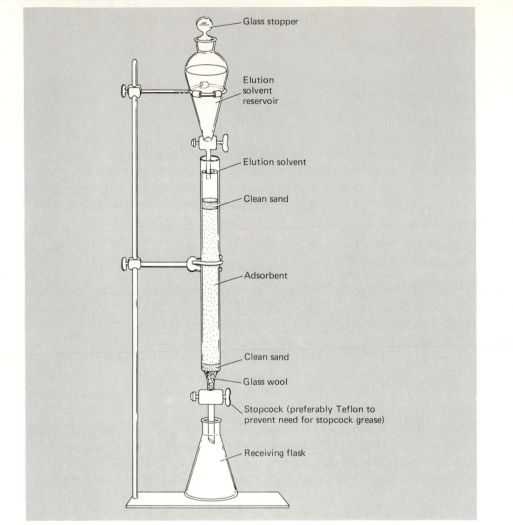

Glass stopper

Elution
solvent
reservoir

Elution solvent

Clean sand

Adsorbent

Clean sand

Glass wool

Stopcock (preferably Teflon to
prevent need for stopcock grease)

Receiving flask

FIGURE 6-1
A column chromatography setup.

a stationary adsorbent such as finely ground silica gel ($SiO_2 \cdot xH_2O$), alumina ($Al_2O_3 \cdot xH_2O$), calcium carbonate, cellulose, or Florisil (fluorinated silicone polymer). A small sample of the mixture, if a liquid, is then poured into the top of the tube and allowed to be adsorbed on the upper portion of adsorbent. If the mixture is a solid, it is first dissolved in a minimum quantity of solvent and then applied.

To separate the mixture, an eluting solvent is next allowed to flow downward through the column. As this solvent passes over the mixture, the components present begin to travel down the column at different rates to form a number of moving bands or zones. Each band will ideally contain only one component of

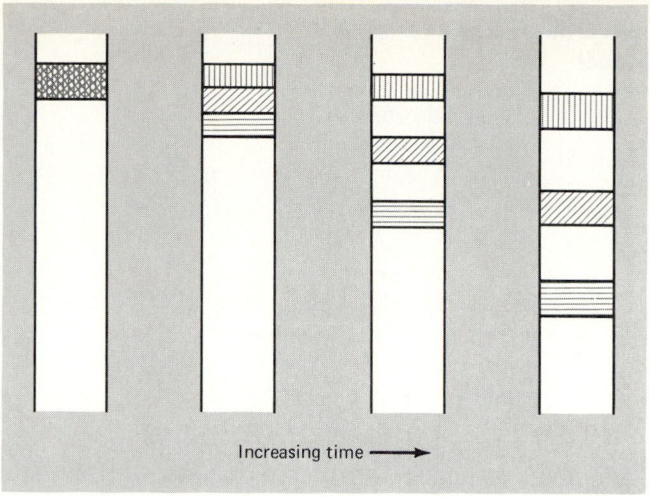

FIGURE 6-2
Development of the chromatogram

the mixture and, as the chromatography progresses, will be spaced at greater and greater distances from the other bands. A complete separation thus results as each band is individually collected as it passes out at the bottom of the column. This progressive separation of the various bands as they travel down the column is illustrated in Figure 6-2.

6.2A. Adsorbents and binding

Table 6-1 lists the stationary adsorbents used in column chromatography, ranked in increasing order of their binding ability. This binding ability depends on both the polarity of the adsorbing material and the polarity of the components in the mixture being separated. Generally, the greater the polarity, the stronger will

TABLE 6-1.

Adsorbents Used in Column Chromatography

Cellulose

Starch

Sugars

Magnesium silicate

Calcium carbonate

Calcium phosphate   Increasing
                           binding
Calcium sulfate   ability

Calcium oxide

Silica gel

Florisil

Charcoal

Alumina

TABLE 6-2.

Adsorbability of Various Classes of Organic Compounds

Alkanes	Least adsorbed, will elute the fastest
Alkenes	
Dienes	
Aromatic hydrocarbons	
Halides	
Ethers	
Esters	Increasing adsorbability
Amides	
Ketones	
Aldehydes	
Alcohols	
Amines	Most adsorbed, will
Carboxylic acids	elute the slowest

be the binding and the more difficult it will be to move each component in the mixture down the column. For components having a similar or identical polarity, molecular weight is the determining factor, with higher-molecular-weight components moving more slowly.

Table 6-2 lists the various classes of organic compounds, ranked in increasing order of their ability to be bound to the adsorbent. Observe how this list confirms the polarity factor, with alkanes being the least strongly bound and carboxylic acids being the most strongly bound. With alkanes, weak van der Waals interactions are the only attractive force present. But with carboxylic acids, very strong ionic bonds are possible through the formation of a salt. Other possibilities are hydrogen bonding, dipole-dipole interactions, and the formation of coordination complexes. These are illustrated in Figure 6-3 for various types of

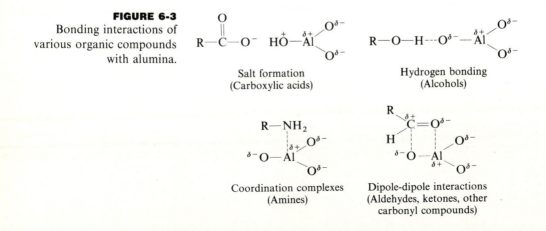

FIGURE 6-3
Bonding interactions of various organic compounds with alumina.

Salt formation
(Carboxylic acids)

Hydrogen bonding
(Alcohols)

Coordination complexes
(Amines)

Dipole-dipole interactions
(Aldehydes, ketones, other carbonyl compounds)

organic compounds being bonded to alumina. For simplicity, only a portion of the alumina molecule is shown. The complete order for the strength of all these bonding interactions is generally the following:

Salt formation > Coordination complexes > Hydrogen bonding
> Dipole-dipole > van der Waals

In choosing any adsorbent for a chromatography, one should remember the following important points:

1. The adsorbent must not react irreversibly with either the compounds being separated or the solvents. It must also not act as a catalyst for their decomposition, dimerization, rearrangement, or isomerization. Otherwise, new and frequently unexpected products will result.

2. The adsorbent should show a large selectivity toward the compounds being separated, so there will be large differences in the rate of migration.

3. To achieve the greatest effectiveness, the adsorbent should have a uniform particle size, small enough to give effective separation and yet large enough to allow a reasonable rate of solvent flow.

4. The adsorbent must be insoluble in the solvents used for the separation.

5. The adsorbent should be nontoxic and relatively inexpensive.

Alumina (Al_2O_3) is a widely used adsorbent in column chromatography and, being strongly adsorbing, is typically utilized for mixtures of nonpolar to medium-polar compounds. Its strong adsorptivity can, however, be lowered by adding small amounts of water. Water binds itself very tightly to alumina, taking up molecular spaces which would otherwise have been used to hold mixture components in place.

Alumina comes in three forms: basic (pH 10), acid-washed (pH 4), and neutral (pH 7). Basic alumina is useful in separating amines and acid-sensitive compounds. Acid-washed alumina is useful for carboxylic acids, amino acids, and base-sensitive compounds. Neutral alumina is commonly used for a wide variety of nonacidic and nonbasic compounds and for compounds sensitive to both acids and bases.

Silica gel and Florisil are not as adsorbent as alumina, and may also be made even less adsorbent by adding small amounts of water. They are widely used for a variety of organic compounds, including alcohols, ketones, esters, azo compounds, amines, and carboxylic acids.

Cellulose, starch, and sugars have the lowest adsorbence and so are useful for mixtures of highly polar compounds which might come off with difficulty if using alumina or some other highly adsorbing material. Cellulose, starch, and sugars are also useful for some natural products very sensitive to acid-base interactions.

6.2B. Solvents

There are many possible eluting solvents for a column chromatography. But in choosing one, the following important points should be remembered.

1. The solvent must not dissolve the adsorbent.

2. The solvent must not react irreversibly with either the adsorbent or the compounds to be separated.

3. For maximum effectiveness, the solvent should show a large selectivity in desorbing the compounds in the mixture from the adsorbent.

4. The solvent must be able to dissolve the compounds being separated.

5. The solvent should, if possible, have a low boiling point to make it easily removed after the separation.

6. The solvent should, if possible, be relatively nontoxic and inexpensive.

Table 6-3 lists some common eluting solvents, ranked in increasing order of their ability to move a substance down a column. Note that this order is only approximate, since the eluting power is somewhat dependent upon the nature of the adsorbent and the compound being moved. However, the order again shows how a more polar solvent generally will elute a compound faster than a less polar solvent. This results from a more polar solvent being more effective than a less polar solvent in

1. Displacing a compound from its site of adsorption

TABLE 6-3.

Eluting Solvents for Chromatography

Petroleum ether (hexane, pentane, heptane)	
Cyclohexane	
Carbon tetrachloride*	
Toluene	
Methylene chloride	
Chloroform*	
Ethyl ether (anhydrous)	Increasing eluting power
Ethyl acetate (anhydrous)	
Acetone (anhydrous)	
Propanol (anhydrous)	
Ethanol (anhydrous)	
Methanol (anhydrous)	
Water	
Acetic acid	

* Is a suspected carcinogen.

2. Occupying these sites, making them unavailable for adsorption of other compounds moving down the column.

The eluting solvent chosen will depend upon how strongly the mixture is held to the adsorbent. If the mixture is only weakly held, then a solvent of low eluting power is necessary. Otherwise, the mixture would move too rapidly down the column, and little separation would occur. But if the mixture is held very strongly, then a solvent of higher eluting power is required to prevent the mixture from moving too slowly or not at all.

Generally, one begins with a solvent of low eluting power to remove the relatively weakly held compounds from the column. One then gradually increases the eluting power to force the more strongly adsorbed components down the column. Rapid changes from one solvent to another are usually avoided by using a solvent mixture containing a gradually increasing percentage of the new liquid being introduced. Otherwise, excessive heat may be generated when the new solvent solvates the adsorbent, forming a weak bond.

$$\text{Solvent} + \text{adsorbent} \longrightarrow \text{Solvent} \cdot \text{adsorbent} + \text{heat}$$

This heat may be enough to vaporize the solvent, with the gas forcing apart the adsorbent and creating cracks or channels. Since the solvent will preferentially flow through the cracks rather than the adsorbent, the bands will become deformed as they travel faster on one side of the column. This results in a poor separation, since the compounds can then be remixed as they are collected at the bottom (see Figure 6-4).

FIGURE 6-4

A crack in the adsorbent, producing poor separation through the formation of deformed bands.

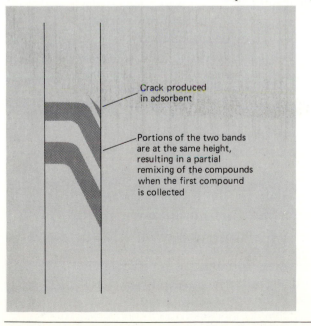

Crack produced in adsorbent

Portions of the two bands are at the same height, resulting in a partial remixing of the compounds when the first compound is collected

TABLE 6-4.

Size of Column and Amount of Adsorbent Needed for Various Sample Sizes

Amount of sample (g)	Amount of adsorbent (g)	Column dimensions alumina (diameter × height, cm)	Column dimensions silica gel (diameter × height, cm)
0.1	3	0.7 × 8	1.0 × 12
1.0	30	1.5 × 18	2.2 × 26
3.0	90	2.1 × 25	3.1 × 37
10.0	300	3.2 × 38	4.7 × 57

One also tries, if possible, to avoid solvents more polar than methylene chloride or ethyl ether in the elutant series. Ethyl acetate, acetone, acetic acid, and the alcohols may all undergo various chemical reactions to give undesired, misleading products. In addition, the alcohols, water, and acetic acid may dissolve and elute some of the adsorbent.

The elution solvents in a column chromatography must be pure. Frequently, commercial-grade solvents will contain small amounts of residue, which will contaminate each component of the separated mixture after evaporation. This is especially true for hydrocarbon solvents. Commercial-grade solvents should all be redistilled and dried before use.

6.2C. Choosing a column

The column size and the amount of adsorbent are very important to the success of a column chromatography. Generally, the amount of adsorbent required is about 25 to 30 times by weight the amount of material to be separated. Even more than this amount of adsorbent should be used for a difficult separation. Less may be used for a relatively easy one.

In addition, the column of adsorbent should usually be about 10 to 12 times as tall as it is wide. Utilizing all these figures and the densities of alumina and silica gel, one arrives at the figures in Table 6-4 for the amounts of adsorbent and the column dimensions when various amounts of a mixture are being separated.

6.2D. Preparing the column

The proper preparation of a chromatography column is critical to its success. If the adsorbent has cracks, channels, or air bubbles, or if it has an irregular or nonhorizontal surface, obtaining a complete separation of a mixture will likely be impossible. As shown in Figure 6-4, a crack in the adsorbent will cause the bands to overlap at the same height, resulting in only a partial separation. Air bubbles and surface irregularies will do the same. Although a nonhorizontal surface probably won't produce this large a deformity, different bands can still be at the same height, as shown in Figure 6-5, thus producing an incomplete separation.

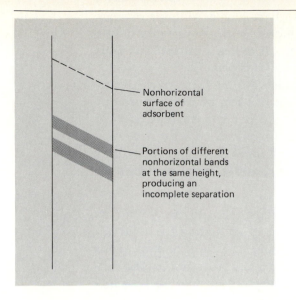

Nonhorizontal surface of adsorbent

Portions of different nonhorizontal bands at the same height, producing an incomplete separation

FIGURE 6-5
A nonhorizontal surface of adsorbent, producing an incomplete separation.

The following steps are used to prepare a column and avoid the problems resulting from uneven packing and column irregularities. Follow them carefully for good results, utilizing Figure 6-1 as a guide.

Step 1: Clamp a 50-mL buret or other chromatography column in an exactly vertical position. Any stopcock present should be made of Teflon, since the stopcock grease used with glass stopcocks may contaminate the product. As an alternative, the bottom may be a piece of flexible tubing and a screw clamp. With this alternative, the tubing should be made of polyethylene, since Tygon and rubber tubing partially dissolves in a number of solvents, contaminating the product.

Step 2: Fill the column three-fourths of the way with an eluting solvent. This solvent will usually be a nonpolar liquid like hexane.

Step 3: Using a long glass rod, place a small wad of glass wool into the bottom of the container. This wad must be thick enough to prevent any solid material from passing through.

Step 4: Drain a small amount of solvent from the column and tamp the glass wool with the glass rod until all entrapped air is forced out. However, do not tamp the glass wool so much that it will totally plug the column.

Step 5: Place a powder funnel on top of the column and add enough clean sand to form a level 1-cm layer on top of the glass wool. Tap the column gently with a piece of pressure tubing to help level the sand. Any sand adhering to the side of the column may be washed down with a small amount of solvent. This sand layer forms a base for the adsorbent and prevents any fine particles of the adsorbent from washing through the stopcock.

Step 6:   Slowly add, in small amounts, an adsorbent such as alumina or silica gel. As you do this, gently tap the column continuously with the pressure tubing so that the adsorbent settles evenly and uniformly. **It is very important that there be no holes, channels, or air bubbles present.** During this time, slowly drain some of the solvent to prevent the column from overflowing. **However, do not drain so much solvent that the top of the adsorbent becomes uncovered.** Add enough adsorbent so that you have at least 25 to 30 times as much adsorbent by weight as the mixture to be separated. Also add enough to make the column at least 10 to 12 times as long as it is wide. **At the end, be sure that the top of the adsorbent is level and has no irregularities.**

Step 6(a):   As an alternative to step 6, add the adsorbent to the column as a thin slurry in the eluting solvent. Prepare the slurry by adding the solid adsorbent, a little at a time, to the solvent with stirring to form a homogeneous mixture, free of lumps and air bubbles. **Do not add the solvent to the adsorbent, since the heat evolved may cause the solvent to boil. In addition, the mixture is more likely to be lumpy.**

After the slurry is prepared, add it to the column slowly with gentle tapping, using the pressure tubing. During this time, drain solvent from the bottom to prevent the column from overflowing. You may mix this solvent with the slurry yet to be added. **Again, it is very important that there be no holes, channels, or air bubbles present. At the end, be sure that the top of the adsorbent is level and has no irregularities.**

Step 7:   Run the collected solvent at the bottom through the column several times to ensure that the settling of the adsorbent is complete and that the column is firmly packed. **Again, do not allow the solvent level to fall below the top of the adsorbent.**

Step 8:   Add a small layer of sand over the adsorbent to protect the surface of the adsorbent. This sand layer must also remain below the surface of the eluting solvent.

6.2E. Applying the sample to the column

Step 1:   Drain the solvent from the column until the solvent level is at the top of the sand layer.

Step 2:   Dissolve the solid mixture to be separated in the **minimum** amount of the same solvent used to prepare the column. If the mixture is not very soluble in this solvent, a very small amount of a more polar solvent may be used. If the original mixture is a liquid, step 2 may be omitted.

Step 3:   Carefully pipette or pour the liquid from step 2 onto the surface of the column, being sure not to disturb the surface.

Step 4: Slowly drain the solvent from the column until the liquid layer is again at the top of the sand layer.

6.2F. Eluting the mixture

Step 1: Place a stoppered separatory funnel, filled with the eluting solvent, in place over the column as shown in Figure 6-1. This funnel **must have** a long stem at the bottom inside the column.

Step 2: With the stopper still in place, open the stopcock of the separatory funnel, and carefully add, with as little turbulence as possible, enough eluting solvent to nearly fill the column. Note that with the stopper in place, it is not necessary to close the stopcock, since as soon as the liquid level in the column rises higher than the bottom of the separatory funnel, the flow of liquid into the column will automatically stop, preventing an overflow.

Step 3: Open the stopcock or screw clamp at the bottom of the column and allow liquid to flow out at a rate of about 2 milliliters per minute. As the liquid level at the top of the column falls, the separatory funnel arrangement prepared in step 2 will automatically refill the column. The setup is thus self-filling.

Step 4: Continue the eluting, changing receivers at regular intervals or as indicated from monitoring the progress of the chromatography (see Section 6.2G). Evaporate the solvent from the receiving flasks to recover the separated compounds, and gradually increase the polarity of the eluting solvent, as indicated by the progress of the chromatography. To prevent the bands from diffusing together, **do not** stop the eluting for long periods. For example, once started, the chromatography may **not** be left overnight or until the next lab period.

6.2G. Monitoring the progress of the chromatography

If all the components of the mixture have color, monitoring the progress of a column chromatography is very easy. Usually, however, this is not the case. One common method with colorless compounds is to collect solvent fractions of constant volume in preweighed flasks. After evaporating the solvent, a plot of the weight versus the fraction number will show when each component is being collected. The resulting graph might look like Figure 6-6. Note that for good results, the volume of solvent in each fraction must not be too large. Otherwise, a fraction could easily contain more than one component. Thin-layer chromatography (see Section 6-3) of the fractions containing compounds can be used to ensure that each fraction does actually have only one compound.

A way to visually monitor the progress with colorless compounds is to initially mix the adsorbent with a small amount of inorganic phosphor. This causes the adsorbent to fluoresce when illuminated with ultraviolet light. Most solutes on the column, however, will suppress the fluorescence and can thus be detected by the resulting dark bands produced.

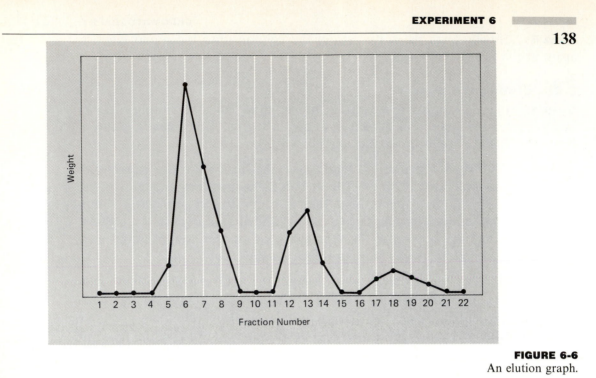

FIGURE 6-6
An elution graph.

6.2H. Dry-column chromatography

Dry-column chromatography is essentially a kind of scaled-up thin-layer chromatography (see Section 6.3). The main differences from regular column chromatography are that the column is packed dry, no solvent is collected at the bottom of the column, and the solvent is not changed as the chromatography progresses.

PROCEDURE

Step 1: Using thin-layer chromatography, discover, if necessary, which solvent will best separate the mixture. This solvent may be a blend of a number of liquids.

Step 2: Clamp a chromatography tube vertically to a ring stand and place a small wad of glass wool at the bottom, pushing it down with a glass rod.

Step 3: Pour the dry adsorbent slowly into the empty tube, tapping the column gently with pressure tubing to help the adsorbent settle evenly in the tube. The top surface of the adsorbent must be level.

Step 4: Dissolve the mixture of compounds being separated in a few milliliters of a volatile polar or nonpolar solvent to make a concentrated solution. Add enough adsorbent to the solution with stirring to make a thick slurry; then evaporate the solvent.

Step 5: Pour the resulting free-flowing and granular adsorbent-mixture combination over the adsorbent in the column, tapping the column gently for even settling. Be sure the top surface is level.

Step 6: Pour 1 cm of adsorbent over the adsorbent-mixture combination, tapping the column gently for even settling. Then add 1 cm of clean sand over this adsorbent, tapping the column gently for even settling.

Step 7: Pour the proper developing solvent gently onto the sand and let it percolate down the column. Maintain at least a 5-cm column of solvent over the sand during this time.

Step 8: When the solvent is near the bottom of the adsorbent, pour off any remaining solvent at the top of the column and lay the column on its side to prevent any further gravity flow.

Step 9: Carefully push the adsorbent column material out of the tube, using gentle air pressure at the tip near the open stopcock. Be certain not to change the order of the material that has come out of the tube.

Step 10: Use a spatula to separate each band in the column. For colorless compounds, an ultraviolet light and the original use of phosphor will be required (see Section 6.2G).

Step 11: Separate the compound in each band from the adsorbent by washing each band three or four times with a volatile, polar solvent. Gravity filter the solvent through fluted filter paper to remove the adsorbent; then evaporate the solvent to isolate each compound.

6.2I. Advantages and disadvantages of column chromatography

Column chromatography has the advantage of being able to effectively purify relatively small amounts of material. Normally this purification is much more complete than either distillation or recrystallization. Its disadvantages include it being quite time consuming, with several trials frequently being necessary to establish acceptable conditions for new mixtures. It is also not normally suitable for the purification of very large quantities, since most columns have only a limited capacity.

6.3 PAPER AND THIN-LAYER CHROMATOGRAPHY

The theory behind the separation of mixtures by paper and thin-layer chromatography is identical to that for column chromatography (see Section 6.2). Again, less polar compounds are held less tightly to the adsorbent than are more polar

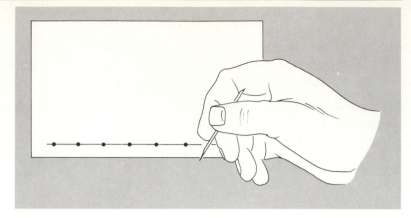

FIGURE 6-7
Spotting a thin-layer or paper chromatography sheet with a
micropipette.

compounds, so they can move faster when the mixture is eluted with a solvent. Again, more polar solvents will move the components in the mixture faster than less polar solvents will.

The main difference between the methods is that in paper and thin-layer chromatography, the adsorbent is used in the form of a thin layer, rather than a column.[1] As a result, not as much of the mixture can be separated, but the number of components in the mixture and the best conditions to separate these components on a larger scale can usually be discovered much faster and with much less effort than with column chromatography. In addition, with paper and thin-layer chromatography, the eluting solvent climbs up the adsorbent by capillary action, instead of flowing downward as with column chromatography.

In paper chromatography, one simply uses a sheet of Whatman No. 1 or similar paper. For thin-layer chromatography, one uses a thin layer of the adsorbent spread over and adhered to a sheet of plastic or glass. The mixture to be separated is then applied, as shown in Figure 6-7, with a micropipette near one edge of the sheet as a small spot of solution. After the solvent has evaporated, the sheet is then placed vertically, as shown in Figure 6-8, in a closed developing chamber with the spotted edge down. The solvent in the bottom of the chamber then climbs up the sheet by capillary action, passes over the spot, and, as it rises, separates the components in the spot. When the solvent has almost reached the top, the developed sheet is then removed from the chamber, and the position of the solvent front and all the separated components is noted and marked with a pencil. When completed, a developed chromatogram might look like Figure 6-9, with a number of spots located at various positions above

---

[1] In paper chromatography, the adsorbent is actually not the paper but the water adsorbed on the cellulose from the solvent and the atmosphere.

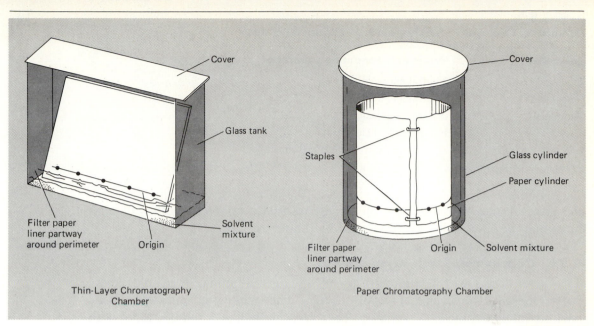

FIGURE 6-8
Development chambers for thin-layer and paper chromatography.

FIGURE 6-9
A typical, developed chromatogram.

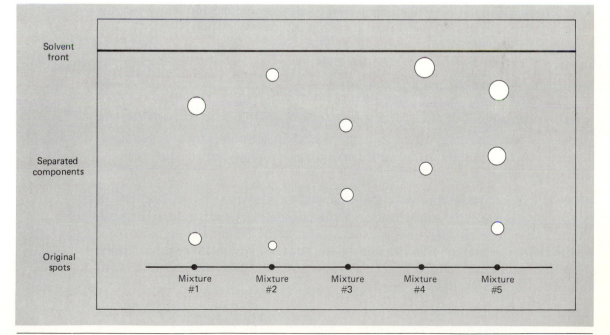

TABLE 6-5.

Adsorbents Commonly Used in Thin-Layer Chromatography

Adsorbent	Types of compounds separated
Silica gel G	Neutral and acidic
Silica gel H	Neutral and acidic
Alumina	Neutral and basic
Cellulose	Polar biochemical substances
Cellulose–silica gel G	Acidic and polar biochemical substances

the original ones. Observe from all the figures that a number of different mixtures may be simultaneously separated by the procedure.

6.3A. Commercial thin-layer sheets

Thin-layer sheets with a durable layer of silica gel, alumina, or cellulose may be purchased from chemical supply houses. The ones with a flexible plastic backing have become very common. They resist flaking, are manufactured with a consistent, uniform thickness of adsorbent, and can be easily cut with a pair of scissors to the needed size. However, they are expensive, buckle in some solvents, such as chloroform, and twist when heated during sulfuric acid charring.

6.3B. Lab preparation of thin-layer sheets

Thin-layer sheets may also be prepared in the laboratory. If glass is the backing, they are also commonly called thin-layer plates; if microscope slides are the backing, they are also commonly called thin-layer slides. Table 6-5 lists some common adsorbents used.

PROCEDURE FOR THIN-LAYER SLIDES

Step 1:   Wash the microscope slides with detergent and water, rinse them with distilled water, and dry. After cleaning, pick them up only at the edges, since fingerprints on the slide surface will make it more difficult for the adsorbent to adhere to the glass.

Step 2:   Stir in a wide-mouth, screw-capped jar a suspension of 40 grams of silica gel G (or another adsorbent) in 85 mL of methylene chloride and 15 mL of methanol. The G stands for gypsum, which is the binder for the adsorbent to the slide. The methanol in the solvent enables the gypsum to set more firmly. This suspension can be used by many people and can be reused later if the container has been kept tightly sealed.

Step 3:   Stir the slurry just before use to ensure thorough mixing. If it has become too thick, add more solvent.

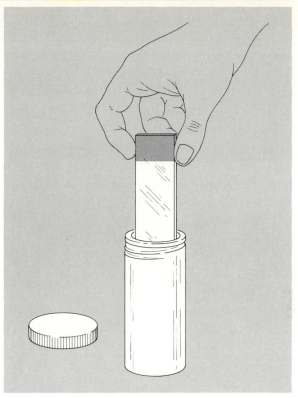

FIGURE 6-10
Preparing thin-layer chromatography slides by dipping.

Step 4: Place two slides together, back to back, and hold them on one end by the edges.

Step 5: Lower the slides as far as possible into the slurry, then remove them about 2 seconds later with a slow and steady motion, as shown in Figure 6-10.

Step 6: Allow the excess slurry to drip back into the container, then carefully separate the slides and place them, face up, on a clean surface to thoroughly dry. Continue to handle them only by the edges or by the portion not covered with adsorbent. The finished slides should have an even coating, without any lumps, streaks, or glass showing through the adsorbent. Thin or streaked slides probably result from the slurry not being thick enough or not being stirred before use.

PROCEDURE FOR LARGER THIN-LAYER PLATES

Step 1: Wash the glass plates, up to about 20 × 25 cm in size, with detergent and water. Rinse with distilled water and dry. After cleaning, pick them up only at the edges, since fingerprints on the surface will make it more difficult for the adsorbent to adhere to the glass.

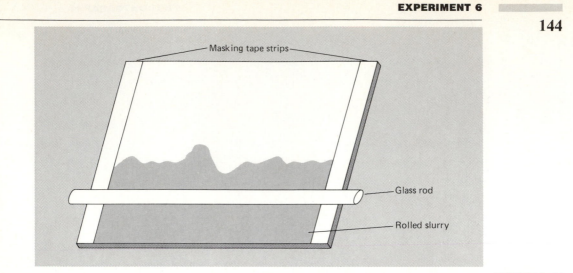

Masking tape strips

Glass rod

Rolled slurry

FIGURE 6-11
Preparing a large thin-layer plate.

Step 2: Place the plates on a sheet of newspaper, and place strips of masking tape on the two opposite sides of each plate alongside the edges. If you want a thicker coating on the plate, put a second, or even a third, layer of masking tape over the first one.

Step 3: Prepare a slurry of the adsorbent in water, using a ratio of about 1 gram of adsorbent to 2 mL of water.

Step 4: Stir the slurry and pour it along one of the untaped edges of the plate. Use a long, thick glass rod to level and spread the slurry over the plate, as shown in Figure 6-11, rolling the rod back and forth, as necessary. Do not delay, since the slurry will thicken in a few minutes.

Step 5: Wait 10 minutes, then carefully remove the strips of masking tape. Dry the plates in a 110° oven for about 30 minutes to 1 hour.

6.3C. Preparing micropipettes

A number of micropipettes must be made in order to apply the liquid sample or solution to the paper or thin-layer chromatography sheet. The number needed is equal to the number of mixtures being analyzed.

PROCEDURE

Step 1: Hold a melting-point capillary tube horizontally by both ends and place the middle portion into a burner flame, as shown in Figure 6-12.

Step 2: Gently rotate the capillary tube in the flame until the middle portion is soft.

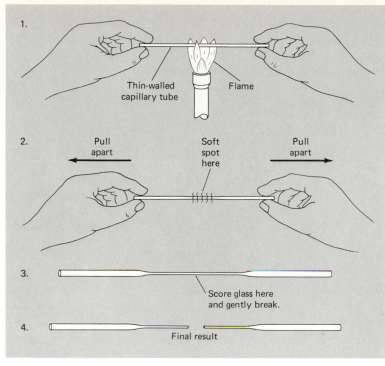

1.

Thin-walled
capillary tube   Flame

2.   Pull
apart   Soft
spot
here   Pull
apart

3.   Score glass here
and gently break.

4.   Final result

FIGURE 6-12
The steps in the preparation of micropipettes.

Step 3: Remove the capillary tube from the flame and immediately pull the two ends apart about 4 to 5 cm until the tube has about one-tenth of its original diameter in the center.

Step 4: Score the tube lightly in the center of the pulled section and carefully break to give two micropipettes. Repeat these steps to produce all the micropipettes needed.

6.3D. Choosing a developing solvent

Success in a paper or thin-layer chromatography is only possible if the proper developing solvent has been chosen. If the solvent is not polar enough to move the components in the mixture, little separation will result. But if the solvent is too polar, all the components will travel almost equally well, and again little separation will result.

If a mixture has been separated previously, a good developing solvent will usually be indicated. For new mixtures, choosing the right solvent is largely a matter of trial and error. For many compounds, successful results can be attained by using hexane or petroleum ether with varying proportions of toluene, methylene chloride, or anhydrous ethyl ether. More polar compounds may require various proportions of ethyl acetate, acetone, methanol, or even water. Frequently, a blend of three or four liquids may be necessary for best results.

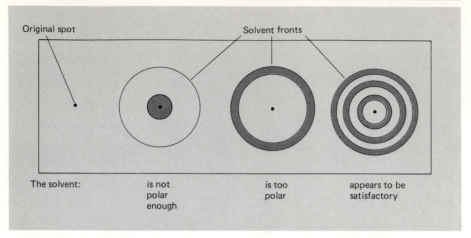

FIGURE 6-13
Testing solvents for chromatography by the concentric ring
method.

To rapidly determine a good solvent, one can apply a number of small spots of the mixture to a thin-layer slide or a small piece of Whatman No. 1 paper. These spots should be at least $1\frac{1}{2}$ cm apart. The possible solvent is then drawn up into a micropipette by capillary action, and gently and quickly touched to one of the spots. As the solvent front expands outward in a circle, the components in the mixture will also move outward to form concentric rings. From the appearance of these rings, the suitability of the solvent can be judged. This process is repeated with a new solvent on another spot until a good solvent is found. Figure 6-13 shows what might be found with various solvents.

6.3E. Preparing the developing chamber

After a satisfactory developing solvent has been found, the developing chamber can be prepared. This should be done before the mixture is applied to the paper or thin-layer sheet to allow the solvent vapor to thoroughly saturate the chamber.

PROCEDURE

Step 1: Place a piece of rectangular filter paper vertically around part of the inside perimeter of the developing chamber. This paper acts as a wick to ensure that the chamber is saturated with solvent vapor for a good and fast chromatographic resolution.

Step 2: Slowly pour the developing solvent over the filter paper, adding enough to produce a 5-mm horizontal layer at the bottom of the chamber.

Step 3: Put the cover on and set the chamber aside to allow the solvent vapor to saturate the inside.

6.3F. Applying the sample

Step 1: Place a piece of Whatman No. 1 paper or a thin-layer sheet or slide onto a larger-size piece of paper. Cleanliness of the surroundings is important for good results.

Step 2: If using a microscope slide, lightly draw a straight pencil line (no pen) across the slide about 1 cm from the bottom, as shown in Figure 6-14. If using a longer thin-layer sheet or a piece of Whatman No. 1 paper, you may draw the line about $1\frac{1}{2}$ to 2 cm from the bottom.

Step 3: If the sample is a solid, dissolve about 1 mg of it in several drops of a volatile solvent such as methylene chloride.

Step 4: Dip the tip of the micropipette into the prepared solution or the liquid to be analyzed. By capillary action, some of the liquid will be drawn up.

Step 5: Quickly touch the tip of the micropipette to the left side of the pencil line, to produce a spot no larger than 1 to 2 mm in diameter. **Do not** hold the micropipette on the paper, sheet, or slide, since its entire contents will be delivered and the spot will be too large. Large spots are unsatisfactory because during development they will tend to streak or tail, giving poor separations. You may want to practice the technique beforehand on a piece of filter paper.

Step 6: If necessary, add material to the first spot by quickly touching the micropipette to the spot one or two more times. Be sure the solvent from the

FIGURE 6-14
Drawing a light pencil line 1 cm from the bottom of a microscope slide.

previous application has completely evaporated before placing more material on the spot.

Step 7: Repeat steps 3 through 6 for every sample being analyzed, evenly spacing the spots along the pencil line as shown in Figure 6-7. To prevent the spots from running together, be sure to place them sufficiently far apart. For example, no more than three samples should be placed on a microscope slide. This distance should be increased to over 1 cm when the solvent has further to travel. Also be sure to use a different micropipette with each new sample to prevent contamination. If a number of samples are being applied, indicate in your notebook the exact sequence to avoid possible confusion later.

6.3G. Developing the chromatogram

Step 1: If performing a paper chromatography, coil the spotted chromatogram to form a cylinder and staple the ends together as shown in Figure 6-8. **Do not allow the ends to touch or overlap after they are stapled.**

Step 2: Remove the cover of the solvent chamber prepared in Section 6.3E, and place the paper or thin-layer chromatogram, spotted-side down, into the chamber, as shown in Figure 6-8. Microscope slides will go into a much smaller chamber. Be sure the level of the spots is above the level of the developing solvent. Otherwise the solvent will dissolve the spots, and you will have to prepare a new chromatogram. Try to have the chromatogram touch the filter paper liner as little as possible. For a paper chromatography, the cylinder should be placed in the middle of the chamber, so that it doesn't have to touch the filter paper liner at all.

Step 3: Put the cover back on the developing chamber and wait until the solvent level has climbed almost to the top of the chromatogram. The solvent level will rise very rapidly at first, but then will rise at a progressively slower rate.

Step 4: Remove the chromatogram from the developing chamber. If performing a paper chromatography, remove the staples from the cylinder. Lay the chromatogram down flat and immediately mark with a pencil the highest level reached by the solvent.

Step 5: Outline with a pencil the location of all the spots. If all the spots are not visible, first make them visible by the procedure in Section 6.3H.

Step 6: Compute the R_f value of each spot by the procedure in Section 6.3I.

6.3H. Visualization methods

If all the components of a mixture have color, their separation by thin-layer or paper chromatography is easily determined visually. Usually, however, this is not the case and visualization methods are necessary.

Many colorless compounds may be detected by illumination with ultra-violet light and will appear as bright spots due to fluorescence. A fluorescent indicator, such as a mixture of zinc and cadmium sulfides, may also be mixed with the adsorbent used to coat the thin-layer sheet. With this technique, any compounds that are fluorescent will appear as bright spots on a light back-ground, while all others will be visible as dark spots due to the absorption of the light.

Another visualization method often used is to treat the developed chro-matogram with iodine vapor. The chromatogram is placed in a container with a few crystals of iodine, and the container is gently warmed on a steam bath. The iodine vapor reacts with most organic compounds to form yellow, brown, or violet complexes. In fact, the only common compounds that don't react are alkanes and alkyl halides. The chromatogram is then removed from the container, and the spots produced are outlined with a pencil. This must be done immediately, since the spots will usually disappear as the iodine sublimes. However, the ultraviolet detection procedure should usually be tried first, since iodine can alter some compounds.

Many other chemical reagents are also available for detecting specific compounds. For example, 2,4-dinitrophenylhydrazine reacts with aldehydes and ketones to form red to yellow spots. Ninhydrin reacts with amino acids to give a purplish-blue color. *p*-Dimethylaminobenzaldehyde can be used to detect amines, and potassium dichromate and potassium permanganate can detect easily oxidized compounds. Ferric chloride will make phenols visible, bromocresol green can be used for carboxylic acids, and silver nitrate can be used for alkyl halides. The silver halides produced decompose with light and yield dark spots of free silver. However, all these detection techniques should be used **only** when one does not mind the destruction or alteration of the orig-inal compounds on the chromatogram.

6.31. The R_f value

Once all the spots have been located and marked, each one is given an R_f value, which is a measure of how much that particular component of the mixture has moved relative to the solvent front. Each R_f value may be computed by the equation

$$R_f = \frac{\text{Distance traveled by the component}}{\text{Distance traveled by the solvent front}}$$

As can be seen, the more a component is able to move with the mobile phase, the larger will be its R_f value. By definition, all R_f values will be in the range of 0.00 to 1.00. In paper chromatography, the distance traveled by each compo-nent is usually measured from the top of the spot; in thin-layer chromatography, from the center of the spot. If there is any extensive tailing or streaking, the R_f value should be reported as a range. The size of the range will indicate the extent of the tailing.

As an example of the calculation, if the solvent front in Figure 6-9 traveled a distance of 145 mm and the two components of mixture 1 traveled 23 and 110 mm, respectively, the R_f values of these compounds would be

$$\text{Component 1} \quad R_f = \frac{23}{145} = 0.16 \qquad \text{Component 2} \quad R_f = \frac{110}{145} = 0.76$$

The R_f value can be very useful in identifying compounds, since it is a constant for a particular substance if the adsorbent, solvent, and other factors such as the thickness of the adsorbent and the amount of material applied are not varied. To avoid any coincidence in the result obtained, the unknown should give the same value as the known when the two are run, side by side, on the same chromatogram for each of several different adsorbents and solvents.

6.3J. Uses of thin-layer and paper chromatography

To achieve complete separation in thin-layer and paper chromatography, it is very important that you apply only a very small quantity of sample, usually much less than a milligram. For this reason, thin-layer and paper chromatography are used almost entirely for qualitative analyses, rather than for preparative separations. They are very valuable for

1. Determining the number of components in a mixture

2. Determining whether certain components are present in a mixture

3. Determining the proper conditions for running a column chromatography

4. Monitoring a separation by column chromatography

5. Determining the effectiveness of the separation achieved by other methods, such as recrystallization or extraction

6. Determining the purity of a commercial product

7. Monitoring the progress of a reaction

Of special value is use 7. For example, assume that a certain reaction converts reactant A into product B with the formation of a small amount of side product C formed from product B. By running thin-layer or paper chromatograms at various times, one can determine what is the optimal time for producing the largest yield of product B without the complication of a large amount of side product C.

After 1.0 hour, for example, one might discover from the chromatogram that a large amount of B had been produced, with no detectable amount of C, but with a significant amount of A still remaining. After 1.5 hours, the situation is even better with more B being present, much less A, and still no detectable amount of side product C. After 2.0 hours, there was no more A present, but a small amount of C was evident. After 2.5 hours, there was still no more A present, but now the amount of C was significantly higher. Therefore, the optimal time for the reaction was either 2.0 hours or slightly less than 2.0 hours.

If desired, one can scale up thin-layer chromatography to preparative separations of half a gram or more by using a larger area and thickness of adsorbent and by applying the mixture as a long thin line rather than as a spot. The separated substances may then be isolated by cutting out or scraping off each individual band produced, extracting each substance from the adsorbent, filtering the solutions, and evaporating the solvent. As an alternative, the dry-column procedure (Section 6.2H) may be used.

6.3K. Advantages and disadvantages of thin-layer and paper chromatography

Both thin-layer and paper chromatography are very valuable for many purposes and have the advantage that only very small quantities of material are needed. Material is not wasted, which can be very important when only a small amount of the mixture is available. Some disadvantages are that

1. The method is only qualitative or, at the most, semiquantitative.

2. Volatile materials may not be a part of the mixture, since they will evaporate from the sheet.

3. It may take some time to find a solvent which will effectively separate all the components in the mixture.

6.4 GAS CHROMATOGRAPHY

Whenever the components of a mixture are volatile or show an appreciable vapor pressure at elevated temperatures, they may be separated and identified by gas chromatography. Developed in 1952, gas chromatography has become very popular due to its simplicity and the excellent results commonly obtained. Besides gas chromatography, the method is also referred to as "GC," vapor-phase chromatography (VPC), and gas-liquid partition chromatography (GLPC).

Gas chromatography is similar in principle to column chromatography, but differs in the following ways:

1. Instead of the stationary phase on the column being a solid adsorbent such as alumina, silica gel, or cellulose, a nonvolatile liquid or solid such as silicone oil, hydrocarbon grease, or polyethylene glycol is used as a coating on an inert, powdered support material, such as crushed firebrick or diatomaceous earth.

2. The mobile phase, instead of being a liquid, is an inert gas, such as helium or nitrogen.

3. The separation of the components in the mixture depends on how strongly these components are adsorbed on the stationary phase and on their relative vapor pressures, which will determine their solubility in the moving gas phase. As a result, if the adsorption of two substances on the stationary phase was about the same, the separation would be primarily the result of their difference in vapor pressure, with the more volatile component moving through faster.

4. The nature of the carrier gas does not affect the solubility and the resulting movement of the components in the mixture. Thus, the polarity or nonpolarity of the mobile phase is **not** a factor in the separation.

5. The temperature of the column can be changed and plays a very important role in gas chromatography, since the column temperature will directly influence the vapor pressures of the mixture components.

6.4A. Gas chromatograph

The instrument used to separate a mixture by GLPC is usually called a **gas chromatograph**. A schematic diagram of a typical gas chromatograph is shown in Figure 6-15. A small sample of the mixture is injected with a syringe through a rubber plug or septum of the heated injection port, where it is immediately vaporized and introduced into the moving stream of helium or nitrogen. The carrier gas then takes the vaporized sample into a column containing the stationary adsorbent, where the various components are separated. When each component leaves the column, a detector sends an electronic signal to a recorder, which plots the output of the detector on a continuously moving roll of chart paper as a function of time. The end result is a gas chromatogram, such as the one in Figure 6-16.

FIGURE 6-15
Schematic diagram of a gas chromatograph.

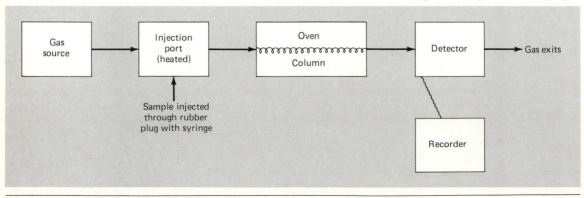

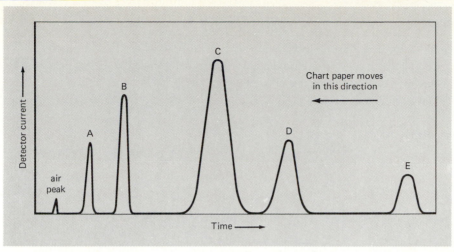

FIGURE 6-16
A typical gas chromatogram.

6.4B. The column

The packed column, where the mixture is separated, is the heart of the gas chromatograph. It is usually made of copper, stainless steel, or glass tubing, is usually $\frac{1}{8}$ to $\frac{1}{4}$ inch in diameter, and may be anywhere from 3 feet to over 300 feet in length. A length of 8 feet to 30 feet is the most common.

To prepare the column, you must first coat the nonvolatile liquid or solid onto the inert, powdered support material. The support material is usually added with stirring to a solution of the nonvolatile liquid or solid in a volatile solvent, such as methylene chloride. The solution is then slowly evaporated, leaving the support material evenly coated. Usually the amount of nonvolatile liquid or solid used will be about 5 to 30 percent by weight of the amount of support material.

The coated support material is then packed into the tubing as evenly as possible, and the tubing is bent or coiled so that it will fit inside the oven compartment of the gas chromatograph. The positions of the ends are then adjusted so that they can be attached to the gas entry and exit ports inside the oven.

6.4C. Support material and the nonvolatile liquid or solid

Some common support materials and nonvolatile liquid and solid adsorbents are given in Tables 6-6 and 6-7. Observe from Table 6-7 that the adsorbents have an upper temperature limit. Raising the temperature above this limit is not advisable, because the adsorbent will begin to "bleed" off the column. This may interfere with the analysis of the less volatile mixture components, and will make the column less effective in future separations.

Also observe from Table 6-7 how the adsorbents vary in polarity and how the choice of the proper adsorbent depends on the type of compounds being

TABLE 6-6.
Common Solid Supports

Chromosorb P: pink diatomaceous earth, highly absorptive, pH 6–7
Chromosorb W: white diatomaceous earth, medium absorptivity, pH 8–10
Chromosorb G: white diatomaceous earth, low absorptivity, pH 8–9
Chromosorb T: 40/60 mesh Teflon 6 beads
Crushed firebrick
Porapak: porous polymer beads
Glass beads
Silica
Alumina
Charcoal

separated. Generally, polar compounds are best separated with a polar adsorbent, while nonpolar compounds are best separated with a nonpolar adsorbent.

6.4D. Separation on the column

When the carrier gas and the mixture enter the heated column, the mixture is adsorbed onto the stationary adsorbent and begins to equilibrate between the liquid and gaseous phases, being liquefied and vaporized numerous times. Each time a compound is vaporized, it can move further through the column in the stream of carrier gas until it is again liquefied and adsorbed onto the column. A more volatile compound will vaporize more readily than a less volatile one and so will spend less time adsorbed on the column and more time moving through the column. This more volatile compound will, therefore, move through the column faster than the less volatile one, and the two will be separated. Under the right conditions, every compound in the mixture can be adsorbed and vaporized at a different rate, with a complete separation resulting.

6.4E. Factors affecting separation

Column temperature.   The right column temperature is critical for a complete separation. If the column temperature is too high, above the boiling point of all the mixture components, then the entire mixture will travel through the column at the same rate as the carrier gas. No separation will occur, because no equilibration has occurred with the liquid phase. On the other hand, if the column temperature is too low, some components of the mixture may remain in the liquid phase, adsorbed on the column. They will never move, since they are never revaporized. If the best column temperature for a particular mixture is not stated, it can be determined by trial and error, utilizing the information about the nature of the mixture.

Rate of flow of the carrier gas.   If the carrier gas is flowing through the column too fast, a poor separation may result because the mixture components

TABLE 6-7.
Common Nonvolatile Liquid and Solid Adsorbents

Name	Type	Composition	Maximum temperature (°C)	Polarity	Types of compounds separated
Apiezon (L, M, N, etc.)	Hydrocarbon greases	Mixture of hydrocarbons	250–300	Nonpolar	Hydrocarbons
Carbowaxes (MW 400–6000)	Polyethylene glycols of varying chain length	$HO\text{---}(CH_2CH_2\text{---}O)_n\,CH_2CH_2OH$	250	Polar	Alcohols, ethers, amines, ketones, aldehydes
DC-200	Silicone oil	$(CH_3)_3Si\text{---}O\text{---}[Si(CH_3)_2\text{---}O]_n$	250	Medium	Low-molecular-weight ketones, aldehydes, and halogen compounds
DC-550	Silicone oil	Like DC-200, but higher numbers for n	275	Medium	Ketones, aldehydes, and halogen compounds
SE-30	Methyl silicone rubber	Like silicone oil, but cross-linked	350–375	Nonpolar	Hydrocarbons, pesticides, and steroids; general for low-polarity compounds
QF-1	Fluoro silicone	Like silicone oil, but fluorinated in the alkyl groups	250	Medium	Polyalcohols, halogen compounds, alkaloids, pesticides, and steroids
Diethylene glycol succinate (DEGS)	Polyester	$\left(CH_2CH_2\text{---}O\text{---}\overset{\displaystyle O}{\underset{}{C}}\text{---}CH_2CH_2\text{---}\overset{\displaystyle O}{\underset{}{C}}\text{---}O\right)_n$	200	Polar	Esters, fatty acids, ethers, ketones, and aldehydes
Butanediol succinate	Polyester	$\left[(CH_2)_4O\text{---}C(\!=\!O)\text{---}CH_2CH_2\text{---}C(\!=\!O)\text{---}O\right]_n$	225	Medium	Esters, fatty acids, ethers, ketones, and aldehydes
Diisodecyl phthalate	Diester	(phthalate ring with —C(=O)—O—isodecyl and —C(=O)—O—isodecyl)	175	Polar	Generally useful

in the gas phase cannot be as effectively readsorbed on the column. But if the rate is too slow, the separation may needlessly take longer, due to some less volatile components being "held up" on the column.

Length of the column.   With other factors being unchanged, a longer column will produce a more effective separation than a shorter one, due to each mixture component undergoing many more adsorptions and revaporizations.

Liquid or solid adsorbent chosen.   Success in a separation depends on choosing the proper adsorbent. Some adsorbents are obviously more suitable than others, depending on the polarity, functional groups, and molecular weights of the mixture components. Table 6-7 may be used as a guide, with even more detailed information coming from a comprehensive text on gas chromatography. Always remember to consider the temperature limit of the adsorbent.

Components in the mixture.   If the components in the mixture are quite different in terms of polarity and volatility, then the mixture will usually be quite easy to separate. But if the components are very much alike, the separation will obviously be much more difficult. A longer column, the choice of the proper packing inside the column, and the correct temperature and carrier gas flow rate are all factors which will have to be utilized for a complete separation.

6.4F.   The detector

The detector is also an important part of the gas chromatograph. How else could one know when a mixture component is leaving the column and how much of that component is present?

The **thermal conductivity detector** is the most common. An electrically heated wire having a constant voltage is positioned in the gas stream at the column exit and is cooled by the passing gas. When only helium is passing by, the amount of cooling will remain constant, and the temperature of the wire and its electrical resistance will therefore also remain constant. But when some gas other than helium passes by, the wire is not cooled as effectively, since helium has a much higher thermal conductivity. Therefore, the temperature and the electrical resistance of the wire will rise. The increase in the electrical resistance is easily measured and is sent electronically to the recorder, which draws a peak proportional to the amount of organic compound present.

The **flame ionization detector** is more sensitive, but is less common. A portion of the gas from the column is sent to a hydrogen and air flame. When an organic compound is present, the electrical conductivity of the hot gas from the flame will increase, due to the formation of ions as the organic compound burns. This increase in conductivity is measured electronically and sent to a recorder, which again draws a peak proportional to the amount of organic compound present.

Flame ionization detectors commonly use nitrogen as the carrier gas. It is much less expensive than the helium gas required with thermal conductivity detectors.

6.4G. Retention time

The retention time is the amount of time required for a compound to pass through the column after the injection of the mixture. It is similar to the R_f value used in thin-layer and paper chromatography, and is a measure of how well a compound is able to move with the carrier gas. Thus, in Figure 6-16, compound A has the lowest retention time, except for air, since it was able to move through the column the fastest. Compound E, on the other hand, has the highest retention time, since it moved through the column the slowest.

The following factors are found to influence the retention time:

1. The nature of the compound: a more polar, less volatile compound is retained on the column longer.

2. The nature of the adsorbent: a more polar material retains the mixture components longer.

3. The concentration of the adsorbent: a higher concentration retains the mixture components longer.

4. The column temperature: a lower temperature retains the mixture components longer.

5. The flow rate of the carrier gas: a lower flow rate retains the mixture components longer.

6. The length of the column: a longer column retains the mixture components longer.

A clock is not necessary to measure the retention time. Most gas chromatograph recorders will move the chart paper at a constant rate, so knowing this rate and the number of divisions the paper has moved since the initial injection are all that is necessary.

The retention time, like the R_f value, can be very useful in identifying compounds, since it is a constant for a particular substance if the adsorbent, column length, temperature, flow rate of the gas, and other factors are not varied. To avoid any coincidence in the result obtained, the unknown should give the same value as the known for several different columns with a variety of adsorbents. To be absolutely certain, you may collect the unknown as it exits from the column and identify it by other means, such as infrared and NMR spectroscopy.

6.4H. Qualitative analysis

Unfortunately, the gas chromatograph gives **no** information about the identities of the compounds it has separated. However, if a certain peak is suspected to be a particular compound, the retention time can be used to help prove it (see Section 6.4G).

One usually determines the identity of compounds separated by a gas chromatograph by collecting separately the compounds as they exit from the column and analyzing them in other ways such as infrared or NMR spectroscopy.

6.4I. Quantitative analysis

After a gas chromatogram has been obtained, the amounts of the components in the mixture can be determined by first measuring the areas under all the peaks. If the recorder has either a mechanical or an electronic integrator, each of these areas is obtained automatically. Otherwise, the area of each peak may be measured either with a planimeter or by counting the number of squares of graph paper under each peak. If the peak is almost perfectly symmetrical, its area may also be obtained mathematically by multiplying 1.08 by the height of the peak by the width of the peak at its half-height. This is illustrated in Figure 6-17.

The areas under the peaks may also be obtained by labeling them, cutting them out carefully with scissors, and then individually weighing the pieces on an **analytical** balance. Since the chart paper has a reasonably constant weight per unit of area, the relative areas of the peaks will be proportional to their weights.

After you obtain all the areas under the peaks, calculate the **approximate** mole percentage composition of the mixture simply by dividing the area of each

FIGURE 6-17
Determining the area of a symmetrical peak.

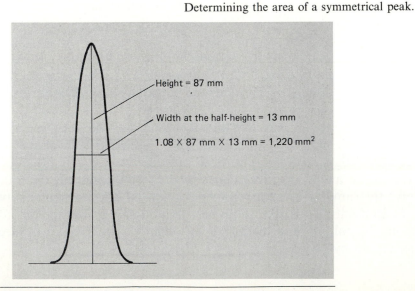

Height = 87 mm

Width at the half-height = 13 mm

1.08 × 87 mm × 13 mm = 1,220 mm^2

peak by the total area and then multiplying the result by 100 percent. For example:

$$
\begin{aligned}
\text{Peak A} &= \quad 395 \text{ mm}^2 \\
\text{Peak B} &= \quad 637 \text{ mm}^2 \\
\text{Peak C} &= 1940 \text{ mm}^2 \\
\text{Peak D} &= \underline{1190 \text{ mm}^2} \\
&\quad\; 4162 \text{ mm}^2
\end{aligned}
$$

$$
\text{Mol \% A} = \frac{395 \text{ mm}^2}{4162 \text{ mm}^2} \times 100\% = \quad 9.5\%
$$

$$
\text{Mol \% B} = \frac{637 \text{ mm}^2}{4162 \text{ mm}^2} \times 100\% = \quad 15.3\%
$$

$$
\text{Mol \% C} = \frac{1940 \text{ mm}^2}{4162 \text{ mm}^2} \times 100\% = \quad 46.6\%
$$

$$
\text{Mol \% D} = \frac{1190 \text{ mm}^2}{4162 \text{ mm}^2} \times 100\% = \underline{\quad 28.6\%}
$$
$$
\phantom{\text{Mol \% D} = \frac{1190 \text{ mm}^2}{4162 \text{ mm}^2} \times 100\% =\;\;} 100.0\%
$$

However, this calculation assumes that the detector is equally sensitive to all the compounds in the mixture. Usually this is not the case, since most compounds have different thermal conductivities. Therefore, for more accurate results, a quantitatively known mixture of the compounds must be prepared and injected. After you obtain this second chromatogram, calculate the relative sensitivity of the detector to each component by dividing the percentage area of each component by its known mole percentage in the mixture. You can then determine the adjusted mole percentage composition of the first chromatogram by dividing the percentage area of each component by its relative sensitivity factor.

For instance, suppose that the four compounds A, B, C, and D in the example are identified. A known mixture, quantitatively containing an equal mole percentage of each compound, is then prepared and injected into the gas chromatograph. Upon analysis, this chromatogram gives peak areas of 21.6% A, 29.2% B, 23.4% C, and 25.8% D. The relative sensitivity of the detector to each component is

$$
\text{Component A} = \frac{21.6\%}{25.0\%} = 0.864
$$

$$
\text{Component B} = \frac{29.2\%}{25.0\%} = 1.17
$$

$$
\text{Component C} = \frac{23.4\%}{25.0\%} = 0.936
$$

$$
\text{Component D} = \frac{25.8\%}{25.0\%} = 1.03
$$

Dividing the percentage area of each component in the original mixture by its respective relative sensitivity, we obtain for the adjusted mole percentages:

$$\text{Component A} = \frac{9.5\%}{0.864} = 11.0\% \times \frac{100.0}{101.7} = 10.8\%$$

$$\text{Component B} = \frac{15.3\%}{1.17} = 13.1\% \times \frac{100.0}{101.7} = 12.9\%$$

$$\text{Component C} = \frac{46.6\%}{0.936} = 49.8\% \times \frac{100.0}{101.7} = 49.0\%$$

$$\text{Component D} = \frac{28.6\%}{1.03} = \frac{27.8\%}{101.7\%} \times \frac{100.0}{101.7} = \frac{27.3\%}{100.0\%}$$

Observe that since the adjusted mole percentages do not initially total 100.0 percent, they are multiplied by whatever factor is necessary to make the total 100.0 percent.

6.4J. Preparative separations

Like paper and thin-layer chromatography, gas chromatography gives best results in separating a mixture when only a very small sample is used, usually around 1 to 20 μL (1 μL = 0.001 mL). For this reason, gas chromatography is used mostly for qualitative and quantitative analyses. However, it may also be used for preparative separations if a larger-size column and samples as large as 0.5 mL are used. The separated substances may then be individually collected in a number of different traps, with the trap being changed each time a new peak appears on the chromatogram. These traps are usually U-shaped tubes, cooled by ice water, liquid nitrogen, or dry ice in acetone.

One can scale a preparative separation up to much more than 0.5 mL by multiple injections. Changing all the traps for each new injection could be very involved and tiresome, but fortunately machines can be attached to the gas chromatograph to do this job automatically.

6.4K. Other specific uses

In addition to qualitative analysis, quantitative analysis, and preparative separations, gas chromatography has a number of other valuable functions. Among them are the following:

1. Determining the purity of a commercial product

2. Determining the effectiveness of the separation achieved by a distillation or extraction

3. Monitoring the progress of a reaction

The last use can be especially important, as demonstrated by the example in Section 6.3J.

6.4L. Advantages and disadvantages

One important advantage of gas chromatography is that only very small quantities of material are needed. Material is not wasted, which can be very important when only a small amount of the mixture is available.

Another advantage is that even trace amounts of compounds can be detected, especially when a flame ionization detector is used. This can be very important during analysis for very toxic materials, such as certain pesticides.

Some disadvantages include the fact that only compounds with an appreciable vapor pressure at higher temperatures may be separated and detected, the compounds may not be thermally unstable, and it may take some time to find the right conditions to effectively separate all the components in the mixture.

6.4M. Procedure for obtaining a gas chromatogram

Do these steps **only** in the presence of your instructor until you have demonstrated competence in the use of the gas chromatograph.

Step 1: Adjust the column temperature and the injection port temperature control as needed or as stated in the experiment. Use low temperatures for compounds having a high volatility, high temperatures for compounds having a low volatility.

Step 2: Turn on the helium carrier gas and adjust it to a flow rate of about 40 mL per minute.

Step 3: Turn on the power supply and allow the equipment to equilibrate for about 2 hours.

Step 4: When the equipment is equilibrated, set the instrument's attenuation (sensitivity scale) as directed or as needed. If the peaks in your chromatogram are too small, decrease the attenuation setting. If the peaks go off the paper, either increase the attenuation setting or decrease the amount of your sample in your next injection.

Step 5: Turn on the recorder, if necessary adjusting the pen to the zero line.

Step 6: Carefully draw from 0.3 to 5.0 μL of the sample or the standard into a 10.0-μL syringe. Be sure no air is trapped in the syringe, and remember to handle the syringe with care since it is fragile and expensive.

If your sample is a solid at room temperature, first dissolve it in a volatile solvent such as methylene chloride or anhydrous ether. Do not confuse the solvent peak with the later sample peaks.

Step 7: Keeping the plunger in place, gently insert the syringe needle as far as it will go through the rubber septum of the injection port.

Step 8: Push the plunger to inject the sample onto the column, remove the syringe, and **immediately** make some type of mark on the chart paper to indicate at what point the sample was injected.

Step 9: After the chromatogram is completed, repeat steps 6 through 8 for all the samples to be analyzed, being sure not to inject a new sample until all of the old sample is off the column.

Step 10: When finished, do **not** turn off the flow of helium until after the instrument has been shut off and the column is completely cooled down. The detector may be ruined if there is not a continuous flow of gas to cool it.

6A. PAPER CHROMATOGRAPHY OF ARTIFICIAL FOOD COLORING AND KOOL-AID POWDERED DRINK MIX

After all, you can't expect men not to judge by appearances.

Ellen Glasgow, *The Sheltered Life*, Harcourt Brace Jovanovich, Inc.

INTRODUCTION

Most of the food people eat today has color. Orange juice is orange. Tomatoes, cherries, raspberries, and strawberries are red. Many vegetables are green. Meat is frequently red before being cooked, and brown afterwards. All these colors give the food eye appeal and make it more appetizing for us.

These colors are all natural ones. Today, however, much of what we eat also contains artificial colors through the addition of synthetic dyes. Now why should artificial colors be added to food at all? This question is more easily answered from the point of view of the food processor, who knows that the eye appeal of a product will influence its sales. For example, a customer is much more likely to buy oranges if they have a bright orange skin, rather than the more natural green and yellow one. Margarine would not sell very well if it did not have artificial yellow color added. A person expects strawberry or cherry soda to be red, orange soda to be orange, and grape soda to be purple. Would you buy dog food that did not have the nice red or brown appearance of meat, even though Rover is color blind and cannot tell the difference?

This practice of adding coloring agents to food is not a recent one. For centuries people have been aware of the importance of eye appeal in making food more appetizing. But until the last century, the needed colors were derived entirely from natural biological sources. For example, green was obtained from chlorophyll, brown from charred sugar and tree bark, blue from grape skins, red from beets, sandalwood, and alkanet roots, and yellow from carrots, annoto seeds, and turmeric.

The era of artificial synthetic dyes began in 1856 when Perkin was successful in synthesizing mauve, the first dye derived from coal tar sources (see Experiment 20). Shortly after, a number of other synthetic dyes were also produced, which gradually found their way into foodstuffs, replacing many of the natural ones. The reasons were greater stability, more intense colors, a wider color range, and a lower cost than the natural coloring agents. Today over 90 percent of the dyes added to foods are synthetic. In fact, since 1950 the per capita consumption of artificial food colors has more than tripled.

Unfortunately, a number of artificial food colors have been found to be unsafe for use. As a result, many of them have been removed from the market. The most recently removed dyes are red dye No. 2, which had been the most widely used food dye in the industry, red dye No. 4, used in maraschino cherries, and carbon black, used in candies. Previously, six once-approved dyes had been banned in 1960, and three others in 1950. Studies of these dyes had revealed that they were either toxic, or could cause birth defects, heart disease, or cancer.

Currently there are eight dyes allowed for food use without any restrictions. Table 6-8 gives their names and structures. The latest member of the group is red dye No. 40, which has taken the place of red dye No. 2. Before being approved, this dye was given the most stringent and extensive testing program ever carried out on a food dye. With the scrutiny now being given to all food additives, it is evident that any future food colors will have to undergo the same rigorous testing before approval for their use is given.

The status of some of these eight dyes, however, is now in limbo. Test results submitted from 1983–1985 have indicated that red dye No. 3 is carcinogenic, while conflicting evidence was attained with blue dye No. 2. As a result, the Public Citizen Health Research Group has sought to ban both of these dyes and has sued the FDA for its refusal to do so. The Public Citizen Health Research Group has also sued the FDA to test more thoroughly yellow dye No. 5 and No. 6. When and how these lawsuits will be resolved is still uncertain.

DISCUSSION

In this experiment, you will analyze by paper chromatography a number of different flavors of Kool-Aid powdered drink mix and some commercial food colors, such as Durkee, Ehlers, or McCormack. From the results, you will be able to determine how complex the food color mixture is in each product and if any dyes are present in more than one product. If pure samples of blue dye No. 1, blue dye No. 2, red dye No. 40, red dye No. 3, yellow dye No. 5, and yellow dye No. 6 are available, you will also, by comparing the R_f values, be

TABLE 6-8.

Eight Food Colors Currently Approved by the Food and Drug Administration

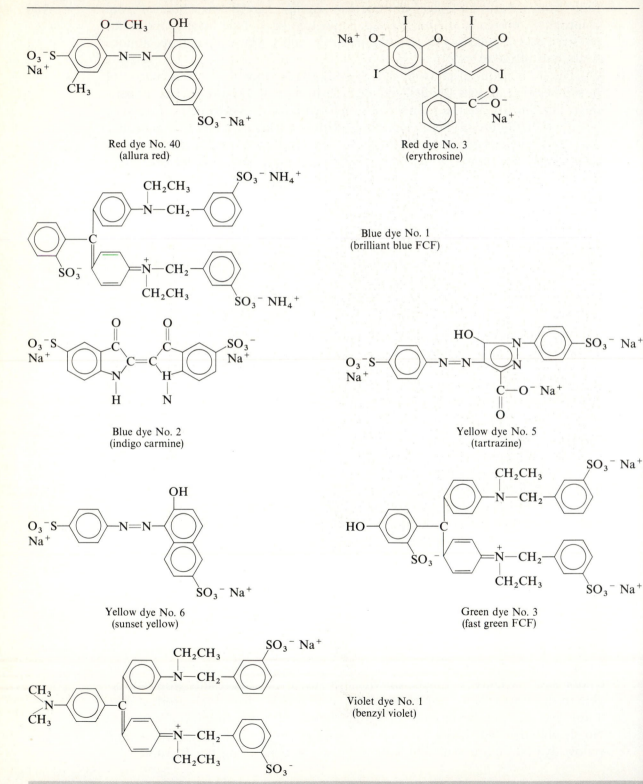

Red dye No. 40
(allura red)

Red dye No. 3
(erythrosine)

Blue dye No. 1
(brilliant blue FCF)

Blue dye No. 2
(indigo carmine)

Yellow dye No. 5
(tartrazine)

Yellow dye No. 6
(sunset yellow)

Green dye No. 3
(fast green FCF)

Violet dye No. 1
(benzyl violet)

able to know specifically which food dyes are present in each of the various products. This is not necessary for green dye No. 3 and violet dye No. 1 because they are the only artificial green and violet dyes permitted. Combinations of the other artificial dyes, of course, could produce green and violet colors.

Three different solvent systems will be used in the chromatography. This will allow you to see how the selection of the solvent affects the R_f value of each dye and determines how effectively the various mixtures are separated. Since it takes at least 40 minutes for each solvent system to climb up the paper, it is advisable during this waiting period to continue with portions of other experiments.

PROCEDURE

To apply the colors, you need 18 micropipettes, which you can prepare by the procedure given in Section 6.3C. You will also need three developing chambers and covers and six rectangular pieces of Whatman No. 1 paper, at least 16×28 cm, which will fit into the developing chambers when curled. Three of these will be used as paper liner wicks, and the other three will be used for the actual chromatograms.

Prepare the three developing chambers, as indicated in Section 6.3E, using the following solvent systems. Be sure to label which solvent system is in each chamber.

Chamber 1	Chamber 2	Chamber 3
Equal volumes of	60% isopropanol	95% ethanol
2 M NH$_4$OH	40% water	
95% ethanol	(by volume)	
Butanol		

Place a small amount of each available flavor of Kool-Aid in separate, small, **labeled**, test tubes. The possible flavors and the way they should be labeled are indicated below. Add warm water dropwise to each tube, mixing until the powder dissolves. Be sure to thoroughly rinse your stirring rod each time.

Black cherry (BC)	Punch (P)
Cherry (C)	Orange (O)
Grape (G)	Raspberry (R)
Lime (L)	Strawberry (S)

Using a different micropipette for each sample, apply each of the Kool-Aid flavors on the three Whatman No. 1 paper chromatograms, using the procedure in Section 6.3F. The spots should be no less than 1.5 cm apart and must be no more than 1 to 2 mm in diameter. Be sure to label underneath in pencil (no pen) which spot corresponds to each flavor.

Next apply the four colors of the commercial Durkee, Ehlers, or McCormack food colors and, if available, 2% aqueous solutions of blue dye No. 1, blue dye No. 2, red dye No. 40, red dye No. 3, yellow dye No. 5, and yellow dye No. 6. Again label these spots underneath in pencil. Although the order of the spots may be different, your final chromatograms before developing should look like the one in Figure 6-18.

When all the spots on each chromatogram are thoroughly dry, develop each chromatogram by the procedure in Section 6.3G. Place the computed R_f values (Section 6.3I) of your Kool-Aid flavor colors, the commercial colors, and the known reference dyes in your notebook. From your results, indicate the complexity of the food color mixture in each flavor and commercial color, which dyes are present in more than one product, and, if you used the reference dyes, state specifically which of the eight approved FDA dyes are present in each product.

Also from your results, indicate which solvent system gave the best results in separating the food colors and which solvent system gave the worst results. What basis did you use for your answer?

When finished, staple your chromatograms inside your notebook.

FIGURE 6-18
The food colors chromatogram before developing.

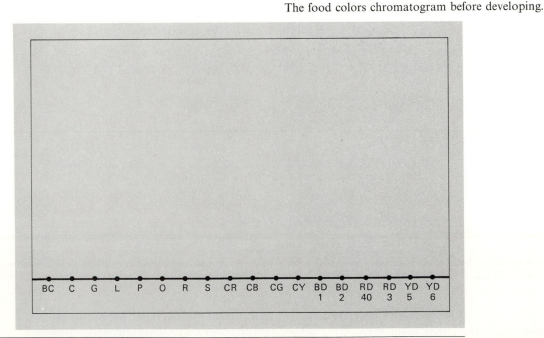

6B. THIN-LAYER CHROMATOGRAPHY OF PEN INKS

In this experiment, you will separate by thin-layer chromatography the dyes in a number of pen inks. This experiment is similar to the paper chromatography of artificial food colors (Experiment 6A), but since there are no FDA restrictions on pen inks, a much wider selection of dyes is possible. Besides black, blue, and red pens, many other colors are available, especially in the felt-tip type. Try to use as many colors as possible to obtain a large variety of results. You will also find it interesting to compare the inks present in the same color but sold by different manufacturers.

PROCEDURE

Obtain from your instructor a commercial, plastic-backed, thin-layer sheet with either a cellulose or an activated silica gel coating. Lay this sheet onto a larger-size piece of paper. (Cleanliness of the surroundings is necessary for good results.) About 1.5 cm from the bottom of the thin-layer sheet, lightly draw a straight **pencil line (no pen)** across the sheet. Then using narrow felt-tip pens of different color and manufacturer, make small spots no more than 1 to 2 mm in diameter and no less than 1.5 cm apart. Large spots are unacceptable because, during development, they will tend to streak or tail, thus giving poor separations or uncertain R_f values. You may want to practice the technique beforehand on a piece of paper to get it perfected. Since the commercial thin-layer plates are not cheap, you may be given only one for the experiment. Therefore, be careful not to make any mistakes. If any spots are too large, another spot of the same ink should be applied elsewhere. Be sure all your spots are labeled in pencil according to their color and the pen manufacturer.

After all the spots have completely dried, place the thin-layer sheet into the developing chamber, to which has been added a solution of 7:2:1 ethyl acetate, formic acid (90 percent), and water by volume. If a filter paper liner is used, be sure it does not touch the thin-layer sheet at any point, since then the solvent will begin to diffuse onto the sheet from that point. Be sure that only about 5 mm of solvent covers the bottom of the chamber to prevent the spots from dissolving in the solvent. Cover the top of the chamber and wait until the solvent rises to near the top of the thin-layer sheet. During this time, portions of other experiments may be continued or completed.

When the solvent is near the top, remove the thin-layer sheet and mark its level with a pencil. Allow the chromatogram to dry. Calculate all the R_f values (see Section 6.3I) and place the results in your notebook. From your results, compare how complex the various inks are, and indicate if any contain the same dyes, especially the ones of the same color by different manufacturers.

When finished, attach your chromatogram, protected inside an envelope, within your notebook.

6C. COLUMN CHROMATOGRAPHY OF TOMATO PASTE: THE ISOLATION OF LYCOPENE AND β-CAROTENE

In this experiment, you will isolate lycopene, the red pigment in tomatoes, by the column chromatography of dehydrated tomato paste. If the separation is done carefully, a small amount of β-carotene, the yellow pigment in carrots, will also be obtained.

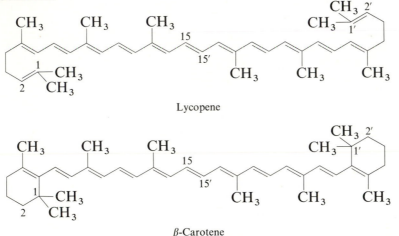

Lycopene

β-Carotene

As you can see from the foregoing structures, lycopene and β-carotene are very similar and consist of a long chain of 11 conjugated double bonds. This extended conjugated system is responsible for the color of each compound. The only difference between the molecules is that at the ends of the chain, the carbons in β-carotene form a six-membered ring, while in lycopene no cyclization occurs and there is an extra unconjugated double bond.

Fresh tomatoes, like all fruits and vegetables, contain mostly water. With tomatoes the amount is about 96 percent, which means that the amounts of the desired organic substances will likely be quite low. This was confirmed in 1907 when R. Willslätter and H. Escher were able to isolate only 0.020 gram of lycopene per 1000 grams of fresh tomatoes. The researchers then discovered that commercial tomato paste was a much more convenient source. With this improvement, Willslätter and Escher were able to obtain 0.150 gram of lycopene per 1000 grams of paste. Because only 10 grams of paste will be used in this experiment, your expected yield will be only about 0.0015 gram. This amount is much too small to be weighed, except on a microbalance.

In the following procedure, the tomato paste will be dehydrated with ethyl alcohol, followed by extraction with methylene chloride. The dehydration is required to make the methylene chloride extraction more efficient. The methylene chloride extracts will then be dried and evaporated to provide a crude residue that will be used for the column chromatography.

PROCEDURE

CAUTION: Lycopene and β-carotene are both very sensitive to photochemical air oxidation. Protect all solutions and solids from undue exposure to light and air.

The toluene, methylene chloride, and petroleum ether used in the experiment are all toxic and should be used with care and evaporated only in the hood. Petroleum ether is also very flammable. Be sure there are no flames near it.

In addition, the packing of the chromatography column and the chromatography itself should be started near the beginning of the lab period so that they may be completed without interruption.

Dehydration and extraction of the tomato paste

Place 10 grams of tomato paste in a 50-mL Erlenmeyer flask. Add 15 mL of 95% ethanol and thoroughly stir the mixture with a spatula for at least 3 minutes until the stickiness of the paste disappears. Filter the liquid into a 50-mL beaker through a loose plug of glass wool held in a small, short-stem funnel. Be sure the glass wool fits snugly in the bottom of the funnel above the stem so that no solid material can get through. When the filtration is almost completed, press the remaining tomato solids in the Erlenmeyer flask with a spatula to squeeze out more liquid.

Place the glass wool and the dehydrated residue held in it into the original Erlenmeyer flask. Add 10 mL of methylene chloride to the solid material and thoroughly stir the mixture for about 3 minutes. Filter the mixture as before into a clean 50-mL beaker through a loose plug of glass wool held in a small, short-stem funnel. Repeat with a second 10-mL portion of methylene chloride and combine the two liquid extracts.

Pour the methylene chloride extracts into a separatory funnel. Add 5 to 10 mL of saturated sodium chloride solution (to aid in layer separation) and shake. After the layers have separated, drain the colored, lower methylene chloride layer into a clean, dry Erlenmeyer flask. Add a small amount of anhydrous magnesium sulfate, swirl gently for a few minutes, and then decant or filter the dried liquid into another clean, dry Erlenmeyer flask. Evaporate the solution to dryness in the hood, using a steam or warm-water bath. **Be careful not to overheat the pigments.** Then dissolve the crude residue in a few milliliters of toluene in a small Erlenmeyer flask or test tube. The resulting solution should be stored in the dark with its container stoppered until it is needed for the column chromatography.

Column chromatography

Read thoroughly Section 6.2 on column chromatography, page 127–139.

Prepare a 15-cm chromatography column by the procedure given in Section 6.2D, using alumina (Fisher No. A-540) as the adsorbent and petroleum ether (35–60°C) as the solvent. Place the toluene solution containing the pigments onto the column by the procedure given in Section 6.2E. Then elute the mixture of pigments by the procedure given in Section 6.2F, using petroleum ether as the eluting solvent. The yellow β-carotene should move rapidly through the column, whereas the red lycopene will move more slowly. Collect the β-carotene fraction in a 50-mL Erlenmeyer flask, separate from any colorless solvent fractions preceding it.

After all the β-carotene has been removed from the column, change receiving flasks and gradually add small portions of toluene to the petroleum ether in the solvent reservoir. The added polarity of toluene will speed up the movement of the remaining lycopene. A highly colored band of oxidation products and other compounds may also be present higher up on the column, but it will hardly move. Also collect the lycopene fraction in a 50-mL Erlenmeyer flask, separate from the colorless solvent fractions preceding it.

Evaporate the colored fractions obtained in the hood. A warm-water or steam bath should be used for the β-carotene, but the lycopene will also later require a hot plate, due to the higher-boiling toluene (111°C). **Again, do not**

FIGURE 6-19

The visible light spectra of isolated lycopene and β-carotene in petroleum ether.

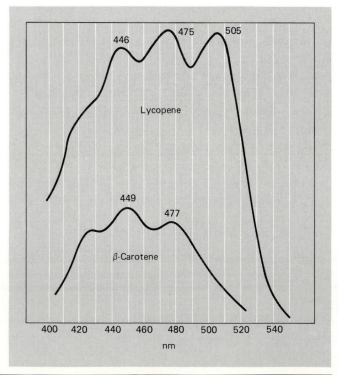

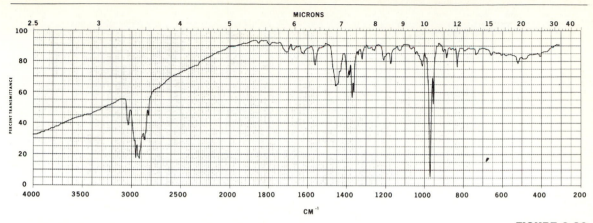

FIGURE 6-20

The infrared spectrum of β-carotene in a KBr pellet. Spectrum courtesy of Sadtler Research Laboratories, division of Bio-Rad Laboratories, Inc., copyrighted 1973.

overheat the pigment solutions. Compare the final products in terms of their quantity, appearance, and intensity of color. Record the results in your notebook. At your instructor's option, separately combine each of your products with those of your neighbors to obtain enough for a visible, infrared, and

FIGURE 6-21

The ¹H NMR spectrum of β-carotene in deuterated chloroform. Spectrum courtesy of Sadtler Research Laboratories, division of Bio-Rad Laboratories, Inc., copyrighted 1978.

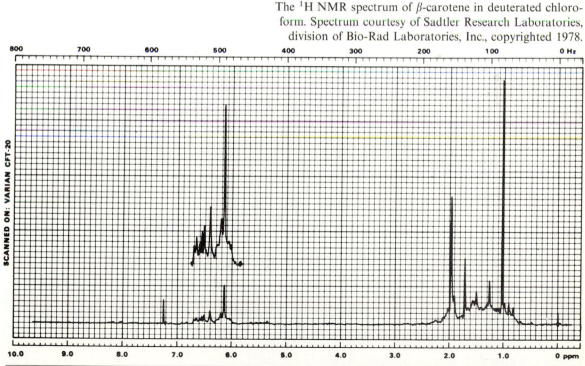

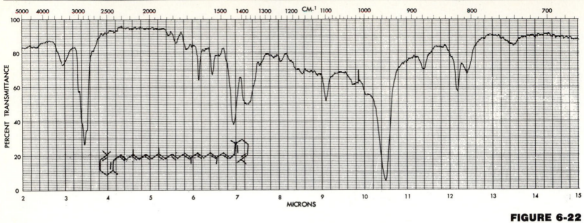

FIGURE 6-22

The infrared spectrum of lycopene in a KBr wafer. Spectrum courtesy of Sadtler Research Laboratories, division of Bio-Rad Laboratories, Inc., copyrighted 1976.

^{1}H NMR spectrum. The visible spectra should be taken in petroleum ether, the infrared spectra in a KBr pellet, and the ^{1}H NMR spectra in deuterated chloroform. Compare your spectra with those shown in Figures 6-19 through 6-22.

6D. COLUMN AND THIN-LAYER CHROMATOGRAPHY OF PAPRIKA PIGMENTS

In this experiment, you will extract paprika with methylene chloride. You will then analyze the crude mixture of extracted pigments by thin-layer chromatography, and separate it on a larger scale by column chromatography.

The major pigment in paprika, responsible for the deep red color, is the fatty acid esters of capsanthin.

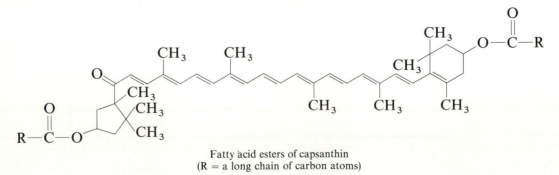

Fatty acid esters of capsanthin
(R = a long chain of carbon atoms)

Paprika also contains minor amounts of the fatty acid esters of capsorbin, which are also red, and β-carotene, which is yellow.

172

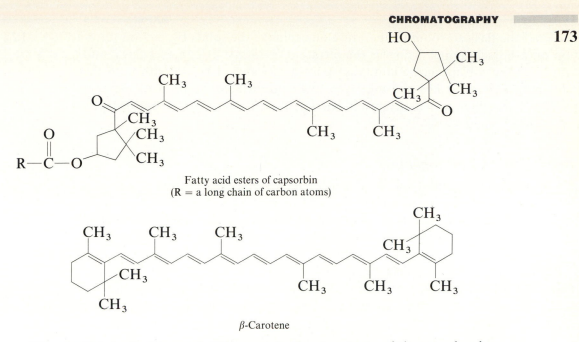

Fatty acid esters of capsorbin
(R = a long chain of carbon atoms)

β-Carotene

Observe the similarities and differences in the structures of these molecules. All of them are tetraterpenes containing eight isoprene units.

Isoprene unit

All of them also contain a long chain of 18 carbons, with 9 conjugated carbon-carbon double bonds included in the chain and 4 methyl groups attached to it. But the rings at the ends of the chain differ in size, mode of attachment to the chain, and whether an ester group is attached. The long conjugated system in all the molecules is responsible for their red or yellow color.

You will note from the chromatographies that only three spots or fractions are typically obtained. The various fatty acid esters of capsanthin are so much alike in their adsorption onto the adsorbent and their polarities that a separation of them does not occur with the experimental conditions used. The same is true with the various fatty acid esters of capsorbin.

PROCEDURE

CAUTION: Methylene chloride is toxic. Do not breathe it or get it on your skin. Solutions of it must be evaporated only in the hood.

Extraction of the pigments

Place 1.5 grams of ground paprika and several boiling chips into a 50-mL round-bottom flask. Add 15 mL of methylene chloride, attach a reflux condenser, and reflux the mixture for 20 to 25 minutes.

Cool the flask to room temperature and remove the solids by suction filtration on a Hirsch funnel. Transfer the filtrate to a 25-mL Erlenmeyer flask, and evaporate the solvent in the hood, using a warm-water or steam bath until the remaining volume is about 1 mL. Use this concentrated solution of the pigments for the thin-layer analysis and the larger-scale column chromatography separation.

Thin-layer chromatography

If necessary, thoroughly read Section 6.3, pages 139–151.

Prepare a developing chamber, capable of holding a 2×10 cm thin-layer sheet, following the procedure given in Section 6.3E. Use methylene chloride as the developing solvent. Also prepare a micropipette by the procedure given in Section 6.3C.

Obtain a commercial, plastic-backed 2×10 cm silica gel G sheet from your instructor. Spot it with the concentrated solution of extracted paprika pigments, using the procedure given in Section 6.3F. Next develop the chromatogram by the procedure in Section 6.3G. Be sure the chromatogram does not touch the liner paper wick inside the developing chamber. Record in your notebook the R_f values of all the spots and indicate which spots correspond to which compounds. At the end of the experiment, attach the chromatogram within your notebook, protected inside an envelope.

Column chromatography

If necessary, thoroughly read Section 6.2, pages 127–139.

Prepare a chromatography column by the procedure given in Section 6.2D, using 12 grams of 60–200 mesh silica gel as the adsorbent and methylene chloride as the eluting solvent.

Place the concentrated solution of extracted paprika pigments onto the column, using the procedure given in Section 6.2E. Then elute the mixture of pigments by the procedure in Section 6.2F. As each colored fraction comes off the column, change receivers, collecting the solution in a preweighed 50-mL Erlenmeyer flask. The mobility of the compounds should be the same as it was in the thin-layer chromatography.

Evaporate the colored fractions in the hood, using a warm-water or steam bath. Compare the final products in terms of their quantity, appearance, and intensity of color, and record the results in your notebook.

At your instructor's option, separately combine each of your products with those of your neighbors to obtain enough for a visible, infrared, and ^{1}H NMR spectrum. The visible spectra should be taken in petroleum ether, the infrared spectra in a KBr pellet, and the ^{1}H NMR spectra in deuterated chloroform. The visible, infrared, and ^{1}H NMR spectra of β-carotene are shown in Figures 6-19 through 6-21.

6E. GAS CHROMATOGRAPHIC ANALYSIS OF GASOLINES

When the oil is exhausted, the lamp goes out.

William Scarborough, *Chinese Proverbs,* No. 928 (1875).

INTRODUCTION

Gasoline is a complex mixture of alkanes, cycloalkanes, aromatic hydrocarbons, and various additives. Well over 100 compounds can be present, but most of the mixture will consist of about 25 compounds. These compounds typically will contain from 5 to 10 carbon atoms and will boil between 40° and 180°C (104° and 356°F). The specific compounds commonly found in gasoline include toluene, benzene, the xylenes, pentane, 2-methylpentane, 3-methylpentane, 2,2,4-trimethylpentane, hexane, 2-methylhexane, 3-methylhexane, heptane, cyclohexane, and methylcyclopentane.

Most gasoline comes from petroleum, as do many other products. Table 6-9 indicates the composition of the various petroleum fractions and the commercial products obtained from them.

Unfortunately, the petroleum fraction useful for gasoline is only about 19% of the entire mixture, whereas the need is 46%. Therefore, after fractional distillation has separated the crude oil components, cracking, polymerization, and alkylation are used to convert other fractions into gasoline. Cracking breaks large hydrocarbon molecules into smaller ones, while polymerization and alkylation convert small molecules into larger ones.

Heat and pressure are required for cracking. In addition, catalysts are frequently used to improve the quality of the gasoline obtained and to allow more moderate reaction conditions. Silica-alumina, silica-magnesia, silica-zirconia-alumina, and bentonitic clays are among the catalysts used. Gaseous hydrogen

TABLE 6-9.

Composition and Commercial Uses of Various Petroleum Fractions

Petroleum fraction	Composition	Commercial uses
Natural gas	C_1—C_4	Heating fuel, synthesis of petrochemicals
Gasoline	C_5—C_{10}	Motor fuel, synthesis of petrochemicals
Kerosene	C_{11}—C_{12}	Jet fuel and heating fuel
Light oil	C_{13}—C_{17}	Heating fuel, diesel engines
Heavy oil	C_{18}—C_{25}	Motor oil, heating fuel, paraffin wax, petroleum jelly
Residuum	C_{26}—C_{60}	Asphalt, tar, road oils

may also be present to convert the alkenes produced into alkanes. For example:

$$C_{16}H_{34} + H_2 \xrightarrow[\text{heat, pressure}]{\text{catalyst}} 2\ C_8H_{18}$$

Various
hexadecanes

Various
octanes

Heat and pressure are also required in polymerization. Natural gas and waste gases produced by cracking and other operations combine to produce liquid products suitable for gasoline.

Alkylation is especially important in the production of high-quality gasolines. The most important example is the production of 2,2,4-trimethylpentane or isooctane by the reaction of isobutylene with isobutane. Isooctane is a high-quality fuel and is a standard used to measure the quality of gasolines.

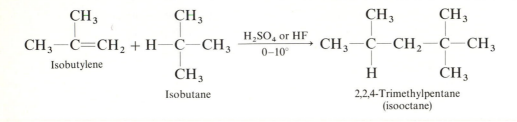

Octane ratings

As you probably know, not all gasolines burn equally well inside an engine. The high temperature and pressure inside an engine cylinder can make some components of gasoline spontaneously ignite before the spark plug has even fired. Some components may also explode rather than burn smoothly. This violent, uncontrolled ignition results in **knocking** or **pinging**. Knocking or pinging is undesirable because the transfer of power to the piston is much less efficient, resulting in a waste of energy and possible engine damage.

To grade all the possible gasolines in terms of their antiknocking ability, researchers developed the octane number system. Engineers determine a gasoline's octane number by comparing its antiknock performance to a **reference fuel**. The reference fuel usually is a mixture of two hydrocarbons, isooctane and heptane.

Isooctane has a high antiknock performance and has been given an octane rating number of 100. Heptane, on the other hand, knocks very badly and has been given an octane rating number of 0. A gasoline's octane number equals the percentage of isooctane in a reference fuel having the same antiknock quality of the gasoline. Therefore, if a certain gasoline knocks like a reference fuel containing 95% isooctane and 5% heptane, the octane number of this gasoline would be 95.

Automobile engineers measure octane numbers in three ways, so every gasoline actually has three different octane numbers.

1. The research octane number (RON) uses a special one-cylinder engine, operating at a mild temperature of 120°F and a moderate engine running speed of 600 revolutions per minute.

2. The motor octane number (MON) also uses a special one-cylinder engine, but operates at 300°F and 900 revolutions per minute. As a result, the MON is usually from 6 to 10 numbers less than the RON.

3. The road octane number is usually the average of the RON and the MON. It is also usually the octane number reading you find at the gas pump. Some of you may have even seen the equation $(R + M)/2$ at the gas pump to describe how the octane number was obtained. Therefore, a gasoline which has an octane rating of 89 on the gas pump may have, for example, a RON of 93 and a MON of 85. Typical road octane numbers for gasolines for cars vary from a low of 86 to a high of 93.

Some specific RONs of various compounds are given in Table 6-10. Note the especially high octane numbers of the aromatic hydrocarbons. In addition, branched-chain alkanes generally have much higher octane numbers than do straight-chain alkanes. Therefore, since the use of lead additives to boost the octane number is being phased out for environmental reasons, oil companies are increasing the proportion of aromatic hydrocarbons and branched-chain alkanes in their gasolines to maintain the octane numbers. This makes the following types of reactions to alter low-octane-number molecules to high-octane-number molecules even more important.

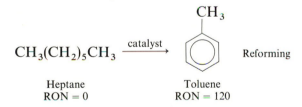

Heptane
RON = 0

Toluene
RON = 120

Reforming

TABLE 6-10.

Research Octane Numbers of Some Organic Compounds

Compound	RON	Compound	RON
Toluene	120	2-Hexene	93
m-Xylene	118	1-Pentene	91
Benzene	106	2,2-Dimethylbutane	89
Methanol	106	Cyclohexane	83
Cyclopentane	101	Pentane	62
2,2,4-Trimethylpentane	100	Hexane	25
Propane	97	Heptane	0
1-Butene	97	Octane	−19
Butane	94		

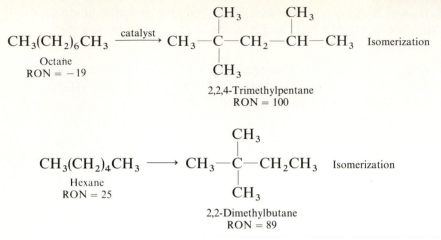

However, since all aromatic hydrocarbons are toxic and benzene is carcinogenic, their use in gasoline presents another hazard, especially for gas station attendants, gasoline truck drivers, and refinery employees. Research is currently being conducted to find economical means of increasing the octane ratings of unleaded gasoline without using high amounts of aromatic hydrocarbons.

Other requirements of gasoline

Burning inside an engine without knocking is only one requirement of gasoline. The gasoline must also

1. Vaporize sufficiently in cold weather to permit easy engine starting

2. Avoid excessive vaporization in hot weather or at high altitudes, which can cause vapor lock—the blockage of the liquid in the fuel line, fuel pump, or carburetor by a buildup of vapor

3. Eliminate carbon or other engine deposits, which can reduce the engine's performance

4. Minimize carburetor icing, spark plug fouling, rusting, or engine corrosion

5. Not oxidize in storage, thus producing undesired gums or other products

To satisfy these requirements, oil companies usually vary the blend of their products as needed and include various additives. To promote easy starting in the winter, refiners commonly use a higher proportion of low-boiling butane or pentane. To avoid vapor lock in the summer or at high altitudes, they use a lower proportion of any low-boiling compounds. They add detergents to prevent harmful engine deposits, alcohols and surface-active agents to prevent icing, corrosion inhibitors to prevent rusting or other corrosion, and aromatic amines or alkyl-substituted phenols to prevent the gasoline from oxidizing. In addition, every gasoline sold contains a dye, whose color identifies both the brand and the grade of the product. The dye also indicates that the product contains poisonous components.

The energy crisis

It is well known that petroleum heats our homes and other buildings, turns the wheels of industry and transportation, and greases these wheels to help them run smoothly. Numerous products, essential to our quality of life, are also obtained from it.

But what may not be as well known is the fact that there is only a limited supply of petroleum and other fossil fuels. And much of what took millions of years to be created has been used up in only a little over a hundred years, never to be available again. As the energy crises of 1973 and 1979, and the conflict in the Persian Gulf clearly indicate, many countries are highly dependent on an adequate supply of energy and organic raw materials. Yet these same countries are using up the remaining supplies, almost as though there were no tomorrow. If there is truly going to be a tomorrow, with a quality of life as high as it is today, then the world must learn to develop other sources of energy and use the remaining fossil fuels only for the numerous products essential in maintaining this quality of life. Otherwise, the ancient Chinese proverb quoted in the beginning may actually come to pass.

DISCUSSION

In this experiment, you will analyze several grades of commercial gasoline by gas chromatography to determine the complexity of the mixtures, the main components present, the amounts of each component, and how the compositions of the grades differ. If several brands are available, you can also see how the compositions of these brands differ. If gasohol is sold in your region, you can also compare its composition to the other gasolines.

You will perform the analyses by first injecting into the gas chromatograph a known mixture which simulates gasoline. By the volumes and boiling points of the components of this mixture, you will be able to identify which component produces each peak in this gas chromatogram. A comparison of the retention times and the sizes of the peaks in this chromatogram with the various gasoline chromatograms will then enable you to identify these components and their percentage of the volume in the various gasolines. It is expected that premium gasoline will contain a higher percentage of the higher-octane-number hydrocarbons, while regular gasoline will contain a higher percentage of the lower-octane-number hydrocarbons. Your analyses should confirm this.

A number of peaks in the gasoline chromatograms cannot be identified with certainty in this experiment. However, by the retention times of these peaks and the boiling points of some other possible compounds, you will at least be able to make an educated guess. You could obtain supporting evidence for your guess by injecting the suspected compound into the gas chromatograph and comparing the retention times.

PROCEDURE

CAUTION: Remember that gasoline and the simulated gasoline mixture are toxic and contain highly volatile and flammable components. Do not breathe them and keep them away from any open flames.

Thoroughly read Section 6.4 on gas chromatography, pages 151–162.

Following the procedure given in Section 6.4M, inject 1 µL of the **simulated** gasoline mixture given in Table 6-11 onto a 12- to 15-foot Chromosorb W column coated with silicone oil (SE-30) as the adsorbent. The carrier gas flow rate should be about 40 mL per minute, the injection port temperature about 150°C, and the column temperature about 105–115°C. The amounts of the components in the simulated mixture have been chosen to allow easy identification when the retention times are correlated with the boiling points. In some cases, you may find that the peaks will partially overlap due to the closeness in the boiling points. In these cases, the size of each peak will have to be estimated. Observe from the table that because of benzene's carcinogenic nature, its volume is only 2.0 percent of the total mixture. If desired, benzene may be eliminated from this mixture and the volume of pentane increased to 14.0 percent.

TABLE 6-11.

Composition of the Simulated Gasoline Mixture

Component	Boiling point (760 mmHg)	% by volume
Pentane	36.1	12.0
2-Methylpentane	60.3	3.0
3-Methylpentane	63.3	7.0
Hexane	69.0	14.0
Benzene	80.1	2.0
Cyclohexane	80.7	12.0
2-Methylhexane	90	3.0
3-Methylhexane	92	6.0
Heptane	98.4	2.0
2,2,4-Trimethylpentane	99.2	8.0
Toluene	110.6	20.0
p-Xylene	138.4	2.0
m-Xylene	139.1	3.0
o-Xylene	144.1	6.0

After the simulated gasoline chromatogram is finished, repeat with a sample of regular gasoline and a sample of premium gasoline. Be sure the gas flow rate, the injection port temperature, and the column temperature all remain the same. Otherwise, an exact comparison of retention times will not be possible. If different brands are available, they should be run by different students and the results compared at the end.

From the percentage volume and the boiling point of each compound in Table 6-11, identify which peak in the simulated gasoline chromatogram corresponds to each compound. Label each peak. Then select and label those peaks in the gasoline chromatograms that have the same retention times as those in the simulated gasoline chromatogram.

Quantitatively obtain the percentage area of each peak in the simulated gasoline chromatogram by one of the methods in Section 6.4I. Then determine the relative sensitivity of the detector to each component by dividing each component's percentage area by its percentage volume in the mixture.

Also quantitatively obtain the percentage area of each peak in the gasoline chromatograms. Then divide the percentage area of each known component by its relative detector sensitivity. This will give you a more accurate percentage by volume of each known component in the gasolines. These percentages by volume will total less than 100 percent because of all the other unidentified components in the gasoline.

By utilizing the information in Table 6-12 and the retention times of your unidentified components, try to determine what these unidentified components

TABLE 6-12.

Some Possible Components in Gasoline

Component	Boiling point (760 mmHg)	Component	Boiling point (760 mmHg)
Butane	−0.5	3,3-Dimethylpentane	86.1
2-Methylpropane	−0.5	2,3-Dimethylpentane	89.8
2-Methylbutane	27.9	**2-Methylhexane**	90
Pentane	36.1	**3-Methylhexane**	92
Cyclopentane	49.3	**Heptane**	98.4
2,2-Dimethylbutane	49.7	**2,2,4-Trimethylpentane**	99.2
2,3-Dimethylbutane	58	Methylcyclohexane	100.9
2-Methylpentane	60.3	**Toluene**	110.6
3-Methylpentane	63.3	2-Methylheptane	117.7
Hexane	69.0	4-Methylheptane	117.8
Methylcyclopentane	71.8	3-Methylheptane	119
2,2-Dimethylpentane	79.2	**p-Xylene**	138.4
Benzene	80.1	**m-Xylene**	139.1
2,4-Dimethylpentane	80.5	**o-Xylene**	144.1
Cyclohexane	80.7		

might be. To assist you, all the components in the simulated gasoline are in boldface type. Your answers many times will only be educated guesses, and there may still be many unidentified peaks. But without isolating and then identifying each component by other means, at least you have done your best.

Quantitatively compare the percentage composition of all the components in the regular and premium gasolines, especially for those known components having very high and very low octane number ratings. If other students analyzed different brands, also compare your percentage compositions with theirs. Record all your findings inside your notebook and place the chromatograms inside your notebook. If you determined the percentage area of each peak by cutting it out and weighing it on an analytical balance, then the pieces for the three chromatograms should be placed in three separate, labeled envelopes and attached to your notebook.

QUESTIONS

1. What would be the result of the following errors in a paper or thin-layer chromatography experiment?
   (a) Too much sample was applied or the spots were made too large.
   (b) Too much solvent was used in the developing chamber, causing the spots to be immersed.
   (c) A solvent with too high an eluting power was used.
   (d) A solvent with too low an eluting power was used.
   (e) The paper or thin-layer sheet was not removed when the solvent system reached the top.
   (f) The solvent system or the exact adsorbent was not specified.
2. In a thin-layer or paper chromatography, how would you expect the R_f values to change if you went from a cyclohexane-toluene solvent system to a methylene chloride–ethyl acetate solvent system? Why?
3. What is meant by the term "capillary action"? How is it able to apparently defy gravity?
4. If colorless compounds are used in a paper, thin-layer, or column chromatography, explain how the spots or bands could be located?
5. In column chromatography, why is it important to pack the column carefully?
6. In column chromatography, why is it important not to let the stationary adsorbent dry out?
7. In column chromatography, why does one begin with a nonpolar eluting solvent and then progress to solvents of higher and higher polarity? Why shouldn't there be an abrupt change in the eluting solvent?
8. Arrange the following compounds in the sequence of their elution from an alumina column: benzyl ethyl ether, *p*-xylene, cyclohexanol, octadecane, butanoic acid. What is the basis for your answer?

9. In gas chromatography, what would be the order of the retention times for a mixture of decane, tetradecane, octane, octadecane, and nonane? What is the basis for your answer?

10. If the various peaks in a gas chromatogram are too close, what experimental changes could you make to separate them more?

11. Suppose you were the manufacturer of inks and needed to compare the components of your products with those of your competitors. Describe how you would carry out the assignment.

12. In the isolation of lycopene and β-carotene from tomato paste, what would be the result in the column chromatography if the paste was not dehydrated with ethanol and extracted with methylene chloride?

13. Circle the separate isoprene units in lycopene, β-carotene, and the fatty acid esters of capsanthin and capsorbin.

14. β-Carotene is manufactured by an efficient synthesis and is in demand as a source of vitamin A. Look up the molecular structure of vitamin A. Compare the structures of β-carotene and vitamin A, and explain specifically why β-carotene is a good source of vitamin A. Why can't lycopene be used as a source of vitamin A?

15. Suppose the Food and Drug Administration banned the use of all green artificial food colors. If natural green food colors proved unsatisfactory, explain an easy way to obtain green food coloring.

16. Suppose you were a police detective and discovered that the suspect in an arson attempt had a can of gasoline in his car. How could you easily prove, without gas chromatography, that this gasoline was the same brand and grade of gasoline as that found at the attempted arson site?

References

1. J. Bobbitt, **Thin-Layer Chromatography,** New York: Van Nostrand Reinhold, 1963.

2. J. Bobbitt, A. Schwarting, and R. Gritter, **Introduction to Chromatography,** 2nd ed., San Francisco: Holden-Day, 1984.

3. D. Browning, **Chromatography,** New York: McGraw-Hill, 1969.

4. S. Dal Nogare and R. Juvet, **Gas-Liquid Chromatography, Theory and Practice,** New York: Wiley Interscience, 1962.

5. Z. Deyl, K. Macek, and J. Janak, **Liquid Column Chromatography,** Amsterdam: Elsevier, 1975.

6. R. Grob, editor, **Modern Practice of Gas Chromatography,** 2nd ed., New York: Wiley, 1985.

7. M. Hais and K. Macek, **Paper Chromatography,** New York: Academic Press, 1963.

8. E. Heftmann, **Chromatography,** 3rd ed., New York: Van Nostrand Reinhold, 1975.

9. F. Helfferich, **Multicomponent Chromatography,** New York: Dekker, 1970.

10. P. Jeffrey and P. Kipping, **Gas Analysis by Gas Chromatography,** 2nd ed., Oxford: Pergamon Press, 1972.

11. C. Kenner, **Instrumental and Separation Analysis,** Columbus, Ohio: Merrill, 1973.

12. J. Kirchner, "Thin-Layer Chromatography," in **Technique of Organic Chemistry,** vol. XII, A. Weissberger, editor, New York: Interscience, 1967.

13. J. Kirchner, **Thin-Layer Chromatography,** New York: Wiley, 1978.

14. J. Kirkland, **Modern Practice of Liquid Chromatography,** New York: Wiley Interscience, 1971.

15. D. Lederer and M. Lederer, **Chromatography,** 2nd ed., Amsterdam: Elsevier, 1957.

16. S. Perry, R. Amos, and P. Brewer, **Practical Liquid Chromatography,** New York: Plenum Press, 1972.

17. C. Poole and S. Schuette, **Contemporary Practice of Chromatography,** New York: Elsevier, 1984.

18. K. Randerath, **Thin-Layer Chromatography,** 2nd ed., New York: Academic Press, 1966.

19. E. Stahl, editor, **Thin-Layer Chromatography, A Laboratory Handbook,** 2nd ed., New York: Springer-Verlag, 1969.

20. J. Schupp in **Technique of Organic Chemistry,** vol. XIII, A. Weissberger, editor, New York: Interscience, 1968.

21. R. Stock and C. Rice, **Chromatographic Methods,** New York: Reinhold, 1963.

7

POLYMERS AND PLASTICS

A chemist setting out to build a giant molecule is in the same position as an architect designing a building. He has a number of building blocks of certain shapes and sizes, and his task is to put them together in a structure to serve a particular purpose . . . . What makes high polymer chemistry still more exciting just now is that almost overnight, within the last few years, there have come discoveries of new ways to put the building blocks together; discoveries which promise a great harvest of materials that have never existed on the earth.

Giulio Natta, "How Giant Molecules Are Made,"
Scientific American, September, 1957, p. 98.

INTRODUCTION

Our world is far different from that of 50 years ago. Atomic energy, computers and miniature calculators, television, VCRs, lasers, open-heart surgery and organ transplants, and rockets to outer space have all been relatively recent developments that have altered and improved our lives. But perhaps the biggest change has been the development of synthetic polymers and plastics. Clothing, carpets,

185

furniture, toys, tires, tools, phonograph records and tapes, bags, electrical insulation, squeeze bottles, floor covering, paints, adhesives, and chewing gum are made from these materials. Words like nylon, Teflon, Formica, Acrilan, Saran, Styrofoam, Lucite, Plexiglas, Orlon, Dacron, silicone, vinyl, polyester, acrylic, polyethylene, and polystyrene have become part of our everyday language.

All of us have obviously benefited from this great harvest of materials, which have made our lives better, easier, and more comfortable. In medicine, the use of plastics has enabled blind people to see and cripples to walk. Damaged or defective heart valves, arteries, tracheas, and larynxes have been repaired or replaced. For many uses, plastics and polymers have replaced cotton, metal, wood, and other materials, because in many ways they have superior properties, can be produced at lower cost, and offer an alternative to other materials in short supply.

Unfortunately, polymers and plastics also have their problems. Because most of them do not decompose when thrown away, the fear has arisen that someday civilization may be buried beneath a mountain of plastic debris. Most plastics burn readily, and a few liberate toxic gases such as hydrogen chloride and hydrogen cyanide to compound the fire hazard. The manufacture of some plastics is hazardous to the health of the production workers and people living near the production area. People complain about how easily certain plastic items break, making replacement necessary. Finally, because the demand for energy has increased at an alarming rate, the indiscriminate use of polymers and plastics, which require petroleum and coal as raw materials, is bringing even closer the day when the world runs out of fossil fuels. The world must not only find other sources of energy but must also learn to use polymers and plastics more wisely and recycle them whenever possible.

7A.   THE SYNTHESIS OF PLEXIGLAS

FREE-RADICAL ADDITION POLYMERIZATION

In this experiment, you will prepare Plexiglas (also known as Lucite) by the free-radical addition polymerization of methyl methacrylate with benzoyl peroxide.

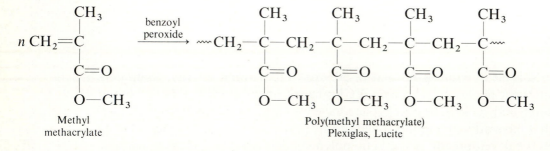

Methyl
methacrylate

Poly(methyl methacrylate)
Plexiglas, Lucite

Benzoyl peroxide functions as the **initiator** for the reaction. With mild heating at 80 to 90°C, the weak oxygen-oxygen bond in the molecule is easily broken to form a pair of benzoyl free radicals. These radicals may then, individually, initiate the reaction or decompose to yield carbon dioxide and phenyl free radicals, which may also start the reaction.

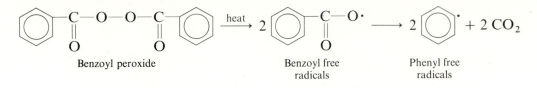

Polymerization begins by the initiating free radical, In·, adding to the double bond of the alkene to form a chemical bond and the generation of a new free radical. This new radical then adds to another molecule of alkene to generate a still larger radical, which then adds to another molecule of alkene, and so on.

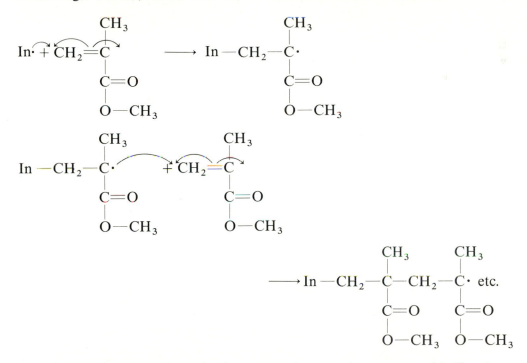

Observe that the polymerization occurs in a regular manner with "head-to-tail" addition to yield a molecule with identical reoccurring groups. This results from the addition forming the more stable of the two possible free radicals.

The chain continues to grow, but eventually is terminated by a number of possible steps, including the combination of two radicals, the loss of a hydrogen atom to form an alkene, and the abstraction of a hydrogen atom from another molecule to form a saturated carbon.

Radical combination

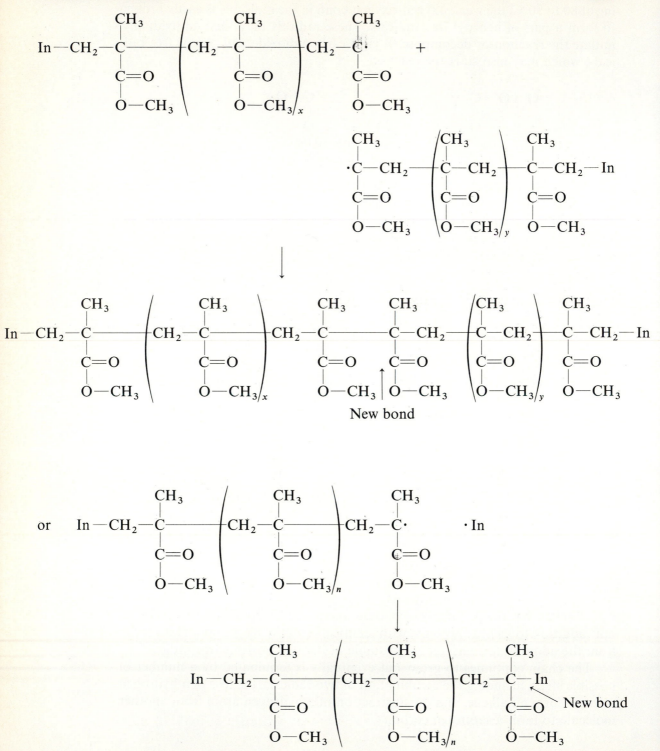

Loss of hydrogen atom

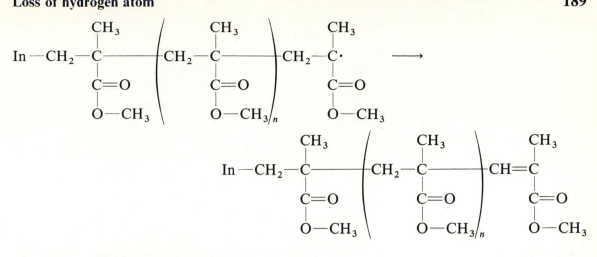

Abstraction of hydrogen atom

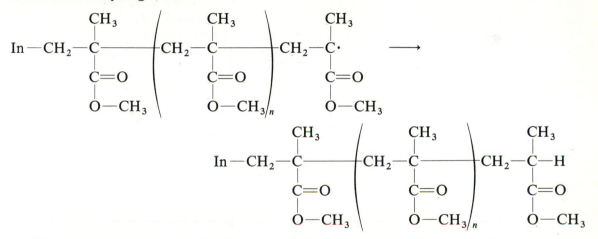

The polymerization can also be initiated by certain acids, bases, metals, metal complexes, light, and heat. If heated to over 300°, the polymer breaks down to again form the monomer, which distills at 102°.

The methyl methacrylate needed for the polymer can be commercially prepared from acetone by the following reactions:

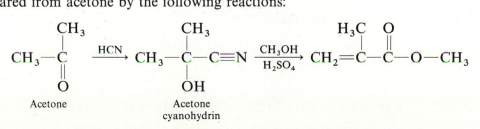

The first reaction involves nucleophilic addition of hydrogen cyanide to acetone to form acetone cyanohydrin. The second reaction involves hydrolysis of the cyano group to yield a carboxylic acid, esterification by the alcohol, and dehydration.

Plexiglas or Lucite has many beneficial uses. Because of its outstanding stability, excellent optical properties, unbreakable nature, and relative lightness, it has in many cases replaced glass in eyeglasses, prisms, and other optical items. Its only drawback with these products is its tendency to be more easily scratched than glass. It is also extensively used as the plastic in screwdriver handles, combs, and so on, and is a major ingredient in latex paints. Its ability to be molded into large, unbreakable, transparent or translucent sheets, which are unusually resistant to sunlight and outdoor weathering and can be drilled and sawed, have also made it valuable for a number of products, including signs and airplane windows.

PROCEDURE

CAUTION: Benzoyl peroxide is a mild explosive that must be handled with care and not subjected to heat, friction, or mechanical shock as a dry powder. It should be carefully manipulated in very small quantities and transferred by a smooth, folded 3 × 5 card. Do not use a metal spatula. Weigh the material on glazed, rather than ordinary, paper. Clean up all spills with water, and rinse the glazed paper with water before discarding. As always, wear safety glasses.

Place about 5 mL of methyl methacrylate in a 6-inch test tube and add 5 to 10 **milligrams** of benzoyl peroxide. If desired, you may also add a small inert object, such as a coin. **Loosely** stopper the tube and heat in a boiling-water bath for 20 minutes or more until the liquid becomes quite viscous. Allow the contents in the tube to solidify, and then remove and examine. To do this, you may have to break the test tube, but first wrap it in a towel. Record the results.

7B.   THE SYNTHESIS OF POLYSTYRENE

FREE-RADICAL ADDITION POLYMERIZATION

In this experiment, you will prepare polystyrene by the free-radical addition polymerization of styrene with benzoyl peroxide.

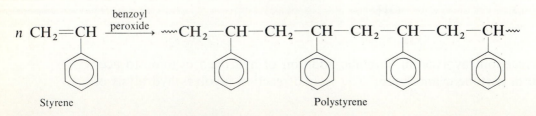

Styrene   Polystyrene

Benzoyl peroxide functions as the **initiator** for the reaction. With mild heating at 80 to 90°C, the weak oxygen-oxygen bond in the molecule is easily broken to form a pair of benzoyl free radicals. These radicals may then, individually, initiate the reaction, or may decompose to yield carbon dioxide and phenyl free radicals, which may also start the reaction.

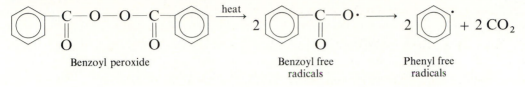

Benzoyl peroxide   Benzoyl free radicals   Phenyl free radicals

Polymerization begins by the initiating free radical, In·, adding to the double bond of the alkene to form a chemical bond and the generation of a new free radical. This new radical then adds to another molecule of alkene to generate a still larger radical, which can then add to another molecule of alkene, and so on.

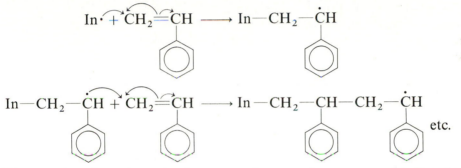

Observe that the polymerization occurs in a regular manner with "head-to-tail" addition to yield a molecule with identical reoccurring groups. This results from the addition forming the more stable of the two possible free radicals.

The chain continues to grow, but eventually is terminated by a number of possible steps, including the combination of two radicals, the loss of a hydrogen atom to form an alkene, and the abstraction of a hydrogen atom from another molecule to form a saturated carbon.

Radical combination

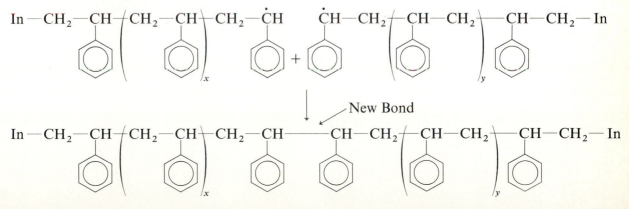

or

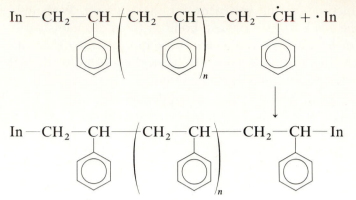

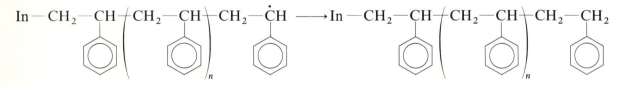

Loss of hydrogen atom

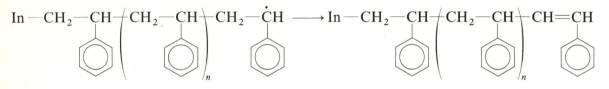

Abstraction of hydrogen atom

Being a conjugated alkenylbenzene, styrene is unusually reactive toward free-radical addition. Just on contact with atmospheric oxygen in the presence of light, the styrene monomer will undergo spontaneous polymerization unless certain **stabilizers** are added. These stabilizers are molecules such as *t*-butyl-catechol, hydroquinone, and diphenylamine, and function by acting as anti-oxidants that remove or neutralize certain free radicals that could initiate the polymerization. Because free-radical polymerization is a chain reaction, it takes

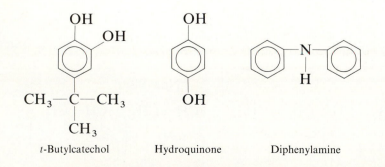

t-Butylcatechol   Hydroquinone   Diphenylamine

only a very small quantity of one of these stabilizers to inhibit the reaction. When polymerization of styrene is desired, the stabilizer can either be removed by extracting with acid or base, or overwhelmed by adding an extra large quantity of benzoyl peroxide.

The polymerization can also be initiated by certain acids, bases, metals, and metal complexes. If heated to very high temperatures, the polymer breaks down to again form the monomer, which distills at 145°. But unless a stabilizer is added, the monomer can again polymerize.

The styrene needed for the reaction is prepared industrially by the Friedel-Crafts alkylation of benzene, followed by dehydrogenation of the ethylbenzene produced.

Ethylbenzene

The laboratory preparation is either by dehydration or dehydrohalogenation.

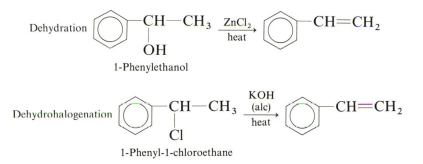

Polystyrene is an important plastic, especially useful as an electrical insulator. As Styrofoam, it is valuable as a packing and cushioning material for household appliances, delicate instruments, and other products. Styrofoam is also useful for many inexpensive, molded household goods, including insulated drinking cups and beverage chests.

PROCEDURE

CAUTION: Benzoyl peroxide is a mild explosive that must be handled with care and not subjected to heat, friction, or mechanical shock as a dry powder. It should be carefully manipulated in small quantities and transferred by a smooth, folded 3 × 5 card. Do not use a metal spatula. Weigh the material on glazed, rather than ordinary, paper. Clean up all spills with water, and rinse the glazed paper with water before discarding. As always, wear safety glasses.

Place about 5 mL of styrene in a 6-inch test tube and add 0.12 gram of benzoyl peroxide. Allow the solid to dissolve and, if desired, add a small, inert object such as a coin. **Loosely** stopper the tube and heat in a boiling-water bath for 60 minutes or more until the liquid becomes quite viscous. During this time, portions of other experiments may be started or completed. Allow the contents in the tube to solidify and then remove and examine. To do this, you may have to break the test tube, but first wrap it in a towel. Record the results.

7C. THE SYNTHESIS OF THIOKOL RUBBER

NUCLEOPHILIC ALIPHATIC SUBSTITUTION

DISCUSSION

The story of rubber goes back at least 2600 years to ancient Egypt and Ethiopia, where it was used to make balls for amusement. This use remained unchanged until 1770, when Joseph Priestley, known for his discovery of oxygen, found that rubber could also be used to erase pencil mistakes. During this time, people also became aware of the waterproofing qualities of rubber and began to use it for manufacturing and covering shoes, boots, raincoats, mailbags, life preservers, and numerous other articles. During the early 1830s, "rubber fever" took hold, with numerous companies springing up in America and Europe to produce the popular new items. But this "fever" was short-lived because customers soon discovered a serious flaw in natural rubber: The substance simply was only usable at mild temperatures; it became bone hard and brittle in the winter and sticky and messy like glue in the summer. Only one company was able to survive more than three years, so investors lost millions.

To most people, it seemed like the end for natural rubber. But a few people, like Charles Goodyear, still had faith in the substance. Obsessed with the mysterious elastic and waterproofing qualities of the material, Goodyear firmly believed that the negative features of natural rubber could be overcome. For over five years, he persistently tried, in vain, to improve the substance. During this time, he and his family frequently lived in extreme poverty, as one promising experiment and venture after another ended in failure.

The necessary breakthrough came by accident in the winter of 1839. Arguing with a number of people about his work, Goodyear, usually calm and mild-mannered, became excited and dropped a large piece of rubber mixed with sulfur onto the hot stove behind him. When he scraped the rubber off, he discovered to his amazement that it had charred like leather. In the accident, an entirely new substance had been formed, which Goodyear discovered was firm and flexible in both heat and cold. After further experimenting, he knew that the problem with rubber had finally been solved.

Goodyear's discovery became known as **vulcanization**, after Vulcan, the Roman god of fire. It is ironic that in vulcanization, Goodyear had used heat, rubber's greatest enemy, to transform it into a material that could withstand that very enemy.

Today we know that natural rubber is a polymer, consisting of many isoprene units linked together in the cis-configuration.

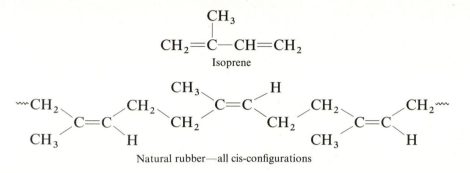

Isoprene

Natural rubber—all cis-configurations

Vulcanization permits the formation of sulfur bridges between different chains of the rubber molecules at both the double bonds and the allylic hydrogens. These bridges or cross-links make the rubber much harder and stronger and able to withstand heat and cold.

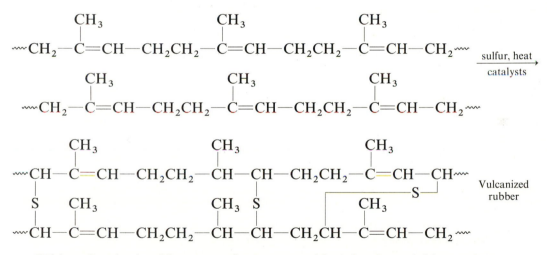

Vulcanized rubber

With vulcanized rubber, manufacturers could make dependable products, and "rubber fever" was again born. Later developments allowed the rubber to have all the desirable qualities of being long-wearing, elastic, pliable, water-resistant, airtight, and shock-absorbing.

Today the basic raw material can be modified or processed to meet almost any requirement. Depending on the intended use, the final product can, ironically, be easily worn away, such as a pencil eraser, or tough enough to withstand the roughest type of abrasion, such as a tire tread. It can be elastic for use in rubber bands, soft and foamy for use as sponge rubber, or as hard as a bowling

TABLE 7-1.

The Structure, Properties, and Uses of Some Synthetic Types of Rubber

Name	Structure and source	Properties and uses
Neoprene	$\left(\!CH_2\!-\!C\!=\!CH\!-\!CH_2\!\right)_n$ with Cl from chloroprene $CH_2\!=\!C\!-\!CH\!=\!CH_2$ with Cl	Resistant to oxygen, sunlight, oil, chemicals, steam, and flames. Used for gasoline hoses, insulation for wire and cables that come into contact with oil, gaskets that come into contact with gas and oil, other goods that come into contact with oil, heat, and weather.
Butyl rubber	$\left(\begin{array}{c}CH_3\\|\\C\!-\!CH_2\\|\\CH_3\end{array}\right)_x\left(CH_2\!-\!CH\!=\!\begin{array}{c}C\\|\\CH_3\end{array}\!-\!CH_2\right)_y$ Copolymer of 98% isobutylene and 2% isoprene	Resistant to aging, heat, acids, solvents, electricity, and, most importantly, the passage of gases. Used in inner tubes, tubeless tire linings, adhesives, and cements.
cis-Polybutadiene	$\left(CH_2\!-\!CH\!=\!CH\!-\!CH_2\right)_n$ from butadiene	Used as a blend with other types of rubber.
Styrene-butadiene rubber (SBR)	$\left(CH_2\!-\!CH\!=\!CH\!-\!CH_2\right)_x\left(CH_2\!-\!CH\right)_y$ with C_6H_5 Copolymer from about 70% butadiene and 30% styrene	Good natural rubber substitute. Especially good for tires.
Buna N nitrile rubber	$\left(CH_2\!-\!CH\!=\!CH\!-\!CH_2\right)_x\left(CH_2\!-\!\begin{array}{c}CH\\|\\C\!\equiv\!N\end{array}\right)_y$ Copolymer of butadiene and acrylonitrile in varying proportions	Especially resistant to oil and solvents and has low swelling properties. Also resistant to abrasion, heat, and aging. Used in gasoline hoses and tank storage linings.
Thiokol	$\left(CH_2CH_2\!-\!\begin{array}{c}S\\\|\|\\S\!-\!S\\\|\|\\S\end{array}\right)_n$ Copolymer of ethylene dichloride and sodium polysulfide	Excellent resistance to oil, fuels, and solvents. Used for cables, gaskets, flexible molds, adhesives, and protective coatings.
Silicone rubbers	$\begin{array}{c}R\qquad R\\|\qquad|\\ \sim\!Si\!-\!O\!-\!Si\!-\!O\!\sim\\|\qquad|\\R\qquad R\end{array}$ from $\begin{array}{c}R\\|\\Cl\!-\!Si\!-\!Cl\\|\\R\end{array}$ with hydrolysis and condensation	Provide maximum resistance to heat and cold. Are also highly resistant to the effects of aging, weather, water, electricity, oil, and other chemicals. Used for high- and low-temperature seals and gaskets, high-temperature electrical insulators, and caulking compounds.

ball. The substance is so versatile that thousands of products are made from it, many of which, such as tires, have no satisfactory substitute. In fact, rubber is so valuable and essential that during times of war, countries have been forced to produce synthetic substitutes in order to survive. This was the case with Germany during World War I and the United States during World War II. During the last 60 years, a number of synthetic rubber compounds have been successfully developed; their structures, properties, and uses are given in Table 7-1.

For two major reasons, most rubber today is synthetic. First, there is simply not enough natural rubber to meet the world's needs. Second, most of the synthetic types have certain special properties that cannot be duplicated with natural rubber, thus making these synthetics even better than natural rubber for certain products. These special properties are indicated in Table 7-1.

EXPERIMENTAL

In this experiment, you will prepare Thiokol rubber by the reaction of sodium polysulfide with ethylene dichloride.

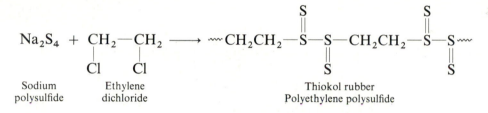

Sodium polysulfide   Ethylene dichloride   Thiokol rubber — Polyethylene polysulfide

The reaction occurs by an S_N2, or a bimolecular nucleophilic substitution mechanism. In this mechanism, the reaction rate depends on the concentration of both reactants, and, as shown in the next reaction, either the polysulfide ion or substituted polysulfide ion attacks either the ethylene dichloride or substituted ethyl chloride molecule from the rear, thus avoiding the departing chloride ion.

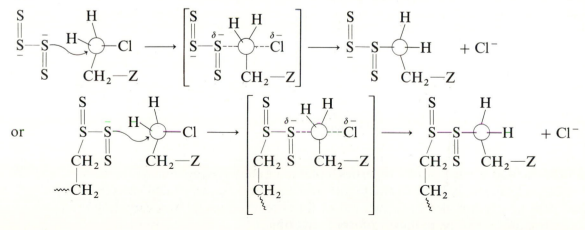

where

$$Z = \text{either Cl or } \overset{\displaystyle S}{\underset{\displaystyle S}{\overset{\|}{\underset{\|}{S - S}}}} \sim\!\!\sim$$

The transition state for the mechanism, as can be seen, involves the sulfur-carbon bond being partially formed and the carbon-chlorine bond being partially broken. The attacking sulfur atom has a diminished negative charge, since it has begun to share its electrons with carbon. On the other hand, the chlorine atom has developed a partial negative charge, since it has partially removed a pair of electrons from carbon. Also notice in the mechanism how the final product has the atoms of the attacked carbon forming a tetrahedral arrangement **opposite** to the carbon in the starting material. A good analogy to this inversion is an umbrella turning inside out on a windy day.

The sodium polysulfide used in the synthesis is prepared by the reaction of dilute sodium hydroxide with elemental powdered sulfur.

PROCEDURE

Place 15 mL of 10% sodium hydroxide solution in a 100-mL beaker and dilute to 40 mL with water. Heat this solution to boiling **in the hood** and add 2.5 grams of powdered sulfur with stirring until the sulfur has dissolved. As the sodium polysulfide is formed, the mixture will change from light yellow to dark brown. Cool the mixture to 70° and add 7 mL of 1,2-dichloroethane (ethylene dichloride) and 5 drops of a liquid dishwashing detergent to promote mixing of the two insoluble layers. Stir the mixture vigorously, maintaining the temperature at 70° until a color change has occurred and a spongy lump of rubber has formed. After the reaction is completed, remove with tongs the white-to-yellow product from the liquid, thoroughly wash with water, and squeeze between paper towels to dry. Test the rubberlike properties of the Thiokol produced by stretching, squeezing, and bouncing it. Note and record your results.

7D. THE SYNTHESIS OF NYLON

CONDENSATION POLYMERIZATION
NUCLEOPHILIC ACYL SUBSTITUTION

In this experiment, you will synthesize the polymer nylon. As a material, nylon is outstanding in most of its properties, combining high dyeing ability, toughness, tensile strength, and elasticity with resistance to moths, molds, mildew, and enzymes. It can also be drawn out in fibers to produce fabrics, carpeting, and tires, and it can be molded to form many objects.

Like other polymers, nylon is made by linking together many smaller molecules. But in contrast to many of these other polymers is the large variety of smaller molecules that can be used to form many different nylon molecules.

For example, nylon 66 may be formed from the reaction of the amine, 1,6-hexanediamine, and the carboxylic acid, adipic acid.

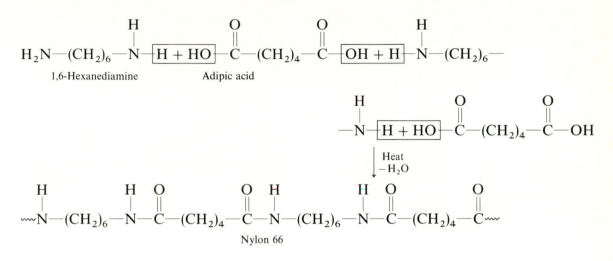

Nylon 66

Observe that each reactant has two functional groups that react with two neighboring molecules of the other reagent to form a long, continuous chain. Nylon 66 gets its name because 1,6-hexanediamine has six carbon atoms, and adipic acid also has six carbon atoms. Many different nylon molecules are named by this system.

Nylon 6 results from the polymerization of the cyclic amide ε-caprolactam.

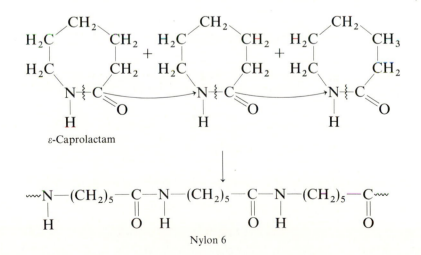

Nylon 6

Observe that there is only one reactant in this polymerization, compared with two before.

Nylon 6-10 can be synthesized from the reaction of the amine, 1,6-hexanediamine, and the acid chloride, sebacoyl chloride.

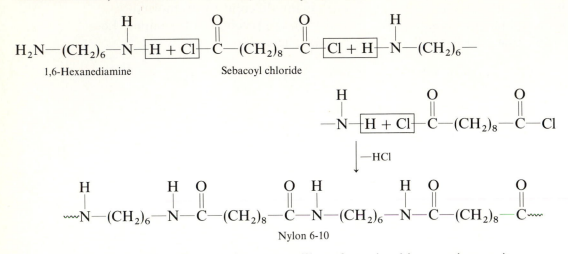

1,6-Hexanediamine

Sebacoyl chloride

Nylon 6-10

This is the equation for the reaction you will perform in this experiment; it occurs readily by simply mixing the two reagents. The acid chloride is dissolved in cyclohexane, and the amine is dissolved in water. The two solvents are insoluble, and the nylon is formed in a continuous strand at the interfacing of the two layers.

But although there are many possible reactants, the synthesis of nylon has just one reaction mechanism: **nucleophilic acyl substitution**. In this mechanism, the amine, acting as a nucleophile, attacks the carbonyl carbon of the acid chloride, carboxylic acid, and so on, to yield a tetrahedral intermediate, which then expels hydrogen chloride, water, and so on, to yield the final product. This is shown in the following reaction for an amine reacting with an acid chloride. The openness of the carbonyl group, due to its flat shape, greatly aids the reaction. So too does the polarity of the carbonyl group and the willingness of the oxygen atom to readily accept a negative charge. The one factor that greatly influences the reaction rate is how easily the leaving group is expelled. Weak bases, such as chloride ion, are found to leave faster than stronger bases.

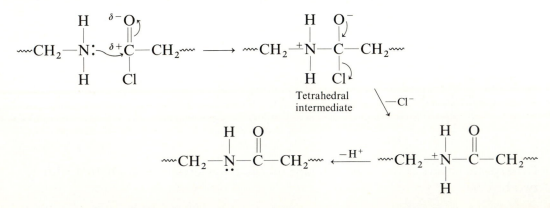

Tetrahedral intermediate

Even though the previous examples show different nylon molecules being formed, every molecule has in common many amide functional groups,

$$\begin{array}{c} H \ O \\ | \ || \\ -N-C- \end{array}$$. As a result, nylon is frequently called a polyamide and undergoes the chemical reactions characteristic of amides. Of these, the most important one is hydrolysis, which can occur under either acidic or alkaline conditions. In this reaction, the polymer is broken apart to yield various forms of the monomers. For example:

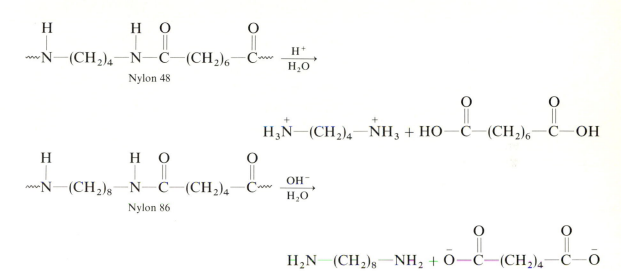

PROCEDURE

▬▬▬▬▬▬

Pour 10 mL of a 5% aqueous solution of 1,6-hexanediamine into a 50-mL beaker, and add 10 drops of 20% sodium hydroxide solution. If you desire your final product to have color, you can also add 1 mL of methyl red, methyl orange, or bromocresol green solution to the aqueous amine.

Very carefully add 10 mL of a 9% solution of sebacoyl chloride in cyclohexane to the aqueous solution by pouring it down the wall of the slightly tilted beaker. Do **not** stir or agitate the liquids, or the polymer formed will be a tangled mess.

Using a thick piece of wire bent at the end to form a hook, grasp and pull out the nylon formed, carefully winding the strands around some object. **Be careful not to touch it with your hands.** If you look closely and wind slowly, you can actually see the thin strands of polymer mysteriously being formed between the two layers and then coming together to form a rope. When completed, wash the product thoroughly with water and allow to dry.

With a stirring rod, vigorously stir the remainder of the two-phase system to see if any additional polymer can be formed. Collect and thoroughly wash.

Record all your results in your notebook. To prevent clogging, do **not** discard any of the nylon in the sink. Use a waste container.

7E.   THE SYNTHESIS OF POLYESTERS

CONDENSATION POLYMERIZATION
NUCLEOPHILIC ACYL SUBSTITUTION

In this experiment, you will synthesize several different polyesters and compare their physical properties. Because many polyesters can be prepared from a variety of different types of reagents, this polymer is similar to nylon (see Experiment 7D).

For example, the polyester poly(ethylene terephthalate) can be prepared by the acid-catalyzed esterification of ethylene glycol and terephthalic acid.

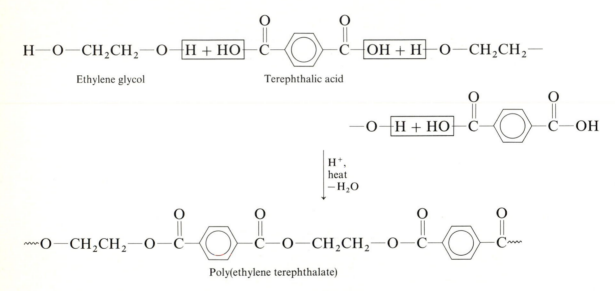

Poly(ethylene terephthalate)

Observe that each reactant has two functional groups that react with two neighboring molecules of the other reagent to form a long, continuous chain. Unfortunately, the polymer produced by this method has a relatively low molecular weight and is not very useful. Furthermore, the high temperature required causes the molecule to undergo some decomposition.

This polymer, however, can be synthesized as a useful product of high molecular weight by **transesterification**, in which an ester reacts with an alcohol to yield another ester and another alcohol. In this case, the ester chosen is dimethyl terephthalate.

The resulting product has many useful applications and can be melt-spun into fibers to produce textiles (Dacron or Terylene), and transparent sheets, films, and coatings of high strength and flexibility (Mylar and Cronar), which

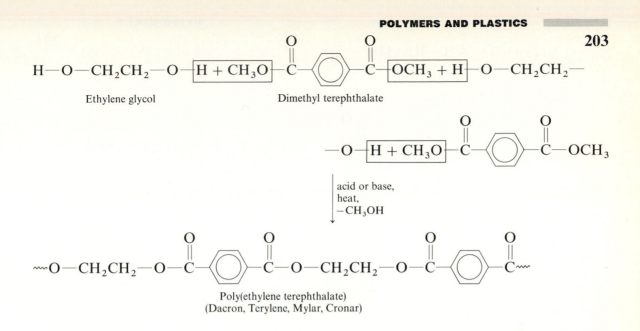

Poly(ethylene terephthalate)
(Dacron, Terylene, Mylar, Cronar)

can be used for many products, including plastic soda bottles and photographic film.

Transesterification is also employed in the production of Kodel, another popular polyester with wide commercial use.

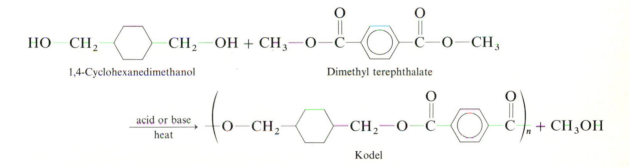

Kodel

Both the alcohol and ester can even be part of the same molecule when they are separated by a sufficient number of carbon atoms to minimize intramolecular ring formation.

$$\text{HO}-(\text{CH}_2)_n-\overset{\overset{\displaystyle O}{\|}}{\text{C}}-\text{O}-\text{CH}_3 \xrightarrow[\text{heat}]{\text{acid or base}} \left[\text{O}-(\text{CH}_2)_n-\overset{\overset{\displaystyle O}{\|}}{\text{C}}\right]_x + \text{CH}_3\text{OH}$$

where $n = 5$ or more.

Cyclic acid anhydrides can also be used instead of esters. For example, maleic anhydride reacts with ethylene glycol to produce an alkyd resin.

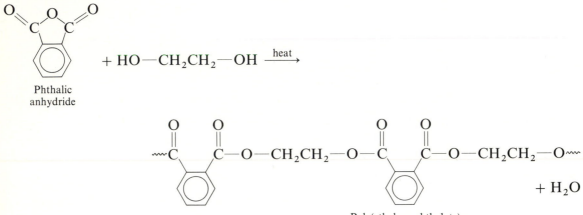

Maleic
anhydride

An alkyd resin

When phthalic anhydride is used with ethylene glycol, the polymer formed is isomeric with Dacron or poly(ethylene terephthalate).

Phthalic
anhydride

Poly(ethylene phthalate)

When glycerol instead of ethylene glycol is reacted with phthalic anhydride, the structure of the polymer formed depends upon the temperature and the proportion of the reagents. At a temperature of 150–160° with a higher proportion of glycerol, only the two more reactive primary hydroxyl groups in glycerol react, and the molecule formed is a linear polyester like all the others up to this point.

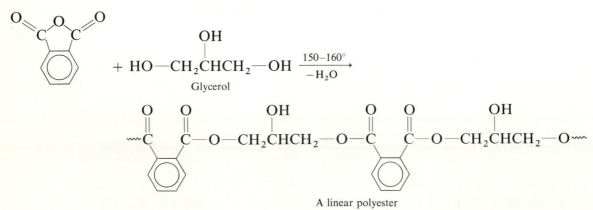

Glycerol

A linear polyester

But with a higher temperature of 200–210° and a higher proportion of phthalic anhydride, all three of the hydroxyl groups in glycerol react to form a molecule, which, instead of being linear, has the polymer chains **cross-linked** to one another to form a three-dimensional polymer of much higher molecular weight. The resulting product, known as Glyptal, is an alkyd resin that, though not suitable for fibers, has a rigid, protective structure, making it useful in paints, lacquers, and similar products.

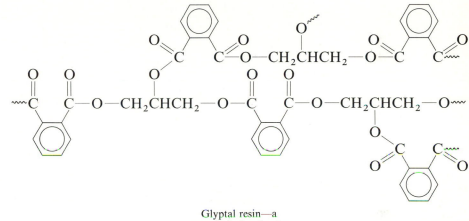

Glyptal resin—a
cross-linked polyester

In this experiment, you will synthesize and compare these last three polyesters, formed by the reaction of phthalic anhydride with ethylene glycol or glycerol.

Although there are many possible reactants, the synthesis of polyesters has one reaction mechanism: **nucleophilic acyl substitution**. In this mechanism, the alcohol, acting as a nucleophile, attacks the carbonyl carbon of the carboxylic acid, ester, or acid anhydride to yield a tetrahedral intermediate, which then expels water, alcohol, or a carboxylate group to yield the product. The following equation shows this for an alcohol reacting with an ester under acidic conditions to give transesterification. Acid speeds up the reaction by protonating the carbonyl oxygen, thus making the carbonyl carbon even more susceptible to nucleophilic attack. The openness of the protonated carbonyl group, due to its flat shape, also greatly aids the reaction.

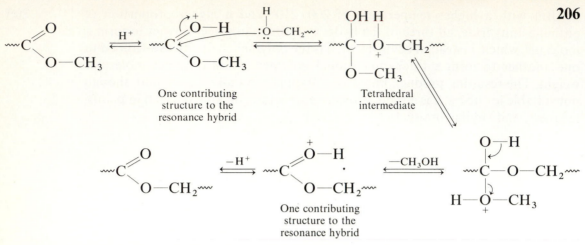

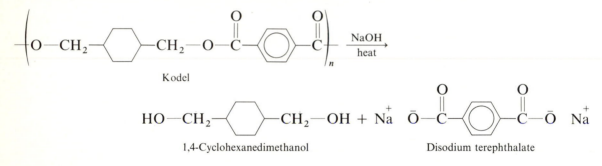

All polyesters will undergo the typical chemical reactions of esters. Of these, the most common one is acidic or alkaline hydrolysis, where the polymer is broken apart to yield the monomers. For example:

Kodel

1,4-Cyclohexanedimethanol   Disodium terephthalate

PROCEDURE

Place 2.0 grams of powdered phthalic anhydride into each of three large test tubes. To the first tube add 0.8 mL of ethylene glycol, to the second add 1.2 mL of glycerol, and to the third add 0.8 mL of glycerol. Thoroughly stir the contents of each tube and with a small flame gently heat the first and second tubes at 150–160° for at least 5 to 10 minutes. During this time, the water formed in the reaction will be removed by distillation. Gently heat the third tube initially at 150–175° until a large amount of the water formed has distilled and then gradually increase the temperature to 200–210° until the mass forms large bubbles that cause it to puff up to a voluminous mass. Remove the thermometer at the completion of each heating.

Allow all three tubes to cool to room temperature. Compare the appearance, viscosity, brittleness, texture, density, and so on, of the polymers formed, and record the results.

Break up the polymers into a powder or very small pieces, and test in small test tubes the solubility of small amounts of each of them in various solvents, such as water, acetone, chloroform, toluene, ether, and ethanol. Perform the tests first in the cold solvent; if necessary, heat the tubes in a hot-water bath. Note and record the results.

Place a small amount of each polymer into separate test tubes. Add several mL of 20% sodium hydroxide solution and heat in a boiling-water bath. Note and record the results.

7F. THE SYNTHESIS OF POLYURETHANE FOAMS

NUCLEOPHILIC ADDITION POLYMERIZATION

In this experiment, you will prepare and compare the physical properties of several different polyurethane foams, synthesized by the reaction of a diisocyanate with a diol or triol. Because a number of polyurethanes can be produced, this polymer is similar in this regard to nylon and polyester (see Experiments 7D and 7E).

The main reaction in the syntheses is the nucleophilic addition of the alcohol across the carbon-nitrogen double bond of the isocyanate.

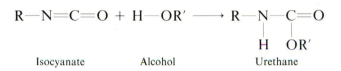

| Isocyanate | Alcohol | Urethane |

With toluene-2,4-diisocyanate and a diol such as triethylene glycol, a linear polymer is produced.

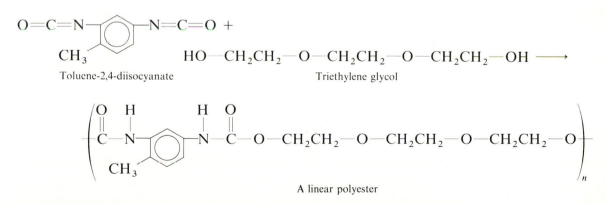

A linear polyester

However, with a triol such as glycerol, a rigid, three-dimensional, **cross-linked** polymer is formed.

Castor oil, a triglyceride or fat (see Experiment 28) of ricinoleic acid, contains a hydroxyl group on each of the ricinoleic acid portions of the molecule. As a result, it can also react as a triol to give a cross-linked polyurethane of much higher molecular weight than can glycerol.

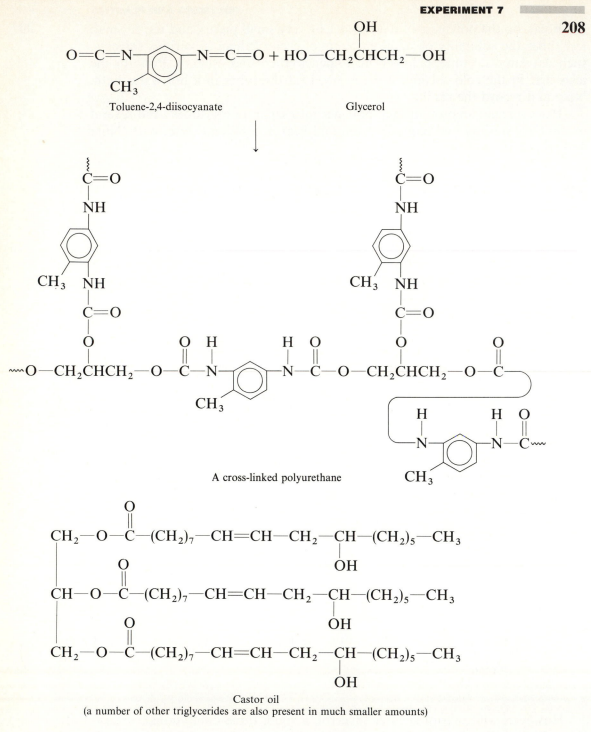

Toluene-2,4-diisocyanate   Glycerol

A cross-linked polyurethane

Castor oil
(a number of other triglycerides are also present in much smaller amounts)

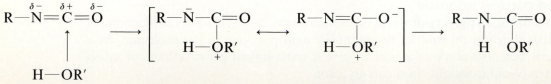

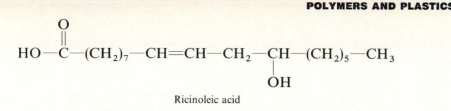

Ricinoleic acid

The reaction mechanism for urethane formation involves the nucleophilic attack by the oxygen atom of the alcohol on the electron-deficient carbon atom of the isocyanate. This produces a resonance-stabilized intermediate, which then transfers a proton to the nitrogen atom to give the final observed product.

With a basic catalyst such as triethylamine, the reaction proceeds more rapidly, due to the hydrogen atom on the alcohol being removed during the formation of the resonance-stabilized intermediate. The transfer of the proton from the catalyst to the intermediate then gives the final product.

$$R-\overset{\delta-}{N}=\overset{\delta+}{C}=\overset{\delta-}{O} \longrightarrow \left[R-\overset{|}{\underset{}{N}}-\overset{|}{\underset{OR'}{C}}=O \longleftrightarrow R-N=\overset{|}{\underset{OR'}{C}}-O^- \right] + (CH_3CH_2)_3\overset{+}{N}-H$$

$$\underset{(CH_3CH_2)_3\ddot{N}}{\overset{H-OR'}{}}$$

$$\longrightarrow \underset{H \quad OR'}{R-\overset{|}{N}-\overset{|}{C}=O} + (CH_3CH_2)_3\ddot{N}$$

The foaming in the reactions is caused by the carbon dioxide produced when added water decomposes a small amount of the isocyanate.

$$R-N=C=O + H_2O \longrightarrow \underset{H \quad OH}{R-\overset{|}{N}-\overset{|}{C}=O} \longrightarrow R-NH_2 + CO_2$$

Analogous to the baking of bread or cake, the carbon dioxide bubbles create pores in the liquid mixture that remain as the foam sets into either a flexible or rigid mass having a much larger volume. Silicone oil is normally added to the mixture to lower the surface tension of the liquid. This results in smaller gas bubbles being produced, leading to a smaller and finer porous structure.

Initially it may appear that the decomposition of an isocyanate group by water to produce carbon dioxide and a primary amine has terminated the growth of the polymer chain at that particular point. However, the primary amine formed may act as a nucleophile, just like the alcohol, and add across the carbon-nitrogen double bond of another molecule of isocyanate. The urea-type linkage formed, along with another isocyanate group in the attacked molecule, allow the chain to be continued. For example:

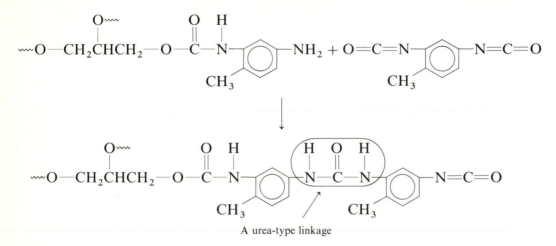

A urea-type linkage

Commercial polyurethane foams are usually prepared from a diisocyanate reacting with a polymeric rather than a simple diol or triol. The flexible types of foams are frequently used in cushions, pillows, mattresses, and rug underliners, where they are superior to rubber foams in resilience and durability. The rigid foams are used to insulate refrigerators and freezers.

Polyurethane fibers, prepared by synthesizing the linear polymer without any water, have considerable strength, elasticity, and flexibility. As a result, they frequently are used as a substitute for rubber fibers in apparel, such as underwear. The three-dimensional, cross-linked polymers, prepared without water, are found in adhesives and surface coatings such as polyurethane floor paints.

PROCEDURE

CAUTION: Toluene diisocyanate is a toxic and hazardous substance that must be used with care. Avoid breathing its vapors and keep it away from your skin and eyes. Plastic gloves are recommended. Use the material only in the hood or in an area with adequate ventilation. Since the compound reacts with the moisture in the air, keep the container tightly closed when not in use. When the experiment is completed, wash your hands thoroughly with water. In addition, the triethylamine catalyst is also quite noxious and must be handled with care.

Weigh in a labeled waxed paper cup 2.0 grams (0.0060 equivalent) of castor oil and 0.50 gram (0.016 equivalent) of glycerol. Add 1 drop each of triethylamine, silicone oil, and water, and stir the mixture well with a glass stirring rod. **In the hood** add 1.5 mL (1.83 grams, 0.021 equivalent) of toluene-2,4-diisocyanate, and stir the resulting warm mixture rapidly until bubbles begin to form. When this occurs, stop stirring the mixture and leave the cup in the hood for the polymerization to proceed. **Do not breathe the evolved vapors.** After the foaming has stopped, allow the sticky material formed to remain undisturbed for several hours to thoroughly set. The paper cup may then be cut away from the mass produced. Note and record the results.

After the above reaction has started, repeat the procedure using 0.70 gram (0.022 equivalent) of glycerol, 1 drop each of triethylamine, silicone oil, and water, and 1.5 mL (0.021 equivalent) of toluene-2,4-diisocyanate. Compare and contrast the foam produced in this experiment with the first one, and record the results.

After the second reaction has begun, repeat the procedure using 1.6 grams (0.022 equivalent) of triethylene glycol, 1 drop each of triethylamine, silicone oil, and water, and 1.5 mL (0.021 equivalent) of toluene-2,4-diisocyanate. Compare and contrast the foam produced in this experiment with the other two, and record the results.

You may also wish to determine the effect of using other reactants, such as ethylene glycol, polyethylene glycol, propylene glycol, polypropylene glycol, or various combinations of all the diols and triols. You can also determine the effect of not using water or silicone oil in the reactions.

QUESTIONS

1. Why couldn't ethylene glycol and methyl benzoate be used as starting materials for the synthesis of a polyester?
2. What are three naturally occurring polymers produced by plants and/or animals?
3. Draw the structure of both the monomer and the polymer for each of the following:
   (a) Teflon
   (b) Orlon
   (c) Saran
   (d) polypropylene
   (e) polyvinyl chloride
4. Other than speeding up the rate of the reactions, what would be the effect of using twice as much benzoyl peroxide initiator in the polymerization of methyl methacrylate and styrene?
5. Why was so much more benzoyl peroxide needed for the polymerization of styrene, compared to the polymerization of methyl methacrylate?

8

THE ESSENTIAL OILS OF PLANTS

As aromatic plants bestow
No spicy fragrance while they grow;
But crushed or trodden to the ground,
Diffuse their balmy sweets around.

Oliver Goldsmith, *The Captivity*, Act I

INTRODUCTION

Practically everyone is familiar with the distinctly pleasant odors produced by many plants. The fragrances of pine and cedar trees, flowers, wintergreen, vanilla, peppermint, spearmint, cinnamon, and cloves are just a few of the examples one can name. All these aromas, of course, are the result of the volatile or **essential oils** produced by the plant, many of which are valuable because of their use as flavorings, perfumes, spices, incense, medicines, insect repellents, and solvents. These essential oils may be found in almost any part of the plant. For example, natural perfumes are obtained from the essential oils found in the petals of flowers. Spearmint, peppermint, and sage all come from the leaves, ginger from the roots, mustard and sesame from the seeds, pepper and nutmeg from the fruit, cinnamon from the bark, and cloves from the buds. When the essential oil comes from a plant used as a flavoring or spice, the oil is valued for both its aroma and its taste.

213

The use of essential oils is as old as civilization, and their importance and value have been so great that they have shaped the course of history. In ancient times, they were used primarily for incense and anointing oils, embalming, sacrificial rites, and funerals. Their ability to add zest and pungency to food was also known, and some were used as medicines, perfumes, and charms. The demand and the resulting monetary incentive was so great that men for centuries fought wars, discovered new worlds, and traveled tremendous distances by ship and caravan to obtain the highly prized treasure. Cities such as Alexandria, Venice, and Genoa prospered from trading in spices, with merchants and distributors becoming very wealthy from their sale. Countries such as Holland, Portugal, Spain, and England fought for control of the spice trade, with discoverers like Columbus, Cabot, Magellan, Frobisher, and Drake making perilous voyages over frequently unknown seas in search of the fabulous spice lands. Frequent references to essential oils and spices are given in the Bible, and frankincense and myrrh are two of the treasures the magi presented to the Christ child.

Frequently, the essential oils are not isolated, but the portion of the plant containing the oil is used as is, or is merely ground up. This is the case with most spices. But essential oils used for flavorings and perfumes are usually isolated from the other components in the plant. This can be done by steam distillation, solvent extraction, enfleurage, and expression methods. You are already familiar with solvent extraction, and steam distillation will be discussed in detail in the background section of Experiment 8A.

The enfleurage process is used mainly with flowers, which either do not yield an appreciable quantity of oil by steam distillation, or whose odor is affected by the high temperature of steam. The flowers are spread over a highly purified mixture of cold tallow and lard, which absorbs the essential oils from the petals. Extraction of the fat with alcohol then transfers the essential oil to the alcohol, which can then be used as is for perfumes or carefully evaporated to yield the concentrated oil.

The expression process is used mainly for the recovery of orange, lemon, and other citrus oils from the peels. After the peels are removed from the fruit, they are squeezed in presses and the oil is decanted or centrifuged to isolate it from peel residue and water.

The essential oils isolated from plants can be either essentially pure compounds or vast mixtures containing up to several hundred components. Wintergreen, orange, and cassia oil are a few examples of essential oils containing mainly one compound. In spices and flavorings, even the minor components of the mixture can be very important, since they give the oil a characteristic and natural odor and taste that can distinguish it from artificial or imitation oils made in the laboratory.

As you will observe in Experiment 9, the flavor of most fruits is due chiefly to organic compounds called esters. But essential oils may consist of a wide variety of other organic molecules such as alkenes, alcohols, aldehydes, ketones,

TABLE 8-1. ▆▆▆▆▆▆▆▆▆▆▆▆▆▆▆▆▆▆

Classification of Terpenes and Terpenoids

Class	Number of Carbon Atoms
Monoterpene or Monoterpenoid	10
Sesquiterpene or Sesquiterpenoid	15
Diterpene or Diterpenoid	20
Triterpene or Triterpenoid	30
Tetraterpene or Tetraterpenoid	40

ethers, lactones, acetals, and phenols. These functional groups, of course, will determine how the oils will react chemically.

Most essential oils may be classified as **terpenes** or **terpenoids**. The word "terpene" is derived from turpentine. In 1818, turpentine was discovered to be a mixture of hydrocarbons having a carbon-hydrogen ratio of 5 to 8. Since many essential oils are also hydrocarbons having this same carbon-hydrogen ratio, they became known as terpenes. Terpenoids are similar in structure to terpenes, but also contain oxygen.

One can also further classify most essential oils according to the number of carbon atoms they contain. This is done in Table 8-1, and the structures of some common terpenes and terpenoids are given in Table 8-2. Terpenes and terpenoids having 5, 25, or 35 carbon atoms either do not exist or are quite rare.

Many of these listed terpenes and terpenoids are a part of the experiments in this book. Menthol is oxidized to menthone in Experiment 8E. Limonene is isolated from orange and grapefruit peels in Experiment 8B and analyzed as a part of turpentine in Experiment 8G. Carvone is isolated from caraway seeds in Experiment 8A. Camphor is synthesized from camphene in Experiment 8C and used to synthesize borneol and isoborneol in Experiment 8D. α-Pinene and β-pinene are analyzed as a part of turpentine in Experiment 8G. Zingiberene may be isolated from ginger, and caryophyllene may be isolated with other compounds from cloves or allspice in Experiment 8A. Finally, β-carotene may be isolated from tomato paste in Experiment 6C, and cholesterol from gallstones and egg yolks in Experiment 30A and 30B. Vitamin A, β-carotene, and cholesterol are not considered to be essential oils, but they nevertheless are either terpenes or terpenoids.

All terpenes or terpenoids contain isoprene units, usually in a head-to-tail sequence.

$$CH_2{=}\underset{\underset{CH_3}{|}}{C}{-}CH{=}CH_2 \quad \text{Isoprene}$$

This is why terpenes or terpenoids usually contain 10, 15, 20, 30, or 40 carbon

TABLE 8-2.

Structures of Some Common Terpenes and Terpenoids

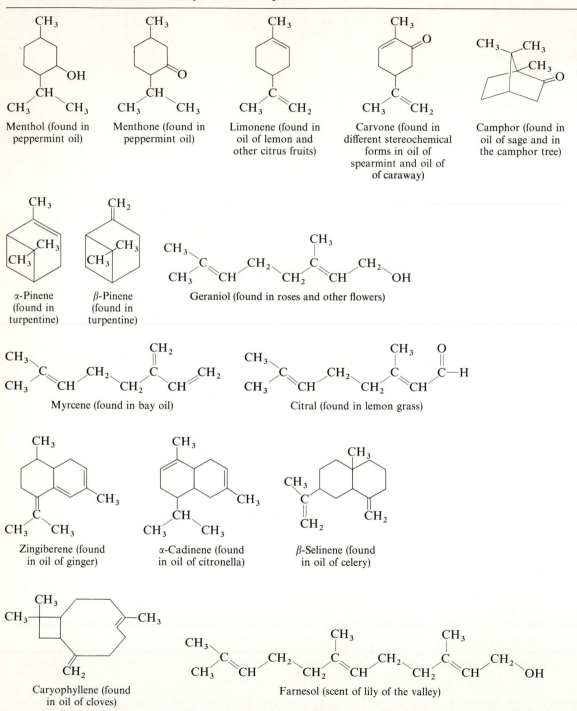

Menthol (found in peppermint oil)

Menthone (found in peppermint oil)

Limonene (found in oil of lemon and other citrus fruits)

Carvone (found in different stereochemical forms in oil of spearmint and oil of of caraway)

Camphor (found in oil of sage and in the camphor tree)

α-Pinene (found in turpentine)

β-Pinene (found in turpentine)

Geraniol (found in roses and other flowers)

Myrcene (found in bay oil)

Citral (found in lemon grass)

Zingiberene (found in oil of ginger)

α-Cadinene (found in oil of citronella)

β-Selinene (found in oil of celery)

Caryophyllene (found in oil of cloves)

Farnesol (scent of lily of the valley)

TABLE 8.2 (continued)

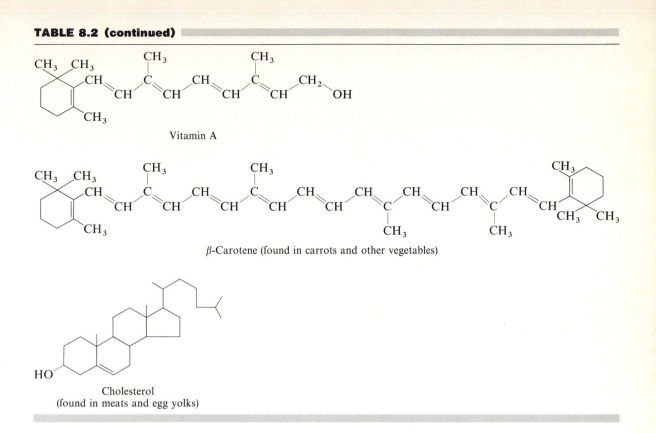

Vitamin A

β-Carotene (found in carrots and other vegetables)

Cholesterol
(found in meats and egg yolks)

atoms. In the terpene or terpenoid, the double bonds of isoprene may be missing or shifted, so one usually just looks for the 5-carbon skeletal structure of each isoprene unit. These structures are circled in the examples shown in Table 8-3.

This common structural unit soon made chemists think that terpenes and terpenoids are all synthesized by the same mode. This mode is now believed to be the following:

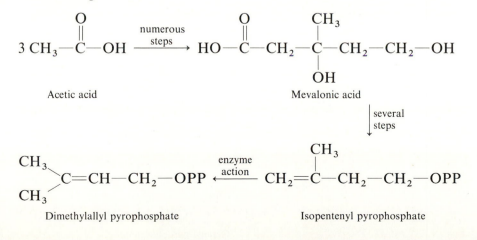

Acetic acid   Mevalonic acid

Dimethylallyl pyrophosphate   Isopentenyl pyrophosphate

TABLE 8-3.

The Isoprene Units of Some Terpenes and Terpenoids

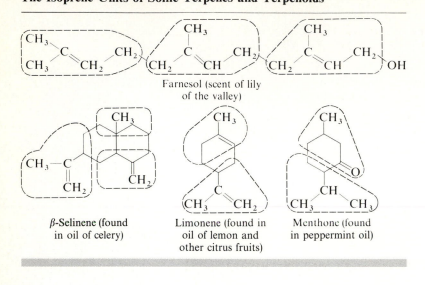

Farnesol (scent of lily of the valley)

β-Selinene (found in oil of celery)

Limonene (found in oil of lemon and other citrus fruits)

Menthone (found in peppermint oil)

$$OPP = pyrophosphate = -O-\overset{\uparrow}{\underset{\underset{OH}{|}}{P}}-O-\overset{\uparrow}{\underset{\underset{OH}{|}}{P}}-OH$$

Acetic acid in a number of steps is converted to mevalonic acid, which is then converted in a number of steps to isopentenyl pyrophosphate. Dimethylallyl pyrophosphate is then formed through an enzyme shifting the double bond.

Once these last two molecules are made, they can react with each other as shown in the next reaction. Using its more reactive π-electrons, isopentenyl

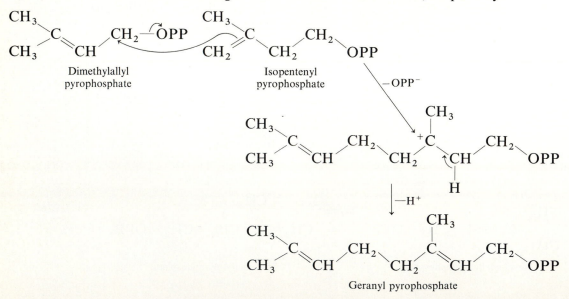

Dimethylallyl pyrophosphate

Isopentenyl pyrophosphate

Geranyl pyrophosphate

pyrophosphate acts as a nucleophile and displaces pyrophosphate from dimethylallyl pyrophosphate to yield a 10-carbon tertiary carbocation. This carbocation then loses a proton to form geranyl pyrophosphate.

Once geranyl pyrophosphate is formed, it can do a number of things, as the following reactions indicate. It can hydrolyze, as shown in route 1, to form geraniol. It can react with another molecule of isopentenyl pyrophosphate, as shown in route 2, to form farnesyl pyrophosphate, which can then be hydrolyzed to form farnesol. Or it can isomerize, as shown in route 3, to form neryl pyrophosphate, which can then form a cyclic tertiary carbocation. Loss of a proton from this cation gives limonene, or hydrolysis gives α-terpineol. Many other reaction sequences exist, and all known terpenes and terpenoids can be shown to be formed from them.

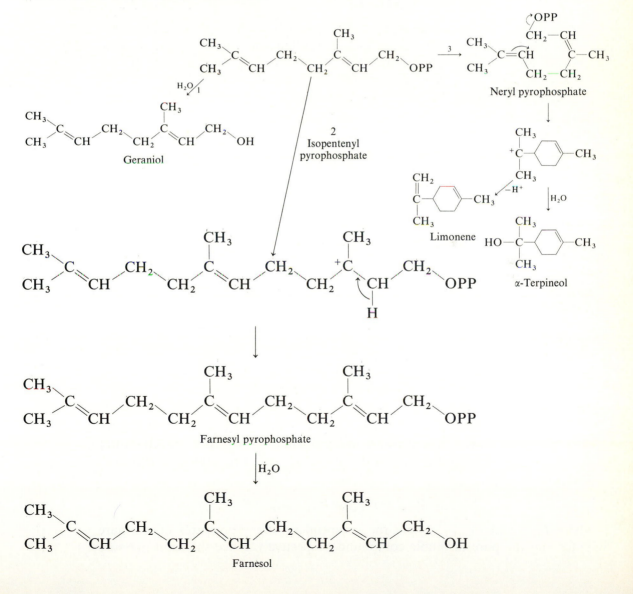

The purpose of essential oils, other than their benefit to people, is not well understood. The odor of flowers may aid their reproduction by acting as attractants to bees. Other oils may act to protect the plant when injured against loss of sap or attack by parasites and disease organisms. Hardly any essential oils are necessary for plant metabolism, and it may be that these materials are just simply side or waste products synthesized by the plant.

8A. STEAM DISTILLATION OF ESSENTIAL OILS FROM SPICES

PREPARATION OF DERIVATIVES

In this experiment, you will separate the essential oil from a spice by steam distillation. Extraction of the oil and water distillate with low-boiling methylene chloride, followed by evaporation of the solvent then allows you to isolate the oil. Because so many spices are available, you may choose which oil you wish to isolate. To compare the results, you may want to do several spices; in certain cases, you can even prepare a derivative from the principal component in the oil.

Solvent extraction can also be used to separate essential oils. However, gums, resins, and fats frequently are also removed, contaminating the desired essential oils. This disadvantage is not common with steam distillation, since these impurities are not usually steam-volatile. Solvent extraction, however, is preferable over steam distillation whenever the essential oils easily decompose with heat, such as with certain flowers.

Steam distillation, however, is not used only for separating essential oils. In a chemical synthesis, for example, it might also be used to separate a steam-volatile reaction product from the other nonvolatile components in the reaction mixture.

A water-insoluble compound is required for any steam distillation, and the compound and water distill together at a lower temperature than the boiling point of either pure compound. If the compound has a very high boiling point but is unstable at this temperature, steam distillation allows the compound to be distilled at a much milder temperature, under 100°C, thereby avoiding thermal decomposition.

This boiling behavior results from the water and the insoluble compound each exerting its own vapor pressure independently of the other, with the total vapor pressure equal to the sum of the individual vapor pressures. In equation form, the result is

$$P_{\text{total}} = P^{\circ}_{\text{H}_2\text{O}} + P^{\circ}_{\text{insol. comp.}}$$

where $P^{\circ}_{\text{H}_2\text{O}}$ and $P^{\circ}_{\text{insol. comp.}}$ are the independent vapor pressures of the pure water and the pure insoluble compound, respectively. Since the total pressure must equal atmospheric pressure at a temperature below the boiling point of either pure substance, the mixture must distill at this lower temperature.

As an example, eugenol, the main component in oil of cloves, is insoluble in water and boils at 252°C at 760 mmHg. However, when steam-distilled, the mixture of eugenol and water together boils at the much lower temperature of 99.86°C at 760 mmHg. At this temperature, the independent vapor pressure of eugenol is 3.8 mmHg, and the independent vapor pressure of water is 756.2 mmHg. Together these pressures equal the atmospheric pressure of 760 mmHg.

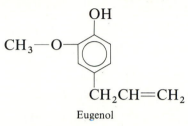

Eugenol

Observe from the above equation that there is no dependence on the mole fraction, X, of either component. This equation therefore contrasts with the following one for Raoult's law (see page 36).

$$P_{total} = P_A^\circ \cdot X_A + P_B^\circ \cdot X_B$$

In this equation, X_A is the mole fraction of component A, and X_B is the mole fraction of component B. Raoult's law applies to soluble, noninteracting liquids that form an ideal solution. Obviously, this is not the case with water and a steam-distillable compound. In fact, in the following ways we may think of a steam distillation as a special type of azeotropic distillation (see page 47).

1. The water and the steam-distillable compound have a higher combined vapor pressure than Raoult's law would predict.

2. The mixture distills at a lower temperature than the boiling point of either pure component.

3. The distillate has a constant composition as long as there is a sufficient amount of the steam-distillable compound available.

STEAM DISTILLATION CALCULATIONS

By knowing the boiling point of a steam distillation, we can determine the vapor pressure of both the water and the insoluble compound. We can then determine the composition of the distillate.

For example, the boiling point of water and eugenol in a steam distillation is 99.86°C at 760 mmHg. From a chemical handbook, we find the vapor pressure of water at this temperature to be 756.2 mmHg. Subtracting this from the atmospheric pressure gives 3.8 mmHg as the vapor pressure of eugenol at this temperature.

Because the number of moles of a substance in its vapor is proportional to its vapor pressure, we can write, for water and eugenol,

$$\frac{\text{Moles of eugenol}}{\text{Moles of water}} = \frac{P^\circ_{\text{eugenol}}}{P^\circ_{\text{water}}}$$

Since moles = weight/molecular weight, we can substitute this equation into the preceding one and obtain

$$\frac{\text{Weight eugenol/molecular weight eugenol}}{\text{Weight water/molecular weight water}} = \frac{P^\circ_{\text{eugenol}}}{P^\circ_{\text{water}}}$$

Rearranging this equation gives

$$\frac{\text{Weight eugenol}}{\text{Weight water}} = \frac{(P^\circ_{\text{eugenol}})(\text{molecular weight eugenol})}{(P^\circ_{\text{water}})(\text{molecular weight water})}$$

Substituting the known values gives

$$\frac{\text{Weight eugenol}}{\text{Weight water}} = \frac{(3.8 \text{ mmHg})(164 \text{ grams/mole})}{(756.2 \text{ mmHg})(18 \text{ grams/mole})}$$

$$= 0.046 \text{ gram eugenol/1.0 gram water}$$

This gives a weight composition of 4.4% eugenol and 95.6% water in the distillate.

As another example, limonene is found in orange, grapefruit, and lemon peels. Its normal boiling point is 178°C at 760 mmHg, but it steam distills at 97.2°C at this pressure. The vapor pressure of water at 97.2°C is found in a handbook to be 687.0 mmHg. Therefore, the vapor pressure of limonene at this temperature must be 73.0 mmHg.

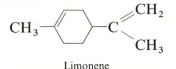

Limonene

The equation for computing the weight composition of the distillate is similar to the one used for eugenol:

$$\frac{\text{Weight limonene}}{\text{Weight water}} = \frac{(P^\circ_{\text{limonene}})(\text{molecular weight limonene})}{(P^\circ_{\text{water}})(\text{molecular weight water})}$$

Substituting the known values gives

$$\frac{\text{Weight limonene}}{\text{Weight water}} = \frac{(73.0 \text{ mmHg})(136 \text{ grams/mole})}{(687.0 \text{ mmHg})(18 \text{ grams/mole})}$$

$$= 0.803 \text{ gram limonene/1.00 gram water}$$

This gives a weight composition of 44.5% limonene and 55.5% water in the distillate.

Observe from these examples how the vapor pressure of a steam-distillable compound directly influences the percentage of this compound in the distillate. Of course, because of the equation used for the calculations, this is what we would expect.

APPARATUS AND TECHNIQUE

Two methods for steam distillation are generally used in the laboratory. One method uses live steam from an outside steam line passed into the distillation flask, and the other one generates the steam directly in the distillation flask itself. The live-steam method has the advantage that the entering steam provides agitation that helps prevent the mixture from bumping. This agitation also proves valuable when bulky or very tarry masses are present. This method is widely used, especially with volatile solids or high-molecular-weight compounds. But it is the more difficult method of the two and should not be used by organic chemistry students unless they are highly proficient in the lab. The apparatus for this method, shown in Figure 8-1, takes longer to set up, requires closer supervision, and can cause mishaps from too much or too little steam pressure.

FIGURE 8-1
Steam distillation using live steam.

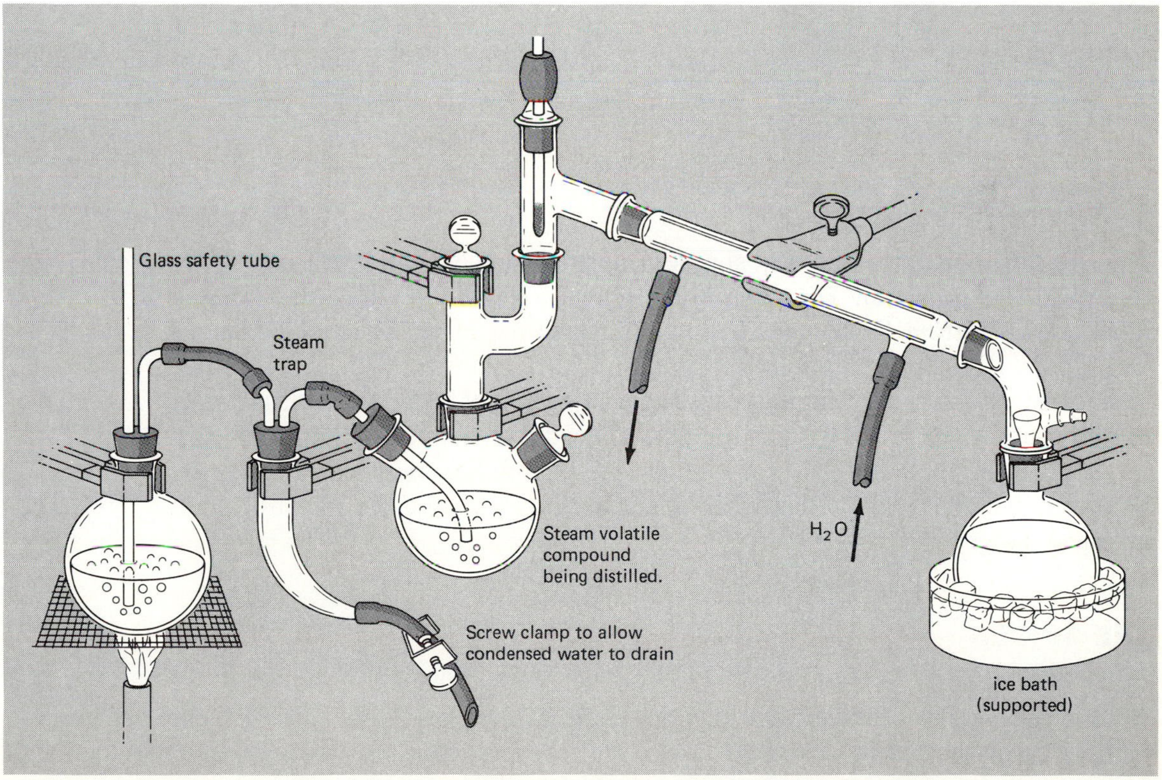

Glass safety tube

Steam trap

Steam volatile compound being distilled.

Screw clamp to allow condensed water to drain

H_2O

ice bath (supported)

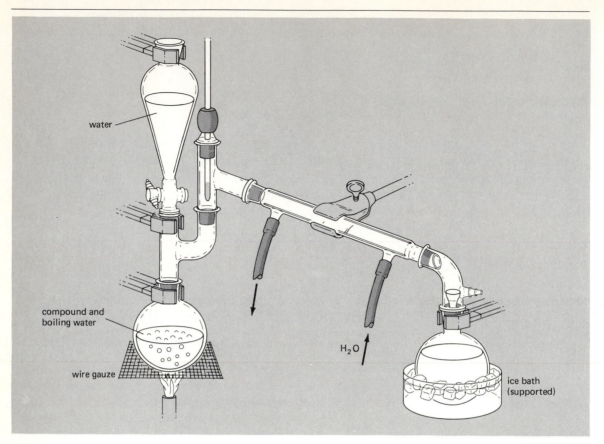

water

compound and
boiling water

wire gauze

H₂O

ice bath
(supported)

FIGURE 8-2
Direct-steam distillation.

The direct-steam method is simpler, more convenient, and will be the one used in this experiment. The apparatus, shown in Figure 8-2, includes a single-neck round-bottom flask, to which is connected a Claisen head. The Claisen head helps reduce the possibility of material being transferred to the receiving flask through excessive bumping. The separatory funnel in the apparatus is used to add more water to the distillation flask. This direct-steam method works well for volatile liquids that have no solids present. Solids may cause excessive bumping, but this can frequently be minimized through the use of a hot plate-magnetic stirrer.

THE SPICES

Among the choices of spices that may be steam distilled are cloves, allspice, cinnamon, caraway, cumin, and anise. If desired, you may also steam distill spices like nutmeg, pepper, ginger, sage, thyme, and oregano. The residue from the distillation contains mainly cellulose and other non-steam-volatile compounds.

Cloves and Allspice

The main constituent in oil of cloves and oil of allspice is eugenol (4-allyl-2-methoxyphenol). Eugenol acetate, caryophyllene, and other compounds are present in small amounts.

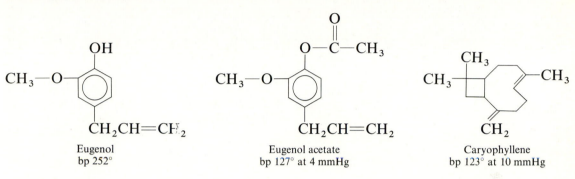

Eugenol
bp 252°

Eugenol acetate
bp 127° at 4 mmHg

Caryophyllene
bp 123° at 10 mmHg

After the oil is isolated, the eugenol may be converted to the benzoate ester by reaction with benzoyl chloride.

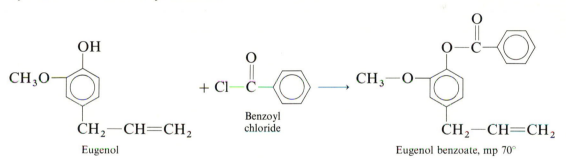

Eugenol

Benzoyl chloride

Eugenol benzoate, mp 70°

The infrared and ^{1}H NMR spectra of pure eugenol are shown in Figures 8-3 and 8-4.

Cinnamon

The principal component of cinnamon oil is cinnamaldehyde (*trans*-3-phenylpropenal).

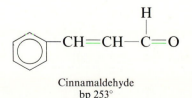

Cinnamaldehyde
bp 253°

After the oil is isolated, the cinnamaldehyde may be converted to its semicarbazone derivative by reaction with semicarbazide hydrochloride.

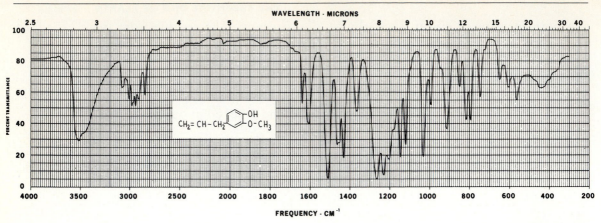

FIGURE 8-3

The infrared spectrum of eugenol, neat. Spectrum courtesy
of Sadtler Research Laboratories, division of Bio-Rad
Laboratories, Inc., copyrighted 1972.

FIGURE 8-4

The ¹H NMR spectrum of eugenol in deuterated
chloroform. Spectrum courtesy of Aldrich Chemical Company,
The Aldrich Library of NMR Spectra, edition II, by Charles
Pouchert.

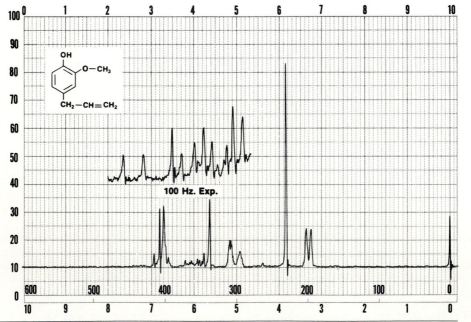

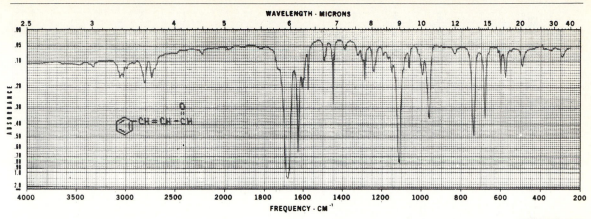

WAVELENGTH · MICRONS

FREQUENCY · CM⁻¹

FIGURE 8-5

The infrared spectrum of cinnamaldehyde, neat. Spectrum courtesy of Sadtler Research Laboratories, division of Bio-Rad Laboratories, Inc., copyrighted 1970.

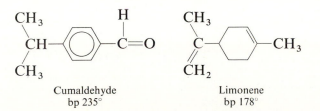

Cinnamaldehyde   Semicarbazide

Cinnamaldehyde
semicarbazone
mp 215°

The infrared and ^{1}H NMR spectra of pure cinnamaldehyde are shown in Figures 8-5 and 8-6.

Cumin

The main constituent in oil of cumin is cumaldehyde (*p*-isopropylbenzaldehyde). The oil also contains a few other compounds, such as limonene, which contribute to the odor.

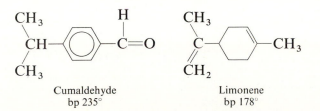

Cumaldehyde
bp 235°

Limonene
bp 178°

227

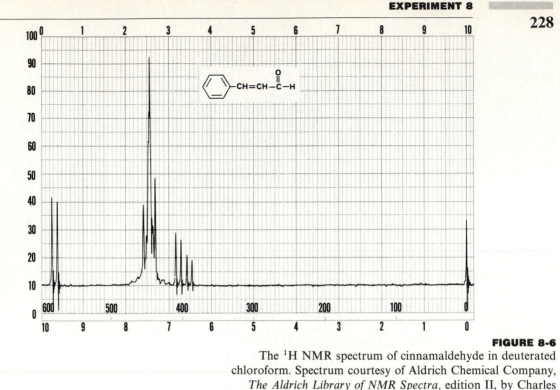

FIGURE 8-6

The ^{1}H NMR spectrum of cinnamaldehyde in deuterated chloroform. Spectrum courtesy of Aldrich Chemical Company, *The Aldrich Library of NMR Spectra,* edition II, by Charles Pouchert.

After the oil is isolated, the cumaldehyde may be converted to its semicarbazone derivative by reaction with semicarbazide hydrochloride. If this is done, you will find it interesting to compare the odor of the oil before and after the derivative is made.

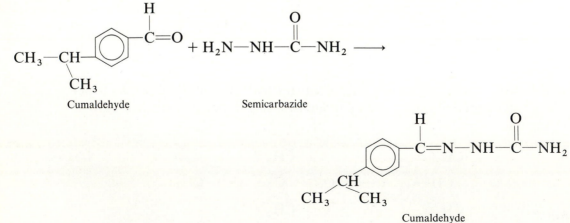

The infrared and ^{1}H NMR spectra of pure cumaldehyde are shown in Figures 8-7 and 8-8.

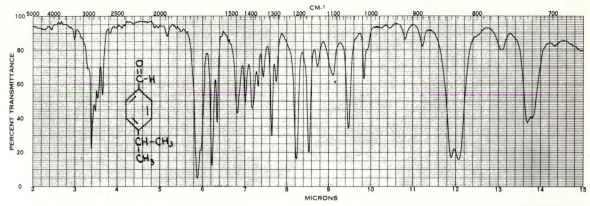

FIGURE 8-7
The infrared spectrum of cumaldehyde, neat. Spectrum
courtesy of Sadtler Research Laboratories, division of Bio-Rad
Laboratories, Inc., copyrighted 1970.

FIGURE 8-8
The ^{1}H NMR spectrum of cumaldehyde in deuterated
chloroform.

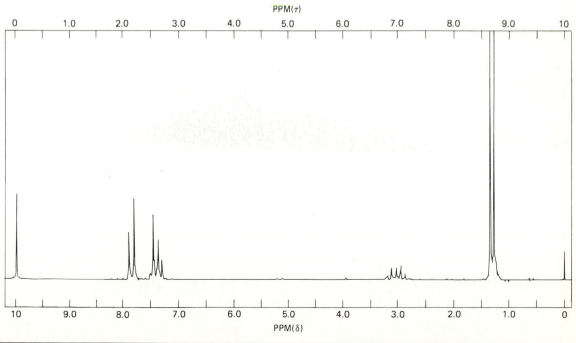

Caraway

The principal components in the essential oil of caraway are *d*-carvone and *d*-limonene. As you can see from the structures below, both of these compounds contain the 1-methyl-4-isopropylcyclohexane skeleton, which is common for many other natural products.

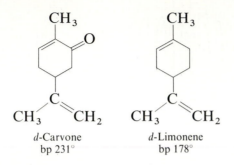

d-Carvone
bp 231°

d-Limonene
bp 178°

After the oil is isolated, the *d*-carvone can be separated almost quantitatively as the 2,4-dinitrophenylhydrazone derivative by reaction with 2,4-dinitrophenylhydrazine. If the yield is high enough to let you accurately weigh the oil and the derivative, then you can estimate the percentage of *d*-carvone in caraway oil.

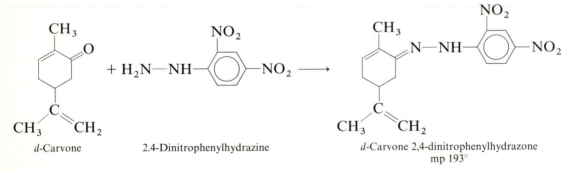

d-Carvone

2.4-Dinitrophenylhydrazine

d-Carvone 2,4-dinitrophenylhydrazone
mp 193°

The infrared and ¹H NMR spectra of pure *d*-carvone and *d*-limonene are shown in Figures 8-9 through 8-12.

Anise

The main constituent in oil of anise is anethole [*trans*-1-(*p*-methoxyphenyl)-propene]. A minor component is *p*-allylanisole, which, as the structures below show, is an isomer of anethole with merely the double bond shifted.

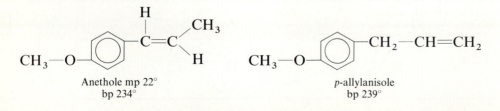

Anethole mp 22°
bp 234°

p-allylanisole
bp 239°

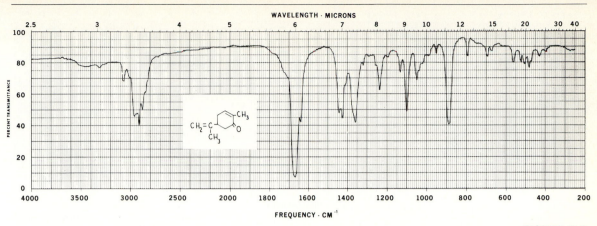

FIGURE 8-9

The infrared spectrum of *d*-carvone, neat. Spectrum courtesy
of Sadtler Research Laboratories, division of Bio-Rad
Laboratories, Inc., copyrighted 1971.

FIGURE 8-10

The ^{1}H NMR spectrum of *d*-carvone in carbon tetrachloride.
Spectrum courtesy of Aldrich Chemical Company, *The Aldrich
Library of NMR Spectra,* edition II, by Charles Pouchert.

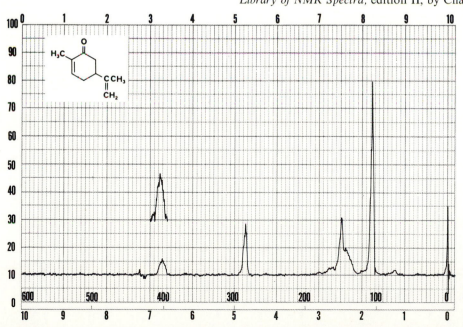

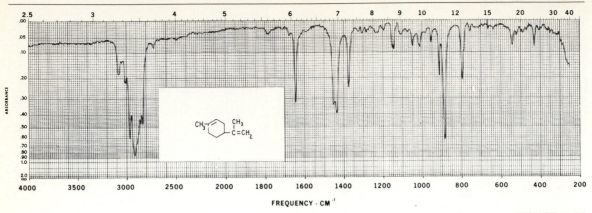

FIGURE 8-11

The infrared spectrum of *d*-limonene, neat. Spectrum courtesy of Sadtler Research Laboratories, division of Bio-Rad Laboratories, Inc., copyrighted 1968.

FIGURE 8-12

The ¹H NMR spectrum of *d*-limonene in carbon tetrachloride. Spectrum courtesy of Aldrich Chemical Company, *The Aldrich Library of NMR Spectra,* edition II, by Charles Pouchert.

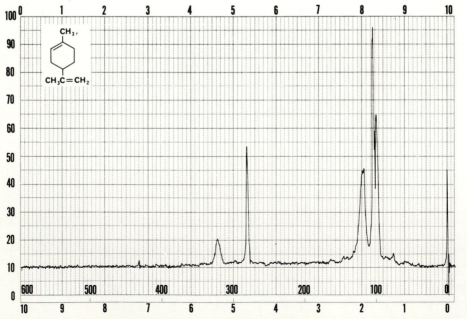

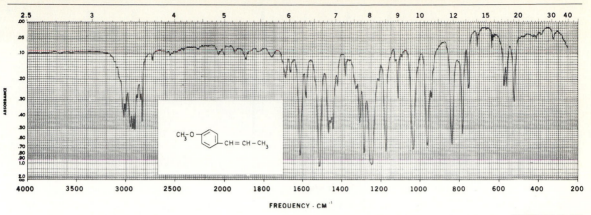

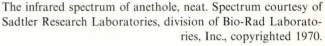

FIGURE 8-13

The infrared spectrum of anethole, neat. Spectrum courtesy of Sadtler Research Laboratories, division of Bio-Rad Laboratories, Inc., copyrighted 1970.

From the observed melting point of anethole, it should be evident that the oil may be easily solidified by cooling in an ice-water bath. The temperature at which this occurs has been used to estimate the percentage of anethole in anise oil. Of course, upon warming to room temperature, the anethole will again liquefy.

The infrared and ^{1}H NMR spectra of pure anethole are shown in Figures 8-13 and 8-14.

PROCEDURE

Place 15 grams of the selected ground spice in a 500-mL round-bottom flask, and add 150 mL of water and several boiling stones. Assemble the apparatus for a direct-steam distillation, as shown in Figure 8-2. Fill the dropping funnel with water and heat the mixture with a Bunsen burner or hot plate-magnetic stirrer. After distillation begins, add water from the dropping funnel to maintain the original volume of liquid in the flask. If heating with a Bunsen burner produces too much bumping, use a hot plate-magnetic stirrer. Continue distilling until no further drops of oil can be seen coming over with the water. At least 100 mL of liquid should be collected.

Pour the distillate into a 250-mL separatory funnel and extract it twice, using 20-mL portions of methylene chloride, and periodically releasing any internal pressure. **(Caution: Methylene chloride is toxic. Do not breathe it excessively or spill it on yourself.)** After shaking the separatory funnel each time, separate the lower methylene chloride layer into a 100-mL beaker. Dry the combined extracts over a small amount of anhydrous magnesium

233

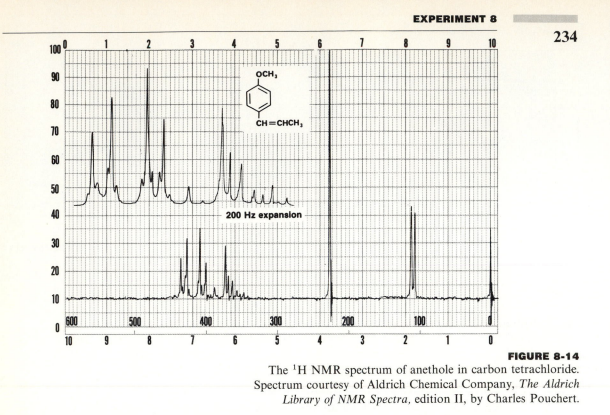

FIGURE 8-14
The ^{1}H NMR spectrum of anethole in carbon tetrachloride.
Spectrum courtesy of Aldrich Chemical Company, *The Aldrich
Library of NMR Spectra*, edition II, by Charles Pouchert.

sulfate, then carefully decant or gravity filter the methylene chloride solution
from the drying agent into a 50-mL Erlenmeyer flask.

Evaporate most of the solvent on a steam bath or hot-water bath in the
hood. When only a small amount remains, transfer the liquid to a previously
weighed test tube and concentrate the liquid further until only an oily residue
remains. Dry the outside of the test tube and determine the weight of essential
oil obtained; record the result in your notebook. Determine the percentage yield
of oil, based upon the original weight of spice used, and show the calculations.

At your instructor's option, obtain the infrared spectrum of your product,
neat between salts, and the ^{1}H NMR spectrum in deuterated chloroform or carbon tetrachloride, using the one specified earlier for your compound. Compare
your spectra to those shown earlier for your compound.

If desired, you may convert the essential oil obtained to a derivative, using
one of the following procedures.

Eugenol benzoate from clove or allspice oil. Place 0.2 mL of the oil in
a small test tube and add 1 mL of water. Add a 1 *M* solution of sodium hydroxide
dropwise until the oil just dissolves. The resulting liquid will not be clear, but
no drops of oil should be visible. To the liquid, carefully add **4 or 5 drops** of
benzoyl chloride; be sure to avoid an excess or the final product will not crystallize. Heat the mixture with a steam bath or hot-water bath for 5 to 10 minutes.

Cool the mixture in an ice bath and scratch the inside of the test tube with a glass rod until crystals form. If the oil does not crystallize, decant the aqueous layer, add a few drops of methanol, and cool the liquid in an ice bath with scratching. Collect the solid by suction filtration, using a Hirsch funnel, and recrystallize from a **minimum** amount of boiling methanol. Collect the crystals on a Hirsch funnel, allow to dry, and determine the melting point; record the result in your notebook. The literature melting point is 70°.

Cinnamaldehyde semicarbazone from cinnamon oil. Dissolve 0.10 gram of semicarbazide hydrochloride and 0.15 gram of anhydrous sodium acetate (which will act as a buffer) in 1 mL of water and add 1.5 mL of ethanol. Add this solution to half of the cinnamon oil and heat the mixture with a steam bath or hot-water bath for 5 minutes. Cool the mixture and allow the resulting cinnamaldehyde semicarbazone to crystallize. Collect the crystals by suction filtration on a Hirsch funnel and recrystallize from the **minimum** quantity of hot methanol. Collect the crystals on a Hirsch funnel, allow to dry, and determine the melting point; record the result in your notebook. The literature melting point is 215°.

Cumaldehyde semicarbazone from cumin oil. Dissolve 0.10 gram of semicarbazide hydrochloride and 0.15 gram of anhydrous sodium acetate (which will act as a buffer) in 1 mL of water and add 1.5 mL of ethanol. Add this solution to half of the cumin oil and heat the mixture with a steam bath or hot-water bath for 5 minutes. Cool the mixture and allow the resulting cumaldehyde semicarbazone to crystallize. Collect the crystals by suction filtration on a Hirsch funnel and recrystallize from the **minimum** quantity of hot methanol. Collect the crystals on a Hirsch funnel, allow to dry, and determine the melting point; record the result in your notebook. The literature melting point is 216°.

Carvone 2,4-dinitrophenylhydrazone from caraway oil. Dilute half of the caraway oil with 1 mL of ethanol and add 0.5 mL of 2,4-dinitrophenylhydrazine solution. Collect the resulting solid by suction filtration on a Hirsch funnel and recrystallize from the **minimum** quantity of hot methanol. Collect the crystals on a Hirsch funnel, allow to dry, and determine the melting point; record the result in your notebook. The literature melting point is 193°.

8B. STEAM DISTILLATION OF *d*-LIMONENE FROM GRAPEFRUIT OR ORANGE PEELS

In this experiment, you will separate by steam distillation *d*-limonene from either grapefruit or orange peels. Extraction of the limonene-water distillate with low-boiling methylene chloride, followed by evaporation of the solvent will then allow you to isolate the limonene. The product obtained is approximately 97 percent pure. The residue from the original distillation contains mainly cellulose and other non-steam-volatile compounds.

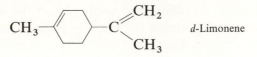

d-Limonene

d-Limonene is a terpene found in many plants, but citrus peels have the greatest amount. Lemons or tangerines could thus also be used in the experiment. When finished, you will find it interesting to compare the odor of your final product with the original peels before distillation.

PROCEDURE

If necessary, read the section on steam distillation in Experiment 8A.

Obtain the peels from one and a half grapefruits or three oranges. Weigh them, cut into $\frac{1}{2}$-cm squares, crush with a mortar and pestle, and place in a 500-mL round-bottom flask with about 200 mL of water. Assemble the direct-steam distillation setup shown in Figure 8-2, but exclude the separatory funnel. Heat the mixture with a Bunsen burner and collect about 75 mL of distillate.

Pour the distillate into a 125-mL separatory funnel and extract it twice, using 15-mL portions of methylene chloride, and periodically releasing any internal pressure. **(Caution: Methylene chloride is toxic. Do not breathe it excessively or spill it on yourself.)** After shaking the separatory funnel each time, separate the lower methylene chloride layer into a 100-mL beaker. Dry the combined extracts over a small amount of anhydrous magnesium sulfate, then carefully decant or gravity filter the methylene chloride solution from the drying agent into a 50-mL Erlenmeyer flask.

Evaporate most of the solvent on a steam bath or hot-water bath in the hood. When only about 10 mL remains, transfer the liquid to a previously weighed 25-mL Erlenmeyer flask and further concentrate the liquid. Dry the outside of the flask and determine the weight of d-limonene obtained; record the result in your notebook. Determine the percentage yield of d-limonene, based upon the original weight of peels used, and show your calculations.

At your instructor's option, obtain the infrared spectrum of your product, neat, and the ^{1}H NMR spectrum in carbon tetrachloride. Compare your spectra with those shown in Figures 8-11 and 8-12.

When finished, hand in your product in a properly labeled container to your instructor.

8C. THE SYNTHESIS OF CAMPHOR FROM CAMPHENE

- **WAGNER-MEERWEIN REARRANGEMENTS**
- **ACETYLATION**
- **OXIDATION USING HOUSEHOLD CHLORINE BLEACH**
- **ELIMINATION, SECOND-ORDER**
- **SUBLIMATION**

In this experiment, you will synthesize camphor from camphene by the following series of reactions involving a Wagner-Meerwein rearrangement:

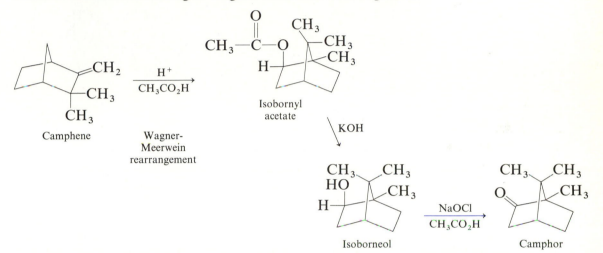

A Wagner-Meerwein rearrangement is any carbocation rearrangement involving a change in the skeletal arrangement of the carbon atoms. Thus the following reaction would be a simple example of a Wagner-Meerwein rearrangement, where the less stable 2° carbocation forms a more stable 3° carbocation by means of an alkyl shift:

However, with bicyclic molecules like camphene, it isn't always apparent that a Wagner-Meerwein rearrangement has occurred, since the product usually retains the same bicyclic structure. Only by a change in the location of the substituents on the bicyclic structure can one tell that a different carbon skeleton is present.

For camphene, as shown below, the rearrangement converts a normally more stable 3° carbocation into a normally less stable 2° carbocation. Evidently, this 2° carbocation must be more stable due to another factor, most likely the new ring system being more stable because the new locations of the three methyl groups produce less steric hindrance. Once formed, the 2° carbocation then reacts with the acetic acid present to yield isobornyl acetate.

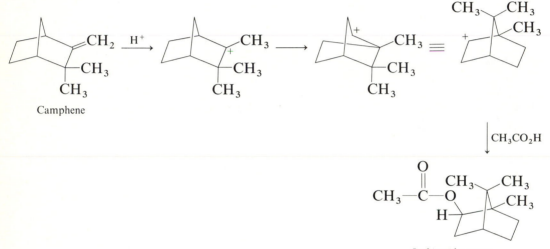

Isobornyl acetate

The rest of the reaction sequence involves nucleophilic acyl substitution by hydroxide ion to yield isoborneol, followed by oxidation with acidic household chlorine bleach to yield camphor. The last reaction is a relatively new way to prepare ketones from secondary alcohols, and appears to occur by the following mechanism:

1.

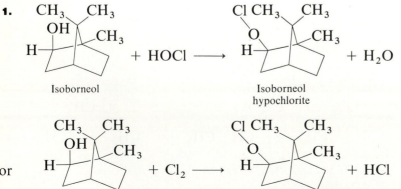

2.

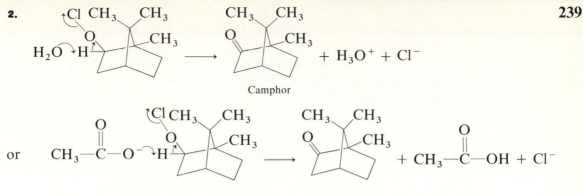

Camphor

In step 1, isoborneol reacts with either the hypochlorous acid or chlorine gas in the reaction mixture to give isoborneol hypochlorite. The isoborneol hypochlorite in step 2 then undergoes an E-2 (elimination, second-order) reaction. As you can see, either water or acetate ion, acting as a base, removes the hydrogen on the hypochlorite-bearing carbon. At the same time, the chlorine atom leaves, taking with it the electron pair, and the carbon-oxygen π bond of the ketone is formed. This mechanism, however, is only tentative, and much more work to establish it needs to be done.

The camphor produced has a familiar penetrating odor. Industrially, it is widely used in manufacturing celluloid plastics and in explosives, pyrotechnics, lacquers and varnishes, embalming fluids, pharmaceuticals, and cosmetics. Its high freezing-point-depression constant of 39.7 degrees per molal also makes it quite useful in molecular-weight determinations of unknown compounds.

PROCEDURE

CAUTION: Be careful using glacial acetic acid, sulfuric acid, and potassium hydroxide. They can all cause severe burns, especially when heated.

Also be careful using household chlorine bleach, since it can damage both your clothes and certain areas of the laboratory.

Also remember that the chlorine gas produced in this experiment must be contained, as indicated in the directions. It is an eye and respiratory irritant, and has an acceptable short-term atmospheric exposure level of 3 parts per million.

Isobornyl acetate

Dissolve 4.1 grams (0.030 mole) of camphene in 10 mL (10 grams, 0.17 mole) of glacial acetic acid in a 50-mL Erlenmeyer flask. Add 1 mL of 6 *M* sulfuric acid and heat the flask on a hot plate or steam bath in the hood for 15 minutes at 90 to 95° with frequent swirling. Then add 8 mL of water, mix, and cool.

Transfer the material to a 125-mL separatory funnel, rinsing the flask with a little water. Separate the ester layer (which one is it?) and wash it in the separatory funnel with water and then with 10% aqueous sodium carbonate, being sure to frequently vent the separatory funnel when using the sodium carbonate. The crude, washed product is suitable for conversion to isoborneol without further purification. The yield is about 5 grams.

Isoborneol

In a 50-mL round-bottom flask, prepare a solution of 2.3 grams (0.035 mole) of 85% pure potassium hydroxide pellets in 12 mL of ethanol and 4 mL of water. Add the isobornyl acetate and heat the mixture under reflux for 1 hour.

Pour the solution slowly over 30 grams of chipped ice in a 100-mL beaker. Stir the mixture until the isoborneol solidifies, then collect the product by suction filtration, and wash it with cold water. The crude isoborneol obtained is sufficiently pure for oxidation to camphor.

Camphor

Place the isoborneol and 10 mL of glacial acetic acid into a clamped three-neck, 500-mL round-bottom flask. Place a clamped condenser in the center neck, a 125-mL separatory funnel in one of the side necks, and a thermometer in the other side neck. Place a piece of glass wool in the bottom of a drying tube and pour in about 3 cm of **moistened but not wet** sodium bisulfite, covered by another piece of glass wool. The sodium bisulfite will react with any chlorine that might escape up the condenser. Be sure that the sodium bisulfite is not so tightly packed that you will have a closed system. If necessary, place a one-hole rubber stopper at the bottom of the drying tube, then insert the drying tube into the top of the condenser.

Carefully add 50 mL (0.041 mole) of 5.25% household chlorine bleach to the separatory funnel. Add about 5 mL of the bleach solution to the isoborneol solution and **carefully** swirl the flask to mix the reactants. Then, keeping the temperature at about 50°, start adding the rest of the bleach dropwise, **carefully** swirling the flask periodically. The addition should take about 15 minutes, and during the early part of the reaction you should see the yellow color of the chlorine disappear as it reacts.

After all the bleach is added, remove the separatory funnel and replace it with a glass stopper. Continue the reaction for another 45 minutes, periodically swirling the flask.

To test for excess oxidant at the end of the reaction period, wet a piece of starch-iodide paper and put a drop of the liquid from the reaction mixture on it. If any oxidant is present, the blue color of the starch-triiodide complex will appear. If this happens, add 1 mL of saturated $NaHSO_3$ solution to the flask and stir. This will remove the excess oxidant by the reaction

$$OCl^- + HSO_3^- \longrightarrow Cl^- + HSO_4^-$$

Test the reaction liquid again with wet starch-iodide paper and, if necessary, continue adding small amounts of NaHSO$_3$ solution until no color change to blue occurs.

Pour the reaction mixture into a 250-mL beaker, rinse the reaction flask with small amounts of water, then gradually add, with stirring, saturated sodium carbonate solution until no more foaming is produced. This will convert the acetic acid to sodium acetate.

Add 50 grams of cracked ice to the beaker, thoroughly cool the mixture in an ice bath, and collect the solid camphor by suction filtration, washing it with a little ice water. Allow the product to thoroughly dry and then purify it by sublimation. Set up the apparatus shown in Figure 8-15, using a 15 × 125 mm, or larger, test tube. The camphor is placed in the bottom of the filter flask, and crushed ice and water go into the test tube. Insert a trap between the apparatus and the water aspirator, and turn on the water.

Heat the sample **carefully** with a small flame, moving the burner back and forth under the filter flask. The camphor will be observed to vaporize from the flask and be deposited onto the cold test tube. If the camphor begins to melt, remove the heat for a few seconds before continuing. When the sublimation is complete, remove the heat and let the apparatus cool.

Turn off the water and slowly release the vacuum by disconnecting the hose at the aspirator or the trap, or, even better, by opening the stopcock or the clamp on the hose coming up from the trap. **Carefully** remove the test tube and try not to have any of the sublimed crystals fall off into the residue. Pour off the remaining ice and water from the tube and scrape the sublimed camphor onto weighing paper.

FIGURE 8-15
Simple sublimation apparatus.

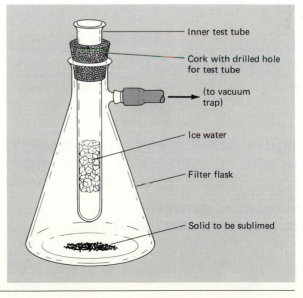

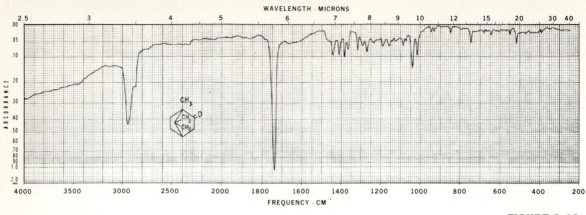

FIGURE 8-16
The infrared spectrum of camphor in a KBr pellet. Spectrum
courtesy of Sadtler Research Laboratories, division of Bio-Rad
Laboratories, Inc., copyrighted 1969.

Determine the yield, the melting point, and the literature melting point of
the camphor, and record the results in your notebook. The melting point must
be taken in a sealed tube. Remember that due to the high freezing-point-
depression constant of camphor, even small amounts of impurities may dramatically affect its melting point. Determine the theoretical and percentage yields
of your product and show the calculations.

At your instructor's option, obtain the infrared spectrum of your product
in a KBr pellet, and the ^{1}H NMR spectrum in carbon tetrachloride. Compare
your spectra with those shown in Figures 8-16 and 8-17. When finished, hand
in your product in a properly labeled container to your instructor.

References

1. J. Mohrig et al., **Journal of Chemical Education, 62,** 519 (1985).

2. R. Stevens, K. Chapman, and H. Weller, **Journal of Organic Chemistry, 45,** 2030
(1980).

8D. SODIUM BOROHYDRIDE REDUCTION
OF CAMPHOR TO BORNEOL AND ISOBORNEOL

▓ NUCLEOPHILIC ADDITION
▓ GAS CHROMATOGRAPHIC ANALYSIS

In this experiment, you will reduce the ketone camphor with sodium borohydride to obtain a mixture of the alcohols borneol and isoborneol. This experiment is therefore almost the opposite of the last portion of Experiment 8C.

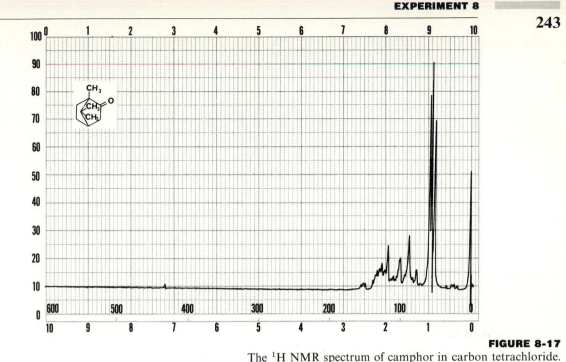

FIGURE 8-17

The ¹H NMR spectrum of camphor in carbon tetrachloride.
Spectrum courtesy of Aldrich Chemical Company, *The Aldrich
Library of NMR Spectra,* edition II, by Charles Pouchert.

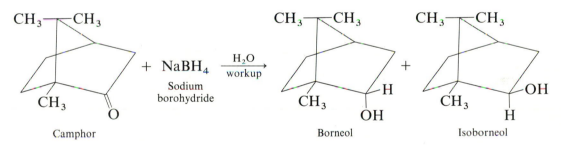

Sodium borohydride is a metal hydride, which by nucleophilic addition
transfers a hydride ion from the boron atom to the electron-deficient carbonyl
carbon of the ketone. The mechanism, as shown below, is believed to occur
through a four-centered transition state, with a bond also being formed between
the boron and the carbonyl oxygen.

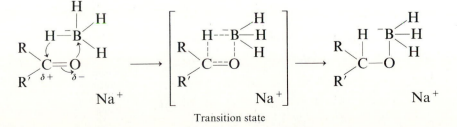

This step can be repeated with three additional molecules of ketone until all the hydrogens on the boron atom have been transferred. Once this has occurred, the boron complex formed can then be decomposed with water to yield the alcohol.

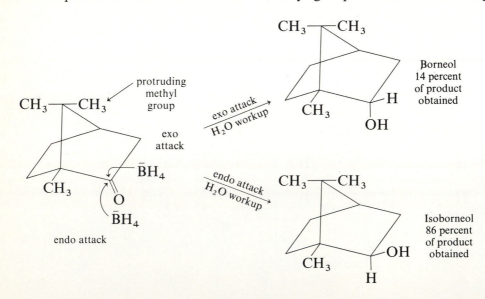

As a mild reducing agent, sodium borohydride will react only with aldehydes, ketones, and acid chlorides. Contrasted with it is the powerful reducing agent lithium aluminum hydride, $LiAlH_4$, which will also react with carboxylic acids, esters, acid anhydrides, and amides. Because lithium aluminum hydride reacts rapidly with water and alcohols, reduction with this reagent must be carried out in anhydrous ether solutions. In contrast, sodium borohydride reacts very slowly with water and alcohol and can easily be used in these solvents without a serious loss of reagent. It is, however, rapidly decomposed in acidic solutions with the evolution of hydrogen gas.

$$NaBH_4 + H_3O^+ + 2 H_2O \longrightarrow 4 H_2 + Na^+ + H_3BO_3$$

Two alcohols, borneol and isoborneol, are formed in the reduction of camphor because the borohydride ion can attack from either the stereochemically different top side or bottom side of the carbonyl group. Attack from the top side is frequently called **exo** attack; attack from the bottom side, **endo** attack. In this example, it is expected that exo attack will not be favored, because of the steric repulsion produced by the one protruding geminal methyl group. In fact, borneol has been found to compose only 14 percent of the final product, versus 86 percent for isoborneol.

The starting material, camphor, can be obtained either synthetically or from the camphor tree; it has a familiar and penetrating odor. Industrially, it is widely used in manufacturing celluloid plastics, in explosives, pyrotechnics, lacquers and varnishes, embalming fluids, pharmaceuticals, and cosmetics. Its high freezing-point-depression constant of 39.7 degrees per molal also makes it quite useful in molecular-weight determinations of unknown compounds.

PROCEDURE

In a 25-mL Erlenmeyer flask dissolve 3.0 grams (0.020 mole) of camphor in 10 mL of ethanol. To this solution cautiously add 1.5 grams (0.039 mole) of sodium borohydride in small portions.[1] Some warming will occur. After all the sodium borohydride has been added, gently heat the mixture in the clamped flask to boiling with a hot-water or steam bath. Allow the mixture to gently boil for 10 minutes, adding extra ethanol as needed to maintain the volume.

Pour the hot reaction mixture into about 50 grams of ice water and rinse out the flask with a small amount of ethanol. After the ice melts, collect the white solid by suction filtration, then dissolve in the minimum amount of hot ethanol in a 50-mL Erlenmeyer flask. Add hot water slowly until the solution becomes turbid; then add hot ethanol to redissolve the material. Cool in an ice bath and collect the purified material by suction filtration.

Allow the product to thoroughly dry and determine the yield and melting point, using a sealed tube. Record the results in your notebook, along with the literature melting points of isoborneol, borneol, and camphor. Also determine the theoretical and percentage yields, showing the calculations.

To analyze your reaction product for the percentages of isoborneol and borneol present, dissolve a small amount of your reaction product in a small amount of reagent-grade acetone or methylene chloride. Inject into a gas chromatograph that has a 5-foot column of 5% Carbowax 4000 on acid-washed firebrick. If necessary, read Section 6.4 on gas chromatography. The column temperature should be 110°C, and the carrier gas flow rate should be 50 mL per minute. The retention time for any unreacted camphor is 6 minutes; the retention times for isoborneol and borneol are 10 and 12 minutes, respectively. Determine the areas and the percentages using one of the methods given in Section 6.4I, page 158.

---

[1] The sodium borohydride should first be checked to see if it is active. Place a small amount in ethanol and heat on a steam bath. Vigorous bubbling should occur if the hydride is active.

Assume that the isoborneol and borneol both affect the detector equally. How do your percentages of isoborneol and borneol compare with the stated amounts?

When finished, attach your chromatogram to your notebook or place the labeled pieces in an envelope attached to your notebook. Submit your product in a properly labeled container to your instructor.

8E. HOUSEHOLD BLEACH
OXIDATION OF MENTHOL TO MENTHONE

ELIMINATION, SECOND-ORDER
VACUUM DISTILLATION

In this experiment, you will oxidize menthol, a principal constituent of peppermint oil, to menthone, also found in peppermint oil. The reaction will be carried out with acidic household chlorine bleach as the oxidizing agent, according to the equation

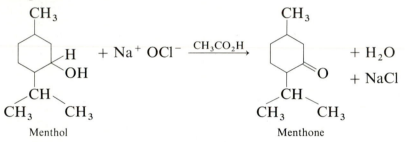

The use of acidic chlorine bleach is a relatively new way to prepare ketones from secondary alcohols, and appears to occur by the following mechanism:

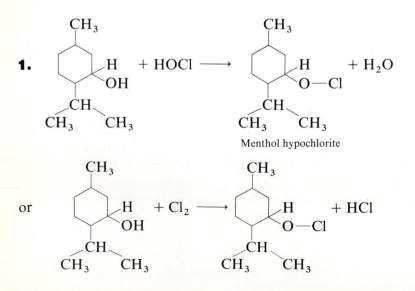

2.

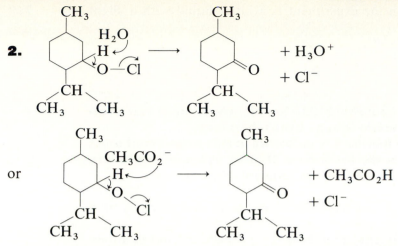

In step 1, menthol reacts with either the hypochlorous acid or chlorine gas in the reaction mixture to give menthol hypochlorite. The menthol hypochlorite in step 2 then undergoes an E-2 (elimination, second-order) reaction. As you can see, either water or acetate ion, acting as a base, removes the hydrogen on the hypochlorite-bearing carbon. At the same time, the chlorine atom leaves, taking with it the electron pair, and the carbon-oxygen π bond of the ketone is formed. This mechanism, however, is only tentative, and much more work to establish it needs to be done.

The menthol used for the experiment has a strong odor and taste of peppermint, and is widely used in liqueurs, confectionary items, cigarettes, cough drops, nasal inhalers, and other products. When necessary, you can prepare the compound synthetically by the hydrogenation of thymol, an effective disinfectant obtained either from thyme oil or the Friedel-Crafts reaction of *meta*-cresol with propylene.

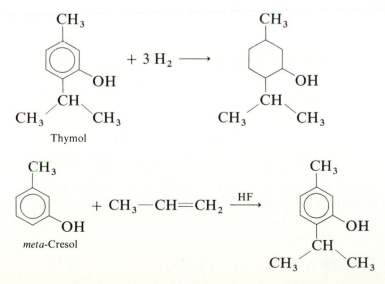

The menthone obtained in the experiment is a bitter liquid with a slight peppermint odor.

PROCEDURE

CAUTION: Be careful using household chlorine bleach, since it can damage both your clothes and certain areas of the laboratory.

Also remember that the chlorine gas produced in this experiment must be contained, as indicated in the procedure. It is an eye and respiratory irritant, and has an acceptable short-term atmospheric exposure level of 3 parts per million.

Be careful not to get glacial acetic acid on your skin, since it can cause severe burns.

Also remember that methylene chloride is toxic. Be careful not to inhale it or get it on your skin.

Place 10.0 mL (0.058 mole) of menthol and 10 mL of glacial acetic acid into a clamped three-neck, 500-mL round-bottom flask. Place a clamped condenser in the center neck, a 125-mL separatory funnel in one of the side necks, and a thermometer in the other side neck. Place a piece of glass wool in the bottom of a drying tube and pour in about 3 cm of **moistened but not wet** sodium bisulfite, covered by another piece of glass wool. The sodium bisulfite will react with any chlorine that might escape up the condenser. Be sure that the sodium bisulfite is not so tightly packed that you will have a closed system. If necessary, place a one-hole rubber stopper at the bottom of the drying tube, then insert the drying tube into the top of the condenser.

Carefully add 90 mL (0.074 mole) of 5.25% household chlorine bleach to the separatory funnel. Add about 10 mL of the bleach solution to the menthol mixture and **carefully** swirl the flask to mix the reactants. Then, keeping the temperature between 45 and 50°, start adding the rest of the bleach dropwise, carefully swirling the flask periodically. The addition should take about 20 minutes, and during the early part of the reaction you should see the yellow color of the chlorine disappear as it reacts.

After all the bleach is added, remove the separatory funnel and replace it with a glass stopper. Continue the reaction for another 40 minutes, periodically swirling the flask.

To test for excess oxidant at the end of the reaction period, wet a piece of starch-iodide paper and put a drop of the liquid from the reaction mixture on it. If any oxidant is present, the blue color of the starch-triiodide complex will appear. If this happens, add 2 mL of saturated $NaHSO_3$ solution to the flask and stir. This will remove the excess oxidant by the reaction

$$OCl^- + HSO_3^- \longrightarrow Cl^- + HSO_4^-$$

Test the reaction liquid again with wet starch-iodide paper and, if necessary, continue adding small amounts of $NaHSO_3$ solution until no color change to blue occurs.

Remove the drying tube, condenser, and thermometer; then gradually add, with stirring, saturated sodium carbonate solution to the reaction mixture until no more foaming is produced. This will convert the acetic acid to sodium acetate.

Cool the reaction mixture in an ice bath; then pour the reaction mixture into a 250-mL separatory funnel, rinsing out the reaction flask with small quantities of water. Extract the reaction mixture twice with 30-mL portions of methylene chloride, periodically releasing in the hood any internal pressure. Combine the methylene chloride extracts, dry them over anhydrous magnesium sulfate, and carefully decant or gravity filter through fluted filter paper the methylene chloride solution into a 100-mL round-bottom flask.

Set up an apparatus for a simple distillation as in Experiment 1, and distill the methylene chloride over a hot-water or steam bath. When no more liquid distills, convert the apparatus and vacuum distill the menthone into a new receiver, using the procedure on page 59. The reason for the change is that severe heat may partially decompose the high-boiling-point menthone. Use a hot plate or heating mantle and collect whatever boils at over 85°. Record the chief boiling range, the yield, and the literature boiling points for menthone at various pressures in your notebook. Also determine the theoretical and percentage yields of your final product, showing the calculations.

At your instructor's option, obtain the infrared spectrum of your product, neat between salts, and the ^1H NMR spectrum in carbon tetrachloride. Compare your spectra to those shown in Figures 8-18 and 8-19. When finished, hand in your product in a properly labeled container to your instructor.

FIGURE 8-18

The infrared spectrum of menthone, neat. Spectrum courtesy of Sadtler Research Laboratories, division of Bio-Rad Laboratories, Inc., copyrighted 1969.

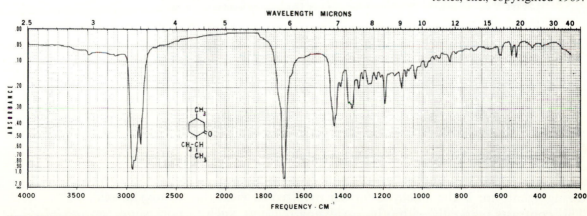

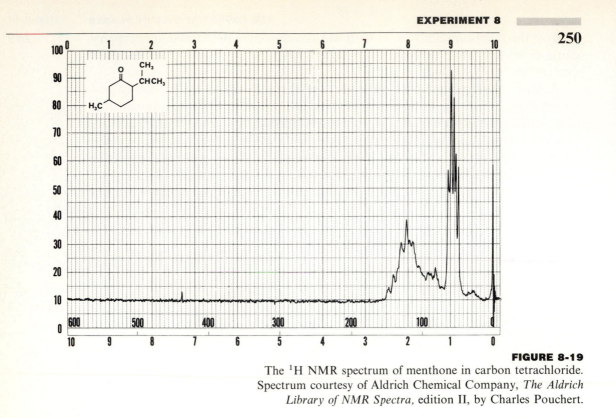

FIGURE 8-19

The ^{1}H NMR spectrum of menthone in carbon tetrachloride. Spectrum courtesy of Aldrich Chemical Company, *The Aldrich Library of NMR Spectra,* edition II, by Charles Pouchert.

References

1. J. Mohrig et al., **Journal of Chemical Education, 62,** 519 (1985).

2. R. Stevens, K. Chapman, and H. Weller, **Journal of Organic Chemistry, 45,** 2030 (1980).

8F. CHROMIUM(VI) OXIDE OXIDATION OF CINNAMYL ALCOHOL TO CINNAMALDEHYDE

In this experiment, you will synthesize cinnamaldehyde, the principal constituent in oil of cinnamon, by the oxidation of cinnamyl alcohol with a pyridine complex of chromium(VI) oxide in methylene chloride (the Collins' reagent).

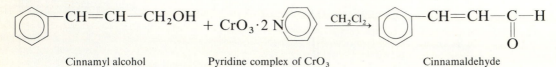

Observe that the carbon-carbon double bond is not attacked by the oxidizing agent, nor is the aldehyde further oxidized to yield the carboxylic acid. This contrasts markedly with other oxidizing reagents, such as potassium permanganate and chromic acid. For a rapid, complete conversion to the aldehyde, it has been discovered that a 6:1 mole-mole ratio of the oxidizing agent to the alcohol is necessary (Ref. 1). Otherwise, prolonged reaction times are necessary.

The reaction occurs simply by allowing the reactants to stand at room temperature for 20 minutes. Then decant the reaction solution, wash the residue with ether, and extract the combined organic liquids with solutions of sodium hydroxide, hydrochloric acid, sodium bicarbonate, and sodium chloride. The first two extractions remove chromium compounds and pyridine, the sodium bicarbonate neutralizes any remaining hydrochloric acid, and the sodium chloride greatly reduces the water content of the ether. After drying, the distillation of the ether and methylene chloride then gives us cinnamaldehyde as a pleasant-smelling oil. If desired, part of the product may be converted to the semicarbazone derivative by the procedure on page 235.

The starting reagent, cinnamyl alcohol, has an odor of hyacinth and is used in the perfume industry. It is found in plants as the ester of cinnamic acid, cinnamyl cinnamate, which is the principal constituent of many balsams.

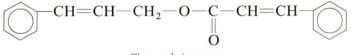

Cinnamyl cinnamate

PROCEDURE

CAUTION: All chromium compounds are toxic, and chromium(VI) compounds are suspected carcinogens. Wear gloves and avoid contact with them.

Also, do not discard chromium compounds or solutions containing them down the drain or into a wastebasket. Use the designated waste disposal jar.

Also try not to inhale pyridine and methylene chloride, and don't get them on your skin. The pyridine is especially hazardous and has a pronounced, unpleasant "fishy" odor.

In addition, never add pyridine to chromium trioxide, since a fire can result. The fire hazard is avoided here by the use of methylene chloride as the solvent.

For a good yield, all the glassware used in the beginning should be flame-dried away from any ether.

In a 250-mL round-bottom flask, dissolve 4.8 mL (4.7 grams, 0.060 mole) of pyridine in 75 mL of methylene chloride, and then add 3.00 grams (0.030 mole) of chromium trioxide. Stopper the flask with a drying tube containing drierite held in place by cotton, and magnetically stir the deep burgundy solution for 15 minutes at room temperature. At the end of this period, add all at once a solution of 0.67 gram (0.0050 mole) of cinnamyl alcohol in a small volume of methylene chloride. A tarry, black deposit should immediately separate.

Stir the mixture for 20 minutes at room temperature, then decant the solution into a 500-mL separatory funnel. Wash the residue in the round-bottom flask with 100 mL of ether, then decant the ether solution into the separatory funnel with the methylene chloride. Extract the combined organic solutions twice with 50-mL portions of 5% aqueous sodium hydroxide solution, twice with 50-mL portions of 5% aqueous hydrochloric acid, once with 50 mL of 5% aqueous sodium bicarbonate solution, and once with 50 mL of saturated sodium chloride solution. **Remember to discard the sodium hydroxide and hydrochloric acid solutions into the designated waste disposal jar and not down the drain.**

Dry the organic solution over anhydrous magnesium sulfate, then decant or gravity filter through fluted filter paper into a 250-mL round-bottom flask. Set up a simple-distillation apparatus and distill the combined ether and methylene chloride, using a hot-water bath. When the volume reaches about 10 to 15 mL, stop the distillation, transfer the remaining liquid to a previously weighed 25-mL Erlenmeyer flask, and rinse out the round-bottom flask with a few milliliters of the distilled liquid. Further concentrate the liquid in a hot-water bath in the hood, then dry the outside of the flask, and determine the weight of cinnamaldehyde obtained; record the result in your notebook. Also determine the theoretical and percentage yields of your product and show the calculations.

At your instructor's option, obtain the infrared spectrum of your product, neat between salts, and the ^{1}H NMR spectrum in deuterated chloroform. Compare your spectra to those shown in Figures 8-5 and 8-6.

If desired, convert part of your product to the semicarbazone derivative, using the procedure on page 235. When finished, hand in your product in a properly labeled container to your instructor.

References

1.   R. Ratcliffe and R. Rodehorst, **Journal of Organic Chemistry, 35,** 4000 (1970).

2.   J. Collins, W. Hess, and F. Frank, **Tetrahedron Letters,** 3363 (1968).

8G.   TERPENES: THE GAS CHROMATOGRAPHIC ANALYSIS OF TURPENTINE

BACKGROUND

Turpentine is a colorless, volatile, highly flammable liquid obtained from various species of pine trees. It has a characteristic pungent odor and has been used for centuries as a solvent and thinner for paints, varnishes, lacquers, and waxes.

Turpentine increases the air hardening of the drying oils and, by thinning the paints and varnishes, permits an easier and more even application.

About 350 million pounds of turpentine are produced annually. Since 1950, its use in paints and varnishes has declined dramatically, it having been replaced by low-cost petroleum solvents and water-miscible paints. However, its overall use has continued to rise. It is used as a solvent for the manufacture of synthetic rubber, and as a reagent for the production of disinfectants, insecticides, medicines, resins, oil additives, and synthetic camphor, pine oil, borneol, terpineol, and terpin hydrate.

Chemically, turpentine is a mixture of monoterpenes, consisting of about 63% α-pinene, 33% β-pinene, and 4% limonene and Δ^3-carene. This composition will vary, however, depending on the species of pine trees used, and the specific gravity, optical rotation, and other physical properties will also vary. In addition, very minor components may be present.

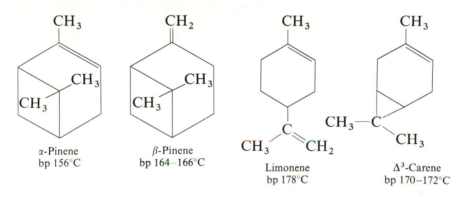

α-Pinene
bp 156°C

β-Pinene
bp 164–166°C

Limonene
bp 178°C

Δ^3-Carene
bp 170–172°C

In this experiment, you will gas chromatograph a sample of turpentine. By comparing this chromatogram with that of a known mixture of terpenes, you will then be able to know what amounts of the various terpenes are present in the turpentine sample.

PROCEDURE

If necessary, reread the section on the gas chromatograph (pages 151–162).

Inject a 1.0-μL standard solution of terpenes containing 50.0% α-pinene, 30.0% β-pinene, 15.0% Δ^3-carene, and 5.0% limonene by volume into a 3-m Chromosorb W column, coated with 10% SE-30 silicone rubber. The column and injection port should both be heated to 160°C, and the carrier gas flow rate should be about 40 mL per minute. After the complete chromatogram has been obtained, repeat with the sample of turpentine.

From the known boiling points of the terpenes and the areas of the peaks, determine which terpene is represented by each peak in the chromatogram from the standard solution. Label these peaks and write down their retention times in your notebook.

Next, determine quantitatively the area of each peak in the chromatogram from the standard solution. Either cut out the labeled peaks with scissors and weigh them on an **analytical** balance, or use the automatic integrator that may be provided with the gas chromatograph. Determine the percentage area of each peak, then divide this percentage by the appropriate known percentage for that compound in the solution. This will give you the relative sensitivity of the detector to each individual compound present. You **cannot** assume that each compound will affect the detector equally.

By comparing the retention times, determine which peaks in the chromatogram from the turpentine sample correspond to the compounds in the known solution. Label these peaks, and record their retention times in your notebook. Remember that any extra minor components in the turpentine may give you more than four peaks. Determine quantitatively the area of each peak and the percentage area of each peak. Then divide the percentage area of each of the four known compounds by that compound's relative detector sensitivity. This will give you the percentage of each of the four compounds in your turpentine sample.

When finished, fold and carefully attach each entire chromatogram to your notebook, or place the individual pieces of each chromatogram in separate labeled envelopes and attach to your notebook.

QUESTIONS

1. Circle or enclose each isoprene unit present in caryophyllene, zingiberene, α-pinene, and vitamin A.
2. What advantage does steam distillation have over solvent extraction in the isolation of the essential oils?
3. What advantage does steam distillation of the essential oils have over the regular distillation of these oils?
4. The steam distillation of any water-insoluble compounds always gives a boiling point of under 100° at atmospheric pressure. What would be the expected boiling point or boiling-point range at atmospheric pressure when water is distilled with a high-boiling, water-soluble compound such as acetic acid?
5. What factor seems to be responsible for the Wagner-Meerwein rearrangement of camphene to yield isobornyl acetate?
6. The aluminum hydride portion of lithium aluminum hydride is even bulkier than the borohydride portion of sodium borohydride. Based upon this information, would you expect the percentage of borneol to increase or decrease if camphor were reduced with lithium aluminum hydride in anhydrous ether solution? Explain why.

7. From the melting point of the recrystallized product from the reduction of camphor, has isoborneol or borneol been the principal product obtained? Explain.

8. *d*-Carvone and *l*-carvone both give the same infrared spectrum and the same ^{1}H NMR spectrum. *d*-α-Pinene and *l*-α-pinene also both give the same infrared spectrum and the same ^{1}H NMR spectrum. In fact, no matter what the terpene or terpenoid is, the *d* and *l* isomers will both give the same infrared spectrum and the same ^{1}H NMR spectrum. Explain why.

9. In the turpentine experiment, how did your percentages of the four known components compare with the approximate values stated in the experiment? How many other components were present and in what amounts?

10. Why were spices so valuable in earlier history, yet are quite inexpensive and commonplace today?

11. For most spices why are the essential oils not isolated for consumer use?

12. Under what conditions should spices be stored to best maintain their flavor and aroma?

ARTIFICIAL FLAVORINGS— THE SYNTHESIS OF ESTERS

NUCLEOPHILIC ACYL SUBSTITUTION
CHEMICAL EQUILIBRIUM

Ah, you flavor everything! You are the vanilla of society.

Sydney Smith, *Lady Holland's Memoir*, Volume I, Chapter 9.

INTRODUCTION

Whenever you eat almost any fruit or many other foods, the flavor you are tasting is due primarily to an organic compound called an ester. Because esters can be synthesized commercially, these flavors do not have to come from nature alone. Food and beverage companies are thoroughly familiar with these esters and frequently use them to enhance the flavor of ice cream, sherbet, gelatin desserts, cakes, soft drinks, candy, chewing gum, and many other products. Frequently the synthetic ester may even be used exclusively.

Now why should artificial flavorings be used at all? There are a number of reasons. Many of the natural flavorings contain ingredients that produce an off-taste when heated, making these flavorings unsuitable for any products requiring high-temperature processing. Some of them also have a poor storage or shelf life, making them unsuitable for certain products. Availability and cost are other factors. The demand for flavorings has increased to the point that without synthetics there would simply not be enough to go around. The cost of producing many natural flavorings has also become prohibitively high. Furthermore, the quality and quantity of many natural products can vary from year to year, depending on the weather, time of picking, and political and labor factors. All these problems can be eliminated through the use of artificial flavorings.

The synthetic flavoring, however, may have a taste or aroma that is not always identical to the natural, since the fruit or other food may also contain a wide array of other ingredients contributing to the flavor. Many of the early artificial flavorings were crude, producing much displeasure with the consumer. But today the public will no longer automatically reject a product containing flavoring labeled as synthetic, artificial, or imitation. Frequently, many of the other ingredients present in the natural product can be included, or the flavor can be closely duplicated by adding related compounds. With these advances, many high-quality products have been developed, some of them so good that even a professional taster can be fooled!

TABLE 9-1.

Some Common Flavors and the Ester Responsible

Flavor	Name of ester	Structure of ester
Banana	Isopentyl acetate	$CH_3-\underset{\underset{CH_3}{|}}{CH}-CH_2CH_2-O-\overset{\overset{O}{\|}}{C}-CH_3$
Peach	Benzyl acetate	$\langle\bigcirc\rangle-CH_2-O-\overset{\overset{O}{\|}}{C}-CH_3$
Pear	Propyl acetate	$CH_3CH_2CH_2-O-\overset{\overset{O}{\|}}{C}-CH_3$
Pineapple	Ethyl butyrate	$CH_3CH_2-O-\overset{\overset{O}{\|}}{C}-CH_2CH_2CH_3$
Raspberry	Isobutyl formate	$CH_3-\underset{\underset{CH_3}{|}}{CH}-CH_2-O-\overset{\overset{O}{\|}}{C}-H$
Wintergreen	Methyl salicylate	$CH_3-O-\overset{\overset{O}{\|}}{C}-\underset{HO}{\bigcirc}$

A list of some common flavors along with the name and structure of the ester responsible is given in Table 9-1. Look over the list and decide which one you would like to make.

BACKGROUND

One method of preparing esters is the reaction of an alcohol and a carboxylic acid in a process called **esterification**. The reaction is reversible and requires a catalyst, usually concentrated sulfuric acid or dry hydrogen chloride, to attain the equilibrium more readily. The equilibrium equation is written as follows:

$$R{-}OH + HO{-}\overset{\displaystyle O}{\overset{\|}{C}}{-}R' \underset{}{\overset{H^+}{\rightleftarrows}} R{-}O{-}\overset{\displaystyle O}{\overset{\|}{C}}{-}R' + H_2O$$

Alcohol   Carboxylic   Ester   Water
            acid

The equilibrium present in the reaction means that the yield of ester frequently may not be as high as desirable. The use of Le Chatelier's principle in the re-action raises the yield to more acceptable levels.

The accepted mechanism of the reaction is shown below and involves (1) the protonation of the carboxylic acid by the acid catalyst to give the resonance-stabilized intermediate I, followed by (2) the nucleophilic attack of the alcohol. The subsequent transfer of a proton (3), loss of water (4), and loss of a proton (5) then give the final observed products. The reverse reaction to yield the car-boxylic acid and alcohol follows the same steps in the opposite order.

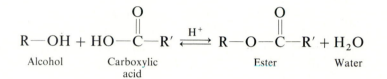

In the reaction workup, the crude product is washed with cold sodium chloride solution and cold sodium bicarbonate solution. The first washing sep-arates from the ester most of the acid catalyst and also perhaps some of the unreacted organic acid and alcohol. The second washing separates completely any remaining acid catalyst and the organic acid by neutralizing each of them to the water-soluble sodium salt. After it is washed and dried, the crude ester is purified by distillation.

PROCEDURE

CAUTION: Concentrated sulfuric acid is extremely hazardous and corrosive. Handle it with care.

Flavor	Alcohol	Carboxylic acid
Banana	Isopentanol 0.10 mole	Acetic acid 0.30 mole
Peach	Benzyl alcohol 0.10 mole	Acetic acid 0.30 mole
Pear	Propanol 0.20 mole	Acetic acid 0.40 mole
Pineapple	Ethanol 0.15 mole	Butyric acid 0.25 mole
Raspberry	Isobutanol 0.15 mole	Formic acid 0.30 mole
Wintergreen	Methanol 1.0 mole	Salicylic acid 0.10 mole

Using the above guide and a handbook of chemistry, fill in the following information in your notebook for the flavor you have selected. The esters may be listed in the handbook under the corresponding carboxylic acid. For example, ethyl butyrate can be listed under butyric acid or butanoic acid, ethyl ester. Have the information checked and approved by your instructor. The stated amounts of alcohol and carboxylic acid were chosen to effectively utilize Le Chatelier's principle and to give you a reasonable yield of product.

Flavor _____ Name of ester _____ Mol wt _____ bp _____ Sp gr _____	Name of alcohol _____ Mol wt _____ bp _____ Sp gr _____ Wt of no. of moles listed _____ Vol of no. of moles listed _____	Name of carboxylic acid _____ Mol wt _____ bp or mp _____ Sp gr (if liquid) _____ Wt of no. of moles listed _____ Vol of no. of moles listed (if liquid) _____

In a 100-mL round-bottom flask, mix thoroughly the calculated amounts of alcohol and carboxylic acid. (**Caution:** Many of the carboxylic acids have especially disagreeable odors and should not be spilled or left uncapped. In certain cases the reaction may even have to be run in the hood.) **Cautiously** add 4 mL of concentrated sulfuric acid, swirl gently, and add several boiling chips. Attach a reflux condenser by placing a water-cooled condenser vertically into the flask. This will allow you to boil the mixture without losing any of the components through vaporization. Allow the mixture to reflux for 2 to 3 hours.

Cool the mixture to room temperature and pour it into a separatory funnel. Rinse the flask with several small portions of cold 15% sodium chloride solution and carefully add them to the separatory funnel. Carefully add additional sodium chloride solution until the aqueous layer has about double the volume of the organic layer. Shake and separate.[1] (**Caution:** Which layer is the organic one?) Now wash the organic layer with an equal volume of cold 10% sodium bicarbonate solution, frequently opening the stopcock to release any internal pressure. Test to see if the organic layer is neutral to blue litmus paper and, if necessary, repeat the washing with fresh sodium bicarbonate solution to ensure the complete removal of acid. Be sure to save the organic layer after each washing.

Transfer the organic layer containing the ester to an Erlenmeyer flask, add a small quantity of anhydrous magnesium sulfate, and swirl gently. Allow the crude ester to stand until the liquid is clear, adding additional small quantities of anhydrous magnesium sulfate, if necessary.

Carefully decant or filter the crude ester into a 50-mL round-bottom flask. Add several boiling chips and set up a distillation apparatus. If your flavor boils at less than 200°, use the simple-distillation setup on page 40 and collect the fraction having a boiling point within about 5° of the listed boiling point. However, if your flavor boils at over 200°, it may partially decompose from the severe heat. Instead, vacuum distill your flavor according to the procedure given on page 59, collecting whatever distills at over 90°. Use a hot plate or heating mantle for both types of distillation. Record the yield and the chief boiling range in your notebook.

At your instructor's option, obtain the infrared spectrum of your product, run neat on salt plates, and the ^{1}H NMR spectrum in carbon tetrachloride. Compare your spectra with those shown for your particular flavor.

Determine the theoretical and percentage yields of your final product, showing the calculations, and submit the product in a properly labeled container to your instructor.

---

[1] In certain cases the ester and water may still not separate effectively. If this occurs, you can obtain a clean separation by adding about 20 mL of methylene chloride to the ester and water mixture. (**Caution: methylene chloride is toxic. If using it, handle it with care and don't breathe it excessively.**) Shake, draw off the lower methylene chloride–ester layer into a clean container and discard the upper, aqueous layer. Pour the methylene chloride–ester layer back into the separatory funnel and wash it with cold sodium bicarbonate solution, as indicated before.

If you added methylene chloride, you can remove it in the final distillation by first heating with a warm-water bath. When all the low-boiling solvent is gone, the remaining ester can then be distilled in the usual manner, using a hot plate or heating mantle.

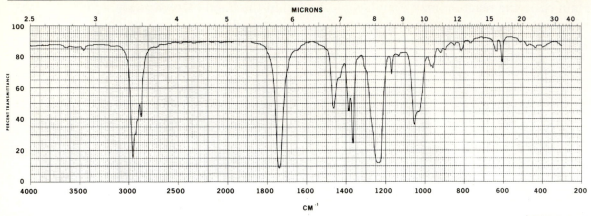

FIGURE 9-1

The infrared spectrum of isopentyl acetate, neat. Spectrum courtesy of Sadtler Research Laboratories, division of Bio-Rad Laboratories, Inc., copyrighted 1973.

FIGURE 9-2

The ^{1}H NMR spectrum of isopentyl acetate in deuterated chloroform. Spectrum courtesy of Sadtler Research Laboratories, division of Bio-Rad Laboratories, Inc., copyrighted 1967.

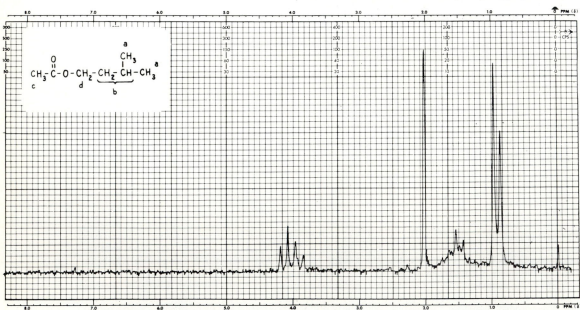

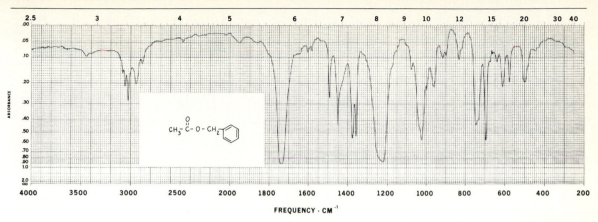

FIGURE 9-3

The infrared spectrum of benzyl acetate, neat. Spectrum
courtesy of Sadtler Research Laboratories, division of Bio-Rad
Laboratories, Inc., copyrighted 1966.

FIGURE 9-4

The ^{1}H NMR spectrum of benzyl acetate in carbon tetrachloride. Spectrum courtesy of Aldrich Chemical Company, *The
Aldrich Library of NMR Spectra,* edition II, by Charles Pouchert.

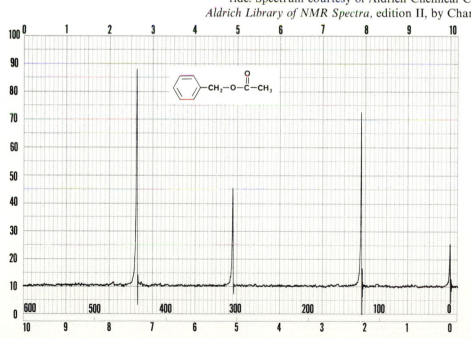

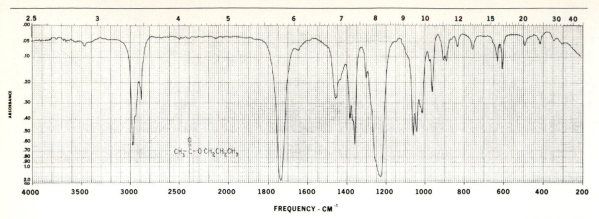

FIGURE 9-5

The infrared spectrum of propyl acetate, neat. Spectrum courtesy of Sadtler Research Laboratories, division of Bio-Rad Laboratories, Inc., copyrighted 1968.

FIGURE 9-6

The ^{1}H NMR spectrum of propyl acetate in carbon tetrachloride. Spectrum courtesy of Sadtler Research Laboratories, division of Bio-Rad Laboratories, Inc., copyrighted 1973.

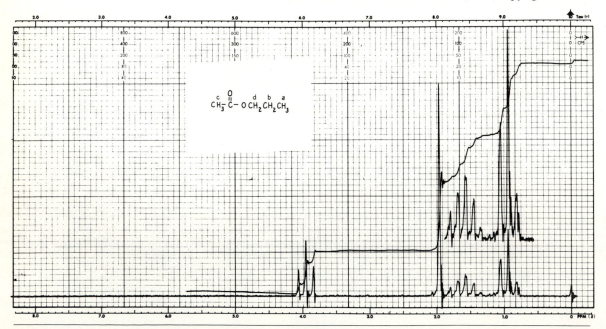

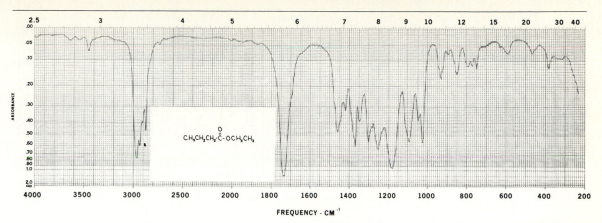

FIGURE 9-7

The infrared spectrum of ethyl butyrate, neat. Spectrum courtesy of Sadtler Research Laboratories, division of Bio-Rad Laboratories, Inc., copyrighted 1966.

FIGURE 9-8

The ^{1}H NMR spectrum of ethyl butyrate in carbon tetrachloride. Spectrum courtesy of Aldrich Chemical Company, *The Aldrich Library of NMR Spectra,* edition II, by Charles Pouchert.

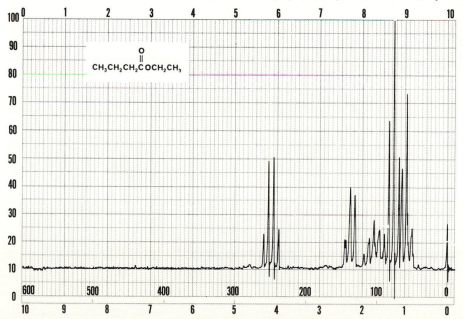

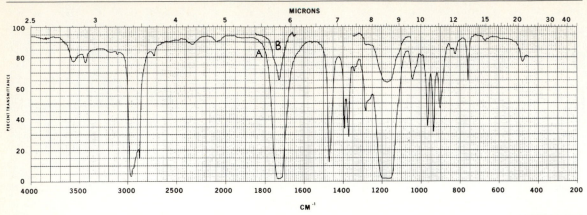

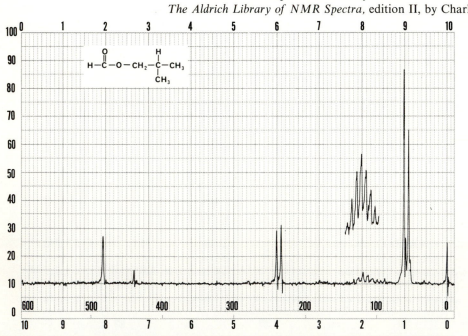

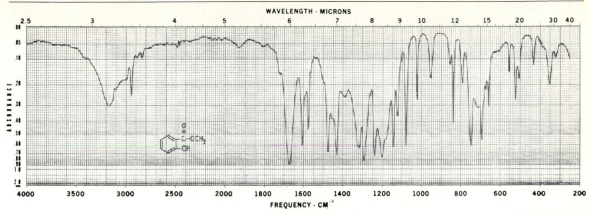

FREQUENCY · CM⁻¹

FIGURE 9-11
The infrared spectrum of methyl salicylate, neat. Spectrum
courtesy of Sadtler Research Laboratories, division of Bio-Rad
Laboratories, Inc., copyrighted 1969.

FIGURE 9-12
The ¹H NMR spectrum of methyl salicylate in carbon tetra-
chloride. Spectrum courtesy of Aldrich Chemical Company, *The
Aldrich Library of NMR Spectra,* edition II, by Charles Pouchert.

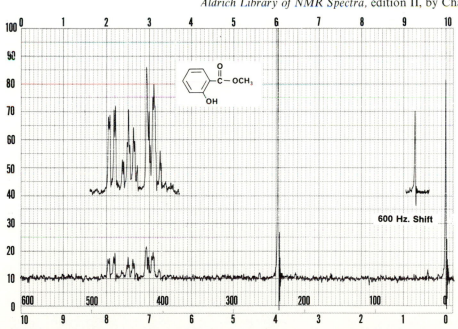

600 Hz. Shift

QUESTIONS

1. Which reagent or reagents did you use as a basis to calculate your theoretical and percentage yields? Why?
2. In what way did you use Le Chatelier's principle in this experiment to raise your yield of product? In what other ways could Le Chatelier's principle have been used?
3. Why was sodium bicarbonate solution used in this experiment? What benefit does it have over sodium hydroxide solution?
4. If you were preparing an ester using ethyl alcohol, why would it be better to use absolute ethyl alcohol rather than 95% ethyl alcohol? If the ethyl alcohol contains a denaturant, why can't the denaturant be methyl alcohol?
5. In the preparation of wintergreen, why is so much more alcohol required than for the other ester syntheses?
6. Why was 15% sodium chloride solution used in this experiment rather than just water? What benefit is there in this solution being cold?
7. What damage would result to the infrared equipment if your sample were not properly dried?
8. If you took an infrared or NMR spectrum of your product, does it show any alcohol contaminant? What regions of the spectrum give you your answer?
9. Why should your final product not contain any carboxylic acid contaminant?
10. Why do manufacturers often prefer to use synthetic rather than natural flavorings in their products? What benefits are there for you?
11. Why might your synthetic flavor not taste exactly like the natural?

10

ARTIFICIAL SWEETENERS: THE SYNTHESIS OF DULCIN

NUCLEOPHILIC ADDITION

Every sweet hath its sour, every evil its good.

Ralph Waldo Emerson, *Compensation*

INTRODUCTION

Eat, drink, and be merry! These words have been looked upon favorably by men and women since the beginning of civilization. Even to just eat and drink is a pleasurable experience for most people. However, one can have too much of a good thing, and many people are overweight. There is obviously a limit to the amount a person can consume without showing the effects.

Fortunately, within the past 40 years, a partial solution to the problem has been found in the widespread use of low-calorie artificial sweeteners. These substances are in soft drinks, desserts, chewing gum, candy, powdered drink mixes, iced tea mixes, jams and preserves, canned fruits, cereals, toothpaste, and pharmaceutical products, and are sold for use in baking, coffee, tea, and other foods and beverages. They have allowed people to "indulge" themselves when consuming certain foods and beverages, and have the added advantages of not promoting tooth decay or affecting diabetics as sugar does. As a result, the average person consumes artificial sweeteners equivalent to over 22 pounds of table sugar per year, up from only 6 pounds in 1975.

The artificial sweeteners which have been used in the United States are saccharin, the cyclamates, and aspartame. Their structures appear in Table 10-1.

Saccharin, discovered in 1879, is about 300 times sweeter than table sugar. It has been used as an artificial sweetener since 1900. In many people, however, it leaves a bitter aftertaste. Cyclamates were first prepared in 1937 and introduced to the public in 1950 under the brand name Sucaryl. The cyclamates are 30 times sweeter than table sugar, and although there is somewhat of an off-taste they do not have the bitter aftertaste of saccharin. Exhaustive testing by experts later determined that this off-taste could be minimized by using a combination of 10 parts of cyclamate to 1 part of saccharin. Due largely to the increased popularity of artificially sweetened soft drinks in the mid-1950s, the market for cyclamates soared and reached almost 20 million pounds per year by the late 1960s.

Then in 1969, the use of cyclamates as a food additive in the United States was banned. This resulted from tests indicating the formation of cancerous

TABLE 10-1.

Artificial Sweeteners that Have Been Used in the United States

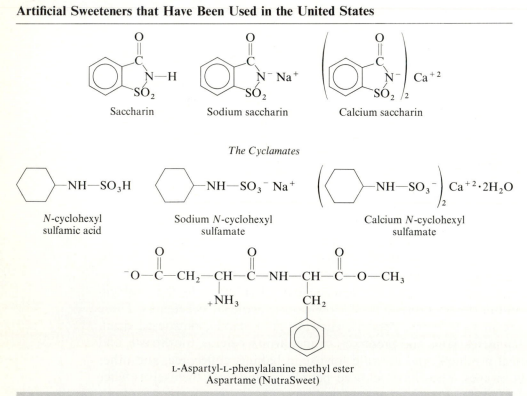

Saccharin   Sodium saccharin   Calcium saccharin

The Cyclamates

N-cyclohexyl sulfamic acid

Sodium N-cyclohexyl sulfamate

Calcium N-cyclohexyl sulfamate

L-Aspartyl-L-phenylalanine methyl ester
Aspartame (NutraSweet)

bladder tumors in rats fed very large quantities of the sweetener. The ban was mandatory due to the 1958 Delaney amendment to the 1938 Food, Drug and Cosmetics Act, which states that "no food additive shall be deemed to be safe if it is found to induce cancer when ingested by man or animal . . . ." With cyclamates no longer allowed, America went back to saccharin.

Then in 1978, even the future of saccharin became uncertain due to a Canadian research study, showing the formation of 13 cancerous bladder tumors in 200 rats fed a diet of 5% saccharin. Earlier studies had also implicated saccharin as a cause of bladder tumors in rats, causing the Food and Drug Administration in 1972 to remove saccharin from the GRAS (generally recognized as safe) list. A Delaney amendment ban was not imposed then due to the smaller number of rats tested and a possible cancer-producing contaminant called *ortho*-toluenesulfonamide in the saccharin. Later Canadian tests used a larger sample of rats and chemically pure saccharin. Furthermore, the results showed that *ortho*-toluenesulfonamide-fed rats had a much lower incidence of cancer than the saccharin group, further strengthening the previous studies.

Because of the Canadian study and the Delaney amendment, the Food and Drug Administration had no choice but to announce that it was starting proceedings to ban the use of saccharin. However, because saccharin was then the only remaining low-calorie sweetener that could be sold, people everywhere bitterly denounced the proposed ban as outrageous, ridiculous, and premature. They questioned the validity of the test results, arguing that the 5% diet of saccharin given the rats would be equivalent to a human consuming between 800 and 1200 cans of diet soda per day, 4000 packets of saccharin per day, or 140 pounds of saccharin per year. In reality, the average consumption of saccharin per human amounted to only 0.026 pound per year or 0.0009 percent of their diet. The level of saccharin fed the rats, therefore, was roughly 5400 times a typical human's intake. Since saccharin is not metabolized by the body, it was conceivable that the sweetener at these very high levels was simply a physical irritant inside the bladder, thereby causing the cancer. This physical irritation would be aggravated for the rats because they concentrate their urine to very high levels, and thus large amounts of saccharin would remain in the bladder for comparatively long periods before being excreted.

In addition, people questioned and proposed updating the Delaney amendment. Although it has served the public well, it is a stringent, all or nothing law that offers no alternatives to an outright ban. Some flexibility or reasonableness, it was argued, should be built into the law for those situations where the medical benefits of an additive greatly exceed the risks. This, it was argued, would be the case with saccharin, where 11 million diabetics and 50 million overweight Americans suffer conditions that are a much more real and immediate danger to their health than the slight possibility of saccharin-induced cancer. Therefore, banning the sweetener would cause great harm and prove to be a hardship to many people while protecting only a theoretical few.

As a result of this outcry, Congress declared a moratorium on the banning of saccharin, pending further studies. This moratorium has been extended periodically ever since 1979, and it will probably continue endlessly. A 1983 Japanese research study confirmed that saccharin promotes the action of a known carcinogen in the bladders in rats, making it a cocarcinogen or a cancer "promoter." However, three studies with humans have shown no real evidence of saccharin causing bladder cancer. For example, a U.S. study of 21,477 diabetics found that they had only 71 percent of the number of bladder cancers expected for an equal number of people from the general population. Diabetics were chosen, of course, because of their high-level intake of saccharin.

Some people believe, however, that saccharin could be a weak carcinogen in humans, and its effects can't be accurately measured by conventional case-control studies. Also, since it has only been widely used since the 1960s, it may be still too early for its cancer-causing potential to register completely. Many known carcinogens, for example, may be ingested for 30 years or more before the cancer becomes evident.

But if saccharin were ever banned, there is fortunately now an approved substitute: aspartame, probably better known as NutraSweet. About 180 times sweeter than table sugar, aspartame, as shown in Table 10-1, is the methyl ester of two amino acids linked together: aspartic acid and phenylalanine. Both of these amino acids are building blocks of proteins and are nutrients needed by the body (see Experiment 34).

The FDA approved aspartame for certain uses in 1981 and for use in soft drinks in 1983. Since 1984, it has largely replaced saccharin for use in many products, primarily because of its much better taste. Not only does it give a more natural sweetness like table sugar, without a bitter aftertaste, but it is a flavor enhancer, making it even more popular with food processors. In fact, the sweetener is so popular that sales have jumped to well over $1 billion per year for its maker, G. D. Searle and Co., now a subsidiary of Monsanto. Over 10,000 tons of aspartame are now produced per year, equivalent to over 2,000,000 tons of table sugar. The familiar red and white swirl emblem of NutraSweet is now found on hundreds of products.

But despite its obvious benefits, aspartame also has its problems. Although it is now widely used in soft drinks, the product in solution breaks down over a period of time, reducing the shelf life. Even more important, aspartame breaks down at high temperatures, making it unsuitable for any kind of baked goods. In addition, people who suffer from a genetic disease called **phenylketonuria** (PKU), detectable at birth, must severely limit their intake of the amino acid phenylalanine, found in aspartame and proteins, or mental retardation occurs.

And, like saccharin, one obvious concern about aspartame is its safety. Just because it comes from two naturally occurring, needed amino acids found in proteins, doesn't necessarily mean it is completely safe, as PKU clearly demonstrates. To prove aspartame's safety, G. D. Searle spent over $70 million for testing. However, despite this, there still has been concern that aspartame causes brain tumors, brain damage, and undesired behavior and mood changes. Certain people have also questioned Searle's test results and the manner in which the FDA approved the sweetener. Nevertheless, aspartame does appear to be acceptably safe when used in moderation. Unlike saccharin and cyclamates, there is no evidence that it either causes or promotes cancer, and, at the moment, it is the most preferred artificial sweetener.

But the future is not necessarily completely sweet for NutraSweet. The patent on aspartame is running out and other artificial sweeteners may also eventually gain widespread acceptance, thus greatly reducing or even eliminating the use of aspartame. Among these are thaumatin, acesulfame K, a French sweetener called superaspartame, L-sugars, and low-calorie sweeteners being developed by the pharmaceutical companies Charles Pfizer and Johnson and Johnson.

Thaumatin is reported to be the sweetest natural substance ever identified, being almost 100,000 times sweeter than table sugar. Also a flavor enhancer, it is extracted from the *Thaumatoccus danilli* plant and has been shown to be a high-molecular-weight protein. But its exact structure is still unknown.

Acesulfame K has the structure below and is an artificial sweetener about 200 times sweeter than table sugar. Currently used in some parts of Europe, it is now before the FDA for approval in the United States; it has been patented by the German company Hoechst.

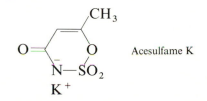

Acesulfame K

Superaspartame is 55,000 times sweeter than table sugar and, unlike aspartame, can be used in baking and frying. L-Sugars are simply ordinary sugars with reverse three-dimensional chemical structures. They might be sweet but, because of their chemical structures, wouldn't be metabolized and thus would have no calories.

These sweeteners could all possibly replace aspartame because of their much lower cost, their greater stability, and, in some cases, their much greater sweetness, which means much less would be needed. This could conceivably make them even safer than aspartame.

In addition, cyclamates are still being used in almost 60 countries, including Canada, and are now being considered for reuse in the United States. A reevaluation of animal tests seems to indicate that cyclamates themselves are not carcinogens, but may be cancer "promoters," thus putting them again in the same category as saccharin. Two benefits of cyclamates are that they taste better than saccharin and can be used in baked goods. However, a breakdown product of cyclamates in the body, called cyclohexylamine, appears to damage chromosomes, so their safety is still questionable.

Other artificial sweeteners have also been discovered, but due to their toxic or carcinogenic effects or the lack of adequate safety testing have not been approved for use in the United States. Included in this group are Dulcin, Perillartine, 5-nitro-2-propoxyaniline, D-tryptophan, stevioside, hernandulcin, naringin dihydrochalcone, and neohesperidin dihydrochalcone, which have the structures shown in Table 10-2.

Dulcin is 250 times sweeter than table sugar, but is not approved in this country because of its toxicity. It has been used in other countries, but is being banned.

Perillartine or perilla sugar is about 2000 times sweeter than table sugar and is utilized to a small extent in Japan, but not elsewhere due to inadequate safety testing. It also has the disadvantages of very low water solubility, appreciable bitterness, and menthol-licorice off-tastes.

5-Nitro-2-propoxyaniline used to be the sweetest substance known, being about 4000 times sweeter than table sugar. Because of its possible carcinogenic effects, it is not used in the United States, but it does have a limited consumption in Europe.

D-Tryptophan is an artificial amino acid 10 times sweeter than table sugar. Like aspartame and thaumatin, it demonstrates that a number of amino acids or amino acid combinations can be sweet.

Stevioside is a natural product found in the Paraguayan shrub *Steviarebaudiana.* Used in Japan, it is about 300 times sweeter than table sugar.

Hernandulcin is a naturally occurring terpene found in the Mexican plant *Lippia dulcis.* About 1200 times sweeter than table sugar, the compound was discovered during an investigation of a sweet herb used by the ancient Aztecs, as described in a monograph written about Mexico in 1570. But although very sweet, hernandulcin also exhibits some bitterness and an unpleasant aftertaste.

Considerable attention has also been given to naringin dihydrochalcone and neohesperidin dihydrochalcone, which are respectively 300 and 2000 times sweeter than table sugar. This attention resulted from both sweeteners being easily obtained as synthetic derivatives of natural products found in citrus fruits. Naringin dihydrochalcone is the hydrogenation product of naringin, a bitter substance found in the rind of the grapefruit; and neohesperidin dihydrochalcone is the hydrogenation product of neohesperidin, a bitter substance in the Seville orange. But both substances will probably be unsatisfactory as sweeteners in foods, since their sweetness is slow in onset, followed by a lingering aftertaste with menthol overtones.

One major problem in the development of new artificial sweeteners is that the relationship between chemical structure and sweetness is still not known. All sugar substitutes, including saccharin, aspartame, and cyclamates, have been discovered by accident, and until more is learned about the taste mechanism, the search for new sweeteners will continue largely on a hit-or-miss basis. As you can see from the structures of sucrose and all the other sweeteners, sweetness can occur in molecules having many variations in their structures. However, just

TABLE 10-2.
Other Discovered Artificial Sweeteners

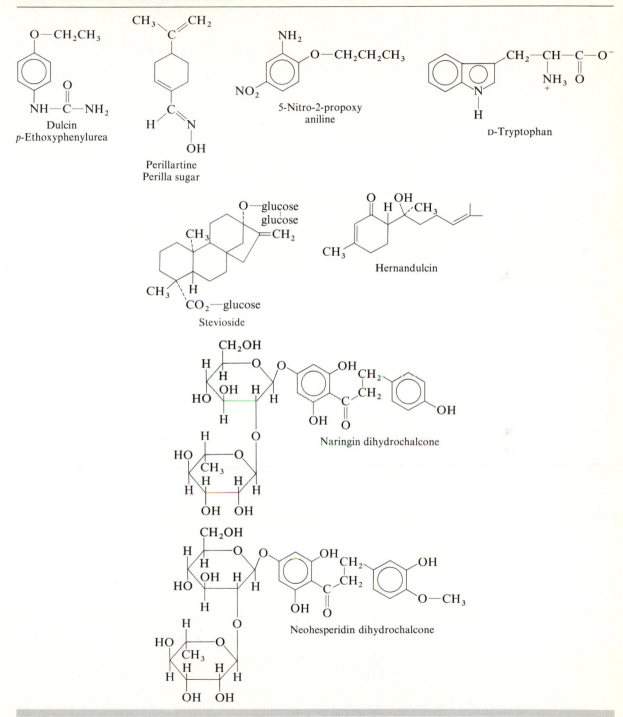

Dulcin
p-Ethoxyphenylurea

Perillartine
Perilla sugar

5-Nitro-2-propoxy
aniline

D-Tryptophan

Stevioside

Hernandulcin

Naringin dihydrochalcone

Neohesperidin dihydrochalcone

slightly altering these structures results in compounds that are either tasteless or bitter. For example, converting saccharin into *N*-methylsaccharin or modifying sodium cyclamate to sodium *N*-phenylsulfamate give compounds that are completely tasteless.

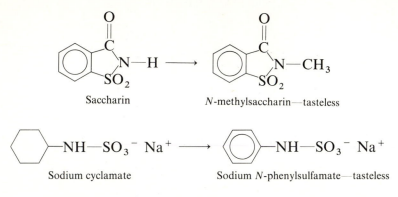

Saccharin   *N*-methylsaccharin—tasteless

Sodium cyclamate   Sodium *N*-phenylsulfamate—tasteless

Lactose or milk sugar (see Experiment 32) is almost tasteless compared to sucrose, even though both are disaccharides.

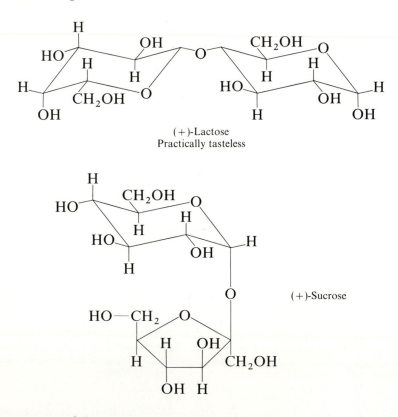

(+)-Lactose
Practically tasteless

(+)-Sucrose

The α anomer of the monosaccharide mannose tastes sweet, but the β anomer is bitter.

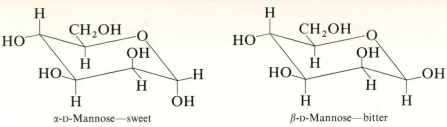

α-D-Mannose—sweet   β-D-Mannose—bitter

D-Tryptophan is sweet, whereas its stereoisomer L-tryptophan is bitter; and D-leucine tastes very sweet, but its stereoisomer L-leucine is slightly bitter.

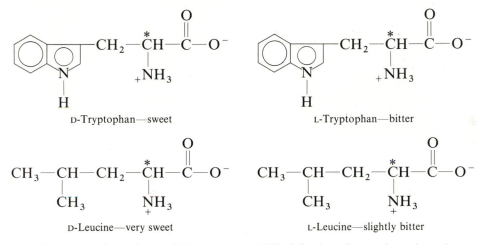

D-Tryptophan—sweet   L-Tryptophan—bitter

D-Leucine—very sweet   L-Leucine—slightly bitter

Finally, changing the —NH$_2$ group of Dulcin-(p-ethoxyphenylurea) to a methyl group results in the compound phenacetin (see Experiment 13), which is not sweet and has been used as an analgesic (pain reliever) and antipyretic (fever reducer).

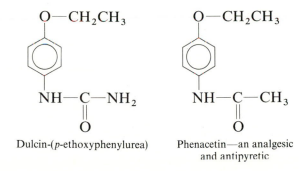

Dulcin-(p-ethoxyphenylurea)   Phenacetin—an analgesic
and antipyretic

BACKGROUND

In this experiment, you will synthesize Dulcin by the reaction of p-ethoxyaniline (p-phenetidine) with isocyanic acid. The isocyanic acid is prepared by the reaction of potassium cyanate with acetic acid.

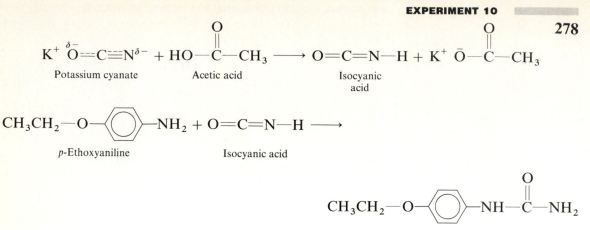

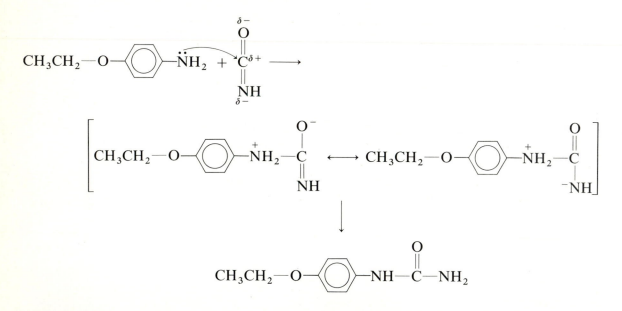

The reaction mechanism involves nucleophilic attack by the amino group of p-ethoxyaniline on the electron-deficient carbon atom of isocyanic acid. This produces a resonance-stabilized intermediate, which then transfers a proton to the other nitrogen atom to give the final observed product.

PROCEDURE

CAUTION: p-Phenetidine is toxic. Avoid breathing it and keep it off your skin.

Also keep glacial acetic acid off your skin, since it causes severe burns.

In addition, although potassium cyanate is nowhere near as toxic as potassium cyanide, keep it off your skin.

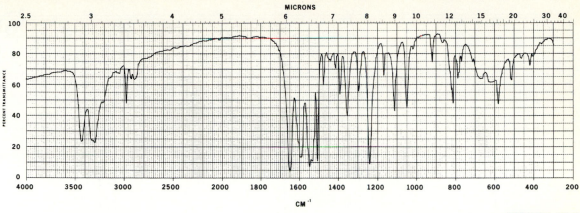

FIGURE 10-1

The infrared spectrum of Dulcin in a KBr pellet. Spectrum
courtesy of Sadtler Research Laboratories, division of Bio-Rad
Laboratories, Inc., copyrighted 1973.

Dissolve 1.4 mL or 1.4 grams (0.010 mole) of **recently distilled** *p*-ethoxyaniline (*p*-phenetidine) in 10 mL of water and 2 mL of glacial acetic acid in a 50-mL Erlenmeyer flask. Then while swirling the solution, add dropwise a solution of 1.6 grams (0.020 mole) of potassium cyanate in 5 mL of water. When the Dulcin begins to precipitate, add all the remaining cyanate solution at once and thoroughly mix. Allow the mixture to stand at room temperature for at least 1 hour, and occasionally stir it. Then add 5 mL of water and cool in an ice bath. Collect the product by suction filtration, wash it with a little ice-cold water, and recrystallize from about 30 mL of hot water, using decolorizing carbon. Allow the purified product to dry until the next laboratory period and determine its weight, melting point, and literature melting point, and record the results in your notebook. Also determine the theoretical and percentage yields of the final product, showing the calculations.

At your instructor's option, obtain the infrared spectrum of your product in a KBr pellet, and the ^{1}H NMR spectrum in polysol-*d*. Compare your spectra with those shown in Figures 10-1 and 10-2. When finished, hand in your product in a properly labeled container to your instructor.

QUESTIONS

1. (a) A pound of table sugar contains 1790 calories. If the average person consumes artificial sweeteners equivalent to 22 pounds of table sugar per year, how many calories has the average person saved per year by using artificial sweeteners instead of table sugar?

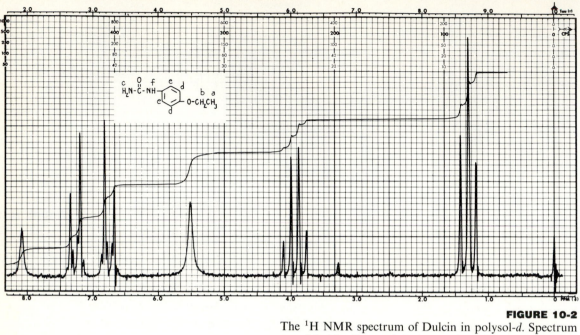

FIGURE 10-2
The ¹H NMR spectrum of Dulcin in polysol-*d*. Spectrum
courtesy of Sadtler Research Laboratories, division of Bio-Rad
Laboratories, Inc., copyrighted 1973.

(b) If the average person consumes 2500 calories daily, what percentage has his or her daily caloric intake been "reduced" by using artificial sweeteners instead of table sugar?

(c) If a pound of weight gained or lost is equivalent to 3500 calories, how many pounds has the average person "lost" each year by using artificial sweeteners?

2. Indicate what factor or factors prevent a large multitude of artificial sweeteners from being indiscriminately sold today. What benefit is there in this for you?

3. Why can't chemists just put their minds to it and decide what organic molecular structure will provide the best artificial sweetener?

11

PERFUMES: THE SYNTHESIS OF NEROLIN

NUCLEOPHILIC ALIPHATIC SUBSTITUTION
WILLIAMSON SYNTHESIS OF ETHERS

Let me have them very well perfumed
For she is sweeter than perfume itself.

Shakespeare, *The Taming of the Shrew,* Act I, Scene 2

INTRODUCTION

Much has been written about the special aura that is a woman, that distinctive feminine quality that reaches out to attract men. Of course, a large part of this attraction is the result of a well-groomed, sweetly scented body. No one can dispute the impact of a personal fragrance that softly whispers something special about the wearer and expresses her femininity. Psychologically it also makes her feel more womanly and more worthy of appreciation.

But what applies to a woman can also apply to a man. In most instances, a woman will be more attracted to and respond more easily to a man who is pleasantly scented with a cologne or after-shave lotion. The essence of cigar or pipe smoke, often offensive, doesn't have to be the only fragrance associated with masculinity.

Today the pleasant fragrances of perfumes have found their way into hundreds of products. Besides being used directly as perfumes, colognes, toilet water, and after-shave lotions, perfumes are now found in soaps, shampoos, bath oils, deodorants, cosmetics, face and body lotions, oils, creams, and powders. In addition to making these products more appealing to the consumer, the scents also provide the psychological lift needed by so many people.

In recent years, the unpleasantness of foul-smelling household products has been greatly reduced with industrial perfumes or odorants. Plastics, deodorizers, antiseptics, glues, insecticides and insect repellents, polishes, and waxes are just some of the items whose odors have been masked or minimized with these odorants.

Originally, all perfumes came from plant or animal sources. But with the tremendous demand for perfumes and the development of organic chemistry, most perfumes today are made synthetically from coal tar or petroleum. The most expensive perfumes contain oils extracted or steam distilled from fresh flowers. The cost is high because the yield may be as little as a pound of oil from a ton of flowers, and is rarely over 1 percent.

Perfume sold today consists of a blend of ingredients, diluted in alcohol and water to a certain concentration. As few as 10 ingredients or as many as 200 may be present. The blending of the right ingredients in the correct amounts is an art requiring much experience and knowledge. Some odors, like musical sounds, harmonize when blended and produce a fragrance richer and fuller or more chaste and delicate than the separate components. But others are antagonistic or incompatible when blended, resulting in a harsh, contrary effect. A trained perfumer knows what the interaction of one scent with another will be, and he or she may even be aware of the chemical structures and possible chemical reactions of the components. The perfumer is able to recognize by odor over a thousand different perfume ingredients, and often their source and quality.

Since good perfume or cologne is quite expensive, care should be taken in its selection. Fragrances should be tested on the skin, not merely sniffed from a bottle. The body chemistries of people differ, and what is good on one person may be unsatisfactory on another. Only a limited number should be tried because after a time it becomes difficult to distinguish the characteristics of the fragrances. Favorable comments from others are a good indication that a proper selection has been made.

BACKGROUND

In this experiment, you will prepare the odoriferous compound nerolin, a synthetic perfume fixative. Nerolin is an important ingredient in many perfumes and, as a fixative, "binds" the other ingredients together, thus diminishing the rate of evaporation of the more volatile components. If a fixative like nerolin were not present, the fragrance of a complex perfume would change with time as the more volatile oils evaporated and left behind the less volatile substances, which have a different scent.

Nerolin chemically is called either 2-ethoxynaphthalene or ethyl β-naphthyl ether and can be prepared by the following Williamson synthesis of ethers.

β-Naphthol   Ethyl iodide   Nerolin

The potassium hydroxide, acting as a base, removes the slightly acidic proton from β-naphthol and converts it into the nucleophile potassium β-naphthoxide, which attacks ethyl iodide and displaces the iodide ion.

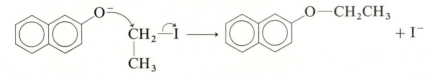

This last reaction occurs by an S_N2 (a bimolecular nucleophilic substitution) mechanism. In this mechanism, the rate of the reaction depends on the concentration of both reactants and involves the β-naphthoxide ion attacking the ethyl iodide molecule from the rear, thus avoiding the departing iodide ion.

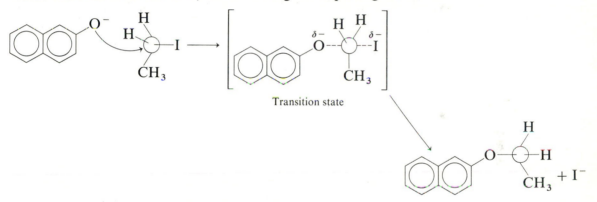

Transition state

The transition state for the mechanism, as you can see, involves the carbon-oxygen bond of the ether being partially formed and the carbon-iodine bond of ethyl iodide being partially broken. The oxygen atom of the β-naphthoxide ion has a diminished negative charge, since it has begun to share its electrons with carbon. On the other hand, the iodine atom has developed a partial negative charge, since it has partially removed a pair of electrons from carbon. Also notice in the mechanism how the final product has the atoms on the attacked carbon forming a tetrahedral arrangement **opposite** to the original one. A good analogy to this inversion is an umbrella turning inside out on a windy day.

In this synthesis, potassium hydroxide is used rather than sodium hydroxide because of its greater solubility in the solvent alcohol. After it is formed, the product nerolin precipitates with the addition of ice-cold water and may be recrystallized from alcohol and water.

PROCEDURE

CAUTION: Potassium hydroxide is caustic. Keep it off of your skin and clothing. Wear protective glasses.

Place 20 mL of methanol, 2.9 grams (0.020 mole) of β-naphthol, and 1.7 grams (0.026 mole) of potassium hydroxide, with stirring, into a 100-mL round-bottom flask. Allow a few minutes for the acid-base reaction to occur, then add 1.8 mL (0.023 mole) of ethyl iodide and attach a reflux condenser by placing an ordinary condensing column vertically into the flask. Boil the mixture for 2 hours.

Cool the hot reaction mixture and add about 50 mL of ice-cold water. The nerolin may at this point form an oil, due to its low melting point and the impurities present, but eventually it will crystallize upon cooling in an ice-salt bath or the freezer compartment of a refrigerator. Collect the crystallized nerolin by suction filtration, place in a 100-mL beaker, dissolve in the minimum amount of hot methanol, then **slowly** add hot water until turbidity is produced. Add enough hot methanol to redissolve the product, treat with charcoal, and then continue with the recrystallization procedure. Collect and dry the crystals and determine the weight, melting point, and literature melting point; record the results in your notebook.

At your instructor's option, obtain the infrared spectrum of your product in a KBr pellet, but be sure all the water has been removed. Also at your instructor's option, obtain the ^{1}H NMR spectrum in carbon tetrachloride. Compare your spectra with Figures 11-1 and 11-2.

FIGURE 11-1
The infrared spectrum of 2-ethoxynaphthalene as a capillary cell melt. Spectrum courtesy of Sadtler Research Laboratories, division of Bio-Rad Laboratories, Inc., copyrighted 1973.

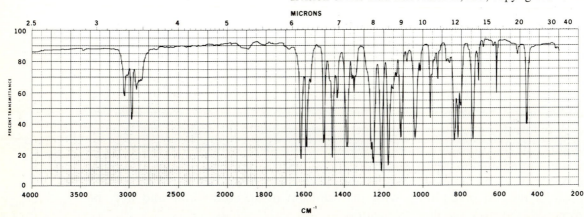

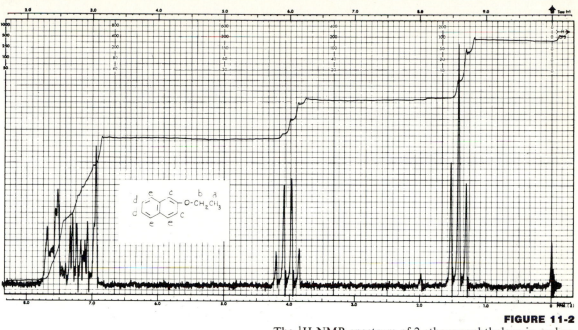

FIGURE 11-2

The ^{1}H NMR spectrum of 2-ethoxynaphthalene in carbon tetrachloride. Spectrum courtesy of Sadtler Research Laboratories, division of Bio-Rad Laboratories, Inc., copyrighted 1973.

Determine the theoretical and percentage yields of your final product, showing the calculations, and submit it in a properly labeled container to your instructor.

QUESTIONS

1. If you suspected that the product nerolin contained unreacted β-naphthol, suggest a way to chemically remove it, based on what you learned in Experiment 4.
2. A possible side reaction in this synthesis is the nucleophilic attack of the hydroxide ion on ethyl iodide to produce ethyl alcohol and the iodide ion. Write the equation for this reaction and explain how the way in which the reagents were added helped to minimize it.
3. Why do we react β-naphthol with ethyl iodide rather than react ethyl alcohol with 2-iodonaphthalene?
4. What is the difference between perfume, cologne, and toilet water?
5. If nerolin is a solid, why are you able to smell it?
6. Why should perfume bottles be stored tightly capped and away from heat and direct sunlight?
7. Why is it a good idea not to place perfume on clothing?

12

LOCAL ANESTHETICS: THE SYNTHESIS OF BENZOCAINE

- **NUCLEOPHILIC ACYL SUBSTITUTION**
- **CHEMICAL EQUILIBRIUM**

For there was never yet philosopher
That could endure the toothache patiently.

Shakespeare, *Much Ado About Nothing,*
Act V, Scene 1

INTRODUCTION

All of us have suffered pain from sunburns or other types of burns, toothaches, earaches, cold sores, insect bites, and minor cuts and abrasions. Such pain can be severe and persistent. What we want is relief until the affliction begins to heal or can be treated by a doctor or dentist.

For many years, products have been marketed to provide this relief. One common ingredient in many of these products is the local anesthetic benzocaine. Go to your local drug store or supermarket and look for all the products that contain benzocaine as a key ingredient. Although no store may carry them all, there are literally dozens of them.

287

But benzocaine is only one of the local anesthetics known today. Cocaine, procaine (Novocain), and lidocaine (Xylocaine) are just a few of the others, some of which are shown in Table 12-1. Observe from the structures that every molecule, with the exception of benzocaine, consists of an aromatic portion and a secondary or tertiary amino portion separated by an intermediate chain. Most local anesthetics are aromatic esters, but some are aromatic amides. Benzocaine is also different in that it does not suffuse well into tissue, which makes it unsuitable for injection but useful as a skin preparation.

The first known local anesthetic is the naturally occurring drug cocaine, found in the leaves of the coca plant *Erythroxylon coca* in Columbia, Peru, and Bolivia. Also known as "coke," "snow," "C," "happy dust," "nose candy," and other names, the drug can be taken by mouth, sniffed, injected, or smoked. But, ironically, although it dulls feeling where it is taken, it is the most potent and the fastest-acting central nervous system stimulant known. Users immediately experience a sense of euphoria and exhilaration. They feel energetic, mentally alert, inspired, talkative, witty, and self-confident. They may also become very productive and have a positive attitude. As a result, cocaine use has become extremely widespread in recent years, and today it is known as the king of the illegal drugs. It has the reputation of being "chic," of being a symbol of wealth and status, and of being able to enhance one's sex appeal and social standing.

Because of its high price, cocaine is also the caviar of illegal drugs. One gram of the pure material can easily cost $500 or more. Frequently, the price appears to be less, but then the drug probably has been diluted or "cut" with lactose or some other sugar to 12% or less of its original potency. The economic incentive is so great that dealers will frequently even add amphetamines and procaine or lidocaine to the highly diluted product to give the illusion of a higher concentration of cocaine. Amphetamines enhance the "energized" effect, and procaine and lidocaine produce a similar numbing effect at the part of the body where they are taken in.

Cocaine advocates talk about it being a relatively risk-free drug. "After all," they say, "there's no hangover, no physical addiction or dependency, no danger of lung cancer, and no necessary holes in the arms. In addition, look at all its positive effects!"

Part of what these people say is true. But experts can easily counter many of these arguments. Because cocaine's psychological and physiological lift lasts for only 20 to 30 minutes, many people soon want another dose to continue the effects. As a result, they can easily become psychologically dependent on the drug and experience many of its negative effects when the drug wears off. These effects usually include extreme depression, despondency, fatigue, nervousness, irritability, and muscle spasms. Hallucinations, paranoia, and a totally "strung out" physical collapse are also common.

TABLE 12-1.

Structures of Local Anesthetics

Local anesthetic	Aromatic portion	Intermediate chain	Amino portion

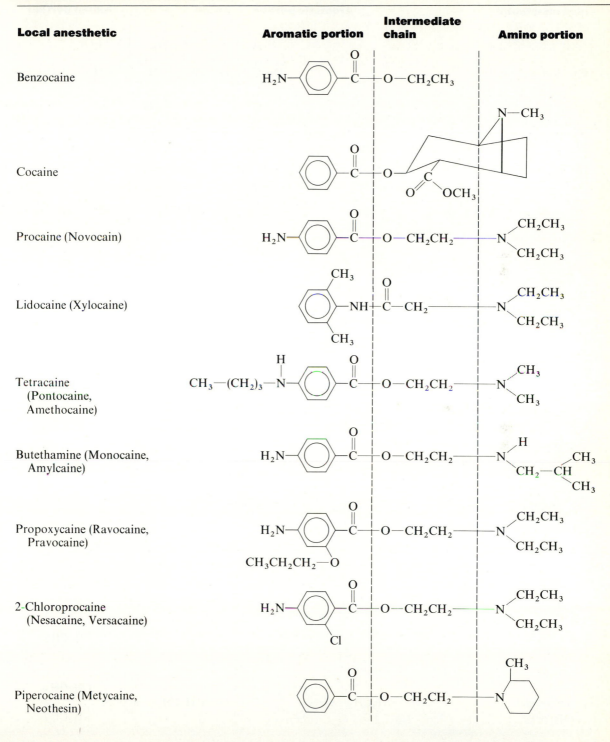

TABLE 12-1. (continued)

Structures of Local Anesthetics

Local anesthetic	Aromatic portion	Intermediate chain	Amino portion

Meprylcaine (Oracaine)

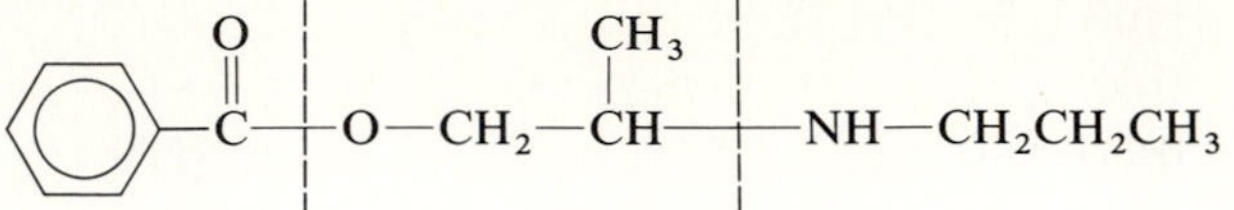

Isobucaine (Kincaine)

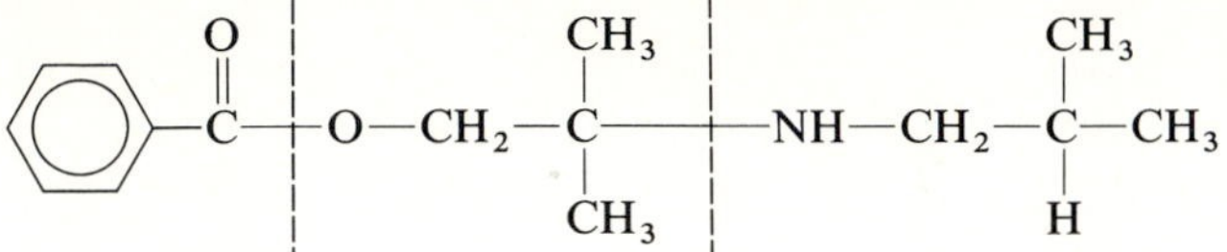

Metabutethamine (Unacaine)

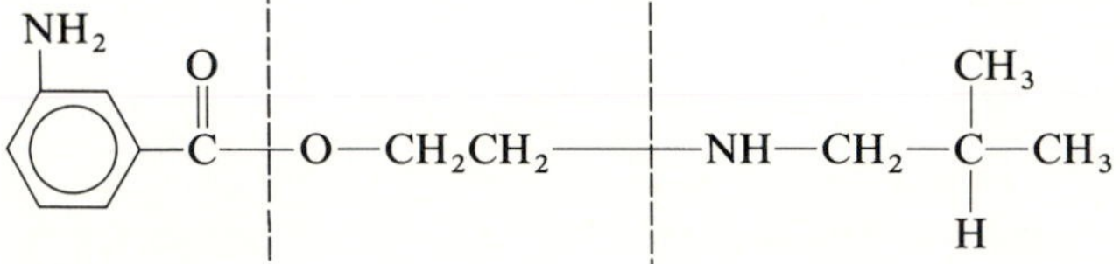

Primacaine

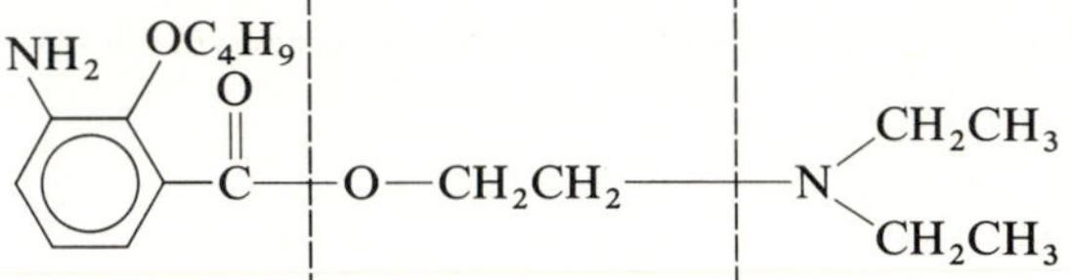

Pyrrocaine (Dynacaine)

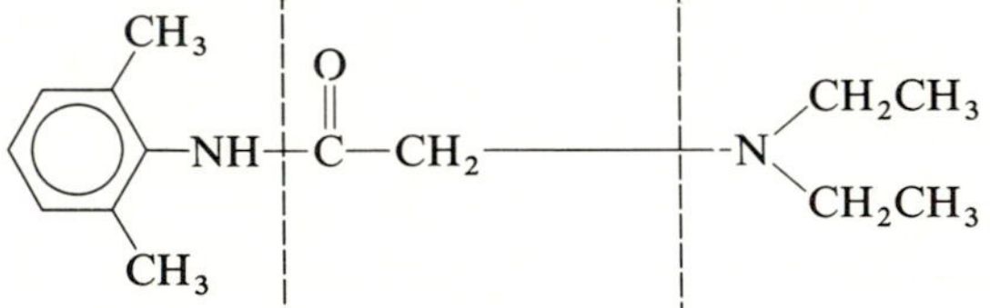

Prilocaine (Citanest)

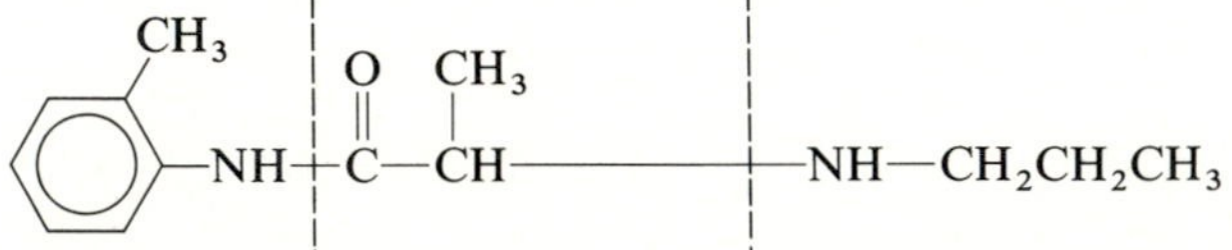

Mepivacaine (Carbocaine)

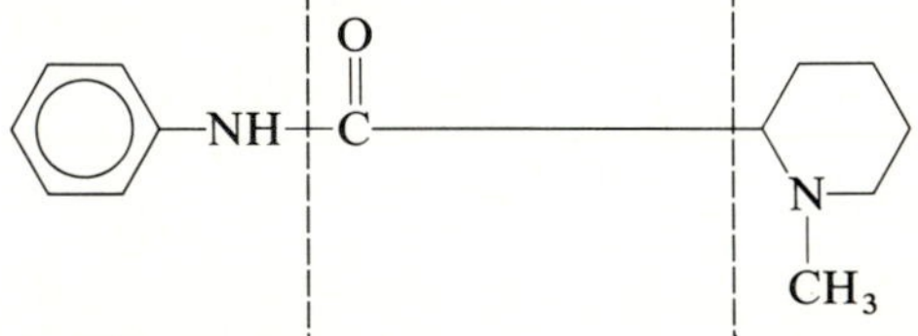

Dibucaine

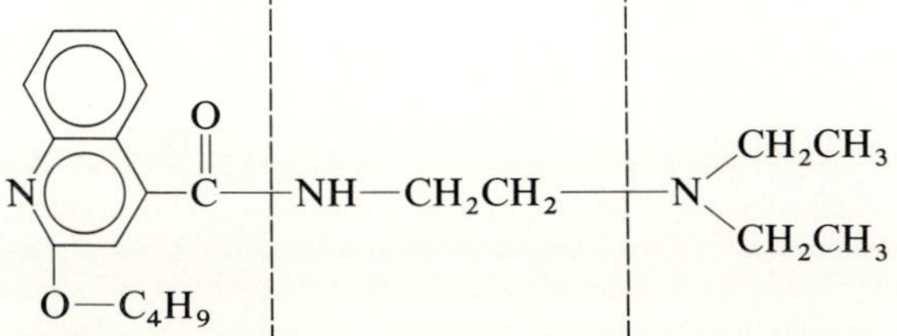

Parethoxycaine (Intracaine)

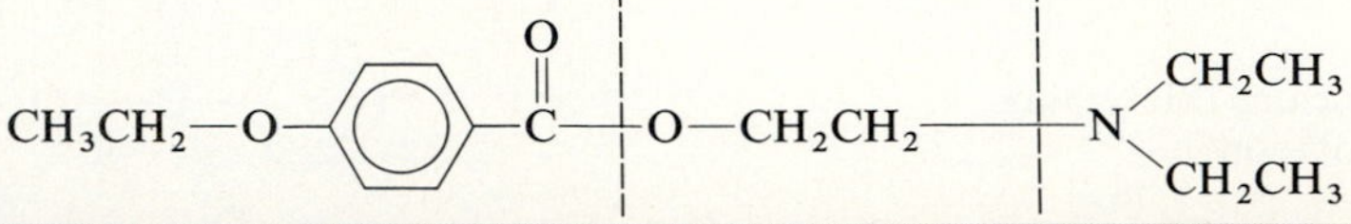

To rebound from these negative effects, users may take more and more of the drug, which puts them on an even larger psychological roller coaster with much steeper drops to even deeper lows. They find they cannot survive without it, thus they are addicted and their craving is excruciating. Like any addict, they will do almost anything to get it—even begging, borrowing, stealing, and selling their bodies and possessions. Some people have even spent over $3,000 a week on the drug for their own use. Test animals addicted to cocaine will even die of starvation if given a choice of being deprived of food or the drug.

But test animals are not the only ones to die from cocaine use. Every year, numerous people are its victims, from suicide during a period of extreme depression or from heart attacks, strokes, and seizures. Even star athletes in the prime of life have died from it, demonstrating that no one is immune. Even small amounts in some people have proven fatal.

As a powerful stimulant, cocaine can set off what is known as the "Casey Jones" reaction. Dramatic physiological increases can occur in a person's blood pressure and in his or her rate of heartbeat and respiration. If these dramatic increases exceed the body's acceptable limits, death can occur. And as people resort to larger and larger amounts of cocaine to get its benefits or to temporarily recover from its negative effects, the chances are even greater that the "Casey Jones" reaction will be set in motion.

But the known structure and physiological properties of cocaine does have one important benefit: From it has come the synthesis and testing of hundreds of other local anesthetics, with the objective of finding compounds that will

1. Produce a reversible action

2. Be nonirritating and nonallergenic

3. Have a low toxicity

4. Have a rapid onset and a sufficient duration

5. Have good penetrating properties

6. Have a potency sufficient to give complete anesthetization without harmful concentrated solutions

7. Have stability in solution, but be readily biotransformed or hydrolyzed in the body to harmless products.

Unfortunately, the perfect local anesthetic to satisfy all these requirements has yet to be found.

An almost unlimited number of local anesthetics could theoretically be synthesized. Any structural changes in the molecule are usually found to alter the toxicity, potency, duration, and diffusibility of the compound. But most of the newer drugs synthesized are never used in actual practice because tests show no advantages over products already in use.

For many years, the most successful synthetic local anesthetic used in injections was procaine (Novocain), which was the standard for comparing other compounds. But the introduction and wide acceptance of nonester local anesthetics have reduced the importance of procaine, and currently lidocaine (Xylocaine) is the product most commonly used for injections.

The method in which local anesthetics prevent pain is still not completely understood. It appears that they compete with calcium ions for a site in the nerve membrane that controls the passage of sodium ions across the membrane. At this site, they are able to prevent the passage of sodium ions through the membrane into the nerve cell and the passage of potassium ions through the membrane and out of the nerve cell. The polarized nerve is, therefore, unable to depolarize and conduct an electrical impulse to signal pain.

BACKGROUND

Benzocaine will be prepared in this experiment by the direct esterification of *p*-aminobenzoic acid with absolute ethanol by the reaction

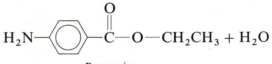

Like the direct esterification reaction to produce artificial flavors (Experiment 9), this reaction gives a lower yield of benzocaine due to chemical equilibrium. But by Le Chatelier's principle, the equilibrium of the reaction is shifted to yield additional product.

The accepted mechanism of the reaction follows and involves (1) the protonation of the carboxylic acid by the acid catalyst to give the resonance-stabilized intermediate I, followed by (2) the nucleophilic attack of the alcohol. The subsequent transfer of a proton (3), loss of water (4), and loss of a proton (5) then give the final observed products. The reverse reaction to yield the carboxylic acid and alcohol follows the same steps in the opposite order.

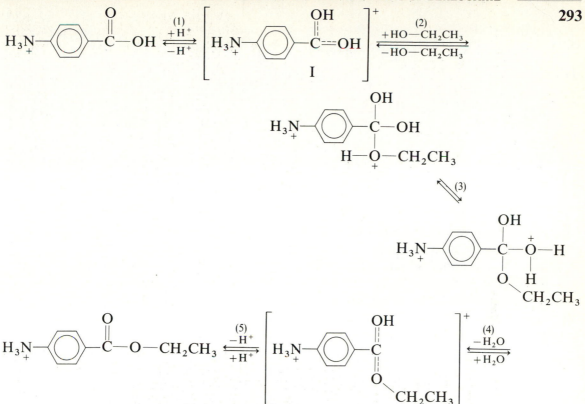

The reaction workup involves first making the reaction mixture alkaline with sodium carbonate solution. With this step, all the following reactions occur:

$$H_2SO_4 + Na_2CO_3 \longrightarrow Na_2SO_4 + H_2O + CO_2$$

Unreacted starting material

Sodium
p-aminobenzoate

$$H_3\overset{+}{N} \underset{HSO_4^-}{\underset{|}{\bigcirc}} \overset{O}{\overset{\|}{C}} - OCH_2CH_3 + Na_2CO_3 \longrightarrow$$

Acid salt of
benzocaine

$$H_2N - \bigcirc - \overset{O}{\overset{\|}{C}} - OCH_2CH_3 + Na_2SO_4 + H_2O + CO_2$$

Benzocaine

Because the amount of sodium carbonate solution added is not very large, roughly an equal volume of water and ethanol results. Under these conditions, much of the nonionic benzocaine formed in the experiment should remain dissolved, and some ionic sodium sulfate and unreacted sodium p-aminobenzoate may precipitate. Extraction of this mixture with ether permits the separation of the benzocaine from the ionic compounds. Then after the ether-ethanol layer is dried and evaporated, crystallization of the resulting oil from an ethanol-water solution further purifies the benzocaine.

As stated in the introduction, benzocaine is unlike the other local anesthetics in not having a secondary or tertiary amino group separated from the aromatic portion of the molecule. It also does not suffuse well into tissue, thus making it unsuitable for injection but useful as a skin preparation. Its low water solubility and consequent slow absorption from the area of application are beneficial in producing a lower level of toxicity and a longer period of anesthetic protection.

PROCEDURE

CAUTION: Concentrated sulfuric acid is extremely hazardous and corrosive. Handle it with care.

Place 2.7 grams (0.020 mole) of p-aminobenzoic acid in a 100-mL round-bottom flask. Add 35 mL (0.60 mole) of **pure, absolute** ethanol, swirling gently to help dissolve most of the solid. Cool the mixture in an ice bath and **slowly add** 2.5 mL (0.045 mole) of concentrated sulfuric acid. A large amount of precipitate will form, but will slowly dissolve when the mixture is refluxed. Attach a reflux condenser and heat the mixture under reflux for 2 hours. During this time, swirl the mixture approximately every 5 to 10 minutes until the solid dissolves.

Pour the contents of the flask into a 250-mL beaker, cool to room temperature, and **slowly add in very small portions** a 20% sodium carbonate solution with stirring. After each addition, an extensive evolution of gas will occur until the acid is nearly neutralized. When additional sodium carbonate solution produces no more gas, check the pH of the solution with pH paper and, if necessary, add an extra small amount of sodium carbonate solution until the pH is 9 or higher. **Do not** dip the pH paper into the reaction mixture, but instead use a stirring rod to transfer a drop of the liquid from the mixture to the paper.

Carefully decant the ethanol-water mixture from any precipitate present into a separatory funnel. Add 25 mL of anhydrous ethyl ether to any precipitate in the beaker and stir. Then carefully decant this ether from any remaining precipitate into the separatory funnel. This ensures that no benzocaine will be lost from the water having been added.

Shake the ether, water, and ethanol mixture in the separatory funnel, and draw off the lower aqueous layer into an Erlenmeyer flask. Draw off the upper ether-ethanol layer into another Erlenmeyer flask and pour the aqueous layer back into the separatory funnel. Extract the aqueous layer with another 25 mL of ether and combine both ether layers.

Dry the ether-ethanol solution with anhydrous magnesium sulfate, then remove the drying agent by gravity filtration. Evaporate the ether and ethanol in the hood, using either a steam or a hot-water bath. When only a few milliliters remain, an oil should appear in the flask. Add hot 95% ethanol **slowly** and heat the mixture until the oil dissolves. Then add hot water **slowly** to the alcohol solution until extensive cloudiness or the oil appears.

Cool the mixture in an ice bath with swirling. During the cooling, any oil initially present will solidify. Collect the crystals of purified benzocaine by suction filtration and thoroughly dry them. Determine their weight, melting point, and literature melting point, and record the results in your notebook.

At your instructor's option, obtain the infrared spectrum of your product in a KBr pellet, and the ^{1}H NMR spectrum in deuterated chloroform. Compare your spectra with those shown in Figures 12-1 and 12-2.

FIGURE 12-1

The infrared spectrum of benzocaine as a capillary cell melt. Spectrum courtesy of Sadtler Research Laboratories, division of Bio-Rad Laboratories, Inc., copyrighted 1967.

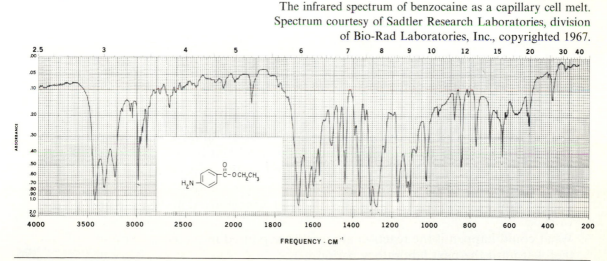

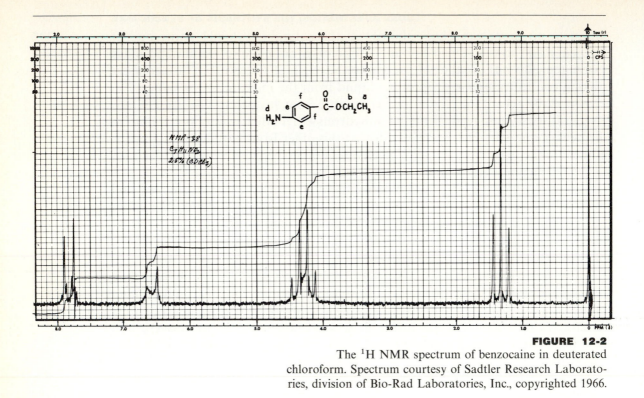

FIGURE 12-2
The ^{1}H NMR spectrum of benzocaine in deuterated chloroform. Spectrum courtesy of Sadtler Research Laboratories, division of Bio-Rad Laboratories, Inc., copyrighted 1966.

Determine the theoretical and percentage yields of your final product, showing the calculations. Submit your product in a properly labeled container to your instructor.

QUESTIONS

1. Outline a synthesis of the starting material *p*-aminobenzoic acid from toluene.
2. How was Le Chatelier's principle used in this experiment to raise the yield of benzocaine?
3. Why were anhydrous ethyl alcohol and concentrated sulfuric acid used in this experiment rather than 95% ethyl alcohol and dilute sulfuric acid? What would be the result if concentrated hydrochloric acid had been used?
4. Why couldn't the anhydrous ethyl alcohol used in this experiment be denatured with methyl alcohol?
5. Explain how any unreacted *p*-aminobenzoic acid was removed in the reaction workup.
6. What could happen if the reaction mixture were poured into cold water and then left until the next laboratory period?

7. What is the structure of the precipitate that formed after the sulfuric acid was added?

8. Why are most local anesthetics such as procaine and lidocaine administered as chloride or sulfate salts? Looking at the structures of these two compounds, which nitrogen atom in each of them will be protonated first? Why?

9. Go to a local drug store or supermarket and write down by brand name and type of product at least five products containing benzocaine.

EXPERIMENT

13

ANALGESICS

Pain is no longer pain when it is past.

Margaret Preston, *Old Songs and New Nature's Lesson*

INTRODUCTION

With the exception of caffeine, analgesics are the most commonly used drug around the world today. In the United States alone, they are a 2.1 billion dollar a year industry, and Americans consume over 45 million pounds annually, the equivalent of over 65 billion tablets, or 270 tablets per person. Their widespread use has made them a popular remedy for many ailments, including headaches, muscular aches, fever, colds, influenza, arthritis, and rheumatism.

One of the most powerful analgesics is morphine. The drug has been used for many years for treating very severe pain, but it has the very negative feature of being highly addictive. As a result, its use is usually very restricted. Other common analgesics that have been used are codeine, salicylic acid, phenyl salicylate, acetanilide, aspirin, salicylamide, phenacetin, acetaminophen, and ibuprofen, the structures of which are given in Table 13-1.

The oldest known pain reliever is salicylic acid. Its use goes back to at least 400 B.C. when Hippocrates, the father of modern medicine, advised women to chew the bitter leaves of willow trees to ease the pains of childbirth. Then in the second century A.D., Galen and other Greek physicians noted that chewing willow leaves also reduced fever and inflammation. No one still knew why, but for the next 1600 years generations of Europeans and American Indians reaped the benefits of the willow leaves' remarkable ability to reduce pain, fever, and inflammation.

Then in 1763, it was discovered that the extract of willow bark was also successful in reducing fever. Almost 100 years later, the same extract was found to be successful in alleviating the symptoms of acute rheumatism. Shortly after,

TABLE 13-1.

Some Common Analgesics

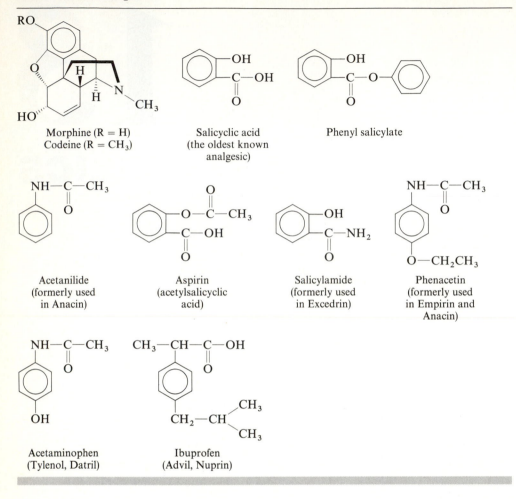

Morphine (R = H)
Codeine (R = CH$_3$)

Salicyclic acid
(the oldest known
analgesic)

Phenyl salicylate

Acetanilide
(formerly used
in Anacin)

Aspirin
(acetylsalicyclic
acid)

Salicylamide
(formerly used
in Excedrin)

Phenacetin
(formerly used
in Empirin and
Anacin)

Acetaminophen
(Tylenol, Datril)

Ibuprofen
(Advil, Nuprin)

organic chemists were able to isolate and identify the active ingredient in the
extract as salicylic acid (see Table 13-1). This identification led in 1875 to the
chemical production of salicylic acid in large quantities. However, it also became
evident that its use as a medical remedy was severely limited. The compound,
because of its acidic properties, caused severe irritation of the mouth, throat,
esophagus, and stomach. This problem was solved by converting the molecule
to its less acidic sodium salt (sodium salicylate), but this substance had a very
objectionable taste and many people would not take it.

As a result, the search continued for other compounds to combat pain and
fever. In 1886, phenyl salicylate and acetanilide (see Table 13-1) were both
introduced. Phenyl salicylate had the advantages of no truly objectionable taste
and no stomach irritation, but, with the hydrolysis to produce sodium salicylate,

it also released phenol, creating the danger of phenol poisoning. Acetanilide was not as effective in relieving pain and fever, and was also somewhat toxic.

The necessary breakthrough came in 1893 when Felix Hofmann, a chemist for the German company Bayer, found a practical way to synthesize acetylsalicylic acid (see Table 13-1). The compound was first made in 1853, but no one ever studied its possible medical benefits. Hofmann did, discovering that it greatly relieved his father's arthritis without causing serious side effects or having a highly objectionable taste. As a result, Bayer introduced its new product to the public in 1899, calling it aspirin. It has been the most popular pain reliever ever since.

Today it is well known that aspirin is hydrolyzed inside the body back to salicylic acid. However, since aspirin is more potent than salicylic acid in its pain-, fever-, and swelling-reducing properties, its medicinal effect cannot be the sole result of its conversion to salicylic acid. In fact, it wasn't until 1971 that anyone knew for certain why aspirin worked. John Vane, a British pharmacologist, then discovered that aspirin blocks the manufacture of prostaglandins, which are important hormonelike chemicals produced in the body. His discovery led to him sharing the 1982 Nobel Prize for medicine. (See the introduction to Experiment 25 for a discussion of the Nobel Prize.)

Prostaglandins are involved in practically every biological function, including blood circulation, blood clotting in response to injury, muscle movement, respiration, digestion, kidney function, and nerve signal transmission. They are thus extremely beneficial, but in excess amounts can trigger pain, fever, and inflammation. How they do this is still not completely clear, however. Nor is it clear exactly how aspirin blocks their manufacture.

One negative feature of aspirin is that it hinders the clotting of blood. But now this negative feature also appears to be a positive one. In short, swallowing an aspirin a day seems to reduce the occurrence of both heart attacks and strokes. Statistical studies indicate that in the United States alone, aspirin could save between 40,000 and 60,000 lives per year from heart attacks and 150,000 lives per year from strokes. The prevention or hindrance of the blood clots that cause many of the heart attacks and strokes seems to be the reason.

As a result, aspirin has become an even more versatile drug and is being called by some a "wonder" or a "miracle" drug. One aspirin company, Bayer, even has a trademark on the slogan, "The wonder drug that works wonders." Aspirin's versatility is also shown in the fact that it delays or slows the formation of cataracts, combats certain complications of diabetes, and reduces by more than 50 percent the death rate from Kawasaki's syndrome, a childhood disease involving abnormal swelling in the arteries to the heart.

But as effective and beneficial as aspirin is, it is not for everyone, as many people know. Despite the buffering that some companies put in their product, aspirin can still be a stomach irritant, and cause ulcers and gastrointestinal bleeding. Therefore, it shouldn't be taken by anyone suffering from these ailments, gout, or any type of bleeding problem. It also causes allergic reactions such as shock, severe asthma, swelling, hives, and abdominal pain in about 5 out of every 1000 people. It can even be fatal to especially sensitive people. Excessive use by people not allergic to it can cause a ringing in the ears, a hearing loss, and kidney and liver damage.

Pregnant women are also advised not to take aspirin during the last three months. It may cause birth defects, a low birth weight, maternal bleeding, and stillbirth.

Just as important, aspirin should be avoided by people under 16 who have the flu or chicken pox. The reason is Reye's syndrome, a severe liver disease which strikes between 600 and 1200 American youngsters each year, killing 300. Symptoms include fever and vomiting, which may progress to convulsions and brain damage before death. Statistical studies show a definite link between Reye's syndrome and aspirin taken by youngsters with the flu or chicken pox. The reason is not yet clear, but aspirin may trigger some of the metabolic and pathological changes accompanying the syndrome.

In cases where aspirin cannot be used, alternative analgesics and antipyretics, such as acetaminophen and ibuprofen, are available (see Table 13-1). Acetaminophen, sold over the counter as Tylenol, Datril, Anacin 3, and Panadol, is especially popular, both in tablet form for adults and a flavored liquid form for children. In fact, acetaminophen is so popular that, despite its higher cost, it holds 44 percent of the analgesic market, only slightly behind the 45 percent of the market held by aspirin. This is a tremendous improvement over 1973, when acetaminophen accounted for only 8 percent of all analgesic sales. It has been available over the counter since 1955.

Acetaminophen is the obvious choice for people who are allergic to aspirin or who suffer from stomach irritation, bleeding, or other ailments aggravated by aspirin. For people under 16 who have the flu or chicken pox, it is a must.

However, acetaminophen also has its limitations and problems. Although it relieves pain and fever as effectively as aspirin, it has absolutely no effect on swelling or inflammation. For this reason, it is not as effective as aspirin for treating arthritis, rheumatism, or certain injuries such as sprains. Furthermore, there is currently no evidence that it helps to prevent strokes, heart attacks, and cataracts as aspirin does. Some people are allergic to acetaminophen, but its biggest danger is severe and permanent liver damage, especially when taken with alcohol. People who take more than eight regular-strength tablets per day are risking this damage, and alcoholics and people with serious liver disease are advised not to take even this many. Aspirin does not cause liver damage in comparable amounts.

Ibuprofen has been available over the counter in the United States since 1984. It had previously been approved only for prescription use under the brand name Motrin, and was the fifth largest-selling prescription drug in the country.

Sold now primarily under the brand names Advil, Nuprin, and Medipren, ibuprofen has captured 11 percent of the analgesic market. Some analysts predict that it could even capture as much as 35 percent of the market, thus giving some aspirin and acetaminophen companies a major headache.

Ibuprofen, like aspirin, relieves pain, fever, and swelling or inflammation. It is thus better than acetaminophen in relieving swelling or inflammation. It is also better than either aspirin or acetaminophen in the amount needed to relieve pain, with one 200-mg ibuprofen tablet providing the same relief as two 325-mg aspirin or acetaminophen tablets. In addition, at a higher dosage of two tablets, it provides significantly more relief than four tablets of either aspirin or acetaminophen. This results from both aspirin and acetaminophen providing only a slight additional benefit at double the dosage, while ibuprofen offers a significant additional benefit. Therefore, one ibuprofen tablet is recommended for moderate pain, and two tablets are recommended for severe pain. However, increasing the number of tablets to three or more has been found to provide no additional relief.

Like aspirin and acetaminophen, ibuprofen relieves pain by lowering the level of prostaglandins. However, the mechanism by which it does this appears to be different. This helps explain why it provides a higher level of pain relief and is more effective than either aspirin or acetaminophen for treating certain kinds of discomfort, such as severe menstrual cramps.

However, ibuprofen also has its limitations and problems. Although it contains no aspirin or salicylates, people allergic to aspirin are also likely to be allergic to ibuprofen. Therefore, they cannot falsely believe they can safely take it. The drug can also aggravate high blood pressure and even cause kidney failure in susceptible people. Anyone taking diuretics or suffering from diabetes, hypertension, or impaired kidney function is advised not to take it. Children under 12 and women in their last three months of pregnancy are also advised not to take the drug, except under the supervision of a doctor. Although ibuprofen is not as likely as aspirin to promote bleeding, it still causes stomach irritation, and should be avoided by anyone suffering from ulcers. Because of the drug's potency, no more than six 200-mg tablets should be taken in one day. And finally, unlike aspirin, there is no evidence that it helps to prevent strokes, heart attacks, or cataracts.

Until 1983, phenacetin (see Table 13-1) was also a popular over-the-counter analgesic, being used in products such as Empirin, Fiorinal, and APC tablets (aspirin-phenacetin-caffeine). It was then banned by the FDA because of studies linking its use to serious kidney disease and kidney and bladder cancer. Chronic abuse, defined as taking seven or more phenacetin tablets per day, seemed to cause most of the health problems associated with the drug.

A list of the major over-the-counter analgesic products is given in Table 13-2. You are familiar with the names of most of these products, having seen them advertised everywhere as well as displayed on the shelves in drugstores and supermarkets. Many of the aspirin products also contain buffering agents, caffeine, and other ingredients. The buffering agents serve to counteract the

TABLE 13-2.

Amounts of the Analgesics and Caffeine in Some Common Products (Milligrams)

Product	Aspirin	Acetaminophen	Ibuprofen	Caffeine
Brand X aspirin	325	—	—	—
Brand X acetaminophen	—	325	—	—
Brand X ibuprofen	—	—	200	—
Advil	—	—	200	—
Anacin	400	—	—	32
Arthritis Pain Formula Anacin	500	—	—	—
Maximum Strength Anacin	500	—	—	32
Anacin 3	—	325	—	—
Maximum Strength Anacin 3	—	500	—	—
Bayer	325	—	—	—
Maximum Bayer	500	—	—	—
Eight-Hour Bayer	650	—	—	—
Children's Bayer	81	—	—	—
Bufferin	325	—	—	—
Extra Strength Bufferin	500	—	—	—
Cope	421	—	—	32
Datril	—	325	—	—
Extra Strength Datril	—	500	—	—
Ecotrin	325	—	—	—
Empirin	325	—	—	—
Excedrin	250	250	—	65
Excedrin P.M.	—	500	—	—
Medipren	—	—	200	—
Nuprin	—	—	200	—
Panadol	—	500	—	—
St. Joseph Aspirin	81	—	—	—
St. Joseph Fever Reducer	—	80	—	—
Trendar	—	—	200	—
Tylenol	—	325	—	—
Extra Strength Tylenol	—	500	—	—
Children's Tylenol	—	80	—	—
Liquid Tylenol	—	160 mg/teaspoon	—	—
Vanquish	227	194	—	33

possible acidic irritation of aspirin inside the stomach. However, a study published in the *New England Journal of Medicine* reported that buffered aspirin failed to prevent severe injury to the stomach lining, even with more antacid than most over-the-counter buffered aspirin contains. Therefore, buffering is of little or no value and only raises the cost of the product, and is perhaps also an enticement to get you to buy it. Caffeine combined with analgesics has been demonstrated to have an effect on migraine headaches. There is, however, no reliable evidence that it aids in relieving other types of pain. All it does is stimulate the user while increasing the cost of the product.

In addition, many of these products are labeled "extra strength," even though it has been demonstrated that the extra 23 to 54 percent of analgesic product provides only a slight additional amount of relief. Therefore, the use of an extra-strength product is questionable, especially when one considers that three regular-strength tablets provide almost as much analgesic as two extra-strength tablets, but at a much lower price. There is frequently a premium price associated with these extra-strength products.

Furthermore, paying more for a brand name analgesic is usually just a waste of money, since all brands sold must meet high USP (United States Pharmacopeia) standards of quality and purity. Brand X is the same analgesic made by one of the big name companies anyway, so you are receiving the same product for much less money. Company profits and the cost of advertising are frequently the reasons for any price differences.

If you are sensitive or allergic to aspirin, you should know that it is found in many other products, such as cold remedies, Alka-Seltzer, and Bromo Seltzer. Read the label on the product so that you know exactly what you are getting.

13A. SYNTHESIS OF ASPIRIN

NUCLEOPHILIC ACYL SUBSTITUTION

In this experiment, you will synthesize aspirin by the acid-catalyzed acetylation of salicylic acid with acetic anhydride.

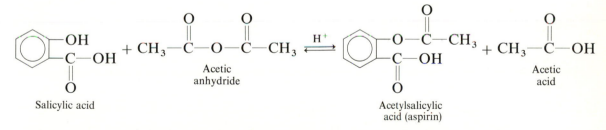

Salicylic acid

Acetic anhydride

Acetylsalicylic acid (aspirin)

Acetic acid

The reaction mechanism is nucleophilic acyl substitution. The phenol group of salicylic acid, acting as a nucleophile, attacks a carbonyl carbon atom of acetic anhydride to yield the tetrahedral intermediate I. Acid speeds up this reaction by protonating the carbonyl oxygen atom, thus making the carbonyl carbon even more susceptible to nucleophilic attack. The openness of the protonated carbonyl group, due to its flat shape, also greatly aids the reaction. Intermediate I then loses a proton to form the tetrahedral intermediate II, which then receives a proton on a different oxygen atom to form the tetrahedral intermediate III. The expulsion of acetic acid and a proton from intermediate

III then yields aspirin as the final product. Observe how the acid protonation to form intermediate III greatly aids the last reaction, since acetic acid is a much better leaving group than the basic acetate ion.

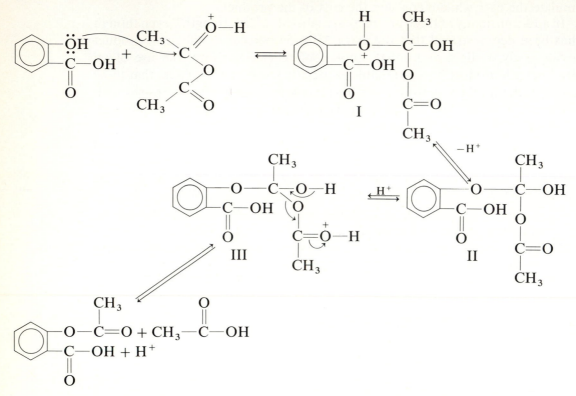

As shown below, acetic acid could be used instead of acetic anhydride for the synthesis. Acetic anhydride, however, is better because it reacts to form acetate esters much more rapidly than acetic acid. This results from the acetic acid formed in the acetic anhydride synthesis being a better leaving group than water in the last step of the mechanism. Using acetic anhydride thereby avoids a long period of refluxing.

Observe that the salicylic acid necessary for the synthesis contains both a hydroxyl and a carboxyl group. Therefore, it can undergo two different types of esterification reactions, acting either as the alcohol or the carboxylic acid. In the synthesis of aspirin the alcohol group is used, but in the presence of excess methanol, the carboxylic acid portion reacts to yield methyl salicylate, which is the flavor wintergreen (see Experiment 9).

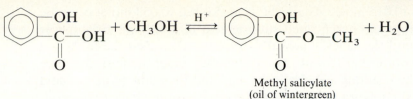

Methyl salicylate
(oil of wintergreen)

The two functional groups in salicylic acid complicate the synthesis of aspirin somewhat because the hydroxyl group of one molecule can react with the carboxyl group of another, resulting in a small amount of undesired polymer.

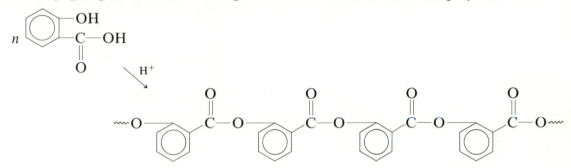

Fortunately, the small amount of polymer is easily removed, since aspirin forms a water-soluble sodium salt in sodium bicarbonate solution, whereas the polymer does not. Acidifying the filtered sodium bicarbonate solution then allows the aspirin to be recovered.

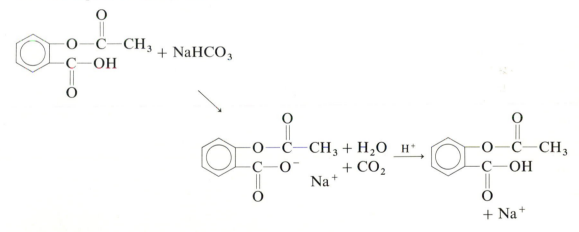

PROCEDURE

CAUTION: Acetic anhydride is a toxic and irritating substance. Avoid breathing it and getting it on your skin.

Also remember that concentrated sulfuric acid is highly corrosive. Handle it with care.

In addition, both ethyl ether and petroleum ether are highly flammable. Avoid any flames when using them.

Place 2.0 grams (0.015 mole) of salicylic acid, 5.0 mL (0.053 mole) of acetic anhydride, and 5 drops of concentrated sulfuric acid in a 125-mL Erlenmeyer flask. Swirl gently until the salicylic acid dissolves and then heat the flask in a steam bath or a hot-water bath for at least 10 minutes. Cool the flask in an ice bath to crystallize the resulting acetylsalicylic acid and then add 50 mL of cold water with continued cooling to complete the crystallization. Collect the product by suction filtration and wash the crystals several times with small portions of **ice-cold** water.

To remove any polymer present, place the crystals in a 150-mL beaker and add 25 mL of saturated sodium bicarbonate solution. Swirl, and when the reaction has ceased, filter the solution with suction. Transfer the filtrate back to the clean 150-mL beaker and slowly add 3 M HCl until the mixture is acid to litmus and the aspirin has precipitated. **Do not dip the litmus paper into the mixture.** Cool the mixture in an ice bath, collect the product by suction filtration, and wash it with a small amount of **ice-cold** water. Allow the crystals to air dry and weigh them. Record the result in your notebook.

Recrystallize the aspirin by dissolving it in ether, adding an equal volume of petroleum ether, and then allow the solution to stand undisturbed in an ice bath. Collect the product by suction filtration, allow to dry, and determine the yield, melting point, and literature melting point. Record the results in your notebook. Also determine the theoretical and percentage yields of the final product, and show the calculations.

Commercial aspirin tablets are pressed together with a small amount of inert binding material. To determine what this material is, boil a crushed aspirin tablet in several milliliters of water in a test tube. Cool the liquid, add a drop of iodine solution, and record the result. From your previous knowledge of

FIGURE 13-1

The infrared spectrum of acetylsalicylic acid in a KBr pellet. Spectrum courtesy of Sadtler Research Laboratories, division of Bio-Rad Laboratories, Inc., copyrighted 1974.

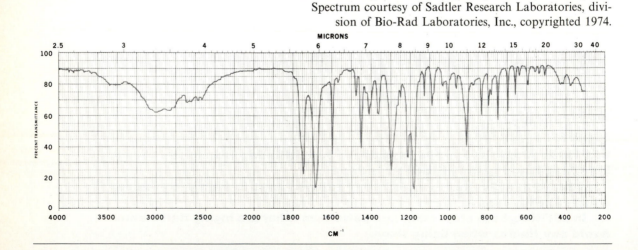

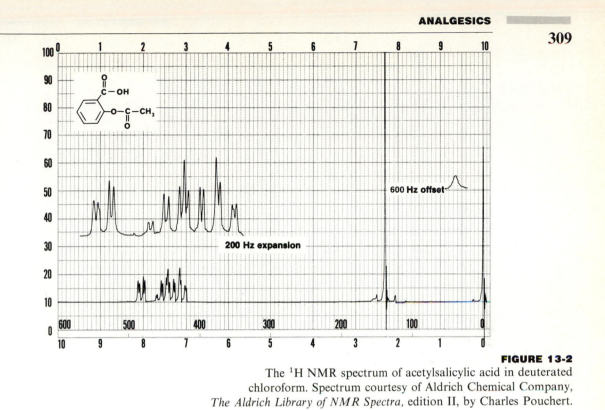

FIGURE 13-2

The ^{1}H NMR spectrum of acetylsalicylic acid in deuterated chloroform. Spectrum courtesy of Aldrich Chemical Company, *The Aldrich Library of NMR Spectra*, edition II, by Charles Pouchert.

chemistry in other courses, what is the binding material? Try the same test on a small quantity of the aspirin you synthesized, and record the result. Explain any difference in the two results.

At your instructor's option, obtain the infrared spectrum of your product in a KBr pellet and the ^{1}H NMR spectrum in deuterated chloroform. Compare your spectra with those shown in Figures 13-1 and 13-2. When finished, hand in your product in a properly labeled container to your instructor.

13B. SYNTHESIS OF PHENACETIN

NUCLEOPHILIC ACYL SUBSTITUTION

In this experiment, you will synthesize the analgesic phenacetin by the acid-catalyzed reaction of *p*-phenetidine with acetic anhydride. However, since large amounts of acid would hinder the reaction by converting *p*-phenetidine to its amine salt, sodium acetate will be added to form a buffer solution.

The reaction mechanism is nucleophilic acyl substitution. The amino group of *p*-phenetidine, acting as a nucleophile, attacks a carbonyl carbon atom of acetic anhydride to yield the tetrahedral intermediate I. Acid speeds up this

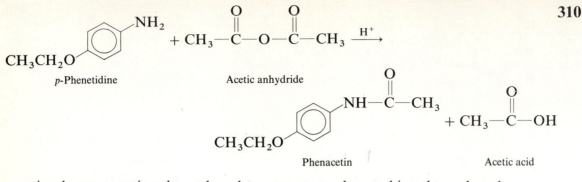

reaction by protonating the carbonyl oxygen atom, thus making the carbonyl carbon even more susceptible to nucleophilic attack. The openness of the protonated carbonyl group, due to its flat shape, also greatly aids the reaction. Intermediate I then loses a proton to form the tetrahedral intermediate II, which then receives a proton on a different oxygen atom to form the tetrahedral intermediate III. The expulsion of acetic acid and a proton from intermediate III then yields phenacetin as the final product. Observe how the acid protonation to form intermediate III greatly aids the last reaction, since acetic acid is a much better leaving group than the basic acetate ion.

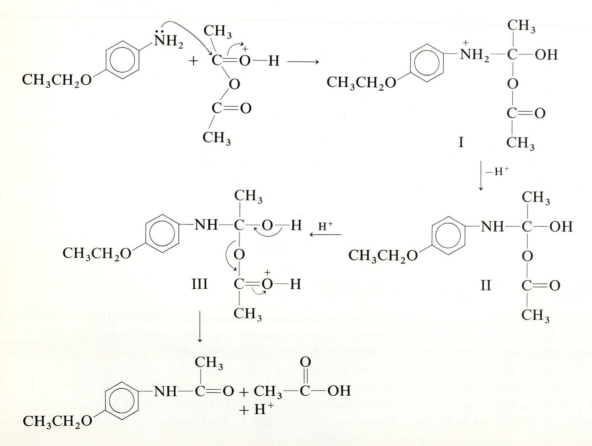

As shown below, acetic acid and acetyl chloride could be used instead of acetic anhydride for the synthesis. Acetic acid, however, reacts with the amine to form the amine salt, which requires severe heat to react further. Acetyl chloride reacts quite vigorously and liberates HCl, which converts half of the amine to its hydrochloride salt, making it incapable of reacting further under the reaction conditions.

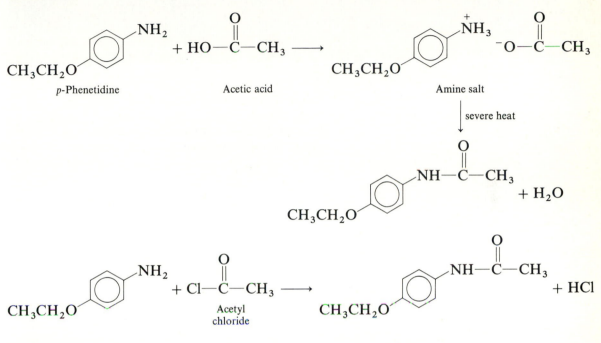

PROCEDURE

CAUTION: *p*-Phenetidine and acetic anhydride are both toxic and irritating substances. Avoid breathing them and getting them on your skin.

Dissolve 2.3 grams (0.017 mole) of sodium acetate trihydrate in 10 mL of water. Set aside for later use.

Place 2.0 mL (0.015 mole) of recently distilled *p*-phenetidine in 30 mL of water in a 100-mL beaker. Add 6.0 mL (0.018 mole) of 3.0 *M* HCl and stir. If some of the amine does not dissolve, add a little more hydrochloric acid. If the *p*-phenetidine has color because of sitting after the distillation, add a small amount of decolorizing carbon to the solution, warm the mixture with stirring on a steam bath or hot plate, then gravity filter the mixture into a 125-mL Erlenmeyer flask, using a fluted filter paper.

Heat the *p*-phenetidine solution on a steam bath or hot plate; then add 2.5 mL (0.026 mole) of acetic anhydride with stirring. Add the sodium acetate solution all at once and continue stirring.

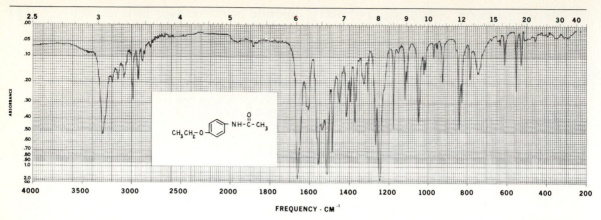

FIGURE 13-3
The infrared spectrum of phenacetin in a KBr pellet. Spectrum courtesy of Sadtler Research Laboratories, division of Bio-Rad Laboratories, Inc., copyrighted 1966.

FIGURE 13-4
The ^{1}H NMR spectrum of phenacetin in a combination of dimethyl sulfoxide-d_6 and deuterated chloroform. Spectrum courtesy of Aldrich Chemical Company, *The Aldrich Library of NMR Spectra,* edition II, by Charles Pouchert.

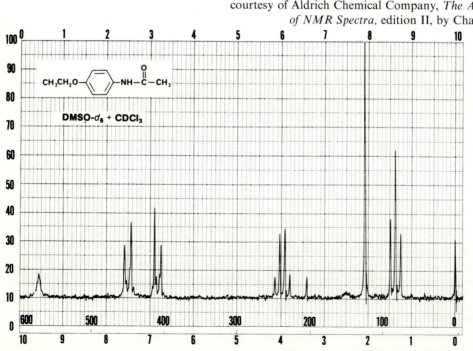

Cool the reaction mixture in an ice bath and stir until the phenacetin crystallizes. Collect the product by suction filtration and wash it with a small amount of ice-cold water.

Recrystallize the product from the minimum amount of boiling water, using decolorizing carbon if necessary. Allow the hot solution to slowly cool and, after crystals form, cool thoroughly in an ice bath. Collect the purified product by suction filtration and wash with a small amount of cold water. Allow the crystals to thoroughly dry and determine the yield, melting point, and literature melting point. Record the results in your notebook. Also determine the theoretical and percentage yields of the final product, and show the calculations.

At your instructor's option, obtain the infrared spectrum of your product in a KBr pellet and the ^{1}H NMR spectrum in either dimethyl sulfoxide-d_6 or a combination of dimethyl sulfoxide-d_6 and deuterated chloroform. Compare your spectra with those shown in Figures 13-3 and 13-4. When finished, hand in your product in a properly labeled container to your instructor.

13C. THIN-LAYER CHROMATOGRAPHIC ANALYSIS OF OVER-THE-COUNTER ANALGESICS

In this experiment, you will determine by thin-layer chromatography the components present in Anacin, Excedrin, Advil, and Tylenol. The active ingredients in these products are listed in Table 13-2, so your results should agree with this table. Your instructor may also ask you to determine the components present in an unknown analgesic, which he or she may have personally created.

Determinations will be made by comparing the positions of the components on the developed thin-layer plate with known samples of pure aspirin, acetaminophen, ibuprofen, and caffeine. You can observe the spots by using an ultraviolet lamp and a fluorescent indicator on the thin-layer plate. The difference in appearance of the spots under ultraviolet light may also aid in their identification.

PROCEDURE

If necessary, read again the section on thin-layer chromatography, page 139.

For application of the analgesic products and the knowns, at least eight to nine micropipets will be required. These can be prepared by softening the middle of a melting-point capillary tube in a Bunsen burner flame and then pulling the ends apart about 4 to 5 cm. The thin portion in the center, which now has a diameter of about one-tenth its previous size, can be carefully broken to yield two micropipets. You will also need a solvent chamber and cover, and a 10 cm × 10 cm commercial, plastic-backed silica gel G plate with a fluorescent indicator.

To analyze the commercial products, you must remove the binder portion and any inorganic buffers. Crush each of the four tablets on pieces of weighing paper and place the powder into four separate, **labeled** test tubes. Add 6 mL of a 50:50 methylene chloride–methanol solution to each tube and **gently** heat the tubes in a steam bath or a warm-water bath for 2 to 3 minutes. The binder portion of the tablets and any inorganic buffers will not dissolve, but the analgesics and any caffeine present will go into solution. Do not allow any solvent to boil away. After the solid settles, the solution is ready for the analysis.

To prepare the chromatogram for developing, place the silica gel G plate onto a larger-size piece of paper. (This is necessary because cleanliness of the surroundings is important for good results.) About 1.5 cm from the bottom of the plate, lightly draw a straight **pencil line (no pen)** across the plate. On this line, mark at least nine evenly spaced dots. Four of these dots will be for the solutions you just prepared, four will be for already prepared 5% standard solutions of pure caffeine, pure aspirin, pure acetaminophen, and pure ibuprofen in 50:50 methylene chloride–methanol, and one will possibly be for a solution of an unknown containing any number of the four ingredients. One possible sequence for arranging the solutions is shown in the diagram. Be sure to indicate in pencil for each dot which solution has been placed there to avoid any possible mixup.

Using a **different** micropipet for each solution, place the tip of the micropipet into the solution to be applied. By capillary action, some of the liquid will be drawn up. Starting at the left side, quickly touch the tip of the pipet to the plate at the appropriate dot so that a spot no larger than 1 to 2 mm in diameter is produced. **Do not** keep the pipet on the plate too long or its entire contents will be delivered and the spot will be too large. Large spots are unsatisfactory, because during development they will tend to streak or tail, giving poor separations or uncertain R_f values. You may want to practice the technique beforehand on a piece of paper to get it perfected. Since the commercial silica gel G plates are not cheap, you may be given only one for the experiment. Therefore, be careful not to make any mistakes. If any spots are too large, another spot of the same solution should be applied elsewhere.

After all the spots have dried completely, place the plate into the solvent chamber to which has been added ethyl acetate. Be sure that not so much solvent has been added that the level is above the level of the spots. Otherwise they will dissolve in the solvent. Cover the top of the solvent chamber and wait until the ethyl acetate rises to near the top of the plate. During this time, other experiments or activities may be continued or completed.

When the ethyl acetate is near the top, remove the plate and mark its level with a pencil. After the chromatogram has dried, observe it under an ultraviolet light in a darkened room or area. **Try not to look directly at the ultraviolet light.** Lightly mark the observed spots with a pencil. Compute the R_f values and determine what ingredients are present in the commercial analgesic products and the unknown. Record the results in your notebook. Do your results

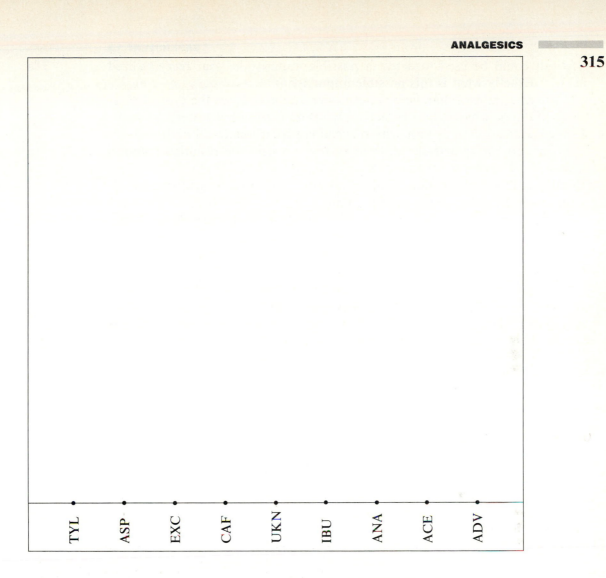

TYL ASP EXC CAF UKN IBU ANA ACE ADV

for the commercial products agree with Table 13-2? You will find that using ethyl acetate will cause the aspirin to "tail" considerably, but no better simple and safe solvent system has been found. One more complex system uses benzene, but this is a known carcinogen. When finished, attach your chromatogram to your notebook.

QUESTIONS

1. In the aspirin experiment, why does the treatment of the crude product with sodium bicarbonate solution remove any polymeric material present but not any unreacted salicylic acid reagent?
2. A general test for phenols is to add to the phenol a 1% ferric chloride solution, which results in a color ranging from red to violet. Explain how

this test could be used to detect a possible impurity in your recrystallized aspirin. Exactly what is this possible impurity?

3. Considering the fact that aspirin is an ester and undergoes the typical reactions of esters, why would one not recrystallize it from hot water?

4. In the purification of aspirin, why was sodium bicarbonate solution used instead of sodium hydroxide solution, and why wasn't the resulting solution heated?

5. What possible complication would there be if you tried to synthesize acetaminophen by the acetylation of p-aminophenol?

6. Explain why p-phenetidine is much more soluble in dilute aqueous acid than phenacetin is.

7. Explain why phenacetin can be recrystallized from hot water in contrast to aspirin.

8. Explain how acetaminophen could be made very water soluble.

9. If a less polar solvent than ethyl acetate had been used for the thin-layer chromatography analysis, how would the R_f values for the various analgesic components have been affected? Why?

EXPERIMENT

14

CHEMOTHERAPY AND THE SULFA DRUGS: THE SYNTHESIS OF SULFANILAMIDE

- ■ **ELECTROPHILIC AROMATIC SUBSTITUTION**
- ■ **PROTECTION OF A GROUP**
- ■ **NUCLEOPHILIC SUBSTITUTION**
- ■ **HYDROLYSIS**

> *Diseases of their own accord,*
> *But cures come difficult and hard.*

Samuel Butler, *The Weakness and Misery of Man,* I, p. 82

INTRODUCTION

People have been plagued by infection and disease ever since the beginning of civilization. From leprosy in biblical times to bubonic plague in Europe to cancer and AIDS today, infection and disease in all their forms have caused pain, suffering, and death. For centuries, people have sought relief and a cure from the afflictions that ail them, but only in recent years through the advancement of modern science have the answers started to come forth.

One area of science that has proven to be quite successful in treating and curing some ailments during the past 50 years and offers even greater promise

317

for the future is **chemotherapy**. This term may be defined as the use of chemical agents or drugs to selectively destroy infectious organisms or diseased areas of the body without simultaneously affecting or damaging the patient. The term is also used more loosely to mean any use of drugs in the treatment of disease. Before chemotherapy, people had to rely on sanitation and immunological methods to defend themselves from the ravages of disease. These methods proved effective against some afflictions but failed miserably against others. All of the possible approaches that could be effectively utilized had not yet been discovered.

The era of modern chemotherapy began with the work of Paul Ehrlich, who, around 1900, began the search for what he called "magic bullets," which would selectively hit and destroy only the infection- and disease-producing components in the body. While working as a medical student, Ehrlich had been impressed by the ability of certain dyes to stain tissues selectively. Thinking this might be the result of a chemical reaction between the dye and the tissue, Ehrlich searched for compounds that would react selectively only with microorganisms. In this way, he hoped to find compounds that would be lethal specifically to infectious microorganisms. In his search, Ehrlich discovered Salvarsan and Neosalvarsan, two organoarsenic compounds used successfully in treating syphilis and sleeping sickness. Before this, the only chemicals effective against disease were quinine (for malaria), the alkaloids present in ipecacuanha (for dysentery), and mercury (for syphilis). For his contributions to medicine, Ehrlich was awarded the Nobel Prize in 1908.

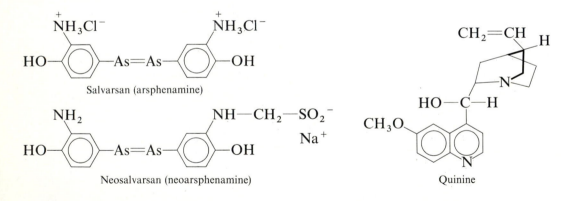

Salvarsan (arsphenamine)

Neosalvarsan (neoarsphenamine)

Quinine

But for the next 30 years, advancement in chemotherapy did not occur. Thousands and thousands of compounds, including many dyes, were tested in a search for the longed-for "magic bullets." But practically all the work was in vain, because very few compounds showed any promising effects.

It took what was almost a personal tragedy to provide the necessary breakthrough. The young daughter of Dr. Gerhard Domagk, a physician employed by a German dye manufacturer, developed a streptococcal infection from a pin prick. As his little girl grew steadily worse and was close to death, Domagk, with very little hope and no medical alternatives, in desperation decided to give her a dose of a red dye called Prontosil. It had been found earlier that Prontosil exhibited remarkable antibacterial action when it was used to dye wool, and tests with mice had shown that the dye inhibited the growth of streptococci. Domagk's gamble paid off and within a short time, his little girl recovered. Later Domagk showed that Prontosil was active against a wide array of bacteria, and for this important discovery he was awarded the Nobel Prize in 1939. Unfortunately, a direct order from Adolf Hitler prevented Domagk from accepting this award.

Prontosil was found to be active against bacteria as a direct result of its breakdown inside the body to produce sulfanilamide, which is the actual active agent.

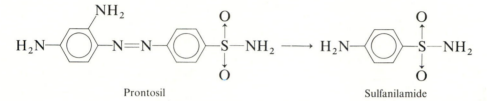

Prontosil   Sulfanilamide

This discovery set in motion the preparation of thousands of sulfanilamide-related compounds in the hope of finding even better antibacterial agents. With the structure of sulfanilamide being modified in every possible way, the best results were attained with compounds in which the amino group bonded to sulfur had one of its hydrogens replaced with an aromatic heterocyclic ring. Among the most successful compounds were the following:

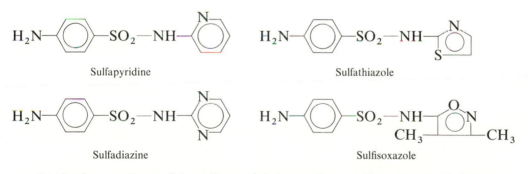

Sulfapyridine   Sulfathiazole

Sulfadiazine   Sulfisoxazole

As the first really useful antibacterial drugs, these sulfa compounds became the wonder drugs of their day. During World War II, sulfathiazole prevented infection in hundreds of thousands of wounded servicemen and civilians. Other sulfa drugs were found to be effective against pneumonia, malaria, tuberculosis, leprosy, scarlet fever, meningitis, and infections of the intestinal and urinary tracts.

The clue as to how sulfa drugs function in the body came in 1940 when it was discovered that *p*-aminobenzoic acid (PABA) inhibits the action of sulfanilamide. Because of the structural similarity between PABA and sulfanilamide, it was concluded that the two compounds must somehow compete with each other in some essential metabolic process. Further study showed that sulfanil-amide does not kill bacteria but inhibits their growth by taking the place of PABA on the active site of a bacterial enzyme that incorporates PABA into the molecule folic acid. The bacterial enzyme thus cannot produce folic acid, which is needed by the bacteria to synthesize nucleic acids essential for growth. As a result, the bacterial growth and reproduction is halted until the body's immune system can rally to destroy the invader.

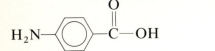

p-Aminobenzoic acid (PABA)

Sulfanilamide

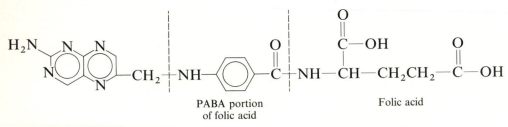

PABA portion
of folic acid

Folic acid

Within the past 40 years, sulfa drugs have been replaced to a large extent by antibiotics such as penicillin, terramycin, chloromycetin, aureomycin, and strep-tomycin, which have the following structures:

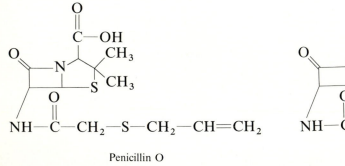

Penicillin O

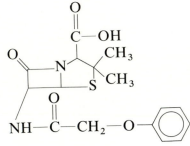

Penicillin V

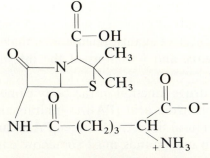

Penicillin N

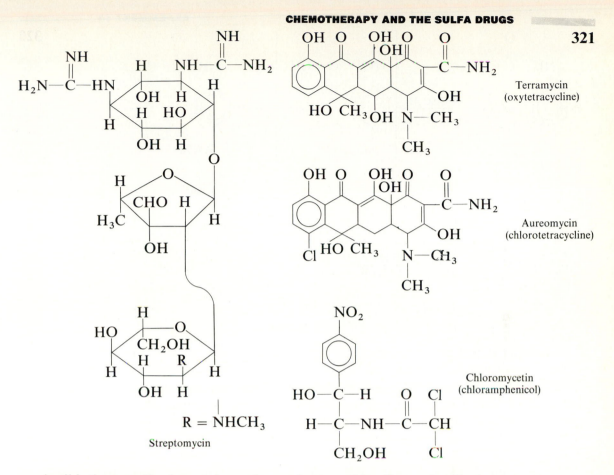

Terramycin
(oxytetracycline)

Aureomycin
(chlorotetracycline)

Chloromycetin
(chloramphenicol)

R = NHCH₃

Streptomycin

Antibiotics usually do not have the unpleasant side effects of sulfa drugs. They seem to operate by preventing the bacteria from synthesizing their rigid cell walls. Since animal cells do not have these rigid cell walls, they are not harmed by these drugs. However, because living things can change, many bacterial strains have developed complete resistance to the antibiotics that once destroyed them. It thus appears that mankind's fight against infectious organisms will continue in the future.

At this time, chemotherapy appears to be mankind's best hope for the future. Cancer, AIDS, arthritis, mental disorders, allergies, hemophilia, gout, cystic fibrosis, muscular dystrophy, and multiple sclerosis are just some of the disorders that may someday become a thing of the past through chemotherapy. Just consider what it has done in the last 45 years. Consider what all of medicine has done in the last 100 years and simply take it from there.

BACKGROUND

In this experiment, you will prepare the simplest and first known sulfa drug, sulfanilamide, from acetanilide. The synthesis is a three-step sequence and can be carried out without purification of intermediates.

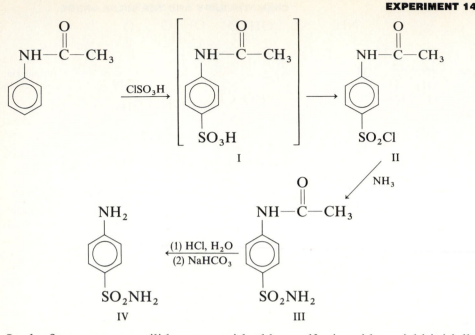

In the first step, acetanilide reacts with chlorosulfonic acid to yield initially
p-acetamidobenzenesulfonic acid (I), which then reacts with additional chlorosulfonic acid to yield p-acetamidobenzenesulfonyl chloride (II). The first reaction
is an example of electrophilic aromatic substitution, where the acetamido group
of acetanilide directs the incoming group almost exclusively to the para position.

Since the final sulfa drug contains an amino group on the benzene ring
rather than an acetamido group, you may be wondering why aniline is not used
in this first step. Three problems would arise if this were the case. First, the
highly acidic chlorosulfonic acid would protonate the amino group of aniline
to make it a **meta** rather than a para director.

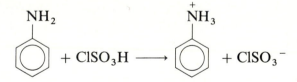

Second, even if the amino group were not protonated, the chlorosulfonic
acid would react preferentially with the amino group rather than with the ring.

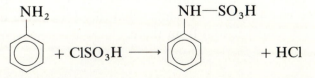

Third, even if the para position could have a sulfonyl chloride group, say
by starting with sulfanilic acid, the sulfonyl group of one molecule could attack the amino group of another to form a sulfonamide linkage with possible
polymerization.

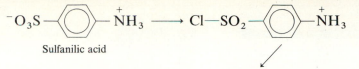

Sulfanilic acid

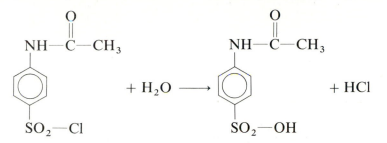

All these difficulties are eliminated if we "protect" the amino group by acetylating it. In a later step, the "protecting" group is then removed after its function is fulfilled.

The intermediate sulfonyl chloride (II) is next converted to the amide (III) by reaction with aqueous ammonia. Hydrogen chloride is also a product in the reaction, but it can be neutralized with excess ammonia. A side reaction in this step is the hydrolysis of the sulfonyl chloride to give the sulfonic acid. This lowers the yield of product but does not cause any difficulty in purification.

In the final step, the "protecting group" on the aromatic amine is removed by acid-catalyzed hydrolysis of the **carboxylic acid amide** to give the sulfa drug as the salt. This reaction proceeds by a nucleophilic acyl substitution mechanism. Notice that of the two amide groups present, **only** the acetamido group is cleaved; the sulfonamide is **not**. This is due to both steric and electronic factors in the transition state of each reaction. The salt of the drug is then easily converted to sulfanilamide itself (IV) by the addition of the base sodium bicarbonate.

PROCEDURE

CAUTION: Chlorosulfonic acid is an extremely hazardous and corrosive chemical that causes severe burns and must be handled with extreme care. Since it reacts violently with water, dry glassware must be used. Plastic gloves are highly recommended and, as always, safety glasses must be worn. If any of the reagent should be spilled on the skin, wash it off immediately with water and notify your instructor.

p-Acetamidobenzenesulfonyl chloride

Prepare a trap as shown in Figure 14-1 to absorb the hydrogen chloride gas formed in the reaction. The funnel must be placed just below the surface of the

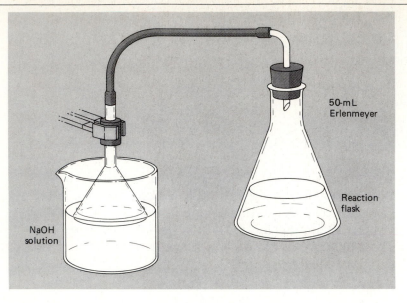

FIGURE 14-1
Hydrogen chloride trap.

sodium hydroxide solution so that all the gas evolved will dissolve in the solution and be neutralized.

Place 4.0 grams (0.030 mole) of dry acetanilide into a **dry** 50-mL Erlenmeyer flask and melt the solid by gentle heating with a flame. Swirl the flask containing the molten liquid so that as it cools the solid will form a thin layer on the lower walls and bottom of the flask. Cool the flask in an ice bath. To the solidified material add, all at once, 10 mL (0.15 mole) of chlorosulfonic acid. Immediately connect the flask to the gas trap and swirl the mixture until part of the solid has dissolved and hydrogen chloride is being evolved rapidly. Remove the flask from the ice bath, but be prepared to again cool it if the reaction should become too vigorous. After about 10 minutes the reaction should subside, and only a small amount of solid should remain. Heat the flask for 10 to 15 minutes with a steam or hot-water bath, keeping the trap in place, and then cool in an ice bath. Remove the trap and then, **in the hood**, slowly and cautiously pour the mixture onto 50–60 grams of crushed ice in a beaker. Stir the precipitated product for a few minutes, and then collect it by suction filtration. Since the product formed reacts slowly with water, this procedure **must** be done as quickly as possible and the next part of the experiment **must** be completed in the same lab period.

p-Acetamidobenzenesulfonamide

Transfer the crude product from above to a 50-mL Erlenmeyer flask and add 14 mL (0.21 mole) of concentrated ammonium hydroxide and 10 mL of water.

Using a hot plate in the hood, heat the mixture and **carefully** maintain it just below the boiling point for 5 minutes with stirring. During this time, you will see a change as the granular sulfonyl chloride is converted to the more pasty sulfonamide. Cool the mixture well in an ice bath and collect the product by suction filtration.

Sulfanilamide

Transfer the moist product from above to a 50-mL round-bottom flask and carefully add 15 mL (0.045 mole) of 3.0 M hydrochloric acid. Attach a reflux condenser and gently boil the mixture for 20 minutes. After 10 minutes, the solid should have dissolved. Cool the mixture to room temperature; if any solid material appears, reheat the mixture to boiling for several minutes so that no solid appears upon cooling. If the cooled solution is not colorless, swirl it with a little decolorizing carbon and filter with suction. Place the solution in a beaker and slowly add a saturated solution of sodium bicarbonate with stirring until the mixture is no longer acid to litmus. Do **not** dip the litmus paper into the mixture. Cool the mixture thoroughly in an ice bath and collect the precipitated sulfanilamide by suction filtration. Recrystallize the product from water, using about 10 to 12 mL of water per gram of crude product. Allow the material to dry until the next lab period. Determine the weight, melting point, and literature melting point, and record the results in your notebook.

At your instructor's option, obtain the infrared spectrum of your product in a KBr pellet, and the ^{1}H NMR spectrum in a 50–50 by volume mixture of deuterated chloroform and dimethyl sulfoxide-d_6. Compare your spectra with those shown in Figures 14-2 and 14-3.

FIGURE 14-2
The infrared spectrum of sulfanilamide in a KBr pellet.

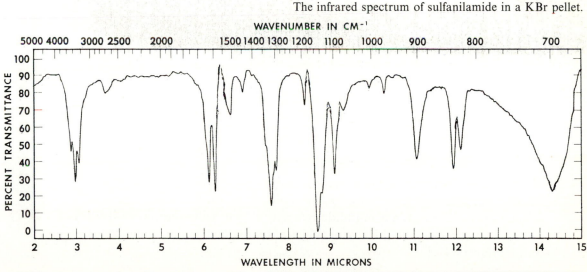

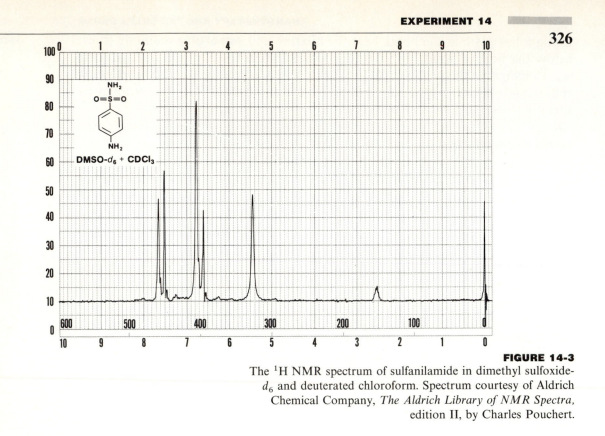

FIGURE 14-3
The ^{1}H NMR spectrum of sulfanilamide in dimethyl sulfoxide-d_6 and deuterated chloroform. Spectrum courtesy of Aldrich Chemical Company, *The Aldrich Library of NMR Spectra,* edition II, by Charles Pouchert.

Determine the theoretical and percentage yields of your final product, showing the calculations, and submit it in a properly labeled container to your instructor.

QUESTIONS

1. Why is more than 1 mole of chlorosulfonic acid required for each mole of acetanilide?
2. Give an equation showing how excess chlorosulfonic acid is decomposed in this experiment.
3. What is the relative rate of reactivity of water, ammonia, and hydroxide ion toward the sulfonyl chloride group?
4. Write the equation for what would happen if the *p*-acetamidobenzenesulfonyl chloride were not combined with ammonia in the same laboratory period.
5. Assume that not all of the acetanilide reacted to form sulfanilamide. Given the workup procedure, could any of this acetanilide contaminate the crude sulfanilamide? Why?

6. Give the complete, step-by-step mechanism for the acid-catalyzed hydrolysis of *p*-acetamidobenzenesulfonamide to form sulfanilamide.
7. Why is taking two aspirin for a headache or bicarbonate of soda for an upset stomach *not* considered to be chemotherapy?
8. Why is mercury or Salvarsan no longer used in the treatment of syphilis?
9. Do you think antibiotics are the ultimate answer for controlling infection and disease? Why or why not?
10. What are several problems that have arisen with drugs used in chemotherapy?

15

BARBITURATES: THE SYNTHESIS OF BARBITURIC ACID

- **HETEROCYCLIC COMPOUNDS**
- **NUCLEOPHILIC ACYL SUBSTITUTION**
- **USE OF SODIUM METAL**

Sleep that knits up the ravell'd sleave of care,
The death of each day's life, sore labour's bath,
Balm of hurt minds, great nature's second course,
Chief nourisher in life's feast.

Shakespeare, *Macbeth,* Act II, Scene 2

INTRODUCTION

Much has been written about sleep, that necessary substance that knits the ravelled mind, revitalizes the weary, and soothes the savage beast. Most people are aware of the importance of sleep in their lives and look forward to a period of peaceful, restful slumber. However, for various reasons many individuals fail to get the amount or the kind of sleep that they desire and need. The end result can be bloodshot eyes that burn or ache, bags or discolored areas beneath the eyes, a fuzzy mind, a short temper, and frayed nerves.

329

For some individuals, the inability to sleep is the result of their daily routine and can be easily corrected. Daytime napping, large meals or stimulants, such as tea or coffee, before bedtime, and strenuous mental and physical activity before bedtime are some things that can be avoided to produce restful sleep.

Unfortunately, for others insomnia is caused by more serious reasons. The person may have a physical disorder such as arthritis, ulcers, or hyperthyroidism, or a psychological disorder such as anxiety or mental depression. The person may have lost a loved one, or may be suffering from financial, marital, or job difficulties.

As a remedy for insomnia, a number of antidepressant and sleeping aids have come onto the market. Among the most effective of these products are the **barbiturates**, which produce sleep by depressing the activity of the brain and the rest of the nervous system. Barbiturates are chemical compounds derived from **barbituric acid**, which, as shown below, is a heterocyclic molecule that can be written either as a triketo compound or as the tautomeric, aromatic compound 2,4,6-trihydroxypyrimidine.

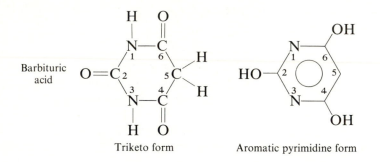

Barbituric acid

Triketo form   Aromatic pyrimidine form

The barbiturates, as derivatives of barbituric acid, have both hydrogens at position 5 replaced by alkyl or aryl groups. They are, therefore, incapable of existing in the aromatic pyrimidine form and are written in the triketo form. Approximately 2500 barbiturates have been synthesized and clinically studied, and about 50 have been placed on the market. Some of the more common ones are listed in Table 15-1 according to their duration of action.

Generally, lengthening the alkyl side chains or introducing double bonds decreases the duration of action. In addition, the depressant activity decreases when the alkyl groups become larger than five to six carbon atoms.

In full medical doses the barbiturates all induce sleep, while in lesser amounts they have a sedative or calming effect. People who have difficulty falling asleep are usually given short acting barbiturates like secobarbital, while those who wake up repeatedly may be given the longer acting type. Some barbiturates such as phenobarbital have other medical benefits by helping to prevent convulsions or epileptic seizures. Others are used by doctors as anesthetics prior to surgery or are given to induce sleep before the introduction of another anesthetic.

TABLE 15-1.
Common Barbiturates

Generic name	Brand name	Structure	Duration of action (hours)
Barbital	Veronal	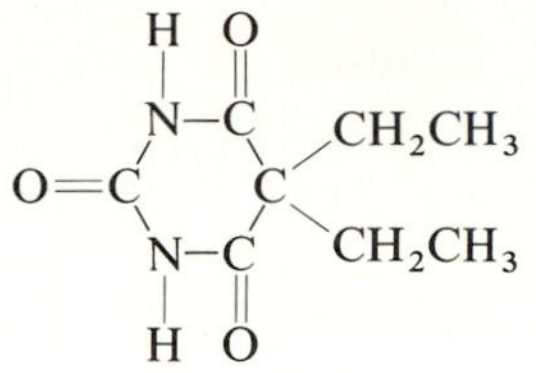	6–24
Phenobarbital	Luminal	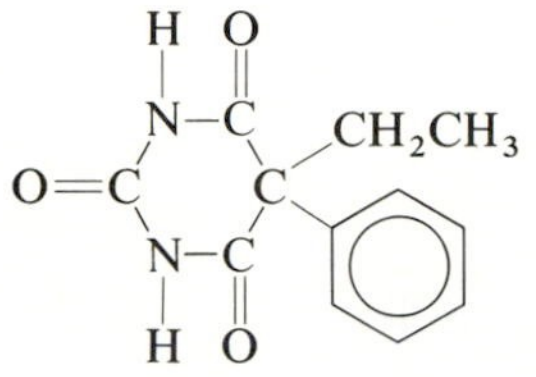	6–24
Amobarbital	Amytal	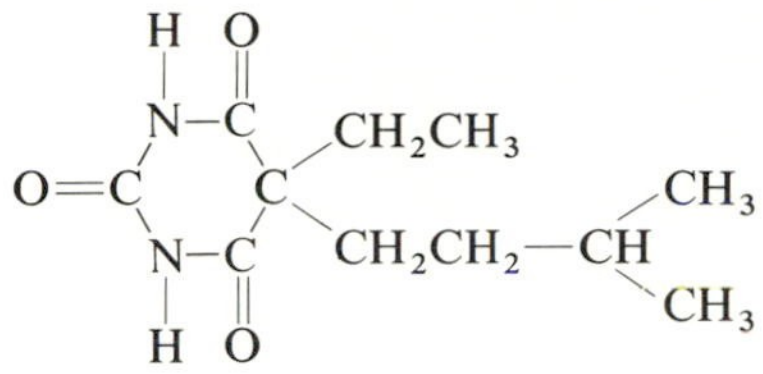	5–12
Pentobarbital	Nembutal	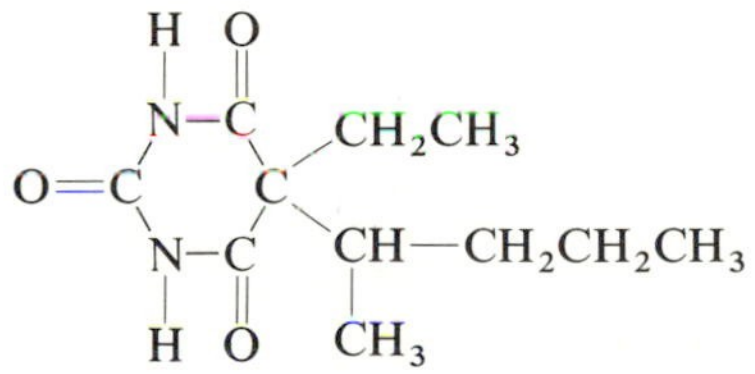	3–8
Secobarbital	Seconal	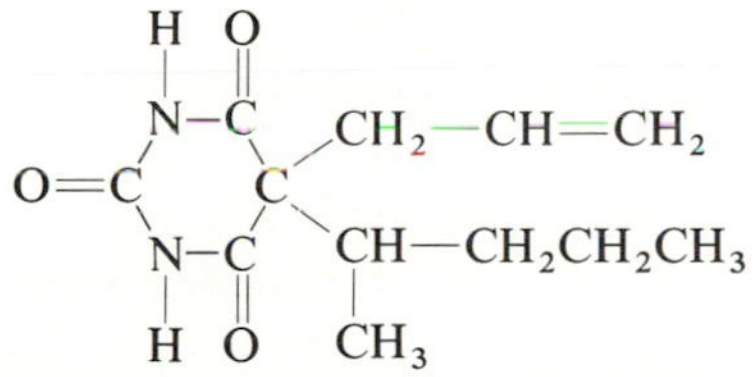	3–8

Useful for this latter purpose is sodium pentothal, the sodium salt of thiopental, which exists as a resonance hydrid of the following structures. At lower dose levels, sodium pentothal has been called the "truth serum." A person with the correct amount of the drug undergoes a type of twilight sleep or hypnosis in which there is very little self-control or willpower. Under these conditions, the subject cannot hide truthful answers, no matter how hard he or she tries. The nonbarbiturate scopolamine has also been used in this way as a truth drug.

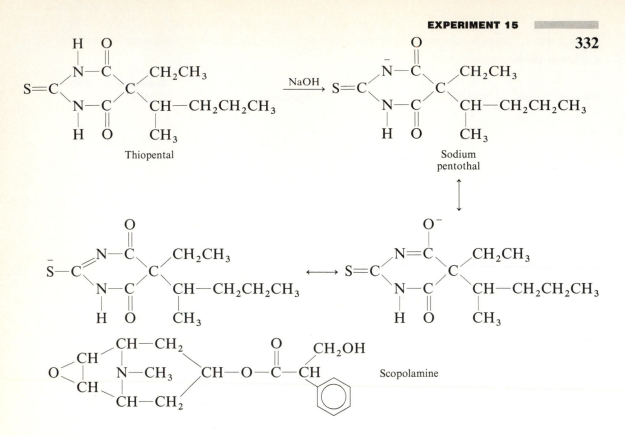

As useful as barbiturates are, they are not a cure for insomnia and should be taken for this purpose only temporarily. Their prolonged consumption can result in a person developing a tolerance, which means that increasingly larger doses are required to achieve the same sleep-producing effect. In these large amounts, the person easily becomes addicted to the drug, and without it suffers withdrawal symptoms of insomnia, nausea, tremors, delirium, convulsions, vomiting, and possible death. Large amounts of the drug also can be fatal, especially when combined with alcohol, and an overdose of sleeping pills is a common form of suicide. Because of these potential dangers, most states have laws forbidding the retail sale of barbiturates without a prescription. But when taken in proper doses, as directed by a doctor, barbiturates are among the safest drugs known.

BACKGROUND

In this experiment, you will synthesize barbituric acid, often regarded as the parent compound of the barbiturates. However, as a parent, barbituric acid is unlike the barbiturates in that it has little sedative or sleep-producing properties, and can therefore be synthesized without encountering legal difficulties. To attain the biological properties of the barbiturates, both hydrogens at position 5 of barbituric acid must be replaced with alkyl or aryl groups.

Barbituric acid doesn't even have to be the starting material or an intermediate in the synthesis of the barbiturates. Most of them can be made by the same method in which you will prepare barbituric acid, with one reactant merely being replaced by another.

Barbituric acid is a valuable compound in the manufacturing of plastics and pharmaceuticals. In particular, it is used in the production of riboflavin or vitamin B_2. That portion of the riboflavin molecule derived from barbituric acid should be evident from its structure.

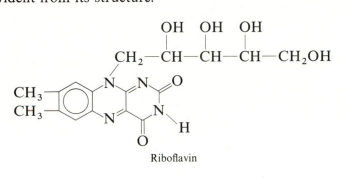

Riboflavin

In this experiment, the synthesis of barbituric acid will be accomplished by the reaction of urea with diethyl malonate in the presence of sodium ethoxide and ethanol.

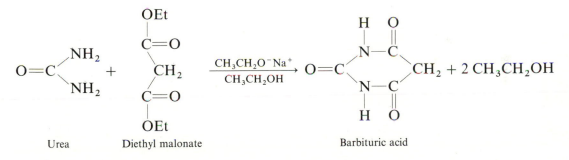

| Urea | Diethyl malonate | Barbituric acid |

The reaction mechanism involves the strongly basic ethoxide ion (**1**) pulling off a weakly acidic hydrogen on urea (**2**) to yield a resonance-stabilized anion of urea (**3**).

$$CH_3CH_2O^- + \underset{\textbf{2}}{NH_2\!-\!\overset{\overset{\displaystyle O}{\|}}{C}\!-\!NH_2} \longrightarrow$$

$$CH_3CH_2OH + \underline{N}H\!-\!\overset{\overset{\displaystyle O}{\|}}{C}\!-\!NH_2 \longleftrightarrow \underset{\textbf{3}}{NH\!=\!\overset{\overset{\displaystyle O^-}{\|}}{C}\!-\!NH_2}$$

The anion of urea, acting as a nucleophile, then attacks a carbonyl carbon atom on diethyl malonate (**4**) to yield intermediate (**5**), which then expels an ethoxide ion to regenerate the catalyst and give the open-chain intermediate (**6**).

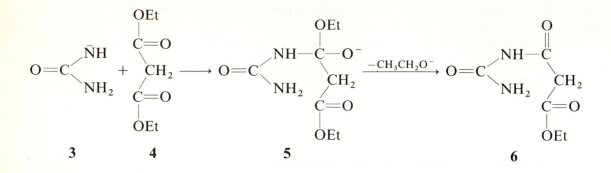

A repeat of this process on the open end of intermediate **6** then gives barbituric acid (**7**) as the final product.

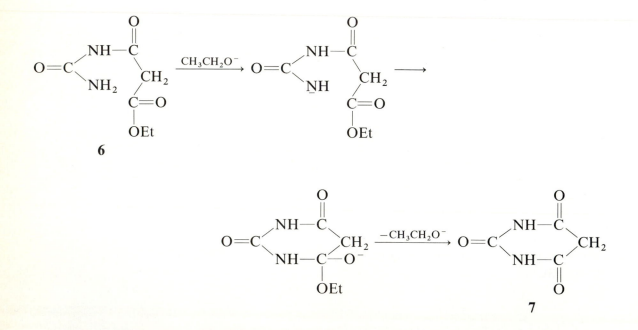

A side reaction in the synthesis is the removal of one of the acidic hydrogens in diethyl malonate by the ethoxide ion or the ion obtained from urea. These hydrogens are acidic due to the resonance-stabilized anion (**8**) formed in the equation.

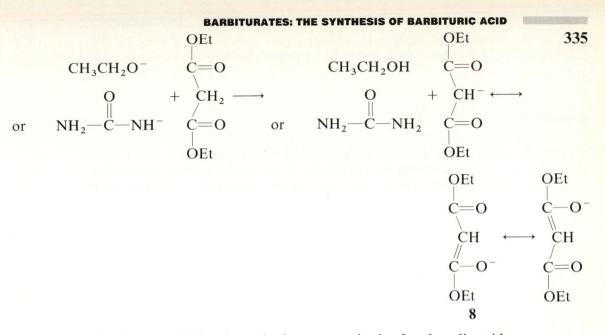

This reaction is essential for the malonic ester synthesis of carboxylic acids and the synthesis of the needed reactants to prepare barbiturates. For example, the barbiturate amobarbital can be prepared by the following synthesis. Each acidic hydrogen in diethyl malonate is individually removed by sodium ethoxide, with the first hydrogen being replaced by an ethyl group and the second one by an isopentyl group. Both of these replacements occur by a S_N2 attack of the anions on the alkyl halides, as indicated. The resulting intermediate (**9**) then reacts with urea to form amobarbital by the same mechanism given earlier for the formation of barbituric acid.

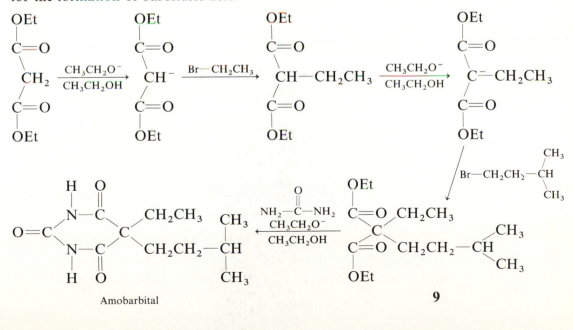

Amobarbital

For the synthesis of barbituric acid, the removal of one of these acidic hydrogens appears to have no benefits. But it doesn't prevent the synthesis, and 1 mole of sodium ethoxide is all that is necessary for every mole of urea and diethyl malonate.

PROCEDURE

CAUTION: Sodium metal reacts violently with water, so keep it away from water or any wet apparatus. Never handle it with your fingers or dispose of it in the sink. Destroy any excess, undesired sodium by reacting it with anhydrous alcohol.

Due to oxidation, sodium tarnishes when exposed to the air, so it is usually kept under xylene, ligroin, or mineral oil. To weigh sodium, transfer the metal with tongs from its container to a paper towel, and cut away the oxide coating with a knife or spatula. Small pea-sized pieces of the metal are then cut and quickly placed (without any oxide coating) into a small, previously weighed beaker of ligroin or xylene. From the increase in weight of the beaker, the amount of sodium added can be determined. When the sodium is needed, use forceps or tweezers to transfer the small pieces of metal from the beaker to the reaction flask.

Place 25 mL of dry, **absolute** ethanol into a **dry** 100-mL round-bottom flask. Attach a **dry** reflux condenser and then add, piece by piece through the top, 0.81 gram (0.035 mole) of sodium metal, prepared by the above procedure. The heat evolved from the reaction of sodium with the ethanol may cause the ethanol to boil. Cool the flask in an ice bath, if necessary, to prevent the reaction from becoming too vigorous.

FIGURE 15-1
The infrared spectrum of barbituric acid in a KBr pellet.
Spectrum courtesy of Sadtler Research Laboratories, division
of Bio-Rad Laboratories, Inc., copyrighted 1967.

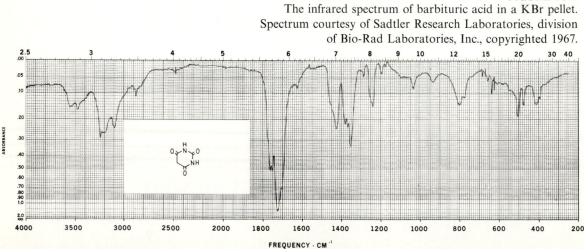

After all the sodium has dissolved, add 5.0 mL (0.032 mole) of diethyl malonate, followed by a warm solution of 1.9 grams (0.032 mole) of **dry** urea in 25 mL of **absolute** ethanol. (The urea should be dried beforehand by heating it for 1 hour in an oven at 105 to 110°.) Attach a **loosely packed** calcium chloride drying tube to the top of the reflux condenser, and heat the flask under reflux for 3 hours or more. (The heating may be conducted over more than one laboratory period; in the interim, portions of other experiments may be completed.)

When the refluxing is completed, pour the solution into 30 mL of cold water in a 125-mL Erlenmeyer flask, and **cautiously** acidify the resulting solution with 4 mL of concentrated hydrochloric acid. Concentrate the solution to about 30 mL, using a hot plate, and cool the solution in an ice bath. Collect the crude precipitate of barbituric acid by suction filtration, and recrystallize the crystals from hot water. After cooling it in an ice bath, collect the product by suction filtration and dry in an oven at 100 to 110° for 4 hours. (The reason is that the barbituric acid obtained from water exists as the dihydrate $C_4H_4N_2O_3 \cdot 2H_2O$). Weigh the dried product and determine its melting point and literature melting point; record the results in your notebook.

At your instructor's option, obtain the infrared spectrum of your product in a KBr pellet, and the ^1H NMR spectrum in dimethyl sulfoxide-d_6. Compare your spectra with those shown in Figures 15-1 and 15-2.

FIGURE 15-2

The ^1H NMR spectrum of barbituric acid in dimethyl sulfoxide-d_6. Spectrum courtesy of Aldrich Chemical Company, *The Aldrich Library of NMR Spectra*, edition II, by Charles Pouchert.

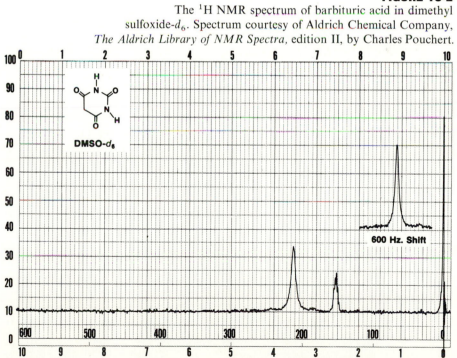

Determine the theoretical and percentage yields of your final product, show-ing the calculations, and submit it in a properly labeled container to your instructor.

QUESTIONS

1. Nonprescription sleeping aids, such as Nytol, Sominex, and Unisom, may be purchased in drug stores and supermarkets. What are the names and chemical structures of the key ingredients in these products?
2. What dangers are there in the excessive use of nonprescription drugs as sleep-ing aids?
3. The presence of water in this experiment is a serious problem that can severely lower the yield of product. Write equations showing how water is responsible for two undesired side reactions in the synthesis.
4. After the reaction of urea with diethyl malonate, why was hydrochloric acid added to the reaction mixture? Write equations for any reactions.
5. Write a reaction sequence for the synthesis of barbital.
6. Write a reaction sequence for the synthesis of sodium pentothal.
7. Draw resonance structures showing why barbituric acid should be acidic. Should it be more or less acidic than urea? Why?
8. From the infrared and ^{1}H NMR spectra of barbituric acid, does the compound exist primarily in the triketo form or the aromatic pyrimidine form? Explain your answer.
9. Predict the relative sedative activity and length of duration of the following compounds. Explain your answer.

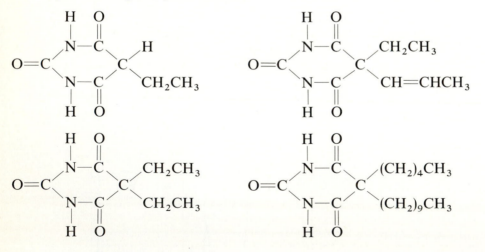

EXPERIMENT

16

CAFFEINE

- **HETEROCYCLIC COMPOUNDS**
- **ALKALOIDS**

Tea! Thou soft, thou sober, sage and venerable liquid, thou female tongue-running, smile-smoothing, heart-opening, wink-tipping cordial, to whose glorious insipidity I owe the happiest moments of my life.

Colley Cibber, *The Lady's Last Stake,* Act I

INTRODUCTION

Caffeine is, by far, consumed more than any other drug on the earth today. Over 260 million pounds of it are consumed each year, an average of 0.05 pound per person. This per capita amount may not seem very high, but remember that we are dealing with a substance that typically has a physiological impact when only 40 milligrams or 0.00009 pound are consumed. In the United States, the annual average is a much higher 0.16 pound per person; in England it is an even higher 0.34 pound per person.

Many products today contain caffeine. Coffee, tea, and soft drinks, especially cola drinks, are the ones people usually think of first. But chocolate also contains it, and so do almost 2000 different brands of nonprescription and prescription drugs. It is found in alertness tablets (NoDoz), analgesics (Anacin, Excedrin), cold and allergy remedies (Dristan), weight-control aids (Dexatrim, Dietac), diuretics—urine producers—(Aqua-Ban), and prescription drugs (Darvon, Fiorinal). A list of the caffeine content of many beverages, foods, and nonprescription and prescription drugs is given in Table 16-1. Notice the large difference that the type of coffee, tea, or soft drink makes in the caffeine content.

339

Caffeine was first isolated in the laboratory in 1820, but has been providing people in many cultures with a stimulating lift for centuries. Its reported use in ancient China goes back to 2737 B.C., where it was consumed in the form of tea. The ancient Arabians consumed it in the form of coffee. The ancient Aztecs found spiritual inspiration from ceremonial drafts of chocolate, and Ethiopians chewed coffee beans mixed with fat to prepare for battle. Even people of the

TABLE 16-1.

The Caffeine Content of Beverages, Foods, and Drugs

Soft drinks

Brand	Milligrams caffeine per 12-oz. serving	Brand	Milligrams caffeine per 12-oz. serving
Mountain Dew	54	Pepsi Light	36
Mellow Yellow	53	RC Cola	36
TAB	47	Diet Rite	36
Coca-Cola	46	Canada Dry Jamaica Cola	30
Diet Coke	46	Canada Dry Diet Cola	1
Shasta Cola	44	7-UP	0
Sunkist Orange	42	Diet 7-UP	0
Mr. PIBB	41	Sprite	0
Dr. Pepper	40	Sunkist Diet Orange	0
Sugar-Free Dr. Pepper	40	Pepsi Free	0
Big Red	38	Patio Orange	0
Pepsi-Cola	38	Fanta Orange	0
Aspen	36	Fresca	0
Diet Pepsi	36	Hires Root Beer	0

Source: National Soft Drink Association

Coffee, tea and foods

Item	Milligrams caffeine	Item	Milligrams caffeine
Coffee (5-oz. cup)		Cocoa beverage (5-oz. cup)	7–20
Brewed, drip method	110–180	Chocolate milk (8 oz.)	4–7
Brewed, percolator	60–170	Milk chocolate (1 oz.)	6–15
Instant	40–120	Dark chocolate, semisweet	
Decaffeinated, brewed	2–5	(1 oz.)	20–35
Decaffeinated, instant	1–5	Baker's chocolate (1 oz.)	26
Tea (5-oz. cup)		Chocolate-flavored syrup	
Brewed, major U.S. brands	20–90	(1 oz.)	4
Brewed, imported brands	25–110		
Instant	25–50		
Iced (12-oz. glass)	67–76		

Source: FDA, Food Additive Chemistry Evaluation Branch

TABLE 16-1 (continued)

The Caffeine Content of Beverages, Foods, and Drugs

Drugs

Prescription drugs (Standard Dose)	Milligrams caffeine	Nonprescription drugs	Milligrams caffeine
		Analgesic/Pain relief (Standard Dose)	
Cagergot (for migraine headaches)	100	Anacin	64
Fiorinal (for tension headaches)	40	Excredrin	130
		Midol	65
Soma Compound (pain relief, muscle relaxant)	32	Vanquish	66
Darvon Compound (pain relief)	32	Plain aspirin	0
Nonprescription drugs		**Diuretics** (Standard Dose)	
Weight-control aids (Daily Dose)		Aqua-Ban	200
		Permathene H_2Off	200
Codexin	200	Pre-Mens Forte	100
Dex-A-Diet II	200	**Cold/Allergy remedies** (Standard Dose)	
Dexatrim	200		
Dietac	200		
Maximum Strength Appedrine	200	Coryban-D capsules	30
Prolamine	280	Triaminicin tablets	30
Alertness tablets (Standard Dose)		Dristan decongestant tablets	32
		Duradyne-Forte	30
NoDoz	200		
Vivarin	200		
Caffedrine	200		

Source: FDA's National Center for Drugs and Biologics

Stone Age are believed to have consumed it in beverages. Today, over 160 plant species are known to produce it.

As a drug, caffeine is a stimulant and causes increased alertness, decreased muscle fatigue, a greater capacity for thinking, and the ability to put off sleep. Studies have shown that it improves driving and typing performance, helps take the monotony out of repetitious activities, and sharpens reactions when using dangerous machinery. Once consumed, its stimulating effect begins in as little as five minutes and typically lasts for three to four hours.

Other effects of caffeine include a faster heartbeat, slightly increased blood pressure, gastric irritation, and increased flow of urine. Because caffeine causes gastric irritation, it can aggravate an ulcer. Some people also suffer nervousness, irritability, sleeplessness, anxiety, dizziness, and a rapid, irregular heartbeat.

Excessive amounts of caffeine can cause a syndrome called **caffeinism**. The symptoms include ringing in the ears, flashes of light, mood changes, diarrhea, mild delirium, rapid breathing, muscle tension, tremors, and heart palpitations. A lethal dose is approximately 10 grams, the amount found in roughly 100 cups of coffee. But for caffeine to be fatal, all of this coffee would have to be consumed in a single sitting. Fortunately, caffeine is quickly metabolized by the body.

Caffeine is not considered to be a truly addictive drug, but it can be habit-forming, and people can easily become physically dependent on it. People who consume the equivalent of five or more cups of coffee per day can suffer from caffeine withdrawal if they suddenly stop its use. The symptoms include feelings of depression, moodiness, lassitude, and irritability, accompanied by severe, throbbing headaches and occasional nausea. Fortunately, all these symptoms usually end within two to three days. Also, the **gradual** reduction of caffeine intake can prevent them entirely.

There is no evidence today linking caffeine to heart disease. Studies have shown that heavy coffee drinking can more than double the rate of heart attacks, but other studies contradict these results. In addition, no studies have found that heavy tea drinking increases the risk of heart attacks. Therefore, caffeine cannot be a culprit. If heavy coffee drinking does indeed increase the risk of heart attacks, some other component or components must be responsible.

In addition, there is no conclusive evidence indicating that caffeine is a carcinogen. One study concluded that drinking five cups of coffee daily tripled a person's risk of getting cancer of the pancreas, and drinking two cups per day doubled the risk. However, these results are suspect because of a bias in choosing subjects for this study. Another study failed to find any link between coffee drinking and pancreatic cancer. Also, if any link is ever found, it must be established that caffeine and not some other substance or substances in the coffee is the actual culprit.

However, caffeine may be responsible for human birth defects, and the Food and Drug Administration has cautioned pregnant women to avoid or at least limit its consumption. Caffeine is known to definitely reach the fetus, and a study has shown harmful effects to the offspring of pregnant rats force-fed caffeine through a tube into the stomach. The offspring often had missing toes and were underweight when the mothers had received the human equivalent of the caffeine in 12 to 24 cups of coffee a day. And when the mothers received the human equivalent of the caffeine in only two cups of coffee, half the offspring had delayed skeletal growth.

Of course, people may not metabolize caffeine the same as rats, and few humans today are missing toes or fingers at birth. However, more subtle, perhaps even unrecognizable defects may result, and caffeine conceivably could be the culprit.

Those people who want to avoid the possible negative effects of caffeine can consume products containing little or none of it. To satisfy these people, coffee companies have been selling decaffeinated coffee for years, and 21 percent of all coffee is now decaffeinated, a sharp rise from a few years ago. Tea is also sold in decaffeinated form, and some types of tea never have had it. In addition, some new soft drinks no longer have caffeine, and others on the market for years have never had it.

Most of the 3 million pounds of caffeine removed from coffee each year is sold to the soft-drink industry and added to cola and some noncola drinks, to give both the consumer and the product sales a lift. No longer does the law say that cola drinks must have caffeine.

The rest of the 3 million pounds goes into prescription and nonprescription drugs, where it frequently counteracts the possible drowsiness produced by some of the other drugs present. There is little or no scientific evidence that caffeine relieves pain, aids weight loss, eases menstrual cramps, helps one fight a cold, or serves any effective nondrowsiness role in these products. Therefore, people who want to avoid the possible negative effects of caffeine should also not consume any analgesic, weight-loss, cold remedy, or other drugs containing it. If caffeine is present in these products, it will be listed on the ingredients label.

When isolated, caffeine is a white, odorless, bitter-tasting solid. It is chemically called 1,3,7-trimethylxanthine and has the structure

1,3,7-Trimethylxanthine
(caffeine)

Observe that caffeine is a dicyclic molecule in which each ring contains two nitrogen atoms. Because nitrogen is an element different from carbon, caffeine as a result is frequently called a heterocyclic compound. It is also called an **alkaloid** (alkali-like), and can be protonated at position 9 to form an acid salt. Many other organic compounds are also called alkaloids, among them nicotine, morphine, cocaine, quinine, and strychnine. Most alkaloids are produced by plants and have certain, specific physiological effects when consumed. These effects can be very different, as you may note by comparing all the alkaloids given above.

16A. THE ISOLATION OF CAFFEINE FROM TEA

In this experiment, you will isolate caffeine from tea. With the exception of plain water, tea is the most common beverage consumed around the world. The annual world production of 4 billion pounds is enough to brew over 800 billion cups, or 160 cups for every human being on the earth. In contrast, the annual world production of coffee is 11 billion pounds, but this is enough to brew only 550 billion cups or 110 cups per person. England is the largest per capita consumer of tea, averaging almost 1800 cups per person each year, about 12 times the world average. The U.S. consumption is a much lower 170 cups per person. This large difference is due to the popularity in the United States of other beverages such as coffee, beer, and soft drinks.

Historically, the first use of tea dates back to 270 A.D. in ancient China, although, according to legend, its use in China may go back as far as 2737 B.C. Although the beverage was very popular in the Orient for many years, it wasn't until 1610 that the use of tea reached Europe. But it caught on quickly here, soon becoming the national drink of England, and by 1650 it was being imported by the American Colonies. Here it was also very popular until the 1770s, when a heavy tax on the substance was responsible for the Boston Tea Party of 1773, which helped spur the American Revolutionary War. In the 1800s, tea played a major role in the development of the fast-sailing clipper ships. Since the flavor of tea deteriorates with time, merchants were willing to pay top prices for tea brought in rapidly.

Today there are over 3000 varieties of tea, grown chiefly in India, Ceylon, Indonesia, China, and Japan. Many of these varieties, like fine wines, take their names from the districts where they are grown. Tea is also like wine in that its quality can vary from year to year. One consequence of this is that almost all tea is blended with as many as 20 varieties being used. This gives each brand its own distinctive taste and helps protect the flavor from any great change in quality of one variety. Professional tea tasters determine the quality of the different varieties and what blend should be used to obtain a particular flavor.

In the isolation of caffeine from tea, the main difficulty is the separation of all the other natural components present. The major substance in tea leaves is cellulose, the principal structural material of all plant cells. But since cellulose is virtually insoluble in water, whereas caffeine is water soluble, there is no difficulty here. However, water-soluble proteins, sugars, essential oils, flavonoid and chlorophyll pigments, tannins (used to "tan" leather), and hydrolysis products of the tannins must also be separated. We shall focus our attention on the tannins and their hydrolysis products.

The tannins that can be hydrolyzed have the general structure

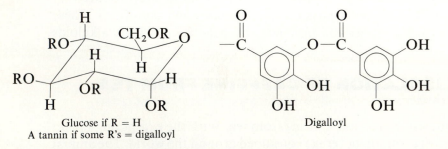

Glucose if R = H
A tannin if some R's = digalloyl

Digalloyl

and yield glucose and gallic acid upon hydrolysis.

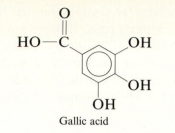
Gallic acid

The tannins that cannot be hydrolyzed are condensation polymers of catechin.

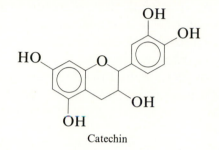
Catechin

Observe in the structures the hydroxyl groups and the carboxyl group on the aromatic rings, making the tannins and gallic acid acidic in nature. Therefore, if a base such as sodium carbonate is present in tea water, both the tannins and gallic acid are converted to their methylene chloride-insoluble sodium salts. The caffeine, however, is soluble in methylene chloride and can thus be separated, since it is nonacidic and won't form a sodium salt. Since most of the other substances present in tea water are also insoluble in methylene chloride, the methylene chloride extraction of the alkaline tea solution removes nearly pure caffeine, which, after distillation of the methylene chloride, may be purified by recrystallization.

$$\text{Tannins and gallic acid} \xrightarrow{\text{Na}_2\text{CO}_3} \text{Methylene chloride-insoluble sodium salts}$$

$$\text{Caffeine} \xrightarrow{\text{Na}_2\text{CO}_3} \text{No reaction, so caffeine remains methylene chloride soluble}$$

PROCEDURE

CAUTION: Methylene chloride is toxic. Be careful when handling it. Do not breathe it excessively or spill it on yourself.

Place 15 grams of tea bags (with the strings cut off) and 11 grams of powdered sodium carbonate into a 500-mL Erlenmeyer flask. Add 175 mL of distilled water and boil the contents for about 20 minutes, being careful not to break any bags.

Cool the liquid and decant into a large separatory funnel. Make sure no tea leaves are carried over due to any broken bags. If loose tea leaves or any residue is present, suction filter first. Add 35 mL of methylene chloride, but do **not** vigorously shake the separatory funnel as you would normally. Otherwise, a possibly troublesome emulsion will be produced. Instead, invert the separatory funnel and swirl the two layers together for 3 to 5 minutes to transfer the caffeine to the methylene chloride.

Remove the lower methylene chloride layer and, if any emulsion is present, allow the methylene chloride to go through a funnel that has a 1-cm layer of anhydrous magnesium sulfate on top of a small wad of cotton in the funnel neck. The drying of the methylene chloride by the magnesium sulfate should break up the emulsion. Repeat the extraction with another 35 mL of methylene chloride and combine both methylene chloride extracts.

Set up an apparatus for simple distillation, and distill the methylene chloride, using a steam or hot-water bath, until about 10 to 15 mL remains. Do not use a stronger source of heat because it may char the caffeine if the caffeine should form a residue in the bottom of the flask.

Transfer the remaining liquid to a weighed 50-mL Erlenmeyer flask. Rinse the round-bottom flask with 5 mL of the distilled methylene chloride and transfer this to the 50-mL Erlenmeyer flask. Evaporate the light green methylene chloride solution to dryness on a steam bath or hot-water bath **in the hood** and determine the weight of crude product obtained.

Recrystallize the solid residue of caffeine from 95% ethanol, using about 5 mL/gram. One recrystallization should be satisfactory, and it yields caffeine in the form of small needles. Collect the crystals by suction filtration on a Hirsch funnel, allow to air dry, and determine the weight, melting point, and literature melting point. Record the results in your notebook.

FIGURE 16-1

The infrared spectrum of caffeine in a KBr pellet. Spectrum courtesy of Sadtler Research Laboratories, division of Bio-Rad Laboratories, Inc., copyrighted 1966.

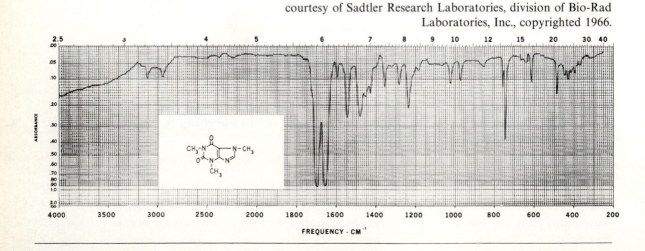

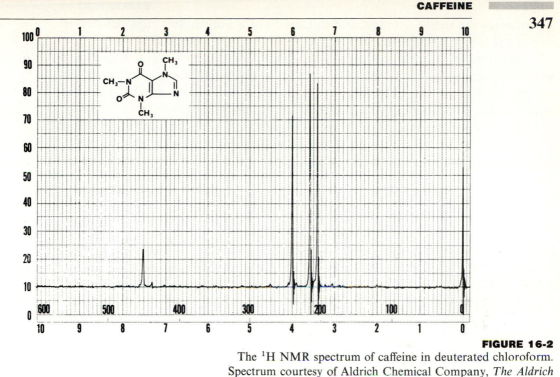

FIGURE 16-2

The ^{1}H NMR spectrum of caffeine in deuterated chloroform.
Spectrum courtesy of Aldrich Chemical Company, *The Aldrich
Library of NMR Spectra*, edition II, by Charles Pouchert.

At your instructor's option, obtain the infrared spectrum of your product
in a KBr pellet, and the ^{1}H NMR spectrum in deuterated chloroform. Compare your spectra with those shown in Figures 16-1 and 16-2.

Determine the percentage yield of the final product, based upon the original
weight of the tea bags. Show the calculations. How would your percentage yield
have been affected if you had used 15 grams of loose tea rather than tea bags?
Why?

When finished, submit your product in a properly labeled container to your
instructor.

16B. THE ISOLATION OF CAFFEINE FROM COFFEE

In this experiment, you will isolate caffeine from instant coffee. Coffee is today
the third most popular beverage in the world, losing out only to plain water
and tea. The annual world production is 11 billion pounds, enough to brew
550 billion cups or about 110 cups per person. In the United States, the annual
consumption is over 700 cups per person, more than six times the world average.
In the Scandinavian countries, it is an even higher 800 cups per person or over

seven times the world average, and in Canada it is almost 400 cups per person or almost four times the world average.

According to legend, coffee was discovered in Ethiopia when a goatherd was surprised to see his animals become unusually frisky after eating the bright red berries from a tree growing wild in the pasture. He also tried the berries, enjoyed their stimulating effect and announced his discovery.

The Arabs cultivated coffee trees as early as 600 A.D. and the first mention of coffee in literature was by an Arab physician about 900 A.D. But for seven centuries, the Arabs' use of coffee was not as we know it today. Typically, the coffee berries were dried, crushed, and then mixed with fat to form a ball that was eaten, either for food or as a medicine. Coffee berries were also added to wine, and wine was even made from the coffee itself by fermenting a mixture of the berry skins with the crushed green beans.

Then in the thirteenth century, it was discovered that a delicious beverage could be made from the roasted beans. As a result, coffee became immensely popular in Arabia and a lucrative article of trade, eventually becoming very popular throughout all of Europe and much of the world. Until the seventeenth century, all coffee came from Arabia. But then the transport of a single plant to the island of Martinique in the West Indies began coffee cultivation in the Western Hemisphere. In fact, most of the coffee trees in the Western Hemisphere today are descendents of this one plant!

Most coffee today comes from Brazil, but Columbia, the Ivory Coast, Mexico, Indonesia, Ethiopia, Uganda, and the Philippines are also large producers. Many countries, in fact, receive from 40 to over 60 percent of all their foreign-trade income by exporting coffee.

Although there are over 60 different species of coffee trees, only two, *Coffea arabica* and *Coffea robusta*, are principally used to produce coffee. *Coffea arabica* is grown almost exclusively in the Western Hemisphere and is used in almost 72 percent of all marketed coffee. Most African and Asian coffee is *Coffea robusta* and comprises almost 28 percent of the market. *Coffea arabica* is favored primarily in the United States, Canada, and Scandinavia; *Coffea robusta* is preferred in France, Italy, Turkey, and many other countries in Europe and the Middle East.

There are, however, many variations of the two main coffee types, which can also be further affected by the country or region where it is grown and the method of cultivation. As a result, most coffee sold today is blended to yield a product of the desired flavor. This flavor not only varies from country to country, but from brand to brand, and can even be different within different regions of the same country. Professional coffee tasters determine what mix of all the varieties should be used.

Coffee attains most of its flavor through roasting, which causes very complex chemical transformations within the beans. The beans are kept at about 220 to 230°C (430 to 445°F) for 16 to 17 minutes. During this time, steam, carbon dioxide, carbon monoxide, and other volatile compounds are released, resulting in a weight loss of between 14 and 23 percent. The internal pressure from all these gases also triples the size of the beans. They become a deep, rich-brown color and develop a porous, crumbly texture.

Green coffee generally keeps well, but after roasting the flavor and aroma are quickly lost if the beans are exposed to the air. This results from the compounds responsible for the flavor and aroma reacting readily with oxygen to yield less desirable products. Therefore, after the coffee is ground, it is immediately packaged to prevent its deterioration, and it is usually either vacuum-packed or pressure-packed, with the air being replaced by an inert gas.

Instant coffee is manufactured by dehydrating a strong brew of regular coffee. Since the resulting crystals absorb moisture readily, they are usually immediately packed in glass jars with airtight tops.

Decaffeinated coffee can be prepared by extracting the caffeine from the green beans with an organic solvent such as methylene chloride. The solvent is then removed from the beans by steaming. However, since trace amounts of the organic solvent still remain, some coffee companies are now instead removing the caffeine by extraction with water.

In this experiment, instant coffee will be used to eliminate the need to extract and filter the coffee from the coffee grinds. Instant coffee contains no water-insoluble components, such as protein, fat, and cellulose, but it still has numerous other substances that must be removed to obtain pure caffeine. Among these are the tannins, glucose, gallic acid, and chlorogenic acid, which have the following structures:

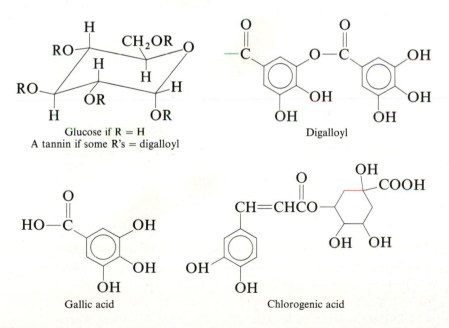

Glucose if R = H
A tannin if some R's = digalloyl

Digalloyl

Gallic acid

Chlorogenic acid

Observe the hydroxyl groups on the aromatic rings and the carboxylic acid groups, making all the molecules except glucose acidic in nature. Therefore, if a base such as sodium carbonate is added to coffee water, the tannins, gallic acid, and chlorogenic acid are all converted to their methylene chloride-insoluble sodium salts. The caffeine, however, is soluble in methylene chloride and can thus be separated, since it is nonacidic and won't form a sodium salt. Since glucose is also insoluble in methylene chloride, the methylene chloride extraction of the alkaline coffee solution removes nearly pure caffeine, which, after distillation of the methylene chloride, may be purified by recrystallization.

$$\text{Tannins, gallic acid,} \atop \text{and chlorogenic acid} \xrightarrow{\text{Na}_2\text{CO}_3} \text{Methylene chloride-insoluble} \atop \text{sodium salts}$$

$$\text{Caffeine} \xrightarrow{\text{Na}_2\text{CO}_3} \text{No reaction, so caffeine remains} \atop \text{methylene chloride soluble}$$

PROCEDURE

CAUTION: Methylene chloride is toxic. Be careful when handling it. Do not breathe it excessively or spill it on yourself.

Heat 10 grams of instant coffee and 10 grams of powdered sodium carbonate together in 175 mL of distilled water for about 5 minutes. Cool the liquid, filter if necessary, and decant into a large separatory funnel. Add 35 mL of methylene chloride, but do **not** vigorously shake the separatory funnel as you would normally. Otherwise, a possibly troublesome emulsion will be produced. Instead, invert the separatory funnel and swirl the two layers together for 3 to 5 minutes to transfer the caffeine to the methylene chloride.

Remove the lower methylene chloride layer and, if any emulsion is present, allow the methylene chloride to go through a funnel that has a 1-cm layer of anhydrous magnesium sulfate on top of a small wad of cotton in the funnel neck. The drying of the methylene chloride by the magnesium sulfate should break up the emulsion. Repeat the extraction with another 35 mL of methylene chloride and combine both methylene chloride extracts.

Set up an apparatus for simple distillation and distill the methylene chloride, using a steam or hot-water bath, until about 10 to 15 mL remains. Do not use a stronger source of heat, since it may char the caffeine if the caffeine should form a residue in the bottom of the flask.

Transfer the remaining liquid to a weighed 50-mL Erlenmeyer flask. Rinse the round-bottom flask with 5 mL of the distilled methylene chloride and transfer this to the 50-mL Erlenmeyer flask. Evaporate the yellow solution to dryness on a steam bath or hot-water bath **in the hood**, and determine the weight of the crude brownish-yellow caffeine obtained.

Recrystallize the solid residue of caffeine from 95% ethanol, using about 5 mL/gram. Collect the crystals by suction filtration on a Hirsch funnel and compare their color now with what it was before. Allow the product to air dry, and determine the weight, melting point, and literature melting point. Record the results in your notebook. Also determine the percentage yield, based upon the original weight of instant coffee, and show the calculations.

Further purifying the caffeine by sublimation

If necessary, caffeine may be further purified by sublimation. Set up the apparatus shown in Figure 16-3, using a 15 × 125 mm test tube inserted first into a #2 neoprene adapter and then into a 20 × 155 mm side-arm test tube. The caffeine is placed in the bottom of the side-arm test tube, and crushed ice and water go into the inner test tube. Insert a trap between the apparatus and the water aspirator, turn on the water, and press the inner tube and the adapter into the outer tube until a good seal is obtained, as noted by a change in the water velocity.

Heat the sample **carefully** with a small flame, moving the burner back and forth under the outer tube and up around its sides. You will see the caffeine vaporize from the outer tube and be deposited onto the cold, inner tube. If the caffeine begins to melt, remove the heat for a few seconds before continuing. When the sublimation is complete, remove the heat and let the apparatus cool.

Turn off the water and slowly release the vacuum by disconnecting the hose at the aspirator or the trap, or, even better, by opening the stopcock or the clamp on the hose coming up from the trap. **Carefully** remove the inner tube of the apparatus, and try not to have any of the sublimed crystals fall off the tube into the residue. Pour off the remaining ice and water from the inner tube, and scrape the sublimed caffeine onto weighing paper. Determine the melting

FIGURE 16-3
Simple sublimation apparatus.

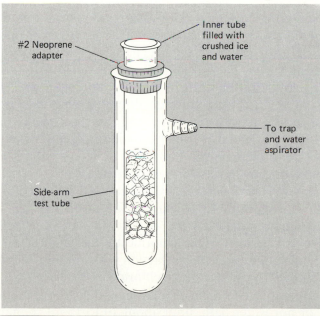

#2 Neoprene adapter

Inner tube filled with crushed ice and water

To trap and water aspirator

Side-arm test tube

point of this purified caffeine and compare its melting point and color with what was obtained after recrystallization. Record the results in your notebook.

At your instructor's option, obtain the infrared spectrum of your product in a KBr pellet, and the ^{1}H NMR spectrum in deuterated chloroform. Compare your spectra with those shown in Figures 16-1 and 16-2.

When finished, submit the product in a properly labeled container to your instructor.

16C. THE ISOLATION OF CAFFEINE FROM NODOZ

In this experiment, you will isolate the caffeine present in NoDoz. Unlike coffee and tea, NoDoz does not have a wide array of other components with caffeine. Only the binder in the tablets prevents NoDoz from being pure caffeine. Since each NoDoz tablet contains 100 milligrams of caffeine, you know from the start how much caffeine can theoretically be obtained, unlike the experiments with coffee and tea. Also, by weighing the NoDoz tablets, you can determine what percentage of the tablet is caffeine and what percentage is binder.

PROCEDURE

CAUTION: Methylene chloride is toxic. Be careful when handling it. Do not breathe it excessively or spill it on yourself.

Weigh 5 NoDoz tablets, then crush them and place the powder in 200 mL of water in a 400-mL beaker. Heat the resulting suspension to boiling to dissolve the caffeine, then cool and filter. (The binder will remain in suspension during the heating.)

Pour the aqueous solution into a large separatory funnel and extract the caffeine with 35 mL of methylene chloride. Briefly shake to determine if a troublesome emulsion results as with the coffee and tea experiments. If not, continue the extraction as you would normally, periodically allowing any vaporized methylene chloride to be released. If any sign of an emulsion is produced, swirl rather than shake the two layers for 3 to 5 minutes.

Remove the lower methylene chloride layer and, if any emulsion is present, allow the methylene chloride to go through a funnel that has a 1-cm layer of anhydrous magnesium sulfate on top of a small wad of cotton in the funnel neck. The drying of the methylene chloride by the magnesium sulfate should break up the emulsion. Repeat the extraction with another 35 mL of methylene chloride and combine both methylene chloride extracts.

Set up an apparatus for simple distillation and distill the methylene chloride, using a steam or hot-water bath, until about 10 to 15 mL remains. Do not use a stronger source of heat, since it may char the caffeine if the caffeine should form a residue in the bottom of the flask.

Transfer the remaining liquid to a weighed 50-mL Erlenmeyer flask. Rinse the round-bottom flask with 5 mL of the distilled methylene chloride and transfer this to the 50-mL Erlenmeyer flask. Evaporate the solution to dryness on a steam bath or hot-water bath **in the hood**, and determine the weight of the caffeine obtained.

Recrystallize the caffeine from 95% ethanol, using about 5 mL/gram. One recrystallization should be satisfactory; it yields caffeine in the form of small needles. Collect the crystals by suction filtration on a Hirsch funnel, allow to air dry, and determine the weight, melting point, and literature melting point. Record the results in your notebook.

At your instructor's option, obtain the infrared spectrum of your product in a KBr pellet, and the ^{1}H NMR spectrum in deuterated chloroform. Compare your spectra with those shown in Figures 16-1 and 16-2.

Determine the percentage yield of the final product, based upon the original known weight of caffeine, and show the calculations. Also determine the percentage of NoDoz that is actually caffeine by weight, and show your calculations. Indicate several reasons why your amount of isolated caffeine should be less than the amount you started with. Spilling the solutions, and similar reasons are *not* acceptable.

When finished, submit your product in a properly labeled container to your instructor.

QUESTIONS

1. Why are the tea, coffee, and NoDoz solutions heated and even boiled for as long as 20 minutes?
2. Why are two extractions with 35 mL of methylene chloride better than one extraction with 70 mL?
3. In the tea experiment, what substance do you know of that could account for the light green color of the methylene chloride solution? Why was this light green color not evident earlier in the experiment?
4. In the tea, coffee, and NoDoz experiments, if the aqueous caffeine solution had been made acidic, could the caffeine have been extracted into the methylene chloride? Why?
5. Write a chemical reaction for the synthesis of gallic acid from the tannins.
6. Which hydroxyl groups in the tannins do you expect to be more acidic? Why?
7. When caffeine is commercially extracted in large quantities from coffee and tea, in what types of products can it be utilized?
8. Why does an emulsion form much more easily in the coffee and tea experiments than in the NoDoz experiment?

17

THE ISOLATION OF NICOTINE FROM TOBACCO

- **HETEROCYCLIC COMPOUNDS**
- **ALKALOIDS**

Tobacco is a dirty weed: I like it.
It satisfies no normal need: I like it.
It makes you thin, it makes you lean,
It takes the hair right off your bean;
It's the worst darn stuff I've ever
seen: I like it.

Graham Hemminger, *Tobacco*

INTRODUCTION

The above quote was selected because it so appropriately expresses the feeling of so many people about tobacco. Most people are aware, or should be aware, of the detrimental effects of smoking to their health, and yet many of these same people continue to engage in the habit. Furthermore, none of the other negative features of the practice, such as the monetary expense, stained teeth, dirty ashtrays, and clothing, homes, and cars that smell of smoke, act as a deterrent to these people.

The reasons why people begin smoking and continue the habit are numerous, complex, and not completely understood. But a large part of it stems from psychological and social drives. The individual begins because others around him or her are smoking, and this is his or her way of becoming "part of the group." This conforming due to peer pressure may begin at a very early age,

355

even before junior high school. Smoking is associated with being "grown up," and a person's desire to be viewed as an adult contributes to the habit, as well as the illusion that smoking is essential to social prominence, business success, and sex appeal. Many individuals also smoke to relieve tension or boredom. Studies have shown that people tend to smoke more in stressful situations or when they have nothing else to do. In addition, a cigarette is often used as a social "prop" or viewed as a reward for a completed task.

Historically, it is unknown when tobacco was first used. In 1492, Columbus observed that Indians were utilizing it for smoking, chewing, and snuffing. By 1550, the smoking habit had been introduced in Europe, and by 1610 it was recognized by most of the known world. The use of tobacco was not universally accepted, however, and in some parts of the world stiff taxes and severe penalties were imposed to prohibit it. All these efforts proved unsuccessful, and its popularity continued to grow. In the American settlement at Jamestown, exports of the leaf to England were substantial. Profits were so large that severe restrictions on tobacco planting had to be imposed to ensure a sufficient quantity of food crops.

As popular as smoking was, cigarettes were not introduced in the United States until after the Civil War. Due largely to advertising, cigarette-making machines, and social acceptability, their use increased sharply, especially between 1920 and 1960. Today over 625 billion cigarettes are smoked annually by Americans, an average of 7.1 cigarettes per person each day. The average for the actual cigarette users is 26 cigarettes per day.

From the beginning, the smoking of tobacco produced controversy over possible health hazards. Much of the early concern, however, was pure conjecture, with very little basis or real evidence. Present concern over the possible ill effects began in the 1930s when a dramatic increase in lung cancer deaths was observed. By the 1950s, the results of a number of studies had been released, showing a definite causal relationship between cigarette smoking and lung cancer, bronchitis, emphysema, cardiovascular diseases, and other health problems. In 1962, the Royal College of Physicians of London reported: "Cigarette smoking is a cause of lung cancer, bronchitis and various other less common diseases. It delays the healing of gastric and duodenal ulcers." This was followed in 1964 by a 150,000 word report by the U.S. Surgeon General's Advisory Committee on Smoking and Health, which concluded that "Cigarette smoking is a health hazard of sufficient importance in the United States to warrant appropriate remedial action." The U.S. Public Health Service in 1967 published a review of 2000 additional research studies, which not only confirmed the Surgeon General's report, but indicated that the extent of the hazard had been underestimated.

Since 1967, the evidence of the hazards of smoking has continued to grow. As a result, at least 35 million Americans are estimated to have quit the habit. In addition, Congress banned the advertising of most tobacco products on television and radio, and has required cigarette packages and the remaining ads to print various warnings about the hazards of smoking. Among the evidence against smoking are the following facts:

1. Cigarette smoking is the most important single cause of preventable death in the United States today. Over 400,000 deaths each year are attributed to smoking, or over eight times the number of deaths caused by automobile accidents. About one-fifth of all deaths in the United States are smoking related. Four of the five leading causes of death are related to cigarette smoking.

2. A smoker is 70 percent more likely to die prematurely than a nonsmoker is. A heavy smoker (more than two packs per day) is 200 percent more likely to die prematurely than a nonsmoker is. A young person who continually smokes two packs per day can expect to live an average of eight years less than a nonsmoker, and a young person who continually smokes half a pack per day can expect to live an average of four years less.

3. Male cigarette smokers have 33 percent more days lost from work and 14 percent more days of bed disability than do nonsmokers. For females, the figures are even higher: 45 percent and 17 percent.

4. Lung cancer kills over 120,000 Americans each year, more than any other form of cancer. At least 85 percent of all lung cancer deaths are the result of cigarette smoking. Smokers are over 10 times more likely to die of lung cancer than nonsmokers are. Heavy smokers are over 25 times more likely to die of lung cancer than nonsmokers are. Almost 70 percent of all lung cancer patients die within one year of diagnosis. Almost 90 percent die within five years. In 1984, lung cancer among women surpassed breast cancer as the leading cause of death by cancer. While the rate of breast cancer deaths in women has been relatively steady since 1950, the lung cancer rate is seven times what it was in 1950, a direct result of increased smoking.

5. Over 30 percent of all cancer deaths are tobacco related. Tobacco, in fact, is responsible for more cancer deaths than all other reliable causes **combined**. Smoking increases the risk of a person developing cancer of the mouth, larynx, pharynx, esophagus, kidney, and bladder.

6. Over 200,000 Americans each year die of coronary heart disease caused by cigarette smoking. This is roughly double the number who die from lung cancer and is about 30 percent of all coronary deaths. A smoker is about 70 percent more likely to die from heart disease than a nonsmoker is. A heavy smoker is about two to three times more likely to die from heart disease than a nonsmoker is.

7. Smoking causes over 50,000, or 85 percent, of all the deaths from chronic bronchitis and emphysema each year.

8. Women who smoke while pregnant have children who weigh, on the average, 7 ounces less at birth than children of women who don't smoke. The risk of miscarriage and stillbirth is almost twice as great as it is for nonsmokers, and the risk of sudden death infant syndrome (SDIS) is 50 percent higher. Smoking during pregnancy also causes more premature births, fetal injuries, and conditions such as immature lungs, bowel problems, infections, and bleeding within the brain. Even the physical and mental development after birth are adversely affected.

9. Over 4000 compounds have been found in tobacco smoke, many of which are either toxic or carcinogenic. A few of these compounds are nicotine, hydrogen cyanide, hydrogen sulfide, carbon monoxide, ammonia, formaldehyde, acetaldehyde, acrolein, and the components of tar. Although cigarette manufacturers have added filters and have reduced the nicotine and tar content of cigarette tobacco, this has been insufficient in completely eliminating the harmful substances. The quantity of them that enter the body still depends on the way and on the amount that one smokes.

10. Smoking can be especially hazardous to people in occupations in the chemical, coal, rubber, textile, and uranium industries, because the substances in the smoke acting together with chemicals from these occupations greatly increase the chances of contracting lung cancer or other ailments.

11. The cost to the nation in terms of medical treatment and time lost from work was over $70 billion in 1987 or $1200 per smoker. Nonsmokers paid most of this cost through higher medical insurance premiums, higher taxes, and the higher cost of goods and services. In addition, life insurance premiums for nonsmokers are higher than they would be if smokers were excluded.

Because of the increased health awareness, since 1965 the percentage of men who smoke has dropped sharply, from 52 percent to 33 percent, while the percentage of women who smoke has dropped moderately, from 34 percent to 28 percent. Cigarette companies are selling less of their product, and many of them have diversified to other products, such as food. In addition, some of them are emphasizing in their ads the greatly reduced tar and nicotine content of their product.

The question of a person's right to smoke has also come under severe attack from nonsmokers who find the smoke offensive and irritating, and are concerned over its ill effects on their health. In many places, such as restaurants, offices, factories, movie houses, trains, buses, and airplanes, smoking is now either banned or restricted to certain designated areas. Precedent-setting court decisions have established the employee's right to a smokefree workplace, and many companies have been compelled to provide it.

A growing number of companies, aware of all the negative effects of smoking, are even paying their employees a bonus for quitting. About 3 percent of American companies and 6 percent of Canadian companies now have various plans. Their rationale is that smoking is not only physically unhealthy for the employees but economically unhealthy for the company. The companies benefit from reduced absenteeism, higher productivity on the job, lower maintenance costs, and sometimes even a rate reduction for employee health care coverage. Many employees have reacted favorably, primarily because the immediate monetary reward is apparently more important than their concern over possibly getting cancer or heart disease 20 years later.

Nicotine is the substance most often associated with smoking and is one of the most toxic drugs known: only 60 mg is lethal. Death occurs within a few minutes from the paralysis of the muscles used in respiration. "Large" doses, in less than the fatal amount, cause extreme nausea, vomiting, convulsions, mental confusion, and evacuation of the bowels and bladder. One cigar contains about 150 mg of nicotine or enough for two lethal doses. Fortunately, less than 10 percent of this amount is actually absorbed by inhaling the smoke, and this occurs over a relatively long time period. The body is also able to metabolize the drug quite rapidly, thus preventing its accumulation. Although nicotine is not considered to be a truly addicting drug, individuals can easily develop a tolerance to it and become dependent upon it, which is one important reason so many people find it difficult to stop smoking. Fortunately, cigarette manufacturers today have somewhat lowered this dependence by greatly reducing the nicotine content of their product.

As detrimental as nicotine can be, it also has its benefits. As the sulfate salt, it is used as an agricultural insecticide, and much of the nicotine removed from cigarette tobacco is utilized for this purpose. As the tartrate salt, it is valuable in counteracting the effects of tetanus and strychnine poisoning. And when used in chewing gum, it has helped almost half of the people in test studies quit smoking, **provided** that these people really wanted to quit and had the proper support group.

BACKGROUND

In this experiment, you will isolate nicotine from tobacco. Observe from its structure below that nicotine is a dicyclic molecule with each ring containing a nitrogen atom. Because nitrogen is different from carbon, nicotine is frequently called a heterocyclic molecule. The nitrogen atoms also make both rings appreciably basic, so nicotine is also called an **alkaloid** ("alkali-like"). Many other organic compounds are also called alkaloids, among them morphine, cocaine, quinine, caffeine, and strychnine. Most alkaloids are produced by plants and contain a chiral (asymmetric) center. Nicotine is no exception and occurs in nature as the (−) or levorotatory compound.

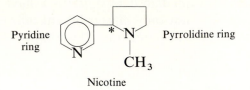

Pyridine ring

Pyrrolidine ring

Nicotine

Although both rings of nicotine are appreciably basic, there is a difference. The K_B value of 10^{-4} for the pyrrolidine ring means that this ring is highly protonated in plain water. This is one reason for nicotine's infinite solubility in water. The K_B value of 10^{-9} for the pyridine ring means that it can only become protonated in dilute acid.

In this experiment, you will remove the nicotine from tobacco by extraction with ordinary water. After the separation of the aqueous solution of nicotine from the tobacco, the solution is then made alkaline with a concentrated solution of sodium hydroxide. This generates the free base, which can be extracted into diethyl ether. Ether was not used originally as the extraction solvent, because of the vast mixture of other compounds that would be obtained.

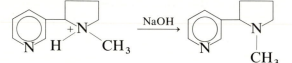

Evaporation of the ether solution leaves nicotine as an oil, which in small amounts is not easily purified. The compound is heat sensitive; even upon standing at room temperature, it slowly decolorizes first to a yellow oil and then to a brown oil. You shall therefore convert the nicotine into its picrate salt by reaction with picric acid. This yields a crystalline material which, after being purified, can be identified by its melting point. Both basic centers of nicotine react with picric acid, and the high molecular weight of the picrate anion results in the salt having a molecular weight almost four times the molecular weight of nicotine. This provides a much larger quantity of material to work with.

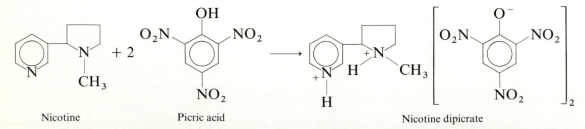

Nicotine   Picric acid   Nicotine dipicrate

Cigarette tobacco could be used in this experiment, but would be a poor source of nicotine since most cigarette manufacturers remove much of the drug. Cigar tobacco is a much better source. Crush it to expose as much surface area as possible to the extraction solvent.

PROCEDURE

CAUTION: Remember that nicotine is extremely toxic; a dose of only 60 mg is lethal. Handle it with care.

Also remember that 50% NaOH solution is extremely caustic. Handle it with care.

In addition, picric acid when dry is a potential explosive. Be sure that it always remains moist or dissolved in solution.

Place 10 grams of crushed cigar tobacco and 100 mL of distilled water in a 400-mL beaker and stir the mixture for 15 minutes. Suction filter the tobacco and cool the filtrate in the filter flask in an ice bath. With the Büchner funnel removed, **carefully** add 15 mL of cooled 50% NaOH solution. Stir or gently swirl the solution, keeping it at room temperature or lower.

Transfer the alkaline filtrate to a 250-mL separatory funnel and extract with 30 mL of ether. After shaking the liquids, collect the lower aqueous layer in a beaker and save for further extraction. The remaining upper ether layer may contain a small amount of a dark oily emulsion at the bottom of the funnel. If this is present, try not to let it contaminate the ether, by carefully decanting the ether from the emulsion through the top of the separatory funnel into a dry beaker.

Transfer the saved aqueous layer back into the separatory funnel and extract with another 30-mL portion of ether. After shaking the liquids, again collect the lower aqueous layer and save for further extraction. Again carefully decant the upper ether layer through the top of the separatory funnel and combine it with the first portion of ether.

Return the aqueous layer to the separatory funnel and extract it with a third 30-mL portion of ether. The lower aqueous layer may now be discarded. Combine the third portion of ether with the other two, being sure there is no emulsion or water present. If the ether is contaminated, the emulsion or water may be removed by pouring the ether back into the **cleaned** separatory funnel, drawing off any water or emulsion present, and carefully decanting the ether again through the top.

The ether is next evaporated under vacuum. No heat should be used, since this will decompose the thermally sensitive nicotine. Using a 125-mL filter flask, assemble the apparatus shown in Figure 17-1. Add about 30 mL of the ether extract to the flask along with a wooden applicator stick, which serves the same purpose as a boiling chip. As the ether evaporates, the filter flask will become cold and may be gently heated with a beaker of lukewarm water. When the volume has been reduced to about 5 or 10 mL, add another 30-mL portion of the ether extract and continue the evaporation. When the volume has been reduced again, add all the remaining ether extract and continue until all the ether has been removed.

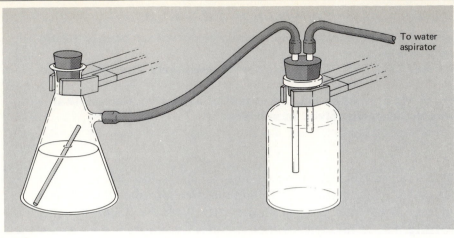

FIGURE 17-1

A vacuum-distillation apparatus for removing the ether from the thermally sensitive nicotine.

Next add 1 mL of water with gentle swirling to dissolve the residue. Add 5 mL of methanol and filter the solution into a small beaker through a **small** piece of glass wool placed loosely at the bottom of a stemless funnel. (Don't let the glass wool touch your skin.) Rinse the filter flask and the glass wool with an additional 5 mL of methanol and combine the two solutions. Be sure the filtered solution contains no suspended particles, or refiltering will be required. To the clear solution add 10 mL of a saturated solution of picric acid dissolved in methanol. A pale yellow precipitate of nicotine dipicrate will immediately appear. Collect the precipitate by suction filtration, using a Hirsch funnel. Any precipitate that forms in the filter flask from the evaporation of the methanol is mostly picric acid and can be discarded. **Do not allow it to become dry.**

To recrystallize the crude nicotine dipicrate, place the crystals into a 50-mL Erlenmeyer flask and slowly add a hot solution of 50% (by volume) ethanol and water until the solid just dissolves. Allow the hot solution to slowly cool, and collect the light yellow crystals by suction filtration, using a Hirsch funnel. The crystallization is slow, and you may want to stopper the flask and allow the solution to stand until the next laboratory period. Cooling the solution in an ice bath after it has reached room temperature will also help. Allow the crystals to dry overnight. Determine the yield and melting point, and record the results in your notebook. The literature melting point is 222–223°C. Also determine the percentage yield of your final product, and show the calculations. Remember that only a portion of the product is nicotine. When finished, hand in your product in a properly labeled container to your instructor. The infrared and the ^{1}H NMR spectra of nicotine are shown in Figures 17-2 and 17-3.

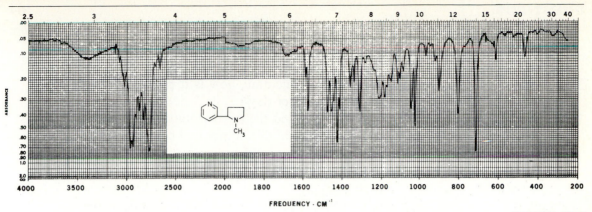

FIGURE 17-2

The infrared spectrum of nicotine, neat. Spectrum courtesy of Sadtler Research Laboratories, division of Bio-Rad Laboratories, Inc., copyrighted 1970.

FIGURE 17-3

The ^{1}H NMR spectrum of nicotine in carbon tetrachloride. Spectrum courtesy of Aldrich Chemical Company, *The Aldrich Library of NMR Spectra*, edition II, by Charles Pouchert.

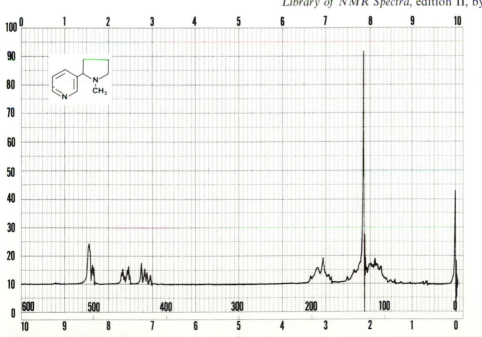

QUESTIONS

1. What possible benefits or uses could the nicotine removed from cigarette tobacco have?
2. If a cigar contains over two fatal doses of nicotine, how is it possible for a person to survive when smoking 10 cigars per day?
3. If you synthetically prepared and then isolated the dextrorotatory isomer of nicotine, how would the physical properties of this compound compare with those of naturally occurring nicotine? If naturally occurring nicotine consisted of both the levorotatory and dextrorotatory isomers, by what process could these isomers be separated?
4. Account for the fact that pure nicotine is colorless and yet the pure nicotine dipicrate is yellow.
5. Could other alkaloids be isolated as the picrate salt as nicotine was? Explain.
6. Briefly compare and contrast the physiological effects that the alkaloids nicotine, morphine, cocaine, quinine, caffeine, and strychnine have upon the body.

18

INSECT CONTROLS

The struggle between man and insects began long before the dawn of civilization, has continued without cessation to the present time, and will continue, no doubt, for as long as the human race endures. We commonly think of ourselves as the lords and conquerors of nature. But insects had thoroughly mastered the world and taken full possession of it long before man began the attempt. They had, consequently, all the advantage of possession of the field when the contest began, and they have disputed every step of our invasion of their original domain so persistently and successfully that we can't even yet scarcely flatter ourselves that we have gained any important advantage over them. . . . We cannot even protect our very persons from their annoying and pestiferous attacks, and since the world began, we have never yet exterminated—we probably shall never exterminate—so much as a single insect specie.

S. A. Forbes, U.S. Entomologist

INTRODUCTION

Although the introductory quote was written 70 years ago, for the most part it is just as true today. Despite humanity's tremendous advances in science and technology, we have made little progress in our battle with bugs. In fact, most experts now agree that if there ever is a winner in this war, it will **not** be humans.

Clearly, the present conditions are not in our favor. Around the world, agricultural pests cause billions of dollars in crop losses each year, producing starvation in many nations and economic ruin for farmers. In the United States alone, the figure is over $10 billion, or about 10 percent of all the crops grown. This percentage is very low compared with some countries, where over 75 percent is lost. In forest areas of the United States, millions of valuable trees are defoliated and killed each year, which could have provided enough lumber to build over a million homes. Insects also attack and annoy people and animals, and are responsible for spreading serious diseases such as yellow fever, malaria, river blindness, and sleeping sickness. These diseases are not very common here, but in other parts of the world they affect hundreds of millions of people, leaving millions of them either dead or sightless.

The situation was not always this severe. After World War II, the development and widespread use of chemical pesticides like DDT, dieldrin, aldrin, chlordane, lindane, heptachlor, and mirex, provided hope that the ultimate weapons against insects had been developed. For a few years, people were indeed winning, as hundreds of damaging species soon came under control.

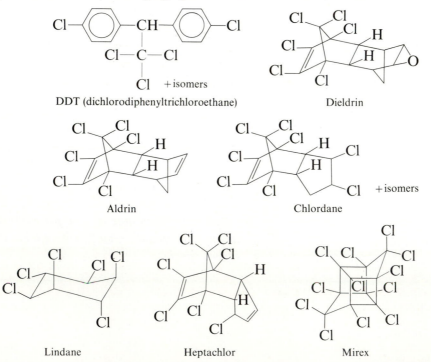

DDT (dichlorodiphenyltrichloroethane)

Dieldrin

Aldrin

Chlordane

Lindane

Heptachlor

Mirex

But because the life cycles of insects are quite short, many species quickly developed an immunity to the insecticides that once destroyed them. In fact, there are today at least 30 known "monster" species that absolutely **no** insecticide will kill! Some insects have even altered themselves so much that they actually thrive on dieldrin and aldrin, which once proved fatal!

Furthermore, DDT, dieldrin, aldrin, and other chlorocarbon insecticides are called "hard" insecticides, because they persist in the environment for a long time without decomposing to harmless materials. This presents an environmental hazard that became quite evident in the 1960s when the population of birds declined sharply after heavy spraying with DDT. The birds, which ate DDT-poisoned insects, accumulated large quantities of the substance, which interfered with their reproductive cycles and caused them to lay eggs that frequently broke before hatching because the shells were too thin. The potential danger to other animals and people has greatly curtailed the use of DDT, dieldrin, and aldrin. In this country, they have been banned by the Environmental Protection Agency, except for certain special purposes.

With the "hard" pesticides being used less and less, people have had to search for and resort to other weapons in their fight against bugs. Among these are "soft" pesticides, insect repellents, sex attractants, insect hormones, insect chemosterilants, physical controls, biological controls, biophysical controls, cultural controls, and legal controls.

"SOFT" PESTICIDES

"Soft" pesticides can be characterized as those that have a short lifetime and are decomposed to harmless products. Among the most common are carbaryl, parathion, malathion, dichlorvos, nicotine, and the pyrethrins, which have the following structures.

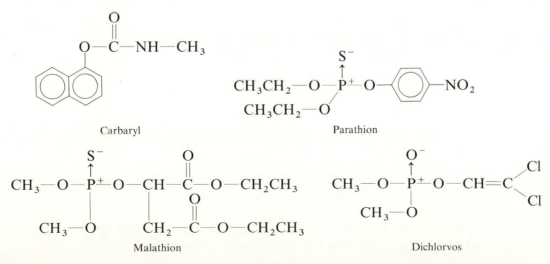

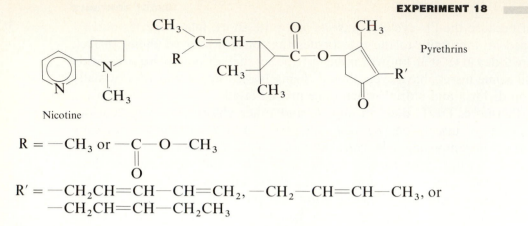

Nicotine

Pyrethrins

$$R = -CH_3 \text{ or } -\overset{\displaystyle O}{\underset{\displaystyle \|}{C}}-O-CH_3$$

$$R' = -CH_2CH=CH-CH=CH_2, -CH_2-CH=CH-CH_3, \text{ or } -CH_2CH=CH-CH_2CH_3$$

The compounds with phosphorus are highly toxic and must be used with stringent safety precautions. They tend to kill many more nontarget animals, such as mammals and birds, than do the chlorocarbon insecticides. Furthermore, insects resistant to the soft pesticides have been observed. Nicotine and the pyrethrins are obtained from plant sources.

INSECT REPELLENTS

Insect repellents are substances that keep insects away from people, animals, plants, food, shelter, and manufactured products. How these substances function is frequently unknown. They may be compounds that give the insects an unpleasant odor or taste, alarm pheromones that warn the insects of imminent danger, or sensor inhibitors that prevent the insects from finding a host. In whatever manner they work, they are an effective but limited method of control.

Ever since the beginning of civilization, people have tried to protect themselves from the biting and stinging attacks of insects. Early methods such as smoke, pitch, and various clays were somewhat effective, but had other obvious drawbacks. Later, plants containing essential oils such as citronellal, citronellol, geraniol, and other terpenes were rubbed over the body.

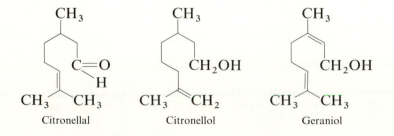

Citronellal   Citronellol   Geraniol

During World War II, the U.S. Department of Agriculture undertook the massive assignment of finding an effective repellent to protect the armed forces.

With over 20,000 compounds being isolated or synthesized and then tested, over 6000 were active against insects, but only 40 of these proved useful. The stringent requirements that eliminated practically all of the compounds were as follows. The product must

1. Be effective against a wide array of insects

2. Have long-lasting effects

3. Be stable to heat and light

4. Have low toxicity

5. Be nonirritating and nonallergenic

6. Be nonobjectionable because of feel, odor, or staining of clothing

7. Be inert or compatible with a wide array of formulation additives

After the war, the search for even better repellents continued, with a number of compounds proving worthwhile. Among those discovered were

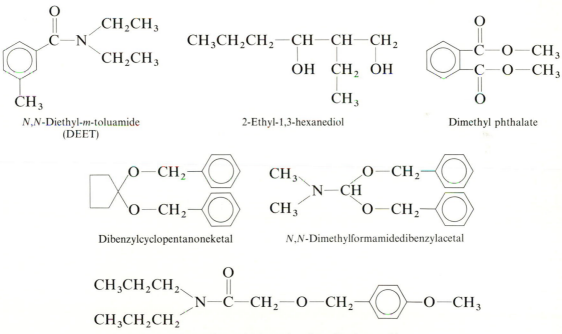

N,N-Diethyl-m-toluamide
(DEET)

2-Ethyl-1,3-hexanediol

Dimethyl phthalate

Dibenzylcyclopentanoneketal

N,N-Dimethylformamidedibenzylacetal

N,N-Dipropyl-2-(p-methoxybenzyloxy)acetamide

Of all the compounds tested, N,N-diethyl-m-toluamide (DEET) is currently considered to be the best all-purpose repellent and is the active ingredient in Off and 6-12 brands of repellents. 2-Ethyl-1,3-hexanediol is also very effective, and until the late 1970s was the active ingredient in 6-12.

To date it is still not known what physical and structural features of a compound are responsible for an effective and useful repellent. Many studies have been made, analyzing every conceivable factor involved, but so far no specific conclusions have been derived. Repellents have been found within practically every class of organic compound, and it now appears that it may be impossible to predict beforehand just how beneficial any particular compounds will be.

SEX ATTRACTANTS

Insect sex attractants are chemical compounds emitted by females in minute amounts to attract males. These compounds offer the possibility of insect control because when sprayed into the air by people, reproduction is dramatically reduced because the males are no longer able to locate the females. The structures of a number of these attractants have been identified and are shown below.

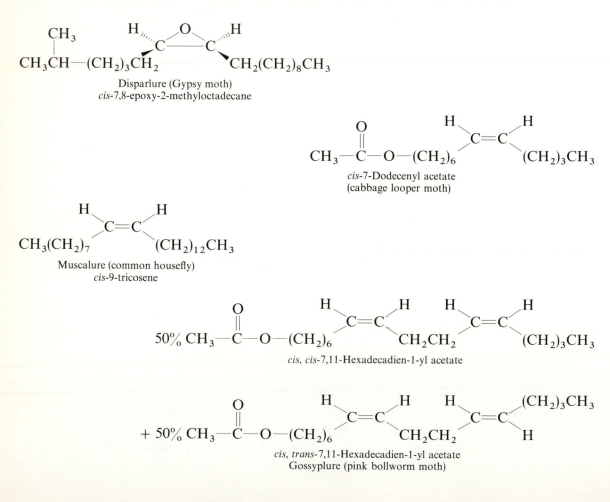

Disparlure (Gypsy moth)
cis-7,8-epoxy-2-methyloctadecane

cis-7-Dodecenyl acetate
(cabbage looper moth)

Muscalure (common housefly)
cis-9-tricosene

cis, cis-7,11-Hexadecadien-1-yl acetate

cis, trans-7,11-Hexadecadien-1-yl acetate
Gossyplure (pink bollworm moth)

As can be seen, a major difficulty with this approach is the complex struc-tures of the molecules, which can make their syntheses quite expensive for any large-scale campaign. In addition, since the insects produce only very minute quantities, determining the exact chemical structures of these sex attractants has been very difficult with the very small samples commonly available. In fact, incorrect structures have been assigned a number of times, and only the biolog-ical inactivity of the synthesized compound has revealed the error.

HORMONES

Insect hormones are substances secreted within the body of the insect to regu-late growth and metamorphosis during the various stages of development. The structures of a number of these hormones are now known, and experiments with one class of them, called **juvenile hormones**, have shown promise. Al-though this class of substances must be secreted in the metamorphosis from larva to pupa, at later times it must be absent or the insect will fail to mature and thus will be unable to reproduce. The structure of the juvenile hormone from the male cecropia moth is shown below, and has been found to prevent the maturing of a number of species.

Although the syntheses of large quantities of the juvenile hormones may be prohibitively expensive for agricultural use, a number of synthetic analogs, such as Entocon, have been prepared and shown to have similar properties and effectiveness. Furthermore, a terpenoid substance, called **paper factor**, found in the American balsam fir, has been discovered to be active against some species.

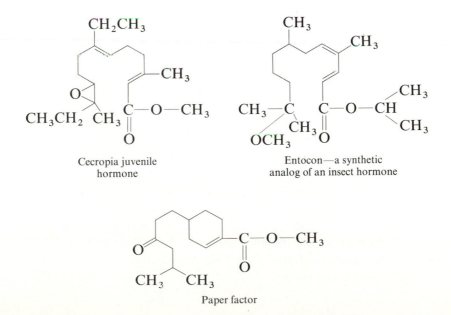

Cecropia juvenile
hormone

Entocon—a synthetic
analog of an insect hormone

Paper factor

CHEMOSTERILANTS

Insect chemosterilants are chemical compounds which reduce or eliminate the ability of an insect to reproduce. With some members made sterile, it can be shown mathematically that the reproductive potential of the population will decrease. In fact, if the ratio of sterile to fertile members becomes and remains sufficiently high, the population will eventually die out. The structures of a few proven chemosterilants are given below. Unfortunately, because of the potential environmental hazards involved, application of chemosterilants to date has been quite limited.

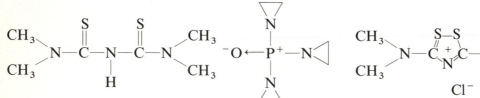

1,1,5,5-Tetramethyl-2,4-dithiobiuret

Tris(1-Aziridinyl)-
phosphine oxide

3,5-*Bis*(Dimethylamino)-
1,2,4-dithiazolium chloride

PHYSICAL CONTROLS

Physical controls include some very simple methods, such as door and window screens, mosquito netting, and the wide furrows that farmers sometimes plow around planted fields to hinder nonflying pests from reaching the crops. Also, since insects become inactive below 40°F, storing grain, vegetables, or other products at low temperatures can be an effective control.

BIOLOGICAL CONTROLS

Biological controls refer to the use of a pest's own natural enemies to bring it under control. Usually these enemies are other insects, birds, or microscopic organisms. They can either be introduced directly into the needed region or the environment of the region can be adjusted to encourage the growth of natural enemies already living there.

A drawback to the method is that the specie introduced to control the pest may become a pest itself. For example, the introduction of certain birds to control an insect problem may prove harmful if the bird population gets out of control and feeds on grains and other crops.

BIOPHYSICAL CONTROLS

Biophysical controls combine the two types given above. A common example is the use of gamma rays to sterilize the males of certain destructive species, which have been captured or raised in laboratories. When these males are released, they compete with fertile males for the available females, and thus reduce the number of offspring produced.

The success of the approach depends upon practical and inexpensive ways of raising, irradiating, and releasing large numbers of the sterile males. Also the pest population must not be so high that there are too many fertile males to compete with the sterile ones.

CULTURAL CONTROLS

Cultural controls are methods of protecting livestock and crops from pests. Included in this category are the removal of shelters and breeding areas for pests, the development of insect-resistant breeds of crops and animals, and crop rotation to control pests that prey on only one type of plant.

LEGAL CONTROLS

Legal controls refer to laws aimed at preventing the introduction of destructive pests from one part of the country or the world to another. This is accomplished by prohibiting the movement of plant or animal products which may carry live insect pests. In the past, at least half a dozen serious accidents have occurred where pests once unknown to a region have become a problem due to their accidental importation. Every state and nation wants to protect itself, and cooperation among them has proven to be very beneficial in this effort.

18A.   SYNTHESIS OF THE REPELLENTS OFF AND 6-12

NUCLEOPHILIC ACYL SUBSTITUTION

In this experiment, you will synthesize *N,N*-diethyl-*m*-toluamide, which is currently the best all-purpose insect repellent and the active ingredient in Off and 6-12. Observe from the structure that *N,N*-diethyl-*m*-toluamide is an aromatic amide, which is disubstituted both on the aromatic ring and the nitrogen atom.

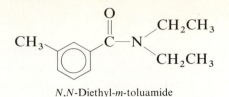

N,N-Diethyl-*m*-toluamide

The synthesis requires two reaction steps. First, *m*-toluic acid will be converted to its acid chloride derivative through the use of thionyl chloride, $SOCl_2$. Other chlorinating agents, such as phosphorus trichloride, PCl_3, and phosphorus pentachloride, PCl_5, could be used, but thionyl chloride has the advantage that the other reaction products are gases and thus easily removed. The acid chloride will then be reacted with diethylamine to yield the final product.

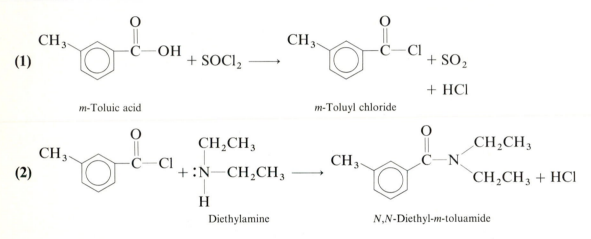

The reaction mechanism follows and involves nucleophilic substitution occurring three times. In step (1), the carbonyl oxygen of the carboxylic acid, acting as a nucleophile, displaces one of the chlorine atoms on thionyl chloride to give the protonated intermediate I. After the proton leaves in step (2) to give intermediate II, the displaced chloride ion from step (1) also acts as a nucleophile and attacks the carbonyl carbon of this intermediate in step (3) to give intermediate III. Sulfur dioxide and the other chlorine atom then leave in step (4) to give the acid chloride IV. The openness of the carbonyl group, due to its flat shape, greatly aids reaction (3). So too does the polarity of the carbonyl group and the willingness of the oxygen atom to readily accept a negative charge. The stability of the departing sulfur dioxide and chloride ion, compared to intermediate III, is critical to reaction (4) and enables it to be energetically favorable.

Once formed, the acid chloride IV easily undergoes nucleophilic attack by diethylamine in step (5) to give intermediate V. Again, the openness of the carbonyl group, due to its flat shape, greatly aids this reaction. So too does the polarity of the carbonyl group and the willingness of the oxygen atom to readily accept a negative charge. Once formed, intermediate V easily expels the weakly basic chloride ion in step (6) to give intermediate VI, which then loses a proton in step (7) to give the final product.

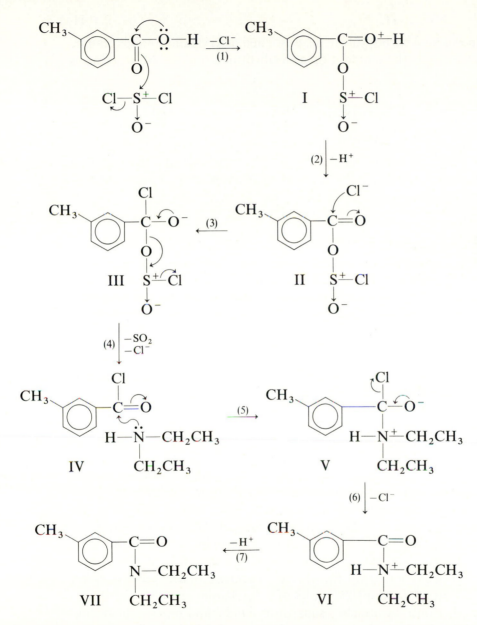

Note that the final product **cannot** be prepared directly by the reaction of *m*-toluic acid with diethylamine. If these two reagents are mixed, an acid-base reaction occurs, yielding the salt of the reactants, which will not react further while in solution.

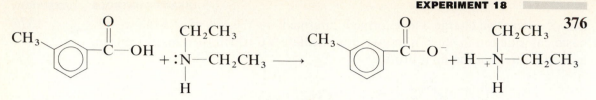

Many amine salts of carboxylic acids, if isolated as crystalline solids, can be strongly heated to yield the amide. For a disubstituted amide, the reaction is

This process is frequently used industrially for the synthesis of amides, but is not a convenient laboratory method. Here the more common and preferred way is the procedure involving acid chlorides.

PROCEDURE

CAUTION: Thionyl chloride is a noxious and corrosive chemical that must be handled with extreme care. Since it reacts with water to liberate hydrogen chloride and sulfur dioxide, all equipment must be dry to avoid unnecessary decomposition of the reagent. As always, safety glasses must be worn. If any of the reagent should be spilled on the skin, wash it off immediately with water and notify your instructor.

Diethylamine, used in the second reaction step, should also be handled with care. It is not only noxious and corrosive, like thionyl chloride, but also quite volatile, boiling at 56°.

Place 4.1 grams (0.030 mole) of *m*-toluic acid (3-methylbenzoic acid) and 5.0 mL (0.069 mole) of thionyl chloride into a **dry** 500-mL three-neck round-bottom flask. Equip the flask, as shown in Figure 18-1, with a reflux condenser, gas trap, and separatory funnel connected to a drying tube. The gas trap is essential to prevent hydrogen chloride and sulfur dioxide from escaping into the room. Add a boiling stone, stopper the unused neck of the flask, and start the flow of water into the reflux condenser. Using either a steam bath or a hot plate, **gently** heat the mixture for 20 to 30 minutes until the evolution of gases stops.

Let the flask cool to room temperature and add 50 mL of **anhydrous** ether. Next pour a solution of 10 mL (0.097 mole) of diethylamine dissolved in 20 mL of **anhydrous** ether into the attached separatory funnel and replace the drying tube. Add the diethylamine solution **dropwise** into the reaction flask over a period of about 30 minutes. Be sure the boiling action doesn't become too vigorous. Each time the diethylamine is added, a white cloud of diethylamine hydrochloride will form in the flask. Be sure the diethylamine solution is added slowly enough to prevent this cloud from becoming too large.

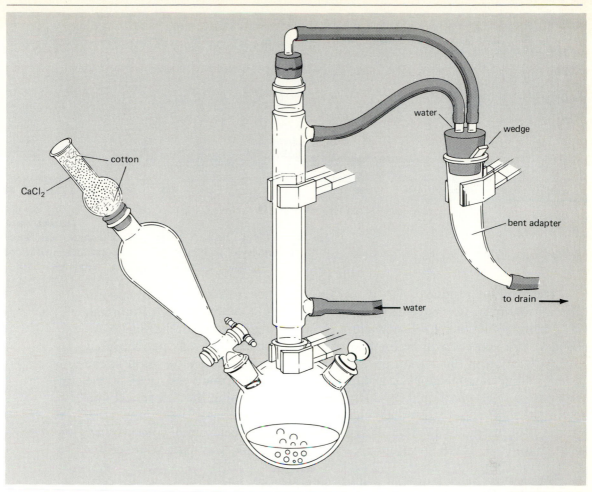

FIGURE 18-1

Apparatus for preparation of *N,N*-diethyl-*m*-toluamide.

After all the diethylamine is added and all signs of reaction have stopped, disconnect the apparatus and pour the contents of the flask into a separatory funnel. Any residue that remains can be washed out with ether and combined with the contents of the separatory funnel. It may be necessary to add additional ether to the funnel, due to evaporation during the reaction. Add 20 mL of 5% NaOH solution and shake to extract any unreacted *m*-toluic acid. Allow the layers to separate and draw off the lower NaOH portion. Repeat with a second 20-mL portion of 5% NaOH solution. To remove the unreacted diethylamine, extract the ether with 20 mL of 10% HCl solution, and after removal of the lower acid layer wash the ether with 20 mL of water. Dry the final ether layer over anhydrous magnesium sulfate, gravity filter the solution through fluted filter paper into a dry flask, and evaporate the ether on a steam bath or hot-water bath in the hood.

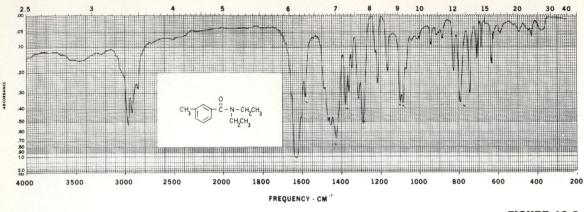

FIGURE 18-2

The infrared spectrum of *N,N*-diethyl-*m*-toluamide, neat. Spectrum courtesy of Sadtler Research Laboratories, division of Bio-Rad Laboratories, Inc., copyrighted 1967.

The resulting residue of *N.N*-diethyl-*m*-toluamide is best purified by column chromatography (see page 127). Prepare a column of 25 grams of alumina in a 25-mm-diameter tube, and elute with petroleum ether or ligroin. *N,N*-Diethyl-*m*-toluamide will be the first compound eluted and may be recovered by evaporation of the solvent on either a steam bath or hot-water bath in the hood. Using a weighed flask will permit you to determine the yield of the colorless to light tan liquid product remaining. Determine the theoretical and percentage yields of the product, and show the calculations.

At your instructor's option, determine the infrared spectrum of your product neat, and the ^{1}H NMR spectrum in carbon tetrachloride. Compare your spectra with those shown in Figures 18-2 and 18-3.

When finished, submit your product in a properly labeled container to your instructor.

18B. PHEROMONES FROM THE GRIGNARD SYNTHESIS: 4-METHYL-3-HEPTANOL AND 4-METHYL-3-HEPTANONE

■ **NUCLEOPHILIC ADDITION**
■ **OXIDATION USING HOUSEHOLD CHLORINE BLEACH**
■ **ELIMINATION, SECOND-ORDER**

In this experiment, you will synthesize two insect pheromones using the Nobel-Prize-winning Grignard reagent (see the introduction to Experiment 25 for a discussion of the Nobel Prize). This reagent, discovered by the French chemist Victor Grignard in 1898, is formed when certain alkyl or aryl halides react with magnesium metal in dry ether.

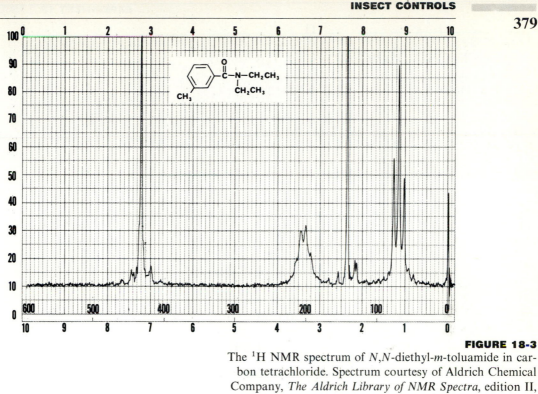

FIGURE 18-3

The ^1H NMR spectrum of *N,N*-diethyl-*m*-toluamide in carbon tetrachloride. Spectrum courtesy of Aldrich Chemical Company, *The Aldrich Library of NMR Spectra*, edition II, by Charles Pouchert.

$$R—X + Mg \xrightarrow{\text{ether}} R—MgX \qquad \text{where R is either an alkyl or aryl group}$$

The structure of the reagent is usually written as RMgX, R—MgX or $\overset{\delta-}{R}—\overset{\delta+}{MgX}$. In reality, however, the exact structure is unknown and is thought to be a mixture of numerous species, some of which may be quite complex. Grignard received the Nobel Prize in 1912 for showing how valuable, versatile, and prolific the Grignard reagent is in organic syntheses. Once formed, the reagent reacts easily with a wide range of molecules, including aldehydes, ketones, esters, epoxides, alkyl halides, acyl halides, nitriles, carbon dioxide, and acids to yield an impressive array of products. For example:

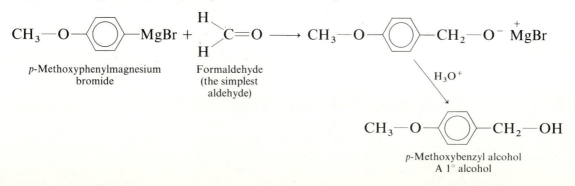

p-Methoxyphenylmagnesium bromide

Formaldehyde (the simplest aldehyde)

p-Methoxybenzyl alcohol
A 1° alcohol

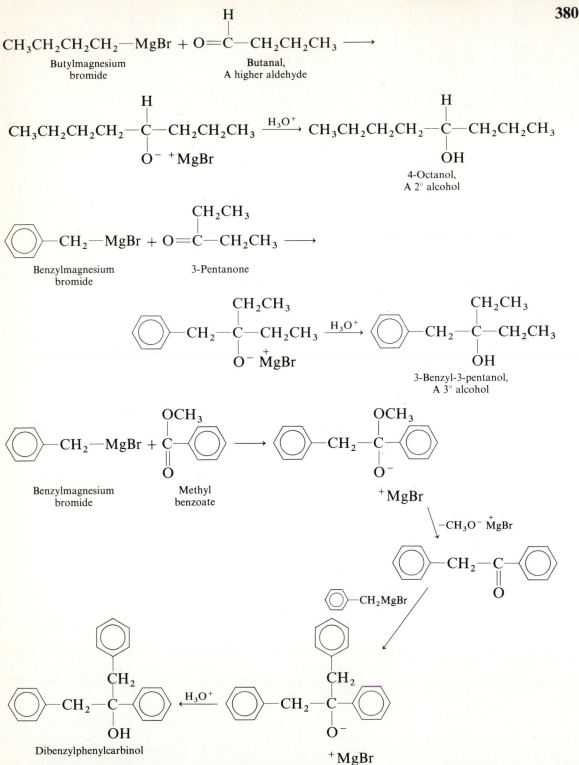

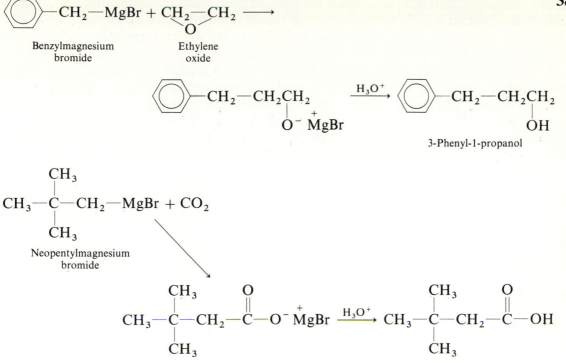

Benzylmagnesium bromide

Ethylene oxide

3-Phenyl-1-propanol

Neopentylmagnesium bromide

3,3-Dimethylbutanoic acid

The value of the reagent can be seen in the fact that when Grignard died in 1935, over 6000 scientific papers dealing with the subject had been published!

The high reactivity of the reagent means that it can also be destroyed by a number of undesired side reactions, unless certain experimental precautions are employed. The reagent, for example, will easily react with water, oxygen, carbon dioxide, and any unreacted organic halide, as shown in the following equations:

$$R—MgX + H_2O \longrightarrow R—H + Mg(OH)X$$

$$R—MgX + O_2 \longrightarrow R—O—O^- \overset{+}{M}gX$$

$$R—MgX + CO_2 \longrightarrow R—\underset{\underset{O}{\|}}{C}—O^- \overset{+}{M}gX$$

$$R—MgX + R—X \longrightarrow R—R + MgX_2$$

Because of the reaction with water, the Grignard reagent must be prepared and used under anhydrous conditions. The glassware, reactants, and solvent must be scrupulously dry. The low molecular weight of water means that it only takes about 2 mL of the liquid to destroy the entire quantity of Grignard reagent prepared in most laboratory procedures. Furthermore, traces of water can also strongly inhibit or even prevent the reaction forming the reagent from even starting.

The reaction with oxygen and carbon dioxide can be avoided by conducting the reaction under an inert atmosphere, such as helium or nitrogen gas. Allowing the ether solvent to reflux can also be somewhat effective, since the solvent vapor will act like a blanket to exclude the air from the surface of the reaction mixture.

The coupling reaction of the Grignard reagent with a second molecule of organic halide cannot be completely prevented, but you can minimize it by avoiding the use of the more reactive organic iodides and by keeping a low concentration of the organic halide in solution. Do this by slowly adding the halide to the magnesium in ether with very efficient stirring.

The order of reactivity of the organic halides with magnesium is RI > RBr > RCl > RF. Very few organomagnesium fluorides have been successfully prepared. Alkyl halides are more reactive than the corresponding aryl or vinyl halides. Usually an organic bromide is the best reagent to choose, since it will usually react reasonably fast without giving extensive unwanted coupling. Aryl and vinyl chlorides are usually avoided, since they react so slowly and, in some cases, don't react at all. Changing the solvent from ethyl ether to the slightly more basic and higher-boiling tetrahydrofuran does allow the reaction to occur more readily. Tetrahydrofuran, however, is completely water soluble, which can make the products of the Grignard reaction more difficult to isolate.

Grignard reactions can frequently be difficult to start, even when the glassware, reagents, solvent, and atmosphere are kept scrupulously dry. The use of magnesium that is pure and relatively free of oxide coating helps. So does the addition of a small crystal of iodine. This induces the reaction by activating the surface of the metal.

In this experiment, you will prepare the insect pheromones 4-methyl-3-heptanol and 4-methyl-3-heptanone by means of the Grignard reaction. 4-Methyl-3-heptanol is produced by female European elm beetles, *Scolytus mulbistriatus,* and attracts large numbers of both male and female beetles when it is emitted with 2,4-dimethyl-5-ethyl-6,8-dioxabicyclo(3.2.1)octane by the female beetles and α-cubebene by the Dutch elm tree. The combination of these three compounds could thus theoretically be used as a lure in traps to destroy

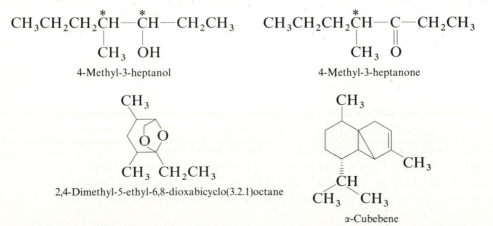

4-Methyl-3-heptanol

4-Methyl-3-heptanone

2,4-Dimethyl-5-ethyl-6,8-dioxabicyclo(3.2.1)octane

α-Cubebene

large numbers of European elm beetles. However, how practical and worthwhile such a method would be is highly questionable.

Observe that 4-methyl-3-heptanol contains two chiral centers. Your synthesized product will therefore be a mixture of two pairs of enantiomers or four optical isomers. Although these isomers can be separated by chromatography and resolution techniques, this will not be done in this experiment.

4-Methyl-3-heptanone is an alarm-response pheromone for a number of ant species, including the Texas leaf-cutting ant *Atta texana* and the harvester ant *Pogonomyrmex barbatus*. When exposed to the compound above ground, the ants sense danger and dig below the surface to escape. The compound thus has the potential of being a method to control these species of ants. Observe that this molecule also has a chiral center and therefore exists as a mixture of two enantiomers. However, of the two, only the S-(+)-isomer is found to have any biological activity with the ants.

4-Methyl-3-heptanol is prepared by the following reactions:

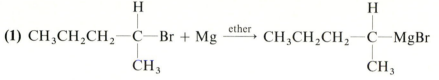

2-Bromopentane

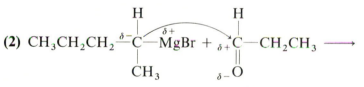

Propanal

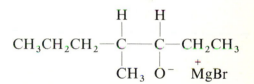

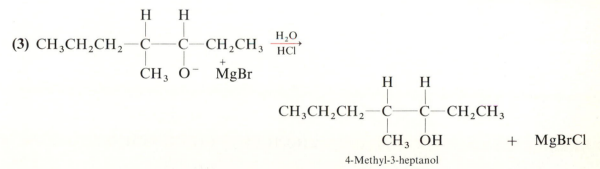

4-Methyl-3-heptanol

In step (1), 2-bromopentane is reacted with magnesium in anhydrous ether to give the Grignard reagent. Because of the differing electronegativity between carbon and magnesium, the Grignard reagent has a partial negative charge on carbon. This carbon atom can therefore act as a nucleophile and attack the electron-deficient carbonyl carbon atom of propanal in step (2) to yield a tetrahedral intermediate by nucleophilic addition. The openness of the carbonyl group, due to its flat shape, also aids the reaction, as does the willingness of the carbonyl oxygen atom to readily accept a negative charge. The workup of the reaction mixture with acid and water in step (3) then gives 4-methyl-3-heptanol as the product.

Once isolated, the 4-methyl-3-heptanol can be oxidized with acidic household chlorine bleach to yield 4-methyl-3-heptanone. Observe that in this reaction, the C-3 chiral center is removed, so the initial mixture of four optical isomers becomes a new mixture of two isomers.

$$CH_3CH_2CH_2-\overset{*}{C}H-\overset{*}{C}H-CH_2CH_3 + Na^+OCl^- \xrightarrow{\ CH_3COOH\ }$$
$$\underset{\underset{CH_3}{|}\ \underset{OH}{|}}{}$$

$$CH_3CH_2CH_2-\overset{*}{C}H-\overset{\ }{C}-CH_2CH_3 + H_2O + Na^+ Cl^-$$
$$\underset{\underset{CH_3}{|}\ \underset{O}{\|}}{}$$

This reaction is also a relatively new way to prepare ketones from secondary alcohols and appears to occur by the following mechanism:

(1) $CH_3CH_2CH_2-\overset{*}{C}H-\overset{H}{\underset{\underset{OH}{|}}{\overset{|}{C}{}^*}}-CH_2CH_3 + HOCl \longrightarrow$
$$\underset{CH_3}{|}$$

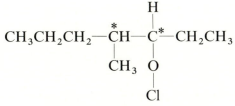

4-Methyl-3-heptanol hypochlorite

or

$$CH_3CH_2CH_2-\overset{*}{C}H-\overset{H}{\underset{\underset{OH}{|}}{\overset{|}{C}{}^*}}-CH_2CH_3 + Cl_2 \longrightarrow$$
$$\underset{CH_3}{|}$$

$$CH_3CH_2CH_2-\overset{*}{C}H-\overset{H}{\underset{\underset{O}{|}}{\overset{|}{C}{}^*}}-CH_2CH_3$$
$$\underset{CH_3}{|}\qquad\underset{Cl}{|}$$

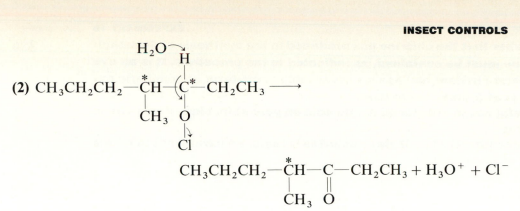

(2)

or

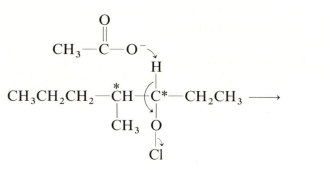

In step (1), the 4-methyl-3-heptanol reacts with either the hypochlorous acid or chlorine gas in the reaction mixture to give 4-methyl-3-heptanol hypochlorite, which then in step (2) undergoes an E-2 or elimination, second-order reaction. As you can see, either water or acetate ion, acting as a base, removes the hydrogen on the hypochlorite-bearing carbon. At the same time, the chlorine atom leaves, taking with it the electron pair, and the carbon-oxygen π bond of the ketone is formed. This mechanism, however, is only tentative, and much more work to establish it still needs to be done.

PROCEDURE

CAUTION: Remember that diethyl ether and propanal are both very flammable. Be certain there are no open flames in your area when you are using them.

Because of the strong odors of the propanal and 2-bromopentane, the synthesis using them must be done primarily in the hood.

Be careful when using household chlorine bleach, since it can damage both your clothes and certain areas of the laboratory.

Remember that the chlorine gas produced in the synthesis of 4-methyl-3-heptanone must be contained, as indicated in the procedure. It is an eye and respiratory irritant and has an acceptable short-term atmospheric exposure level of 3 parts per million.

Be careful not to get glacial acetic acid on your skin, since it can cause severe burns.

Also remember that methylene chloride is toxic. Be careful not to inhale it or get it on your skin.

All the glassware and the magnesium used in the Grignard synthesis of 4-methyl-3-heptanol must be scrupulously dry. Even apparently dry glassware contains moist air and a surprisingly large amount of water adhered to the walls, which will greatly lower the yield and can even prevent the reaction from starting.

All the glassware and the magnesium may be dried by leaving them in a drying oven overnight. But if a drying oven is unavailable or cannot hold everything, gently heat all the surfaces of the glassware with a burner for about 5 minutes. Be sure there is no ether or propanal nearby.

To ensure dryness, use only a recently opened, sealed metal can of anhydrous or absolute diethyl ether.

In addition, the propanal used in this experiment must contain no traces of propanoic acid, which would also destroy the Grignard reagent. To remove the propanoic acid, the propanal must be freshly distilled.

Also remember that the Grignard reagent cannot be stored, so the experiment cannot be left until after the reagent has reacted with the propanal.

4-Methyl-3-heptanol

Put 5.0 grams (0.21 mole) of oven-dried magnesium turnings, 75 mL of anhydrous ether, an iodine crystal, and 2.0 mL of 2-bromopentane into a dried 250-mL round-bottom flask. Attach a dried Claisen head, a dried reflux condenser, a dried separatory funnel, and two drying tubes filled with anhydrous calcium chloride, as shown in Figure 18-4. Be sure the apparatus is securely clamped in the places indicated, using a single stand to hold all the clamps. Also do not tightly pack the drying tubes. Otherwise, you may have a closed system which cannot easily release any pressure buildup.

Warm the reaction mixture with a warm-water bath until the reaction begins. About 2 to 5 minutes should be needed. Also have present a cold-water bath in case the reaction shows any signs of getting out of control.

After the reaction has started, remove the drying tube from the separatory funnel, be sure the stopcock is closed, then add to the separatory funnel 17.0 mL (0.152 total mole) of 2-bromopentane. Replace the drying tube; then gradually add the 2-bromopentane to the reaction mixture over a 30-minute period. Holding the stand, carefully and gently swirl the reaction mixture periodi-

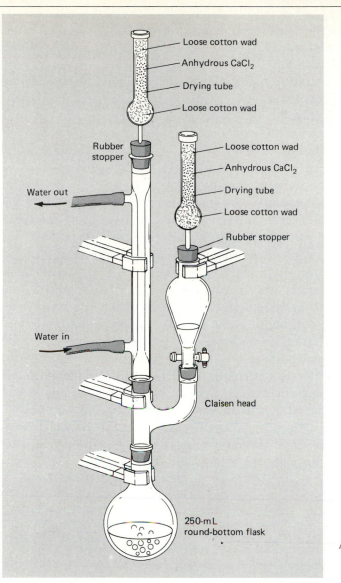

FIGURE 18-4
Apparatus for Grignard
preparation of
4-methyl-3-heptanol.

cally during the addition, **being certain each apparatus piece remains secure**. The reaction should proceed with moderate refluxing, and may be controlled by using the cold-water and the warm-water baths as needed.

After all the 2-bromopentane has been added, continue the refluxing for another 20 minutes with occasional **careful** and **gentle** swirling. Cool the mixture to room temperature, remove the drying tube from the separatory funnel, be sure the stopcock is closed, then add 10.9 mL (0.152 mole) of the predistilled propanal to the separatory funnel. Replace the drying tube, then gradually add the propanal over a 10- to 15-minute period. **Carefully** and

gently swirl the reaction mixture periodically during the addition, and be prepared to cool it in the cold-water bath if it begins to get out of control. After all the propanal has been added, reflux the mixture for 20 minutes with periodic **careful** and **gentle** swirling.

Cool the reaction mixture to less than 15°C, remove the two drying tubes and the separatory funnel, and stopper the one opening in the Claisen head. Then **carefully** and **slowly** add 35 mL of distilled water in 5- to 10-mL portions through the top of the condenser. Keeping the drying tube off, swirl the flask after each addition, then add enough 10% HCL solution to dissolve the magnesium salts. Some magnesium metal may still remain.

Remove the reflux condenser and the Claisen head, and gravity filter the reaction mixture through a small wad of cotton into a 250-mL separatory funnel. Remove the lower aqueous layer and wash the upper ether layer with two 35-mL portions of 10% NaOH solution and once with 25 mL of saturated salt solution. Dry the ether layer over anhydrous magnesium sulfate and decant into a clean, dry 250-mL round-bottom flask, filtering if necessary through a small piece of cotton in a stemless funnel.

Set up a simple distillation apparatus and distill the ether, using a steam bath or a warm-water bath, until about 25 mL of liquid remains. Transfer the remaining liquid to a 50-mL round-bottom flask, set up a simple distillation apparatus, and continue the distillation, using a steam bath or a warm-water bath. When no more ether distills, change receivers and continue the distillation, using a hot plate or a heating mantle. Collect the fraction boiling between 150 and 165°C in a preweighed 25-mL Erlenmeyer flask, and determine the yield. Record the result and the chief boiling range in your notebook. The literature boiling point is 160–161°C. Also determine the theoretical and percentage yields of your product, and show the calculations.

FIGURE 18-5

The infrared spectrum of 4-methyl-3-heptanol, neat. Spectrum courtesy of Sadtler Research Laboratories, division of Bio-Rad Laboratories, Inc., copyrighted 1966.

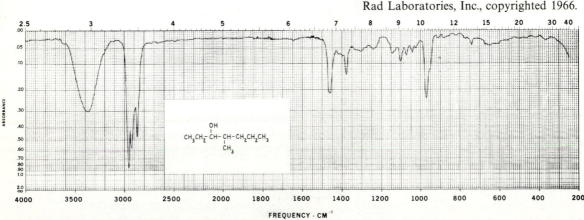

At your instructor's option, determine the infrared spectrum of your product, run neat between salt plates, and the ^{1}H NMR spectrum in deuterated chloroform. Compare your spectra with those shown in Figures 18-5 and 18-6. Save 7.0 mL of your product for the oxidation to 4-methyl-3-heptanone, and hand in the remainder to your instructor in a properly labeled container at the end of the experiment.

4-Methyl-3-heptanone

Place 7.0 mL (0.040 mole) of 4-methyl-3-heptanol and 10 mL of glacial acetic acid into a clamped three-neck, 500-mL round-bottom flask. Place a clamped condenser in the center neck, a 125-mL separatory funnel in one of the side necks, and a thermometer in the other side neck. Place a piece of glass wool in the bottom of a drying tube and pour in about 3 cm of **moistened but not wet** sodium bisulfite, covered by another piece of glass wool. The sodium bisulfite will react with any chlorine that might escape up the condenser. Be sure the sodium bisulfite is not so tightly packed that you will have a closed system. Place the drying tube into the top of the condenser.

Carefully add 70 mL (0.057 mole) of 5.25% household chlorine bleach to the separatory funnel. Add about 10 mL of the bleach solution to the 4-methyl-3-heptanol mixture and **carefully** swirl the flask to mix the reactants. Then

FIGURE 18-6
The ^{1}H NMR spectrum of 4-methyl-3-heptanol in deuterated chloroform. Spectrum courtesy of Aldrich Chemical Company, *The Aldrich Library of NMR Spectra*, edition II, by Charles Pouchert.

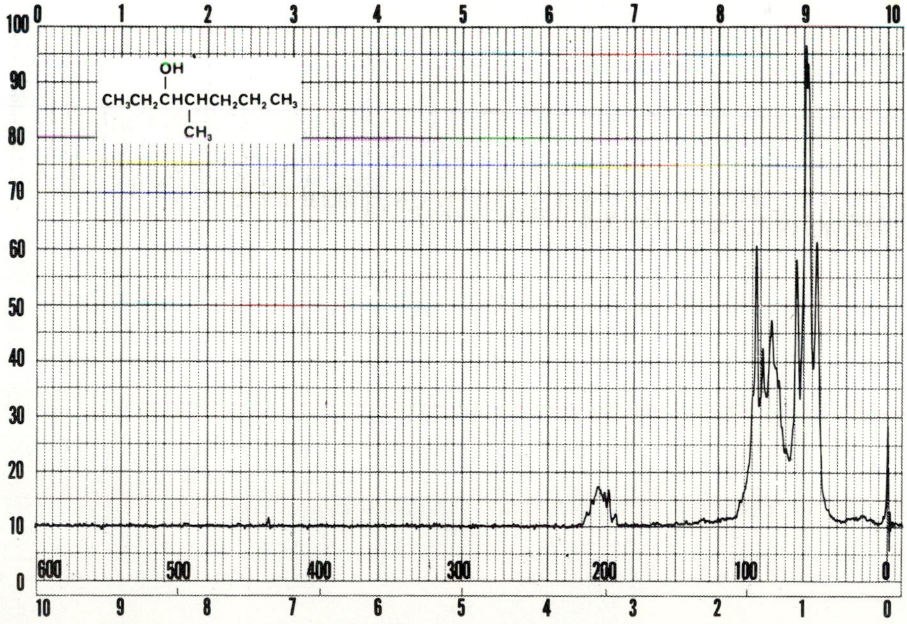

keeping the temperature at about 50°, start adding the rest of the bleach drop-wise, carefully swirling the flask periodically. The addition should take about 20 minutes, and during the early part of the reaction you should see the yellow color of the chlorine disappear as it reacts.

After all the bleach is added, remove the separatory funnel and replace it with a glass stopper. Continue the reaction for another 40 minutes, periodically swirling the flask.

To test for excess oxidant at the end of the reaction period, wet a piece of starch-iodide paper and put a drop of the liquid from the reaction mixture on it. If any oxidant is present, the blue color of the starch-triiodide complex will appear. If this happens, add 2 mL of saturated $NaHSO_3$ solution to the flask and stir. This will remove the excess oxidant by the reaction

$$OCl^- + HSO_3{}^- \longrightarrow Cl^- + HSO_4{}^-$$

Test the reaction liquid again with wet starch-iodide paper and, if necessary, continue adding small amounts of $NaHSO_3$ solution until no color change to blue occurs.

Remove the drying tube, condenser, and thermometer. Then gradually add, with stirring, saturated sodium carbonate solution to the reaction mixture until no more foaming is produced. This will convert the acetic acid to sodium acetate.

Cool the reaction mixture in an ice bath; then pour the reaction mixture into a 250-mL separatory funnel, rinsing out the reaction flask with small quantities of water. Extract the reaction mixture twice with 30-mL portions of methylene chloride, periodically releasing in the hood any internal pressure.

FIGURE 18-7

The infrared spectrum of 4-methyl-3-heptanone, neat. Spectrum courtesy of Sadtler Research Laboratories, division of Bio-Rad Laboratories, Inc., copyrighted 1975.

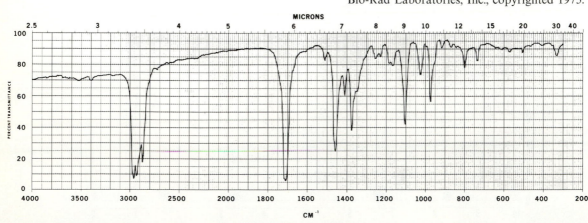

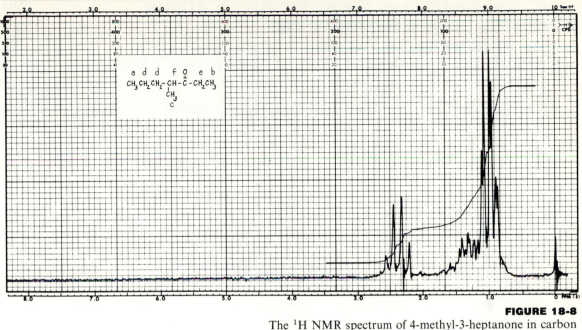

FIGURE 18-8

The ¹H NMR spectrum of 4-methyl-3-heptanone in carbon tetrachloride. Spectrum courtesy of Sadtler Research Laboratories, division of Bio-Rad Laboratories, Inc., copyrighted 1975.

Combine the methylene chloride extracts, dry them over anhydrous magnesium sulfate, then carefully decant or gravity filter through fluted filter paper the methylene chloride solution into a 100-mL round-bottom flask.

Set up a simple distillation apparatus and distill the methylene chloride over a hot-water or steam bath. When no more liquid distills, change receivers and continue the distillation, using a hot plate or heating mantle. Collect the fraction boiling between 150 and 165°C in a preweighed 25-mL Erlenmeyer flask, and determine the yield. Record the result and the chief boiling range in your notebook. The literature boiling point is 156–157°C at 745 mm Hg. Also determine the theoretical and percentage yields of your product, showing the calculations.

At your instructor's option, obtain the infrared spectrum of your product, run neat between salt plates, and the ¹H NMR spectrum in carbon tetrachloride. Compare your spectra with those shown in Figures 18-7 and 18-8. When finished, hand in both of your products in properly labeled containers to your instructor.

References

1. R. Einterz, J. Ponder, and R. Lenox, **Journal of Chemical Education, 54,** 382 (1977).

2. J. Mohrig et al., **Journal of Chemical Education, 62,** 519 (1985).

3. R. Stevens, K. Chapman, and H. Weller, **Journal of Organic Chemistry, 45,** 2030 (1980).

18C. INSECTICIDES AND THE DIELS-ALDER REACTION, THE SYNTHESIS OF *CIS*-NORBORNENE-5,6-*ENDO*-DICARBOXYLIC ANHYDRIDE, A DIELS-ALDER INSECTICIDE ANALOG

In 1928, Otto Diels and Kurt Alder discovered that alkenes and alkynes add to conjugated dienes to produce six-membered rings. The reaction proved to be such a valuable and versatile synthetic method that Diels and Alder were awarded the Nobel Prize in 1950. (See the introduction to Experiment 25 for a discussion of the Nobel Prize.)

The Diels-Alder reaction is called a **cycloaddition reaction** because a cyclic compound is formed from the addition of two frequently noncyclic molecules. As shown below, the six-membered ring is produced from the shifting of three pairs of π electrons. Three π bonds are broken, but two σ bonds and a new π bond are formed. Since these σ bonds are stronger than the broken π bonds, the Diels-Alder reaction is usually favored energetically. However, it is also usually reversible. And since the alkene or alkyne is frequently eager to react with the diene, it is called a **dienophile**, meaning a "lover of dienes."

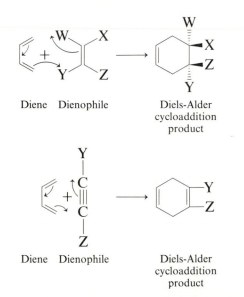

Diene	Dienophile		Diels-Alder cycloaddition product

Almost any diene or dienophile can undergo the Diels-Alder reaction. But for the reaction to occur readily, there must be a marked difference between the electron densities of the two. The dienophile must contain strong electron-withdrawing groups such as cyano ($-C{\equiv}N$) or carbonyl ($>C{=}O$), while the diene reacts best when electron-releasing groups such as alkyl and alkoxy ($-OR$) are present.

Some specific examples of the Diels-Alder reaction follow. Observe that when the dienophile is cyclic, a bicyclic product results. When the diene is cyclic, a bridged bicyclic product results.

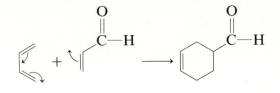

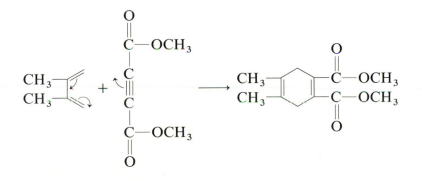

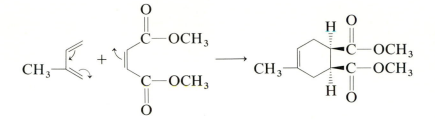

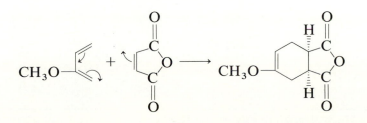

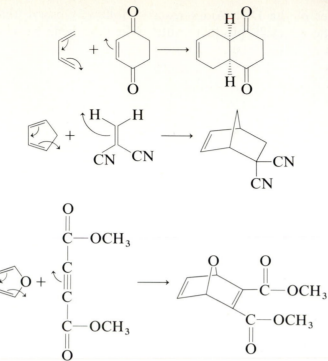

The Diels-Alder reaction is believed to have a concerted mechanism, with all the bond making and bond breaking occurring in a single step. There is a simultaneous cyclic movement of the four π electrons in the diene and two π electrons in the dienophile. This is believed to occur as shown in Figure 18-9, with the diene and dienophile approaching each other in parallel planes. Bonding occurs through the overlap of the π-electron clouds of both molecules.

FIGURE 18-9

Bond formation in the Diels-Alder reaction by overlapping of the π-electron clouds in the diene and dienophile.

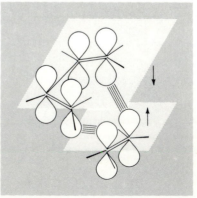

Observe from Figure 18-9 how the diene has the s-cis rather than the s-trans conformation. Only in this way can the two end *p* orbitals of the diene overlap in the transition state with the two *p* orbitals of the dienophile. In fact, if a diene cannot form the s-cis conformation, it will not undergo the Diels-Alder reaction. For example, both of the following dienes fail to react, since their double bonds are "locked" into the s-trans arrangement.

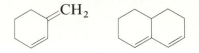

On the other hand, if the diene is "locked" into the s-cis arrangement, it will usually be even more reactive than the corresponding open-chain diene. Some examples of "locked" s-cis dienes are

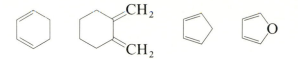

This concerted Diels-Alder mechanism also explains why the reaction gives only **syn addition**. With all the bonds being made or broken in the same step, there simply is no chance for any of the substituents on the diene or the dienophile to change their stereochemical positions during the reaction. Therefore, the arrangement before the reaction is the same as the arrangement afterward, and one can easily predict exactly what will be formed. For example:

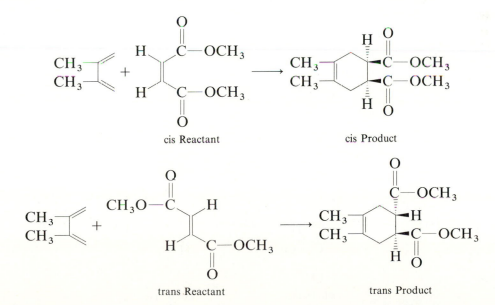

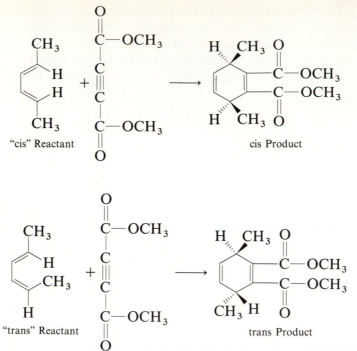

Another feature of the Diels-Alder reaction is the formation of primarily the *endo* rather than the *exo* product when a cyclic diene reacts. For example:

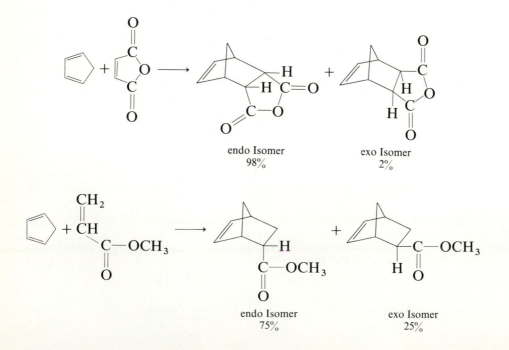

This is **not** what we would have predicted, since the **exo** product is usually more stable than the **endo**. An explanation seems to be the secondary overlap between the *p* orbital of the electron-withdrawing group of the dienophile and one of the inner *p* orbitals of the diene. This is illustrated in Figure 18-10. As a result, there is additional stabilization of the transition state for the endo isomer not possible for the exo.

In cases where both the diene and the dienophile are unsymmetrical, the Diels-Alder product is usually only a single isomer. For example:

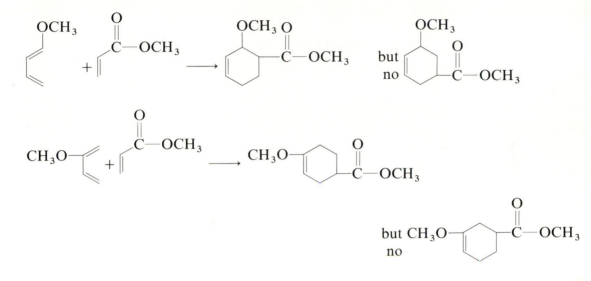

Observe that when the unsymmetrical diene has a group at the end of the molecule, the final product has the two substituents on adjacent carbons. But when the unsymmetrical diene has the group not on the end, the final product has the two groups at positions 1 and 4 on opposite sides of the molecule. An explanation is to use resonance structures of both starting materials and to have

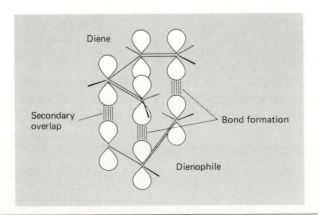

FIGURE 18-10
Secondary overlap between the *p* orbital of an electron-withdrawing group of the dienophile and one of the inner *p* orbitals of the diene.

the molecules line up so that partial positive charges are attracted to partial negative charges. For example:

(1)

$$
\begin{array}{ccc}
CH_3 & CH_3 & CH_3 \\
| & | & | \\
O & O^+ & O^{\delta+} \\
\end{array}
\qquad \longleftrightarrow \qquad \text{or} \qquad {}^{\delta-}
$$

$$
\underset{C-OCH_3}{\overset{O}{\|}} \quad \longleftrightarrow \quad \underset{C-OCH_3}{\overset{O^-}{\|}}_{+} \quad \text{or} \quad \underset{C-OCH_3}{\overset{O^{\delta-}}{\|}}_{\delta+}
$$

$$
\underset{\delta-}{\overset{CH_3}{\underset{|}{O^{\delta+}}}} \;+\; \underset{\delta+}{\overset{O^{\delta-}}{\underset{C-OCH_3}{\|}}} \quad \longrightarrow \quad \text{OCH}_3\;\; \overset{O}{\underset{\|}{C}}-OCH_3
$$

(2) $CH_3-O=\!\!\!\longleftrightarrow\; CH_3-\overset{+}{O}=\!\!\!\!\underset{-}{} \quad \text{or} \quad CH_3-\overset{\delta+}{O}=\!\!\!\cdots\;^{\delta-}$

$$
\underset{O}{\overset{\|}{\underset{C-OCH_3}{}}} \quad \longleftrightarrow \quad \underset{O^-}{\overset{+}{\underset{C-OCH_3}{}}} \quad \text{or} \quad \underset{O^{\delta-}}{\overset{\delta+}{\underset{C-OCH_3}{}}}
$$

$$
CH_3-\overset{\delta+}{O}\cdots\,^{\delta-} \;+\; \underset{O^{\delta-}}{\overset{\delta+}{\underset{C-OCH_3}{}}} \quad \longrightarrow \quad CH_3-O\!\!\!-\!\!\!\overset{}{\underset{C-OCH_3}{\overset{\|}{O}}}
$$

To predict what starting materials are needed to synthesize a certain Diels-Alder product, simply write down the structure of the product and then reverse the electron flow that would have created this product. Be certain that you end up with a conjugated diene and an alkene or alkyne, and that the alkene or alkyne has the electron-withdrawing group, if any are present. For example:

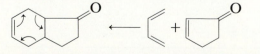

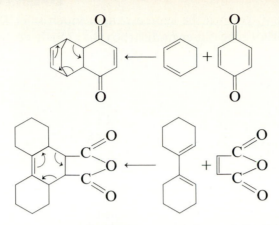

One very important practical and commercial use of the Diels-Alder reaction has been the synthesis of various insecticides, such as dieldrin, aldrin, chlordane, and heptachlor. All of these insecticides, as you can see, are structurally similar and are synthesized from the same cyclic diene. Two of the insecticides, in fact, are even named after Diels and Alder.

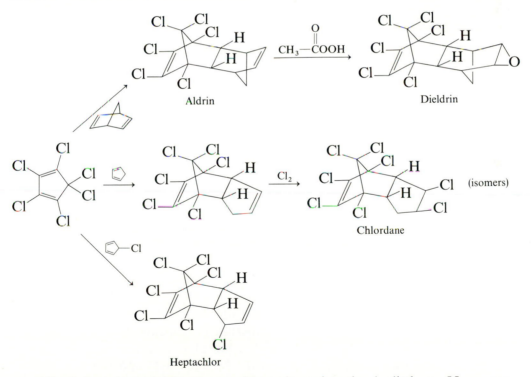

All of these insecticides are carcinogenic and toxic. As little as 55 mg per kilogram of body weight can be fatal, and they may be breathed in, eaten, or absorbed through the skin. In humans, poisoning causes headaches, dizziness, nausea, vomiting, and rapid breathing. Convulsions followed by a coma can also occur in severe cases.

Therefore, in this experiment, you will be synthesizing a much safer Diels-Alder insecticide analog, *cis*-norbornene-5,6-*endo*-dicarboxylic anhydride, by the following reaction:

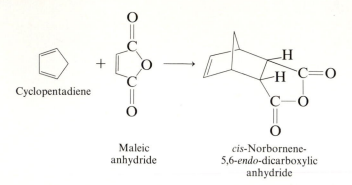

Cyclopentadiene   Maleic
                    anhydride

cis-Norbornene-
5,6-*endo*-dicarboxylic
anhydride

The cyclopentadiene is prepared when needed by the thermal decomposition of *endo*-dicyclopentadiene. This reaction, you will observe, is an example of the reverse Diels-Alder reaction. Note that cyclopentadiene cannot just be poured out of a reagent bottle because it readily dimerizes by the Diels-Alder reaction at room temperature to yield *endo*-dicyclopentadiene.

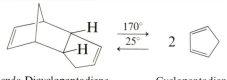

endo-Dicyclopentadiene   Cyclopentadiene

PROCEDURE

CAUTION: Cyclopentadiene is extremely flammable and volatile. Handle it with care, being sure there are no flames in the area.

Cyclopentadiene

Place 10 mL of dicyclopentadiene in a 25-mL round-bottom flask. Set up a fractional distillation apparatus **in the hood** as in Experiment 1, using an ice-cooled receiver for the distillate. Heat the dimer with a hot plate or a heating mantle until brisk refluxing occurs (about 170°C in the boiling flask) and the monomer begins to distill in the region of 40–42°C. Do not allow the temperature of the distilling vapor to rise above 44–45°C. At least 4 mL of cyclopentadiene should be obtained within 25 minutes. Keep the distillate cold, adding a small amount of anhydrous calcium chloride if it should become cloudy from moisture. Use it as soon as possible.

cis-Norbornene-5,6-*endo*-dicarboxylic anhydride

Place 2.5 grams (0.025 mole) of maleic anhydride in a 50-mL Erlenmeyer flask, dissolve it in 10 mL of ethyl acetate, and cool the solution in an ice bath.

Dissolve 2.5 mL (0.030 mole) of the cold, dry cyclopentadiene in 10 mL of cold ligroin, bp 60–90°C. Add this solution to the cold solution of maleic anhydride in ethyl acetate, and swirl the combined liquids in an ice bath for a few minutes until a white solid of the adduct is formed. Then heat the mixture on a steam bath or in a hot-water bath with stirring until all the solid has dissolved. Allow the solution to slowly cool, and collect the large crystals by suction filtration. Allow them to thoroughly dry and determine the yield and melting point. Record the results in your notebook. The literature melting point is 160–161°C. Also determine the theoretical and percentage yields of your product, and show the calculations.

At your instructor's option, obtain the infrared spectrum of your product in a KBr pellet and the ^{1}H NMR spectrum in deuterated chloroform. When finished, hand in your product in a properly labeled container to your instructor.

QUESTIONS

1. Write a balanced equation for the reaction of thionyl chloride with water.
2. Write a balanced equation for the reaction that would occur in the DEET experiment if the acid chloride formed in step 1 of the synthesis were mixed with water.
3. Write the equation for the formation of the white cloud that appeared in the DEET experiment.
4. In the DEET experiment, why was the number of moles of diethylamine added so much larger than the number of moles of the other reagents?
5. In the DEET experiment, why is the final reaction mixture extracted with sodium hydroxide solution?
6. Indicate what active ingredients are present in at least three common, household insecticidal products.
7. Write the equation for what would occur in the synthesis of 4-methyl-3-heptanol if the Grignard reagent had moisture present and if the propanal had propanoic acid present.
8. Why is 2-bromopentane slowly added in portions over a 30-minute period to a mixture of magnesium in ether?
9. Why wouldn't 2-chloropentane be as good as 2-bromopentane in a Grignard synthesis?
10. What reaction would occur in the distillation of 4-methyl-3-heptanol if the added acid was not completely removed or neutralized? Write the equation, showing all the products.
11. Give two reasons why the distilled cyclopentadiene must be kept cold.

12. What reactants would give the following Diels-Alder products?

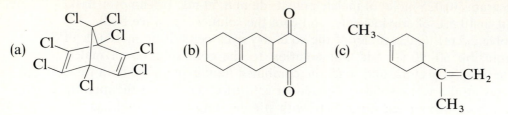

(a) (b) (c)

13. What Diels-Alder products will be formed with the following reactants? Include the stereochemistry where appropriate.

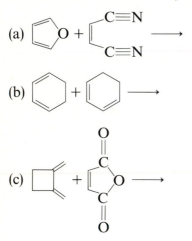

(a)

(b)

(c)

PHOTOCHEMICAL REACTIONS

And God said, Let there be light: and there was light. And God saw that the light was good and separated it from the darkness.

Genesis, Chapter 1, Verses 3 and 4

DISCUSSION

Photochemical reactions occur all around us and are vitally important to us. Without photochemical reactions occurring within our eyes, we could not see. Without photosynthesis, plants could not grow, which would mean the end of life on earth. Photochemical reactions enable us to take photographs and movies and to photocopy books, magazines, and reports. Ozone, which shields us from harmful ultraviolet light, is produced by a photochemical reaction of oxygen gas in the stratosphere. Solar cells use photochemical reactions to convert light energy into electrical energy and are vitally important in powering satellites and space vehicles—perhaps someday they may supply power for our entire society.

As the name implies, photochemistry is the chemistry of reactions initiated by light. A molecule can absorb light energy and can then isomerize, rearrange, dimerize, dissociate, and be oxidized or reduced. For example:

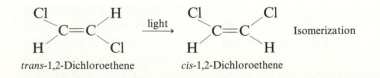

trans-1,2-Dichloroethene   *cis*-1,2-Dichloroethene

403

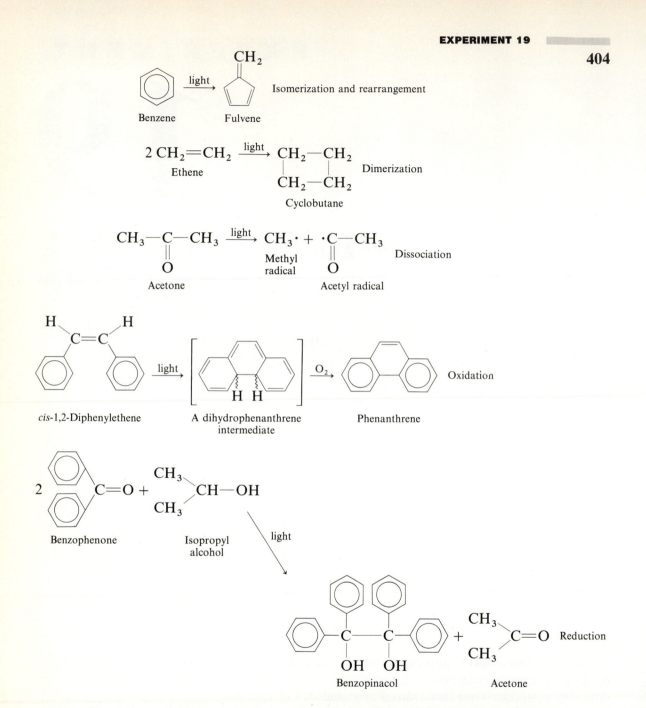

Usually only a specific wavelength of light is absorbed for a particular process; all other wavelengths have no effect. In addition, only one molecule can react for each photon of light absorbed. However, once this molecule has reacted, it can sometimes initiate or trigger other reactions, causing a process to occur spontaneously. For example, chlorine gas and hydrogen gas can remain together for years at room temperature in the dark without reacting. But if ultraviolet light reaches the mixture, the two substances react explosively in a chain reaction, releasing 44 kcal/mole of chemical energy. Each photon of light absorbed allows the formation of two chlorine free radicals to begin the chain reaction.

$$Cl—Cl \xrightarrow{\text{light}} 2\ Cl\cdot$$

$$Cl\cdot + H—H \longrightarrow H—Cl + H\cdot$$

$$H\cdot + Cl—Cl \longrightarrow H—Cl + Cl\cdot \qquad \text{etc.}$$

In a typical photochemical reaction, the absorbed light excites one or two key electrons to a higher energy level, thus **activating** the molecule. In this activated state, the molecule then can isomerize, dimerize, and so on. Usually the reaction occurs in the extremely short time of 10^{-15} to 10^{-8} seconds, but sometimes a few seconds are necessary. Ultraviolet or visible light is usually used, with ultraviolet light of shorter wavelengths providing more energy per photon. Thus ultraviolet light with a wavelength of 2859 Å will provide 100 kcal of energy per mole of photons, whereas blue-violet light with a wavelength of 4000 Å will provide 71 kcal of energy per mole of photons. These amounts of energy are far greater than what the molecules could gain from other methods, such as ordinary heating. As a result, photochemistry is the only way that many reactions can ever occur.

In addition, photochemistry frequently provides us with complex products, obtainable only with difficulty by other synthetic procedures. In most cases, photochemically allowed reactions are thermally forbidden reactions. Thus, usually one cannot just heat the reactant or reactants and hope that the same product will result. It won't. The tremendous synthetic advantage of photochemistry therefore makes it even more valuable in chemistry.

However, the absorption of light by a molecule doesn't necessarily mean that a reaction has to occur. The molecule can frequently just release the energy from its excited state as light and return to its ground state. This process in general is called **luminescence**. It has three specific types:

1. **Chemiluminescence** is the emission of light produced by a chemical reaction. As an example, fireflies produce light when a compound called luciferin reacts with oxygen through the action of an enzyme called luciferase (see Experiment 19A).

2. **Fluorescence** usually involves light emitted for only a brief period after the source of excitation is removed. Some organic molecules exhibit strong fluorescence. A good example is the dye fluorescein.

3. **Phosphorescence** involves light emitted for a relatively long time after the source of excitation is removed. Anthracene and zinc sulfide are good examples of compounds exhibiting phosphorescence.

19A. CHEMILUMINESCENCE: THE LIGHT OF THE FIREFLY

- **NUCLEOPHILIC ACYL SUBSTITUTION**
- **REDUCTION OF A NITRO GROUP TO AN AMINE**
- **OXIDATION AND CLEAVAGE OF A HETEROCYCLIC RING**

Before us, beside us, and above
The firefly lights his lamp of love.

Reginald Heber, *Tour Through Ceylon*

INTRODUCTION

On a summer night, probably most of you have observed in a yard or field a twinkling array of flashing lights produced by an insect appropriately called the firefly or "lightning bug." This blinking light perhaps provides the most enjoyment and fascination to children, who like to catch the insects with their hands and store them in glass jars in order to watch them more closely.

Much study has now established that the light produced by the firefly is a mating device. From dusk until about midnight, the male will fly several feet above the ground, emitting flashes of light at regular intervals to attract a female. The female, remaining stationary on the ground, if interested then flashes a response. This process of the male flashing and the female responding occurs a few more times until he reaches her. Fireflies can even distinguish one another by their characteristic flash patterns, which will vary in number, rate, and duration from one firefly to another.

The light produced is caused by the almost instantaneous oxidation of a substance called luciferin through the action of molecular oxygen, magnesium ion, adenosine triphosphate (ATP), and an enzyme called luciferase according to the equation

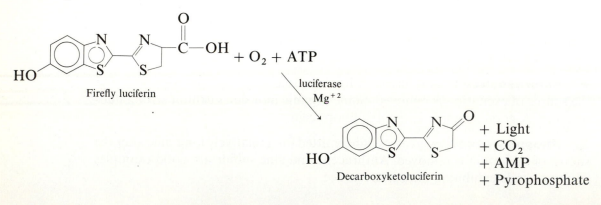

Firefly luciferin

luciferase
Mg^{+2}

Decarboxyketoluciferin

+ Light
+ CO_2
+ AMP
+ Pyrophosphate

In this equation, ATP and AMP (adenosine monophosphate) are both nucleotides and have the structures

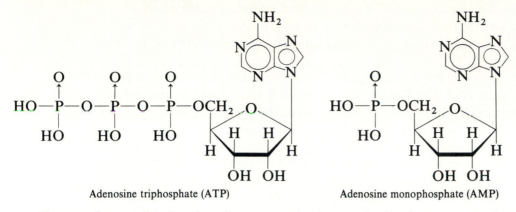

Adenosine triphosphate (ATP)   Adenosine monophosphate (AMP)

The reaction mechanism is still not completely certain, but it appears to be the following:

$$\text{Luciferase} + \text{ATP} \xrightarrow{1} \text{Luciferase-ATP}$$

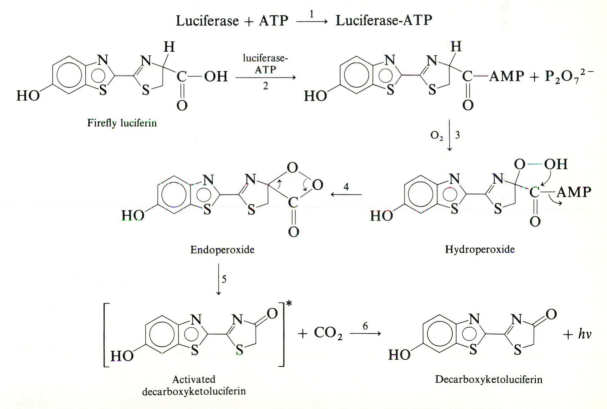

Firefly luciferin

Endoperoxide   Hydroperoxide

Activated
decarboxyketoluciferin   Decarboxyketoluciferin

In step 1, the luciferase, acting as the catalyst, complexes with ATP. The firefly luciferin then reacts with this complex in step 2. The AMP becomes bonded to the carboxyl carbon, displacing the hydroxyl group, with a pyrophosphate ion being produced as a side product. In step 3, the luciferin-AMP complex then reacts with molecular oxygen to form a hydroperoxide, which in step 4 reacts with the former carboxyl carbon, expelling AMP and producing the cyclic endoperoxide. The endoperoxide is unstable and decarboxylates easily in step 5, producing an electronically excited state of decarboxyketoluciferin. In step 6, a photon of light is emitted, and stable decarboxyketoluciferin is the final product.

Partially supporting this mechanism is that when decarboxyketoluciferin is excited photochemically by photon absorption, it gives a fluorescence emission spectrum identical to the emission spectrum of firefly luciferin reacting with O_2 and ATP in the presence of Mg^{+2} and firefly luciferase.

Although luciferase catalyzes the reaction, its structure is still unknown. Also, a bountiful supply of luciferin will apparently never be obtainable from the firefly itself, since one research study was able to isolate only 0.009 gram of the compound from 15,000 fireflies. The luminescence of the reaction contains few ultraviolet or infrared rays and is called "cold light," since it is produced by an almost complete conversion of potential energy into light.

But the firefly is not the only organism that emits light. Much marine life—certain species of clams, fish, squids, coral, crustaceans, and sponges—also have the ability, as well as certain bacteria and fungi. The luminescent compounds from these organisms are all given the general name of luciferins, and any necessary enzymes are all called luciferases. But none of these other luminescent organisms has generated as much interest as the firefly.

BACKGROUND

Firefly luciferin is too complex in structure to permit its easy synthesis in the laboratory. However, the compound luminol (3-aminophthalhydrazide) emits the same kind of light and is readily synthesized by the reaction of hydrazine and 3-nitrophthalic acid, followed by reduction of the resulting intermediate.

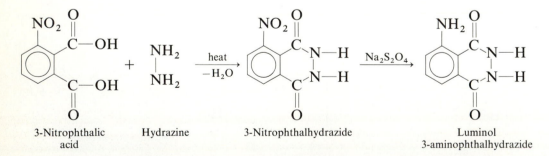

| 3-Nitrophthalic acid | Hydrazine | 3-Nitrophthalhydrazide | Luminol 3-aminophthalhydrazide |

The first reaction involves hydrazine displacing two molecules of water by nucleophilic acyl substitution on the dicarboxylic acid to produce a cyclic diamide. The second reaction involves the reduction of the nitro group on the benzene ring through the use of sodium hydrosulfite (sodium dithionite).

Due to the acidity of the hydrazide hydrogens, luminol in neutral solution exists largely as a dipolar ion or zwitterion.

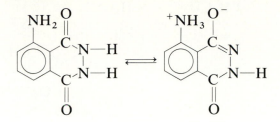

But in alkaline solution, both of these acidic hydrogens are removed, thus converting luminol to a dianion.

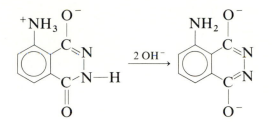

This dianion in the presence of molecular oxygen is oxidized to yield nitrogen gas and an excited state of the 3-aminophthalate dianion, which then emits a photon of visible light.

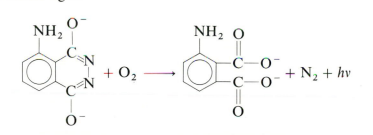

This same oxidation reaction can also be accomplished with potassium ferricyanide and hydrogen peroxide in alkaline solution.

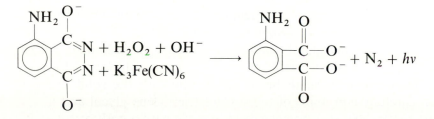

The energy of the light that is emitted in the above reactions can also be transferred to fluorescent dyes, resulting in lights of different colors and intensities. Examples of fluorescent dyes that can be mixed with luminol are fluorescein, rhodamine B, eosin, and phenolphthalein, which have the structures

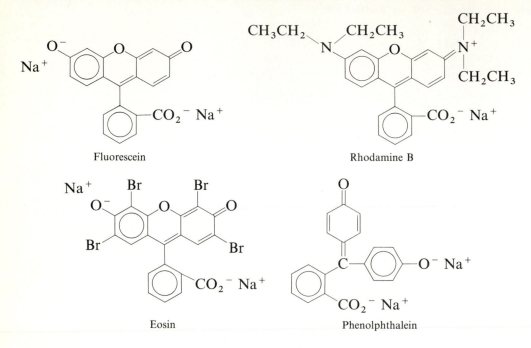

Fluorescein

Rhodamine B

Eosin

Phenolphthalein

PROCEDURE

CAUTION: Hydrazine is very toxic and should not be inhaled or spilled on the skin. Wear gloves when using it.

Dimethyl sulfoxide, if used, is also potentially dangerous, since anything mixed with it is rapidly absorbed through the skin with possible harmful results.

In addition, potassium hydroxide, sodium hydroxide, and glacial acetic acid may all cause severe burns. Use them with care.

3-Nitrophthalhydrazide

Heat a flask containing 20 mL of water on a steam bath or hot plate. (This hot water will be needed shortly in the experiment.) Next clamp a 25×200 side-arm test tube upright on a ring stand and add 1.3 grams (0.0062 mole) of 3-nitrophthalic acid and 2 mL (0.006 mole) of a 10% aqueous solution of hydrazine. Heat the mixture over a small burner flame until all the solid has dissolved, and then carefully add 4 mL of the solvent triethylene glycol and a boiling stone. Insert a 250–300° thermometer held in place with a cork or rubber stopper, and attach to the side arm a piece of pressure tubing connected to a water aspirator (use a trap). Heat the solution with a small burner flame until the liquid boils vigorously and the water is drawn off by the aspirator. Continue heating and allow the temperature to rise rapidly for 5 minutes until it has reached 210–220°. Remove the burner briefly and, by intermittent gentle heating, maintain the temperature at 210–220° for another 2 minutes. Allow the test tube to air cool to around 100°, add the 20 mL of previously prepared hot water, and further cool the test tube to room temperature. Collect the light yellow crystals of 3-nitrophthalhydrazide by suction filtration, and with no further drying of the crystals proceed to the next reaction step.

Luminol (3-aminophthalhydrazide)

Transfer the damp product back to the cleaned test tube in which it was prepared. Add 6.5 mL (0.018 mole) of 10% sodium hydroxide solution and stir the mixture until the hydrazide dissolves. To the deep red-brown solution add 4.0 grams (0.019 mole) of sodium hydrosulfite dihydrate (sodium dithionite) and wash the solid off the walls of the test tube with a little water. Heat the mixture to boiling with stirring and keep the liquid hot without boiling for 5 minutes. **Carefully** add 2.6 mL (0.045 mole) of glacial acetic acid and cool the mixture thoroughly in an ice bath. Collect the resulting precipitate of light yellow luminol by suction filtration. The damp product is satisfactory for the following chemiluminescence experiments.

CHEMILUMINESCENCE EXPERIMENTS

Using oxygen and potassium hydroxide

In a 125-mL Erlenmeyer flask, place approximately 10 grams of potassium hydroxide pellets, 25 mL of dimethyl sulfoxide, and 0.15 gram of moist luminol. In a dark room, stopper the flask, shake it, and observe the color and intensity of the light produced. When the light emitted decreases in intensity, remove the stopper, swirl the flask, restopper the flask, shake it, and observe what happens. Repeat this process for as long as the light continues to be emitted and record your results.

Using potassium ferricyanide, hydrogen peroxide, and sodium hydroxide

As an **alternative** method, dissolve 0.15 gram of moist luminol in 5 mL of 10% sodium hydroxide solution and 45 mL of water. Prepare a second solution by mixing 15 mL of 3% aqueous potassium ferricyanide, 15 mL of 3% hydrogen peroxide, and 70 mL of water. Next dilute 15 mL of the first solution with 85 mL of water and, in a dark room, pour the diluted first solution and the second solution simultaneously into a large beaker. Stir the contents and, to increase the brilliance, gradually add further small quantities of sodium hydroxide solution. Observe the color and intensity of the emitted light and record the results.

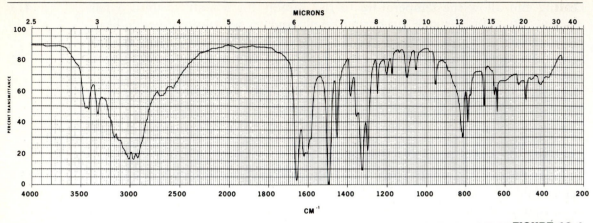

FIGURE 19-1

Infrared spectrum of luminol in a KBr pellet. Spectrum courtesy of Sadtler Research Laboratories, division of Bio-Rad Laboratories, Inc., copyrighted 1973.

FIGURE 19-2

^{1}H NMR spectrum of luminol in dimethyl sulfoxide-d_6. Spectrum courtesy of Aldrich Chemical Company, *The Aldrich Library of NMR Spectra*, edition II, by Charles Pouchert.

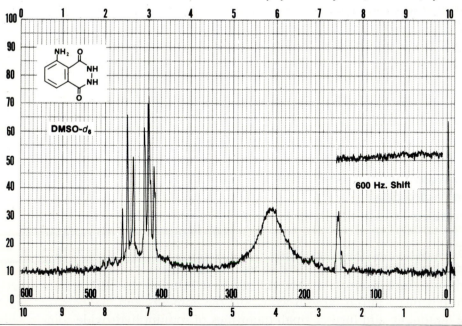

Colors using fluorescent dyes

To observe the energy transfer to a fluorescent dye, dissolve a few crystals (5–10 mg) of the dye in a few milliliters of dilute base and add to a fresh, unused sample of either the dimethyl sulfoxide solution or the aqueous sodium hydroxide solution of luminol. Repeat the procedure given earlier and observe and record the color and intensity of the light produced. Perform this test on fluorescein, rhodamine B, eosin, phenolphthalein, and any other dyes you would like to try.

At your instructor's option, wash the remaining material with ice-cold water, **thoroughly** dry, and obtain the infrared spectrum in a KBr pellet and the ^{1}H NMR spectrum in dimethyl sulfoxide-d_6. Compare your spectra with those shown in Figures 19-1 and 19-2. The literature melting point of luminol is 315°.

When finished, hand in the remaining material in a properly labeled container to your instructor.

19B. A PHOTOCHROMIC COMPOUND: SYNTHESIS OF 2-(2,4-DINITROBENZYL)PYRIDINE

ELECTROPHILIC AROMATIC SUBSTITUTION
NITRATION

In this experiment, you will synthesize 2-(2,4-dinitrobenzyl)pyridine by the reaction of 2-benzylpyridine with fuming nitric acid in the presence of sulfuric acid.

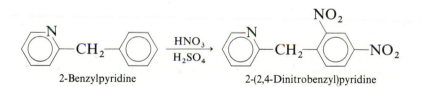

2-Benzylpyridine   2-(2,4-Dinitrobenzyl)pyridine

The product formed has the very unusual property of turning deep blue when exposed to sunlight. If it is then stored in the dark, the product will return to its original sandy color. Formation of the blue color only takes a few minutes in bright sunlight, but the loss of color takes about a day. Changing the color back and forth may be repeated many times.

The phenomenon occurs when light converts 2-(2,4-dinitrobenzyl)pyridine into a tautomer by the transfer of a benzyl hydrogen to the doubly bonded oxygen atom of the ortho nitro group. The transfer back of this hydrogen then regenerates the original compound.

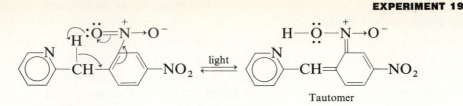

Tautomer

The nitration is another example of electrophilic aromatic substitution and has the following mechanism:

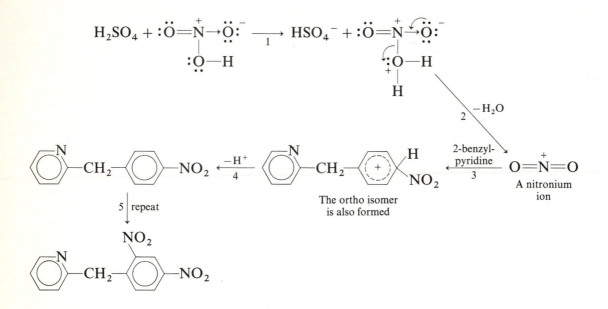

The ortho isomer is also formed

A nitronium ion

In step 1, nitric acid acts as a base and receives a proton from the sulfuric acid, which is the stronger acid. A loss of water then occurs in step 2 to form the nitronium ion, which, needing electrons, finds them available in the π cloud of the slightly activated benzene ring of 2-benzylpyridine. A covalent bond is formed at either the ortho or para position, producing a resonance-stabilized arenium ion, which then loses a proton in step 4 to form the mononitration product. Repeating these steps then gives 2-(2,4-dinitrobenzyl)pyridine as the final product.

Note that the intermediate nitronium ion prefers **not** to attack the pyridine ring. The highly electronegative nitrogen atom in pyridine greatly decreases the availability of the π-electrons, thus making pyridine much less reactive than benzene. In addition, the highly acidic medium easily protonates the basic nitrogen atom of pyridine, which makes this ring even more deactivated. This deactivation is analogous to what occurs when aniline is reacted in a highly acidic medium or in the Friedel-Crafts reaction.

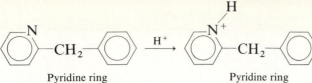

Pyridine ring
deactivated compared
to benzene

Pyridine ring
even more
deactivated

PROCEDURE

CAUTION: Concentrated sulfuric acid and fuming nitric acid are both extremely corrosive and must be used with great care. If you spill either of them on your skin, immediately flush the area with plenty of cold water and notify your instructor. Also, because nitric acid reacts with an amino acid in the skin, your skin will turn yellow. This yellow color will not wash off; it can only wear off.

Place 10 mL (0.18 mole) of concentrated sulfuric acid into a 50-mL Erlenmeyer flask. Cool to below 5°C in an ice bath and gradually add, with careful swirling to the ice bath, 2.6 mL (2.7 grams, 0.016 mole) of 2-benzylpyridine. To the still cooled, still swirled mixture, add slowly, over a 2-minute period, 1.5 mL (2.3 grams, 0.035 mole) of red fuming nitric acid. The addition of the nitric acid will initially cause a color change to deep brown, but the color will become lighter as the rest of the acid is added. After all the nitric acid is added, heat the mixture for about 30 minutes on a steam bath or in a boiling-water bath.

When the heating is completed, **carefully** pour the hot mixture onto about 150 grams of cracked ice in a 400-mL beaker. **Carefully** rinse out the Erlenmeyer flask with a small amount of ice-cold water, and add the water to the beaker. Next, **carefully** and **slowly** add with stirring 90 mL (0.45 mole) of ice-cold 20% sodium hydroxide solution. Toward the end, the change in pH will cause the product to separate to give a milky yellow suspension.

Pour the suspension into a 500-mL separatory funnel and extract with three 40-mL portions of ether. If a large amount of suspension still remains after the first extraction, remove most of the upper ether layer from the separatory funnel with a pipet. **(Caution: Do not mouth-pipet.)**

Combine the ether extracts, pour into the clean, empty, separatory funnel, and shake with 25 mL of saturated salt solution to reduce the water content of the ether. Separate the ether solution, dry it over anhydrous magnesium sulfate, filter or decant, then distill from a 250-mL round-bottom flask, using a warm-water bath and collecting the distillate in an ice-water-cooled receiver. When the volume of ether solution remaining is about 20 mL, stop the distillation and cool the mixture in an ice bath.

Collect the large sandy prisms of 2-(2,4-dinitrobenzyl)pyridine by suction filtration and recrystallize from 95% ethanol, using about 10 mL per gram. Allow

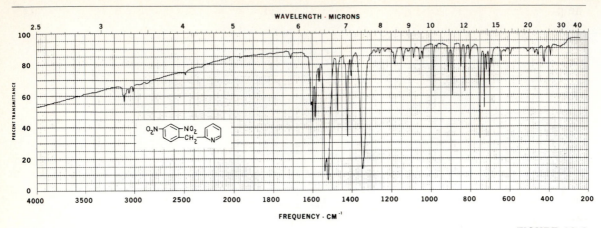

FIGURE 19-3

Infrared spectrum of 2-(2,4-dinitrobenzyl)pyridine in a KBr pellet. Spectrum courtesy of Sadtler Research Laboratories, division of Bio-Rad Laboratories, Inc., copyrighted 1972.

FIGURE 19-4

^{1}H NMR spectrum of 2-(2,4-dinitrobenzyl)pyridine in deuterated chloroform. Spectrum courtesy of Sadtler Research Laboratories, division of Bio-Rad Laboratories, Inc., copyrighted 1971.

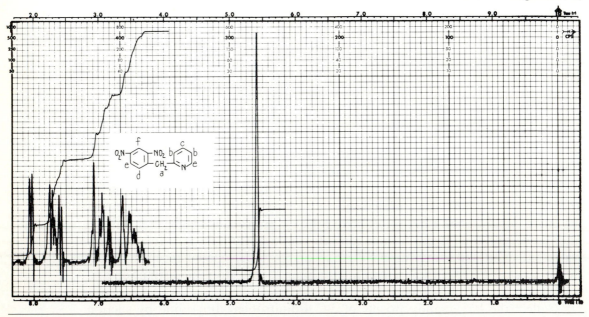

the product to thoroughly dry and determine the yield and melting point. Record the results in your notebook. The literature melting point is 91–93°C. Also determine the theoretical and percentage yields of your final products, and show the calculations.

Place about half of your product in bright sunlight. Note and record what happens. Let it stand in the dark, and note and record what has happened after 1 hour, 2 hours, and overnight.

At your instructor's option, obtain the infrared spectrum of your sandy-colored product in a KBr pellet and the ^{1}H NMR spectrum in deuterated chloroform. Compare your spectra with those shown in Figures 19-3 and 19-4. Also obtain the infrared spectrum of your blue product in a KBr pellet, and the ^{1}H NMR spectrum in deuterated chloroform. Compare the results with your first two spectra and see if these last two spectra support the tautomer explanation for the color change. Explain specifically how.

When finished, hand in your product(s) in a properly labeled container(s) to your instructor.

References

1. A. Ault and C. Kouba, **Journal of Chemical Education, 51,** 395 (1974).

2. A. Blumn, J. Weinstein, and J. Sousa, **Journal of Organic Chemistry, 28,** 1989 (1963).

3. J. Sousa and J. Weinstein, **Journal of Organic Chemistry, 27,** 3155 (1962).

4. A. Nunn and K. Schofield, **Journal of the Chemical Society, 1952,** 586.

5. K. Schofield, **Journal of the Chemical Society, 1949,** 2411.

19C.   A THERMOCHROMIC COMPOUND FROM A PHOTOCHEMICAL REACTION: SYNTHESIS OF DIXANTHYLENE

DEHYDRATION OF AN ALCOHOL

In this experiment, you will synthesize dixanthylene by first reacting xanthene and xanthone in the presence of sunlight to yield 9-hydroxydixanthyl, followed by dehydration of this intermediate alcohol.

The product formed is unusual because it is **thermochromic**—its color depends on the temperature. At room temperature, dixanthylene is a very pale yellow-green solid, but it becomes dark blue when either heated in solution or melted. A loss of color then occurs as the hot compound is cooled, and it even has no color at −200°C.

The highly conjugated system in dixanthylene appears to be responsible for this thermochromic effect. As evidence of this, dianthraquinone, which has a

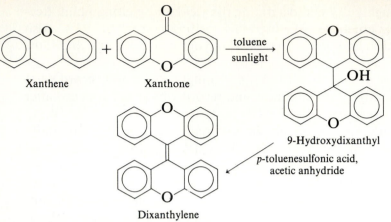

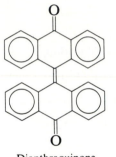

Dianthraquinone

similar highly conjugated system, is also thermochromic, being a bright yellow at room temperature and a brilliant light green when heated.

The first reaction in the synthesis involves the photochemical addition of xanthene to the carbonyl group of xanthone to yield 9-hydroxydixanthyl.

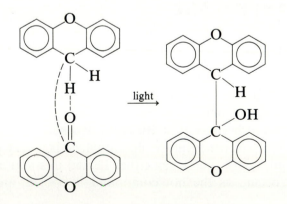

The second reaction involves the hydroxyl group of 9-hydroxydixanthyl being protonated by the strongly acidic p-toluenesulfonic acid. Loss of a water molecule to form the highly stable benzylic carbocation, followed by the loss of a proton then yields the final product. Observe that the acetic anhydride is not just a solvent, but a reactant to remove all the water present.

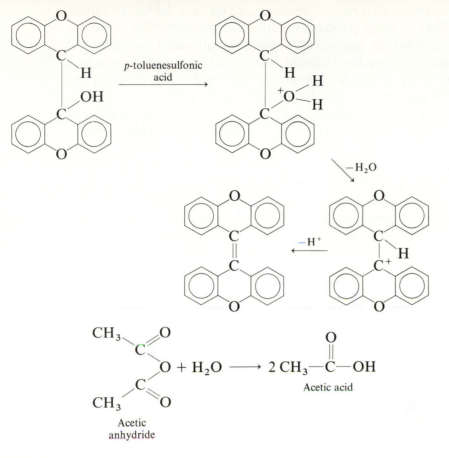

PROCEDURE

CAUTION: Use care when using all the reactants and solvents, being sure you don't breathe them or get them on your skin.

9-Hydroxydixanthyl

Place 0.65 gram (0.0036 mole) of xanthene and 0.70 gram (0.0036 mole) of xanthone into a 50-mL Erlenmeyer flask. Add 20 mL of toluene and dissolve the solids by swirling and mild heating. Stopper and label the flask, and leave it in bright sunlight for 5 to 14 days until the formation of the coarse crystalline material has stopped. Overexposure does not harm the results.

Collect the crystalline material by suction filtration, thoroughly dry a few crystals, and take their melting point. Record the result in your notebook. The literature melting point is 194°C.

Dixanthylene

Place all of the 9-hydroxydixanthyl into a 50-mL round-bottom flask. Add 16 mL of acetic anhydride and reflux the mixture until all the solid has dissolved. Add 0.80 gram (0.0042 mole) of *p*-toluenesulfonic acid monohydrate down the reflux condenser and reflux the mixture for about 3 minutes until no more solid appears to form and the color has changed from a red or a dark brown to a dark green. Let the mixture cool for 1 hour and then collect the dixanthylene by suction filtration, washing it with a little glacial acetic acid.

Recrystallize the product by refluxing it with about 10 mL of mesitylene and allowing the blue solution to cool to room temperature. Note the color change as the solution cools. Collect the pale yellow-green crystals by suction filtration, allow to thoroughly dry, and determine the yield. Record the result in your notebook. Also determine the theoretical and percentage yields of your final product, and show the calculations. The literature melting point is 315°C.

At your instructor's option, obtain the infrared spectrum of your product in a KBr pellet and the ^{1}H NMR spectrum in deuterated chloroform. When finished, hand in your product in a properly labeled container to your instructor.

References

1. A. Ault, R. Kopet, and A. Serianz, **Journal of Chemical Education, 48,** 410 (1971).

2. A. Schonberg and A. Mustafa, **Journal of the Chemical Society, 1944,** 67.

3. G. Gurgenjanz and S. von Kostanecki, **Chemische Berichte, 28,** 2310 (1895).

19D. PHOTOREDUCTION OF BENZOPHENONE TO BENZOPINACOL

- **ALKALINE CLEAVAGE OF BENZOPINACOL**
- **PINACOL REARRANGEMENT**

In this experiment, you will reduce benzophenone in the presence of isopropyl alcohol and ultraviolet light to obtain benzopinacol. Acetone is also obtained as a by-product.

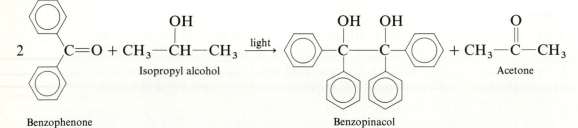

Benzophenone   Benzopinacol

This reaction is one of the oldest and most thoroughly studied photochemical reactions. It all began when it was discovered that benzophenone was unstable to light in certain solvents that could act as hydrogen donors. Isopropyl alcohol is one of those solvents.

The reaction mechanism is shown below. In step (1), benzophenone is activated by a photon of ultraviolet light to an electronically excited triplet state. This activated ketone then abstracts a hydrogen atom from isopropyl alcohol in step (2) to form two radicals: benzhydrol and hydroxyisopropyl. In step (3), the hydroxyisopropyl radical then transfers a hydrogen atom to unactivated benzophenone, producing another benzhydrol radical and a molecule of acetone. Combination of the two benzhydrol radicals in step (4) then produces benzopinacol, which crystallizes from the solution.

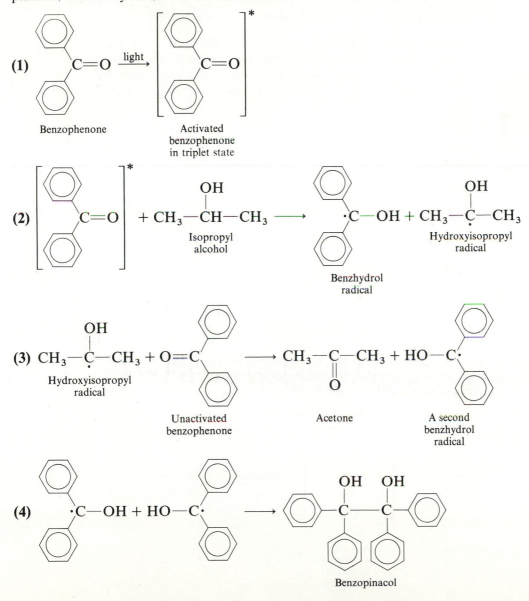

The reaction must be carried out in a very slightly acidic solution. This counteracts the traces of alkali present from the glass, which would cleave the benzopinacol to yield benzhydrol and benzophenone.

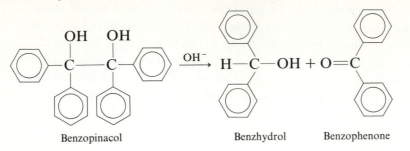

Benzopinacol   Benzhydrol   Benzophenone

The benzophenone produced, of course, is one of the reactants in the initial photochemical reaction. It can thus be converted into more benzopinacol, which can be cleaved by the base to yield even more benzhydrol and benzophenone. With a continuation of this process, benzhydrol can be obtained in almost 100% yield. When benzhydrol is desired, sodium isopropoxide, formed from sodium metal and isopropyl alcohol, is a more effective base for the cleavage reaction.

Benzopinacol can also react if heated with acid. This reaction, which is one example of the pinacol rearrangement, gives benzopinacolone as the product.

Benzopinacol   Benzopinacolone

The initial photoreduction reaction to produce benzopinacol can also be stopped by simply adding a small quantity of naphthalene. The naphthalene is said to **quench** the reaction by accepting the excess energy of the excited triplet state of benzophenone before the triplet state has a chance to react with isopropyl alcohol to form the two free radicals.

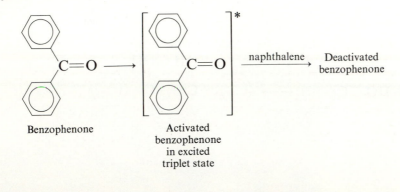

Benzophenone   Activated
                      benzophenone
                      in excited
                      triplet state

CAUTION: Glacial acetic acid can cause severe burns, especially when hot. Be careful when using it. If any should get on your skin, wash it off with water immediately.

Also remember that sodium metal reacts violently with water, so keep it away from water or any wet apparatus. Never handle it with your fingers nor dispose of it in the sink. Destroy any excess, undesired sodium by reacting it with anhydrous alcohol.

Since sodium tarnishes due to oxidation when exposed to the air, it is usually kept under xylene, ligroin, or mineral oil. To weigh sodium, transfer the metal with tongs from its container to a paper towel and cut away the oxide coating with a knife or spatula. Cut very small pieces of the metal and quickly place them into a small, previously weighed beaker of ligroin or xylene. From the increase in weight of the beaker, you can determine the amount of added sodium. When you need them, transfer the small pieces of metal from the beaker to the reaction flask or test tube, using forceps or tweezers.

Place 2.9 grams (0.016 mole) of benzophenone into each of three 20 × 150 mm test tubes, labeled 1, 2, and 3. Add 0.25 gram of napththalene to tube 2, then add about 15 mL of isopropyl alcohol to each of the three tubes and warm on a steam bath to dissolve the solids. Add **1 drop** of glacial acetic acid to tube 1 and tube 2 and 0.1 gram of sodium metal to tube 3. Fill each tube to near the top with more isopropyl alcohol, stopper the tubes tightly with rubber stoppers wrapped in plastic wrap, shake the tubes well, then place them in a beaker on a window ledge for direct irradiation by the sun. A southern exposure is best. Periodically follow the progress of the reaction by noting the amount of crystals forming in tube 1. Compare with tubes 2 and 3, and record the results in your notebook. Continue to irradiate the three tubes until no more product seems to form in tube 1 (about 5 to 10 days, depending on the amount of sunlight received.)

Benzopinacol

Collect the crystals formed in tube 1 by suction filtration, wash with a little ice-cold isopropyl alcohol, and allow to dry. No recrystallization should be necessary. Determine the yield, melting point, and literature melting point of the product, and record the results in your notebook. Also determine the theoretical and percentage yields, showing your calculations. Remember that 2 moles of benzophenone are needed to produce 1 mole of benzopinacol.

At your instructor's option, obtain the infrared spectrum of your product in a KBr pellet and the ^{1}H NMR spectrum in deuterated chloroform. Compare your spectra with those shown in Figures 19-5 and 19-6.

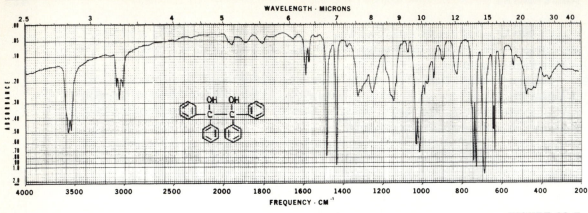

WAVELENGTH · MICRONS

FREQUENCY · CM⁻¹

FIGURE 19-5

Infrared spectrum of benzopinacol in a KBr pellet. Spectrum courtesy of Sadtler Research Laboratories, division of Bio-Rad Laboratories, Inc., copyrighted 1970.

FIGURE 19-6

¹H NMR spectrum of benzopinacol in deuterated chloroform. Spectrum courtesy of Aldrich Chemical Company, *The Aldrich Library of NMR Spectra,* edition II, by Charles Pouchert.

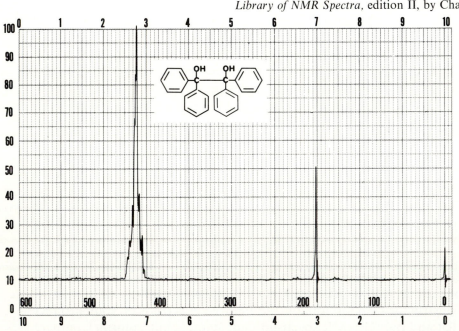

Benzhydrol

Pour the contents of tube 3 into a 125-mL Erlenmeyer flask and dilute with water. Acidify with about 5 to 7 mL of 1.0 M HCl, then evaporate the isopropyl alcohol in the hood, using a hot-water bath. Cool the mixture and collect the crystals of benzhydrol by suction filtration, allow to thoroughly dry, and determine the yield, melting point, and literature melting point. Record the results in your notebook. Also determine the theoretical and percentage yields, showing your calculations.

At your instructor's option, obtain the infrared spectrum of your product in a KBr pellet and the ^{1}H NMR spectrum in deuterated chloroform. Compare these spectra with those obtained from benzopinacol and benzopinacolone.

Benzopinacolone

Place 1.3 grams (0.0035 mole) of benzopinacol, 7 mL of glacial acetic acid, and a *small* crystal of iodine into a 50-mL round-bottom flask. Attach a reflux condenser and reflux the red solution for 10 minutes. Some benzopinacolone may separate from the solution. Slightly cool the mixture, add 7 mL of ethanol, thoroughly swirl the mixture, and allow it to cool.

Collect the crystals by suction filtration, wash with ice-cold ethanol to remove the iodine, and thoroughly dry. No recrystallization should be necessary. Determine the yield, melting point, and literature melting point of your product, and record the results in your notebook. Since the melting point of the product is close to the melting point of the starting material, you should also do a mixed melting point of the two substances and record the result. Also determine the theoretical and percentage yields of your product, and show the calculations.

At your instructor's option, obtain the infrared spectrum of your benzopinacolone in a KBr pellet and the ^{1}H NMR spectrum in deuterated chloroform. Compare these spectra with those obtained from benzopinacol and benzhydrol.

When finished, hand in all three products in properly labeled containers to your instructor.

19E. PHOTOCHEMICAL DIMERIZATION OF *TRANS*-CINNAMIC ACID

In this experiment, you will dimerize *trans*-cinnamic acid photochemically to obtain α-truxillic acid (cyclobutane-*trans*-1,3-dicarboxylic acid, *trans*-2,4-diphenyl).

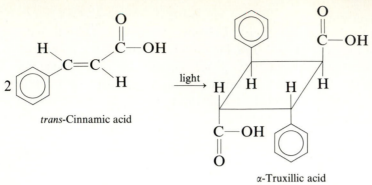

trans-Cinnamic acid

α-Truxillic acid

For unsymmetrical alkenes like *trans*-cinnamic acid, two modes of reaction are possible, giving us truxinic acids by head-to-head dimerization or truxillic acids by head-to-tail dimerization.

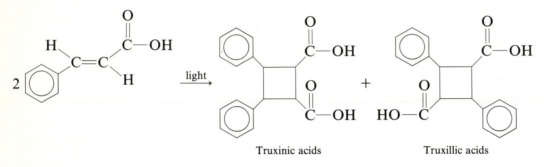

Truxinic acids   Truxillic acids

Since only α-truxillic acid is produced in this reaction, only head-to-tail dimerization is occurring. Furthermore, there are five possible stereoisomers of truxillic acid. Therefore, like most photochemical reactions of this type, this one is stereospecific, giving us just the one reaction product.

PROCEDURE

Place 1.5 grams (0.010 mole) of *trans*-cinnamic acid in a 125-mL Erlenmeyer flask. Add 2 mL of tetrahydrofuran (THF) and dissolve the acid by heating the flask on a steam bath. Remove the flask and rotate it to coat the sides and bottom with the crystallizing acid. If the coating obtained is very uneven, reheat with a little additional THF to redissolve the acid, and repeat the rotating. When the coating is sufficiently dry, clamp the flask upside down **in the hood** for 15 minutes to allow all the solvent vapor to escape. Stopper and label the flask, and place it upside down in a beaker on a window sill for irradiation by the sun. A southern exposure is best. Leave the flask there for two weeks, periodically rotating it to expose all sides.

After the second week, add 20 mL of toluene and swirl to dissolve any un-reacted cinnamic acid. Collect the crude α-truxillic acid by suction filtration and recrystallize from 95% ethanol, using a little decolorizing carbon. Collect the purified product by suction filtration, allow to thoroughly air dry, and determine the yield and melting point. Record the results in your notebook. The literature melting point is 290–292°C. Also determine the theoretical and percentage yields of your final product, showing the calculations. Remember that the yield is based upon 2 moles of *trans*-cinnamic acid producing 1 mole of α-truxillic acid.

At your instructor's option, obtain the infrared spectrum of your product in a KBr pellet and the ^{1}H NMR spectrum in deuterated chloroform. When finished, hand in your product in a properly labeled container to your instructor.

Reference

Klemm, Gopinath, Dooley, and Klopfenstein, **Journal of Organic Chemistry, 31,** 3003 (1966).

QUESTIONS

1. List all the possible ways that energy may be released in a chemical reaction.
2. In the luminol experiment, why would unstoppering the flask of luminol in dimethyl sulfoxide cause the light to become more intense? Why would this effect not be as pronounced in the potassium ferricyanide–hydrogen peroxide solution of luminol?
3. In the luminol experiment, what physical properties of triethylene glycol caused it to be used as the solvent in the synthesis of 3-nitrophthalhydrazide? Why wasn't heavy mineral oil used as an alternative solvent?
4. Is there any possible commercial use of luminol as a source of light? (See H. Schneider, **Journal of Chemical Education, 47,** 519 (1970).)
5. The photochemical reduction of benzophenone occurs less efficiently when isopropyl alcohol is replaced by ethyl alcohol, and the reaction doesn't even occur at all when *t*-butyl alcohol is used. Looking at the mechanism of the reaction, explain why.
6. The excited triplet state of benzophenone is 68.5 kcal/mole above the ground state. Calculate the wavelength of its phosphorescence.
7. Give the mechanism for the formation of benzhydrol from benzopinacol in the presence of the isopropoxide ion.
8. Assuming no alkali is present, could benzophenone react with benzhydrol in the presence of light to yield benzopinacol? Write the mechanism for this reaction.
9. Benzhydrol can be easily dehydrated in the presence of acid and heat, due to the highly stable conjugated system formed. Why doesn't this reaction predominate in the reaction workup of benzhydrol?

10. Give the mechanism for the formation of benzopinacolone from benzopinacol.

11. Give the structures of the other possible stereoisomers of α-truxillic acid.

12. α-Truxillic acid will form a dimethyl ester, but will not form a cyclic acid anhydride unless epimerization occurs. Explain why.

20

DYES: THE SYNTHESIS OF PARA RED— AMERICAN FLAG RED

- **DIAZONIUM IONS**
- **ELECTROPHILIC AROMATIC SUBSTITUTION**

The purest and most thoughtful minds are those which love colour the most.

John Ruskin, *The Stones of Venice*

INTRODUCTION

The use of dyes is an ancient art that goes back over 4000 years. Early dyes were obtained from natural plant or animal sources, where fruits, berries, flowers, roots, bark, and so on, were crushed and then extracted with water. After perhaps being boiled and strained, the extract would be brushed onto the fabric, or the fabric would be dipped into the extract. Some of the ancient dyes used were indigo, Tyrian purple, madder, and henna. Ironically, indigo is still used for the dyeing of denim to produce blue jeans.

429

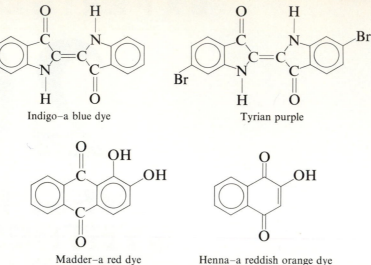

Indigo–a blue dye   Tyrian purple

Madder–a red dye   Henna–a reddish orange dye

A significant discovery occurred in 1856, when the first synthetic dye was made. A young British chemist, William Perkin, was attempting to synthesize the drug quinine. Unfortunately, the molecular structure of quinine was then unknown, and with the approach he was using, Perkin never could have succeeded. Nevertheless, in one of his experiments, he oxidized aniline sulfate with potassium dichromate and obtained a black, tarry material. Just as he was throwing the substance away as another failure, he noticed dark purple crystals, which upon extraction with methanol gave a beautiful purple solution. This solution was found to be an excellent dye for fabrics, and Perkin, after overcoming numerous obstacles, became a famous and wealthy industrialist. The dye he synthesized became known as **mauve**, and subsequently was found to have the structure

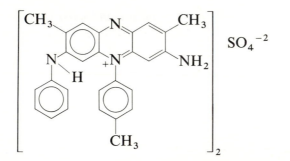

From the methyl groups on the aromatic rings, it should be evident that Perkin had oxidized impure aniline, which contained a mixture of toluidines. Ironically, if he had used pure aniline, he might never have succeeded!

Because of Perkin's success, many renowned chemists were attracted to the exciting and expanding new field, leading to the discovery of many other synthetic dyes, including malachite green, crystal violet, butter yellow, orange II, diamine green B, and indanthrene blue.

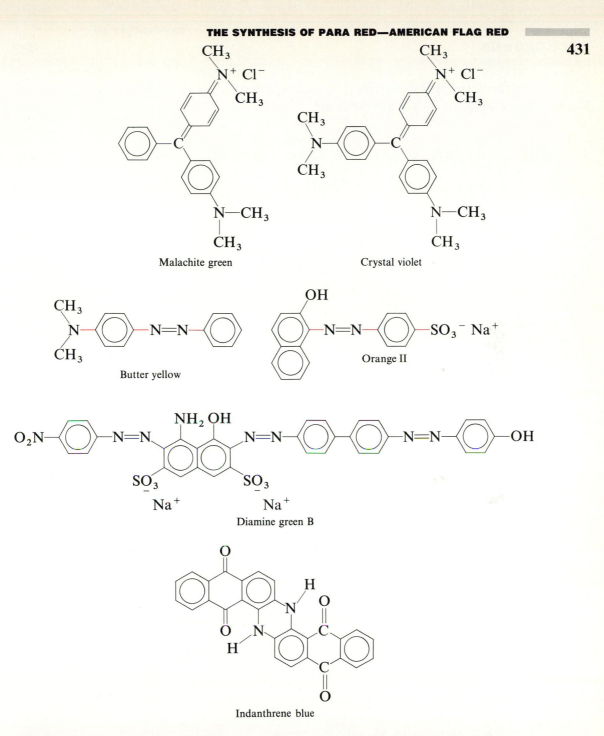

Malachite green

Crystal violet

Butter yellow

Orange II

Diamine green B

Indanthrene blue

This has led today to over 5000 synthetic dyes being produced, in what has become a multibillion dollar industry. The synthetic dyes generally are found to cover a wider range of colors and to give more "brilliant" colors than the natural ones. As a result, they usually are more popular and preferred.

BACKGROUND

In this experiment, you will synthesize the azo dye para red, also called American flag red. The equations for the synthesis are

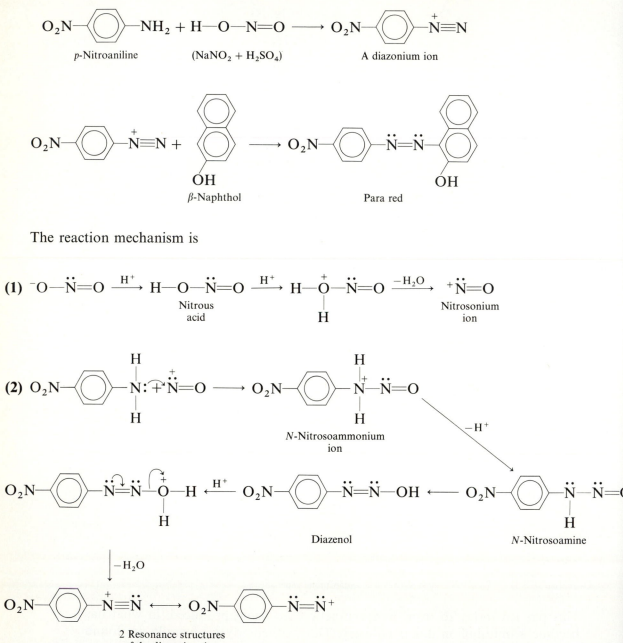

p-Nitroaniline   (NaNO$_2$ + H$_2$SO$_4$)   A diazonium ion

β-Naphthol   Para red

The reaction mechanism is

(1) Nitrous acid → Nitrosonium ion

(2) *N*-Nitrosoammonium ion

Diazenol ← *N*-Nitrosoamine

2 Resonance structures of the diazonium ion

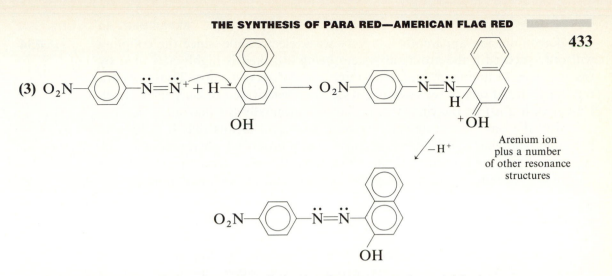

(3)

Arenium ion
plus a number
of other resonance
structures

In the first series of reactions, the nitrite ion is protonated to yield nitrous acid, which then also becomes protonated and loses a molecule of water to yield the nitrosonium ion. The second protonation is necessary because water is a much better leaving group than hydroxide ion, which would be the only alternative.

Then in the second series of reactions, the amino group of p-nitroaniline acts as a nucleophile and forms a chemical bond to the nitrosonium ion to form an N-nitrosoammonium ion. The loss of a proton forms an N-nitrosoamine, which then transfers a proton to the oxygen atom to form a diazenol. This proton transferal is analogous to what occurs in keto-enol tautomerism. Protonation of the diazenol, followed by a loss of water then forms the diazonium ion, which exists as a resonance hybrid of a number of structures, two of which are shown. In fact, resonance gives this diazonium ion its moderate stability, since nonaromatic diazonium ions usually decompose immediately after being formed. Again in the formation of this ion, water is a much better leaving group than hydroxide ion.

Then in the third series of reactions, the diazonium ion, acting as an electrophile, forms a chemical bond to the β-naphthol at position 1. Position 1 is favored because the intermediate arenium ion formed is the most stable, due to the especially stable resonance structure shown, allowing the delocalization of the positive charge onto the oxygen atom of the hydroxyl group. The loss of a proton from this arenium ion then gives us the final product.

Observe that para red contains two doubly bonded nitrogen atoms between two aromatic ring systems. This is the general structure of all azo dyes, which today constitute over 3000 of all synthetic dyes. These dyes receive their color because the azo linkage —N=N— brings the two aromatic ring systems into conjugation. This results in an extended system of delocalized π-electrons, allowing the absorption of light in visible regions.

As observed in the reaction mechanism, the azo linkage is formed from the coupling of the two aromatic ring systems by electrophilic aromatic substitution.

The diazonium ion apparently is a very weak electrophile, since the coupling will only succeed if the aromatic system being attacked is highly activated by the presence of amino, substituted amino, or hydroxyl groups. Thus, as para red illustrates, if one of the two ring systems is deactivated or only moderately activated, it must be the ring system that was originally the diazonium ion.

Also, the only moderate stability of nitrous acid means that it cannot be prepared beforehand. It is therefore conveniently produced when needed by the reaction of acid with sodium nitrite. Once formed, the diazonium ion can also unwantedly decompose, so it is prepared in a cold solution and used immediately.

PROCEDURE

CAUTION: Aromatic diazonium ions may explode if they are isolated and allowed to dry. Keep your prepared diazonium ion cold and in solution. React it as soon as possible with the cold solution of sodium β-naphthoxide.

In a 50-mL Erlenmeyer flask place 0.83 gram (0.0060 mole) of *p*-nitroaniline and 10 mL (0.030 mole) of 3 *M* sulfuric acid. If necessary, gently warm the mixture to dissolve the amine as much as possible and then cool the mixture to below 10° in an ice bath. Small pieces of ice may be added to the flask, if desired, to raise the efficiency of cooling. To the resulting suspension of *p*-nitroanilinium bisulfate, slowly add with cooling and stirring a solution of 0.41 gram (0.0060 mole) of sodium nitrite in 5 mL of cold water. Do not allow the temperature to rise above 10°.

FIGURE 20-1

Infrared spectrum of para red in a KBr pellet. Spectrum courtesy of Sadtler Research Laboratories, division of Bio-Rad Laboratories, Inc., copyrighted 1982.

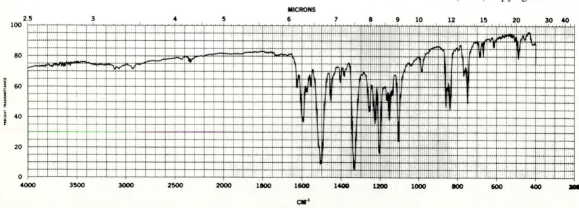

Prepare a solution of 0.86 gram (0.0060 mole) of β-naphthol in 15 mL of 10% sodium hydroxide solution, warming gently if necessary. Cool the solution to below 10° and slowly pour it into the still-cooled diazonium ion solution with stirring. Acidify the mixture, collect the red product by suction filtration, and allow to dry. The product may be recrystallized from toluene. Allow the purified product to dry thoroughly and determine the weight but **not** the melting point, since para red decomposes upon melting. Also determine the theoretical and percentage yields of the final product, showing all calculations.

At your instructor's option, obtain the infrared spectrum of your product in a KBr pellet. Compare your spectrum with the one shown in Figure 20-1.

When finished, submit your product in a properly labeled container to your instructor.

QUESTIONS

1. What is the limiting reagent, if any, in this experiment? Why?
2. If the double bond in the azo linkage were converted into a single bond, what would be the result? Why?
3. What would be the decomposition product if the p-nitrobenzenediazonium ion were heated in the aqueous solution or allowed to stand in the aqueous solution without any β-naphthol being added?
4. Why is p-nitroaniline more soluble in an acidic aqueous solution than in a neutral aqueous solution?
5. Why is β-naphthol more soluble in an alkaline aqueous solution than in a neutral aqueous solution?
6. Why can't the following azo dye be prepared directly by a coupling reaction?

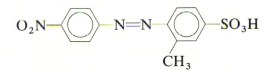

7. Which dyes, if any, in the introduction section could be classified as azo dyes? Why?
8. Draw the structures of the azo dyes that would be formed from the following organic starting materials:

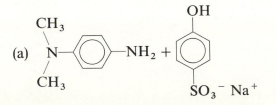

(b)

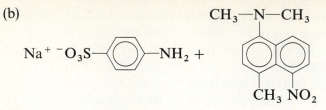

9. Draw the structures of the organic starting materials that would be required to make the following azo dyes:

(a)

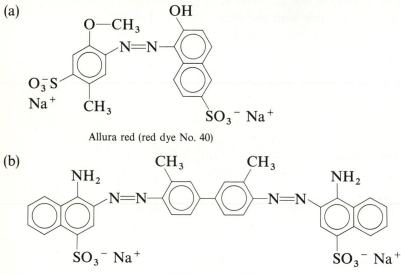

Allura red (red dye No. 40)

(b)

Benzopurpurin 4B

EXPERIMENT

21

DYEING OF FABRICS

It is a bad clothe that will take no colour.

John Heywood, *Proverbs,* Part I, Chapter 9 (1546)

INTRODUCTION

At some time in your life, you or some member of your family may have dyed some fabric or clothing. You went to the store, selected the color you liked, and at home followed the relatively simple directions included with the dye. The dye was dissolved in a specified amount of hot water, and the article was stirred in the dye bath until a shade darker than the desired color was obtained. The article was then rinsed thoroughly in cool or lukewarm water, the excess water squeezed out, and the article dried. When finished, you may have even thought or remarked how easy the whole operation was.

But in reality, the dyeing process is more complex than you may have realized. As you may have discovered, not all dyes are suitable for every type of fabric. Dyeing is an interaction between the dye and fiber, and a dye that is good for wool may not dye cotton at all, or a dye that works well with cotton and linen may prove to be unsatisfactory with polyester or nylon.

This fact was perhaps most strikingly demonstrated to the public at the DuPont exhibition at the 1964 New York World's Fair. When an ordinary looking white towel was dipped into a dye bath and then rinsed, the end result was a colorful array of stripes! Actually the towel was not ordinary, but consisted of many different kinds of fabrics. When dipped into the dye bath, each fabric had accepted only one dye and had rejected all the others!

Even though a dye may appear to have adhered to a fabric, it must be demonstrated that the bonding is permanent and that the dye can withstand

437

washing, bleaching, ironing, steaming, light, air oxidation, perspiration, dry cleaning, and friction. If tests show no appreciable bleeding or fading of the dye after treatment with the above, then the dye may be said to be colorfast to washing, perspiration, light, etc., on the given fabric.

Although most dyes on a given fabric are not equally colorfast to all these destructive agents, they must be colorfast to those with which they will come in contact. Thus the dyes used for upholstery or drapery fabrics must be able to withstand sunlight, but it is not as important for them to withstand washing or perspiration. The dyes for an evening dress must be colorfast to perspiration and dry cleaning, but not necessarily colorfast to sunlight. But dyes in sportswear fabrics must be colorfast to sunlight, washing, and perspiration. Any dyes intended for use in swimming suits must also be colorfast to chlorine bleach.

In the manufacturing process, other important factors must also be considered. The shade and depth of color of the dye on the yarn or fabric must be precisely reproducible and not change from day to day or lot to lot. A level, even color must be attained, and, for economic reasons, the dye must be utilized as thoroughly as possible. In the dyeing of blended and mixed fibers, the process frequently requires a mixture of dyes or more than one step. As demonstrated in the DuPont exhibition, the end result of this can be patterned colors and special effects, as well as textured colors.

It must be emphasized that although the dyeing process is determined by the type of dye and fabric used, the end result will depend on a number of variables. Frequently the best conditions are found entirely by experimenting, with skill and long experience providing the best guidelines for success.

BACKGROUND

In this experiment, you will dye small samples of cotton, wool, nylon, polyester, acetate, and acrylic with picric acid, indigo, methylene blue, para red, Congo red, malachite green, eosin, and alizarin. The structures of these fibers and dyes are given in Tables 21-1 and 21-2. The purpose of the experiment is to compare the ability of the dyes to color the fabrics, and to test the durability or colorfastness of the dyes to some destructive agents.

CLASSES OF DYES

It has been found that the dyes in Table 21-2 and many others may be divided into the following six classes by the method in which they bond themselves to the fiber.

1. Direct dyes
2. Disazo dyes
3. Ingrain or developed dyes

TABLE 21-1.

Structures of Some Natural and Synthetic Fibers

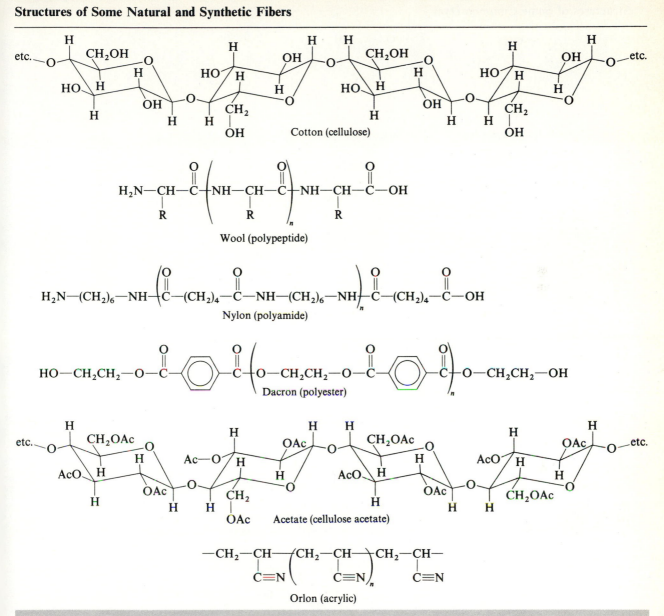

4. Vat dyes
5. Mordant dyes
6. Fiber-Reactive dyes

From these classifications, it is frequently possible to predict the effectiveness of the bonding of a certain dye to a given fiber.

TABLE 21-2.
Structures of Some Common Dyes

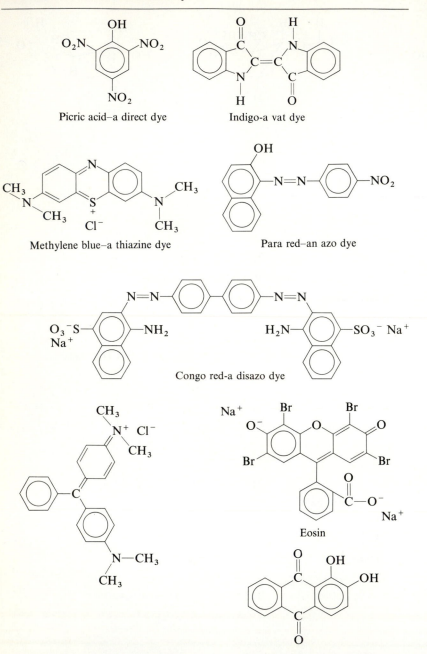

Picric acid–a direct dye

Indigo–a vat dye

Methylene blue–a thiazine dye

Para red–an azo dye

Congo red–a disazo dye

Malachite green–a triphenylmethane dye

Eosin

Alizarin–an anthraquinone dye

Direct dyes

This type of dye attaches itself to the fiber through the formation of a strong ionic bond. The dye either contains or is capable of forming a positive or negative charge, which is electrostatically attracted to a negative or positive charge on the fiber. For example, picric acid can easily lose a proton to an amino group in wool to form an ionic bond.

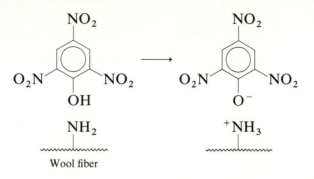

Wool fiber

Disazo dyes

This type of dye contains two azo linkages, as seen in the structure for Congo red given earlier. Through hydrogen bonding, this type of dye can bond itself to the hydroxyl groups in cotton, although this attachment is not effective for simple azo dyes.

Ingrain dyes

If a dye such as an azo dye will not effectively bond itself to a fiber such as cotton, the problem can be overcome by using an ingrain or developed dye. The dye is simply synthesized right inside the fiber. Although the dye molecule itself is too large to fit inside the spaces between the fibers, the two reactants making up the dye are small enough to enter. When they react to form the dye, the larger molecule becomes trapped inside. This process is depicted for para red in Figure 21-1.

FIGURE 21-1
Synthesis of the dye para red inside the fiber.

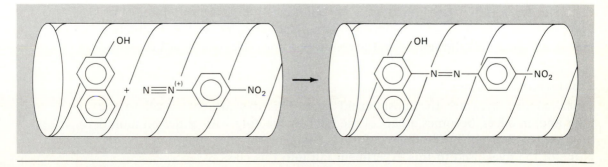

Vat dyes

Vat dyes can be used for all fibers. The dye is introduced into the fiber in a colorless, soluble form, and is then "developed" into a colored, insoluble form, which precipitates both on the inside and outside of the fabric fibers. Since the dye is insoluble in water, the color is fast.

A good example of a vat dye is indigo, used to dye blue jeans. Initially, indigo is in the insoluble blue form, but can be reduced by sodium hydrosulfite to soluble, colorless leucoindigo. Air oxidation then regenerates the original blue dye.

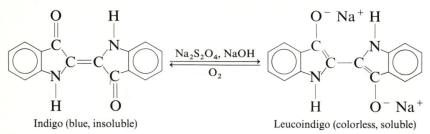

Indigo (blue, insoluble)   Leucoindigo (colorless, soluble)

Mordant dyes

Certain dyes and the fibers wool, silk, cotton, and linen are found to form chelated complexes with **mordants** or the salts of heavy metals such as tin, iron, chromium, and copper. Since the chelated complex is, in effect, a bond to the metal ion, then the metal can be used as a link to bond these dyes to the fibers. This is depicted below for alizarin bonding itself to cotton through the mordant chromium.

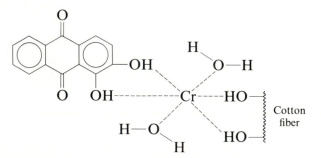

In order for the mordant dye to be colorfast, the complex of the mordanting metal, the dye, and the fiber must be insoluble and stable. If different metals are used with the same dye, the resulting color is frequently not the same.

Fiber-reactive dyes

Fiber-reactive dyes are the most recent type. This dye attaches itself to the fiber and, in effect, becomes a part of the fiber by reacting with either an amino or a hydroxyl group in the fiber to form a covalent bond, as illustrated below. None of the dyes used in this experiment fall into this class.

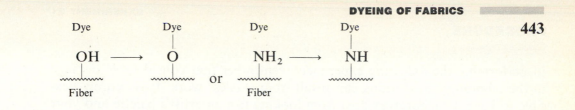

COLORFASTNESS TO DESTRUCTIVE AGENTS

Laundering

The testing of a dye for fastness to washing uses a soap stronger than that used normally. One or two washings with strong soap generally give the same results as many washings with mild soap. Any appreciable bleeding or fading means that a dye on the given fabric is not fast to laundering.

Dry cleaning

The dyed fabric in this test is placed in a dry cleaning solution composed of dry cleaning soap in tetrachloroethylene or carbon tetrachloride for 25 minutes. The fabric is then removed, rinsed with clean solvent containing no soap, and any discoloration or bleeding of color is noted.

Bleaching

The dyed fabric in this test is placed in undiluted chlorine bleach solution, and any destruction of the color is noted.

Perspiration

To test for fastness to perspiration, the most satisfactory method is to subject the dyed fabric to actual wear. A quicker test is to attach a swatch of the dyed fabric to a white cloth of the same material and immerse them in 5% acetic acid. If the white fabric becomes stained, the dyed fabric is not fast to perspiration. The tested sample can also be compared with an untested piece for any change in color. Acetic acid solution is used because it comes closest chemically to perspiration.

Friction

Dyed fabrics used for clothing and other items subjected to wear must be able to withstand a great deal of friction without the color rubbing off or crocking. Not only will the color fade as the dye comes off, but other fabrics rubbed against it may be discolored. To test for friction, the dyed sample is rubbed briskly against a white cloth, and any transfer of color is noted. To be absolutely fast to friction, the color should not rub off when the fabric is either dry or wet.

PROCEDURE

In performing the following experiment, remember that considerable damage to the laboratory or clothing can result from careless work. It is entirely possible to keep the laboratory area from looking like an artist's palette and your clothes from looking like painter's dungarees. Furthermore, you may find it socially uncomfortable to sport almost every color of the rainbow around the school or anywhere else you go. Stirring rods and tongs can be used almost as conveniently as fingers for handling dyed cloth and are more easily cleaned. Paper towels are a good and inexpensive way to keep the laboratory area clean, and much of the work can be done on them.

To save time, you can do several parts of the experiment simultaneously. Since each part has a certain amount of waiting, this time can be efficiently utilized for starting a later section.

DYEING THE FABRICS

The nine dyes in this experiment will be used to color wool, acrylic, polyester, nylon, cotton, and acetate fibers. Indicate in chart form the effectiveness of each dye in coloring each fiber. To save time, use nine pieces of approximately 2-inch by 4-inch multifiber fabric. One such fabric, called Multifiber Fabric 10A,[1] has the six needed fibers woven in a striped sequence. **Be sure you know which stripes are which fibers!**

Picric acid

Place about 30 mL of a 0.5% solution of picric acid in a 150-mL beaker. **(CAUTION: PICRIC ACID IS A POTENTIAL EXPLOSIVE WHEN DRY. IT MUST EITHER BE KEPT UNDER WATER OR COMPLETELY DISSOLVED IN SOLUTION.) Carefully** add 5 drops of concentrated sulfuric acid and heat the solution on a steam bath or hot plate. Immerse one piece of the multifiber fabric in the hot solution. After a few minutes, remove the piece with tongs and transfer to another beaker of warm rinse water. Rinse it well, changing the water when necessary, and repeat this procedure until the dye no longer runs. Note and record the results in your chart.

Indigo

Place about 0.05 gram of indigo into a 150-mL beaker along with 0.1 gram of sodium hydrosulfite (sodium dithionite) and a sodium hydroxide pellet. Add 30 mL of hot water and stir until the indigo is dissolved and bleached. Add a

---

[1] Available from Testfabrics, Inc., P.O. Box 118, 200 Blackford Ave., Middlesex, New Jersey 08846

little more sodium hydrosulfite if the indigo does not dissolve completely. To the resulting green leucoindigo solution, place a piece of the multifiber fabric. After a few minutes, remove the fabric and press it between paper towels to remove the excess dye solution. Hang the fabric to air dry, and compare and record the results as the oxygen oxidizes the leucoindigo.

Methylene blue

Dissolve 0.05 gram of methylene blue in about 30 mL of hot water in a 150-mL beaker. Immerse a piece of multifiber fabric. After a few minutes, remove the fabric and thoroughly rinse. Note and record the results in your chart.

Para red

Dissolve about 0.05 gram of the para red dye prepared in the last experiment in about 30 mL of hot water in a 150-mL beaker. **Carefully** add a few drops of concentrated sulfuric acid and immerse a piece of multifiber fabric. After about 5 minutes, remove the fabric and thoroughly rinse. Note and record the results.

Para red developed

Solution A.   Place 0.25 gram of *p*-nitroaniline and 10 mL of 2% hydrochloric acid in a small beaker and heat until most of the *p*-nitroaniline dissolves. Cool the solution to below 5° in an ice bath and add a cold solution of 0.12 gram of sodium nitrite dissolved in 5 mL of water. Stir the solution well, keeping it in the ice bath, and prepare Solution B.

Solution B.   Place 0.1 gram of *β*-naphthol in a small beaker containing 25 mL of water and, while stirring, add a solution of 10% sodium hydroxide **drop-wise** until **most** of the *β*-naphthol has dissolved. It is important **not** to add too much base, because a number of the fibers will disintegrate under strongly alkaline conditions.

Place a piece of multifiber fabric into Solution B and allow it to soak for 2 to 3 minutes. Remove the fabric and lightly pat dry on paper towels. Next dilute Solution A with about 20 mL of ice-cold water and immerse the fabric removed from Solution B. After several minutes, remove the fabric and rinse well. Note and record the results, comparing them in particular with the para red used in the previous dyeing step.

Congo red

Dissolve about 0.05 gram of Congo red in 35 mL of hot water in a 150-mL beaker and then add 0.5 mL each of 10% sodium carbonate and 10% sodium sulfate solution. Immerse a piece of multifiber fabric. After 10 minutes, remove the fabric and rinse in warm water. Note and record the results.

Malachite green

Dissolve about 0.025 gram of malachite green in 35 mL of hot water, and immerse the fabric for about 10 minutes. Remove the fabric, thoroughly rinse it, and note and record the results.

Eosin

Dissolve about 0.025 gram of eosin in 35 mL of hot water, and immerse the fabric for about 10 minutes. Remove and thoroughly rinse the fabric, and note and record the results.

Alizarin

Dissolve about 0.025 gram of alizarin and 0.025 gram of sodium carbonate in 35 mL of hot water. Immerse the fabric for about 10 minutes, then remove the fabric and thoroughly rinse it. Note and record the results.

DESTRUCTIVE AGENTS

Each dyed fabric will now be tested with the following destructive agents. With a pair of scissors, cut each piece of cloth into four pieces; be sure that all six fibers are still present on each piece. Three of these pieces will be used for the destructive agents, and the fourth will serve as a control. **It is very important that you continue to know which fibers are which so that no mixup occurs.**

Laundering

Separately immerse each of the nine different, smaller-size fabrics in a hot solution of a strong laundry detergent for about 15 minutes, rinse, and look for any bleeding or fading. Each color is done separately to avoid one color bleeding onto another. Record the results in chart form.

Bleaching

Immerse each of the nine different, smaller-size fabrics in undiluted chlorine bleach solution for 15 minutes, and note and record any fading of the color.

Perspiration

Attach each of the nine fabrics to a piece of laundered, white cotton fabric. Immerse them separately in 5% acetic acid solution, and note and record any staining of the white fabric. Also compare the tested sample to the control for any change in color.

1. What dyes were best able to dye cotton before being treated with the destructive agents? After being treated with the destructive agents? Repeat your answer for each of the other five fibers.
2. What dyes were affected most by bleaching? Look at the molecular structures of these dyes and determine what functional groups, if any, might be responsible.
3. Most household dyes sold in the store contain sodium chloride. What is the purpose of this added salt? (Hint: See page 106. This will not exactly answer the question for you, but it should hopefully guide your mind to the correct answer).
4. Look up and describe the molecular structures of the fibers linen, silk, and rayon. Is there any similarity between the structures of these fibers and the structures of any of the fibers used in this experiment?
5. Why wouldn't you use a vat dye or developed dye at home if you were desiring a certain shade of color in an item you might be dyeing?

FRIEDEL-CRAFTS ALKYLATION AND ACYLATION

ELECTROPHILIC AROMATIC SUBSTITUTION

DISCUSSION

Of all the methods used to introduce alkyl and acyl groups onto aromatic rings, the most important and versatile is the Friedel-Crafts reaction. Discovered in 1877 by Charles Friedel and James Crafts, this reaction today has been extensively studied and is frequently considered a classic example of electrophilic aromatic substitution. The typical electrophile in the reaction is a carbocation or acylium ion, prepared respectively from an alkyl halide or acid chloride losing a halogen ion to the Lewis acid $AlCl_3$. Once formed, the electrophile then attacks the aromatic ring to form a resonance-stabilized intermediate, which then loses a proton to give the observed product. For example:

Carbocation Electrophile

$$(1) \quad \underset{\substack{| \\ CH_3}}{\overset{\substack{CH_3 \\ |}}{Cl-C-CH_3}} + AlCl_3 \;\underset{}{\overset{\text{fast}}{\rightleftharpoons}}\; \underset{\substack{| \\ CH_3}}{\overset{\substack{CH_3 \\ |}}{{}^+C-CH_3}} + \bar{A}lCl_4$$

tert-Butyl chloride

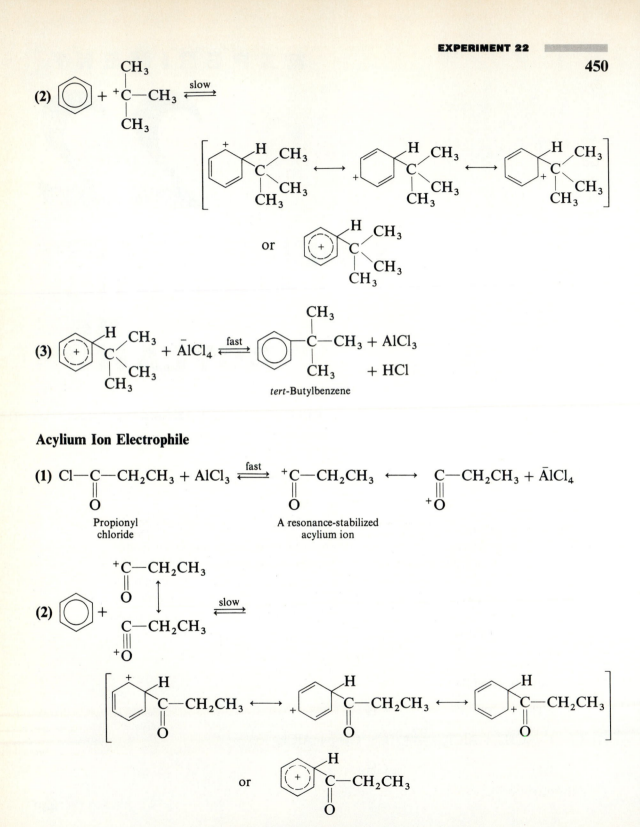

Acylium Ion Electrophile

tert-Butylbenzene

Propionyl
chloride

A resonance-stabilized
acylium ion

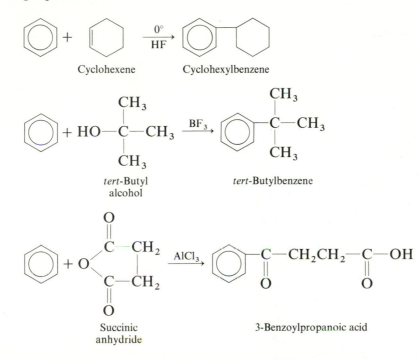

(3) A phenyl–C(=O)–CH$_2$CH$_3$ cation $+ \bar{\text{A}}\text{lCl}_4 \xrightarrow{\text{fast}}$ phenyl–C(=O)–CH$_2$CH$_3$ + AlCl$_3$ + HCl

Ethyl phenyl ketone

The Friedel-Crafts reaction, however, is not limited to an alkyl halide or acid chloride forming the electrophile through the use of AlCl$_3$. Alkenes, alcohols, carboxylic acids, and acid anhydrides form similar electrophiles with the aid of other Lewis acids such as BF$_3$, HF, SnCl$_4$, and FeCl$_3$, thus expanding the scope of the reaction. A few examples of these variations are shown in the following equations:

Cyclohexene + benzene $\xrightarrow[\text{HF}]{0°}$ Cyclohexylbenzene

tert-Butyl alcohol + benzene $\xrightarrow{\text{BF}_3}$ *tert*-Butylbenzene

Succinic anhydride + benzene $\xrightarrow{\text{AlCl}_3}$ 3-Benzoylpropanoic acid

Evidence indicates that the attacking electrophile may not always be a free carbocation or acylium ion. Instead, the starting material may just form a highly polarized complex with the Lewis acid, which then attacks the aromatic ring. For an alkyl halide or acid chloride reacting with AlCl$_3$, this polar complex would have the structures

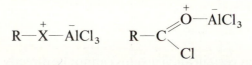

$$R\text{---}\overset{+}{X}\text{---}\overset{-}{\text{A}}\text{lCl}_3 \qquad R\text{---}C\overset{\overset{+}{O}\text{---}\overset{-}{\text{A}}\text{lCl}_3}{\diagdown_{\text{Cl}}}$$

But the mechanism can also be considered to be acid-catalyzed nucleophilic substitution and nucleophilic acyl substitution, with the aromatic ring acting as the nucleophile.

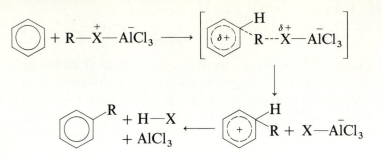

Nucleophilic Acyl Substitution

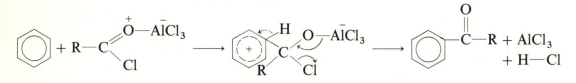

Even though the Friedel-Crafts reaction is very important and versatile in organic syntheses, a number of restrictions limit its usefulness.

1. Since a carbocation intermediate is frequently involved, rearrangements may occur in the side chain attaching itself to the aromatic ring. As the following example demonstrates, sometimes the rearranged product is the only monosubstituted product formed:

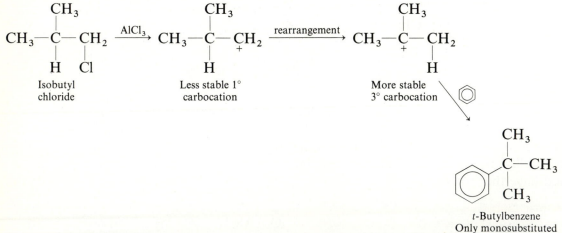

This rearrangement means that certain alkyl groups cannot be directly placed on the aromatic ring by Friedel-Crafts alkylation. But this limitation frequently can be overcome by using Friedel-Crafts acylation, followed by the Clemmensen reduction of the carbonyl group with amalgamated zinc and hydrochloric acid.

For example:

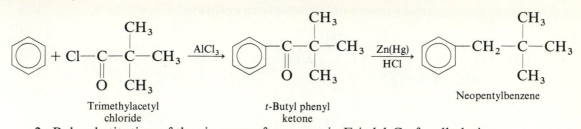

Trimethylacetyl chloride — *t*-Butyl phenyl ketone — Neopentylbenzene

2. Polysubstitution of the ring can often occur in Friedel-Crafts alkylations. This results from the alkyl groups already on the ring releasing electrons, thereby making the product more reactive than the starting material. For example:

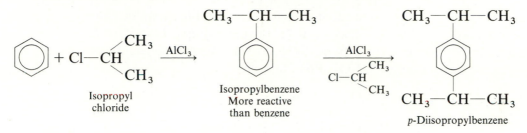

Isopropyl chloride — Isopropylbenzene More reactive than benzene — *p*-Diisopropylbenzene

This limitation can be minimized by using a large excess of the starting aromatic compound. What the starting compound thus lacks in reactivity is partially compensated for by the much higher probability of it colliding with the carbocation or the highly polarized complex.

Polysubstitution, however, is not a problem with Friedel-Crafts acylation. This results from the acyl group withdrawing electrons from the aromatic ring, thereby deactivating the ring from further substitution. For example:

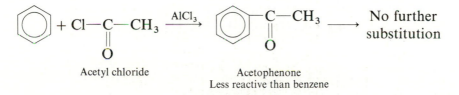

Acetyl chloride — Acetophenone Less reactive than benzene

3. Because of the low reactivity of halogen atoms attached to aromatic rings and doubly-bonded carbons, aryl and vinyl halides cannot be used in place of alkyl halides. For example:

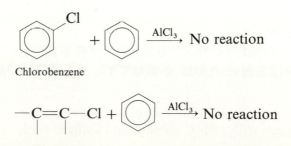

Chlorobenzene

4. Migration or "scrambling" of the alkyl groups on the ring can occur whenever a thermodynamically more stable product can be formed. For example:

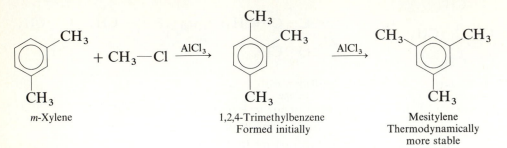

m-Xylene   1,2,4-Trimethylbenzene   Mesitylene
    Formed initially   Thermodynamically
              more stable

5. Finally, Friedel-Crafts reactions do not occur whenever the aromatic ring is highly deactivated by powerful electron-withdrawing groups. Evidently a Friedel-Crafts electrophile is not as powerful as other electron-deficient reagents, such as the nitronium ion $^+NO_2$. For example:

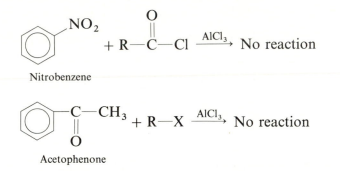

Nitrobenzene

Acetophenone

Aniline is highly activated and therefore should undergo the reaction. But it doesn't, because the free pair of electrons on nitrogen readily react with the Lewis acid needed in the reaction to form a highly deactivated complex.

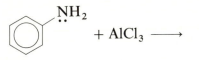

Aniline   Incapable of undergoing
         the Friedel-Crafts reaction

22A. SYNTHESIS OF 1,4-DI-t-BUTYL-2,5-DIMETHOXYBENZENE FROM 1,4-DIMETHOXYBENZENE AND t-BUTYL ALCOHOL

In this experiment, you will prepare 1,4-di-t-butyl-2,5-dimethoxybenzene by the reaction of 1,4-dimethoxybenzene with t-butyl alcohol and sulfuric acid.

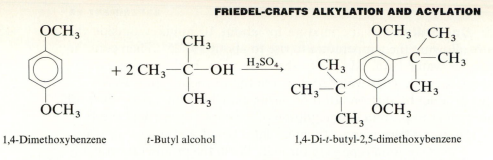

1,4-Dimethoxybenzene   *t*-Butyl alcohol   1,4-Di-*t*-butyl-2,5-dimethoxybenzene

The reaction occurs quite readily, due to the two methoxy groups activating the aromatic ring toward electrophilic aromatic substitution. As in the reaction of *t*-butyl chloride and benzene, the electrophile is the *t*-butyl carbocation, but this time it is produced by *t*-butyl alcohol reacting with sulfuric acid.

The final product of 1,4-di-*t*-butyl-2,5-dimethoxybenzene predominates from the excess quantity of *t*-butyl alcohol used and the bulkiness of the *t*-butyl group preventing two *t*-butyl groups from readily occupying two adjacent positions on the aromatic ring. There is also more steric hindrance when both *t*-butyl groups are placed ortho to one methoxy group.

PROCEDURE

CAUTION: Concentrated sulfuric acid and glacial acetic acid cause severe burns. Use them very carefully. The use of plastic gloves is recommended. Also, methylene chloride is toxic. Do not breathe it excessively or spill it on yourself.

Place 4.0 grams (0.029 mole) of 1,4-dimethoxybenzene into a 125-mL Erlenmeyer flask. Add 7.0 mL (5.5 grams, 0.075 mole) of *t*-butyl alcohol and 13 mL of glacial acetic acid and put the flask into an ice-water bath. **Carefully** place 20 mL of concentrated sulfuric acid into a 50-mL Erlenmeyer flask and also cool it in an ice-water bath, being certain the flask is properly supported.

When the contents of both flasks have reached a temperature of 0–3°, clamp a 125-mL separatory funnel over the 125-mL Erlenmeyer flask. **Carefully** transfer the cold sulfuric acid to the separatory funnel and then add the cold sulfuric acid dropwise to the still chilled 125-mL Erlenmeyer flask over a period of about 5 minutes. The flask should be periodically swirled during this time.

After all the sulfuric acid has been added, a considerable amount of solid reaction product should have separated and the temperature should have risen to about 15°. **Carefully** swirl the mixture for about 10 minutes outside the ice-water bath, allowing the temperature to rise to about 20–25°. Then cool the mixture in the ice-water bath, add ice to the mixture to both cool and dilute the sulfuric acid, and add cold water to near the top of the reaction flask.

Collect the product by suction filtration, using glass fiber filter paper instead of regular filter paper to prevent disintegration of the paper by the acid solution. As a second choice, use **two** pieces of regular filter paper and only very gentle suction when the acid solution is passing through. Wash the product thoroughly with water and allow it to drain well.

In the meantime, cool 20 mL of methanol, which will be used to further wash the crystals. Release the suction, cover the crystals with about 7 mL of the cold methanol, and then reapply the suction. Repeat this washing procedure twice more.

Place the moist product into a 125-mL Erlenmeyer, add enough methylene chloride to dissolve the crystals, then add anhydrous magnesium sulfate. If the drying agent clumps together, use a little more and swirl the mixture. Allow the drying to occur for 10 minutes, then carefully decant the liquid into another 125-mL Erlenmeyer, filtering through a small wad of cotton in the neck of a funnel, if necessary. Add 20 mL of methanol to the solution and evaporate on a steam bath or a hot water bath **in the hood** to remove most of the lower-boiling methylene chloride. When the volume is about 20 mL, let the solution stand for recrystallization. When recrystallization is complete, cool in ice and collect by suction filtration. Allow the product to thoroughly dry and determine the yield, melting point, and literature melting point. Record the results in your notebook. Also determine the theoretical and percentage yields, and show the calculations.

FIGURE 22-1
Infrared spectrum of 1,4-di-*t*-butyl-2,5-dimethoxybenzene in a KBr pellet. Spectrum courtesy of Sadtler Research Laboratories, division of Bio-Rad Laboratories, Inc., copyrighted 1973.

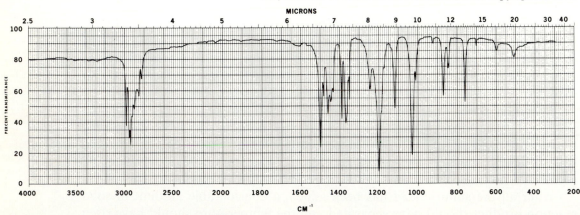

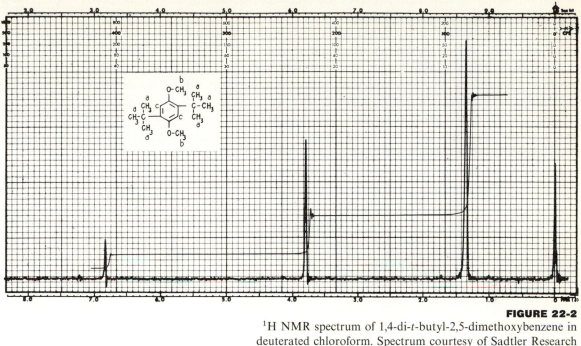

At your instructor's option, obtain the infrared spectrum of your product in a KBr pellet and the ^{1}H NMR spectrum in deuterated chloroform. Compare your spectra with those shown in Figures 22-1 and 22-2. When finished, hand in your product in a properly labeled container to your instructor.

22B. SYNTHESIS OF RESACETOPHENONE FROM RESORCINOL AND ACETIC ACID

In this experiment, you will synthesize resacetophenone from resorcinol and acetic acid by Friedel-Crafts acylation. Because the aromatic ring in resorcinol is highly activated, the very reactive acetyl chloride is not needed.

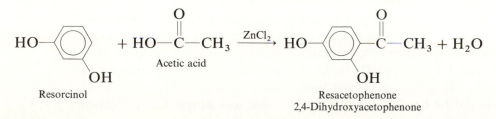

Resorcinol

Acetic acid

Resacetophenone
2,4-Dihydroxyacetophenone

457

The reaction mechanism involves the zinc chloride acting as a weak Lewis acid catalyst and forming a chemical bond to the carbonyl oxygen of acetic acid in step 1. This makes the carbonyl carbon atom of acetic acid even more electron deficient, and it is readily attracted to the loosely held π electrons of resorcinol, forming a highly stable arenium ion (I) in step 2. Since the hydroxyl groups of resorcinol are both powerful ortho, para directors, substitution occurs predominantly at the indicated position. For steric reasons, little substitution occurs at the ortho position between the two hydroxyl groups. Once formed, the arenium ion then loses a proton in step 3 to form intermediate II, which then expels another proton and the oxygen–zinc chloride complex in step 4 to form resacetophenone as the final product. Notice from the mechanism that no intermediate acylium ion is formed.

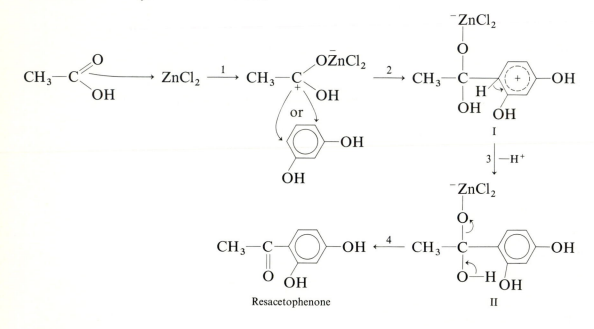

Resacetophenone

PROCEDURE

CAUTION: Be careful when using hydrochloric acid, glacial acetic acid, and resorcinol. Keep them off your skin, especially when they are hot. Severe burns may result.

Place 6.0 mL (0.11 mole) of glacial acetic acid into a 50-mL Erlenmeyer flask **in the hood**. Add 6.0 grams (0.044 mole) of finely powdered, anhydrous $ZnCl_2$, and dissolve by heating the mixture on a hot plate. When all the $ZnCl_2$ has dissolved, add 4.0 grams (0.036 mole) of resorcinol in small amounts with constant stirring. The addition should take a few minutes, during which time the solution should be kept heated at 140–145°. **Do not allow the solution**

to boil. Continue heating the solution for another 25 minutes with occasional stirring, keeping it between 140–145°.

Allow the mixture to cool and then **carefully** and slowly add, with stirring, 25 mL of 6 *M* HCl to dissolve the zinc salts. Cool the mixture in an ice bath to 10°. If a thick reddish paste forms, **carefully** add another 5 mL of 6 *M* HCl with cooling and stirring.

Suction filter the mixture and wash the collected reddish crystals with 25 mL of cold 1 *M* HCl. Place the collected crystals in a 125-mL Erlenmeyer flask and recrystallize, using 1 *M* HCl solution. After the crystals dissolve, remove the tarry impurities by suction filtering the hot solution. Cool the filtrate in an ice bath and collect the resulting crystals by suction filtration.

Recrystallize the reddish-tan crystals a second time from hot water, using decolorizing carbon. Pure resacetophenone has no color, but, in the literature, it has also been obtained as light tan crystals. Allow your purified product to thoroughly dry and determine the yield, melting point, and literature melting point. Record the results in your notebook. Also determine the theoretical and percentage yields, and show the calculations.

At your instructor's option, obtain the infrared spectrum of your product in a KBr pellet and the ^{1}H NMR spectrum in deuterated chloroform. When finished, hand in your product in a properly labeled container to your instructor.

QUESTIONS

1. Why can acetic acid rather than acetyl chloride be used in the synthesis of resacetophenone?
2. Why does polysubstitution frequently occur with Friedel-Crafts alkylation but not with Friedel-Crafts acylation?
3. If benzene is reacted with *t*-butyl chloride, explain why the chief disubstituted product is the para isomer rather than the ortho isomer.
4. In Friedel-Crafts acylation reactions, why must you use slightly more than 1 mole of catalyst per mole of the acid chloride? Why is this not necessary in Friedel-Crafts alkylation reactions?
5. Why may anhydrous calcium chloride be the drying agent in the reaction workup for many Friedel-Crafts alkylation reactions but not for Friedel-Crafts acylation reactions?
6. Outline a Friedel-Crafts synthesis of the compound 4-*n*-hexylresorcinol, the antiseptic found in Sucrets and Listerine throat lozenges. You may use resorcinol and any other needed reagents for the synthesis.

4-*n*-Hexylresorcinol

7. Starting with benzene and succinic anhydride, outline a procedure for the synthesis of α-tetralone, utilizing the Friedel-Crafts reaction twice.

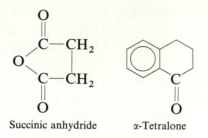

Succinic anhydride   α-Tetralone

8. Explain the following reaction, remembering that an ethyl group is an ortho, para director in electrophilic aromatic substitution.

23

ALDOL CONDENSATION

- **NUCLEOPHILIC ADDITION**
- **CARBANIONS**
- **ELIMINATION**

DISCUSSION

The two fundamental types of reactions generally undergone by aldehydes and ketones are

1. Nucleophilic addition

2. Substitution reactions due to the slight acidity of α-hydrogens on the carbons adjacent to the carbonyl group

The reaction that combines both of these reaction types is the **aldol condensation**, which is very useful synthetically for generating new carbon-carbon bonds, and is also used biologically for building many complex naturally occurring compounds.

The simplest example of an aldol condensation is the reaction of acetaldehyde with itself in the presence of dilute base.

$$CH_3-\overset{\overset{\text{O}}{\|}}{C}-H + CH_3-\overset{\overset{\text{O}}{\|}}{C}-H \xrightarrow{\text{10\% NaOH}}$$

Acetaldehyde

$$CH_3-\overset{\overset{\text{OH}}{|}}{CH}-CH_2-\overset{\overset{\text{O}}{\|}}{C}-H \xrightarrow[-H_2O]{\text{heat}} CH_3-CH=CH-\overset{\overset{\text{O}}{\|}}{C}-H$$

3-Hydroxybutanal
"aldol"

Crotonaldehyde

461

Observe that the product formed initially contains both an aldehyde functional group and an alcohol functional group. Combining the first three letters of the word **ald**ehyde and the last two letters of alcoh**ol**, we obtain the word **aldol**, which can generally mean any compound containing both these functional groups. Because both molecules of aldehyde have added together, this reaction may be called an "aldol addition." The same type of self-addition reaction undergone by ketones is also called an aldol addition.

The initial aldol addition product, however, is frequently not isolated, since water can be readily eliminated to form a stabilized conjugated system. Thus 3-hydroxybutanal can react further to yield crotonaldehyde. Because the loss of water in a reaction is called a "condensation," the loss of water in this particular type of reaction gives us the name "aldol condensation," even though the initial product may not lose water at all.

An aldol condensation may be acid, base, or enzyme catalyzed. In the laboratory, catalysis by bases is most commonly used, whereas in biology nature obviously uses enzymes to speed up the reaction. One common example of a biological aldol condensation is the synthesis of D-fructose-1,6-diphosphate from D-glyceraldehyde-3-phosphate and dihydroxyacetone phosphate. Note that the enzyme catalyzing this reaction is appropriately called aldolase.

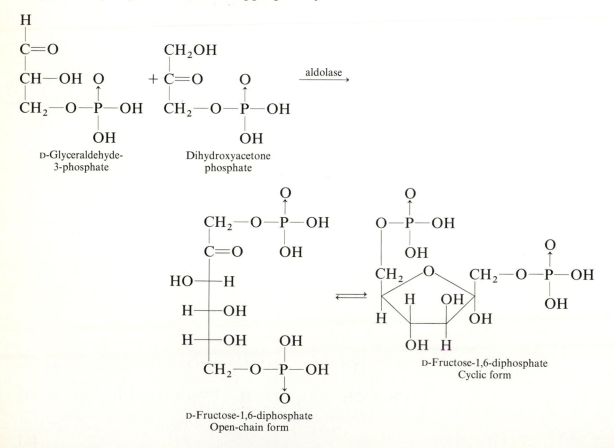

D-Glyceraldehyde-3-phosphate

Dihydroxyacetone phosphate

D-Fructose-1,6-diphosphate
Open-chain form

D-Fructose-1,6-diphosphate
Cyclic form

The generally accepted mechanism for the base-catalyzed reaction involves the following steps, with acetaldehyde being used as an example.

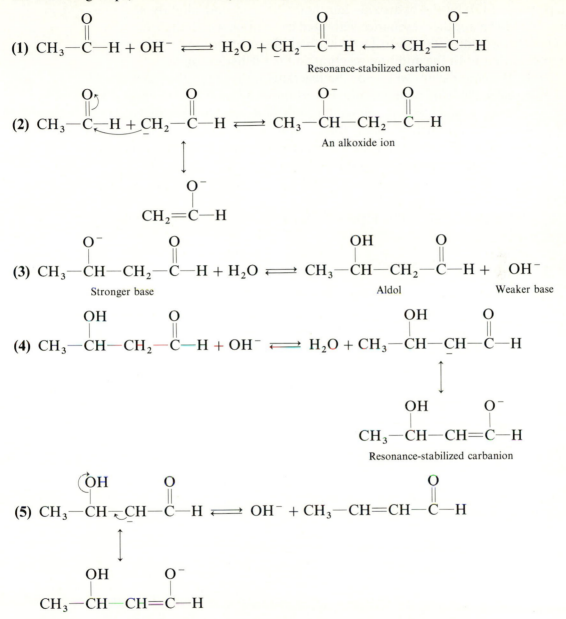

In step (1), the hydroxide ion abstracts a proton from the α-carbon of the aldehyde to form a resonance-stabilized carbanion. In fact, the stabilization of this ion is the reason for the slight acidity of the α-hydrogen. In step (2), this carbanion, acting as a nucleophile, then attacks the carbonyl carbon of a second molecule of aldehyde to form an alkoxide ion. In step (3), this alkoxide ion, being a stronger base than hydroxide ion, then abstracts a proton from water to yield aldol as the initially formed product. This product may be isolated or the reaction mixture may be heated to eliminate water. This occurs, as shown in steps (4) and (5), by the hydroxide ion removing a second, slightly acidic α-hydrogen to form another resonance-stabilized carbanion, which then expels the OH group as a hydroxide ion to form the final observed product.

Observe that both the addition reaction and the dehydration are reversible. Frequently, the equilibrium will favor the reactants rather than the aldol product. This is especially true with ketones. For example, when acetone reacts with itself to form diacetone alcohol, the equilibrium is so unfavorable that only a small percentage of product is formed. But by using a special reaction apparatus that continuously separates the product from the reactant and catalyst, you can obtain over a 70 percent yield of product.

$$CH_3-\overset{\overset{\displaystyle O}{\|}}{C}-CH_3 + CH_3-\overset{\overset{\displaystyle O}{\|}}{C}-CH_3 \underset{}{\overset{Ba(OH)_2}{\rightleftharpoons}} CH_3-\overset{\overset{\displaystyle OH}{|}}{\underset{\underset{\displaystyle CH_3}{|}}{C}}-CH_2-\overset{\overset{\displaystyle O}{\|}}{C}-CH_3$$

Acetone   Diacetone alcohol

If two different carbonyl compounds are used, a mixed or crossed aldol condensation occurs, giving a complex mixture of as many as four or more reaction products. For example, the reaction of acetaldehyde with propionaldehyde gives the following four products and, possibly, some dehydration products as well.

$$CH_3-\overset{\overset{\displaystyle O}{\|}}{C}-H + CH_3CH_2-\overset{\overset{\displaystyle O}{\|}}{C}-H \overset{OH^-}{\longrightarrow}$$

$$CH_3-\overset{\overset{\displaystyle OH}{|}}{CH}-CH_2-\overset{\overset{\displaystyle O}{\|}}{C}-H$$

3-Hydroxybutanal (from 2 molecules of acetaldehyde)

$$+ CH_3CH_2-\overset{\overset{\displaystyle OH}{|}}{CH}-\overset{\overset{\displaystyle}{\underset{\underset{\displaystyle CH_3}{|}}{CH}}}-\overset{\overset{\displaystyle O}{\|}}{C}-H$$

2-Methyl-3-hydroxypentanal (from 2 molecules of propionaldehyde)

$$+ CH_3-\overset{\overset{\displaystyle OH}{|}}{CH}-\overset{\overset{\displaystyle}{\underset{\underset{\displaystyle CH_3}{|}}{CH}}}-\overset{\overset{\displaystyle O}{\|}}{C}-H$$

2-Methyl-3-hydroxybutanal (from the carbanion of propionaldehyde adding to acetaldehyde)

$$+ CH_3CH_2-\overset{\overset{\displaystyle OH}{|}}{CH}-CH_2-\overset{\overset{\displaystyle O}{\|}}{C}-H$$

3-Hydroxypentanal (from the carbanion of acetaldehyde adding to propionaldehyde)

But a crossed aldol condensation can be worthwhile if one reactant has no α-hydrogens and thus cannot react with itself. In such cases, the two possible reactions become the desired mixed condensation and the self-condensation of the other compound. The self-condensation is undesired but you can minimize it by maintaining a low concentration of the other compound. The following synthesis of benzalacetophenone illustrates the effective use of a crossed aldol condensation.

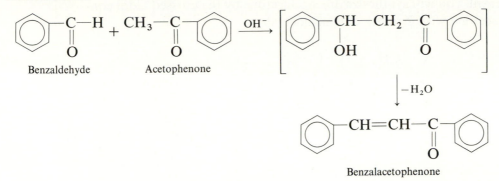

The use of brackets in this reaction indicates that the initial reaction product cannot be isolated. In this case, spontaneous dehydration occurs from the very stable and extended conjugated system being formed with benzalacetophenone.

Once an aldol condensation product has been formed, it can undergo other reactions to give a variety of useful products. For example:

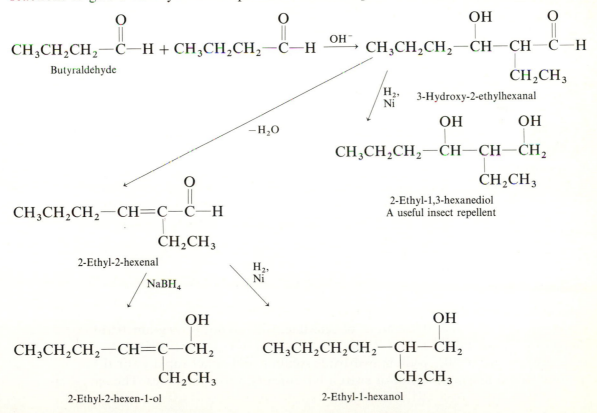

23A. PREPARATION OF DIBENZALACETONE FROM BENZALDEHYDE AND ACETONE

In this experiment, you will synthesize dibenzalacetone by the crossed aldol condensation of acetone with 2 moles of benzaldehyde.

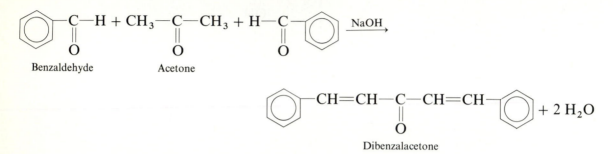

Benzaldehyde   Acetone

Dibenzalacetone

The reaction occurs readily by merely stirring the reactants and catalyst. Water and ethyl alcohol are used as the solvent, with enough alcohol being used to keep the initially formed benzalacetone in solution until it has reacted with a second molecule of benzaldehyde.

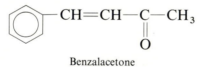

Benzalacetone

Although this reaction is a crossed aldol condensation, a complex mixture of products is not formed because benzaldehyde has no α-hydrogens. Furthermore, the equilibrium is very unfavorable for the acetone reacting with itself to yield diacetonealcohol. As a result, the concentration of acetone does not have to be kept small.

None of the initial aldol addition product is ever obtained in the reaction, due to the especially stable, extended conjugated system causing water to be eliminated from the molecule.

PROCEDURE

Place 3.0 mL (3.2 grams, 0.030 mole) of benzaldehyde, 1.0 mL (0.80 gram, 0.014 mole) of acetone, and 25 mL of 95% ethyl alcohol into a 125-mL Erlenmeyer flask. Add 30 mL of 10% sodium hydroxide solution and either rapidly stir the mixture with a magnetic stirrer or swirl it for at least 15 to 20 minutes. The reaction mixture should at first be clear and homogeneous, but should shortly become milky with yellow solid particles of product visible.

Collect the yellow solid by suction filtration, using cold water to completely transfer and thoroughly wash the product. After washing, press the solid onto the filter to remove as much water as possible. Turn off the suction, break up the crystalline mass with a spatula, and add to the solid on the filter an ice-cold solution of 0.3 mL of glacial acetic acid in 5 mL of 95% ethanol. Allow to stand for 30 to 60 seconds and then remove the liquid by reapplying the suction. Turn off the suction, break up the crystalline mass, add another ice-cold solution of glacial acetic acid and 95% ethanol, allow to stand, and then remove the liquid by reapplying the suction. These two washings neutralize any remaining base and also remove any impurities that might cause the product to oil out during recrystallization.

Recrystallize the product from either 2-propanol or 95% ethanol, allowing the hot mixture to slowly cool to room temperature; then cool it further in an ice bath. Collect the purified yellow crystals by suction filtration and wash with a very small quantity of fresh, ice-cold solvent. Allow the product to thoroughly dry and determine the yield, melting point, and literature melting point. Record the results in your notebook. Also determine the theoretical and percentage yields of the final product and show the calculations.

At your instructor's option, obtain the infrared spectrum of your product in a KBr pellet and the ^{1}H NMR spectrum in deuterated chloroform. Compare your spectra with those shown in Figures 23-1 and 23-2. When finished, hand in your product in a properly labeled container to your instructor.

FIGURE 23-1

Infrared spectrum of 1,5-diphenyl-1,4-pentadiene-3-one in a K Br pellet. Spectrum courtesy of Sadtler Research Laboratories, division of Bio-Rad Laboratories, Inc., copyrighted 1970.

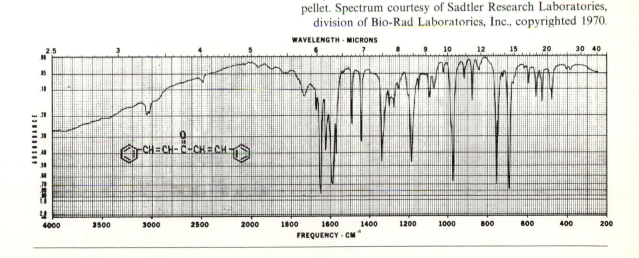

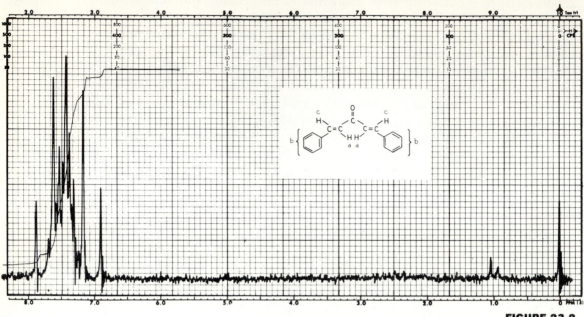

FIGURE 23-2

^{1}H NMR spectrum of 1,5-diphenyl-1,4-pentadiene-3-one in deuterated chloroform. Spectrum courtesy of Sadtler Research Laboratories, division of Bio-Rad Laboratories, Inc., copyrighted 1969.

23B. PREPARATION OF 2-METHYL-2-PENTENAL FROM PROPIONALDEHYDE

In this experiment, you will prepare 2-methyl-2-pentenal by the aldol condensation of propionaldehyde with sodium hydroxide.

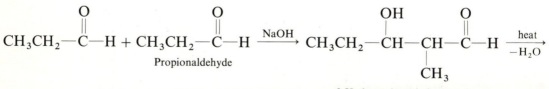

Propionaldehyde

3-Hydroxy-2-methylpentanal

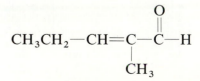

2-Methyl-2-pentenal

468

Although the 3-hydroxy-2-methylpentanal initially formed can be isolated, you will heat the reaction mixture, causing this intermediate to dehydrate to a conjugated enal as the final product.

PROCEDURE

Clamp a 50-mL round-bottom flask over a magnetic stirrer and add a stirring bar and 5 mL of 10% sodium hydroxide. Fit the flask with a Claisen head and place a clamped separatory funnel into the center neck of the head and a clamped reflux condenser into the other neck. Place 20 mL (16 grams, 0.28 mole) of recently distilled propionaldehyde, free of polymer and propionic acid, into the separatory funnel and add it slowly with stirring to the flask. The temperature will rise during the addition, and the low-boiling propionaldehyde may reflux. After all the reagent has been added, continue stirring the reaction mixture until it has cooled to nearly room temperature.

Remove the reflux condenser, separatory funnel, and Claisen head. Remove as much of the aqueous layer as possible with a pipet and suction bulb. Assemble an apparatus for fractional distillation as in Experiment 1 and distill the residue, keeping the temperature at the still head below 140°. When the temperature falls to below 120°, discontinue the distillation and remove the water present in the distillate with a pipet and suction bulb. Dry the distillate over anhydrous calcium chloride, removing the aqueous phase with a pipet as it forms. Add more calcium chloride, if necessary, and continue the drying process for about 15 minutes.

Decant the dried liquid into another flask and again distill. Collect the material boiling between 133 and 137° and determine the yield and literature boiling point, recording the results in your notebook. Also determine the theoretical and percentage yields of the final product, showing the calculations. Remember that 2 moles of propionaldehyde yield 1 mole of product.

At your instructor's option, obtain the infrared spectrum of your product, neat, and the ^{1}H NMR spectrum in carbon tetrachloride. Compare your spectra with those shown in Figures 23-3 and 23-4. When finished, hand in your product in a properly labeled container to your instructor.

QUESTIONS

1. Give the structures of two key side products formed in the synthesis of dibenzalacetone. How were these side products removed in the purification step?
2. Why should the intermediates in the formation of dibenzalacetone dehydrate so much more readily than the intermediate in the formation of 2-methyl-2-pentenal?

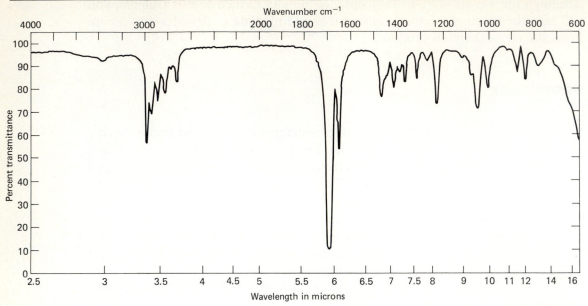

FIGURE 23-3
Infrared spectrum of 2-methyl-2-pentenal, neat.

FIGURE 23-4
¹H NMR spectrum of 2-methyl-2-pentenal in carbon
tetrachloride.

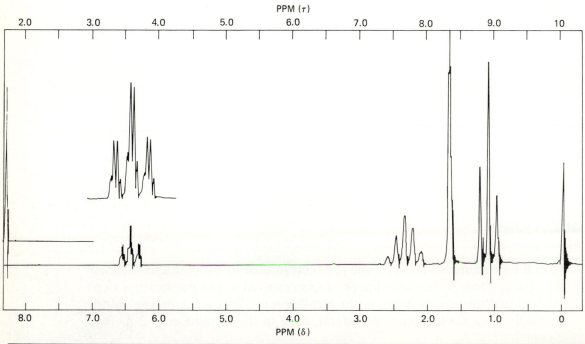

3. Why are the α-hydrogens in an aldehyde or ketone considerably more acidic than any of the other hydrogens?

4. The γ-hydrogen in the compound

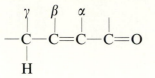

is also quite acidic. Explain why this should be true.

5. Related to the aldol condensation is the Cope reaction, an example of which is

Give a detailed mechanism for the formation of the product in this reaction.

6. Ketones with nonequivalent α-hydrogens usually are not as valuable in the aldol condensation. Using 2-butanone as an example, explain why this is true.

7. Write structural formulas of all the possible geometric isomers of dibenzalacetone. Which isomer would you expect to be the most stable? Why? The least stable? Why?

8. How would you change the procedure for Experiment 23A if you wanted to synthesize benzalacetone, $C_6H_5CH\!=\!CHCOCH_3$? Benzalacetophenone, $C_6H_5CH\!=\!CHCOC_6H_5$?

EXPERIMENT
24

THE CANNIZZARO REACTION: CUMIC ALCOHOL AND CUMIC ACID FROM CUMALDEHYDE

- OXIDATION OF AN ALDEHYDE
- REDUCTION OF AN ALDEHYDE
- NUCLEOPHILIC ADDITION
- INTERMOLECULAR HYDRIDE TRANSFER

Alike—but oh how different!

William Wordsworth, *Yes, It Was the Mountain Echo*

DISCUSSION

It isn't often in organic chemistry that two identical molecules will react with each other and have the principal functional group in each molecule altered differently as a result. Yet this is the case in the Cannizzaro reaction, where a compound such as benzaldehyde or formaldehyde reacts with itself in the

473

presence of a strong base to yield products where both aldehyde groups have become converted to two different functional groups.

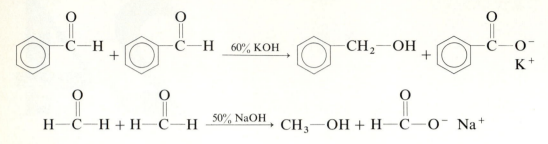

Observe that one molecule of aldehyde becomes reduced to the alcohol while the other one becomes oxidized to the salt of the carboxylic acid. The Cannizzaro reaction is, therefore, a self-oxidation and reduction reaction. As shown in the following examples, both functional groups can even be part of the same molecule to give an **internal** Cannizzaro reaction.

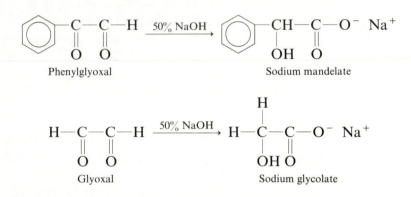

One common feature of all these examples is that the carbonyl group must have no adjacent carbon atoms or no hydrogens on the adjacent carbon atoms. If there is an adjacent carbon atom with a hydrogen on it, the Cannizzaro reaction does not occur, due to the compound preferentially undergoing the aldol condensation (see Experiment 23). Without an α-hydrogen, the compound cannot experience the aldol condensation and so undergoes self-oxidation and reduction as a second choice.

A likely mechanism for the Cannizzaro reaction is

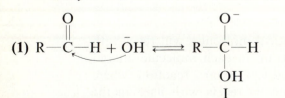

where R is a hydrogen, aromatic ring, or a carbon with no α-hydrogens

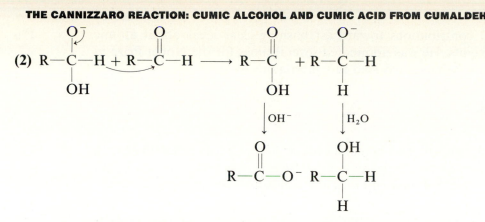

In step (1), a hydroxide ion attacks the electron-deficient carbonyl carbon atom of one molecule of aldehyde by nucleophilic addition to give intermediate I. This intermediate in step (2) then transfers a hydride ion as a nucleophile to the electron-deficient carbonyl carbon atom of a second molecule of aldehyde. The presence of the negative charge on I, plus the willingness of the second molecule to receive a nucleophile, aid this hydride transfer. A carboxylic acid and the anion of an alcohol are the resulting products. But because of the highly alkaline media, the carboxylic acid is then converted to its carboxylate salt. And the anion of the alcohol, because it is a stronger base than hydroxide ion, receives a proton from the solvent to become the alcohol.

In general, a mixture of two aldehydes undergoes the Cannizzaro reaction to yield high amounts of all four possible products. But if one of the aldehydes is formaldehyde in an excess amount, the reaction yields almost exclusively sodium formate and the alcohol formed from the reduction of the other aldehyde. This use of formaldehyde makes the **crossed Cannizzaro reaction** a useful synthetic method. For example:

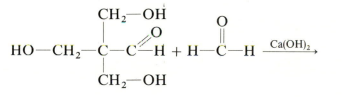

Pentaerythritol
(used in preparing
explosives)

The Cannizzaro reaction is just one of the achievements of Stanislao Cannizzaro, an Italian researcher and professor of chemistry. He is also recognized for his important contributions toward establishing clear ideas about atomic and molecular weights. He was nominated several times for the Nobel Prize in the early 1900s, but he never received this prestigious award.

BACKGROUND

In this experiment, you will react cumaldehyde or 4-isopropylbenzaldehyde with itself in the presence of alcoholic potassium hydroxide to produce cumic alcohol (4-isopropylbenzyl alcohol) and cumic acid (4-isopropylbenzoic acid).

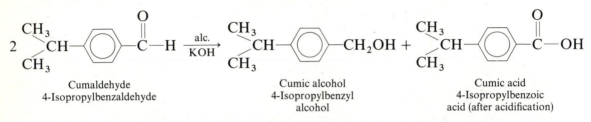

Cumaldehyde
4-Isopropylbenzaldehyde

Cumic alcohol
4-Isopropylbenzyl
alcohol

Cumic acid
4-Isopropylbenzoic
acid (after acidification)

The reaction proceeds by simply allowing the reactants to stand together at room temperature overnight or until the next laboratory period. The two products are easily separated because the potassium salt of cumic acid is water-soluble but not ether-soluble in contrast to cumic alcohol. Therefore, after the ethanol solvent is evaporated, the addition of water to the reaction mixture, followed by extraction with ether, is a key part of the reaction workup.

The ether layer in the extraction may also contain a small amount of unreacted cumaldehyde. But this impurity is easily removed because aldehydes and unhindered ketones react with sodium bisulfite to give a water-soluble bisulfite addition product. For cumaldehyde, the equation is

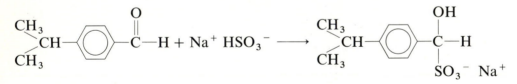

If desired, you can recover any cumaldehyde by adding an acid or a base, either of which destroys the bisulfite addition product by the reactions

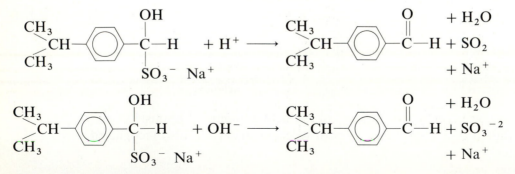

Cumaldehyde is the main component in oil of cumin, obtained from the herb cumin. The herb has a sharp taste and is used to flavor soups, eggs, meats, chili, and other foods, and is also used in certain perfumes and medicines.

Cumic alcohol is found in caraway seeds and has an odor like caraway and an aromatic taste. It is also used as a flavoring.

PROCEDURE

CAUTION: Potassium hydroxide is very caustic and generates heat when dissolved in solution. Use it with care.

In a 50-mL Erlenmeyer flask, dissolve 4.0 grams (0.061 mole) of 85% potassium hydroxide pellets in the minimum amount of 95% ethanol at room temperature. Add 5.0 mL (0.033 mole) of cumaldehyde and stir the resulting mixture. Then stopper the flask and allow the reaction mixture to stand overnight or until the next laboratory period. After this time, the odor of cumaldehyde should have distinctly changed.

Carefully evaporate the solution in a steam or hot-water bath to remove all the ethanol. Add 20 mL of distilled water and transfer the resulting solution to a separatory funnel. Then add 15 mL of ether to the Erlenmeyer flask to dissolve the remaining liquid, and transfer the ether solution to the separatory funnel. **(CAUTION: BE SURE THERE ARE NO FLAMES IN THE AREA.)**

Extract the aqueous alkaline solution with two separate 15-mL portions of ether to remove the cumic alcohol and any unreacted cumaldehyde. Save both the aqueous and ether layers and obtain each of the final products by the following procedures.

Cumic alcohol

Combine the ether extracts and extract them with 10 mL of 20% sodium bisulfite to remove any cumaldehyde. Next wash the ether with 10 mL of water and dry it over a small quantity of anhydrous magnesium sulfate. Carefully decant or filter the ether solution into a 50-mL Erlenmeyer flask, and evaporate the ether **in the hood**, using a steam or a hot-water bath.

When about 10 mL of ether remains, transfer the solution to a tared 25-mL Erlenmeyer flask. Continue the evaporation **in the hood** until a residual oil of cumic alcohol is obtained. **Carefully** compare its odor with the starting material and record the result.

Cool the oil in an ice bath or in a freezer until a solid is formed. Allow the solid to warm and determine its melting point in the flask, the yield, and the literature melting point. Record the results in your notebook. Also determine the theoretical and percentage yields, showing the calculations. Remember that the yield is based upon only half the cumaldehyde giving cumic alcohol as a product.

At your instructor's option, obtain the infrared spectrum of your product in a KBr pellet and the ^{1}H NMR spectrum in deuterated chloroform. When finished, submit your product in a properly labeled container to your instructor.

Cumic acid

Mildly heat the alkaline aqueous layer from the ether extractions with a steam or warm-water bath **in the hood**. This will remove the small quantity of ether present. Cool the solution in an ice-water bath and **carefully** add with stirring 3 M HCl until the solution is strongly acidic to pH paper. Test the pH by transferring a drop of the solution to the paper with a stirring rod. Collect the precipitated cumic acid by suction filtration and wash twice with small quantities of cold water.

Recrystallize the crystals from 95% ethanol, using charcoal if the solution is colored, and collect the purified product by suction filtration. Allow the product to thoroughly dry and determine the yield, melting point, and literature melting point. Record the results in your notebook. Also determine the theoretical and percentage yields, showing the calculations. Remember that the yield is based upon only half of the cumaldehyde giving cumic acid as a product.

At your instructor's option, obtain the infrared spectrum of your product in a KBr pellet and the ^{1}H NMR spectrum in deuterated chloroform. When finished, submit your product in a properly labeled container to your instructor.

QUESTIONS

1. Explain why the Cannizzaro reaction is limited to aldehydes having no α-hydrogens.
2. Is the potassium hydroxide a catalyst or a reactant in the Cannizzaro reaction? Explain.
3. When the Cannizzaro reaction is carried out in D_2O solution, no deuterium is found attached to the carbon atom of the alcohol. Explain how this supports the mechanism given for the reaction. Would you expect deuterium to be attached to the oxygen atom of the alcohol? Why?
4. If a mixture of 4-methylbenzaldehyde and 4-nitrobenzaldehyde undergoes the Cannizzaro reaction, what would be the products expected?
5. Write equations for the synthesis of benzaldehyde from (a) benzoic acid; (b) benzyl alcohol.
6. Write the structure of the product or products formed in the reaction

$$\text{C}_6\text{H}_5-\overset{\displaystyle O}{\overset{\displaystyle \|}{\text{C}}}-\overset{\displaystyle O}{\overset{\displaystyle \|}{\text{C}}}-\text{H} + {}^-\text{O}-\text{CH}_3 \xrightarrow{\text{CH}_3\text{OH}}$$

25

THE WITTIG REACTION: THE SYNTHESIS OF E,E-1,4-DIPHENYL-1,3-BUTADIENE

- **CARBANIONS**
- **ZWITTERIONS**
- **NUCLEOPHILIC BIMOLECULAR SUBSTITUTION**
- **NUCLEOPHILIC ADDITION**
- **USE OF SODIUM METHOXIDE AND PHOSPHORUS COMPOUNDS**

The whole of my remaining realizable estate shall be dealt with in the following way: The capital shall be invested by my executors in safe securities and shall constitute a fund, the interest on which shall be annually distributed in the form of prizes to those who, during the preceding year, shall have conferred the greatest benefit on mankind.... It is my express wish that in awarding the prizes, no consideration whatever shall be given to the nationality of the candidates, so that the most worthy shall receive the prize ....

From the will of Alfred Bernhard Nobel, November 27, 1895

TABLE 25-1.

Some Notable Nobel Prize Winners in Chemistry and Physics

Wilhelm Roentgen	Discovery of x-rays	1901
Emil Fischer	Work on sugars and purines	1902
Svante Arrhenius	Theory of electrolytic dissociation	1903
Antoine-Henri Becquerel	Discovery of spontaneous radiation	1903
Pierre Curie Marie Curie	Investigations in radiation	1903
Sir William Ramsay	Discovery of the inert gases and their place in the periodic table	1904
Adolf von Baeyer	Work on organic dyes and hydroaromatic compounds	1906
Lord Ernest Rutherford	Investigations in the chemistry and disintegration of radioactive substances	1908
Guglielmo Marconi	Development of wireless telegraphy	1909
J. van der Waals	The equation of state of gases and liquids	1910
Marie Curie	Discovery of radium and polonium; isolation of radium	1911
Victor Grignard	Discovery and development of the Grignard reagent	1912
Sir William Bragg Sir Lawrence Bragg	Analysis of crystal structures by x-rays	1915
Max Planck	Discovery of the elemental quanta	1918
Walther Nernst	Work in thermochemistry	1920
Albert Einstein	Many accomplishments in theoretical physics	1921
Niels Bohr	Investigations in atomic structure and radiation	1922
Prince Louis de Broglie	Discovery of the wave nature of electrons	1929
Sir James Chadwick	Discovery of the neutron	1935
Peter Debye	Studies of dipole moments and the diffraction of x-rays and electrons in gases	1936
Enrico Fermi	Disclosure of artificial radioactive elements produced by neutron irradiation	1938
Wolfgang Pauli	Discovery of the Pauli exclusion principle	1945
Otto Diels Kurt Alder	Discovery and development of diene syntheses	1950
Linus Pauling	Study of the nature of the chemical bond	1954
Frederick Sanger	Determining the entire amino acid sequence in insulin	1958
Giulio Natta and Karl Ziegler	Structure and syntheses of polymers in the field of plastics	1963
Robert Woodward	Outstanding contributions to synthetic organic chemistry	1965
Derek Barton	Demonstrating that functional group reactivity depends on the axial and equatorial positions on a ring	1969
Herbert Brown	Development and synthetic uses of organoboranes	1979
Georg Wittig	Development of Wittig reaction	1979
Kenichi Fukui and Roald Hoffman	Developing frontier orbital methods to understand organic reactions	1981

INTRODUCTION

Of all the prizes and awards given for excellence and achievement in science, none has received more prestige or acclaim than the Nobel Prize. These prizes are given annually to those people who have demonstrated the most outstanding work in chemistry, physics, physiology and medicine, economics, literature, and world peace. The prize consists of a gold medal, a diploma, and a monetary gift that has ranged from $40,000 to $385,000. In some years, the award in one field has been shared by several people.

Of course, Nobel could not have foreseen that his prizes would become the ultimate symbol of excellence. He was only interested in rewarding those who had done the most for mankind. His fortune of $8.5 million (over $100 million in today's dollars) was derived from his invention and manufacture of dynamite, blasting gelatin, smokeless powder, and detonators. Having lost a younger brother to a nitroglycerine explosion, Nobel was determined to make the explosive safe for others. He succeeded when he discovered that the substance could be stabilized substantially by combining it with the inert powder kieselguhr. The resulting product, dynamite, is still used today. Completely aware of the potential military uses of his inventions, Nobel became interested in the preservation of peace, which ultimately led to the prizes being established.

A list of some of the prize winners in chemistry and physics is given in Table 25-1. You may already have read about many of these people and their accomplishments. This partial list, of course, is not intended to reduce the value and importance of the tremendous achievements of the other winners.

DISCUSSION

One of the most common Nobel Prize winning accomplishments, seen in practically all undergraduate organic chemistry books, is the Wittig reaction. Discovered in 1954 by Georg Wittig, the tremendous synthetic usefulness of the reaction won for Dr. Wittig the Prize in 1979. The reaction allows the carbonyl group of aldehydes and ketones to be converted into the carbon-carbon double bond of an alkene without the complications of rearrangements, mixtures of non-cis-trans isomers, or uncertainty over the location of the double bond. These advantages plus the relatively mild reaction conditions make it a very valuable, versatile procedure for synthesizing a wide variety of alkenes and alkene derivatives. In fact, if an organic chemist has to synthesize a new alkene, the Wittig reaction will frequently be the preferred method. As a result, if you were to peruse a recent copy of an organic chemistry journal, the chances are excellent of finding another example of the Wittig reaction. Literally thousands of new alkenes and their derivatives are being made by it each year.

The Wittig reaction begins with triphenylphosphine, $(C_6H_5)_3P$, and a methyl, primary, or secondary alkyl halide. As shown below, the phosphorus atom of triphenylphosphine acts as a nucleophile and displaces a halide anion from the alkyl halide in step (1) to form an alkyltriphenylphosphonium salt.

In step (2), this salt is then treated with a strong base, such as phenyllithium, butyllithium, or sodium ethoxide, to form a phosphorus **ylide** (pronounced "ill'-id"). This ylide exists as a resonance hybrid of the two structures shown, with the charged one making the major contribution. This occurs because the phosphorus-carbon π bond in the other structure is quite weak.

Then in step (3), the strongly nucleophilic carbon atom of the ylide attacks the carbonyl carbon of an aldehyde or ketone to form an intermediate called a **betaine** (pronounced "bay'-tuh-ene"). The opposite charges on the betaine produce a chemical bond between phosphorus and oxygen to form an **oxaphosphetane**, which then decomposes to yield the desired alkene and triphenylphosphine oxide. The exceptional stability of triphenylphosphine oxide and the instability of a four-membered ring provide the driving force for the last reaction.

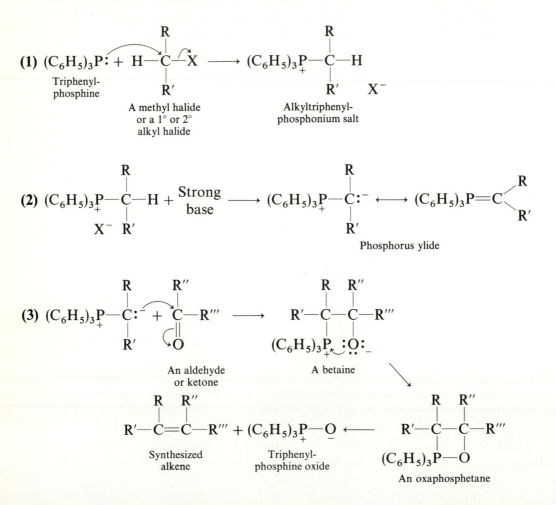

Some specific examples of the Wittig reaction are

$(C_6H_5)_3P: + \underset{\underset{Br}{|}}{CH_2}CH_2CH_2CH_3 \longrightarrow (C_6H_5)_3\overset{+}{P}-CH_2CH_2CH_2CH_3$ $\quad Br^-$

$\searrow$ R-Li

⬡$=CHCH_2CH_2CH_3 \xleftarrow{\text{⬡}=O} (C_6H_5)_3\overset{+}{\underset{-}{P}}-CHCH_2CH_2CH_3$

$(C_6H_5)_3P: + \underset{\underset{Br}{|}}{CH_2}CH_2-⬡ \longrightarrow (C_6H_5)_3\overset{+}{P}-CH_2CH_2-⬡$ $\quad Br^-$

$\searrow$ C_6H_5Li

$\underset{CH_3CH_2}{\overset{CH_3}{>}}C=CHCH_2-⬡ \xleftarrow{CH_3-\overset{O}{\overset{||}{C}}-CH_2CH_3} (C_6H_5)_3-\overset{+}{\underset{-}{P}}-CHCH_2-⬡$

One problem with the Wittig reaction is that mixtures of cis-trans isomers frequently cannot be avoided. Another is the low yield in the first reaction step if secondary alkyl halides are used. Thus if one has a choice, primary alkyl halides are usually preferred in the first step. For example, in the synthesis of 3-methyl-3-heptene, the first combination of reagents is better than the second.

$(C_6H_5)_3P: + \underset{\underset{Br}{|}}{CH_2}CH_2CH_2CH_3 \longrightarrow (C_6H_5)_3\overset{+}{P}-CH_2CH_2CH_2CH_3$ $\quad Br^-$

$\searrow$ R-Li

$\underset{CH_3}{\overset{CH_3CH_2}{>}}C=CHCH_2CH_2CH_3 \xleftarrow{CH_3-\overset{O}{\overset{||}{C}}-CH_2CH_3} (C_6H_5)_3\overset{+}{\underset{-}{P}}-CHCH_2CH_2CH_3$

3-Methyl-3-heptene

$CH_3CH_2\underset{\underset{CH_3}{|}}{CH}-Br + :P(C_6H_5)_3 \longrightarrow CH_3CH_2\underset{\underset{CH_3}{|}}{CH}-\overset{+}{P}(C_6H_5)_3$ $\quad Br^-$

$\searrow$ R-Li

$\underset{CH_3}{\overset{CH_3CH_2}{>}}C=CHCH_2CH_2CH_3 \xleftarrow{H-\overset{O}{\overset{||}{C}}-CH_2CH_2CH_3} CH_3CH_2\underset{\underset{CH_3}{|}}{\overset{-}{C}}-\overset{+}{P}(C_6H_5)_3$

3-Methyl-3-heptene

To plan a Wittig synthesis, you must first identify the carbonyl compound and the ylide needed to prepare the desired alkene. Mentally disconnect the double bond so that one of its carbons originally came from the aldehyde or ketone and the other one came from the ylide. Write the structures of both of these starting materials. This will give you one possible combination. Then repeat this process by switching the doubly bonded carbon that originally came from the aldehyde or ketone with that which originally came from the ylide. This will give you the second possible combination. Next work backward to obtain the structures of the two possible alkyl halides, and decide which combination, if any, will more likely give better results. For example:

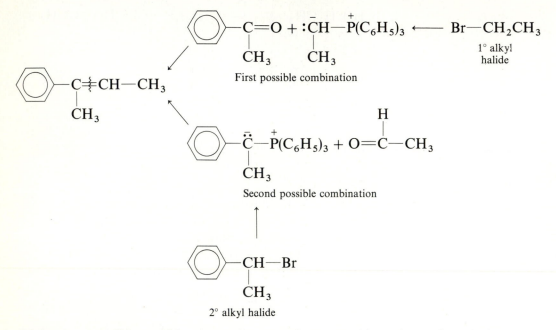

Of the two possible combinations, the second one would not be preferred, since it involves the use of a secondary alkyl halide.

Many other functional groups can also be present in the aldehyde, ketone, or alkyl halide, giving the Wittig reaction even more versatility. The alkene produced can then be converted to some other desired product. For example:

Various modifications of the original Wittig synthesis have been made. One of these, the Horner procedure, uses triethyl phosphite, $(CH_3CH_2O)_3P$, instead of triphenylphosphine. This procedure works whenever activated alkyl halides such as benzyl halides, allyl halides, and α-halo esters are used. For allyl chloride, the reactions are

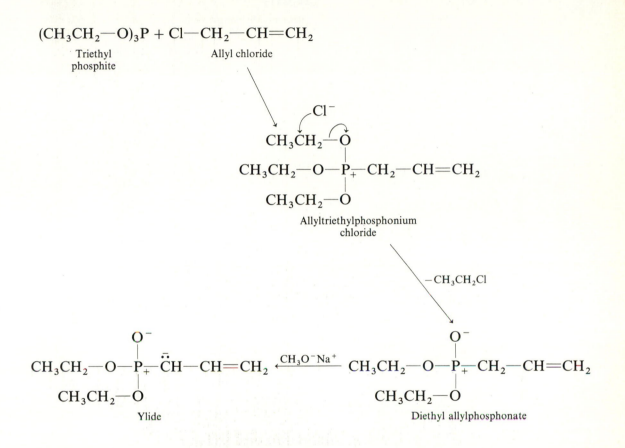

The reaction of the ylide with the aldehyde or ketone then gives the alkene as in the regular Wittig procedure.

BACKGROUND

In this experiment, you will synthesize *E,E*-1,4-diphenyl-1,3-butadiene by means of the Horner modification of the Wittig reaction. The reaction sequence is

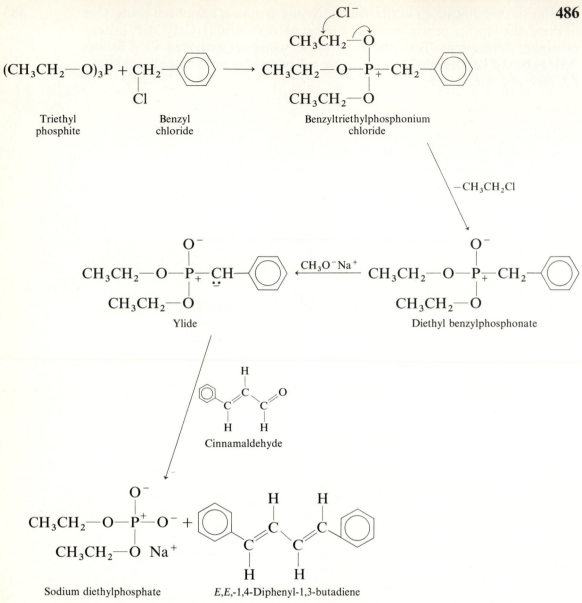

Triethyl phosphite

Benzyl chloride

Benzyltriethylphosphonium chloride

$-CH_3CH_2Cl$

Ylide

$CH_3O^- Na^+$

Diethyl benzylphosphonate

Cinnamaldehyde

Sodium diethylphosphate

E,E,-1,4-Diphenyl-1,3-butadiene

Observe that although the final reaction could produce either a cis or trans arrangement for the second carbon-carbon double bond, hardly any of the cis product is formed. This occurs because the trans, trans product is more stable due to less steric hindrance.

In addition, the product formed is colorless despite the extended conjugated system. Two more conjugated double bonds are needed in order for the molecule to absorb visible light.

CAUTION: Organophosphorus compounds are toxic, and benzyl chloride is both a skin irritant and a lachrymator (it produces tears). Do not breathe them or get them on your skin. You must use a hood and plastic gloves!

In addition, for best results the sodium methoxide should be freshly opened and the jar kept tightly closed. Do not leave it out for long periods, since it reacts with the moisture in the air to produce methanol and sodium hydroxide. It is also an even stronger base than sodium hydroxide. Treat it accordingly, and keep it off your skin.

Place 3.7 mL (0.032 mole) of benzyl chloride, 5.7 mL (0.032 mole) of triethyl phosphite and a boiling stone into a 25-mL or a 50-mL round-bottom flask. Attach a water-cooled reflux condenser and gently reflux the mixture for 1 hour with a small flame. Ethyl chloride will begin to be eliminated from the reaction intermediate at about 130°C, and the temperature will ultimately climb to over 190°C.

Cool the phosphonate ester to room temperature. In the meantime, place 1.8 grams (0.032 mole) of sodium methoxide in a dry 125-mL Erlenmeyer flask and immediately add 15 mL of N,N-dimethylformamide. Pour the cooled phosphonate ester mixture into the Erlenmeyer flask and rinse the round-bottom flask with another 15 mL of N,N-dimethylformamide, adding this rinse solution to the Erlenmeyer flask.

Thoroughly cool the Erlenmeyer flask in an ice bath and slowly add 3.8 mL (0.032 mole) of cinnamaldehyde with swirling and continued cooling. The mixture quickly becomes deep red, and crystals of product begin to form. When there is no further change, remove the flask from the ice bath and let it stand at room temperature for about 10 minutes.

Add 15 mL of water and 7 mL of methanol to the mixture, swirl it, and collect the product by suction filtration, rinsing the flask with a small amount of water. Wash the crystals with water until the red color on them is removed; then wash with methanol until there is no yellow color in the methanol filtrate.

Recrystallize the crude, faint yellow product from methylcyclohexane or absolute ethanol. Allow the purified, colorless crystals to thoroughly dry, and determine the yield and melting point. Record the results in your notebook. The literature melting point is 152–153°C. Also determine the theoretical and the percentage yields of your product, showing the calculations.

At your instructor's option, obtain the infrared spectrum of your product in a KBr pellet and the ^{1}H NMR spectrum in deuterated chloroform. Compare your spectra with those shown in Figures 25-1 and 25-2. When finished, hand in your product in a properly labeled container to your instructor.

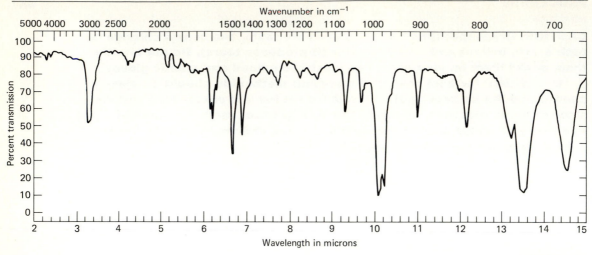

Wavenumber in cm^{-1}

Wavelength in microns

FIGURE 25-1
Infrared spectrum of *E,E*-1,4-diphenyl-1,3-butadiene in a KBr pellet.

FIGURE 25-2
^{1}H NMR spectrum of *E,E*-1,4-diphenyl-1,3-butadiene in deuterated chloroform. Spectrum courtesy of Aldrich Chemical Company, *The Aldrich Library of NMR Spectra*, edition II, by Charles Pouchert.

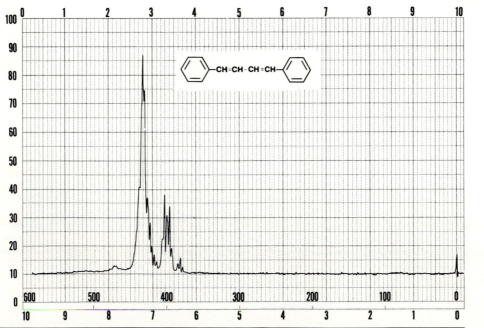

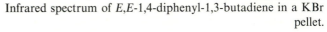

1. What is the limiting reagent, if any, in the experiment? Why?
2. Explain specifically why secondary alkyl halides give lower yields than primary alkyl halides in the Wittig reaction.
3. Write the structures for all the possible stereoisomers of 1,4-diphenyl-1,3-butadiene.
4. Methylenecyclohexane can be prepared in high yield by the Wittig reaction. What would be wrong with reacting cyclohexanone with methylmagnesium bromide, followed by the acid-catalyzed dehydration of the resulting alcohol?

Methylenecyclohexane

5. Outline a synthesis of each of the following alkenes by the Wittig reaction. In each case, show each of the possible combinations of reactants. Decide for each synthesis which possible combination of reactants, if any, would be the better choice and why.

(a) $\left\langle\right\rangle$—CH$_2$—CH=CH—CH$_2$—$\left\langle\right\rangle$

(b) $\left\langle\right\rangle$—$\overset{\displaystyle CH_3}{\underset{\displaystyle |}{C}}$=CHCH$_2$—$\left\langle\right\rangle$

BENZYNE AS AN INTERMEDIATE: THE SYNTHESIS OF TRIPTYCENE, A CAGE-RING HYDROCARBON

DIELS-ALDER ADDITION

DISCUSSION

For a long time, the reaction of sodium amide with certain aryl halides in liquid ammonia presented organic chemists with an interesting puzzle. With some aryl halides, the amide ion simply replaced the halogen on the same carbon atom. For example:

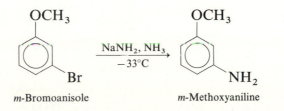

m-Bromoanisole   *m*-Methoxyaniline

But other similar aryl halides yielded products having the amino group attached

only to the adjacent carbon atom of the departing halogen. For example:

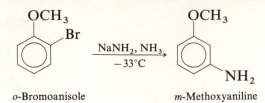

o-Bromoanisole   m-Methoxyaniline

And still other similar aryl halides gave an equal mixture of products formed by substitution at both the position of the departing halogen and any adjacent positions. For example:

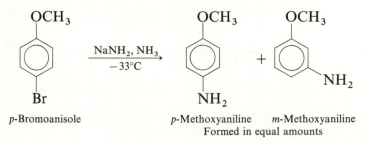

p-Bromoanisole   p-Methoxyaniline   m-Methoxyaniline
Formed in equal amounts

It took an imaginative experiment by J. D. Roberts in 1953 to solve the puzzle. Roberts synthesized a sample of chlorobenzene having the chlorine attached **not** to an ordinary carbon atom, but to a carbon-14 isotope. When the reaction with sodium amide in liquid ammonia was completed, the aniline obtained had almost exactly half of the carbon-14 at the amino group position and half at the adjacent position. Labeled bromobenzene gave similar results.

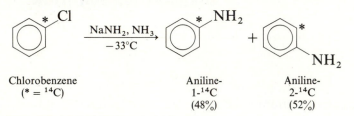

Chlorobenzene   Aniline-   Aniline-
(* = ^{14}C)   1-^{14}C 2-^{14}C
                            (48%)   (52%)

The following elimination-addition mechanism, involving the formation of an intermediate called **benzyne**, seemed to best explain Roberts' results.

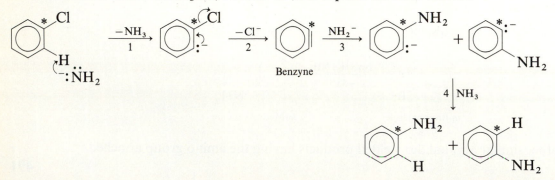

Benzyne

In step 1, an amide ion initiates the elimination by acting as a very strong base and abstracting the weakly acidic hydrogen on the carbon atom ortho to the chlorine. The negative charge that develops on this carbon is partially stabilized by the inductive effect of the chlorine. Then, in step 2, the elimination is completed by the loss of a chloride ion to give benzyne.

As an intermediate, benzyne is highly reactive because the bond formed is not very strong. As shown below, both electrons of this bond lie in sp^2 hybrid orbitals in the same plane as the aromatic ring, and the overlap between these orbitals is not very large. As a result, an amide ion can easily break this bond in step 3 to yield a carbanion addition product. The abstraction of a proton by this carbanion in step 4 then yields aniline.

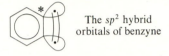

The sp^2 hybrid orbitals of benzyne

Observe from the mechanism that the amide ion in step 3 may attach itself equally well to either of the two available carbons of benzyne. Thus, the mechanism explains why the two products from carbon-14 chlorobenzene or bromobenzene are obtained in almost equal amounts.

After the benzyne mechanism became established, the reason for certain products being formed or not formed in various aryl halide reactions could now be understood. For example, in the reaction of *o*-bromoanisole with amide ion in liquid ammonia to yield *m*-methoxyaniline, the mechanism is

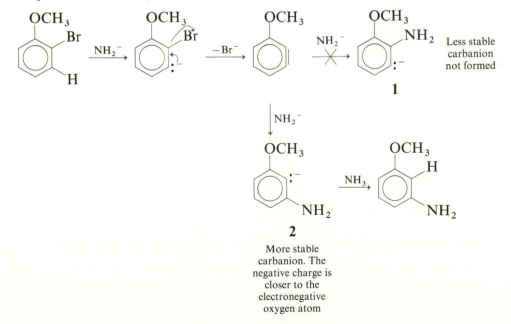

Observe that the reaction occurs in the way that produces the more stable carbanion **2** rather than the less stable carbanion **1**.

Carbanion **2** is more stable than carbanion **1** due to the negative charge being closer to the highly electronegative oxygen atom, which stabilizes the charge by its electron-withdrawing inductive effect. The resonance effect does not operate here because the sp^2 hybrid orbital containing the electron pair does not overlap with the p orbitals of the aromatic system.

For the reaction of p-bromoanisole with amide ion in liquid ammonia, the mechanism is

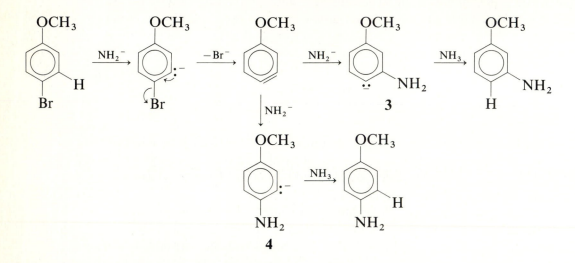

Here, the two products are formed in equal amounts because there is no difference in the stability of carbanion **3** versus carbanion **4**. Evidently the electronegative oxygen's inductive effect is of no value because of the greater distance. In addition, the resonance effect again cannot operate here.

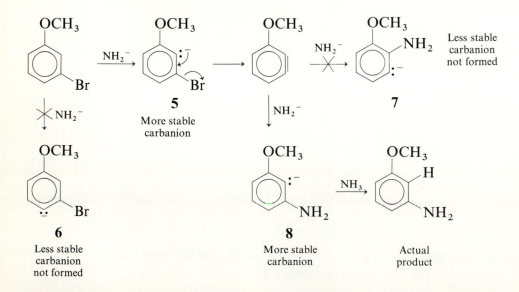

The formation of *m*-methoxyaniline from *m*-bromoanisole seemed perfectly normal prior to the benzyne mechanism. But with this mechanism, we discover that this reaction is actually more complex than was originally thought. As shown on page 494, the starting material has a choice of forming either carbanion **5** or carbanion **6**. Of the two, carbanion **5** is more stable and so is the one actually made. Then after benzyne is produced, there is again a choice of two carbanions that could be made. Carbanion **8** is more stable than carbanion **7**, and so is the one actually formed.

It has also been discovered that the benzyne mechanism can operate even when the reactant is not sodium amide in liquid ammonia. For example:

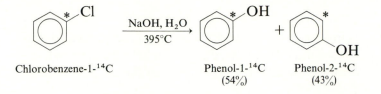

Chlorobenzene-1-^{14}C

Phenol-1-^{14}C
(54%)

Phenol-2-^{14}C
(43%)

The somewhat larger than expected amount of phenol-1-^{14}C can be explained if a nucleophilic aromatic substitution mechanism also partially occurs.

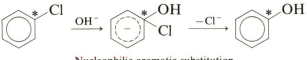

Nucleophilic aromatic substitution

Benzyne can even be produced under conditions that are not strongly alkaline. Although the intermediate is too reactive to isolate, further evidence of its existence comes from "trapping it." The following example illustrates both.

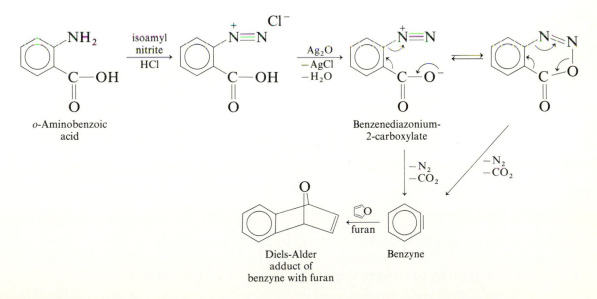

o-Aminobenzoic
acid

Benzenediazonium-
2-carboxylate

Diels-Alder
adduct of
benzyne with furan

furan

Benzyne

If no furan is present, the benzyne formed can also be "trapped" by dimerization. For example:

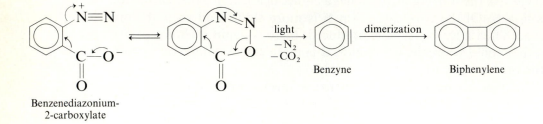

Benzenediazonium-
2-carboxylate

When generated by this flash photolysis procedure, the benzyne produced was found to last less than 10^{-4} seconds. It is indeed a very reactive intermediate.

Benzyne can also be produced in other ways. For example:

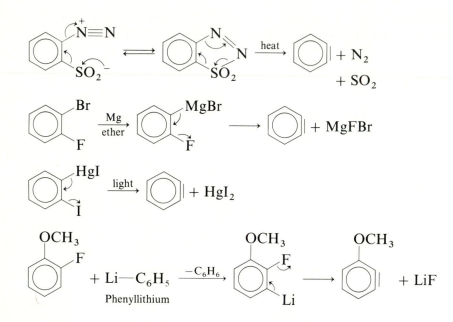

All these methods together provide even more evidence for its existence.

BACKGROUND

In this experiment, you will synthesize benzyne by the decomposition of benzenediazonium-2-carboxylate. The benzyne will then be "trapped" by its Diels-Alder addition to anthracene to form triptycene, a cage-ring hydrocarbon.

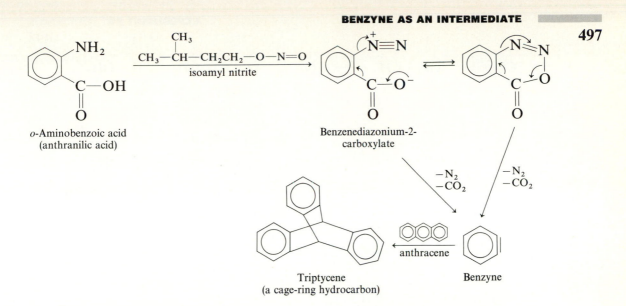

o-Aminobenzoic acid
(anthranilic acid)

isoamyl nitrite

Benzenediazonium-2-
carboxylate

$-N_2$
$-CO_2$

$-N_2$
$-CO_2$

anthracene

Triptycene
(a cage-ring hydrocarbon)

Benzyne

Observe that this Diels-Alder addition (see Experiment 18C) occurs at positions 9 and 10 of anthracene. The addition could also conceivably occur at positions 1 and 4 to form compound **9** as an alternative product.

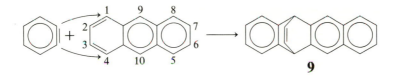

9

However, since anthracene has a resonance stabilization of 84 kcal/mole, and the naphthalene ring of compound **9** has a resonance stabilization of only 61 kcal/mole, this means that 23 kcal/mole of resonance stabilization would have to be sacrificed in the formation of compound **9**. Triptycene, on the other hand, has the two nonfused aromatic rings from anthracene having 2×36 or 72 kcal/mole of resonance stabilization. As a result, only 12 kcal/mole of resonance stabilization is sacrificed, making triptycene the preferred product. In fact, anthracene readily undergoes a number of addition reactions at positions 9 and 10, showing that 12 kcal/mole of resonance energy is a relatively small sacrifice.

A possible reaction mechanism follows. In step 1, isoamyl nitrite is protonated. It is then attacked by the nucleophilic amino group of anthranilic acid in step 2 to form intermediate **10**. A proton transfer occurs in step 3, and then, in step 4, isoamyl alcohol is expelled to yield intermediate **11**. After a proton transfer in step 5, water is expelled in step 6 to yield the diazonium ion **12**. The loss of a proton in step 7 yields benzenediazonium-2-carboxylate, which can exist in equilibrium with another possible cyclic structure. The loss of carbon dioxide and nitrogen gas in step 8 then yields benzyne, which is "trapped" by anthracene in step 9 to give triptycene as the final product.

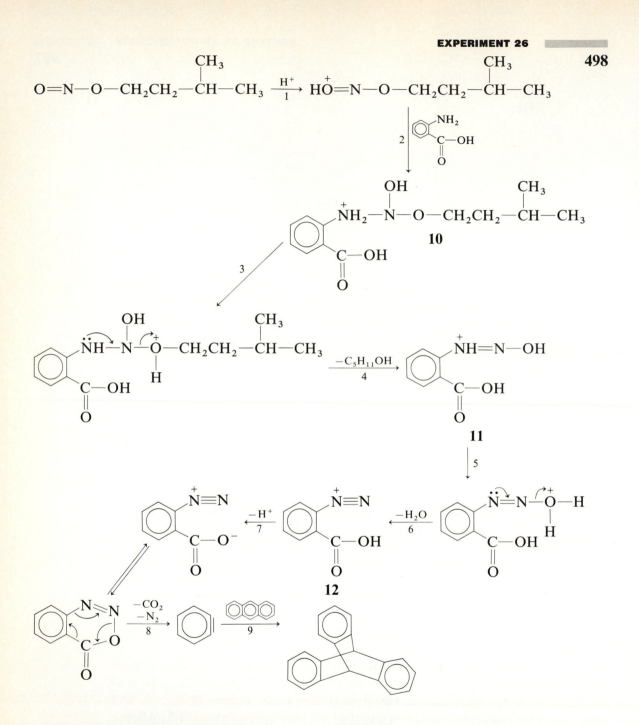

The triptycene produced will initially not be pure due to the presence of unreacted anthracene. Both compounds have similar solubilities in various solvents, so recrystallization and extraction are both unsatisfactory purification techniques. Anthracene can, however, be separated by treating the crude reaction mixture with maleic anhydride, followed by aqueous sodium hydroxide.

As shown below, anthracene forms a Diels-Alder adduct with maleic anhydride. Alkaline hydrolysis of this adduct then yields a water-soluble salt. Triptycene, however, fails to react with either maleic anhydride or aqueous sodium hydroxide, and so remains water-insoluble.

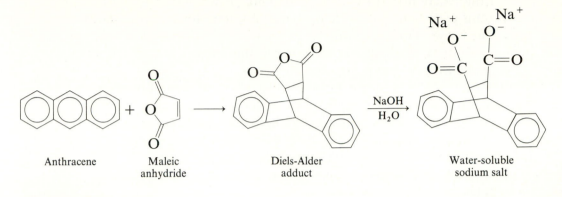

| Anthracene | Maleic anhydride | Diels-Alder adduct | Water-soluble sodium salt |

PROCEDURE

CAUTION: Isoamyl nitrite is a powerful heart stimulant and must not be inhaled. The use of a hood is highly recommended for the first portion of the experiment. The compound also causes flushing of the skin, a rapid pulse, headaches, and a drop in blood pressure. When not in use, store the reagent in the refrigerator in a brown, tightly closed bottle. During the experiment, keep the reagent bottle in the hood, tightly closing it after the liquid has been removed. These precautions also help prevent the decomposition of the liquid by light and air. For best results, the compound should be recently purchased.

Also remember that sodium hydroxide is very caustic. Handle it with care. If you get it on your skin, rinse immediately with water.

The benzenediazonium-2-carboxylate prepared in this experiment presents no danger when reacted immediately with anthracene, following the stated procedure. However, if isolated and dried, it becomes a potential explosive. Be sure you don't alter the procedure in this manner.

Place 2.0 grams (0.011 mole) of reagent-grade anthracene, 2.0 mL (0.015 mole) of isoamyl nitrite, and 20 mL of 1,2-dimethoxyethane solvent (bp 84°C) into a 100-mL round-bottom flask. Attach a water-cooled reflux condenser and heat the mixture with a hot plate or steam bath until gentle refluxing occurs. Dissolve 5.2 grams (0.038 mole) of reagent-grade anthranilic acid in 20 mL of 1,2-dimethoxyethane. Place 10 mL of this solution in a dropping funnel (separatory funnel), place the funnel on top of the reflux condenser, and add this solution dropwise through the condenser into the refluxing mixture over a period of about 20 minutes. The mixture will become ruby red as the reaction proceeds.

Add an additional 2.0 mL (0.015 mole) of isoamyl nitrite by using the dropping funnel; then add the remaining 10 mL of anthranilic acid solution dropwise over a period of about 20 minutes with refluxing. After the addition, continue the refluxing for another 10 minutes.

Slightly cool the reaction mixture, then add 10 mL of 95% ethanol and a solution of 3.0 grams of sodium hydroxide in 40 mL of water. Stir and cool the resulting solid and brown liquid thoroughly in an ice bath, then collect the solid by suction filtration, washing it with a small amount of ice-cold 80% methanol in water solution to remove the brown liquid.

Transfer the moist, nearly colorless solid to a clean 100-mL round-bottom flask, and remove any liquid by heating the flask under reduced pressure on a steam bath for about 15 minutes. Disconnect the water aspirator and the vacuum adapter, then add to the flask 1.0 gram (0.010 mole) of maleic anhydride and 20 mL of triethylene glycol dimethyl ether solvent (triglyme, bp 222°C). Add a boiling chip, attach a reflux condenser, and reflux the mixture with a flame for 10 minutes.

Allow the flask to cool to about 100°C, and add with stirring 10 mL of 95% ethanol and a solution of 3.0 grams of sodium hydroxide in 40 mL of water. Cool the mixture in an ice bath and collect the nearly colorless crystals of triptycene by suction filtration, washing them with a small amount of ice-cold 80% methanol in water.

Dry the crystals as much as possible between sheets of filter paper and then recrystallize from methylcyclohexane. Collect the colorless crystals by suction filtration, allow to thoroughly dry, and determine the yield and melting point. Record the results in your notebook. The literature melting point is 252–254°C. Also determine the theoretical and percentage yields of your product, and show the calculations.

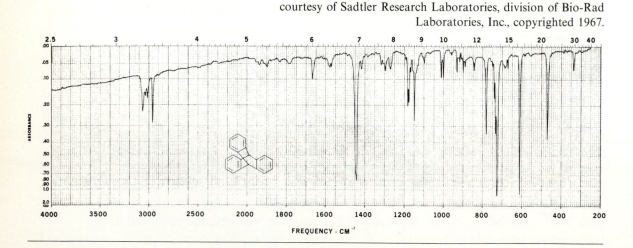

FIGURE 26-1

Infrared spectrum of triptycene in a KBr pellet. Spectrum courtesy of Sadtler Research Laboratories, division of Bio-Rad Laboratories, Inc., copyrighted 1967.

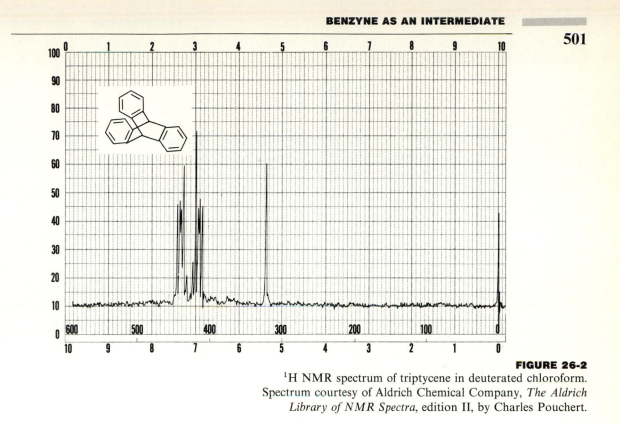

FIGURE 26-2
^{1}H NMR spectrum of triptycene in deuterated chloroform.
Spectrum courtesy of Aldrich Chemical Company, *The Aldrich
Library of NMR Spectra*, edition II, by Charles Pouchert.

At your instructor's option, obtain the infrared spectrum of your product
in a KBr pellet and the ^{1}H NMR spectrum in deuterated chloroform. Compare
your spectra with those in Figures 26-1 and 26-2. When finished, hand in your
product in a properly labeled container to your instructor.

QUESTIONS

1. If no anthracene is used in this experiment, then biphenylene is the reaction
   product. Indicate how it may be isolated.

Biphenylene

2. What was the purpose of adding sodium hydroxide solution after the first
   refluxing? After the second refluxing?
3. Explain why 2-bromo-3-methylanisole does not react when placed with
   sodium amide in liquid ammonia.

4. In the following reaction, diphenyl ether is a side product. Indicate a mechanism for its formation.

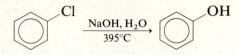

5. The following reaction occurs by the benzyne mechanism. Explain the results, specifically stating why the percentages of the products obtained are different from those obtained when *o*-bromoanisole is used.

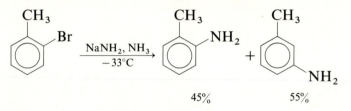

NONBENZENOID AROMATIC COMPOUNDS: THE SYNTHESIS OF FERROCENE

- CARBANIONS
- REVERSE DIELS-ALDER

DISCUSSION

When one thinks of aromaticity, benzene and benzene derivatives are the examples that most readily come to mind. But a benzene ring does not have to be present in order for a molecule to be aromatic. Pyridine, pyrrole, furan, and thiophene are just some of the heterocyclic aromatic compounds that do not have a benzene ring.

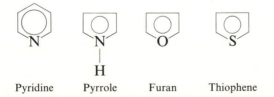

Pyridine   Pyrrole   Furan   Thiophene

But as the following examples demonstrate, a nonbenzenoid aromatic compound can also have only carbon atoms within the ring system. Note that many of these compounds are ions.

503

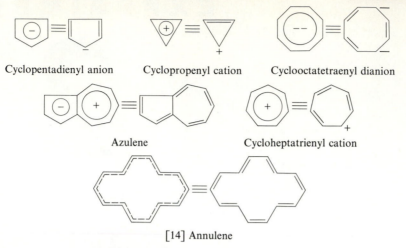

Cyclopentadienyl anion   Cyclopropenyl cation   Cyclooctatetraenyl dianion

Azulene   Cycloheptatrienyl cation

[14] Annulene

Because of all the known nonbenzenoid aromatic compounds and other similar nonaromatic compounds, our understanding of what constitutes aromaticity has been expanded and refined. From this, four structural requirements for aromaticity have been derived.

1. The molecule must be cyclic.

2. The molecule must be completely unsaturated within the ring system. This allows the complete delocalization of the π electrons within the ring system.

3. The ring system must be flat or able to easily become flat. This allows the p orbitals containing the π electrons to be properly aligned for the complete delocalization of the π electrons.

4. The ring system must contain Hückel's number of $(4n + 2)$ π electrons (2, 6, 10, 14, 18, etc.). This rule is based upon the molecule having the π electrons only in the bonding orbitals, with none in the antibonding orbitals.

Because of these rules, the following molecules are **not** aromatic. In addition, many other examples exist.

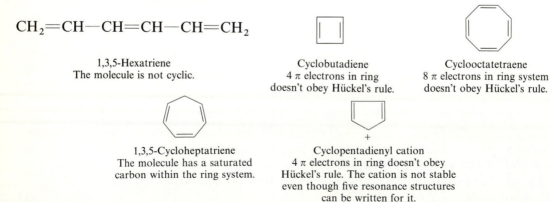

$CH_2{=}CH{-}CH{=}CH{-}CH{=}CH_2$

1,3,5-Hexatriene
The molecule is not cyclic.

Cyclobutadiene
4 π electrons in ring doesn't obey Hückel's rule.

Cyclooctatetraene
8 π electrons in ring system doesn't obey Hückel's rule.

1,3,5-Cycloheptatriene
The molecule has a saturated carbon within the ring system.

Cyclopentadienyl cation
4 π electrons in ring doesn't obey Hückel's rule. The cation is not stable even though five resonance structures can be written for it.

As time passes, even more "exotic" nonbenzenoid aromatic compounds will likely be synthesized, thus increasing our understanding of the subject.

BACKGROUND

In this experiment, you will synthesize the nonbenzenoid aromatic compound ferrocene. As shown in Figure 27-1, ferrocene contains an iron(II) cation between two cyclopentadienyl anions. As a result, ferrocene is frequently called a "sandwich" compound. Also observe that each carbon atom in the upper ring is exactly staggered between two carbons in the lower ring. This arrangement allows the carbon atoms in the two rings to stay as far away from each other as possible.

The chemical bonding involves the six π electrons of each anion binding every carbon atom equally to the iron atom. Thus, as expected, every carbon-iron bond length is the same, and the iron atom is exactly in the center of the molecule. Taking the 12 π electrons from these aromatic rings and adding the 6 electrons from the outer orbit of iron, we arrive at the same electron arrangement for iron as the inert gas krypton has. As a result, ferrocene should be relatively stable and exhibit the chemical properties of an aromatic compound. Usually it does, undergoing electrophilic aromatic substitution even more readily than benzene and being resistant to chemical addition and most acids and bases. The compound, however, is sensitive to oxidizing agents because the iron atom can lose an electron and go from the $+2$ to the $+3$ oxidation

FIGURE 27-1
Ferrocene.

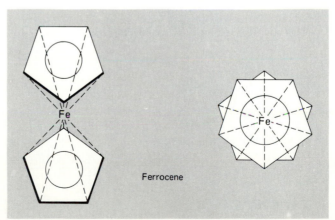

Ferrocene

state. Now the compound is much less stable because the iron no longer has the same electronic structure as krypton.

The procedure for synthesizing ferrocene is outlined below. In step (1), cyclopentadiene is prepared when needed by the thermal decomposition of dicyclopentadiene. This reaction is an example of the reverse Diels-Alder reaction (see Experiment 18C). Note that cyclopentadiene cannot just be poured out of a reagent bottle, because it readily dimerizes by the Diels-Alder reaction at room temperature to yield dicyclopentadiene.

In step (2), cyclopentadiene then loses one of its two moderately acidic hydrogens through the use of potassium hydroxide and becomes an aromatic cyclopentadienyl anion. Reaction of this anion with iron(II) chloride in step (3) then gives us ferrocene.

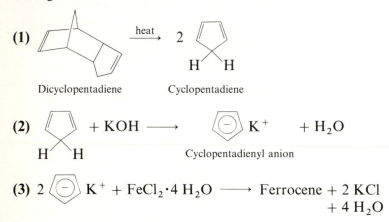

(1) Dicyclopentadiene $\xrightarrow{\text{heat}}$ 2 Cyclopentadiene

(2) + KOH $\longrightarrow$ Cyclopentadienyl anion K^+ + H_2O

(3) 2 K^+ + $FeCl_2 \cdot 4 H_2O \longrightarrow$ Ferrocene + 2 KCl + 4 H_2O

Although aromatic, the cyclopentadienyl anion is rapidly decomposed by oxygen in the air. Therefore, during its formation in step (2), nitrogen gas should be used to flush all the air from the system.

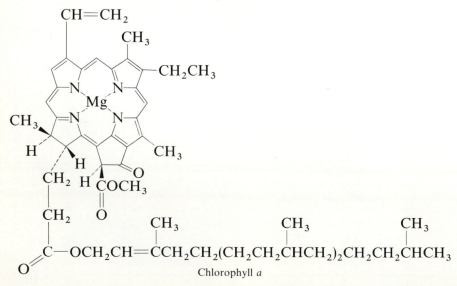

Chlorophyll *a*

Ferrocene is just one of the many thousands of known organometallic compounds. Some of the other ones important to you are chlorophyll *a*, heme, and coenzyme B_{12}, which have the indicated structures. Chlorophyll *a* is the green substance in plants and is necessary for photosynthesis. Heme is an important part of hemoglobin, the red substance in the blood that is necessary to carry oxygen. Coenzyme B_{12} is necessary to prevent pernicious anemia.

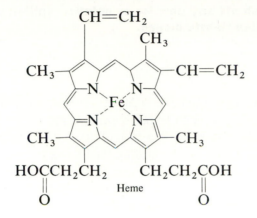

Heme

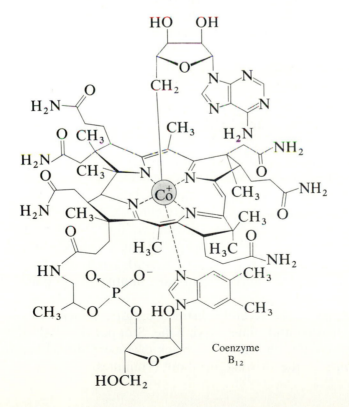

Coenzyme B_{12}

PROCEDURE

CAUTION: Cyclopentadiene is extremely flammable and volatile. Handle it with care, being sure there are no flames in the area.

In addition, dimethyl sulfoxide is potentially dangerous because anything mixed with it is rapidly absorbed through the skin, with possible harmful results. Disposable gloves are recommended when using it.

Also remember that potassium hydroxide is extremely corrosive and hygroscopic. Immediately wash off any powder or solution spilled on the skin and keep all reagent bottles tightly closed.

Cyclopentadiene

Place 20 mL (0.16 mole) of dicyclopentadiene in a 50-mL round-bottom flask. Set up a fractional-distillation apparatus as in Experiment 1, using an ice-cooled receiver for the distillate. Heat the dimer with a hot plate or heating mantle until brisk refluxing occurs (about 170°C in the boiling flask) and the monomer begins to distill in the region of 40–42°C. About 5 minutes of refluxing will be necessary before the monomer begins to be collected. Distill as rapidly as possible, but do not allow the temperature of the distilling vapor to rise above 44–45°C. At least 7 mL of cyclopentadiene should be obtained within 40 minutes. Keep the distillate cold, adding about 1 gram of anhydrous calcium chloride if the distillate should become cloudy from moisture.

While distillation is occurring, you can measure the reagents for the remainder of the experiment and set up the apparatus. Everything should be ready before the reactants are mixed together in the later steps.

Cyclopentadienyl anion

Place 25 grams (0.44 mole) of potassium hydroxide flakes or pellets into a 250-mL round-bottom flask. Add 70 mL of 1,2-dimethoxyethane solvent and swirl the mixture to dissolve as much of the potassium hydroxide as possible. Cool the mixture in an ice bath; then bubble nitrogen gas through the solution for about 2 minutes. When finished, quickly stopper the flask.

Also dissolve 6.8 grams (0.034 mole) of iron(II) chloride tetrahydrate in 30 mL of dimethyl sulfoxide solvent in a 50-mL Erlenmeyer flask, gently warming the flask on a steam bath to speed up the rate of dissolving. Cool the solution to 15–20°C in an ice bath, transfer the solution to a 125-mL separatory funnel, and bubble nitrogen gas through the solution for about 2 minutes. Then quickly stopper the funnel.

Add 5.5 mL (0.067 mole) of the recently distilled cyclopentadiene to the mixture of potassium hydroxide and dimethoxyethane. Stopper the flask and swirl the mixture, noting the color change as the black cyclopentadienyl anion is formed in solution. Continue the swirling for about 5 minutes.

Ferrocene

Unstopper the reaction flask and place the bottom end of the separatory funnel into it. Open the stopcock and allow all the ferrous chloride solution to drip into the black solution with swirling over a period of about 10 to 15 minutes. Disconnect the reaction flask, stopper, and swirl the mixture another 10 minutes.

Carefully pour the dark mixture onto a mixture of 90 mL of 6 M HCl and 100 grams of ice in a 600-mL beaker. Rinse the reaction flask with about 20 mL of water, pour this into the beaker, and thoroughly stir the contents of the beaker to neutralize all the potassium hydroxide. Collect the crystalline orange ferrocene by suction filtration, wash with water, and allow the product to thoroughly dry. Any blue color is the result of the ferrocene being oxidized to the cation, $Fe(C_5H_5)_2{}^+$.

Dissolve about 1 gram of tin(II) chloride in 10 mL of water, and add this solution to the filtrate. Note and record any color change that occurs.

Recrystallize the crude ferrocene from hexane or ligroin, allow to dry thoroughly, and determine the yield and melting point. Record the results in your notebook. The melting point should be taken with the upper end of the capillary tube sealed after the sample has been introduced. The literature melting point is 172–174°C. Also determine the theoretical and percentage yields of your product, and show the calculations. Remember that the balanced equations for this synthesis do not always have the same number of moles of the various reactants.

At your instructor's option, obtain the infrared spectrum of your product in a KBr pellet and the ^{1}H NMR spectrum in deuterated chloroform. Compare your spectra with those shown in Figures 27-2 and 27-3. When finished, hand in your product in a properly labeled container to your instructor.

QUESTIONS

1. Account for any color change that occurred when the tin(II) chloride solution was added to the filtrate from the ferrocene.
2. What reaction would occur faster if the recently distilled cyclopentadiene were not kept cold? What other reason is there to keep it cold?
3. Why does the ^{1}H NMR spectrum of ferrocene have only one peak?
4. Why were 1,2-dimethoxyethane and dimethyl sulfoxide used as solvents in this experiment rather than diethyl ether?
5. If ferrocene is acetylated with excess acetic anhydride, why does the second acetyl group prefer not to go to the same ring as the first one?
6. Ferrocene can be nitrated with nitronium fluoroborate. What would be wrong with using nitric acid?

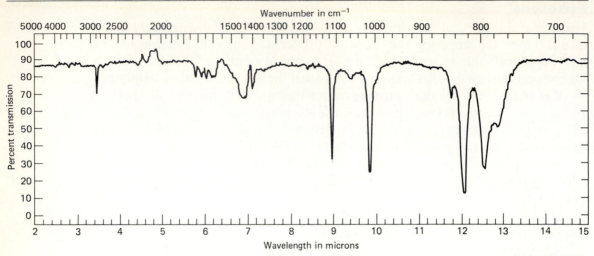

Wavenumber in cm⁻¹

FIGURE 27-2
Infrared spectrum of ferrocene in a KBr pellet.

FIGURE 27-3
¹H NMR spectrum of ferrocene in deuterated chloroform.

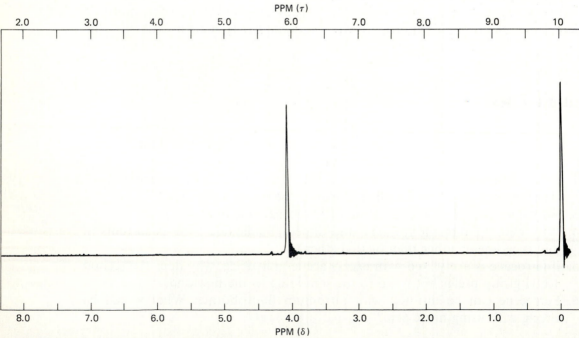

7. Even though [10] annulene and [14] annulene satisfy Hückel's rule, they are much less stable than [18] annulene. Considering the geometries of all these compounds, explain why.

8. Predict whether the following molecules will be aromatic. Explain your answer in each case.

(a)   (b)   (c)

EXPERIMENT

28

FATS AND OILS

*Our Garrick's a salad, for in him we see
Oil, vinegar, sugar and saltness agree!*

Oliver Goldsmith, *Retaliation*

INTRODUCTION

Fats and oils in their various forms are familiar to all of us. We eat meat containing fat, dairy products containing butterfat, and margarine, mayonnaise, salad dressings, and baked goods containing vegetable oil. We fry chicken, fish, potatoes, onions, rice, and doughnuts in shortening or vegetable oil. We wash ourselves with soap made from fats and oils, use paints and varnishes containing drying oils, and, unfortunately, must wash greasy pots and pans. Words such as saturated, polyunsaturated, corn oil, peanut oil, olive oil, and fatty acids have become a part of our everyday language, along with brand names such as Wesson, Crisco, Puritan, Blue Bonnet, Promise, Chiffon, and Parkay.

Generally, the terms "fat" and "oil" are used interchangeably. Technically, however, "fat" refers to fats that are solids at room temperature, whereas "oil" refers to liquid fats.

Depending on the person, between 20 and 60 percent of all calories consumed comes from fats and oils. They are the most concentrated form of food energy in our diet and produce almost 4100 calories of energy per pound, more than double the amount of 1800 calories per pound produced by carbohydrates and proteins. Reducing the intake of fats and oils can thus be a very effective way of losing weight.

Some fats and oils are necessary in the diet to provide several essential fatty acids and to absorb and transport the fat-soluble vitamins A, D, E, and K. The specific functions of the essential fatty acids are still unknown, but infants and young animals fail to grow at a normal rate if these nutrients are missing. Sparse hair, scaly skin, and poor healing of wounds are some other symptoms.

Fats and oils also make food more palatable and satisfying, and are valuable in the diet because they are slow to be digested. Slow digestion prevents a person from feeling hungry for a longer time than after eating foods without fats and oils. The digestion takes place not in the stomach, but in the small intestine.

Any excessive amount of food in the diet is converted inside the body to fat, which is stored as a reserve source of energy. A certain amount of fat is necessary to insulate the body from excessive heat loss and to form a protective layer and support around some of the vital organs. But excessive quantities can result in obesity and put an unnecessary burden on the heart and other organs.

The chemical nature of fats and oils was first investigated between 1810 and 1820 by the French chemist Michel-Eugene Chevreul. He discovered that, when hydrolyzed, fats and oils produce fatty acids and glycerol. Fats and oils, therefore, are esters of glycerol and are frequently called **glycerides** or **triglycerides**. The structural relationship between the fatty acids, glycerol, and the triglycerides is shown below.

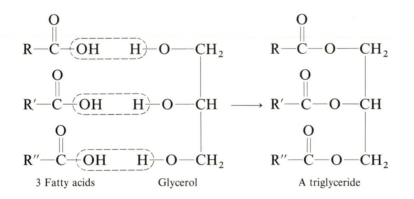

3 Fatty acids   Glycerol   A triglyceride

More than 100 different fatty acids have been found in fats and oils, giving a possible grand mixture of over 500,000 triglycerides! Most of these fatty acids, however, are present in only trace amounts, and only about 10 occur as principal components in the triglycerides. The names and structures of these fatty acids are given in Table 28-1.

Observe from the table that the most common fatty acids have 12, 14, 16, 18, or 20 carbon atoms, and have all the carbons forming a straight chain. The acids with 12 and 14 carbons are both saturated, but with 16, 18, and 20 carbons, unsaturation can occur. As many as four double bonds may be present, usually in the cis configuration. The fact that all these fatty acids contain an even number of carbons suggests that they apparently are built up of two-carbon units, and all present evidence indicates that these two-carbon units are acetyl coenzyme A groups.

TABLE 28-1.

Common Fatty Acids

Acid	Name	Structure	Comments
C_{12}	Lauric	$CH_3(CH_2)_{10}COOH$	Saturated
C_{14}	Myristic	$CH_3(CH_2)_{12}COOH$	Saturated
C_{16}	Palmitic	$CH_3(CH_2)_{14}COOH$	Saturated
	Palmitoleic	$CH_3(CH_2)_5CH=CH(CH_2)_7COOH$	Unsaturated: 1 cis double bond
C_{18}	Stearic	$CH_3(CH_2)_{16}COOH$	Saturated
	Oleic	$CH_3(CH_2)_7CH=CH(CH_2)_7COOH$	Unsaturated: 1 cis double bond
	Ricinoleic	$CH_3(CH_2)_5CH(OH)CH_2CH=CH(CH_2)_7COOH$	Contains —OH + 1 cis double bond
	Linoleic	$CH_3(CH_2)_4CH=CHCH_2CH=CH(CH_2)_7COOH$	Unsaturated: 2 cis double bonds
	Linolenic	$CH_3CH_2CH=CHCH_2CH=CHCH_2CH=CH(CH_2)_7—COOH$	Unsaturated: 3 cis double bonds
	Eleostearic	$CH_3(CH_2)_3CH=CHCH=CHCH=CH(CH_2)_7—COOH$	Unsaturated: 1 cis and 2 trans double bonds
C_{20}	Arachidonic	$CH_3(CH_2)_4(CH=CHCH_2)_4CH_2CH_2COOH$	Unsaturated: 4 cis double bonds

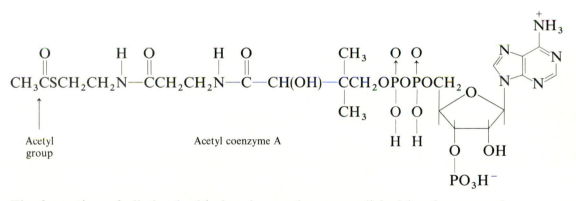

Acetyl group

Acetyl coenzyme A

The formation of all the double bonds may be accomplished by the removal of hydrogen atoms through the action of other enzymes.

In recent years, people have become even more aware of the importance of the right kinds of fats in their diet. In general, vegetable fats are better than animal fats, and polyunsaturated fats are better than saturated fats. The reason is cholesterol (see Experiment 30), which in excessive amounts in the blood can lead to **atherosclerosis**, or a hardening and narrowing of arteries due to the buildup of cholesterol deposits. Sometimes the deposits become so thick that blood flow is severely restricted or even cut off. If this reduced blood flow is to vital organs such as the brain or heart, a stroke or heart attack results.

515

Vegetable fats are generally considered better than animal fats because they contain no cholesterol to contribute to atherosclerosis. However, certain highly saturated fats such as palm and coconut oil actually increase the level of blood cholesterol. Their consumption may, therefore, be actually worse for the body than consuming certain types of animal fats. Palm and coconut oil are commonly used for commercially made cookies, cakes, and other baked goods because of the lower cost. Economics obviously takes priority over health.

Vegetable fats containing high amounts of polyunsaturated fatty acids are considered best for the body, since they actually lower the level of blood cholesterol, and, as we might expect, vegetable fats containing mostly monounsaturated fatty acids have little or no effect on the level of blood cholesterol.

Table 28-2 gives the percentages of fatty acids in various types of animal and vegetable fats. Observe from the table the comparatively high amounts of saturated fatty acids in most animal fats and the comparatively high amounts of unsaturated fatty acids in vegetable fats. Of course, there are exceptions. Note how the fats from fish tend to be more unsaturated, which helps explain why a diet higher in fish and lower in other meats helps to prevent heart disease.

TABLE 28-2.

Mole Percentages of Fatty Acids in Various Animal and Vegetable Fats

Animal	Saturated fatty acids					Unsaturated fatty acids			
	C_4–C_{12}	Myristic	Palmitic	Stearic	Total	Oleic (1 d.b.)	Linoleic (2 d.b.)	Linolenic (3 d.b.)	Total
Beef fat	1	6	27	14	48	49	2	0	52
Butter	9–13	7–9	23–26	10–13	43–52	30–40	4–5	0	39–50
Lard	0	1–2	24–30	12–18	41–50	41–48	10	0	51–58
Cod oil	0	5–7	8–10	0–1	13–18	27–33	27–32	0	82–88
Herring oil	0	5	14	3	22	0	0	30	78
Sardine oil	0	6–8	10–16	1–2	17–26	6–10	27–11	8–12	78–85
Vegetable									
Coconut	55–73	17–20	4–10	1–5	77–97	2–10	0–2	0	3–13
Corn	0	0–2	7–11	3–4	10–17	43–50	34–42	0	77–93
Cottonseed	0	0–3	17–23	1–3	18–29	23–44	34–55	0	71–82
Linseed	0	0	4–7	2–5	6–12	9–38	3–43	25–58	88–94
Olive	0	0	5–15	1–4	6–19	69–84	4–12	2	81–94
Palm	0	1–6	32–47	3–6	36–59	38–42	5–11	0	43–58
Peanut	0	0–1	6–9	2–6	8–15	50–70	13–26	2	85–92
Safflower	0	1	3	4	8	15	76	1	92
Sunflower	0	1	6	5	12	21	66	1	88
Soybean	0	0–1	10–13	2–5	12–19	21–29	50–59	2–10	81–88
Tung	0	0	1–3	1–3	2–6	4–16	0–1	0	94–98

A high percentage of the fatty acids in fish even contains three or more double bonds. And, as stated before, both coconut and palm oil have very high amounts of saturated fatty acids. Coconut oil even has a higher percentage of saturated fatty acids than all the animal fats! Also note from the table how safflower, sunflower, and soybean oils are especially good sources of polyunsaturated fats.

Increasing the unsaturation of a fat lowers its melting point. This explains why most unsaturated fats are liquids and most saturated fats are solids. To obtain margarine or shortening, vegetable oils are partially hydrogenated until the desired semisoft consistency is attained. Although the resulting products are not as good as the original oils in lowering blood cholesterol, they still are much better in this regard than butter or lard. To make margarine, the resulting semi-soft solid is mixed with β-carotene to give it color, with milk to form an emulsion, and with salt, biacetyl, and acetoin to give it flavor.

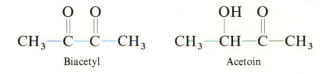

$$\underset{\text{Biacetyl}}{CH_3-\overset{\displaystyle O}{\overset{\|}{C}}-\overset{\displaystyle O}{\overset{\|}{C}}-CH_3} \qquad \underset{\text{Acetoin}}{CH_3-\overset{\displaystyle OH}{\overset{|}{C}H}-\overset{\displaystyle O}{\overset{\|}{C}}-CH_3}$$

Some fats turn rancid and have an unpleasant taste and odor when left unrefrigerated and exposed to the air. One reason why this occurs is that the moisture in the air hydrolyzes the fat to give foul-smelling and foul-tasting fatty acids such as butyric acid from rancid butter. The other reason is that air oxidation occurs at the very reactive allylic positions to give hydroperoxides. These hydroperoxides can then decompose to yield foul-smelling and foul-tasting low-molecular-weight aldehydes.

$$R-CH=CH-CH_2-CH=CH+(CH_2)_7COOH \xrightarrow{O_2}$$

$$R-CH=CH-\overset{\displaystyle OOH}{\overset{|}{C}H}-CH=CH+(CH_2)_7COOH$$

$$\downarrow$$

$$R-CH=CH-\overset{\displaystyle O}{\overset{\|}{C}}-H + H-\overset{\displaystyle O}{\overset{\|}{C}}+(CH_2)_8COOH$$

Highly unsaturated fats such as linseed and tung oil thicken when exposed to air and eventually harden to give a clear, smooth finish. As a result, these oils are used as drying oils in the manufacture of some varnishes and oil-based paints. Oxidation and polymerization at the double bonds are evidently occurring during this exposure.

Two numbers of value when analyzing fats and oils are the **saponification number** and the **iodine number**. The saponification number is defined as the number of milligrams of KOH required to completely convert 1 gram of a fat to glycerol and soap. A high saponification number therefore indicates a low-molecular-weight fat. For example, coconut oil has a higher saponification number than olive oil, since coconut oil contains much greater amounts of low-molecular-weight fatty acids.

The iodine number is a quantitative measure of the amount of unsaturation present in a fat. It is defined as the number of grams of iodine required to completely saturate 100 grams of the fat. Therefore, a higher iodine number means a more unsaturated fat. Iodine numbers can be as low as 8 to 10 for a highly saturated fat, such as coconut oil, and as high as 175 to 205 for a highly unsaturated fat, such as linseed oil.

Vegetable fats and oils are isolated by three principal methods: cold-pressing, hot-pressing, and solvent extraction. Cold-pressing merely involves squeezing out the oil from the plant in a hydraulic press. Hot-pressing is like cold-pressing, but is done at a higher temperature. The hot-pressing method usually gives a higher yield than cold-pressing, but a somewhat lower quality product due to undesirable components in the mixture. Solvent extraction is the most expensive method, but gives the best yields and the highest grade of oils.

Animal fats are usually obtained by cooking the fat out of the tissue at high temperatures. Alternatively, the fatty tissue may be placed in boiling hot water. The fat floats to the surface and can thus be easily separated.

28A. CHEMICAL AND PHYSICAL PROPERTIES OF FATS AND OILS

Solubility of vegetable oils

Place 3 drops of a liquid fat such as cottonseed oil, olive oil, Wesson, Crisco, or Mazola oil into each of six small, labeled, dry test tubes. Add 1 mL of water to the first tube, 1 mL of ethyl alcohol to the second, 1 mL of ethyl ether to the third, 1 mL of acetone to the fourth, 1 mL of methylene chloride to the fifth, and 1 mL of toluene to the sixth. Stopper and shake the tubes, wait to allow any settling to occur, then record the observed results in your notebook. Which solvents completely dissolved the vegetable oil? Which solvent was the poorest?

Place 3 more drops of the vegetable oil into the solvents that completely dissolved the first 3 drops. Stopper, shake, and allow any settling to occur. Record which solvents, if any, continue to completely dissolve the oil. Repeat this procedure as long as necessary, and rank the solvents in their ability to dissolve vegetable oil.

Spot test

Place a drop of vegetable oil and a drop of water on a piece of filter paper. Observe the spots. Warm the filter paper by **carefully** passing it back and forth over a flame, being sure it does not ignite. Record what happens to the spots. This crude test for a fat has been used for centuries.

Acrolein test

When heated with $KHSO_4$, glycerol is dehydrated and oxidized to **acrolein**, which is detectable by its sharp, irritating odor. The acrolein test can thus possibly be used to determine the presence of glycerol in fats.

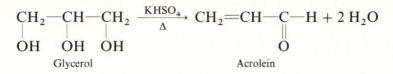

Glycerol   Acrolein

Add a few crystals of potassium bisulfate to each of two small dry test tubes. To one tube add 2 drops of glycerol, and to the other add 2 drops of vegetable oil. **Gently** heat both tubes with a small flame until the contents **begin** to turn yellowish brown, and then **cautiously** smell. Record the results, noting if there is any similarity between the two odors.

Unsaturation tests

Dissolve 1 mL of each of the following oils or fats in 4 mL of methylene chloride in separate **labeled** test tubes: cottonseed oil, olive oil, corn oil, linseed oil, margarine, butter, lard, and tallow. Add, drop by drop, a 0.1 *M* solution of bromine in methylene chloride to each tube with gentle agitation, counting the number of drops it takes for the red color to persist. **(CAUTION: BROMINE SOLUTIONS CAUSE SEVERE BURNS. WEAR PLASTIC GLOVES WHEN USING IT AND BE VERY CAREFUL TO KEEP THE REAGENT OFF YOUR SKIN. IF YOU SPILL ANY BROMINE ON YOUR SKIN, IMMEDIATELY RINSE IT OFF WITH WATER AND TREAT WITH A WET DRESSING OF 10% SODIUM THIOSULFATE. NOTIFY YOUR IN-STRUCTOR OF THE INJURY.)** Record and explain your test results in your notebook.

28B. GAS CHROMATOGRAPHIC ANALYSIS OF FATS AND OILS

- **SAPONIFICATION**
- **ESTERIFICATION**
- **CHEMICAL EQUILIBRIUM**

In this experiment, you will quantitatively determine the amounts of various fatty acids in a fat or oil. The possible fats and oils include Wesson, Crisco, Mazola, Puritan, shortening, lard, cottonseed oil, and linseed oil. If you use a brand name cooking oil or shortening, inspect the label to see if the source of the oil is indicated (e.g., corn oil, safflower oil, soybean oil). Remember that if shortening is used, some of the fatty acids have been changed from the partial hydrogenation.

One problem with analyzing all the fatty acids in a fat is that they have combined together in many possible ways to form hundreds or even thousands of different triglyceride molecules. Injecting the fat into the gas chromatograph would thus give a very complex chromatogram. But even more important, since these peaks are for the triglycerides and not the fatty acids, we wouldn't know which fatty acids were present in a triglyceride that produced a certain peak. Analysis of the tryglyceride producing each peak could be done, but this would be a very long and laborious project.

Obviously then, the fatty acids in the fat or oil must be separated from the triglyceride molecules before the analysis. We could use saponification in the same way that soap is prepared (see Experiment 29). Acidifying the soap mixture would then yield the fatty acids.

$$\underset{\substack{O \\ \parallel}}{R-C-O^- \, Na^+} + H^+ \longrightarrow R-C-OH + Na^+$$

Unfortunately, most of the fatty acids have boiling points of around 300°C and higher. Some of them even decompose when strongly heated. Therefore, gas chromatographic analysis is not very feasible, and saponification alone is clearly not a very good approach.

However, if the fatty acids are converted into their methyl esters, the method succeeds. Methyl esters have lower boiling points than fatty acids, do not decompose, and are easily analyzed by gas chromatography. The reactions are

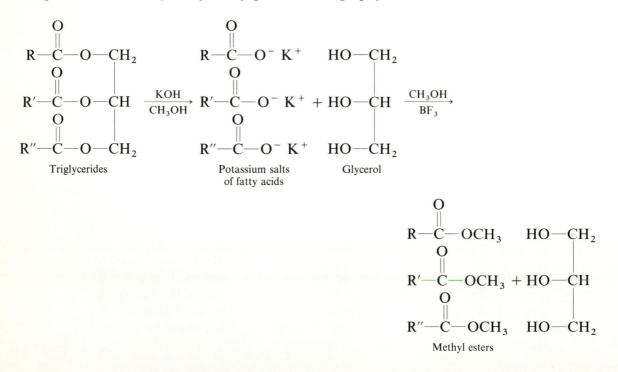

Triglycerides   Potassium salts   Glycerol
                       of fatty acids

Methyl esters

In the first reaction, KOH saponifies the triglycerides to give a mixture of the potassium salts of the fatty acids. A solution of BF_3 in methanol then esterifies the fatty acids. Boron trifluoride is a Lewis acid catalyst for the esterification, and a large excess of methanol ensures that the equilibrium of the second reaction is shifted completely to the right.

The mechanism of the first reaction involves nucleophilic acyl substitution, with hydroxide ion, acting as a nucleophile, attacking one of the carbonyl carbons of the triglyceride to give a tetrahedral intermediate. The openness of the carbonyl group, due to its flat shape, greatly aids this reaction. So too does the polarity of the carbonyl group and the willingness of the oxygen atom to readily accept a negative charge. The oxygen atom of glycerol is then expelled from this intermediate to yield a fatty acid and the anion of a diglyceride. A proton exchange then produces the fatty acid as its potassium salt and a diglyceride. This mechanism is repeated until all the fatty acids are released from the fat.

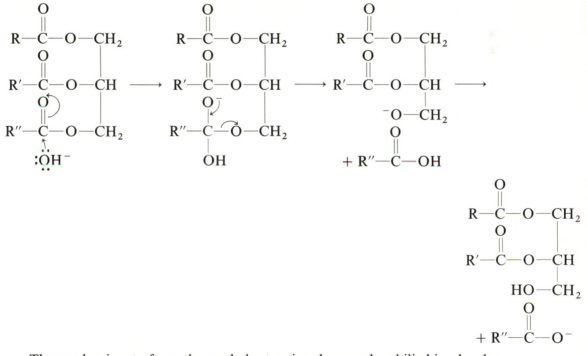

The mechanism to form the methyl esters involves nucleophilic bimolecular substitution by the anion of a fatty acid on the boron trifluoride–methanol complex. The function of the boron trifluoride is therefore to make the hydroxyl group of the alcohol a much better leaving group for this reaction.

CAUTION: Boron trifluoride is toxic. Avoid breathing it and store it in the hood. Also be certain there are no flames around when using petroleum ether because it is very flammable.

If necessary, reread the section on the gas chromatograph (pages 151–162).

Formation of the methyl esters of the fatty acids

Place 10 drops of the vegetable oil or an equivalent amount of the solid into a large 50-mL test tube containing 5 mL of a 0.5 M solution of KOH in methanol. Place the test tube in a boiling-water bath **in the hood**. After the fat or oil dissolves, add 10 mL of a 12% BF_3-methanol solution. Heat in the boiling-water bath for another 3 minutes.

Cool the solution and add 35 mL of petroleum ether. Place the solution in a 125-mL separatory funnel and wash twice with 30-mL portions of saturated salt solution. Save the upper petroleum ether layer each time.

Wash the petroleum ether with 30 mL of water, then dry it in a 50-mL Erlenmeyer flask with a small amount of anhydrous magnesium sulfate. Decant the petroleum ether from the magnesium sulfate into another 50-mL Erlenmeyer flask, filtering if necessary. Place this flask in a steam bath or hot-water bath **in the hood** and concentrate the liquid down to a volume of about 1 mL or less. This liquid will be used for the gas chromatographic analysis.

Gas chromatographic analysis

To obtain the retention times, inject separately about 0.3 μL of the methyl esters of myristic acid, palmitic acid, palmitoleic acid, stearic acid, oleic acid, linoleic acid, and linolenic acid into a 9-foot Chromosorb W column coated with 15% diethyleneglycol succinate. The column temperature should be 195°C, the injection port temperature should be 245°C, and the carrier gas flow rate should be about 40 mL per minute. Do **not** inject a new sample if there is a sample still on the column. To save time, your instructor may permit you to inject only one sample and obtain the other retention times from other students. Be certain in this case, however, that the same gas chromatograph is being used, to avoid errors from a different gas flow rate, a different length of column, and so on.

After the retention times are known, inject about 1 μL of your mixture of fatty acid esters. After the chromatogram is obtained, identify as many of the peaks as possible from the known retention times of the given methyl esters. Also indicate which peaks have not been identified.

Next determine quantitatively the area of each peak in the chromatogram. You may cut out the **labeled** peaks with scissors and weigh them on an **analytical** balance, use the automatic integrator that may be provided with the gas chromatograph, or use one of the other techniques described in Section

TABLE 28-3.

Detector Sensitivity Factors for the Methyl
Esters of Various Fatty Acids

Fatty acid	Detector sensitivity factor
Myristic	0.90
Palmitic	0.91
Palmitoleic	0.91
Stearic	1.01
Oleic	0.98
Linoleic	1.07
Linolenic	1.17

6.4I. Divide the area of each known peak by the appropriate detector sensitivity factor in Table 28-3 to obtain the corrected area for each known peak. Obviously, the various methyl esters do **not** affect the detector equally.

From the corrected areas of the known methyl esters and the uncorrected areas of the unknown peaks, determine the percentage composition of the known fatty acids and the other fatty acids in your fat or oil. If you know the source of your fat or oil (corn oil, safflower oil, etc.), compare your percentage composition with that given in Table 28-2 and record the results.

When finished, either fold and carefully attach your entire chromatogram to your notebook, or place the individual pieces of your chromatogram in a labeled envelope and attach to your notebook.

28C. ISOLATION OF TRIMYRISTIN AND MYRISTIC ACID FROM NUTMEG

SAPONIFICATION

In this experiment, you will isolate the fat trimyristin or glyceryl trimyristate from nutmeg. You will then saponify part of the product to obtain myristic acid. Observe from the structure of trimyristin how all the fatty acid units are the same. This contrasts with the usual situation of the fatty acid units being different. Trimyristin can be found in many different fats and oils, but only in nutmeg and coconut oil is it found in high amounts. Normally, only trace amounts are present.

The isolation of trimyristin first involves the extraction of trimyristin and other organic compounds from nutmeg powder with ether. The ether is then distilled, and the residue from the distillation is recrystallized from acetone. All the other compounds, in contrast to trimyristin, are very soluble in cold acetone, so a separation occurs.

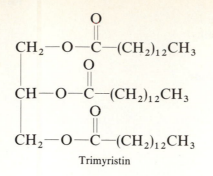

Trimyristin

The saponification of trimyristin yields glycerol and the sodium salt of myristic acid. Acidification of the alkaline solution then produces myristic acid, which may be collected by suction filtration.

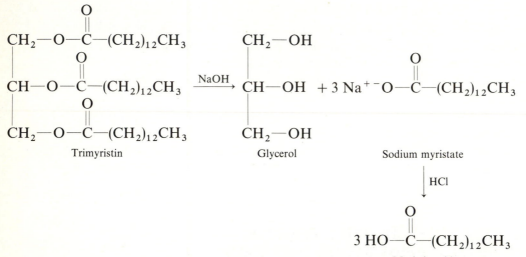

PROCEDURE

CAUTION: Use extreme care when handling sodium hydroxide and concentrated hydrochloric acid. They are both very caustic. Also be certain there are no flames around when using diethyl ether.

Isolation of trimyristin

Place 8 grams of powdered nutmeg and 75 mL of diethyl ether in a 250-mL round-bottom flask. Add a boiling chip, attach the flask to a reflux condenser, and gently reflux the mixture for 15 minutes, using a steam bath or a warm-water bath. Gravity filter the warm mixture through a fluted filter paper into a 100-mL round-bottom flask. Add a boiling chip, set up a simple-distillation apparatus, and distill the ether, using a steam bath or a warm-water bath. Collect the ether in a receiver cooled in an ice bath.

Slowly add hot acetone to the residue in the round-bottom flask to dissolve the material, gravity filter the hot solution through fluted filter paper, and allow the filtered liquid to cool to room temperature. Cool the liquid further in an ice-water bath, collect the white crystals of trimyristin by suction filtration, and allow them to dry thoroughly. Weigh the product, determine its melting point, and record the results in your notebook. The literature melting point is 56–57°C. Also determine the percentage yield of the product and show your calculations.

At your instructor's option, obtain the infrared spectrum of your product in a KBr pellet and the ^{1}H NMR spectrum in deuterated chloroform. When finished, save 0.8 gram of the trimyristin for the saponification part of the experiment and place the remainder in a properly labeled container.

Saponification of trimyristin to obtain myristic acid

Place 0.80 gram (0.0011 mole) of trimyristin, 12 mL of 6 M sodium hydroxide, and 12 mL of ethanol in a 50-mL round-bottom flask. Add a boiling chip, attach a reflux condenser, and reflux the mixture for 1 hour.

Pour the solution into a 250-mL beaker containing 100 mL of ice water and **cautiously** and **slowly** add 12 mL of concentrated hydrochloric acid with stirring. The addition of the acid will cause the myristic acid to precipitate. Cool the beaker in an ice-water bath, collect the product by suction filtration, and wash the crystals with about 10 mL of cold water. Allow the product to dry thoroughly, and determine its weight, melting point, and literature melting point. Record the results in your notebook. Also determine the theoretical and percentage yields of your product, and show your calculations. Remember that 1 mole of trimyristin yields 3 moles of myristic acid.

At your instructor's option, obtain the infrared spectrum of your product in a KBr pellet and the ^{1}H NMR spectrum in deuterated chloroform. When finished, hand in both of your products in properly labeled containers to your instructor.

28D. ISOLATION OF OLEIC ACID FROM OLIVE OIL

- **SAPONIFICATION**
- **COLD TEMPERATURE FILTRATION**
- **FORMATION OF AN UREA COMPLEX**

In this experiment, you will isolate oleic acid from olive oil. Although Table 28-4 indicates that olive oil contains mostly oleic acid, there are other fatty acids that must be removed. To accomplish this, you will use saponification to free the fatty acids from the fat molecules. This is the same reaction used to produce

TABLE 28-4.

Fatty Acids Found in Olive Oil

Fatty acid	Structure	%	mp (°C)
Palmitic	$CH_3(CH_2)_{14}COOH$	5–15	63
Stearic	$CH_3(CH_2)_{16}COOH$	1–4	70
Oleic	$CH_3(CH_2)_7CH=CH(CH_2)_7COOH$	69–84	16
Linoleic	$CH_3(CH_2)_4CH=CHCH_2CH=CH(CH_2)_7COOH$	4–12	−5
Linolenic	$CH_3CH_2CH=CHCH_2CH=CHCH_2CH=CH(CH_2)_7COOH$	2	−11

soap (see Experiment 29). After cleavage by the hydroxide ion, the mixture of fatty acids is then obtained by acidifying the alkaline solution.

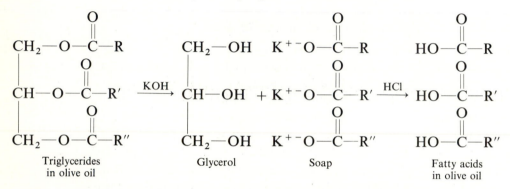

Triglycerides in olive oil → Glycerol + Soap → Fatty acids in olive oil

To partially purify the freed oleic acid, the mixture of fatty acids is dissolved in acetone and cooled to −15°C with dry ice or an ice-salt bath. Stearic acid and palmitic acid are both saturated and thus have much higher melting points and lower solubilities in acetone than do the other unsaturated fatty acids. As a result, they both crystallize from the solution and can be separated by suction filtration.

The unsaturated fatty acids are then recovered from the filtrate and treated with urea in methanol. A large variety of long-chain fatty acids, alkanes, alkynes, esters, and alcohols form inclusion complexes with urea and fit inside a cylindrical channel, surrounded by a number of urea molecules arranged in a spiral pattern. However, this channel has a limited diameter of slightly over 5 Å. Therefore, only thin molecules, having no skeletal branching or other widening protrusions, are able to fit inside. Among these are oleic acid, whose urea inclusion complex is shown in Figure 28-1. For simplicity, the spiral arrangement of 14 urea molecules enclosing the oleic acid is indicated as a spiral line.

Linoleic acid and linolenic acid might also be expected to form urea inclusion complexes. However, they don't because the cis arrangement of two or three double bonds requires more room than one cis double bond, and they are therefore unable to fit inside the thin channel. As a result, oleic acid can be separated from linoleic and linolenic acid. Its urea complex will crystallize when the hot

526

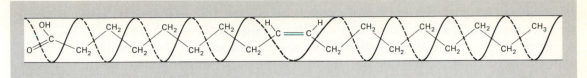

FIGURE 28-1
Urea inclusion complex of oleic acid.

methanol solution is cooled, but both linoleic acid and linolenic acid remained dissolved and can be separated by suction filtration. Placing the isolated urea complex in water and extracting with ether then allows the pure oleic acid to be recovered.

PROCEDURE

CAUTION: Do not touch dry ice because it causes severe frostbite. Also handle potassium hydroxide, concentrated hydrochloric acid, and hot triethylene glycol with care.

Weigh 10.0 grams of olive oil in a preweighed 125-mL Erlenmeyer flask. Add 2.3 grams of potassium hydroxide pellets and 20 mL of triethylene glycol, then heat the mixture with stirring at 160°C for 5 to 7 minutes, using a hot plate. The two layers initially observed will soon combine, and a thick yellow syrup containing the potassium salts of the fatty acids will be formed.

Cool the syrup to room temperature, add 50 mL of ice water and **carefully** and **slowly** acidify the soap solution with 10 mL of concentrated hydrochloric acid. Cool to room temperature and extract the oil twice with 35-mL portions of ether in a 250-mL separatory funnel. Wash the ether with 25 mL of saturated salt solution, then dry the ether solution over a small amount of anhydrous magnesium sulfate. Carefully decant the ether into a 100-mL round-bottom flask, filtering if necessary, and set up a simple-distillation apparatus. Distill the ether, using a steam bath or a warm-water bath, until an oily residue of about 9 grams of the fatty acid mixture remains. The distilled ether should be collected in a receiver cooled in an ice bath.

Add 75 mL of acetone to the round-bottom flask to dissolve the residue. Cool the solution to −15°C in a dry ice bath of ethanol or isopropanol for about 15 minutes. Add crushed dry ice as needed to maintain the temperature at −15°C. As an alternative, the cold temperature may be attained by using an ice-salt bath.

After the crystals of stearic acid and palmitic acid have completely formed, suction filter the mixture. **Do it quickly, using a Büchner funnel that has**

527

been thoroughly chilled in a refrigerator freezing compartment. Otherwise, part of the solid may redissolve, and the separation will be incomplete. Do not allow the unfiltered liquid to warm up.

Pour the filtered acetone solution back into the cleaned round-bottom flask and set up a simple-distillation apparatus. Distill the acetone, using a steam bath or a hot-water bath, until only a residue remains. Collect the distillate in a receiver cooled in an ice bath. In the meantime, dissolve 11 grams of urea in 50 mL of warm methanol in a 125-mL Erlenmeyer flask. Pour this solution into the round-bottom flask containing the unsaturated fatty acids and heat to dissolve all the material. Pour the hot solution into the 125-mL Erlenmeyer flask, rinse out the round-bottom flask with 5 mL of hot methanol, and add to the Erlenmeyer flask.

Allow the hot urea-methanol solution to cool until a large crop of crystals have formed. Then cool further in an ice bath with swirling and collect the product by suction filtration, washing it with a small amount of cold methanol.

Place the crystals of the oleic acid-urea complex into a 50-mL beaker. Add 30 mL of water and record what happens. Transfer the mixture to a 125-mL separatory funnel and extract with two 25-mL portions of ether. Wash the combined ether extracts with 25 mL of saturated salt solution; then dry the ether over a small amount of anhydrous magnesium sulfate.

Decant the ether into a clean, dry 100-mL round-bottom flask, filtering if necessary, and set up a simple-distillation apparatus. Distill the ether, using a steam bath or a warm-water bath, until about 10 mL remains. Collect the distillate in a receiver cooled in an ice bath. Transfer the remaining ether into a preweighed 25-mL Erlenmeyer flask and continue the evaporation of ether **in the hood** until a constant weight is obtained. Cool the oil of oleic acid in an ice-

FIGURE 28-2

Infrared spectrum of oleic acid, neat. Spectrum courtesy of Sadtler Research Laboratories, division of Bio-Rad Laboratories, Inc., copyrighted 1973.

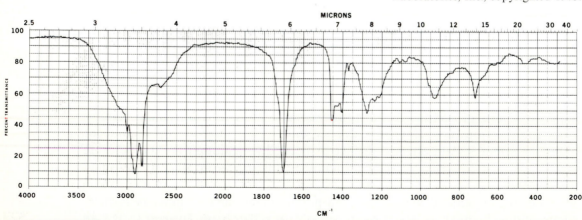

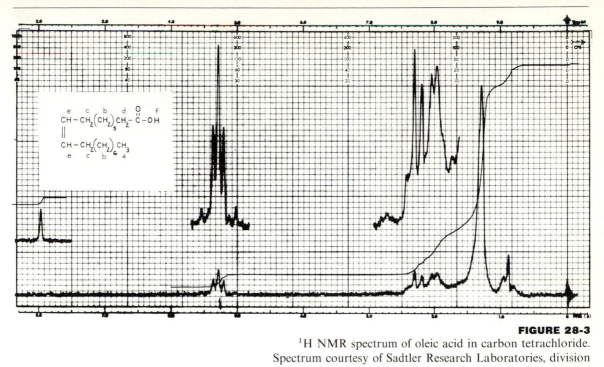

FIGURE 28-3

^{1}H NMR spectrum of oleic acid in carbon tetrachloride.
Spectrum courtesy of Sadtler Research Laboratories, division
of Bio-Rad Laboratories, Inc., copyrighted 1967.

water bath until it solidifies; then slowly warm and determine the yield, melting point, and literature melting point, and record the results in your notebook. Also determine the percentage yield of the oleic acid, showing the calculations.

At your instructor's option, obtain the infrared spectrum of your product, neat between salts, and the ^{1}H NMR spectrum in carbon tetrachloride. Compare your spectra with those shown in Figures 28-2 and 28-3. When finished, hand in your product in a properly labeled container to your instructor.

28E. SAPONIFICATION NUMBER OF FATS AND OILS

The saponification number of a fat is the number of milligrams of KOH required to completely convert 1 gram of the fat to glycerol and soap. A higher saponification number therefore indicates a lower-molecular-weight fat having lower-molecular-weight fatty acids.

In this experiment, you will determine the saponification number and the average molecular weight of coconut oil, butter, and a vegetable oil such as corn oil, olive oil, or safflower oil. You will do this by saponifying each fat or

529

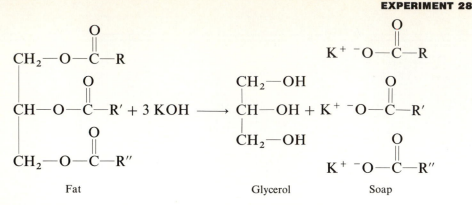

oil with an excess of KOH, then titrating the remaining KOH with HCl until the endpoint is reached. Another identical excess amount of KOH **not** used for the saponification will also be titrated. The difference in the amounts of HCl needed to neutralize the KOH used in the saponification and not used in the saponification will then indicate how much KOH was consumed during the saponification.

PROCEDURE

To ensure a satisfactory amount of equipment, the students may either share their equipment or their results. As an alternative, the experiment may be done in stages, with the refluxing of one fat being conducted while the remaining KOH from another fat is being titrated.

Accurately weigh to the nearest 0.001 gram about 1–1.5 grams of coconut oil, butter, and vegetable oil in separate, **labeled**, preweighed 50-mL beakers. Dissolve each sample in hot ethanol and transfer to separate, **labeled** 100-mL round-bottom flasks. Rinse each beaker with a few milliliters of hot ethanol to ensure a complete transfer. Then carefully deliver from a buret 30.0 mL of approximately 0.5 M KOH into each flask.

Add a boiling stone to each flask, attach a reflux condenser, and reflux the mixtures on a steam bath or hot plate for 30 minutes. Allow the solutions to cool to room temperature, add 2 to 3 drops of phenolphthalein indicator, and titrate with standardized 0.500 M HCl until the pink color disappears. Also titrate a separate 30.0-mL portion of the 0.5 M KOH with the standardized HCl until the endpoint is reached.

Subtract the amount of HCl used in each titrated refluxed KOH sample from the amount used in the unrefluxed KOH sample. The difference gives the amount of HCl **equivalent to** the KOH used in the saponification: 1.0 mL of 0.500 M HCl = 28 mg of KOH.

Determine the saponification number of the coconut oil, butter, and the vegetable oil by dividing the calculated number of milligrams of KOH used for that fat sample by the number of grams of the fat. Then knowing that

$56 \times 3 = 168$ grams of KOH is needed to saponify 1 mole of oil or fat, calculate the average molecular weight of the coconut oil, butter, and the vegetable oil. Show all of your calculations.

28F. IODINE NUMBER OF FATS AND OILS

The iodine number of a fat or oil is the number of grams of iodine required to completely saturate 100 grams of the fat. It is a quantitative measure of the amount of unsaturation present in a fat.

In this experiment, you will determine the iodine number of butter, olive oil, and safflower oil by placing a known amount of the fat or oil with an excessive amount of iodine. After the reaction is completed, the unreacted iodine is titrated with sodium thiosulfate solution until the blue color from added starch disappears. The same titration is also done with an identical amount of iodine that was not placed with a fat or oil. The difference in the amounts of sodium thiosulfate needed to reach the endpoints will then indicate how much iodine was consumed by the fat. The reaction of sodium thiosulfate with iodine is

$$2\,Na_2S_2O_3 + I_2 \longrightarrow 2\,NaI + Na_2S_4O_6$$

PROCEDURE

CAUTION: Do not breathe methylene chloride and glacial acetic acid nor get them on your skin. Do not mouth-pipet.

Also, for good results, do not allow the iodine solutions to stand for over 1–1.5 hours before titrating.

Accurately weigh to the nearest 0.001 gram about 1–1.5 grams of butter, olive oil, and safflower oil in separate, **labeled**, preweighed 50-mL beakers. Dissolve each fat in about 15 mL of methylene chloride and transfer to separate, **labeled**, 50-mL volumetric flasks. Rinse each beaker with additional methylene chloride and add to the appropriate volumetric flask to ensure a complete transfer. Add methylene chloride to bring each volumetric flask up to the 50-mL mark and mix thoroughly.

Using a **clean** 10-mL pipet, transfer 10.0 mL of each solution to separate, **labeled**, 250-mL glass-stoppered bottles. Also put 10.0 mL of methylene chloride in a fourth **labeled** bottle. Add 25.0 mL of 0.15 M I_2 solution in glacial acetic acid to each of the four bottles, stopper all four, and then store them in a dark place, such as your locker, for 1 hour.

Add 10 mL of 5% KI solution and 50 mL of water to each of the four bottles, and titrate each solution with standardized 0.250 M sodium thiosulfate until the solution is pale brown. Periodically, stopper each bottle and vigorously shake the solution throughout the titration to ensure that all the iodine from the lower methylene chloride layer is transferred to the upper aqueous layer. When the solution becomes pale brown, add about 1 mL of 1% starch solution and continue the titration and shaking until the blue color disappears.

Subtract the volume of sodium thiosulfate needed to titrate each fat from the volume needed to titrate the fourth bottle, which had no fat. The difference measures the amount of iodine that reacted with each fat. One milliliter of 0.250 M sodium thiosulfate is equivalent to 0.0318 gram of iodine.

Divide the weight of each fat used by 5, since only 10 mL out of 50 mL of solution in the volumetric flask was used for the determination. Knowing the number of grams of iodine used and the weight of the fat, calculate the iodine number of butter, olive oil, and safflower oil, showing all of your calculations. Compare with the known values of 26–40, 79–90, and 130–150, respectively. In terms of blood cholesterol, which fat or oil is best for you?

QUESTIONS

1. In the gas chromatographic analysis of fats and oils, why do methanol and glycerol not give peaks in the gas chromatogram? What other liquid does give a peak or peaks? Why are these extra peaks not a problem in the analysis?
2. Why was boron trifluoride rather than a protonic acid used as a catalyst in Experiment 28B?
3. The trimyristin and myristic acid obtained from Experiment 28C both have the same color and almost the same melting point. Give two ways to prove they are not the same compound.
4. Why did the saponification of olive oil in Experiment 28D take much less time than the saponification in some other experiments, such as 28C?
5. Draw all the different triglycerides that could possibly be formed from only three different fatty acids. For convenience, label these fatty acids A, B, and C. Be on the lookout for possible duplication of structures.
6. Why are vegetable oils frequently partially hydrogenated by food manufacturers?
7. Why would people not as likely buy a margarine or shortening product made by blending vegetable oil with lard or beef fat?
8. Why do fats and oils turn rancid?
9. Why are fish oils considered to be healthier than other animal fats?

EXPERIMENT

29

SOAPS AND DETERGENTS

Beauty will fade and perish, but personal cleanliness is practically undying, for it can be renewed whenever it discovers symptoms of decay.

W. S. Gilbert, *The Sorcerer*, Act II

INTRODUCTION

An old proverb states that "cleanliness is next to godliness." Although this statement may not have much significance in religion, the fact remains that we live in a society where cleanliness is a very fundamental part of our lives. We wash our bodies with soap, our hair with shampoo, and our clothes and dishes with detergent. Practically everything around us from floors to walls and windows to curtains gets dirty, so the average American uses over 40 pounds of soap and detergents each year. This figure is higher than for any other country and reflects our higher standard of living as well as our awareness of the value of cleanliness in personal hygiene and aesthetics. The sale of soap and detergents is a very competitive field, and we are continually saturated with advertisements about these products.

The making of soap is one of the oldest organic syntheses known, ranking second only to the production of ethyl alcohol. Credit for its manufacture is usually given to the ancient Phoenicians, who, around 600 B.C. prepared it from goats' tallow and wood ashes. Soap was widely known during the Roman Empire, and the ruins of a soap factory dating back 2000 years were unearthed at Pompeii.

But the importance of soap for washing and cleaning was apparently not recognized, or at least not reported, until the second century after Christ. Later

533

writings continued to mention its manufacture and its value as a cleaning agent, but for hundreds of years the use of soap remained very limited. Apparently, cleanliness of the body and clothing was not considered important, and, for those who could afford it, perfumes were used to conceal body odor.

The widespread production and use of soap did not occur until the 1700s. The discovery by Nicolas Leblanc of a method for producing caustic soda from salt brine changed the manufacture of soap from a handicraft to an industry. This enabled it to be made and sold at prices that most people could afford and was the beginning of the modern soap industry today.

The manufacture of soap involves the alkaline hydrolysis or saponification of a fat or oil to yield soap and glycerol. As discussed in Experiment 28, fat consists of a mixture of organic molecules called **triglycerides**, which are esters of long-chain fatty acids, usually containing from 12 to 18 carbon atoms in a straight chain. Soap consists of a mixture of the salts of the long-chain carboxylic acids and is formed when a hydroxide ion cleaves the triglyceride.

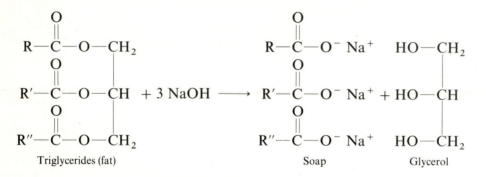

Triglycerides (fat)　　　　　　　　　　Soap　　　　　　Glycerol

Even though this reaction has remained unchanged for hundreds of years, soap is much more attractive and appealing to the consumer than it was even 50 years ago. Dyes, perfumes, germicides, and lotions have been added to change ordinary soap into a colorful, creamy, and pleasant-smelling product, capable of fighting bacteria and body odor. Soap may also vary in its composition and method of processing. Air can be blown in to make it float. Alcohol may be added to make it transparent. If the potassium rather than the sodium salt is made, the soap is soft.

Even with all of these variations, soap still chemically remains the same and cleans in the same manner. A soap molecule has a **polar, water-soluble** head, $-COO^-$, and a long, **nonpolar, oil-soluble** tail, the long chain of the remaining carbon atoms. Ordinarily oil and water tend to separate, forming two distinct layers, but the presence of soap changes this. As shown in Figure 29-1, the long nonpolar ends of soap molecules dissolve in the oil droplets, leaving the polar, water-soluble ends projected outward into the surrounding water. Each individual oil droplet thus becomes surrounded by negative charges. Repulsion between these charges keeps the oil droplets from coalescing, and a stable emulsion of oil in water occurs. Since the oil is what holds the dirt, emulsifying or

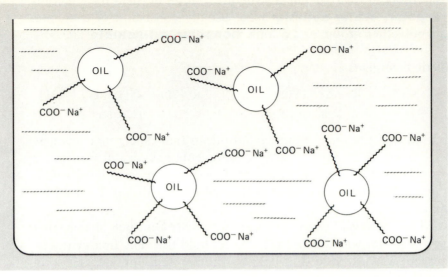

FIGURE 29-1
Emulsifying of an oil in water by soap.

"cutting" the oil also "emulsifies" or "frees" the dirt and allows it to be washed away with rinse water.

The biodegradable nature of soap is an advantage for the environment. The long, linear soap molecule can easily be converted by microorganisms into carbon dioxide and water, thus not becoming a pollution problem in lakes and streams.

However, soap has the disadvantage of being a less effective cleaner in hard water, due to the presence of calcium, magnesium, and iron salts that react with soap to yield insoluble precipitates frequently called **soap curd** or **soap scum**, the familiar "bathtub ring" in sinks and tubs. This precipitate, formed as shown in the reaction below, means that part of the soap has been wasted, and an even greater amount is necessary for effective cleaning.

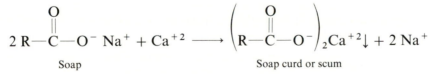

This problem can be lessened for laundry soaps through the addition of a water softener, such as sodium carbonate, that precipitates the hard-water ions as shown below. Unfortunately, this precipitate may become lodged in the laundry fabric and cause discoloration.

$$Na_2CO_3 + Ca^{+2} \longrightarrow CaCO_3\downarrow + 2\,Na^+$$

Hard water is not a problem, however, with detergents, which is one reason detergents have replaced soap for many household uses. Synthetic detergents may vary considerably in their chemical structure, but all of them have a common feature: a **polar, water-soluble** head, usually consisting of a sulfate, sulfonic acid or polyether group, and a long, **nonpolar, oil-soluble** tail.

Types of synthetic detergents

$$R—O—SO_3^- \ Na^+ \qquad R—Ar—SO_3^- \ Na^+ \qquad R(O—CH_2CH_2)_n OH$$

Sulfate   Sulfonic acid   Polyether

The polar head and the long, nonpolar tail mean that detergents clean by the same process as soap. However, because detergents have a different polar head than soap does, they do not form an insoluble precipitate in hard water. This results in greater cleaning effectiveness in hard water without the need for water-softening agents.

The first synthetic detergent was made in 1916 by Fritz Gunther. His product was used industrially, but was too harsh for household use. The first synthetic household detergent was introduced in 1933 by Procter and Gamble, and soon other soap companies began marketing similar products. These cleaning agents, however, were relatively expensive, and it wasn't until 1950 that the first of the inexpensive detergents appeared. This product, called an alkylbenzenesulfonate, can be prepared from inexpensive petroleum sources by the reaction sequence

This type of detergent became very popular and rapidly replaced soap as the most widely used cleaning agent. The highly branched side chain in the molecule, however, created a problem because bacterial enzymes in sewage treatment plants and septic tanks were unable to biodegrade the molecule after use. Lakes and waterways all over the country became polluted with detergent foam that even found its way into the drinking water sources of many cities.

Fortunately, this problem was solved by replacing the branched alkylbenzenesulfonate detergent with a linear alkylbenzenesulfonate. This detergent, as shown below, contains only one branch on the side chain, which makes it biodegradable.

A linear alkylbenzenesulfonate

Laundry detergents and some other detergents are not sold as pure compounds. As little as 10 to 25% of the product may actually be detergent, with the remainder consisting of builders, foam stabilizers, bleaches, optical brighteners or whiteners, antiredeposit agents, coloring agents, and corrosion inhibitors. Builders such as sodium carbonate, sodium silicate, and sodium perborate enhance the cleaning ability of detergents and also act as cheap fillers. These builders have replaced the use of phosphates in detergents, which promote the growth of algae in lakes and streams. Excess quantities of algae are undesirable because they consume so much dissolved oxygen from the water when they die and decompose that no animal life can exist in the water. Optical brighteners or whiteners absorb ultraviolet light and reemit it as visible light, giving the laundry a "whiter" and "cleaner" look. Antiredeposit agents help to keep the dirt suspended and prevent it from returning to the cleaned materials.

29A. SYNTHESIS OF SOAP

In this experiment, you will prepare soap by the alkaline hydrolysis or saponification of a fat or oil.

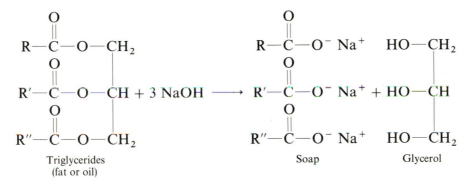

Triglycerides (fat or oil) + 3 NaOH → Soap + Glycerol

The fat or oil used may come from any of various products, such as Crisco or Wesson oil, shortening, lard, and olive oil. Since the amounts of the various fatty acids in these products will differ, it is expected that the soap obtained will vary in terms of yield, composition, and physical properties. If different students use a variety of all the possible fats and oils, you may find it interesting to compare the feel, odor, solubility, and lathering ability of the various products obtained.

PROCEDURE

CAUTION: Sodium hydroxide is very caustic. Handle it with care, remembering that it will also produce a tremendous quantity of heat when dissolved in solution.

To a solution of 10 mL of water and 10 mL of 95% ethanol in a 100-mL round-bottom flask, dissolve 3 grams of sodium hydroxide pellets. Add 3 grams of a fat or oil and a few boiling chips; then attach a reflux condenser and reflux the mixture for at least 30 minutes.

In the meantime, prepare a solution of 15 grams of sodium chloride in 50 mL of water in a 100-mL beaker. Do **not** use table salt, due to the other components present. Heat the solution if necessary to dissolve the salt, but allow to cool to room temperature. To this salt solution, add the hot refluxed mixture with stirring and cool to room temperature in an ice bath. Collect the precipitated soap by suction filtration and wash twice with small quantities of ice-cold water. Use part of your soap for the tests in Experiment 29C, dry the remainder, and submit it in a properly labeled container to your instructor.

29B. SYNTHESIS OF A DETERGENT

In this experiment, the detergent sodium lauryl sulfate will be prepared by the reaction of lauryl or dodecyl alcohol with chlorosulfonic acid to yield the lauryl ester of sulfuric acid. This will then be neutralized with sodium carbonate solution to yield the detergent.

$$CH_3-(CH_2)_{10}-CH_2OH + Cl-\overset{\overset{O}{\uparrow}}{\underset{\underset{O}{\downarrow}}{S}}-OH \longrightarrow$$

Lauryl or dodecyl   Chlorosulfonic
alcohol   acid

$$CH_3-(CH_2)_{10}-CH_2-O-\overset{\overset{O}{\uparrow}}{\underset{\underset{O}{\downarrow}}{S}}-OH + HCl$$

Lauryl ester of sulfuric acid

$$2\ CH_3-(CH_2)_{10}-CH_2-O-\overset{\overset{O}{\uparrow}}{\underset{\underset{O}{\downarrow}}{S}}-OH + Na_2CO_3 \longrightarrow$$

$$2\ CH_3-(CH_2)_{10}-CH_2-O-\overset{\overset{O}{\uparrow}}{\underset{\underset{O}{\downarrow}}{S}}-O^-\ Na^+ + H_2O + CO_2$$

Sodium lauryl sulfate

Sodium lauryl sulfate is used as a detergent in many household products, including toothpastes and shampoos. When a mixture of sodium alkyl sulfates is satisfactory for the intended product, fats and oils may be the starting material, just as in the manufacture of soap. One industrial method of preparation is shown in the following equations.

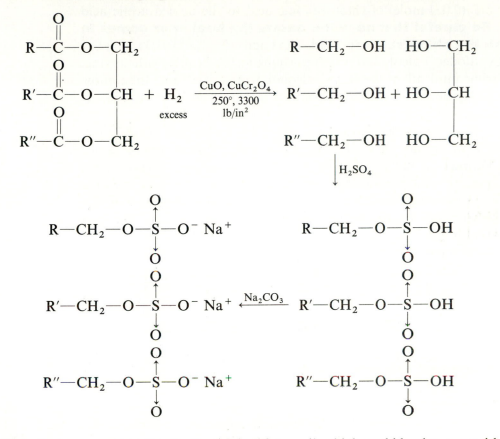

Note that any unsaturation in the original fat or oil, which could lead to a rancid product, is removed in the first hydrogenation step.

PROCEDURE

CAUTION: Chlorosulfonic acid is an extremely hazardous and corrosive chemical that causes severe burns and must be handled with extreme care. Since it reacts violently with water, you must use dry glassware. Plastic gloves are highly recommended, and, as always, wear safety glasses. If you spill any of the reagent on your skin, wash if off immediately with water and notify your instructor.

Glacial acetic acid also causes burns. Use it carefully.

Dry all the glassware for this experiment by thoroughly heating all the surfaces with a burner flame. Even apparently dry glassware has small amounts of moisture on it which will react with chlorosulfonic acid and lower your yield.

Add 5 mL of glacial acetic acid to a 50-mL beaker and moderately cool in an ice bath so that it remains a liquid. **In the hood**, slowly add, in **small** portions, 2.0 mL (0.031 mole) of chlorosulfonic acid to the cooled acetic acid with stirring. **Be careful that no water enters the beaker or comes in contact with the chlorosulfonic acid.** Next add 5.8 grams (0.031 mole) of lauryl alcohol[1] (dodecyl alcohol) in small portions to the beaker with stirring. Continue stirring until all of the lauryl alcohol is dissolved and then allow to stand with occasional stirring for about 20 minutes until the reaction is completed.

In the meantime, prepare a solution of 7 grams of sodium chloride in 25 mL of water in a 100-mL beaker, and add cracked ice until the total volume is not quite 50 mL. Do **not** use table salt, due to the other components present. Pour the reaction mixture with stirring into this ice-salt mixture and slowly add a saturated sodium carbonate solution in 1-mL portions with stirring until the mixture is basic to litmus. Collect the precipitated detergent by suction filtration and wash twice with small quantities of ice-cold water. Use part of your detergent for the tests in Experiment 29C, dry and weigh the remainder, and submit it in a properly labeled container to your instructor. Also determine the theoretical and percentage yields of the final product, showing the calculations.

29C. TESTS ON SOAPS AND DETERGENTS

Effect of calcium and magnesium salts

Dissolve 0.35 gram of your prepared soap in 25 mL of distilled water and place 5 mL of the solution in a test tube. Shake the solution to build a foam, allow the solution to stand for 30 seconds, observe the level of the foam, and record the result in your notebook. Add 3 drops of a 5% calcium chloride solution, shake, and allow to stand. Observe the effect of the calcium chloride on the foam and record the result. Did you notice anything else? Add 0.3 gram of trisodium phosphate or sodium carbonate and shake again. Allow the solution to stand, and note and record the result.

Repeat the above tests with a second 5-mL portion of fresh soap solution, using 5% magnesium chloride solution. Note and record all the results.

Dissolve 0.35 gram of your prepared detergent or a commercial detergent in 25 mL of distilled water and place 5 mL of the solution in a test tube. Shake the solution to build a foam, allow the solution to stand for 30 seconds, observe

---

[1] The lauryl alcohol, usually supplied in a narrow-mouth bottle, has a melting point of 24–27° and may be a solid at room temperature. To facilitate handling, melt it beforehand, pour it into a wide-mouth bottle, and break it into small pieces after crystallizing.

the level of the foam, and record the result. Add 3 drops of a 5% calcium chloride solution, shake, and allow to stand. Observe the effect of the calcium chloride on the foam and record the result. Did you notice anything else? Is there a need for the trisodium phosphate or sodium carbonate? Why?

Repeat the above tests with a second 5-mL portion of fresh detergent solution, using 5% magnesium chloride solution. Note and record all the results.

Effect of oil

Place 5 drops of a vegetable oil into each of three test tubes. To the first add 5 mL of distilled water, to the second add 5 mL of fresh soap solution, and to the third add 5 mL of fresh detergent solution. Shake each tube vigorously, let stand a few minutes, and compare and record the results.

Effect of dilute acid

To the remaining 10 mL of soap solution and 10 mL of detergent solution, add dilute sulfuric acid until the solutions are acid to litmus. Note and record the result in each case.

QUESTIONS

1. Why does soap dissolved in water produce an alkaline solution?
2. In Experiment 29A, why was a mixture of alcohol and water, rather than simply water itself, used to saponify the fat or oil to produce soap?
3. In Experiments 29A and 29B, why does the addition of a salt solution cause the soap or detergent to precipitate?
4. From the information given in Experiment 29A, why is it impossible to calculate an exact theoretical yield of soap?
5. In Experiment 29B, why was glacial acetic acid and not acetic acid and water used to dissolve the chlorosulfonic acid? Give two reasons.
6. In Experiment 29B, why wasn't the amount in moles of chlorosulfonic acid greater than the amount in moles of lauryl alcohol?
7. In Experiment 29C, explain the results of adding calcium and magnesium salts to a solution of your prepared soap. Explain what happened after the trisodium phosphate or sodium carbonate was added.
8. In Experiment 29C, explain the results of adding calcium and magnesium salts to a solution of your detergent.
9. In Experiment 29C, explain what happened to the vegetable oil when it was shaken with soap and detergent solutions.
10. In Experiment 29C, explain the results of adding dilute sulfuric acid to your soap and detergent solutions.
11. Even though soap in the eye may be an unpleasant experience, it is not as harmful or irritating as laundry detergent. What is present in laundry detergent that could account for this difference?

12. Some companies manufacture baby shampoos, which do not sting the eyes at all. What nonirritating cleansing ingredients are present in these products?

13. Some shampoos on the market contain special antidandruff formulas. What key ingredient or ingredients are present in these formulas?

14. Carbona and K2r have been on the market for many years to remove spots and stains without washing. What are the key ingredients in Carbona and K2r?

15. Several organic chemistry books describe various detergents as being either "hard" or "soft." These terms do **not** refer to how the detergents feel to the skin, nor to their solubility, nor to the type of water used. What are the authors talking about when they use these terms?

16. Very often, manufacturers will deliberately add a detergent to a chemical herbicide or pesticide that is to be mixed with water before application. What possible benefit could this have?

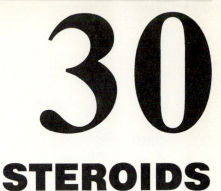

STEROIDS

INTRODUCTION

One of the most important classes of organic compounds found in humans, plants, and animals is **steroids**. This group includes the **female sex hormones** estrone, progesterone, estriol, and estradiol, the **male sex hormones** androsterone and testosterone, the **synthetic birth control agents** Norlutin and Norethynodrel, **vitamin D, cortisone, cholesterol**, and the **bile acids** cholic acid and deoxycholic acid, the structures of which are given in Table 30-1.

Observe from the structures that the steroids have in common the four-ring carbon skeleton of perhydrocyclopentanophenanthrene, where three six-membered rings and one five-membered ring are fused together.

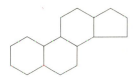

Perhydrocyclopentanophenanthrene

The structures of the steroids, however, are much more complex than indicated, due to the three-dimensional nature of the rings. As shown below, any

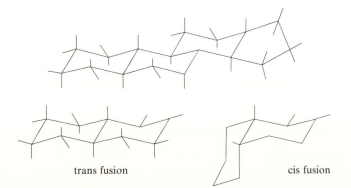

trans fusion   cis fusion

543

TABLE 30-1.

Common Steroids

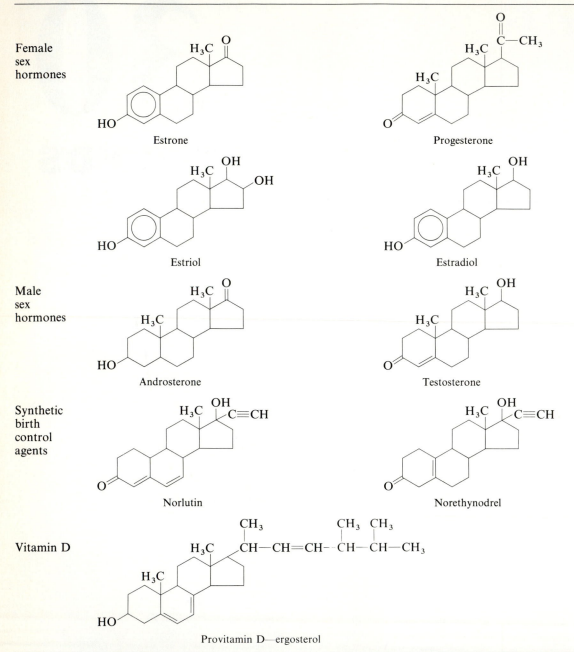

Female sex hormones

Estrone

Progesterone

Estriol

Estradiol

Male sex hormones

Androsterone

Testosterone

Synthetic birth control agents

Norlutin

Norethynodrel

Vitamin D

Provitamin D—ergosterol

TABLE 30-1. (continued)

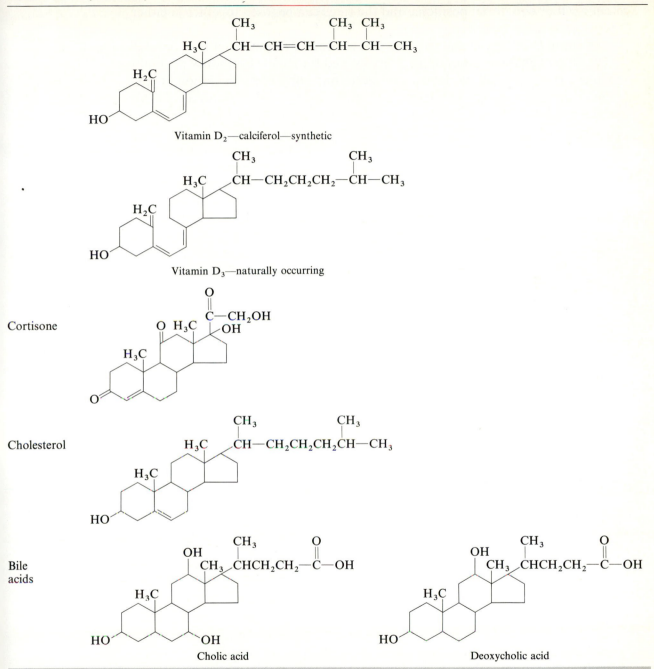

Vitamin D₂—calciferol—synthetic

Vitamin D₃—naturally occurring

Cortisone

Cholesterol

Bile acids

Cholic acid

Deoxycholic acid

substituents on these rings can be located in either the stereochemically different axial or equatorial positions, and the rings can be fused together in either a trans or a cis arrangement.

As steroids, the male and female sex hormones are responsible for the secondary sexual characteristics of each species, and the female hormones maintain the menstrual cycle in the female. Progesterone suppresses ovulation in the female during pregnancy and therefore can be used at other times as a birth control agent. Norlutin and Norethynodrel are synthetic steroids similar in structure to progesterone. As birth control agents, they are more effective orally, since they are better absorbed from the stomach and intestines than is progesterone.

Vitamin D is a group of vitamins essential for the proper formation of bone tissue, and whose absence leads to rickets. The first clue leading to the structure of the D vitamins was the discovery that ultraviolet light was beneficial to patients suffering from rickets. When it was next realized that food rather than the patient could be irradiated with the same beneficial results, investigation eventually led to the compound ergosterol. Although this steroid had no beneficial activity itself, ultraviolet light converted it into vitamin D_2 or calciferol, having a very high activity.

At first it was thought that the naturally occurring vitamin D in food was identical in structure to vitamin D_2. But subsequent investigation revealed that the naturally occurring vitamin had a slightly different structure, now known as vitamin D_3.

Cortisone is one of about 30 adrenal hormones, which as steroids govern a wide variety of metabolic processes. Cortisone is used medically in the treatment of rheumatism to reduce the swelling of inflamed joints.

In recent years, anabolic steroids or synthetic derivatives of the male hormone testosterone have achieved prominence because of their use by male and female athletes. On the positive side, these steroids produce improved athletic performance due to increased muscle size, gains in strength, and a shorter workout recovery time. However, the picture isn't all positive, and among the negative effects are increased aggressive behavior, compulsive eating, high blood pressure, and breast growth in males. Long-term side effects also include strokes, heart disease, liver damage, liver cancer, and sterility. For teenagers, steroid use can also stunt growth in those who haven't fully matured. Thus those people who feel compelled to use steroids are playing a dangerous game and should fully ask themselves if the immediate benefits are worth all the risks.

Cholesterol is the most abundant steroid in humans and other animals, being found in all bodily tissues, but in the highest concentrations in the brain, spinal cord, and nerves. The total amount of cholesterol in an average human adult is roughly half a pound. Because of its wide availability, cholesterol was the first steroid to be isolated and identified. But because of its complex structure, over 150 years passed between its isolation in 1775 and its structure determination in 1932. But since cholesterol also has eight asymmetric centers and could be any one of 256 possible stereoisomers, it required another 23 years to ascertain its complete three-dimensional structure.

The human body receives part of its cholesterol from the consumption of foods such as dairy products, eggs, meat, and animal fat. But most of it is manufactured by the body itself, primarily in the liver.

Cholesterol is essential to the body for the production of the male and female sex hormones, the adrenal hormones, and the bile acids which are natural detergentlike substances that emulsify fats and oils to make them more digestible. The high concentration of cholesterol in all nerve tissue suggests that it may also play a vital role in nerve conduction or some other important nerve function.

But as important as cholesterol is, too much of it, from eating the wrong kinds of food, can lead to **atherosclerosis** and **cholelithiasis**. Atherosclerosis, or hardening of the arteries due to cholesterol deposits, produces a loss of elasticity and a narrowing of the arterial channel, which can lead to heart strain and heart disease. Sometimes the deposits become so thick that the blood flow is severely restricted or even cut off. If this blockage is to vital organs such as the brain or heart, a stroke or heart attack occurs.

Cholelithiasis is the formation of gallstones in the bile duct or gallbladder. The stones consist of cholesterol and other steroids, bile pigment, protein, calcium, and phosphate, and may range in size from tiny particles to nuggets the size of eggs. The formation of stones occurs when there is an impairment in the bile salt production by the liver, caused by liver disease, infection, or chronic intoxication.

Cholelithiasis may produce no symptoms at all, but indigestion is a common sign. Severe abdominal pain or jaundice can also occur when a stone travels through a duct or obstructs a duct. After diagnosis by x-rays, the stones may be removed by surgery, lithotripsy (ultrasonic "crushing"), or by dissolving in methyl *t*-butyl ether via a catheter. These last two procedures are relatively new, but they may soon eliminate the need for surgery.

30A. ISOLATION OF CHOLESTEROL FROM GALLSTONES

■ **ELECTROPHILIC ADDITION OF BROMINE TO A DOUBLE BOND**
■ **DEBROMINATION**

In this experiment, you will isolate cholesterol from human gallstones (obtained from the department of surgery or pathology of a hospital). The gallstones are crushed with a hammer and towel, blender, or a mortar and pestle, and the cholesterol is extracted from the bile pigment with hot diethyl ether. The dirty yellow appearance of the bile pigment is primarily from bilirubin, a metabolite of hemoglobin.

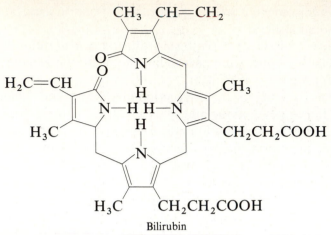

Bilirubin

Unfortunately, recrystallization is not sufficient to purify cholesterol, because of the presence of small amounts (0.1 to 3%) of three other steroids, which are very similar to cholesterol in their solubility properties.

But cholesterol can be separated from these steroids by chemical means. Cholesterol is converted to cholesterol dibromide by the electrophilic addition of bromine across the double bond. The impurities, having no double bond or less active ones, fail to react. Since the cholesterol dibromide has different solubility properties than the impurities, it can be separated and then converted back to cholesterol by debromination with zinc metal in acetic acid. Recrystallization then yields pure cholesterol.

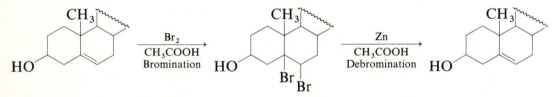

PROCEDURE

CAUTION: Bromine solutions cause severe burns. Wear plastic gloves when using it and be very careful to keep the reagent off your skin. If you spill any bromine on your skin, immediately rinse it off with water and treat with a wet dressing of 10% sodium thiosulfate. Notify your instructor of the injury.

Glacial acetic acid can also cause severe burns, especially when hot. Use it with care.

Isolation

Place 2.0 grams of crushed gallstones and 10 mL of diethyl ether in a 50-mL Erlenmeyer flask. While stirring, heat the mixture with a steam or hot-water bath until all the solid has disintegrated and the cholesterol has dissolved. While the brown-yellow mixture is still hot, remove the brown residue of bile pigment by gravity filtration through a small fluted filter, and collect the filtrate in another 50-mL Erlenmeyer flask. Rinse the original flask with 2 to 3 mL of additional heated ether, and filter.

Add 10 mL of methanol and a little decolorizing carbon to the filtrate. Heat the mixture with a steam or hot-water bath, filter the still hot mixture through a small, preheated fluted filter, and collect the filtrate in a 50-mL Erlenmeyer flask. Reheat the greenish yellow filtrate and add just enough hot water dropwise to make the solution cloudy. The solution is now saturated, and cholesterol will crystallize on cooling. Collect the crystals by suction filtration, wash with a very small quantity of ice-cold methanol, and leave the suction on to dry the crystals. Weigh the product, determine its melting point, and record the results in your notebook.

Bromination

Dissolve 1.0 gram of the crude cholesterol from above in 10 mL of gently warmed **anhydrous** ether in a 50-mL Erlenmeyer flask. Then **slowly and carefully** add 5 mL of a solution of bromine and sodium acetate in glacial acetic acid. The cholesterol dibromide should begin to crystallize in a few minutes. Cool the flask in an ice bath to obtain complete crystallization and collect the crystals by suction filtration. Wash the crystals with an ice-cold mixture of 3 mL of ether and 7 mL of glacial acetic acid, and a second time with 5 to 10 mL of ice-cold methanol.

Debromination

Transfer the cholesterol dibromide crystals to a 50-mL Erlenmeyer flask and add 20 mL of **anhydrous** ether, 5 mL of **glacial** acetic acid, and, with careful swirling, 0.2 gram of **fresh** zinc dust. The dibromide should dissolve within a few minutes, and zinc acetate should separate as a white paste after 5 to 10 minutes. If the reaction is slow, add a little more zinc dust and warm gently. Swirl for an extra 5 minutes and then slowly add, dropwise, enough water to dissolve the white solid and give a clear solution.

Decant the ether solution from any unreacted zinc into a separatory funnel and extract twice with equal volumes of water to remove the zinc acetate and zinc bromide. (You may have to add a little extra ether to bring the solution up to its original volume.) Next, remove the acetic acid by extracting the ether solution with 10% sodium hydroxide solution until a drop of the ether solution no longer turns blue litmus red. Reduce the water content in the ether solution by next shaking it with an equal volume of saturated sodium chloride solution. Separate the ether layer and complete the drying process by having the ether layer stand over a small amount of anhydrous magnesium sulfate, followed by decanting or gravity filtering the ether into a dry 50-mL Erlenmeyer flask.

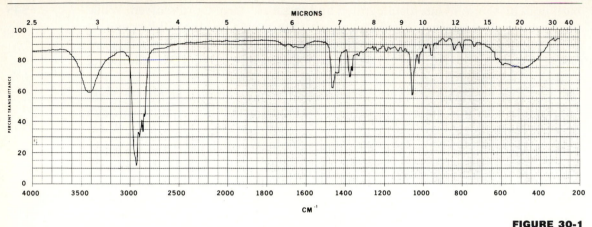

FIGURE 30-1

Infrared spectrum of cholesterol in KBr pellet. Spectrum courtesy of Sadtler Research Laboratories, division of Bio-Rad Laboratories, Inc., copyrighted 1973.

Recrystallization

Add 10 mL of methanol to the dry ether solution and evaporate the solution on a steam bath or hot-water bath **in the hood** until the purified cholesterol begins to crystallize. Let the solution cool to room temperature, then place it in an ice bath to complete the crystallization. Collect the crystals by suction filtration, wash them with a little ice-cold methanol, and allow them to dry. Weigh the product and determine the melting point and literature melting point. Record the results in your notebook. Also determine the percentage yield of the pure cholesterol from the gallstones and show the calculations. Remember that only a portion of the crystals was used after the initial isolation.

At your instructor's option, obtain the infrared spectrum of your product in a KBr pellet and the ^{1}H NMR spectrum in deuterated chloroform. Compare your spectra with those shown in Figures 30-1 and 30-2. When finished, hand in your product in a properly labeled container to your instructor.

30B. ISOLATION OF CHOLESTEROL FROM EGG YOLKS

In this experiment, you will isolate cholesterol from egg yolks. Many other compounds are also present in egg yolks, including lecithins and fats (see Experiment 28), which will also be isolated. In fact, as shown below, lecithins are structurally related to fats, having one of the fatty acid groups replaced by the phosphate ester of choline.

The procedure for isolating the cholesterol, lecithins, and fats is outlined in Figure 30-3. Cholesterol, lecithins, and fats are all soluble in a combination of

550

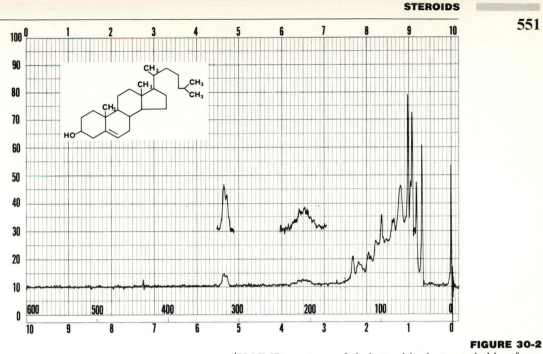

FIGURE 30-2

^{1}H NMR spectrum of cholesterol in deuterated chloroform.
Spectrum courtesy of Aldrich Chemical Company, *The Aldrich
Library of NMR Spectra,* edition II, by Charles Pouchert.

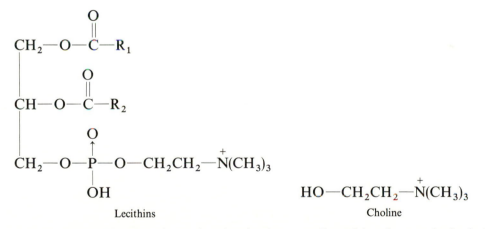

Lecithins

Choline

ether and alcohol. Extraction of pulverized egg yolks with ether and alcohol
therefore separates these compounds and some others from all the insoluble
compounds in egg yolks.

To isolate the lecithins, the ether-alcohol solution is evaporated and the
residue is redissolved in ether. Acetone is then added, which precipitates the
lecithins into a gummy bulk that is then separated.

The remaining acetone-ether solution is evaporated to yield a residue of
cholesterol, fats, and other compounds. A solution of potassium hydroxide in

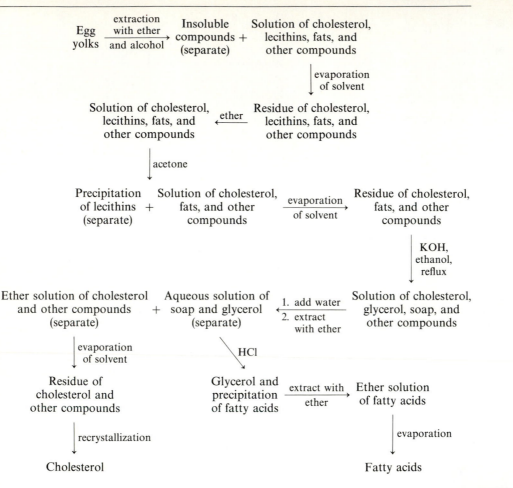

FIGURE 30-3
Outline of procedure for isolating cholesterol, lecithins, and
fatty acids from egg yolks.

ethanol is added, and refluxing saponifies the fats to yield glycerol and the
potassium salts of the fatty acids (soap, see Experiment 29). The equation is

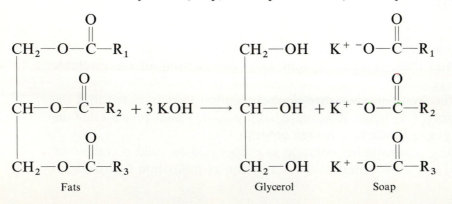

Water is then added, and the alcohol-water solution is extracted with ether. Cholesterol and other compounds are dissolved by the ether, but glycerol and the potassium salts of the fatty acids remain in the water. Evaporation of the ether yields crude cholesterol, which is then purified by recrystallization. Acidification of the aqueous alkaline solution precipitates the fatty acids, which are then removed from the water and glycerol by extraction with ether. Evaporation of the ether then yields the isolated fatty acids.

PROCEDURE

Removal of insoluble material

Heat two medium-size eggs in boiling water for 10 minutes, remove, and cool. **(CAUTION: Do not use a flame if ether is present.)** Separate the rest of the eggs from the yolks; then pulverize the yolks and weigh them.

Place the pulverized yolks into a 250-mL beaker and add 40 mL of methanol and 50 mL of diethyl ether. Mix well for 5 minutes, then gravity filter into another 250-mL beaker while keeping the insoluble material in the first beaker. Add another 20 mL of ether to the insoluble material in the first beaker, mix well, and filter. The insoluble material may now be discarded.

Using a steam bath or a hot-water bath in the hood, evaporate the alcohol-ether solution to dryness. A thin yellow residue should appear.

Isolation of lecithins

Add 25 mL of ether to the residue and slightly warm it to dissolve all the material. Next add acetone slowly with constant stirring until the lecithins precipitate as a gummy bulk. As much as 150 mL of acetone may be needed.

Gravity filter the solution into a 250-mL round-bottom flask, using a loose wad of cotton (no filter paper). Wash the lecithins with 10 mL of acetone and filter into the round-bottom flask, using the same cotton wad. Allow the lecithins to thoroughly dry, weigh them, and determine the percentage yield; be sure to show your calculations. Place the material in a properly labeled container and submit it to your instructor at the end of the experiment with your other products.

Isolation of cholesterol

Using a steam bath or a hot-water bath, distill the acetone-ether solution to obtain a liquid oil residue. The distillate should be collected in a receiver cooled in an ice bath. Remove the flask from the heating bath and add 25 mL of a 15% potassium hydroxide solution in ethanol. Attach a reflux condenser and reflux the mixture for 25 minutes.

Cool the solution to room temperature and pour it into a 250-mL separatory funnel. Add 35 mL of water to the round-bottom flask, swirl, and add the wash water to the separatory funnel. Repeat with another 35 mL of water.

Extract the water-alcohol solution twice with 35-mL portions of ether. Combine the two extracts, then wash the ether with 10 mL of water. Add this water to the alcohol-water solution and save for the isolation of the fatty acids.

Shake the ether solution with 15 mL of saturated salt solution to lower the water content, separate the solutions, and further dry the ether solution over a small amount of anhydrous magnesium sulfate. Filter or decant into a tared 150-mL beaker and evaporate the solution to dryness in the hood, using a steam bath or hot-water bath. Determine the weight and the percentage yield of the crude cholesterol obtained; show the calculations.

Recrystallize the cholesterol from methanol, using decolorizing carbon. Allow the purified product to thoroughly dry and determine its weight, melting point, and literature melting point. Record the results in your notebook. Also determine the percentage yield of the purified product, and show the calculations.

At your instructor's option, obtain the infrared spectrum of your product in a KBr pellet and the ^{1}H NMR spectrum in deuterated chloroform. Compare your spectra with those shown in Figures 30-1 and 30-2. When finished, hand in your product in a properly labeled container along with the other products to your instructor.

Isolation of the fatty acids

Acidify the aqueous solution containing the soap by **carefully** adding concentrated HCl with stirring until a drop of the solution turns blue litmus paper red and the fatty acids have precipitated. Then place the entire mixture into a 250-mL separatory funnel and extract with two 25-mL portions of ether. Part of the ether used in the first extraction should be used to transfer the precipitated fatty acids to the separatory funnel. Combine the ether extracts, shake with 15 mL of saturated salt solution to reduce the water content, separate, then further dry the ether over a small amount of anhydrous magnesium sulfate.

Decant or filter the dried ether into a 125-mL Erlenmeyer flask and evaporate the solvent on a steam bath or a hot-water bath in the hood until about 10 to 15 mL remains. Transfer the ether to a tared 25-mL Erlenmeyer flask, and complete the solvent evaporation to obtain the fatty acids. Determine the yield and the percentage yield of the fatty acids, and show the calculations. When finished, hand in your product in a properly labeled container to your instructor.

30C.   CONCENTRATION OF CHOLESTEROL IN BLOOD

In this experiment, you will determine the concentration of cholesterol in human blood by using a visible spectrometer. Outdated blood from a blood bank may be used as the source. Human blood normally contains between 160 and 200 mg of cholesterol per 100 mL. About 25% of the cholesterol is free, and the rest is in the form of its fatty acid esters (cholesteryl stearate, etc.).

The absorption of visible light by cholesterol is the result of the cholesterol reacting with concentrated sulfuric acid and acetic anhydride (the Liebermann-Burchard reagent) to produce a transient purple color changing to blue and then green. The exact chemical structure of the colored species is still not known, but this color reaction is the method generally used to analyze cholesterol. Both free and esterified cholesterol react to give the color, so the intensity of the color is directly proportional to the concentration of **all** the cholesterol present.

In the experiment, the amount of absorbance on the spectrometer is determined by Beer's law,

$$A = alc$$

where A = absorbance on spectrometer

a = molar extinction coefficient of absorbing species, also given by ε; it will be a constant at a specified wavelength of light

l = path length in centimeters of light beam through solution of absorbing species

c = concentration of absorbing species in moles per liter

Note from the equation that if the path length and the specified wavelength of the light are held constant, then the absorbance of a solution should be directly proportional to the concentration of the absorbing species in the solution.

$$A = (\text{constant})(c)$$

A plot of the absorbance at a specified wavelength versus the concentration is therefore expected to yield a straight line.

In the experiment, the absorption of light by blood cholesterol will be graphed with the absorption of some standard cholesterol solutions. From its position on the graph, the concentration of the blood cholesterol can then be determined. To save time and labor, students should share in the preparation of the standard solutions and the reagents.

PROCEDURE

CAUTION: Traces of water can interfere with the results of this experiment. Be careful to use completely dry apparatus, solvents, and reagents, except where water is already present.

In addition, be very careful when using concentrated sulfuric acid, acetic anhydride, glacial acetic acid, and potassium hydroxide. Do not get them on your skin or breathe those with an odor.

Preparation of blood cholesterol

Centrifuge 2 mL of human blood for 10 minutes to separate the various components. Then transfer 1.0 mL of the blood plasma (the fluid portion) to a 25-mL volumetric flask containing 5.0 mL of freshly prepared alcoholic potassium hydroxide solution, made by adding 1.0 mL of 33% aqueous potassium hydroxide solution to 16 mL of absolute ethanol. Stopper, shake well, and incubate in a water bath at 37°C for 55 minutes.

Cool the solution to room temperature, add 10.0 mL of hexane and 5 mL of water, and vigorously shake for 1 minute. If an emulsion forms, centrifuge the liquid at slow speed until clear layers are produced. Transfer 2.0 mL of the upper hexane layer to a **labeled** dry test tube and evaporate the solvent in a steam bath or a hot-water bath.

Preparation of the standards

Place 5.0 mL of a standard solution containing 200 mg of cholesterol per liter of absolute ethanol into a 25-mL volumetric flask. Add 0.3 mL of aqueous 33% potassium hydroxide solution, 10.0 mL of hexane, and 6 mL of water. Vigorously shake the flask for 1 minute. If an emulsion forms, centrifuge as indicated above. After complete separation occurs, transfer **one** of the following amounts of the upper hexane layer to a **labeled** dry test tube and evaporate the solvent by heating in a steam bath or a hot-water bath: 2.0 mL, 3.0 mL, 4.0 mL, 5.0 mL, or 6.0 mL. These volumes correspond to 100 mg, 150 mg, 200 mg, 250 mg, and 300 mg of cholesterol per 100 mL of blood, respectively.

Color development

The Liebermann-Burchard reagent is prepared by adding 1 volume of concentrated sulfuric acid to 20 volumes of acetic anhydride, chilled to 10°C. The mixture is carefully stirred and kept cold for 10 minutes; then 10 volumes of glacial acetic acid are added with stirring. The reagent is allowed to warm up to room temperature, but it should be used within 1 hour.

Add 6.0 mL of the Liebermann-Burchard reagent to each of the test tubes containing the blood cholesterol and the five standards. Do this in a known sequence with a regular time interval of 1 minute. Stir the contents in each tube with a clean stirring rod and place the tubes in a 25° constant-temperature bath for 30 minutes.

Warm up the visible spectrophotometer and zero it for a wavelength of 620 nm. For the zeroing, place the Liebermann-Burchard reagent in the sample cell. Measure the absorbance of each cholesterol solution at 1-minute intervals in the same sequence in which they were prepared. Plot the absorbance versus the concentration of the five standard solutions on regular graph paper. Draw the best line possible for the five standard solutions, and from the absorbance of the blood cholesterol determine its concentration. How does this concentration compare with the normal concentration of cholesterol in blood?

Reference

Abell, Levy, Brodie, and Kendall, **Journal of Biological Chemistry, 195,** 357 (1952).

1. Why weren't melting points taken for the lecithins or the fatty acids after these compounds were isolated from egg yolks?
2. List the important classes of organic compounds found in the human body or necessary for the human body?
3. What organic functional groups are present in each of the following steroids?
   (a) Cholesterol (b) Androsterone
   (c) Testosterone (d) Estrone
   (e) Progesterone (f) Norlutin
   (g) Vitamin D_3 (h) Cortisone
   (i) Cholic acid
4. Which general types of steroids have the long aliphatic side chain? Which types do not?
5. How can people benefit when a hormone such as a sex or adrenal hormone is isolated from the body, has its structure determined with certainty, and is capable of being synthesized at reasonable cost in the laboratory?

EXPERIMENT

31

VITAMINS: THE ANALYSIS OF VITAMIN C IN COMMERCIAL TABLETS AND FRUIT JUICE

All the vital mechanisms, varied as they are, have only one object, that of preserving constant the conditions of life in the internal environment.

Claude Bernard, *Lessons on Reactions Common to Animals and Plants*

INTRODUCTION

Vitamins are organic compounds that must be supplied in very small amounts in the diets of animals and humans. If any vitamins are missing, normal growth and development are not possible and certain diseases, such as scurvy, rickets, and beriberi, can result.

These diseases have all been known for many centuries. But the idea that they might be due to a deficiency in the diet is relatively new. The discovery that the consumption of small amounts of citrus fruits could prevent scurvy didn't occur until 1757. And over 120 years passed before it was discovered that adding barley to a diet of polished rice could prevent beriberi. But these

559

discoveries still did not reveal the real reason for the successes, and a number of people all reached the wrong conclusions.

Then in 1911, Casimir Funk, a Polish biochemist, originated the "vitamine" theory of food substances essential for preventing certain diseases. He believed he had isolated the pure antiberiberi material from food and called it "vitamine," because it was an amine vital to life. Many of the vitamins known today are not amines, but the name has persisted, minus the final *e*. And even more important, Funk got people to start thinking the right way about food-related diseases.

In the beginning, the exact chemical names and structures of practically all the vitamins were unknown. In order to avoid the use of cumbersome terms such as "scurvy preventive factor," researchers used letters to designate the various vitamins. This system became so popular that it has persisted, despite the fact that the exact chemical names and structures of all the known vitamins have been completely established. In some cases, what was thought to be one vitamin was later shown to be a mixture of several vitamins. Thus the use of subscripts (e.g., vitamin B_1, vitamin B_2, etc.) was born.

All vitamins may be classified as either fat-soluble or water-soluble. Vitamins A, D, E, and K are fat-soluble, whereas all the B vitamins and vitamin C are water-soluble. Excess fat-soluble vitamins are stored in the liver of the body, but excess water-soluble vitamins cannot be stored. As a result, the B vitamins and vitamin C must be supplied more regularily in the diet.

All the functions of most of the vitamins are still not completely known or understood. But it is known that most vitamins become a component of various enzyme systems and thus are needed to catalyze certain specific reactions within the body. As a result, one vitamin cannot replace or act for another one. Because enzymes are not consumed by the reactions they promote, large amounts of vitamins are not needed and there is no advantage in giving extra amounts. However, consuming extra amounts from either vitamin-rich foods or one multi-vitamin tablet per day has been shown to be harmless. This is especially true of the water-soluble vitamins, which are eliminated by the body after it has taken what it needs. However, megadoses of the vitamins should be avoided.

Of course, all foods are not equally good sources of the various vitamins. Thus either a "balanced" diet or a multivitamin tablet is needed to avoid a vitamin deficiency. In addition, the commercial processing of food frequently either destroys or removes considerable amounts of vitamins. But the vitamins lost are frequently replaced by the processor. Some foods are also "fortified" with vitamins not normally present in them. Thus vitamin D is usually added to milk, and a mixture of vitamins may be added to certain breakfast cereals.

But even foods not processed commercially can lose their vitamin content. Many foods are rich in vitamins A and C, but cooking these foods will destroy these vitamins. In addition, water-soluble vitamins may be extracted from the food into the cooking water and thereby lost.

Table 31-1 gives the names, chemical structures, properties, effects, deficiency symptoms, and good sources of the commonly known vitamins. Study this table well; it has much valuable information.

TABLE 31-1.

Commonly Known Vitamins

Name	Chemical structure	Properties	Effects	Deficiency symptoms	Good sources
Vitamin A (retinol)	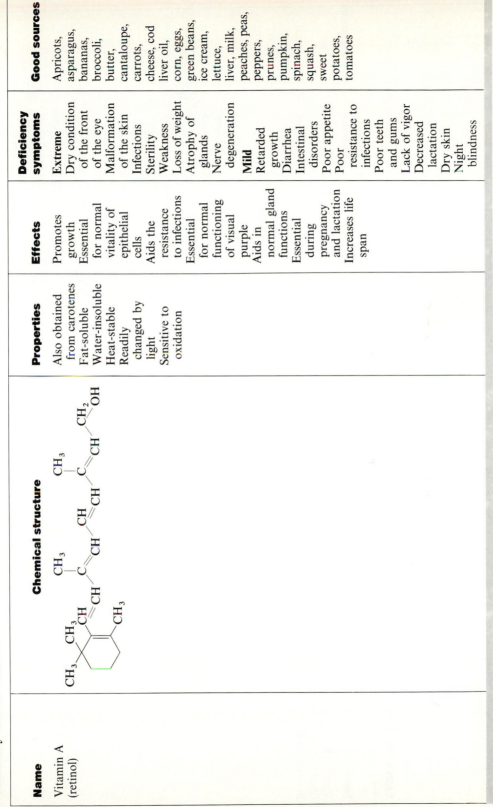	Also obtained from carotenes Fat-soluble Water-insoluble Heat-stable Readily changed by light Sensitive to oxidation	Promotes growth Essential for normal vitality of epithelial cells Aids the resistance to infections Essential for normal functioning of visual purple Aids in normal gland functions Essential during pregnancy and lactation Increases life span	**Extreme** Dry condition of the front of the eye Malformation of the skin Infections Sterility Weakness Loss of weight Atrophy of glands Nerve degeneration **Mild** Retarded growth Diarrhea Intestinal disorders Poor appetite Poor resistance to infections Poor teeth and gums Lack of vigor Decreased lactation Dry skin Night blindness	Apricots, asparagus, bananas, broccoli, butter, cantaloupe, carrots, cheese, cod liver oil, corn, eggs, green beans, ice cream, lettuce, liver, milk, peaches, peas, peppers, prunes, pumpkin, spinach, squash, sweet potatoes, tomatoes

TABLE 31-1 (continued)

Name	Chemical structure	Properties	Effects	Deficiency symptoms	Good sources
Vitamin B_1 (thiamine hydrochloride)		Water-soluble Fat-insoluble Somewhat stable to dry heat	Promotes growth Stimulates appetite Related to carbohydrate and fat metabolism Essential for normal condition and functioning of nerve tissue	**Extreme** Beriberi Polyneuritis Paralysis Muscle atrophy Gland atrophy Weight loss Convulsions Head retraction **Mild** Retarded growth Weakness Slow heart-beat Heart disturbances Poor appetite Digestive disturbances Decreased peristalsis Poor lactation Glandular disfunction Nervousness	Asparagus, bacon, bran, bread, cauliflower, corn, eggs, fish, grains, lettuce, liver, meat, milk, nuts, oranges, peas, pineapples, potatoes, prunes, soybeans, spinach, tomatoes

Vitamin	Properties	Function	Deficiency symptoms	Sources
Vitamin B_2 (riboflavin)	Water-soluble Heat-stable Fairly stable to oxidation Light-sensitive	Improves growth Promotes general health Prolongs active life span Essential in nerve tissues Essential to cell respiration	**Extreme** Weakness Loss of body weight Loss of intestinal function Breakdown of central nervous system Loss of hair Cataracts Seborrheic dermatitis **Mild** Disturbances to digestive processes Impaired growth Lack of vigor Shortened life span Poor lactation Impairment of tissue respiration	Asparagus, bacon, bananas, bran, broccoli, cauliflower, corn, cheese, eggs, fish, grains, grapefruit, legumes, lettuce, meat, milk, nuts, onions, peas, peppers, prunes, raisins, strawberries, spinach, tomatoes
Vitamin B_6 (pyridoxine)	Water- and alcohol-soluble Heat-stable Destroyed by light	Essential for utilization of fatty acids	**Extreme** Dermatitis Tissue lesions Ophthalmia Abscesses Diarrhea Epileptiform fits **Mild** Growth retardation Lack of muscle coordination	Cabbage, egg yolks, fish, legumes, meat, milk, wheat germ, whole grains

Structural formula for Vitamin B_2 (riboflavin):

$$CH_3,\ CH_3 \text{ (ring substituents)};\quad \text{side chain: } CH_2-(CHOH)_3-CH_2OH;\quad \text{with } O,\ NH,\ N \text{ ring atoms}$$

Structural formula for Vitamin B_6 (pyridoxine):

$$\text{pyridine ring with } CH_2OH,\ CH_2OH,\ HO,\ CH_3,\ N$$

TABLE 31-1 (continued)

Name	Chemical structure	Properties	Effects	Deficiency symptoms	Good sources
Vitamin B_{12} (cyanocobalamin)		Water-soluble	Necessary for the production of DNA	Pernicious anemia	Eggs, meat, milk, milk products
Biotin		Water-soluble Synthesized by intestinal bacteria, so deficiencies in humans are usually not seen	Necessary for healthy circulatory system and for maintaining healthy skin	Skin and nerve lesions	Eggs, organ meats, vegetables

Vitamin	Structure	Properties	Function	Deficiency symptoms	Sources
Folic acid		Water-soluble. Synthesized by intestinal bacteria, so deficiencies in humans are usually not seen	Essential for the manufacture of important structural components of body cells	Anemia. Impaired absorption of nutrients through the intestinal wall	Green leafy vegetables, meat, yeast
Niacin (nicotinic acid)		Soluble in alcohol and hot water	Promotes health. Promotes growth. Helps skin function normally. Helps gastrointestinal tract function normally	**Extreme** Pellagra (rough skin) Dermatitis Redness of tongue Wasting away of bone marrow Diarrhea Vomiting Insanity Salivation **Mild** Bright red skin color Mouth soreness Nervousness Indigestion Constipation Loss of appetite Nausea Headache Loss of weight	Bran, eggs, fish, meat, milk, oatmeal, peanuts, peas, peppers, potatoes, spinach, tomatoes, wheat germ, whole wheat bread
Pantothenic acid		Water-soluble. Deficiencies in people are unlikely	Necessary for the production of energy and the synthesis of several hormones	Graying of the hair Hemorrhages Skin inflammation Disturbances in growth	Egg yolks, grains, milk, milk products, organ meats, vegetables, yeast

TABLE 31-1 (continued)

Name	Chemical structure	Properties	Effects	Deficiency symptoms	Good sources
Vitamin C (ascorbic acid)		Water-soluble Fat-insoluble Destroyed by drying and by cooking in air, but not by cooking in steam Very sensitive to oxidation	Prevents scurvy Helps produce good teeth Stimulates growth Improves appetite Essential for glandular functions Protects vascular system Involved in defense against bacterial toxins Essential to tissue respiration	**Extreme** Scurvy Anemia Hemorrhages Swollen joints Swollen gums Loose teeth Fragile bones Sterility Respiratory and intestinal infections Muscle atrophy Gastric ulcers **Mild** Retarded growth Defective teeth Poor bone knitting Tender joints Headaches Poor resistance to infections Digestive disturbances Weakened blood capillaries Restlessness Weakness Poor lactation	Asparagus, bananas, broccoli, beans, cantaloupe, cauliflower, citrus fruits, clams, corn, lettuce, peaches, peas, pineapples, potatoes, raspberries, rhubarb, spinach, squash, strawberries, tomatoes

Vitamin	Structure	Properties	Functions	Deficiency	Sources
Vitamin D	Provitamin D—ergosterol Vitamin D_2—calciferol—synthetic Vitamin D_3—naturally occurring	Fat-soluble Water-insoluble Stable to heat and oxidation	Regulates metabolism of calcium and phosphorus Essential to normal bone growth and tooth development	**Extreme** Rickets Enlarged joints Curved spine Softened bones Beaded ribs Porosity of bones Severe loss of calcium and phosphorus Lesions in bones and teeth Retarded growth **Mild** Poor assimilation of calcium and phosphorus Low levels of blood calcium, phosphate, and phosphatase Poor deposition of calcium and phosphorus in teeth and bones "Bow legs" Lack of vigor Tooth cavities Delayed closing of fontanelles Restlessness Poor growth	Beef, butter, cheese, chocolate, coconut, corn oil, egg yolks, fish, ice cream, vitamin D milk

TABLE 31-1 (continued)

Name	Chemical structure	Properties	Effects	Deficiency symptoms	Good sources
Vitamin E (α-tocopherol)		Fat-soluble Water-insoluble Stable to heat	Deficiencies detected only in laboratory animals fed a special diet	**Extreme** Sterility Growth retardation Muscle dystrophy Degenerative diseases of nervous system **Mild** Low fertility Poor lactation Muscular weakness	Eggs, meat, milk, nearly all green leaves and whole grains, vegetable oils
Vitamin K (phylloquinone)		Fat-soluble Water-insoluble Heat-stable Easily changed by strong acids and oxidizers	Essential for producing clotting of blood and preventing hemorrhage	**Extreme** Hemorrhage Anemia **Mild** Prolonged bleeding from delayed coagulating of blood	Cauliflower, egg yolks, rice, spinach, soybean oil, tomatoes

BACKGROUND

In this experiment, you will determine the amount of vitamin C present in a commercial tablet and in fruit juice. You will also determine how exposure to air and heat affects the amount of vitamin C present. This is accomplished by the following titration reaction:

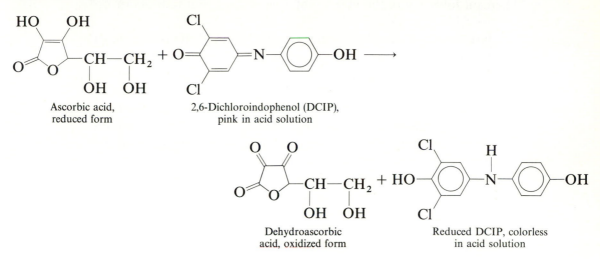

Ascorbic acid,
reduced form

2,6-Dichloroindophenol (DCIP),
pink in acid solution

Dehydroascorbic
acid, oxidized form

Reduced DCIP, colorless
in acid solution

Observe that 2,6-dichloroindophenol (DCIP) is pink in acid solution, but becomes colorless in acid solution when reduced. The titration can thus be monitored by the presence or lack of the pink color, with the disappearance of the pink color indicating the endpoint. The acidity is provided by mixing the crushed vitamin C tablet and the fruit juice with 1% oxalic acid solution. Oxalic acid is used over other possible acids because it retards the oxidation of ascorbic acid by oxygen in the air.

In the beginning, the DCIP solution will be blue, since the compound is also an acid-base indicator, having a blue color above pH 3. However, the acidity of the added vitamin C solution soon changes the color to pink by the following reaction:

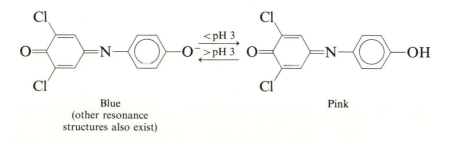

Blue
(other resonance
structures also exist)

Pink

Thus the total overall color change in the titration will be from blue to pink to colorless.

First isolated in 1928, vitamin C for many years has been widely known as the necessary substance for the prevention of scurvy. In addition, it received wide publicity beginning in 1970 when Nobel Prize laureate Linus Pauling advocated large doses of it to curb the frequency and severity of the common cold. Dr. Pauling appears to be partially correct. Although controlled studies have shown that large doses of vitamin C do not reduce the incidence of colds, there is a 30 percent lowering of the number of days either spent indoors or on sick leave from work.

In addition, large doses of vitamin C seem to reduce the amount of cholesterol in the blood. It can therefore help to prevent heart attacks and strokes by alleviating atherosclerosis, a narrowing and hardening of the inner walls of arteries.

Furthermore, the vitamin has been shown to play a key role in the prevention of wound reopening after surgery. In fact, patients deficient in vitamin C have a wound reopening rate eight times greater than normal. The exact reason is still not clear, but it is known that there is a 20 percent average drop in the vitamin level immediately after surgery.

Vitamin C toxicity is not a problem, but doses of 1 to 5 grams per day can produce some negative effects. Among these are the destruction of vitamin B_{12}, the production of kidney stones, acidosis, hyperglycemia, and gastrointestinal disturbances. A moderate intake of from 0.2 to no more than 1.0 gram per day therefore seems to be called for.

PROCEDURE

31A.   ANALYSIS OF A COMMERCIAL VITAMIN C TABLET

Crush either two 100-mg or one 250-mg vitamin C tablet on a small piece of waxed paper with a spatula. Transfer the powder to a 250-mL Erlenmeyer flask and add about 200 mL of a 1% solution of oxalic acid in distilled water. Stir to dissolve the solid as much as possible, then filter into a 250-mL volumetric flask. Wash any remaining residue in the Erlenmeyer flask with additional 1% oxalic acid solution and filter. Bring the volume in the volumetric flask up to the 250-mL mark by adding more 1% oxalic acid solution, stopper the flask, and mix the entire solution.

Rinse a clean buret with a small amount of the vitamin C solution; then discard the rinse solution. Also rinse a second clean buret with a small amount of the provided DCIP solution; then discard this rinse solution. Fill each buret with the respective, appropriate solution and add 60.0 mL of the DCIP solution to a clean 250-mL Erlenmeyer flask, refilling the DCIP buret as needed. Place this Erlenmeyer flask on a piece of white paper; then titrate the DCIP solution by adding your vitamin C solution until the pink color disappears. Be sure to adequately swirl the Erlenmeyer flask after each vitamin C solution addition and to slow down the rate of addition as the endpoint is approached.

One milliliter of the DCIP solution is equivalent to about 0.30 mg of vitamin C. The reagent bottle will tell you the exact figure. From the amount of DCIP solution and vitamin C solution used in the titration, compute the concentration of the vitamin C solution in milligrams per milliliter. Multiply by the total volume of the vitamin C solution to determine the total amount of vitamin C present in the tablet or tablets. Compare this with the stated amount that is supposed to be present. Record your result.

31B. ANALYSIS OF VITAMIN C IN FRUIT JUICE

Filter 200 mL of a citrus juice, such as orange or grapefruit juice, to remove the pulp. The juice may come from either the fresh fruit itself, the frozen concentrate, or bottled or canned juice. Juice that is freshly prepared or recently opened and kept cold provides the best results. Alternatively, a fruit drink that is not red, such as Hi-C apple drink, may be used without filtration.

To test the effect of air oxidation over a period of time, stir 80 mL of the juice for an hour or longer in an open beaker, using a magnetic stirrer. Also, to test the effect of heat, warm another 80 mL of the juice at 70°C or more in a hot-water bath. Do not allow the solution to boil away.

Transfer 20.0 mL of the unstirred, unheated juice to a 100-mL volumetric flask. Dilute to the 100-mL mark with 1% oxalic acid solution and titrate exactly 5.00 mL of the standardized DCIP solution with the diluted fruit juice in the same manner as in Part A. From your results, calculate the concentration of vitamin C in the fruit juice in milligrams per milliliter in the same manner as in Part A. From this concentration, calculate the amount of vitamin C present in a 6-fluid-ounce glass (177 mL) of the fruit juice. What percentage is this of the minimum recommended daily allowance (RDA) of 60 mg per day of vitamin C?

Next transfer 40.0 mL of the heated fruit juice to a 100-mL volumetric flask. Dilute to the 100-mL mark with 1% oxalic acid solution and titrate exactly 1.00 mL of the standardized DCIP solution with this diluted fruit juice in the same manner as in Part A. From your results, calculate the concentration of vitamin C remaining in the heated fruit juice in milligrams per milliliter. What percentage of the original vitamin C has been destroyed by the heating?

Next transfer 20.0 mL of the stirred fruit juice to a 100-mL volumetric flask. Dilute to the 100-mL mark with 1% oxalic acid solution and titrate 2.00 mL of the standardized DCIP solution with this diluted fruit juice. From your results, calculate the concentration in milligrams per milliliter of vitamin C remaining in the stirred fruit juice. What percentage of the original vitamin C in the fruit juice has been destroyed by the stirring?

QUESTIONS

1. Why would raw rather than cooked tomatoes be better for you nutritionally? Be specific.
2. How would your results have changed if you had diluted the vitamin C in the tablet and the fruit juice with distilled water instead of 1% oxalic acid solution?
3. What was the purpose of the white paper used in all the titrations?
4. Why does a 250-mg vitamin C tablet actually weigh more than 250 mg?

EXPERIMENT

32

THE CHEMISTRY OF MILK

Things are seldom what they seem;
Skim milk masquerades as cream.

W. S. Gilbert, *H. M. S. Pinafore,* Act II

INTRODUCTION

Milk is often called the almost perfect food, because it is the most nutritionally complete food found in nature. This property is vital, since milk is the only food consumed by most young mammals during the nutritionally critical period following birth.

All kinds of milk, human or animal, contain vitamins, minerals, proteins, carbohydrates, and fats. But the amounts of these nutrients differ greatly, as can be seen from Table 32-1. Cows' milk and goats' milk are almost identical in every respect. But human milk contains less than half of the proteins and minerals of cows' or goats' milk, but almost $1\frac{1}{2}$ times as much sugar. Horses' milk is quite low in proteins and fats, compared with the others, whereas reindeer milk is very high in proteins, fats, and minerals, but quite low in carbohydrates.

TABLE 32-1.

Average Percentage of Nutrients in the Milk of Various Mammals

Nutrient	Human	Cow	Goat	Sheep	Horse	Reindeer
Proteins	1.4	3.5	3.3	5.5	2.0	9.8
Fats	4.0	3.8	4.2	6.5	1.1	17.0
Carbohydrates	7.0	4.8	4.8	4.5	5.9	2.5
Minerals	0.2	0.7	0.7	0.9	0.4	1.2

One quart of milk supplies about 660 calories or roughly one-fifth of the calories required daily by an average size man, one-fourth of the calories required daily by an average size woman, and one-half the calories needed by a two-year-old child. A quart of milk supplies adults with their complete daily requirement of riboflavin, calcium, and vitamin D, and with well over half of their daily requirement of protein, phosphorus, and vitamin B_{12}. The only important nutrients lacking in milk are iron and vitamin C, but these can be obtained from other foods or specially fortified milk.

In the United States, the average consumption per person of fluid milk and cream is over 150 quarts, or 300 pounds, per year. In addition to this is 250 pounds of dairy products, consisting of cheese, butter, ice cream, yogurt, sour cream, and whipping cream, giving an almost unbelievable total of 130 billion pounds each year.

Almost all milk consumed in the United States today is pasteurized and homogenized. During pasteurization, which kills disease-causing bacteria, the milk is heated to at least 161°F for at least 15 seconds, and then immediately cooled. Homogenization breaks up the fat globules and distributes the fat throughout the milk so that it doesn't rise to the top. The process involves forcing the milk through tiny openings at a pressure of 2000 pounds per square inch, and gives homogenized milk the same cream content throughout. Most milk is also fortified with vitamin D, and some of it is also fortified with other vitamins and minerals. All these added nutrients are indicated on the container.

People who don't care for the taste of milk can still receive its nutritional benefits by using it in cooking or by consuming dairy products such as cheese, butter, ice cream, yogurt, sour cream, and whipping cream.

Literally hundreds of different types of cheese are enjoyed around the world. This great variety is possible due to the many different microorganisms used to separate the **curd** (the solid part of the milk) from the **whey** (the liquid part), along with variations in the type of milk used, the aging time, seasonings, the temperature, and the amount of salt added or the amount of moisture drained from the curd. Cottage cheese is made by adding the enzyme rennin (found in the fourth stomach of young calves) to pasteurized skim milk.

Butter is made by separating cream from milk through centrifugal action, and then pasteurizing, tempering, and churning or mechanically agitating the cream. Churning causes the fat globules in the cream to cluster together in masses and separate from the liquid. The separated butter is then washed in cold water, drained, and lightly salted. The original liquid from the separated butter is drained and sold as buttermilk.

Yogurt is a smooth, semisolid product with a high acid content and a thick curd. It is commonly made by adding to milk the bacteria *Streptococcus lactis* and *Lactobacillus bulgaricus,* which cause the milk to thicken and ferment by changing the milk sugar into lactic acid. The resulting product can either be eaten as is or flavored with fruit.

32A. ISOLATION OF CASEIN FROM MILK

In this experiment, you will isolate casein from milk. Casein is the principal protein found in milk: about 80% of all milk protein is casein. The other proteins are lactalbumin, which resembles blood albumin, and lactoglobulin, which is believed to be the same as serum globulin in the blood. Nutritionists call these proteins "complete proteins" because they contain all the amino acids essential for building blood and tissue and can sustain life and provide normal growth even if they are the only proteins in the diet (see Experiment 34). These proteins not only contain more amino acids than plant proteins, but they contain greater amounts of amino acids than the proteins in eggs and meats.

Casein, which as a protein is found only in milk, consists of many different amino acids linked together to form a large polypeptide structure. It exists in milk in the form of a water soluble calcium salt, which can be easily precipitated with acids according to the equation

$$\text{Ca—caseinate} + 2\,H^+ \longrightarrow \text{casein}\downarrow + Ca^{+2}$$

The other proteins remain water-soluble in acidic solution, making the isolation of casein a simple task. They can also be precipitated and isolated by merely heating the acidic solution and filtering. The isolated casein is insoluble in water, alcohol, and ether, but dissolves in alkaline and some acidic solutions.

Casein is isolated from milk commercially and is industrially important because after dissolving in alkaline solutions and drying, it becomes a sticky substance that can be used in glues, the coating of paper, and the binding of colors in paints and wallpaper. It is also used as a coating for fine leather, and is cured with rennet to produce a plastic material used for buttons. When isolated under sanitary conditions and dissolved in alkaline solutions, casein also is employed in the manufacture of pharmaceutical and nutritional products. As research progresses, the list of applications will certainly continue to grow.

PROCEDURE

Dissolve 10 grams of powdered nonfat dry milk in 40 mL of warm water in a 150-mL beaker. With the temperature kept at 40–45°C, add **dropwise** a solution of 10% acetic acid, with stirring. Do **not** add the acid too rapidly. Continue the addition until the liquid changes from milky to clear and the casein no longer separates. It is important **not** to add too much acid, because it may hydrolyze some of the lactose in the milk to glucose and galactose. Stir the casein until it forms a large amorphous mass; then remove the mass with a stirring rod or tongs and place it in another beaker. Immediately add 1.5 grams of powdered calcium carbonate to the original beaker containing the remaining liquid, stir for a few minutes, and save the resulting mixture for the later separation of lactose in Experiment 32B. The separation of lactose should be done as soon as possible during the same laboratory period.

Suction filter the mass of casein to remove as much water as possible, pressing the solid with a spatula. Place the casein in a small beaker and add 10 mL of ethyl ether. Stir the casein in the ether for a few minutes, decant the ether, and repeat the process with a second 10-mL portion of ether. This removes any small quantities of fat that may have precipitated with the casein. Place the casein between several layers of paper towels to help dry the product and allow it to dry for several days. Submit the dried, weighed casein in a labeled bottle to your instructor, but save part of it for the protein experiment, 34A. Determine the percentage yield of casein from the powdered milk, and show the calculations.

32B.   ISOLATION OF LACTOSE FROM MILK

In this experiment, you will isolate lactose from the whey of the casein experiment. Lactose, frequently called milk sugar, is the only carbohydrate synthesized by mammals and, unlike other sugars, is practically tasteless. The compound, as shown in Figure 32-1, is a disaccharide composed of one molecule of glucose and one molecule of galactose and is capable of existing in three forms: α-lactose, β-lactose, and the free aldehyde. These structures differ in the stereochemistry and the functional carbon group at position 1 of the glucose portion. The α anomer is easily obtained by crystallization from a water-ethanol mixture at room temperature. But to obtain the β anomer, crystallization must occur from concentrated solutions at temperatures over 94°C.

The free aldehyde exists in very small amounts as an intermediate whenever there is an interconversion of the α and β anomers in solution. The interconversion, called a **mutarotation**, can be monitored by the change in optical rotation of the two anomers (see Experiment 33E). For example, at 20°C α-lactose is converted largely to β-lactose in aqueous solution, with a change in optical rotation from $+92.6$ to $+52.3°$. Since the specific rotation of pure β-lactose is $+34.2°$, the equilibrium mixture of the two in aqueous solution must contain 69% of β-lactose at this temperature.

Glucose is frequently referred to as blood sugar or grape sugar. Galactose, whose only nutritional source is lactose, is found in brain and nerve tissue and is apparently needed by the developing infant to build brain and nerve tissue.

Although practically all infants have the ability to digest lactose, unfortunately some adults lose this ability later in life. An enzyme called lactase is necessary for the digestion of lactose. Lactase cleaves the molecule into glucose and galactose. When this enzyme is no longer present, the lactose remains in the digestive tract, where it causes cramps and diarrhea. People inflicted with this condition must limit their intake of milk to no more than one glass per day.

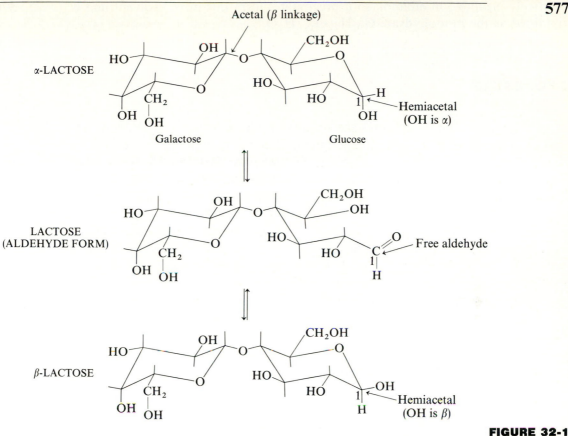

FIGURE 32-1
The three structural forms of lactose and their interconversions.

Lactose and bacteria called lactobacilli cause milk to sour. The lactose molecule is hydrolyzed by microorganisms into glucose and galactose. Once galactose has been formed, lactobacilli convert it to the sour-tasting lactic acid. This process, shown in the following equations, can be slowed by keeping the milk cold or can be stopped entirely by converting the milk into powder form. But once redissolved in water, the milk can again sour.

$$C_{12}H_{22}O_{11} + H_2O \longrightarrow C_6H_{12}O_6 + C_6H_{12}O_6$$

Lactose   Galactose   Glucose

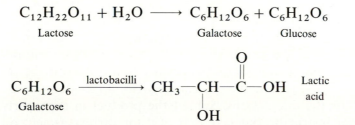

In this experiment, the α anomer of lactose will be isolated by the easier crystallization from a water-ethanol mixture at room temperature. Lactose is relatively insoluble in ethanol, so it precipitates from this mixed solvent and is obtained as the monohydrate $C_{12}H_{22}O_{11} \cdot H_2O$.

PROCEDURE

Gently boil for about 5 minutes the original liquid to which the calcium carbonate was added in the casein experiment. This will precipitate the remaining proteins lactalbumin and lactoglobulin. **(CAUTION: BUMPING AND THE RESULTING LOSS OF LIQUID WILL OCCUR IF THE LIQUID IS NOT CONSTANTLY AND VIGOROUSLY STIRRED. DO NOT ALLOW THE HOT LIQUID TO BUMP ONTO YOU AND BURN YOU.)** Suction filter the hot mixture to remove the proteins and calcium carbonate, and transfer the hot filtrate to a 125-mL Erlenmeyer flask. If necessary, concentrate the filtrate to a volume of about 10 mL by heating with stirring, again being careful that no hot liquid bumps onto you. Vigorous boiling may cause the mixture to foam out of the flask. You can control this foaming by heating the liquid less vigorously and gently blowing onto it. **Do not put your face directly over the boiling liquid.**

Turn off the heat. To the hot, concentrated solution add 50 mL of hot 95% ethanol **(heated without a flame)** and about 0.4 gram of decolorizing carbon. Put this mixture aside and prepare a slurry of about 2 grams of filter aid (Celite) and 15 mL of 95% ethanol. Suction filter the slurry in a Hirsch funnel to obtain a filter pad, and discard the alcohol in the filter flask. This filter pad prevents fine particles in the hot ethanol mixture from either passing through the filter paper or clogging the filter paper and stopping the filtration.

Stir the **slightly** cooled ethanol mixture and then suction filter it through the Celite filter pad, making sure the filtrate is clear. Otherwise, the lactose will likely not crystallize later. If necessary, reheat the solution, add 0.3 gram of Celite, and refilter by gravity. Rapidly occurring cloudiness, however, may result from premature crystallization of the lactose. If this happens, add a small amount of water to the remaining unfiltered mixture and reheat before continuing the filtration.

Pour the filtrate into a 125-mL Erlenmeyer flask, stopper it, and allow it to stand overnight or until the next laboratory period for crystallization.

Collect the crystals of lactose by suction filtration, and wash the product with a small amount of cold 95% ethanol. Thoroughly dry the lactose and determine its weight, melting point, and literature melting point. Record the results in your notebook. Also determine the percentage yield of lactose from the powdered milk, and show the calculations. Submit the product in a properly labeled container to your instructor, but save part of it for the experiment on carbohydrates.

QUESTIONS

1. Give two reasons why it is best to use powdered, nonfat milk rather than homogenized whole milk in this experiment. (Cost is not one of the answers.)
2. Why does homogenized whole milk or skim milk sour much more rapidly than powdered milk? (The same powdered milk dissolved in water is also susceptible to souring quickly.)
3. From the experiment, what component of milk is responsible for its white color?
4. Why is milk not considered to be a solution?
5. Most people think of bacteria as causing disease, and milk products are pasteurized to kill these disease-causing bacteria. Yet certain milk products like cheese and yogurt are made by deliberately adding certain types of bacteria. How can you explain this apparent contradiction?
6. Lactose has only a slightly sweet taste. Why is this fact important for infants who consume mostly milk during the early part of their life?
7. Why is lactose much more soluble in water than in ethanol?

EXPERIMENT

33

CARBOHYDRATES

A loaf of bread, the Walrus said
Is what we chiefly need.
Pepper and vinegar besides
Are very good indeed.

Lewis Carroll, *Through the Looking Glass*, Chapter 4

INTRODUCTION

All of us owe our existence to plants and a process occurring within the plant, called **photosynthesis**. In this process, carbon dioxide and water are converted into oxygen and **sugars, starches**, and **cellulose**, which are all known as **carbohydrates**. As sugars and starches, carbohydrates are the fundamental source of much of our food. Sugar cane and sugar beets provide us with **sucrose** or table sugar. Grapes, bananas, apples, and other fruits supply us with other sugars such as **glucose** and **fructose**. Wheat, corn, rice, potatoes, and other grains and vegetables are our primary source of **starch**. The meat, milk, and fat we obtain from animals originally came from the carbohydrates in plants. Even the food value of alcoholic beverages originates from the sugars and starches in plants.

But carbohydrates furnish us with much more than food. Much of the fabric produced for clothing, sheets and pillowcases, draperies, and tablecloths comes from **cellulose** in the form of cotton, linen, rayon, and cellulose acetate. **Cellophane**, to wrap many of the goods we buy, comes from cellulose, as well as all the paper we use for books, newspapers, writing, currency, bags, and boxes. A large portion of our homes and furniture is made from cellulose in the form of wood, which can also keep us warm and cook our food when used in fireplaces. Starch, in addition to being necessary for food, is useful in the laundry and in the production of many glues and adhesives.

581

Carbohydrates may be defined as **polyhydroxy aldehydes, polyhydroxy ketones, or compounds that can be hydrolyzed to them**. But this definition is not entirely satisfactory, since we shall later find that they exist primarily as **hemiacetals, acetals, hemiketals**, and **ketals**. All carbohydrates contain the elements carbon, hydrogen, and oxygen in a ratio of roughly 1:2:1. The 2:1 ratio of hydrogen to oxygen is the same as in water, hence the name carbo**hydrate**. Although carbohydrates contain no actual water in the molecule, most of the carbons have a hydrogen and a hydroxyl group attached, and the elements of water can be removed as water by a reagent such as concentrated sulfuric acid. Black carbon is left as a residue. Many of you perhaps have seen this dramatic, volcanolike reaction when sulfuric acid is poured over a mound of sugar.

Carbohydrates can be simple organic compounds having as few as three carbons, or complex polymeric substances with over a million carbons and molecular weights up to 100,000,000. But even the simple compounds can prove complex in structure determination, because of numerous chiral centers that give a vast array of possible stereoisomers.

Whenever possible, practically all carbohydrates prefer to have the atoms arranged in five- or six-membered rings. If only one ring is present, the carbohydrate is called a **monosaccharide**. If there are two rings, the carbohydrate is a **disaccharide**, and with many rings it becomes a **polysaccharide**. Nearly all sugars are examples of monosaccharides and disaccharides, whereas starch and cellulose are polysaccharides. The three-dimensional structures of some of the most important carbohydrates are given in Table 33-1.

As can be seen, **(+)-glucose** (also called dextrose) is a monosaccharide, existing primarily as the cyclic hemiacetal. It is by far the most abundant and the most important monosaccharide, since starch, cellulose, and glycogen all contain many (+)-glucose units linked together.

(−)-Fructose is a hemiketal, which occurs widely in fruits and combined with (+)-glucose in the disaccharide (+)-sucrose.

Of all organic chemicals, **(+)-sucrose** as table sugar is produced in the largest amount in pure form.

(+)-Lactose, discussed in the previous experiment as the carbohydrate present in milk, is a disaccharide consisting of (+)-glucose and (+)-galactose.

(+)-Maltose as a disaccharide consists of two molecules of (+)-glucose and is formed in one stage of the fermentation of starch to ethyl alcohol.

Cellulose is the chief component of wood and plant fibers and is the most abundant organic compound of natural origin on the earth. As a polysaccharide, it contains many (+)-glucose units with β anomer linkages. The term **anomer** refers to the stereochemistry or the direction in space of the oxygen atom at the carbon atom where cyclization has occurred (see Table 33-1 and Figure 33-1). Thus for (+)-glucose, the α anomer has this oxygen pointing downward, and the β anomer has it pointing to the right.

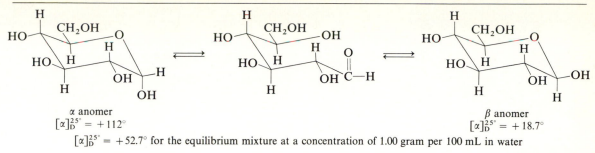

α anomer
$[\alpha]_D^{25°} = +112°$

β anomer
$[\alpha]_D^{25°} = +18.7°$

$[\alpha]_D^{25°} = +52.7°$ for the equilibrium mixture at a concentration of 1.00 gram per 100 mL in water

FIGURE 33-1
Mutarotation of glucose in water.

Starch is a mixture of two carbohydrates. About 20 percent of the mixture is a water-soluble fraction called **amylose**, and the other 80 percent is a water-insoluble fraction called **amylopectin**. Both these molecules contain many (+)-glucose units with α anomer linkages. Amylose has all the units forming a straight chain, but in amylopectin there are a number of branched chains, with a new branch occurring about once every 20 to 30 glucose units.

Structurally related to amylopectin is **glycogen**, which is animal starch commonly found in the liver and muscle cells. Glycogen serves as a form of reserve energy for the animal. Glycogen also contains branched chains of (+)-glucose units with α anomer linkages, but the branches occur about two to three times more frequently than in amylopectin.

Comparing the structures of cellulose and starch, some of you may be wondering why humans can digest starch but not cellulose, as termites do. The answer lies in the way all the (+)-glucose units are linked together. In every form of starch, the stereochemistry of the linkages is the α anomer, whereas in cellulose the stereochemistry is the β anomer. In order for starch or cellulose to be digested, the linkages must be broken. And although we possess an enzyme in our body capable of breaking the α anomer linkage of starch, no enzyme within us can break the β anomer linkage of cellulose. Termites, however, possess in their intestinal tracts microorganisms with an enzyme capable of breaking this linkage.

You may have noted that the structures of all the carbohydrates had the designation of each ring as an α or β anomer. This is necessary because, as a hemiacetal or hemiketal, the ring is capable of opening to yield a small amount of the hydroxy aldehyde or hydroxy ketone and then reclosing to yield a ring of differing stereochemistry in a reaction called **mutarotation**. For example, if a pure sample of the α anomer of glucose is dissolved in water, as shown in Figure 33-1, the β anomer will be formed through ring opening to yield an equilibrium mixture of both. The same equilibrium mixture would also be formed if the pure β anomer were dissolved in water. This possible ring open-

TABLE 33-1.

Structures of Various Carbohydrates

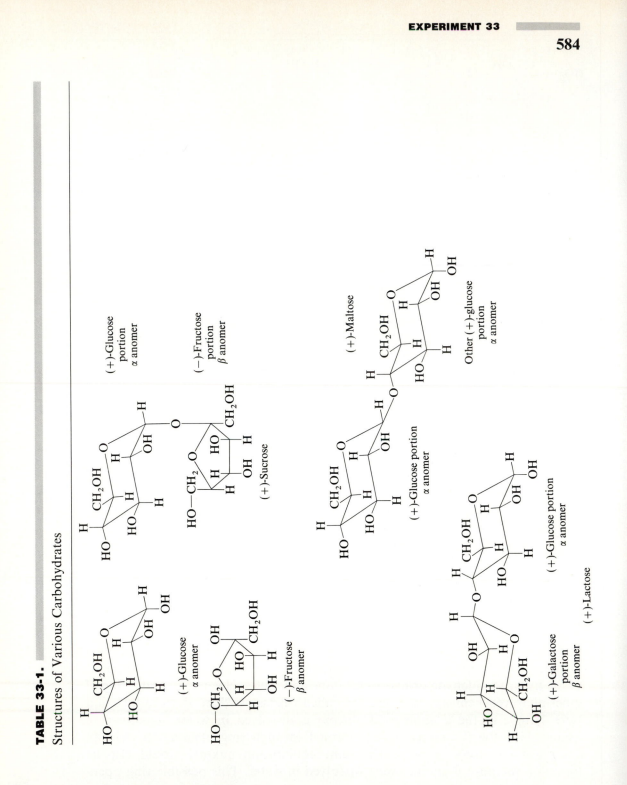

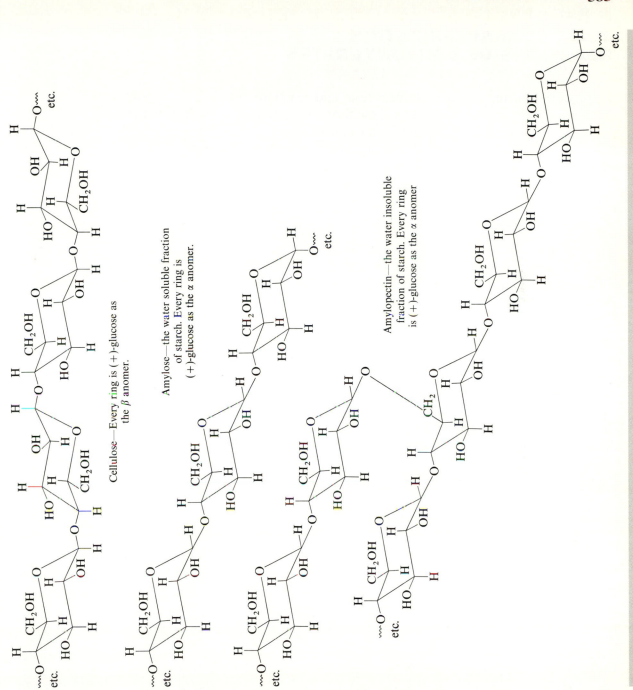

Cellulose—Every ring is (+)-glucose as the β anomer.

Amylose—the water soluble fraction of starch. Every ring is (+)-glucose as the α anomer.

Amylopectin—the water insoluble fraction of starch. Every ring is (+)-glucose as the α anomer

ing of carbohydrates is responsible for several of their chemical reactions, as will be observed later.

33A. CHARACTERIZATION REACTIONS OF CARBOHYDRATES

In this experiment, you will perform tests and reactions on various carbohydrates and food products. Many of these tests will be found to be characteristic for only certain types of carbohydrates and therefore will enable you to distinguish the different structural features of the molecules. Hopefully, you will be able from the results to determine which carbohydrates are present and absent in the various food products.

1. MOLISCH TEST

The Molisch test is a general test for **all** monosaccharides having either five or six carbons. The test is based on the ability of the monosaccharides to undergo acid-catalyzed dehydration to yield either furfural or 5-hydroxymethylfurfural.

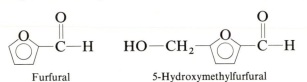

Furfural   5-Hydroxymethylfurfural

Even disaccharides and polysaccharides will give the test when they are hydrolyzed with acid to yield 5- or 6-carbon monosaccharides. The Molisch reagent is a solution of α-naphthol in alcohol, and yields with the carbohydrate a purple color by the following reactions:

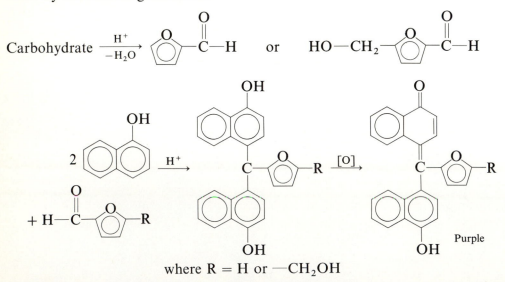

where R = H or —CH$_2$OH

Procedure

Place 2 mL of each of the following 1% carbohydrate solutions in 10 separate, **labeled** test tubes: glucose, fructose, xylose, arabinose, galactose, sucrose, lactose, maltose, starch (shaken), and glycogen. Also place 2 mL of distilled water in another tube to serve as a control. Add 1 drop of the Molisch reagent to each tube and thoroughly mix. Tilt each tube slightly and pour, slowly and carefully, about 2 mL of concentrated sulfuric acid down the side to form a layer below the carbohydrate solution. **(CAUTION: CONCENTRATED SULFURIC ACID IS VERY CORROSIVE. HANDLE WITH CARE.)** Note and record in chart form the color at the interface between the two layers for each solution. A purple color indicates a positive test.

2. BIAL'S TEST

Bial's test differentiates **5**-carbon from **6**-carbon sugars. Furfural is formed from the dehydration of a **5**-carbon sugar and reacts, as shown below, with orcinol and ferric chloride to yield a blue condensation product. 5-Hydroxymethylfurfural is formed from the dehydration of a **6**-carbon sugar and reacts, as shown below, to yield a brown, reddish-brown, or green condensation product. Disaccharides and polysaccharides will also give the test after acid hydrolysis.

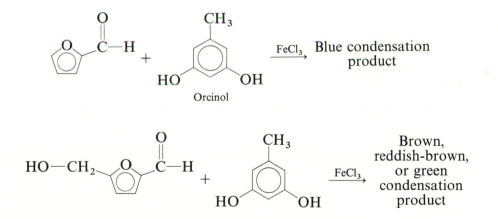

Procedure

Place 1 mL of each of the following 1% carbohydrate solutions in 10 separate, **labeled** test tubes: glucose, fructose, xylose, arabinose, galactose, sucrose, lactose, maltose, starch (shaken), and glycogen. Also place 1 mL of distilled water in another tube to serve as a control. Add 1 mL of Bial's orcinol reagent to each tube and heat the tubes in a boiling-water bath until a color develops. Note and record for each solution the color formed, if any, the time required, and the information obtained from the test about the structure of the carbohydrate.

3. SELIWANOFF'S TEST

Seliwanoff's test differentiates a **ketohexose** (a 6-carbon sugar containing a ketone in the open-chain form) from an **aldohexose** (a 6-carbon sugar containing an aldehyde in the open form). A ketohexose dehydrates rapidly to yield 5-hydroxymethylfurfural, whereas an aldohexose dehydrates more slowly. Once 5-hydroxymethylfurfural is produced, it reacts, as shown below, with resorcinol to yield a deep red condensation product.

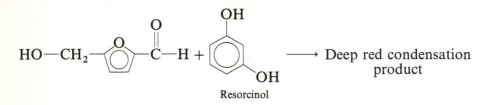

Resorcinol

Procedure

Place 1 mL of each of the following 1% carbohydrate solutions in 10 separate, **labeled** test tubes: glucose, fructose, xylose, arabinose, galactose, sucrose, lactose, maltose, starch (shaken), and glycogen. Also place 1 mL of distilled water in another tube to serve as a control. Add 2 mL of Seliwanoff's reagent to each tube and heat the tubes in a boiling-water bath until a color develops. Note and record for each solution the color formed, if any, the time required, and the information obtained from the test about the structure of the carbohydrate.

4. BENEDICT'S TEST

Benedict's test determines whether a monosaccharide or disaccharide contains a potential aldehyde group that can be oxidized to a carboxylic acid. The sugar in turn reduces the cupric ion in Benedict's reagent to a red precipitate of cuprous oxide by the equation

$$\underset{\text{}}{R-\overset{\overset{\text{O}}{\|}}{C}-H} + 2\,Cu^{+2} + 4\,OH^- \longrightarrow R-\overset{\overset{\text{O}}{\|}}{C}-OH + \underset{\substack{\text{Red}\\\text{precipitate}}}{Cu_2O} + 2\,H_2O$$

The sugar is therefore called a **reducing sugar**. To give a positive test, the sugar must contain a hemiacetal, which upon ring opening will give the aldehyde. For example:

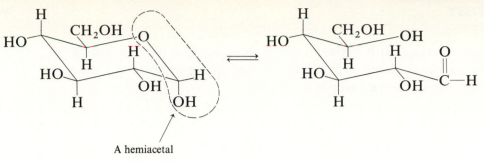

A hemiacetal

Procedure

Place 1 mL of each of the following 1% carbohydrate solutions in 10 separate, **labeled** test tubes: glucose, fructose, xylose, arabinose, galactose, sucrose, lactose, maltose, starch (shaken), and glycogen. Also place 1 mL of distilled water in another tube to serve as a control. To each tube, add 5 mL of Benedict's reagent and heat the tubes in a boiling-water bath for 5 minutes. Remove the tubes and note and record the results. A precipitate, varying in color from red to brown or yellow, constitutes a positive test. Any color change with no precipitate is a negative test. Also record for each carbohydrate the information obtained from the test about the structure of the carbohydrate.

5. BARFOED'S TEST

Barfoed's test is similar to Benedict's test, but determines if a reducing sugar is a monosaccharide or a disaccharide. Barfoed's reagent reacts with monosaccharides to produce cuprous oxide at a rate faster than disaccharides do.

$$
\underset{\substack{\| \\ O}}{R-C-H} + 2\,Cu^{+2} + 2\,H_2O \longrightarrow \underset{\substack{\| \\ O}}{R-C-OH} + Cu_2O + 4\,H^+
$$

Procedure

Place 1 mL of each of the following 1% carbohydrate solutions in 10 separate, **labeled** test tubes: glucose, fructose, xylose, arabinose, galactose, sucrose, lactose, maltose, starch (shaken), and glycogen. Add 5 mL of Barfoed's reagent to each tube and heat the tubes in a boiling-water bath for 10 minutes. Remove the tubes and note and record the results. Compare these results with those obtained with Benedict's reagent, especially for the carbohydrates that gave a positive Benedict's test. Record for each carbohydrate the information obtained from the test about the structure of the carbohydrate.

6. IODINE TEST

Some carbohydrates react with iodine to give a characteristic blue, or red to violet color. The colored product is believed to be due to the trapping of iodine within the open spaces of some carbohydrates.

Procedure

Place 1 mL of each of the following 1% carbohydrate solutions in 10 separate, **labeled** test tubes: glucose, fructose, xylose, arabinose, galactose, sucrose, lactose, maltose, starch (shaken), and glycogen. Also place 1 mL of distilled water in another tube to serve as a control. To each tube add 1 drop of iodine solution and heat if necessary. Note and record the results.

7. TES-TAPE TEST FOR GLUCOSE

A commercially available product called Tes-Tape is used to detect glucose. Available at most drug stores, the tape contains the enzymes glucose oxidase and peroxidase, as well as *ortho*-toluidine. The glucose oxidase oxidizes glucose, as shown below, to give gluconic acid and hydrogen peroxide. Once formed, the hydrogen peroxide then reacts with peroxidase to produce oxygen, which oxidizes the *ortho*-toluidine to give products ranging in color from light green to blue-black.

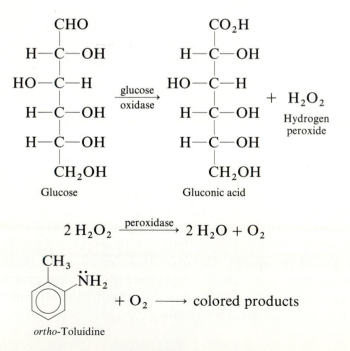

$$2\ H_2O_2 \xrightarrow{\text{peroxidase}} 2\ H_2O + O_2$$

Procedure

Place several drops of each of the following 1% carbohydrate solutions on separate 0.5-cm pieces of Tes-Tape: glucose, fructose, xylose, arabinose, galactose, sucrose, lactose, maltose, starch (shaken), and glycogen. Also place a few drops of distilled water on another piece to serve as a control. Note and record the color of the tape pieces after 1 minute and after 2 minutes.

8. HYDROLYSIS TEST

Disaccharides and polysaccharides can be hydrolyzed in acid solution into their component parts, which can then be tested by some of the previous reagents.

Procedure

Place 5 mL of each of the following 1% carbohydrate solutions in five separate, **labeled** test tubes: sucrose, lactose, maltose, starch (shaken), and glycogen. Add 2 to 3 drops of concentrated hydrochloric acid to each tube and heat the tubes in a boiling-water bath for 10 minutes. Cool the tubes and neutralize the contents by adding approximately 10 drops of 10% sodium hydroxide solution to each tube until the mixtures are just basic to litmus. Remember **not** to dip the litmus paper into the carbohydrate solutions or they will become contaminated by the litmus dye. Test each solution with Tes-Tape and Benedict's reagent. Record the results and compare them with those obtained earlier with each carbohydrate before hydrolysis. Record the reason for any difference obtained.

9. CARAMELIZATION

When sucrose is heated, caramel is formed by a complex series of reactions. In cooking, sugar is caramelized until a light brown, pleasant-tasting product is obtained. Further reaction produces acids that cause a sour, unpleasant flavor.

Procedure

Place approximately a teaspoon of sucrose into an evaporating dish, and place the dish on a wire gauze on a ring stand. Grasp the dish with tongs and slowly heat the dish with a Bunsen burner while stirring the sugar with a stirring rod. Notice that the sugar melts and begins to turn brown. Heat the dish only long enough to change the sugar color to a light brown. Allow the dish to cool, examine the product formed, and record the results.

To a portion of the product, add 5 mL of distilled water and test a few drops of the solution on red and blue litmus paper. Record the results.

Test a small amount of the solution with Tes-Tape and Benedict's reagent, and record the results.

10. DEHYDRATION

Sucrose can be dehydrated by concentrated sulfuric acid to yield carbon. Because of the spectacular nature of the reaction, it is being included as a part of the experiment.

Procedure

Place a quantity of sucrose in a large evaporating dish and indent the top of the sucrose to form a depression. The end result should look like a small mountain. **While standing back, cautiously** and slowly pour concentrated sulfuric acid into the depression until smoke begins to rise. **Stand back** and note and record the results as the sugar reacts with the acid.

11. FOOD PRODUCT TESTS

Perform the Bial, Seliwanoff, Benedict, Barfoed, Tes-Tape, and iodine tests on diluted solutions of brown sugar, honey, molasses, pure maple syrup, white corn syrup, canned fruit syrup, unsweetened apple juice, and unsweetened white grape juice. Record your results in chart form and from your results try to determine the carbohydrates that are present, absent, or uncertain in each of these food products. Indicate how you arrived at your conclusions. If you do Experiment 33B, you can also compare your results here with those obtained from the thin-layer chromatography of carbohydrates and food products.

33B.   THIN-LAYER CHROMATOGRAPHY OF SUGARS AND FOOD PRODUCTS

In this experiment, you will try to determine by thin-layer chromatography which sugars are present in brown sugar, honey, molasses, maple syrup, corn syrup, canned fruit syrup, unsweetened apple juice, and unsweetened white grape juice. With your results, you can then substantiate your conclusions in part 11 of Experiment 33A, and you may be able to make a decision on any uncertain sugars in part 11.

Since sugars are extremely polar compounds and adhere quite strongly to adsorbents, a high-eluting-power solvent is needed. A combination of 1-butanol, acetic acid, ethyl ether, and water in a volume ratio of 10:6:3:1 has been found to give good results.

You can see the sugars on the developed chromatogram by spraying it with a fine mist of p-anisidine and phthalic acid in ethanol or with aniline and phthalic acid in ethanol. These aromatic amines and phthalic acid react with the sugars to give colors after warming. Nonreducing sugars such as sucrose are usually more difficult to observe, due to much weaker color reactions.

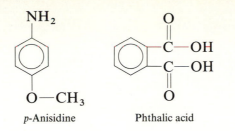

p-Anisidine   Phthalic acid

Procedure

If necessary, read pages 139–151 on thin-layer chromatography. Use commercial silica gel G plates on plastic backing. Plates prepared in the laboratory by dipping are generally unsatisfactory for sugars.

For application of the sugars and food products, at least 15 micropipets will be required. Prepare them by softening the middle of a melting-point capillary tube in a Bunsen burner flame and then pulling the ends apart about 4 to 5 cm. The thin portion in the center, which now has a diameter of about one-tenth its previous size, can be carefully broken to yield two micropipets. You will also need a rectangular solvent chamber and cover, and a 20×20 cm silica gel G plate.

To prepare the chromatogram for developing, place the silica gel G plate onto a larger piece of paper. (This is necessary because cleanliness of the surroundings is important for good results.) About 2 cm from the bottom of the plate, lightly draw a straight **pencil line (no pen)** across the plate. On this line, mark at least 15 evenly spaced dots. Seven of these dots will be for 5% solutions of glucose, fructose, xylose, arabinose, galactose, sucrose, and maltose. The other eight will be for 5% solutions of brown sugar, honey, molasses, maple syrup, corn syrup, 25 percent canned fruit syrup solution, undiluted and unsweetened apple juice, and undiluted and unsweetened white grape juice. One possible sequence for arranging the solutions is shown in the diagram. Be sure to indicate in pencil for each dot which solution has been placed there to avoid any possible mixup.

Using a **different** micropipet for each solution, place the tip of the micropipet into the solution to be applied. By capillary action, some of the liquid will be drawn up. Starting at the left side, quickly touch the tip of the pipet to the plate at the appropriate dot so that a spot no larger than 1 to 2 mm in diameter is produced. **Do not** keep the pipet on the plate too long or its entire contents will be delivered and the spot will be too large. Large spots are unsatisfactory, because during development they will tend to streak or tail, giving poor separations or uncertain R_f values. You may want to practice the technique beforehand on a piece of paper to get it perfected. Since the commercial silica gel G plates are not cheap, you may be given only one for the experiment. Therefore, be careful not to make any mistakes. If any spots are too large, another spot of the same solution should be applied elsewhere.

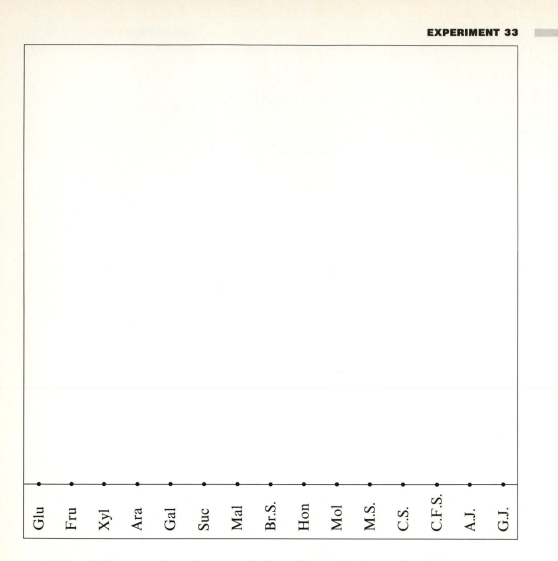

After all the spots have been applied and have dried completely, make a second application of the respective solutions at the same individual points. Allow these spots to dry and then place the plate into the solvent chamber to which has been added the 1-butanol, acetic acid, ethyl ether, and water mixture. Be sure that not so much solvent has been added that the level is above the level of the spots. Otherwise they will dissolve in the solvent. Cover the top of the solvent chamber and wait until the solvent rises to near the top of the plate. During this time, other experiments or activities may be continued or completed.

When the solvent is near the top, remove the plate and mark the solvent level with a pencil. After the chromatogram has dried, place it on a large piece of cardboard **in the hood**. Spray the chromatogram with **a fine mist** of the prepared solution of *p*-anisidine and phthalic acid in ethanol or with aniline and phthalic acid in ethanol. **Do not soak the plate!** A commercial spray bottle can be used for the spraying, but check it first to ensure that only a fine mist is being produced. If necessary, adjust the nozzle using plastic gloves, **not your bare skin**.

Place the plate in an 80° oven until the spots appear. **Wear plastic gloves to protect your skin from the spraying solution.** Remove the plate and compute the R_f values of all the spots. Record the results in your notebook. Compare the results obtained with the various sugars and food products and determine which of these sugars are present in the food products. Do your results substantiate those obtained in the last part of Experiment 33A? State specifically where the results agree, disagree, or are inconclusive. Also indicate the basis for your decisions. Are you able now to identify specifically any of the uncertain sugars in the food products in Experiment 33A? If so, which ones? Do any of the food products have any spots not accounted for by the sugars used in this experiment? If so, which food products? Where are the spots?

33C. OPTICAL ROTATION OF CARBOHYDRATES

In this experiment, you will determine with a polarimeter the optical rotation of carbohydrate solutions of various concentrations. From the values obtained, you will then calculate the specific rotation of each carbohydrate. Because some carbohydrates exhibit **mutarotation** in solution, in some cases the values obtained will be those of the equilibrium mixture of the α and β anomers after the solutions have been allowed to stand for some time (see page 583). The specific rotation of the pure α or β anomer could be obtained if the readings were taken immediately after the carbohydrate was dissolved in solution.

Although there are different makes and models of polarimeters, all of them have certain principal parts in common, which are illustrated in Figure 33-2. Light passes from the light source into the **polarizer**, which is a fixed Nicol prism or Polaroid lens that allows only **polarized light, vibrating in only one plane**, to pass through. The polarized light then passes through the **sample**, which, if optically active, will rotate the plane of the light either clockwise ($+$) or counterclockwise ($-$) a certain number of degrees. The exact amount of rotation is determined by the **analyzer**, which is a second Nicol prism or Polaroid lens that is rotated to whatever angle allows either the maximum or minimum quantity of light from the sample to pass through. The amount of rotation and the direction are then read in degrees on a circular scale with the aid of an auxiliary eyepiece and vernier.

The amount of observed rotation depends on several variables, which must be recorded along with the observed rotation.

Concentration

It is reasonable to expect that the amount of observed rotation should be proportional to the concentration of the solution for a sample tube of a given length.

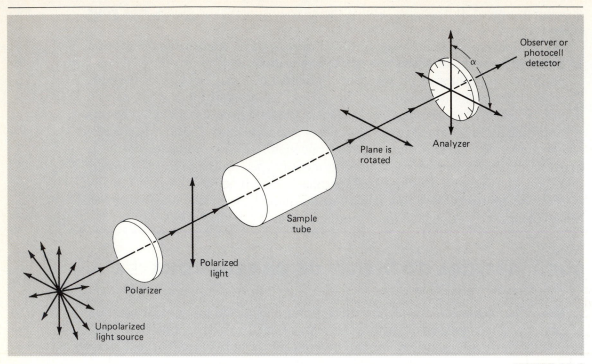

FIGURE 33-2
Schematic representation of a polarimeter.

Thus, doubling the concentration should double the amount of rotation, since the plane-polarized light will then encounter twice as many solute molecules. But this proportionality is not always true, especially at higher concentrations. This deviation can be explained by increased solute-solute interactions at higher concentrations, which presumably affect the rotation differently than solute-solvent interactions. As a result, dilute solutions are generally preferred unless the observed rotation is very low.

Sample-tube length

Other factors being unchanged, the amount of rotation will be directly proportional to the length of the sample tube. Thus, doubling the tube length will double the amount of rotation. This results from the plane-polarized light encountering and being affected by exactly twice the number of molecules in the double-length tube.

Solvent

Changing the solvent can have a very pronounced effect on the magnitude of the observed rotation. In many cases, even the direction of rotation can be changed!

596

Wavelength of light

The wavelength of light used can also determine the magnitude and direction of the observed rotation, with the magnitude of rotation generally increasing at shorter wavelengths. Most rotations in the chemical literature today are reported using the D line of sodium light at 5893 Å.

Temperature

A rather large change in the observed rotation can occur with a change in temperature. Among the factors responsible are the following:

1. The change with temperature of the relative proportions of certain conformational isomers present, each of which produces a different observed rotation.

2. The change with temperature of the relative proportions of certain compounds existing together in equilibrium, such as the α and β anomers of carbohydrates, each of which produces a different observed rotation.

3. The change with temperature of the concentration of the solution, caused by the expansion or contraction of the solvent.

The temperature of the sample can, whenever necessary, be controlled by placing the sample tube in a jacket through which water is passed from a constant-temperature bath.

The observed rotation of compounds can be determined under a variety of concentration and sample-tube-length conditions. In order to standardize the results to a definite concentration and sample-tube length, the **specific rotation** of the substance is calculated by the equation

$$[\alpha]_\lambda^t = \frac{\alpha}{l \cdot c}$$

where $[\alpha]$ = specific rotation in degrees

α = observed optical rotation in degrees

l = length of sample tube in decimeters

c = concentration of solution in grams/mL or density of pure liquid in grams/mL

t = temperature in degrees Celsius

λ = wavelength of light used (usually indicated as "D" for the D line of sodium)

From the equation, note that using a recommended concentration of 2 grams of solute per 100 mL of solution gives an observed rotation of only one-fiftieth of the specific rotation, necessitating a very accurate reading of the circular scale and exact positioning of the analyzer for good results.

When you indicate the specific rotation of a solution, it is good practice to include the concentration of the solution, especially if a high concentration was necessary. This is illustrated in the following calculation of the specific rotation, where a sample of 6.79 grams of cholesterol dissolved in chloroform to make 100 mL of solution gives an observed rotation of $-5.3°$ in a 20-cm tube.

$$[\alpha] = \frac{\alpha}{l \cdot c} = \frac{-5.3°}{2.0 \text{ dm} \cdot 6.79 \text{ grams/100 mL}} = -39°$$

Therefore, the specific rotation of cholesterol is

$$[\alpha]_D^{25} = -39° \qquad (c = 6.79 \text{ grams/100 mL in chloroform})$$

PROCEDURE

The procedure for filling the sample tube and measuring the optical rotation will depend upon the type of polarimeter used. Your instructor will demonstrate or explain to you the method to be used for the particular instrument available.

No matter what instrument is used, it is important that the sodium lamp be turned on at least 10 minutes beforehand, so it is sufficiently warmed up. Also, the sample tube must be perfectly clean, with the solution containing no suspended particles that might disperse the polarized light. Any air bubbles in the liquid must be removed by adding more liquid or by tilting and tapping the tube until the bubbles are out of the path of the light. In cleaning, filling, and handling the sample tube, be careful not to get any fingerprints on the ends of the tube, because they could also interfere with the light.

It is important that you check the zero-degree calibration mark on the instrument by making a zero reading with the sample tube filled only with solvent. Any difference between the two must then be used to correct all subsequent readings. For example, if the zero position is actually $-0.3°$, you must subtract this number from the actual reading of any optically active substance.

Example 1: $\alpha_D^{20} = +10.9°$ (uncorrected)

$\alpha_D^{20} = +10.9° - (-0.3°) = +11.2°$ (corrected)

Example 2: $\alpha_D^{20} = -7.6°$ (uncorrected)

$\alpha_D^{20} = -7.6° - (-0.3°) = -7.3°$ (corrected)

Determine the readings by laying the sample tube in the holder, closing the cover, and turning the knob for the analyzer until the proper angle is reached. Most polarimeters are the double-field type, where a number of sections must be exactly matched in light intensity to attain the correct angle. Figure 33-3 shows what the view through the eyepiece would look like for correct and incorrect adjustments. **CAUTION**: In order to avoid any false rotation values, you **must** be consistent in looking **only** for all sections equally dark **or** all sections equally light. **Do not** use a combination of both.

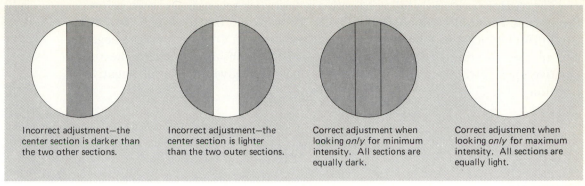

FIGURE 33-3
Various split-field images in the eyepiece of the polarimeter.

When you make any optical rotation determinations, it is best to take at least four readings, with two approaching the actual value from one side, and two approaching it from the other. Discard any value far away from the others, and take another reading. After four reasonably close values have been obtained, they may then be averaged and, if necessary, corrected to obtain the observed rotation.

Determine the observed rotation of distilled water and 2%, 5%, and 10% aqueous solutions of the individual sugars glucose, fructose, sucrose, and lactose. Be sure with each new set of readings that the previous substance has been completely removed by rinsing with a small quantity of the new carbohydrate solution. From the corrected average of the values obtained for each solution, compute the specific rotation for each carbohydrate, showing the calculations. Do your results agree with the literature values for these carbohydrates? Remember that some carbohydrates may exist in solution as an equilibrium mixture of more than one specie, and you must make your comparison with this equilibrium mixture. Also look at your results and determine if the specific rotation for each carbohydrate solution depends on the concentration. Write your conclusion in your notebook and explain how you arrived at your answer.

33D. KINETICS OF THE ENZYMATIC AND ACID-CATALYZED HYDROLYSIS OF SUCROSE

POLARIMETRY

In this experiment, you will use a polarimeter to determine the rates of reaction for the acid-catalyzed and the invertase-enzyme-catalyzed hydrolysis of sucrose to yield glucose and fructose. From your results, you will then determine the comparative efficiencies of these two catalysts in this reaction.

The hydrolysis is easily followed with a polarimeter because, as shown below, the specific rotation changes from $+66.5°$ for pure sucrose to $-19.9°$ for the equilibrium mixture of the glucose and fructose isomers. Observe that the negative specific rotation of the fructose isomers is high enough to overcome the positive specific rotation of the glucose isomers and give an overall negative rotation.

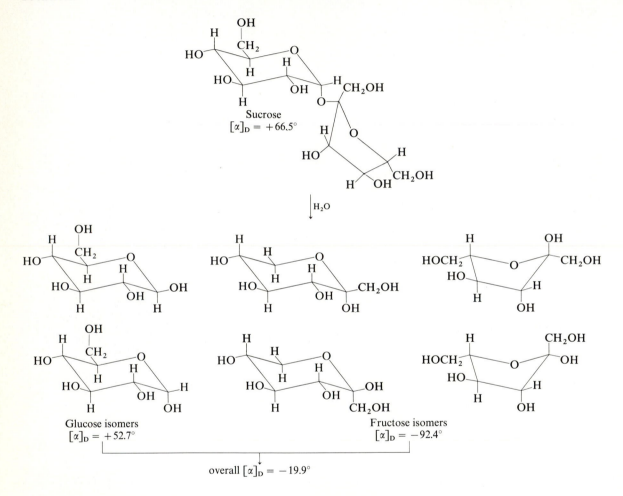

The concentration of sucrose can be determined at any time during the hydrolysis from the observed optical rotation of the solution. This rotation, α_t, is directly proportional to the specific rotations, $[\alpha]$, of all the sugars present and their concentrations. From the specific rotations of $+66.5°$ for pure sucrose and $-19.9°$ for the final equilibrium mixture of all the glucose and fructose isomers, the rotation after complete hydrolysis, α_f, is -0.299 times as large as the initial rotation, α_i. In equation form, the result is

(1)
$$\alpha_f = -0.299\alpha_i$$

The concentration of sucrose at any time, $[\text{sucrose}]_t$, compared to its initial concentration, $[\text{sucrose}]_i$, can be quantitatively expressed in terms of rotations by

(2)
$$\frac{[\text{sucrose}]_t}{[\text{sucrose}]_i} = \frac{\alpha_t - \alpha_f}{\alpha_i - \alpha_f}$$

Substituting equation (1) into equation (2), we obtain

(3)
$$\frac{[\text{sucrose}]_t}{[\text{sucrose}]_i} = \frac{\alpha_t + 0.299\alpha_i}{1.299\alpha_i}$$

Thus, using equation (3), we can easily calculate the concentration of sucrose at any time compared with its initial concentration.

ACID-CATALYZED HYDROLYSIS OF SUCROSE

The acid-catalyzed hydrolysis of sucrose is a second-order reaction, being proportional to the concentration of both sucrose and H_3O^+. But since the acid is not consumed by the reaction, its concentration remains constant and the hydrolysis kinetically becomes a pseudo first-order reaction with the rate equation

(4)
$$\log \frac{[\text{sucrose}]_t}{[\text{sucrose}]_i} = \frac{-k_1 t}{2.303}$$

Substituting equation (3) into equation (4), we obtain

(5)
$$\log \frac{\alpha_t + 0.299\alpha_i}{1.299\alpha_i} = \frac{-k_1 t}{2.303}$$

or

(6)
$$\log(\alpha_t + 0.299\alpha_i) = \log(1.299\alpha_i) - \frac{k_1 t}{2.303}$$

We can find the value of k_1, the pseudo first-order rate constant, by first plotting $\log(\alpha_t + 0.299\alpha_i)$ versus the time. Multiplying the slope of this plot by -2.303 then gives k_1.

(7)
$$k_1 = -2.303 \times \text{slope}$$

Once k_1 is known, the second-order rate constant k_2 can be obtained by equation (8):

(8)
$$k_2 = \frac{k_1}{[H_3O^+]}$$

Procedure

If necessary, read pages 595–599 on optical rotation and the use of a polarimeter. It is very important that the sodium lamp be already warmed up and that you check the zero calibration mark on the instrument by making a zero reading with the sample tube filled with distilled water.

Dissolve 20.0 grams of sucrose in about 30 mL of distilled water in a 50-mL volumetric flask. To this solution, add 10.0 mL of previously measured 6.0 M HCl solution. Rinse the container that held the 10.0 mL of HCl with a few milliliters of distilled water, add to the volumetric flask, and then fill this flask up to the 50-mL line with additional distilled water. Record the exact time and thoroughly mix the sucrose and acid by shaking the stoppered volumetric flask for a few seconds.

As quickly as possible, rinse the empty polarimeter tube with a few milliliters of the sucrose-acid solution, discard the solution, and then completely fill the polarimeter tube with the sucrose-acid solution, being sure there are no air bubbles or particles that could interfere with the path of the light. Measure and record in table form, as shown below, the exact rotation of the light, and the time. Repeat at approximately 5-minute intervals for the next 40 to 50 minutes. At the end, measure and record the temperature of the solution.

Time	t (min) = time − time$_i$	α_t	$\alpha_t + 0.299\alpha_i$	$\log(\alpha_t + 0.299\alpha_i)$

To calculate the values in the last two columns, you must know the initial rotation α_i. Calculate it by multiplying the specific rotation of sucrose, $[\alpha_D] = +66.5°$, by the initial concentration of sucrose, $c_i = 0.40$ gram/mL, by the path length in decimeters of the light in the sample tube:

$$\alpha_i = [\alpha]_D c_i l$$

After all the values in the table are determined, plot on **regular** graph paper $\log(\alpha_t + 0.299\alpha_i)$ on the vertical axis versus the time. Draw a straight line that best approximates the locations of all the points and calculate the slope of this line, $s = (y_2 - y_1)/(t_2 - t_1)$, using two points on opposite sides of the paper.

Using this slope and equation (7), calculate the value of the pseudo first-order rate constant k_1 in units of min^{-1}. But since rate constants are commonly expressed in units of sec^{-1}, divide k_1 by 60. Next, use the **diluted** concentration of the HCl and equation (8) to obtain the value of k_2 in units of molarity^{-1} sec^{-1}.

ENZYMATIC HYDROLYSIS OF SUCROSE

The enzymatic-catalyzed hydrolysis of sucrose in this experiment by invertase is the very reaction that occurs within you when sucrose is digested. The first step in the reaction is, as shown next, the formation of a complex between sucrose and invertase. This complex then breaks down to yield the products glucose and fructose plus the regeneration of invertase.

$$\text{Sucrose} + \text{invertase} \underset{k_{-1}}{\overset{k_1}{\rightleftharpoons}} \underset{\text{complex}}{\text{Sucrose-invertase}} \overset{k_2}{\longrightarrow} \underset{+ \text{ invertase}}{\text{Glucose} + \text{fructose}}$$

The rate of sucrose disappearance depends on the total invertase concentration [I] and the rate constants k_1, k_{-1}, and k_2. But when the sucrose concentration is quite high, it can be shown that essentially all the invertase is complexed with the sucrose, and the rate of hydrolysis becomes simply the rate of decomposition, k_2, of this complex to give the products. In equation form, the kinetic result is

(9)
$$[\text{sucrose}]_t = [\text{sucrose}]_i - k_2[\text{I}]t$$

Substituting equation (3) into equation (9), we obtain

(10)
$$\frac{\alpha_t + 0.299\alpha_i}{1.299\alpha_i} = 1 - \frac{k_2[\text{I}]}{[\text{sucrose}]_i}$$

This can be rearranged to give

(11)
$$\alpha_t = \alpha_i - \frac{1.299\alpha_i k_2[\text{I}]}{[\text{sucrose}]_i} t$$

From this equation, a plot of the rotation α_t versus time should give a straight line with slope

(12)
$$-\frac{1.299\alpha_i k_2[\text{I}]}{[\text{sucrose}]_i}$$

If the initial sucrose concentration, the total invertase concentration, the initial rotation, and the slope are known, we can calculate k_2 from

(13)
$$k_2 = \frac{-\text{slope}[\text{sucrose}]_i}{1.299\alpha_i[\text{I}]}$$

In this experiment, the invertase solution will have a concentration of 5 mg/mL. From its approximate molecular weight of 135,000 grams/mole, the molarity of the **undiluted** invertase is thus 0.000,037 M.

Procedure

Place 20.0 grams of sucrose in a 50-mL volumetric flask. Add enough distilled water to dissolve the sucrose, fill the flask up to the 50-mL line with additional distilled water, and mix.

Rinse the empty polarimeter tube with a few milliliters of the sucrose solution, discard the solution, and then almost completely fill the polarimeter tube with the sucrose solution, quantitatively measuring and recording how much is added. Be sure there are no air bubbles or particles, because they could inter-fere with the path of the light. Measure the optical rotation α_i of the sucrose solution, record the result and the temperature, and then add, using a calibrated Mohr pipet, 0.20 mL of the invertase solution, containing 5 mg of invertase per milliliter. Close the tube and shake it for a few seconds to mix.

Measure the optical rotation of the sucrose-invertase solution for the next 40 to 50 minutes at approximately 5-minute intervals, recording the time and the optical rotation in table form. On a sheet of regular graph paper, plot the rotation versus the time after the invertase was added. Draw the straight line that best approximates the locations of all the points, and calculate the slope of this line, $s = (\alpha_2 - \alpha_1)/(t_2 - t_1)$, using two points on opposite sides of the paper.

Calculate the molarity of the **diluted** invertase, using its initial concentra-tion, the volume added, and the total volume of the sucrose-invertase solution. Also calculate the molarity of the slightly diluted sucrose. Then using equation (13), calculate the rate constant k_2 for the decomposition of the sucrose-invertase complex. Divide the result by 60 to convert it to units of sec^{-1}, which is the standard way to express rate constants.

Compare this value of k_2 with the one obtained in the acid-catalyzed hydrol-ysis experiment. For this comparison, use the initial sucrose concentration of 1.17 M for the acid-catalyzed hydrolysis. Which is the more effective catalyst, acid or invertase? What is the relative ratio of their reactivities under these conditions? At the beginning of the reaction, how many molecules of sucrose can be hydrolyzed per second by one molecule of acid? How many molecules of sucrose can be hydrolyzed per second by one molecule of invertase?

33E. KINETICS OF THE MUTAROTATION OF D-GLUCOSE

POLARIMETRY

In this experiment, you will determine with a polarimeter the rate of mutarotation of α-D-glucose to give β-D-glucose. As the equations show, this reaction occurs in aqueous solution by means of an open-chain, free-aldehyde intermediate, and eventually all three isomers will be in equilibrium with each other in solution. However, very little of the open-chain, free aldehyde actually exists in this equilibrium mixture.

α-D-Glucose has a specific rotation at 20° of $+112.2°$. But in aqueous solu-tion, the specific rotation of α-D-glucose gradually declines to $+52.7°$. On the other hand, β-D-glucose has a specific rotation at 20° of $+18.7°$. In aqueous solution, the specific rotation of β-D-glucose gradually increases to $+52.7°$. Obviously, $+52.7°$ must therefore be the specific rotation of the equilibrium mixture of these isomers at 20°C.

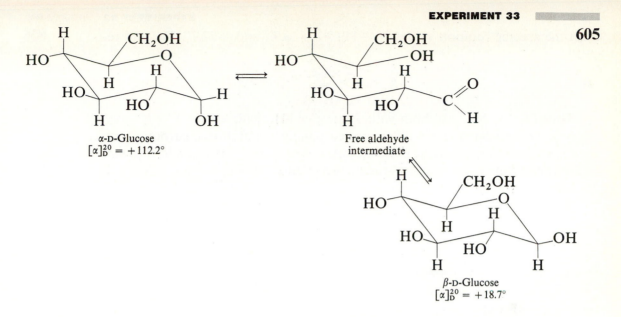

α-D-Glucose
$[\alpha]_D^{20} = +112.2°$

Free aldehyde
intermediate

β-D-Glucose
$[\alpha]_D^{20} = +18.7°$

We can determine the rate of mutarotation by following the rate of change in optical rotation of either α- or β-D-glucose. Equation (1) represents this type of reaction, and equation (2) relates the reaction rates and the time in terms of the concentrations.

(1)
$$A \underset{k_2}{\overset{k_1}{\rightleftharpoons}} B$$

(2)
$$(k_1 + k_2)t = 2.303 \log \frac{A_e - A_i}{A_e - A_t}$$

In equation (2), k_1 and k_2 are the rates of the forward and reverse reactions, t is the time, A_e is the concentration of A at equilibrium, A_i is the initial concentration of A at $t = 0$, and A_t is the concentration of A at any other time during the reaction.

In this experiment, you can easily determine these concentrations by measuring the optical rotations at the beginning, at the end, and at various other times during the reaction. In fact, since the optical rotation is linearly related to the concentration, the optical rotation measured at the beginning (α_i), at the end (α_e), and at various other times (α_t), may all be substituted in equation (2) to give

(3)
$$(k_1 + k_2)t = 2.303 \log \frac{\alpha_e - \alpha_i}{\alpha_e - \alpha_t}$$

From this equation, if the various values of $2.303 \log (\alpha_e - \alpha_i)/(\alpha_e - \alpha_t)$ are plotted versus the time, the slope of the resulting straight line will give the value of $k_1 + k_2$. The individual values of k_1 and k_2 are still not known, but they can be calculated from the equilibrium constant K_e for α- and β-D-glucose, obtained from the specific rotation values of $+112.2°$, $+18.7°$, and $+52.7°$. The end result is

(4)
$$K_e = 1.75 = \frac{k_1}{k_2}$$

Thus $k_1 = 1.75k_2$, and both values can now be calculated.

Since α- and β-D-glucose differ in their configuration at one carbon atom, they are diastereomers having different physical properties. As a result, they can be separated from each other by specific recrystallization techniques, which will actually be done in this experiment to obtain the needed α-D-glucose.

PROCEDURE

Preparation of α-D-glucose

α-D-Glucose is obtained when D-glucose is slowly recrystallized from a mixture of acetic acid and water at room temperature. Place 10 grams of anhydrous D-glucose (dextrose) in 5 mL of distilled water in a 125-mL Erlenmeyer flask. Heat the mixture with continuous stirring on a steam bath until a quite viscous, syrupy solution with no crystals is produced. Remove from the heat and add 20 mL of chilled, **glacial** acetic acid. **(CAUTION: GLACIAL ACETIC ACID CAUSES BURNS. HANDLE WITH CARE.)** Swirl the flask until a homogeneous mixture is attained. Then stopper the flask and allow it to remain undisturbed until the next laboratory period.

Collect the α-D-glucose by suction filtration. Thoroughly wash the crystals with 10 to 15 mL of 95% ethanol and then with 10 to 15 mL of absolute ethanol. Allow them to thoroughly air dry before continuing with the rest of the experiment.

Rate of mutarotation

If necessary, read pages 595–599 on optical rotation and the use of a polarimeter. It is very important that the sodium lamp be already warmed up and that you check the zero calibration mark on the instrument by making a zero reading with the tube filled with distilled water.

Weigh to the nearest 0.01 gram about 3 grams of your α-D-glucose in a dry 50-mL volumetric flask. Fill the flask to near the volumetric line with distilled water, stopper the flask, and shake it. Note the time to the nearest minute. When the solid has completely dissolved, fill the volumetric flask with additional distilled water up to the volumetric line, stopper, and briefly shake.

Working carefully but rapidly, rinse the empty polarimeter tube with a few milliliters of the α-D-glucose solution, discard, and then completely fill the polarimeter tube with the α-D-glucose solution, being sure there are no air bubbles or particles that could interfere with the path of the light. Measure and record in table form, as shown below, the exact rotation of the light and the time. Repeat every minute for the next 5 minutes. Then take additional readings every 2 minutes for the next 10 minutes, and then every 3 to 4 minutes until about 50 total minutes have passed. At the end, record the temperature of both this solution and any glucose solution remaining in the volumetric flask.

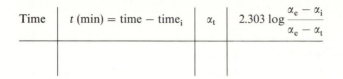

Time	t (min) = time − time$_i$	α_t	$2.303 \log \dfrac{\alpha_e - \alpha_i}{\alpha_e - \alpha_t}$

To obtain the values of the last column in your table, you must know the initial optical rotation and the equilibrium optical rotation. You can obtain the initial optical rotation at the time of mixing by plotting on ordinary graph paper α_t on the vertical axis versus the time on the horizontal axis. Draw the best curved line through these points and extrapolate this curve back to the time of the initial mixing.

Approximately 4 to 5 hours will be required for the optical rotation to drop to its equilibrium value. At your instructor's option, either place the polarimeter tube in your locker until the next laboratory period, or pour the D-glucose solution into your volumetric flask, stopper the flask, and place it in your locker until the next laboratory period. No matter which alternative is used, be sure you use the same polarimeter for measuring the equilibrium optical rotation. Your instructor will have placed identifying numbers on each polarimeter, and you should record the appropriate number in your notebook.

Calculate the values in the last column of your table, and then plot on ordinary graph paper the results of the last column on the vertical axis versus the time on the horizontal axis. Draw the straight line that best approximates the locations of all the points, and calculate the slope of this line, $s = (y_2 - y_1)/(t_2 - t_1)$, using two points on opposite sides of the paper. The value of this slope in $\min^{-1}$ again equals $k_1 + k_2$. Calculate the individual values of k_1 and k_2 and divide the results by 60 to obtain these rate constants in units of $\sec^{-1}$, the standard way to express rate constants.

Also, using your initial optical rotation value and the concentration of your solution, calculate the specific rotation of your α-D-glucose. Repeat the calculation using your equilibrium optical rotation value to obtain the specific rotation of the equilibrium mixture. Compare your results with the known specific rotation values. Is there a significant difference between either of them? If so, comment on possible reasons for these differences.

QUESTIONS

1. Why does a disaccharide molecule not have exactly twice the molecular formula of a monosaccharide molecule?
2. What is meant by the term "invert sugar"?

3. Glucose and fructose together are over twice as sweet as sucrose. Explain why boiling a sucrose solution with a trace of acid is important in candy making.

4. A number of sugars are frequently called "reducing sugars." Yet a diet of them will not cause one to lose weight. Why are they then called reducing sugars?

5. One carbohydrate not mentioned in this experiment was cellobiose. What is cellobiose and how do you think its name was derived?

6. An experimentally observed optical rotation of $+20°$ could also be correctly interpreted as a rotation of $+380°$, $-340°$, or other values. How could you be certain which value is correct?

7. Why are dilute solutions preferred in optical rotation determinations? When might it be necessary to use a concentrated solution?

8. When the pure β anomer of lactose is dissolved in water, the observed optical rotation is found to increase over a period of time, but then stops and remains at a fixed value. From Experiment 33E, explain exactly what is occurring.

9. Why can humans digest starch but not cellulose, like the termite?

10. Give the step-by-step mechanism for the acid-catalyzed hydrolysis of sucrose.

11. Write the step-by-step mechanism for the mutarotation of α-D-glucose to give β-D-glucose. What would be the mechanism for the mutarotation of β-D-glucose to give α-D-glucose?

12. From the specific rotation values given in Experiment 33E, calculate the percentage of α-D-glucose and β-D-glucose present at equilibrium. Looking at how these two anomers differ in structure, specifically explain in terms of stability why one anomer should predominate over the other.

PROTEINS
AND AMINO
ACIDS

INTRODUCTION

It has often been said that man shall not live by bread alone. Although this statement is used primarily in religion, it also has significant application in nutrition. Even though bread and related foods may satisfy the body's energy needs, the other nutritional requirements, especially that of **proteins**, cannot be fulfilled with this food alone.

Proteins are a group of organic molecules found in all living cells. The name **protein** originated in 1839 by the Dutch chemist Gerardus Mulder, who derived the term from the Greek word **proteious**, meaning first. Protein certainly was first to Mulder, who described it as "unquestionably the most important of all known substances—without it no life appears possible—through its means the chief phenomena of life are produced." Today only the nucleic acids RNA and DNA, which control heredity and direct the synthesis of proteins, can challenge proteins as being number one in importance. Proteins, as we shall see, also constitute the structures of enzymes, antibodies, and many hormones.

Chemically, proteins are extremely complex, high-molecular-weight polymers made by the chemical combination of α-amino carboxylic acids, commonly called the **amino acids**. You remember that a typical reaction of a carboxylic acid is its joining with an amine to form a larger molecule, called an amide.

$$R-\overset{\overset{\textstyle O}{\|}}{C}-OH + H_2N-R' \longrightarrow R-\overset{\overset{\textstyle O}{\|}}{C}-NH-R' + H_2O$$

Amino acids contain both the carboxylic acid and amino functional groups within the same molecule. Two amino acid molecules can therefore react to

609

produce an amide that still possesses a free carboxyl group at one end and a free amino group at the other.

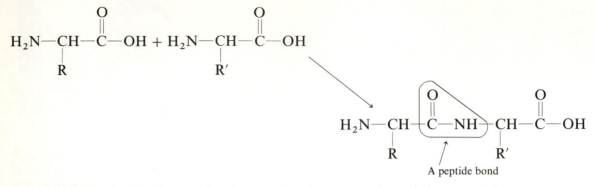

A peptide bond

The amide linkage formed in the reaction is commonly called a **peptide bond**, and the resulting product formed from two amino acids is known as a **dipeptide**. Because the dipeptide can react with other amino acid units, the molecule can get progressively larger and form what is called a **polypeptide**. Proteins therefore are also known as polypeptides.

Table 34-1 gives the names, abbreviations, and structures of 22 amino acids found in proteins. Observe that the table is given in two parts, with the first 8 amino acids being listed as **essential** and the remaining 14 listed as **non-essential**. All the amino acids are needed by the body, **but the essential ones are those the body cannot make itself and must be obtained directly from the food consumed**.

All foods containing protein are **not** all alike, and they can be rated on both the **amount** and the **quality** of protein present. Thus peanuts, although they contain a high percentage of protein, are not a good source due to the low number of essential amino acids. Milk, on the other hand, has a lower percentage of protein, but is **nutritionally complete** because it contains all the essential amino acids. Generally a combination of foods such as milk and dairy products, eggs, meat, and soybeans provide nutritionally complete proteins and should be included in the diet. Foods like bread and cereal that contain low-quality proteins, however, can be satisfactory when supplemented with milk or similar foods.

Removing any of the eight essential amino acids from the diet means that normal growth and health are not possible. In severe cases, the result can be weight loss, weakness, fatigue, poor appetite, anemia, edema, and mental retardation. Protein deficiency also results in a greater susceptibility to disease and infection, and a longer period of recovery and healing. An explanation is that the necessary antibodies to fight disease and infection originally came from the amino acids in proteins.

In the table you may have noted that only the L-amino acids are found in proteins. All the amino acids (except glycine) contain a chiral center and may exist in either one of two enantiomeric forms, D and L.

TABLE 34-1.

L-Amino Acids Found in Proteins

Name	Abbreviation	Structural formula
Essential		

1. Isoleucine — Ile

$$CH_3CH_2-\underset{\underset{CH_3}{|}}{CH}-\underset{\underset{\overset{+}{NH_3}}{|}}{CH}-\overset{\overset{O}{\|}}{C}-O^-$$

2. Leucine — Leu

$$CH_3-\underset{\underset{CH_3}{|}}{CH}-CH_2-\underset{\underset{\overset{+}{NH_3}}{|}}{CH}-\overset{\overset{O}{\|}}{C}-O^-$$

3. Lysine — Lys

$$^+NH_3-CH_2CH_2CH_2CH_2-\underset{\underset{NH_2}{|}}{CH}-\overset{\overset{O}{\|}}{C}-O^-$$

4. Methionine — Met

$$CH_3-S-CH_2CH_2-\underset{\underset{_+NH_3}{|}}{CH}-\overset{\overset{O}{\|}}{C}-O^-$$

5. Phenylalanine — Phe

$$-CH_2-\underset{\underset{_+NH_3}{|}}{CH}-\overset{\overset{O}{\|}}{C}-O^-$$

6. Threonine — Thr

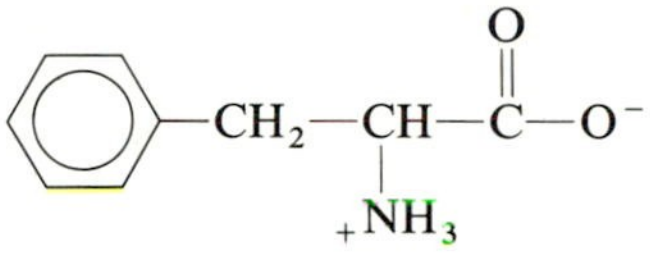

$$CH_3-\underset{\underset{OH}{|}}{CH}-\underset{\underset{_+NH_3}{|}}{CH}-\overset{\overset{O}{\|}}{C}-O^-$$

7. Tryptophan — Trp

$$-CH_2-\underset{\underset{_+NH_3}{|}}{CH}-\overset{\overset{O}{\|}}{C}-O^-$$

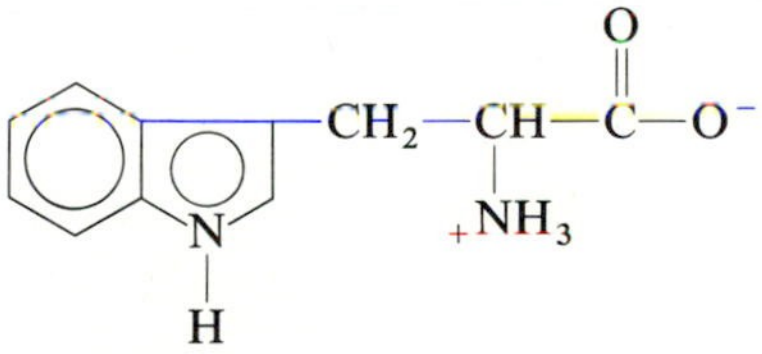

8. Valine — Val

$$CH_3-\underset{\underset{CH_3}{|}}{CH}-\underset{\underset{\overset{+}{NH_3}}{|}}{CH}-\overset{\overset{O}{\|}}{C}-O^-$$

| **Nonessential** | | |

1. Alanine — Ala

$$CH_3-\underset{\underset{_+NH_3}{|}}{CH}-\overset{\overset{O}{\|}}{C}-O^-$$

TABLE 34-1. (continued)

Name	Abbreviation	Structural formula
2. Arginine	Arg	$NH_2-C-NH-CH_2CH_2CH_2-CH-C-O^-$ with $+NH_2$ below C, O above C, NH_2 below CH
3. Asparagine	Asn	$NH_2-C-CH_2-CH-C-O^-$ with O below first C, O above last C, $+NH_3$ below CH
4. Aspartic acid	Asp	$HO-C-CH_2-CH-C-O^-$ with O above both C's, $+NH_3$ below CH
5. Cysteine	Cys	$HS-CH_2-CH-C-O^-$ with O above C, $+NH_3$ below CH
6. Glutamic acid	Glu	$HO-C-CH_2CH_2-CH-C-O^-$ with O above both C's, $+NH_3$ below CH
7. Glutamine	Gln	$NH_2-C-CH_2CH_2-CH-C-O^-$ with O below first C, O above last C, $+NH_3$ below CH
8. Glycine	Gly	CH_2-C-O^- with O above C, $+NH_3$ below CH$_2$
9. Histidine	His	imidazole ring $-CH_2-CH-C-O^-$ with O above C, $+NH_3$ below CH
10. Hydroxylysine	Hyl	$+NH_3-CH_2-CH-CH_2CH_2-CH-C-O^-$ with OH below second CH, O above last C, NH_2 below last CH
11. Hydroxyproline	Hyp	pyrrolidine ring with $HO-$ and $-C-O^-$, O above C, $+N$ with H, H

TABLE 34-1. (continued)

Name	Abbreviation	Structural formula
12. Proline	Pro	
13. Serine	Ser	$HO-CH_2-CH-\overset{\displaystyle O}{\overset{\|}{C}}-O^-$
		$_+NH_3$
14. Tyrosine	Tyr	$HO-\bigcirc-CH_2-CH-\overset{\displaystyle O}{\overset{\|}{C}}-O^-$
		$_+NH_3$

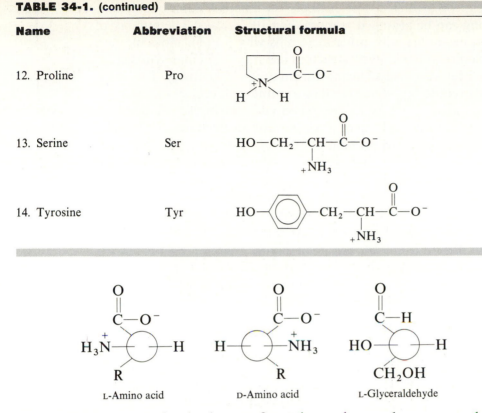

L-Amino acid   D-Amino acid   L-Glyceraldehyde

These letters represent the absolute configuration at the α-carbon atom and were chosen based upon the relative configuration with optically active glyceraldehyde. Thus the L-amino acids in proteins have the same relative configuration at the chiral center as L-glyceraldehyde. A few D-amino acids have been found in the cell walls of bacteria and in the hydrolysis products of certain antibiotics.

The structures of all the amino acids in the table are given as a dipolar ion or **zwitterion**. This results from the moderately acidic and basic properties of the carboxyl and amino groups present. The melting point, solubility, and high dipole moment of amino acids are exactly what would be predicted from such a dipolar ion. In acidic and basic solutions, the zwitterion no longer exists, due to the conversion of the amino acids into their cationic and anionic forms.

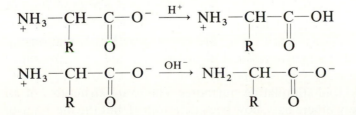

Even though there are only 22 amino acids found in proteins, the large possible size of a protein results in an enormous number of possible ways in which the amino acids can be arranged. For example, 20 different amino acids in a chain 100 units long can be arranged in more than 10^{130} ways! The number of possible combinations, along with other factors, is why protein molecules are so complex in structure and why their structure determination and synthesis in the laboratory has been such a challenging area in organic chemistry. This theoretical number far exceeds the actual number of proteins found in nature.

The proteins in the body frequently are compared to the bricks of a building. Although this is a good analogy, there is an important difference. Unlike the bricks, which remain in place, the proteins of the body are continuously broken down into their individual amino acids and then rebuilt into whatever proteins the body needs. This reassembly process is regulated by the nucleic acids, which place the amino acids into whatever sequence is required. Because the nucleic acids vary from animal to animal and individual to individual, it follows that the proteins formed will never be identical in every respect. For example, casein from cow's milk is different from casein in human milk, although it is chemically impossible to distinguish between them. The same amino acids are present in the same proportions, and only by immunological tests can a distinction be made. The difference between these proteins is undoubtedly due to the arrangement of the amino acids.

As stated previously, proteins are the principal material in enzymes, antibodies, and many hormones. Enzymes are necessary for catalyzing all the essential chemical reactions within the body. A few of the known ones include **pepsin, trypsin**, and **chymotrypsin**, which function as digestive agents, **thrombin**, used to promote the healing of wounds, and **acetylcholinesterase**, used for the conduction of nerve impulses.

Antibodies are formed within the body as agents to neutralize foreign invaders such as antigens, toxins, and certain bacteria and viruses. In fact, antigens, bacteria, and viruses are also proteins, so the neutralizing effect of antibodies is a complex, highly specific reaction of one protein with another.

Hormones are chemical substances, formed in the endocrine glands, that regulate specific bodily functions. A few of the important protein hormones include **thyroglobulin, calcitonin, parathormone, insulin, glucagon, vasopressin**, and **thyrotropin**. Thyroglobulin, calcitonin, and parathormone are all produced by the thyroid or parathyroid glands. The amount of thyroglobulin present has been demonstrated to be related to the rate of body metabolism, and calcitonin and parathormone have been found to control the amount of calcium in the blood. Insulin and glucagon are both pancreas hormones that control the level of blood sugar. All of you are familiar with the relationship between insulin and diabetes. Vasopressin and thyrotropin are both produced in the pituitary gland. Vasopressin is responsible for raising the blood pressure, and thyrotropin is a thyroid-stimulating hormone. The exact structures of all these hormones and many others have now been determined, but the mechanism by which they control the various bodily processes is still not yet fully understood.

34A. PAPER CHROMATOGRAPHY OF AMINO ACIDS

In this experiment, you will hydrolyze a number of proteins into their individual amino acids and, through paper chromatography, attempt to identify many of these amino acids. The hydrolysis involves the refluxing for 1 hour of the proteins with 6 M HCl by the reaction

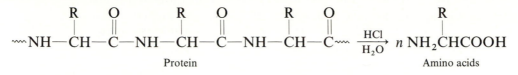

After the chromatographic separation, the individual amino acids are made visible with ninhydrin, which reacts with amino acids to produce a deep blue color. The reactions involved in the production of the color are

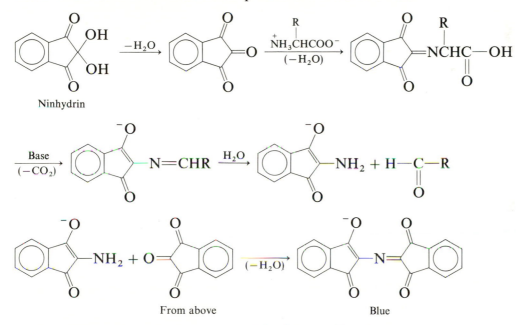

Some amino acids, such as proline and hydroxyproline, however, react to yield a pale yellow color.

PROCEDURE

Since it takes quite some time for the protein hydrolysis and chromatography separation, you will find it advantageous to start this experiment and then proceed to some other experiment during the waiting period.

Hydrolysis of proteins

The proteins of hair, undyed wool, gelatin, and casein (from the milk experiment) will be hydrolyzed. To save time and effort, each student should hydrolyze only one of the four sources of protein, and obtain from other students the other needed hydrolyzed proteins.

Place 20 mL of 6 M HCl and 0.5 gram of the protein in a 100-mL round-bottom flask with a few boiling chips. Attach a reflux condenser and heat the mixture under reflux for at least 1 hour. Then **carefully** add about 0.5 gram of decolorizing carbon to the partially cooled mixture and swirl. Gravity filter the hydrolyzed protein into a labeled 50-mL Erlenmeyer flask. Allow the colorless to pale yellow liquid to cool, stopper it, and save for the paper chromatography analysis.

Separation of amino acids by paper chromatography

If necessary, read pages 139–151 on paper chromatography. Obtain a solvent chamber, cover, and Whatman No. 1 paper that will fit into the solvent chamber when curled. The paper should be at least 15 to 20 cm high and 20 to 25 cm wide. Handle the paper by the **top edge only** so that your fingerprints won't appear as colored spots on the developed chromatogram, or wear plastic gloves. The top edge will be identified by the letter T, placed there when the paper was cut. The paper's thickness may vary, and marking it ensures that the direction of the solvent flow on each piece will be the same for all students. More reproducible R_f values are attained when chromatograms are developed in only one direction.

Place the paper onto a larger piece of paper, since cleanliness of the surroundings is important for good results. About 2 cm from the bottom of the paper, lightly draw a straight **pencil line (no pen)** across the paper. On this line mark at least 16 evenly spaced dots, leaving extra space for both the right and left margins. Four of these dots will be for the hydrolysates from hair, wool, gelatin, and casein. The other 12 will be for 0.1 M solutions of the essential amino acids leucine (Leu), lysine (Lys), threonine (Thr), tryptophan (Try), and valine (Val), and the nonessential amino acids alanine (Ala), aspartic acid (Asp), glutamic acid (Glu), glycine (Gly), proline (Pro), serine (Ser), and tyrosine (Tyr). Each of these solutions has been acidified with 1 drop of 6 M HCl for each milliliter of solution. One possible sequence for arranging the solutions is shown in the diagram. Be sure to indicate for each dot which solution has been placed there to avoid any possible mixup.

For application of the amino acids and protein hydrolysates, you need at least 16 micropipets, which you can prepare by softening the middle of a melting-point capillary tube in a Bunsen burner flame and pulling the ends apart about 4 to 5 cm. The thin portion in the center, which now has a diameter of about one-tenth its previous size, can be carefully broken to yield two micropipets.

Using a **different** micropipet for each solution, place the tip of the micropipet into the solution to be applied. By capillary action, some of the liquid

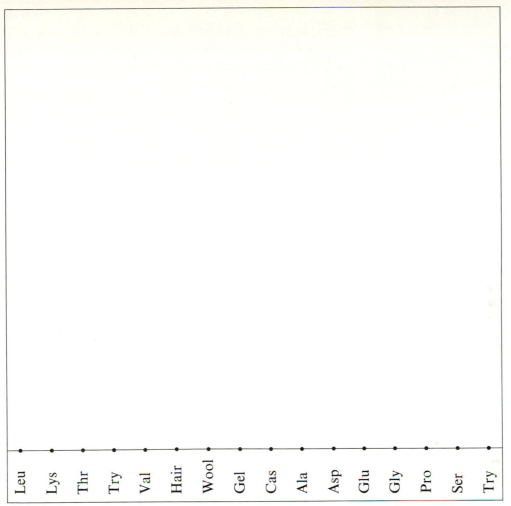

will be drawn up. Starting at the left margin, quickly touch the tip of the pipet to the paper at the appropriate dot so that a spot no larger than 1 to 2 mm in diameter is produced. **Do not** keep the pipet on the paper too long or its entire contents will be delivered and the spot will be too large. Large spots are unsatisfactory, because during development they will tend to streak or tail, thus giving poor separations or uncertain R_f values. You may want to practice the technique beforehand on another piece of paper to get it perfected.

After all the spots have been applied and have dried completely, make a second application of the respective solutions at the same point. The protein hydrolysates should be applied a third time. After these spots have dried, coil the sheet into a cylinder (no fingers please), fasten it together with staples at the top and bottom so that the edges do **not** overlap, and place the cylinder into the solvent chamber to which has been added **one** of the following solvent systems (all measured by volume):

Solvent 1	Solvent 2	Solvent 3[1]
70% Isopropyl alcohol	70% Ethyl acetate	80% Phenol
10% Formic acid	20% Formic acid	20% Water
20% Water	10% Water	

Be sure that not so much solvent has been added that the level is above the level of the spots. Otherwise they will dissolve in the solvent. Cover the top of the chamber and perform other activities until the solvent ascends to near the top of the paper.

When the solvent has reached the top, remove the cylinder (no fingers please), open it, and mark the solvent level with a pencil. After the chromatogram has dried, spray it uniformly but lightly with ninhydrin, allowing the paper to become moist but not soaked. Place the chromatogram in a 110° oven and, within a few minutes, colored spots will appear. Remove the chromatogram and **immediately** outline all the spots with a pencil. (They tend to fade in the light.) Calculate the R_f values and place the results in table form in your notebook, noting which solvent system was used. From other students, obtain the R_f values from using the other two solvent systems and place these results in your table, again noting which solvent system was responsible.

From the R_f values of the known amino acids and the R_f values for each of the hydrolysates, try to determine which of the 12 amino acids are definitely present and absent in hair, wool, gelatin, and casein. Remember that amino acids with similar R_f values may not separate well enough to be distinguished by the chromatogram. Also remember that some amino acids not used in this experiment may give R_f values very close to those of the knowns. Place your results in table form, indicating for hair, wool, gelatin, and casein the amino acids that are definitely present, definitely absent, uncertain, and the R_f values for any extra spots not corresponding to any of the knowns. You should also indicate the basis for each decision.

34B. CHEMICAL TESTS OF PROTEINS AND AMINO ACIDS

In this experiment, you will perform chemical tests on proteins and amino acids. Some of these tests are characteristic for only certain amino acids and will enable you to verify the presence or absence of these amino acids in a protein. Others

---

[1] This solution must have a protective layer of ligroin to exclude air from the phenol. When needed, insert a pipet **below** the layer of ligroin and remove the necessary amount of aqueous phenol with a **bulb**. **DO NOT MOUTH-PIPET OR GET ANY PHENOL ON YOUR SKIN BECAUSE IT CAUSES SEVERE BURNS.**

are general tests given by practically all proteins, and they find many useful applications for everyday life. Some of these applications are presented in the questions at the end.

1. MILLON'S TEST

Millon's test is given by any compound containing a phenolic hydroxy group. Consequently, any protein containing tyrosine will give a positive test of a pink to dark red color. The Millon reagent is a solution of mercuric and mercurous ions in nitric and nitrous acids. The red color is probably due to a mercury salt of nitrated tyrosine. Since the chloride and ammonium ions in urine tend to interfere with the test, it is unreliable for urinalysis.

Procedure

Place 1 mL of 2% gelatin, 2% egg albumin, 2% casein, and 0.1 M tyrosine into separate, **labeled** test tubes. Add 3 drops of Millon's reagent and immerse the tubes in a boiling-water bath for 5 minutes. Cool the tubes and record the colors formed.

Do the results of this test for gelatin and casein confirm the results obtained in the paper chromatography? How?

2. HOPKINS-COLE TEST

The Hopkins-Cole test involves the reaction of the indole portion of the amino acid tryptophan with glyoxylic acid in the presence of sulfuric acid. As shown below, a violet-colored complex is formed between the amino acid and the aldehyde group.

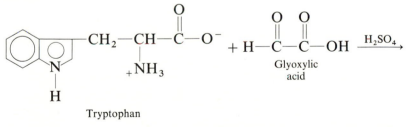

Tryptophan

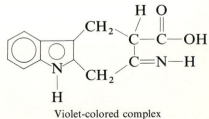

Violet-colored complex

Procedure

Place 1 mL of 2% gelatin, 2% egg albumin, 2% casein, and 0.1 M tryptophan into separate, **labeled** test tubes. Add an equal volume of the Hopkins-Cole reagent and mix well. Tilt each tube and **carefully** pour 2 mL of concentrated sulfuric acid down the side so that it forms a layer below the protein or amino acid mixture. **(CAUTION: CONCENTRATED SULFURIC ACID CAUSES SEVERE BURNS. HANDLE WITH CARE.)** A color should develop at the interface of the two layers. If the color does not appear, gently tap the tube to provide a slight mixing of the two layers. Record the results.

Do the results of this test for gelatin and casein confirm the results obtained in the paper chromatography? How?

3. BIURET TEST

The biuret test is a general test for proteins containing two or more peptide bonds. It is named after the compound **biuret**, which gives the same violet color in the test as proteins.

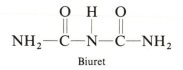

Biuret

The color is thought to result from a coordination complex between the cupric ion in the reagent and the amide groups in the peptides. This color also forms when the reagent is placed with any compound (protein or otherwise) that contains **two or more** of the following groups:

$$-\overset{\overset{\displaystyle O}{\|}}{C}-NH- \qquad -\overset{\overset{\displaystyle O}{\|}}{C}-NH_2 \qquad -\overset{\overset{\displaystyle NH}{\|}}{C}-NH_2 \qquad -\underset{\underset{\displaystyle OH}{|}}{CH}-CH_2-NH_2 \qquad -CH_2NH_2$$

Thus some individual amino acids can also give a positive test.

Procedure

Place 1 mL of 2% gelatin, 2% egg albumin, and 2% casein into separate, **labeled** test tubes along with 1 mL of 10% sodium hydroxide solution. After mixing the two reagents, add 2 to 3 drops of 0.5% copper sulfate solution. Record the results.

4. NINHYDRIN TEST

The ninhydrin reaction is used to detect the presence of α-amino acids and proteins containing free amino groups. When heated with ninhydrin, these molecules give characteristic deep blue colors (or occasionally pale yellow). The reactions involved in the production of the color are given on page 615.

Procedure

Place 1 mL of 2% gelatin, 2% egg albumin, 2% casein, and 0.1 M glycine into separate, **labeled** test tubes along with 0.2 mL of 0.1% ninhydrin solution. Mix the solutions and heat to boiling. Record the results.

Why is this test used to detect the presence of amino acids in paper chromatography?

5. XANTHOPROTEIC TEST

Some amino acids such as tyrosine and tryptophan contain aromatic rings that can easily undergo nitration to yield yellow products. Any proteins containing these amino acids will therefore become yellow in the presence of nitric acid. After the nitrated product is formed, the color may be intensified through the addition of a base, which removes the slightly acidic phenolic hydrogen on tyrosine to give the deeply colored anion.

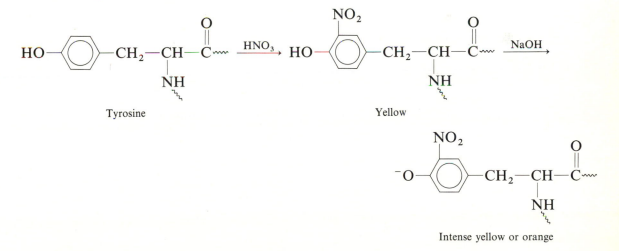

Tyrosine   Yellow

Intense yellow or orange

Procedure

Place 1 mL of 2% gelatin, 2% egg albumin, 2% casein, 0.1 M tyrosine, and 0.1 M tryptophan into separate, **labeled** test tubes and add 5 drops of concentrated nitric acid. **(CAUTION: CONCENTRATED NITRIC ACID CAUSES SE-VERE BURNS. HANDLE WITH CARE.)** Gently warm the mixtures in a water bath and observe any change in color. Cool the mixtures and carefully add 10% sodium hydroxide solution dropwise until basic to litmus. **Do not** place the litmus paper into the liquid. Record the results.

Do the results of this test for gelatin and casein confirm the results obtained in the paper chromatography? How?

6. SULFUR TEST

You can determine the presence of sulfur-containing amino acids, such as cysteine in proteins, by converting the sulfur to an inorganic sulfide through the cleavage by base. When the resulting solution is combined with lead acetate, a black precipitate of lead sulfide results.

$$\text{Sulfur-containing protein} \xrightarrow{\text{NaOH}} S^{-2} \xrightarrow{Pb^{+2}} \underset{\text{Black precipitate}}{\text{PbS}}$$

Procedure

Place 1 mL of 2% gelatin, 2% egg albumin, 2% casein, and 0.1 M cysteine into separate, **labeled** test tubes along with 2 mL of 10% sodium hydroxide solution and 2 drops of 5% lead acetate solution. Heat the mixtures in a boiling-water bath for 5 minutes and record the results.

7. HEAVY-METAL IONS TEST

Heavy-metal ions such as mercury, lead, and silver react with the free carboxylate groups of proteins to form insoluble heavy-metal salts of the protein. The germicidal properties of silver nitrate and mercuric chloride solutions are the results of the precipitation of the proteins in bacteria.

$$\underset{}{\text{Protein}-\overset{\overset{\text{O}}{\|}}{\text{C}}-\text{O}^-} \xrightarrow{Ag^+} \underset{}{\text{protein}-\overset{\overset{\text{O}}{\|}}{\text{C}}-\text{O}^-\,Ag^+} \quad \underset{\text{precipitate}}{\text{Heavy-metal}}$$

Procedure

Place about 1 mL of 2% gelatin, 2% egg albumin, and 2% casein into separate, **labeled** test tubes. Add 5% mercuric chloride solution dropwise and record the results. Repeat the above tests with 5% lead nitrate solution, using new samples of the gelatin, egg albumin, and casein solutions. Record the results.

8. PROTEIN COAGULATION TEST

Proteins may also be precipitated or **coagulated** through the use of intense heat, alcohol, and other reagents. Many proteins possess certain three-dimensional shapes as the result of weak attractions between different portions of the molecule. These weak linkages may easily be disturbed or broken by physical and chemical means, resulting in a change of shape that affects the solubility and other physical properties of the protein. When this occurs, the protein is said to be **denatured** and frequently precipitates from solution.

Procedure

A. Effect of heat.   Place about 1 mL of 2% gelatin, 2% egg albumin, and 2% casein into separate, **labeled** test tubes. Immerse the tubes in a water bath and slowly raise the temperature. Observe the temperature on a thermometer immersed in the bath and note the temperature at which any changes occur. Record these changes and the temperature.

B. Effect of alcohol.   Place about 1 mL of 2% gelatin, 2% egg albumin, and 2% casein into separate, **labeled** test tubes. Add about 3 mL of 95% ethanol to each of the tubes and record the results.

C. Effect of tannic acid.   Place about 1 mL of 2% gelatin, 2% egg albumin, and 2% casein into separate, **labeled** test tubes. Add about 0.5 mL of 10% tannic acid solution to each of the tubes and record the results.

D. Effect of picric acid.   Place about 1 mL of 2% gelatin, 2% egg albumin, and 2% casein into separate, **labeled** test tubes. Add about 0.5 mL of saturated picric acid solution to each of the tubes and record the results.

34C. ENZYMATIC RESOLUTION OF D,L-METHIONINE

- **ACETYLATION**
- **POLARIMETRY**

α-Amino acids may be obtained either from proteins or from a chemical synthesis. One chemical synthesis approach, as shown below, involves the appropriate

$$R-CH_2-\overset{\overset{\displaystyle O}{\|}}{C}-OH \xrightarrow[Br_2]{P} R-\underset{\underset{\displaystyle Br}{|}}{CH}-\overset{\overset{\displaystyle O}{\|}}{C}-OH \xrightarrow{NH_3} R-\underset{\underset{\displaystyle NH_2}{|}}{CH}-\overset{\overset{\displaystyle O}{\|}}{C}-O^-$$

carboxylic acid reacting with phosphorus and bromine (the Hell-Volhard-Zelinsky reaction) to give the α-bromo product. Reaction with ammonia then gives the α-amino acid.

If the amino acid is capable of optical activity, one almost always obtains **only** the L-amino acid when proteins are the source. This results from the amino acid being formed in the asymmetric environment of an enzyme.

But with the above synthesis, an equal mixture of D- and L-amino acids must result. The Hell-Volhard-Zelinsky procedure uses no optically active reagents, and either α-hydrogen on the carboxylic acid is equally capable of being replaced by a bromine.

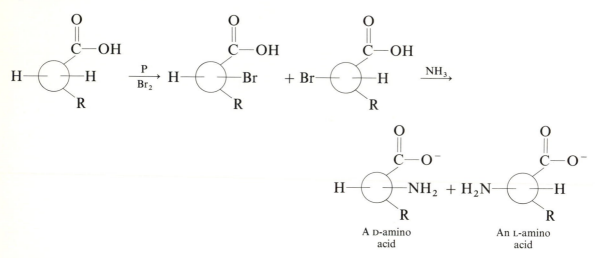

A D-amino acid   An L-amino acid

In fact, **any** synthetic procedure of α-amino acids involving ordinary chemical reagents **must** give an equal mixture of D- and L-isomers. Therefore, if either of these isomers is desired in pure form, some type of resolution is needed to separate them.

One especially interesting way to accomplish this uses an enzyme called **acylase** (isolated from kidney tissue), which catalyzes the hydrolysis of N-acylamino acids in living organisms. The active site of acylase is chiral, and only N-acylamino acids having the L configuration react. Thus, if a mixture of a D- and L-N-acylamino acid is present, the D-isomer is unaffected and the two products are easily separated, due to their different physical properties. The reaction sequence is

$$\text{D,L-RCHCO}_2^- \xrightarrow[\text{anhydride}]{\text{acetic}} \text{D,L-RCHCO}_2^- \xrightarrow{\text{acylase}} \text{L-RCHCO}_2^- + \text{D-RCHCO}_2^-$$

$$\underset{+}{\text{NH}_3} \qquad\qquad \text{NHCOCH}_3 \qquad\qquad \underset{+}{\text{NH}_3} \qquad \text{NHCOCH}_3$$

Easily separated

In this experiment, you will separate by resolution the essential amino acid L-methionine from a commercial mixture of D,L-methionine. Methionine is chosen because the N-acyl derivatives of L-methionine react the fastest of all α-amino acids with acylase. In fact, commercial acylase preparations are standardized on the basis of the rate of hydrolysis of *N*-acetyl-L-methionine. The specific reaction sequence for the separation is

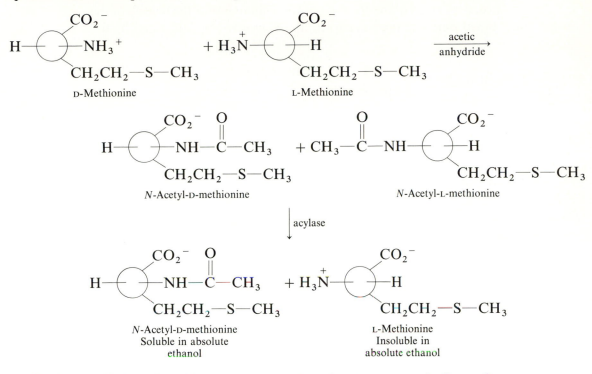

In the acetylation of methionine, care must be taken to prevent the formation of an azlactone. As shown below, *N*-acetyl derivatives of amino acids can tautomerize to form the enol, which then undergoes ring closure with the carboxyl group to form the azlactone.

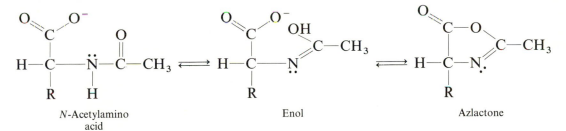

In addition, the methionine may not be completely acetylated. Fortunately, unreacted methionine can be detected by ninhydrin, which forms a characteristic blue color with most amino acids (see page 615). Any unreacted methionine can be removed by its insolubility in acetone or absolute ethanol. *N*-Acetylmethionine, on the other hand, is readily soluble in both solvents. This difference in solubility also allows one to separate L-methionine from *N*-acetyl-D-methionine after treatment with acylase.

Once isolated, L-methionine can be analyzed by its specific rotation. Pure L-methionine has a specific rotation of $[\alpha]_D^{25} = -8.2°$ in water ($c = 1.00$ gram/ 100 mL water) or $[\alpha]_D^{25} = +22.5°$ in either 0.2 M or 1.0 M HCl. Observe how both the magnitude and sign of the specific rotation change with the acidity of the solvent. The optical purity of the isolated product can then be calculated by the equation

$$\text{Optical purity} = \frac{\text{observed specific rotation}}{\text{specific rotation of pure substance}} \times 100\%$$

PROCEDURE

N-Acetylmethionine

Dissolve 2.0 grams (0.013 mole) of D,L-methionine in 7.0 mL of 2.0 M NaOH solution in a 100-mL round-bottom flask. Add 5.0 mL (0.053 mole) of acetic anhydride over a period of 5 to 10 minutes, stoppering and shaking the flask after each addition. Record any observed change in appearance each time. Let the mixture stand another 10 minutes, then add 7.0 mL of 1.0 M H$_2$SO$_4$.

To isolate the N-acetylmethionine, you must evaporate the water and acetic acid from the mixture under reduced pressure. Assemble the apparatus in Figure 34-1 and evaporate the liquid slowly over a steam bath with swirling until a moist solid residue is obtained.

Add to the residue an equal volume of anhydrous magnesium sulfate powder, mix thoroughly and then extract the N-acetylmethionine from the solid by stirring with 15 mL of warm ethyl acetate. Carefully decant the ethyl acetate into a dry 125-mL Erlenmeyer flask through a small cotton plug in a funnel. Repeat the extraction with 10 mL of warm ethyl acetate and carefully decant through the cotton plug. The resulting ethyl acetate solution should be completely clear with no solid particles. If the solution is cloudy, add a little more anhydrous magnesium sulfate, allow to stand, and then decant again through a cotton plug.

Transfer the solution to a dry 50-mL Erlenmeyer flask and evaporate the liquid to about 5 mL on a steam bath. Place the flask in an ice bath and, if necessary, induce crystallization by scratching the oily liquid. Collect the product by suction filtration on a Hirsch funnel.

To see if all the methionine has been acetylated, dissolve about 5 mg of the product in 0.5 mL of water, add 3 drops of a 1% ninhydrin solution in ethanol, and heat for 30 to 60 seconds in a steam bath. **If** a blue-violet color develops, dissolve the N-acetylmethionine in about 5 to 10 mL of hot acetone and filter to remove the unreacted, insoluble methionine. Concentrate the acetone solution over a steam bath, cool, and add a few milliliters of anhydrous ethyl ether. Stratch the solution, if necessary, collect the purified crystals of N-acetyl-methionine, and air dry.

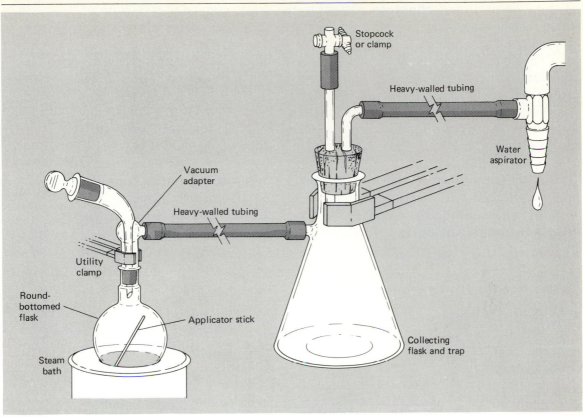

FIGURE 34-1
Reduced-pressure evaporation of water and acetic acid.

No matter whether the product was recrystallized or not, determine its theoretical and percentage yields, showing the calculations, and its melting point. Record the results in your notebook. The literature melting point is 114–115°C.

L-Methionine

Dissolve the *N*-acetylmethionine in 50 mL of distilled water in a clean 125-mL Erlenmeyer flask. Add 5 mL of 5% NaOH solution. Check the pH of the resulting solution with Hydrion paper for a pH of 7 to 8, making sure the paper is **not** dipped into the solution. Then add either 5% sodium hydroxide solution or glacial acetic acid **dropwise** until the pH is between 7 and 8.

Add 10 mg of acylase enzyme powder and mix well. Stopper the **labeled** flask and let it stand at room temperature for 2 to 7 days.

To isolate the hydrolyzed L-methionine, add glacial acetic acid to the reaction mixture **dropwise** until the pH is 4 or less when tested with Hydrion paper. Filter the solution through medium-porosity filter paper into a **labeled** 150-mL beaker and allow it to evaporate to dryness by placing it in an 80° drying oven overnight.

Add 20 mL of hot **absolute** ethanol to the dried residue for extraction of the unreacted *N*-acetylmethionine. Stir well and then place the entire contents in a large centrifuge tube.

Centrifuge the alcohol mixture, decant the alcohol, and wash the precipitate with another 15 mL of absolute ethanol. Centrifuge and decant the second alcohol wash.

Dissolve the white residue of L-methionine in hot distilled water, adding the water a few drops at a time. Keeping the solution hot, add hot absolute ethanol until precipitation begins. Then add 5 mL of additional absolute ethanol and cool.

Collect the purified crystals of L-methionine by suction filtration, then dry and weigh them. Determine the theoretical and percentage yields of the L-methionine, showing the calculations. Remember **not** to include the weight of the original D-methionine.

Optical rotation of L-methionine

If necessary, read pages 595–599 on optical rotation and the use of a polarimeter. It is very important that the sodium lamp be already warmed up and that you check the zero calibration mark on the instrument by making a zero reading with the sample tube filled with 1.0 *M* HCl. (Distilled water is **not** going to be the solvent this time.)

To ensure that adequate L-methionine is used, three to four students should pool all but about 50 mg of their product and collaborate on the determination. Weigh the collected samples to the nearest 0.001 gram in a 25-mL volumetric flask and dissolve them in 1.0 *M* HCl to make 25.0 mL of solution.

Rinse the empty polarimeter tube with a few milliliters of the L-methionine solution, discard it, and then fill the polarimeter tube with the L-methionine solution. Be sure there are no air bubbles or particles that could interfere with the path of the light. Measure the optical rotation of the solution and calculate the specific rotation and optical purity of the L-methionine, showing your calculations. Hand in the small amount of saved L-methionine in a properly labeled container to your instructor.

QUESTIONS

1. Why are amino acids shown as **zwitterions**, or ions that have both a positive and negative charge?
2. Why should an optically active amino acid **not** be expected to have the same specific rotation in acidic, alkaline, and neutral aqueous solutions? As L-methionine illustrates, even the sign of rotation can be different. Remember that only the acidity is being changed, not the solvent.

3. In the L-methionine experiment, how could the unreacted *N*-acetyl-D-methionine be isolated? If pure D-methionine were desired, how could you obtain it from the *N*-acetyl-D-methionine?

4. Why would you expect L-methionine to be soluble in acidic, alkaline, and neutral aqueous solutions but insoluble in acetone and absolute ethanol?

5. If the infrared and ^{1}H NMR spectra of a pure D and a pure L amino acid were taken, would there be any difference in the two infrared spectra? In the two ^{1}H NMR spectra? Explain your answers.

6. Even though bread, cereal, and nuts all contain protein, why aren't these foods capable of providing all the necessary protein the body needs? Why are meat, eggs, and dairy products satisfactory?

7. Why is egg white or milk used as an antidote for patients who have swallowed heavy-metal salts of lead, silver, or mercury? Why must the treatment with the above antidotes be followed by an emetic?

8. Why is alcohol successful as a skin disinfectant?

9. What do the alkaloidal reagents tannic acid and picric acid accomplish in the treatment of burns?

10. What would happen if a urine specimen containing albumin were heated to boiling? Why?

11. Why does a yellow stain appear when nitric acid is spilled on the skin? Why doesn't the stain wash off?

12. Why do all proteins give a positive biuret test?

13. Which individual amino acids contain two or more of the necessary functional groups to give a positive biuret test?

14. Look up the Strecker synthesis of amino acids in another book and explain specifically why at two different points in the synthesis a mixture of equal amounts of D and L isomers is expected. Do **not** just merely say that optically inactive reagents give optically inactive products.

35

INFRARED SPECTROSCOPY: PREPARING THE SAMPLES AND INTERPRETING THE SPECTRA

35.1 INTRODUCTION

If a chemist were given a small sample of a pure organic compound to identify, he or she would certainly take its infrared spectrum. With it, he or she would know much about the structure of the compound by specifically knowing what functional groups were present or missing. After other information about the compound was known, proof of the identity (except for optical isomers) could then be established by comparing the infrared spectrum of the compound with that of an actual known sample. If the two spectra were identical, when taken under identical conditions, then the structure of the compound must be the same as that of the known.

An infrared spectrum is also frequently valuable in telling where the substituents are located on an aromatic ring, if there is conjugation present, and if there is extensive hydrogen bonding. It can even frequently give a quantitative analysis of a mixture with an accuracy of $\pm 1-5$ percent. Other benefits are the very small quantity of sample needed, the affordability of the instrument, the speed with which the spectrum can be taken, and the fact that the sample can be recovered and used for other purposes, if necessary.

To understand the principle behind an infrared spectrum, consider the chemical bonds between atoms to be analogous to a spring with a weight attached to each end. Just as these weights will move from side to side or up and down after the spring is stretched or bent, so will the atoms forming a chemical bond move in the same manner. With a spring, the energy to cause this movement comes from some outside force such as a human hand. But with atoms, the stretching or bending results from the absorption of infrared light, which excites the atoms into motion. For a nonlinear molecule, the number of possible ways this can occur is $3n - 6$, where n is the number of atoms. Thus, ethane with eight atoms could stretch or bend in $3(8) - 6$ or 18 ways. Combinations or multiples of these ways are also possible. The types of stretching and bending in a methylene unit are depicted in Figure 35-1.

One restriction is that in order for the light to be absorbed, the stretching or bending must produce a change in the dipole moment in that region of the molecule. Furthermore, the greater the change in the dipole moment, the greater will be the amount of light absorbed. Thus atoms with highly differing electronegativities such as carbon and oxygen will generally cause light to be more strongly absorbed with the stretching or bending.

The amount of energy needed to produce the stretching or bending varies from about 2 to 11 kcal/mole. This means that the infrared region, having a wavelength from 2.5 to 16 microns or a wave number from 4000 to 625 cm^{-1},

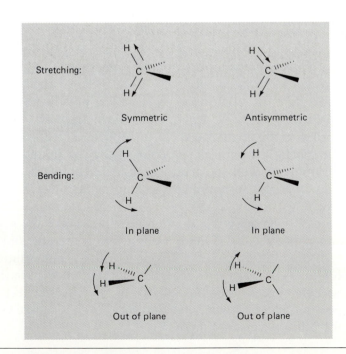

FIGURE 35-1
The types of stretching and bending movements in a methylene unit.

has the greatest practical value in identifying functional groups or other structural features of a molecule. The micron is symbolized by μ and has a value of 1.00×10^{-4} cm. It is related to the wave number by the equation

$$\mu = \frac{10,000}{\text{Wave number in cm}^{-1}}$$

Thus the wavelength in microns and the wave number in cm^{-1} are inversely proportional to each other, and one of them can always be calculated by dividing the other one into 10,000. The wave number is also frequently regarded as the frequency, even though, strictly speaking, it doesn't have frequency units. Also, from the equation $E = hc/\lambda$, a higher wavelength means that a smaller amount of energy is being absorbed per photon. Conversely, a higher wave number means that a greater amount of energy is being absorbed per photon. Figure 35-2 shows a typical infrared spectrum.

In any infrared spectrum, the frequency of the light absorbed in cycles per second equals the frequency of the resulting molecular vibration. With springs, it is well known that the frequency of the vibration increases quantitatively with lighter weights and with the spring's stiffness. It is the same way with atoms. Thus, lighter atoms will have a higher frequency of vibration than will heavier atoms in a chemical bond, and they will absorb infrared light having a higher frequency or a lower wavelength than heavier atoms. And stronger chemical bonds will have a higher frequency of vibration than will weaker chemical bonds, and will absorb infrared light having a higher frequency or a lower wavelength than weaker chemical bonds. Thus the stronger carbon-carbon triple bond absorbs light of a higher frequency than does the weaker carbon-carbon double

FIGURE 35-2
The infrared spectrum of acetylsalicylic acid in a KBr pellet.
Spectrum courtesy of Sadtler Research Laboratories, division
of Bio-Rad Laboratories, Inc., copyrighted 1974.

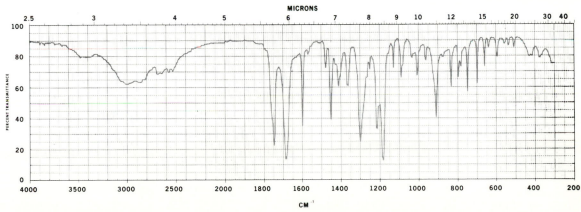

bond, which in turn absorbs light of a higher frequency than does the still weaker carbon-carbon single bond.

From studies of numerous infrared spectra, it has been well established exactly how strongly each type of chemical bond will absorb light and what will be the exact wavelength or wave number range. For example, the carbon-oxygen double bond in a saturated, aliphatic aldehyde is known to strongly absorb light in the 5.75–5.81 μ or 1740–1720 cm^{-1} range. All this information has been tabulated, and by looking at these tables, a person can know which functional groups are responsible for many of the peaks in an infrared spectrum. In fact, you will be learning how to do this later in the interpretation section of this experiment. A few of the peaks appearing above 6.5 μ or below 1540 cm^{-1} are characteristic of certain functional groups, but most of these peaks result from the complex vibrations of the whole molecule. These peaks are frequently difficult to interpret, but are highly characteristic of each individual compound. As a result, this region is called the "fingerprint region," and this is where two

FIGURE 35-3
A schematic diagram of a typical infrared spectrophotometer.

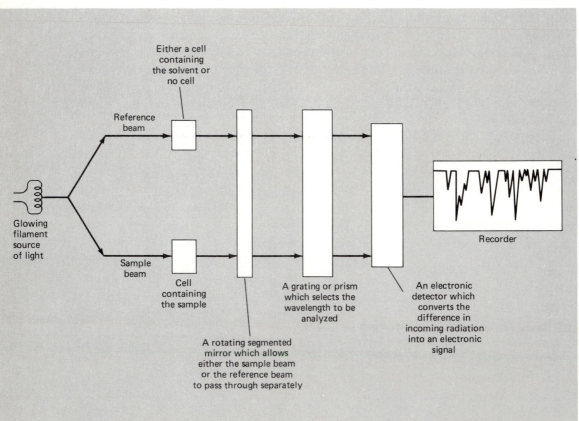

Either a cell containing the solvent or no cell

Reference beam

Glowing filament source of light

Sample beam

Cell containing the sample

A rotating segmented mirror which allows either the sample beam or the reference beam to pass through separately

A grating or prism which selects the wavelength to be analyzed

An electronic detector which converts the difference in incoming radiation into an electronic signal

Recorder

similar compounds would primarily give different spectra. As a student, you would **not** be asked to interpret most of the peaks within this region.

35.3 INFRARED SPECTROPHOTOMETER

The schematic diagram of a typical infrared spectrophotometer is shown in Figure 35-3. An electrically heated filament serves as the source of the infrared light, which passes through both the sample cell and either a reference cell containing the solvent, if any, or no cell. The two beams then pass through a rotating segmented mirror, which continuously allows one beam and then the other to separately pass through. The beams then travel to either a prism or grating, which selects which wavelength will be read at any particular time. From here the beams travel to the electronic detector, which compares the absorbed beam with the reference beam at each individual wavelength, and signals the difference electronically to a recorder.

Because of the numerous makes and models of infrared spectrophotometers available, your instructor will explain how to operate the particular one you will be using. Remember to always immediately write on the spectrum sheet the name of the sample, the type of cell or method used, the solvent, if any, and any other pertinent information. Otherwise, all this information may be forgotten.

35.4 SAMPLE PREPARATION

To take an infrared spectrum, first place the sample in a cell or mount it between plates. Unfortunately, glass, quartz, and plastics are all unsatisfactory for use as a cell or plates, since they absorb strongly throughout the entire infrared region. Ionic substances such as sodium chloride and silver chloride are transparent to infrared light and are actually used. However, they present the following problems:

1. WATER AND LOWER-MOLECULAR-WEIGHT ALCOHOLS WILL DISSOLVE SODIUM CHLORIDE. THEREFORE, TO PREVENT AN EXPENSIVE CELL OR A PAIR OF PLATES FROM BEING RUINED, ALL WATER AND LOWER-MOLECULAR-WEIGHT ALCOHOLS MUST BE SCRUPULOUSLY KEPT AWAY. THE CELLS OR THE PLATES MUST BE STORED IN AIRTIGHT DESICCATORS, NEVER HANDLED WITH THE FINGERS EXCEPT ON THE EDGES, AND USED ONLY WITH SAMPLES CONTAINING ABSOLUTELY NO WATER OR LOWER-MOLECULAR-WEIGHT ALCOHOLS. The reason these sample cells and plates are so expensive is that they have been cut and polished from single large crystals of sodium chloride.

2. SODIUM CHLORIDE AND SILVER CHLORIDE CELLS AND PLATES ARE BREAKABLE AND CAN BE EASILY SNAPPED IN TWO. NO EXCESSIVE PRESSURE SHOULD EVER BE PLACED ON THEM.

3. SILVER CHLORIDE CELLS AND PLATES ARE PHOTOSENSI-TIVE AND MUST BE STORED IN THE DARK TO PREVENT THEM FROM TURNING BLACK. However, since silver chloride is insoluble in water and lower-molecular-weight alcohols, it can be used when these substances are present. But realize that they will also absorb in the infrared, and so could interfere with the infrared spectrum of another compound.

35.5 PREPARING A LIQUID SAMPLE BETWEEN SALT PLATES

The easiest way to obtain an infrared spectrum of a liquid is to place a drop of the liquid between two salt plates to form a thin film. The plates are then placed in the holder shown in Figure 35-4, and the spectrum is taken. The word **neat,** meaning without solvent present, is frequently used to describe this procedure. Highly volatile or low-viscosity liquids, however, do not make good films. These liquids should instead be injected into a solution cell as described in Section 35.6.

PROCEDURE

1. Remove the salt plates from the desiccator and, if necessary, carefully clean their surfaces with methylene chloride and a soft tissue. Be sure they are thoroughly dry before use. Be sure to handle the plates only by their edges.

2. Place 1 drop of the liquid on the surface of one plate and then place the second plate on top. The weight of the second plate will cause the liquid to spread out and form a thin film between the plates. Be sure the sample is dis-tributed evenly and that there are no air pockets present.

3. Place the plates in the holder as indicated in Figure 35-4. **Be sure not to tighten the cinch nuts excessively, since the pressure could snap the plates.** Also the liquid film could become very thin, resulting in all of the peaks in the spectrum being too short. Unlike other instruments, an infrared spectrophotometer usually has no knob to control the intensity of the absorption. This control is possible only by adjusting the sample concentration or the path length through the sample.

4. Place the sample holder in the first sample compartment toward the front of the spectrophotometer. Leave the second compartment empty.

5. Run the spectrum, being sure the paper has been accurately placed in its holder.

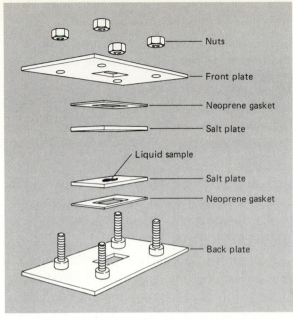

FIGURE 35-4
An infrared salt-plate holder assembly.

6. Remove the sample holder from the instrument and, with the spectrum paper still in place, run a portion of the spectrum of polystyrene film to calibrate the spectrum as described in Section 35.12. Completely label the spectrum.

7. Remove the salt plates from the sample holder. Carefully clean them with methylene chloride, using a soft tissue. Dry them and return them to the storage desiccator.

35.6 PREPARING A LIQUID SAMPLE NEAT IN A SOLUTION CELL

A highly volatile or a low-viscosity liquid can have its infrared spectrum taken in a solution cell, shown in Figure 35-5. Observe that a Teflon spacer is placed between the salt plates to control the thickness of the sample. The top salt plate also has two holes drilled in it to introduce the sample. These holes are below two tubular extensions, which are then plugged to seal the internal chamber and prevent evaporation.

PROCEDURE

1. Remove the solution cell from the desiccator and, if necessary, carefully clean the salt plates with methylene chloride and a soft tissue, being sure they are thoroughly dry before use. Be sure to handle the plates **only** by their edges.

2. Assemble the parts of the solution cell as indicated in Figure 35-5, but leave the tubular extensions unplugged.

3. Using a syringe, fill the cell with the sample liquid. Usually this is done by holding the cell somewhat upright and filling the bottom entrance port. Plug the ports when finished.

4. Place the solution cell in the first sample compartment toward the front of the spectrophotometer. Leave the second compartment empty.

5. Run the spectrum, being sure the paper has been accurately placed in its holder.

6. If necessary, use a different Teflon spacer between the salt plates to get the right path length of the light through the sample and the proper length of the peaks. Unlike other instruments, an infrared spectrophotometer usually has no knob to control the intensity of the absorption.

7. Remove the solution cell from the instrument and, with the spectrum paper still in place, run a portion of the spectrum of polystyrene film to calibrate the spectrum as described in Section 35.12. Completely label the spectrum.

8. Remove the salt plates from the solution cell. If a highly volatile liquid sample was used, it is only necessary to disassemble the cell, dry the salt plates,

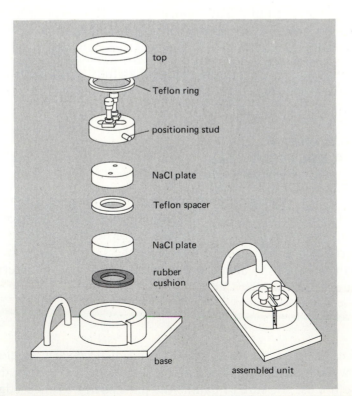

FIGURE 35-5
A solution cell for highly volatile or low-viscosity liquids.

top

Teflon ring

positioning stud

NaCl plate

Teflon spacer

NaCl plate

rubber cushion

base

assembled unit

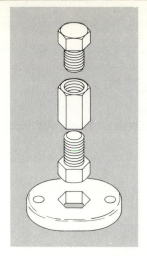

FIGURE 35-6
A minipress for preparing KBr pellets.

and return them to the storage desiccator. Otherwise, carefully clean the plates and the rest of the cell with methylene chloride, dry them, and return them to the storage desiccator.

35.7 PREPARING A SOLID SAMPLE IN A KBr PELLET OR DISK

With this method, a finely ground mixture of the solid sample and spectral-grade KBr is pressed to form a disk or pellet. Both the sample and the KBr must be anhydrous to avoid any absorption of light by water. Since KBr absorbs moisture on standing, it must be previously dried in an oven and then stored in a desiccator. A special minipress to both create the disk and hold it in place in the spectrophotometer is shown in Figure 35-6. It is available from the Wilkes Scientific Company, 80 Republic Drive, North Haven, Conn. 06473.

Some people prefer to use KI rather than KBr for preparing pellets or disks. It is much less hygroscopic.

PROCEDURE

1. Screw one of the bolts into the minipress until only two turns remain.

2. Weigh out about 1–2 mg of the anhydrous solid sample on an analytical balance and thoroughly mix it with about 100 mg of anhydrous, spectral-grade KBr.

3. Grind the mixture with a mortar and pestle for at least 3 minutes to form a very fine powder. Scattering of the infrared light will occur if the powder is not ground fine enough.

4. Carefully place no more than half of the powder into the open end of the minipress. Tap the minipress on the desk top to give an even layer of filling.

5. Put the second bolt into the open end of the minipress and screw by hand until firm.

6. Place one end of the minipress into the holder bolted to the table. Then turn the other end with a torque wrench until a loud click is heard. **Do not apply more than 20 foot-pounds of force or you may break the minipress.**

7. Leave the bolts in for about 1 minute, then remove. But leave the compressed KBr disk in the minipress. The disk when examined with light should be either clear or translucent. A cloudy disk is unacceptable because it will excessively scatter the beam of infrared light. If the disk is cloudy, the following may have caused it:
(a) Either the sample or the KBr may not have been completely dry.
(b) The disk may be too thick from too much of the mixture being put into the minipress.
(c) The mixture may not have been ground finely enough.
(d) Too much sample may have been put in with the KBr.
(e) The sample may have a low melting point and may have melted under pressure.
(f) A torque wrench was not used, and as a result the minipress was not tightened enough.

8. Place the minipress into its special holder and place it in the first sample compartment toward the front of the spectrophotometer. Leave the second compartment empty.

9. Run the spectrum, being sure the paper has been accurately placed in its holder.

10. Remove the minipress and its holder from the instrument and, with the spectrum paper still in place, run a portion of the spectrum of polystyrene film to calibrate the spectrum as described in Section 35.12. Completely label the spectrum.

11. Remove the KBr pellet from the minipress and thoroughly wash and dry the minipress.

35.8 PREPARING A SOLID SAMPLE IN A NUJOL MULL

If a satisfactory KBr pellet cannot be obtained, the finely ground anhydrous powder can be mixed with Nujol mineral oil to form a mull. However, since Nujol absorbs infrared light at 3.4–3.6 μ (2900–2850 cm^{-1}), 6.8 μ (1460 cm^{-1}), and 7.3 μ (1380 cm^{-1}), the spectrum is worthless in these regions. Furthermore, it is frequently difficult to get a consistency in the mull that will not scatter the light beam. Therefore, one usually tries the KBr pellet method first.

PROCEDURE

1. Weigh 5 mg of the sample on an analytical balance and grind the sample with a mortar and pestle.

2. Add 1 to 2 drops of Nujol and grind until the mixture has a very fine dispersion and is about the consistency of milk of magnesia.

3. Remove the salt plates from the desiccator. Carefully clean them, if necessary, with methylene chloride, using a soft tissue. Be sure to handle the plates **only** by their edges.

4. Place the mull in the middle of one salt plate with a rubber policeman; then cover the first plate with the second one. The weight of the second plate will cause the mull to spread. Be certain the sample is distributed evenly and that there are no air pockets.

5. Place the plates in the holder as shown in Figure 35-4. **Be sure not to tighten the cinch nuts excessively, since the pressure could snap the plates.** Also, too much pressure could spread out the liquid too much, resulting in the peaks of the spectrum being too short.

6. Place the sample holder in the first sample compartment toward the front of the spectrophotometer. Leave the second compartment empty.

7. Run the spectrum, being sure the paper has been accurately placed in its holder.

8. Remove the sample holder from the instrument and, with the spectrum paper still in place, run a portion of the spectrum of polystyrene film to calibrate the spectrum as described in Section 35.12. Completely label the spectrum.

9. Remove the salt plates from the sample holder and carefully wipe them clean with a soft tissue. Then carefully clean them with methylene chloride, using a soft tissue. Dry them and return them to the storage desiccator.

35.9 PREPARING A SOLID SAMPLE DISSOLVED IN A SOLUTION

If a satisfactory KBr pellet cannot be obtained, a solid can also be dissolved in a solvent and placed in the solution cell shown in Figure 35-7. Solvents which may be used are carbon tetrachloride and carbon disulfide. However,

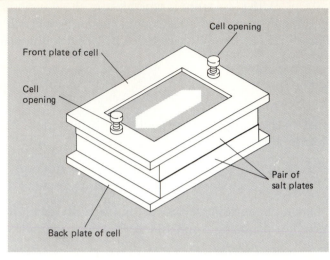

Cell opening

Front plate of cell

Cell opening

Pair of salt plates

Back plate of cell

FIGURE 35-7
A solution cell for a solid sample dissolved in a solvent.

both these solvents have the disadvantage of strongly absorbing in various regions of the infrared spectrum. Carbon tetrachloride absorbs weakly at 6.3–6.4 μ (1590–1565 cm^{-1}) and strongly from 12–16 μ (835–625 cm^{-1}), while carbon disulfide absorbs weakly at 4.4 μ (2270 cm^{-1}), moderately at 4.7 μ (2130 cm^{-1}), and strongly from 6.3–7.0 μ (1590–1430 cm^{-1}). Thus, unless two spectra with two separate solvents are taken, any peaks from the sample compound in these regions could easily be lost. Other possible solvents such as chloroform and methylene chloride have too many interfering peaks and should therefore not be used.

In addition, these solvents must both be used with care and not inhaled or spilled on the skin. Carbon disulfide is very toxic, and carbon tetrachloride is toxic, causes severe liver damage, and is suspected to be carcinogenic. Rubber or plastic gloves are advisable, and as much work as possible should be done in the hood.

If this method is used, a second cell containing only the solvent is placed in the second compartment of the spectrophotometer. The purpose is to cancel out as much as possible the absorbency caused by the solvent.

PROCEDURE

1. Dissolve the sample in the solvent **in the hood** to make a 5 to 10% solution (weight-volume). Only a few drops are actually needed.

2. Remove the cells from the storage desiccator. If necessary, clean them by the procedure given later. Be sure to handle the cells **only** by their edges.

3. Remove the plugs at the cell openings and, with the cell slightly inclined, place about 3 drops of the solution into the lower opening with a capillary dropper. You should be able to see through the window the solution rising between the salt plates. When finished, be certain the cell window contains only the liquid and no air. Be careful not to spill any solution on the outside of the cell windows. Stopper the openings.

4. By the same procedure, using a clean capillary dropper, fill the second cell with the solvent used to dissolve the sample.

5. Place the solvent cell in the far compartment of the spectrophotometer, and put the sample cell in the near compartment.

6. Run the spectrum, being sure the paper has been accurately placed in its holder.

7. Remove both cells from the instrument and, with the spectrum paper still in place, run a portion of the spectrum of polystyrene film to calibrate the spectrum as described in Section 35.12. Completely label the spectrum.

8. Remove the plugs at the openings to the cells. Clean out the sample cell by flushing it with excess methylene chloride. **Dry, clean**, compressed air can then be used to dry both cells. Return the cells to the storage desiccator.

35.10   PREPARING A SOLID SAMPLE NEAT

Many solids will form a translucent residue on a salt plate if a concentrated solution of the solid is evaporated. This residue will not scatter infrared light, and thus a spectrum may be taken. An opaque residue, however, will scatter the light and be unsatisfactory. Highly volatile solvents such as anhydrous ether, methylene chloride, or pentane are preferred.

PROCEDURE

1. Remove the salt plates from the storage desiccator. If necessary, carefully clean them with methylene chloride, using a soft tissue. Be sure to handle the plates **only** by their edges.

2. Prepare a highly concentrated solution of the solid in 4 or 5 drops of the best highly volatile solvent. If ether, methylene chloride, or pentane do not dissolve enough material, a somewhat higher-boiling solvent may be used, but the evaporation period will be longer.

3. Place the 4 or 5 drops of the solution onto the middle of one salt plate, 1 drop at a time. Allow the solvent to evaporate. If the residue obtained is opaque, place 1 or 2 drops of the solvent used on the salt plate to make the original solution. Allow the solvent to evaporate. If the residue is still opaque, **stop**. Some other method will have to be used.

4. If the residue obtained is translucent, continue by placing the second salt plate onto the first.

5. Place the pair of salt plates in the holder, as illustrated in Figure 35-4. **Be sure not to tighten the cinch nuts excessively, since the pressure could snap the plates.**

6. Place the sample holder in the first sample compartment toward the front of the spectrophotometer. Leave the second compartment empty.

7. Run the spectrum, being sure the paper has been accurately placed in its holder.

8. Remove the sample holder from the instrument and, with the spectrum paper still in place, run a portion of the spectrum of polystyrene film to calibrate the spectrum as described in Section 35.12. Completely label the spectrum.

9. Remove the salt plates from the sample holder and carefully wipe them clean, using a soft tissue. Then carefully clean them with methylene chloride, using a soft tissue. Dry them and return them to the storage desiccator.

35.11 PREPARING A LOW-MELTING-POINT SOLID SAMPLE NEAT

Many solids having low melting points up to about 60°C will form a translucent thin film on a salt plate after they are melted and then resolidified. This film will not scatter infrared light, and thus a spectrum may be taken. An opaque film, however, will scatter the light and be unsatisfactory. Because of the melting-point restriction and the difficulty in frequently getting a satisfactory film to form, the method is not as common as the others for preparing solid samples. Higher-melting-point compounds are generally not used for the method because of the more severe heating required.

PROCEDURE

1. Remove the salt plates from the storage desiccator. If necessary, carefully clean them with methylene chloride, using a soft tissue. Be sure to handle the plates **only** by their edges.

2. Melt a sample of the compound in a small test tube in a hot-water bath. **Keeping the salt plates away from any water**, place 1 drop of the liquid sample onto the surface of one plate, and then place the second plate on top. The weight of the second plate will cause the liquid to spread out and form a thin film between the plates. Be sure the sample is distributed evenly and that there are no air pockets present. If the residue becomes opaque upon cooling, **stop**. Some other method will have to be used.

As an alternative method, no second salt plate is used and several drops of the melted compound are allowed to flow in a thin layer over the first salt plate. Again be certain the sample is distributed evenly and does not become opaque upon cooling.

Another method is to place a small amount of the solid sample on one salt plate, which is then placed on a clean watch glass on a **warm** hot plate. The sample is melted and allowed to flow in a thin layer over the salt plate. Again, be certain the sample is distributed evenly. The second warmed salt plate may now be placed over the first, or just the one salt plate may be used. Again, be certain the sample does not become opaque upon cooling.

3. If the film obtained remains translucent, place the salt plate or plates in the holder, as illustrated in Figure 35-4. **Be sure not to tighten the cinch nuts excessively, since the pressure could snap the plate or plates.**

4. Place the sample holder in the first sample compartment toward the front of the spectrometer. Leave the second compartment empty.

5. Run the spectrum, being sure the paper has been accurately placed in its holder.

6. Remove the sample holder from the instrument and, with the spectrum paper still in place, run a portion of the spectrum of polystyrene film to calibrate the spectrum as described in Section 35.12. Completely label the spectrum.

7. Remove the salt plate or plates from the sample holder and carefully wipe them clean, using a soft tissue. Then carefully clean them with methylene chloride, using a soft tissue. Dry them and return them to the storage desiccator.

35.12   CALIBRATING THE SPECTRUM

As a check that the spectrum paper has been accurately placed, the spectrum is usually calibrated afterward with polystyrene film. The peak in polystyrene most often used is at 6.24 μ (1602 cm^{-1}), but the ones at 3.51 μ (2850 cm^{-1}) and 11.04 μ (906 cm^{-1}) are also valuable. The card holding this polystyrene film is placed in the front sample cell holder, and the **tips** (not the entire spectrum) of the desired peaks are recorded on the spectrum of the other compound. Any deviation between the known wavelength or wave number of the polystyrene peaks and what the spectral paper says is then noted.

35.13   INTERPRETING AN INFRARED SPECTRUM

One determines which functional groups or other structural features are present in an organic compound by correlating their known wavelength or wave-number values with the actual peaks in the infrared spectrum. But with so many peaks frequently present in the spectrum, how can one know where to begin? The following guide should be immensely helpful. Remember that, as a student, you are **not** expected to be able to interpret every peak.

1. Check for a carbonyl group. The C=O group gives strong absorption in the region 5.5–6.1 μ (1820–1660 cm^{-1}). The peak is often the strongest absorbing one in the spectrum and has a medium width.

2. If a carbonyl group is present, check for the following peaks. If it is absent, continue on to step 3.
(a) **Acids:** O—H absorption: Check for a **broad, strong** peak in the range 3.0–4.0 μ (3300–2500 cm^{-1}). This peak usually overlaps the C—H absorption peak.
(b) **Aldehydes:** C—H absorptions: Check for **two** medium to weak peaks at 3.50 μ (2850 cm^{-1}) and 3.65 μ (2750 cm^{-1}). The peak at 3.50 μ is often obscure.
(c) **Amides:** N—H absorption: Check for a narrow or broad medium peak at 2.85 μ (3500 cm^{-1}). The peak will sometimes be two equivalent halves.
(d) **Anhydrides:** C=O absorptions: Check for **two** peaks near 5.5 μ (1810 cm^{-1}) and 5.7 μ (1760 cm^{-1}). Anhydrides are much less common than the other functional groups.
(e) **Esters:** C—O absorption: Check for **two** medium-intensity peaks in the region 7.7–10.0 μ (1300–1000 cm^{-1}). One peak will be symmetrical, the other an asymmetrical stretch.
(f) **Ketones:** Are usually assigned by eliminating possibilities (a)–(e).

3. (a) **Alcohols and phenols:** O—H absorption: Check for a strong, broad peak in the region 2.8–3.2 μ (3600–3200 cm^{-1}). Confirm by finding a C—O peak near 7.7–10.0 μ (1300–1000 cm^{-1}).
(b) **Amines:** N—H absorption: Check for one or two medium, possibly broad peaks in the region 2.8–3.0 μ (3550–3060 cm^{-1}).
(c) **Alkyl ethers:** C—O absorption: Check for a peak in the region 8.7–9.2 μ (1150–1085 cm^{-1}). Be sure there is no O—H absorption.
(d) **Aryl-alkyl ethers:** C—O absorption: Check for one peak in the region 7.8–8.3 μ (1275–1200 cm^{-1}) and one in the region 9.3–9.8 μ (1075–1020 cm^{-1}). Be sure there is no O—H absorption.

4. Alkenes: C=C absorption: Check for a weak-to-medium peak near 6.1 μ (1650 cm^{-1}). The spectrum may also have a weak vinyl C—H peak to the left of 3.33 μ (3000 cm^{-1}). Aliphatic C—H peaks are to the right of 3.33 μ.

5. Aromatic rings: C—C bonds in ring absorption: Check for a medium-to-strong peak near 6.25 μ (1600 cm^{-1}) and one peak at 6.9 μ (1450 cm^{-1}) or two peaks at 6.7 μ (1500 cm^{-1}) and 6.9 μ (1450 cm^{-1}). Also check for a weak-to-medium aromatic C—H peak to the left of 3.33 μ (3000 cm^{-1}).

6.   Alkynes: C≡C absorption: Check for a sharp, weak peak near 4.65 μ (2150 cm^{-1}). Also check for a medium acetylenic C—H peak near 3.0 μ (3300 cm^{-1}).

7.   Nitro groups: N=O absorption: Check for **two** strong peaks in the region 6.3–6.7 μ (1600–1500 cm^{-1}) and 7.1–7.7 μ (1390–1300 cm^{-1}).

8.   Nitriles: C≡N absorption: Check for a weak, sharp peak at 4.42–4.65 μ (2260–2150 cm^{-1}).

9.   Halogens:
(a) **Fluorine:** C—F absorption: Check for a strong peak in the region of 7.1–10.0 μ (1400–1000 cm^{-1}).
(b) **Chlorine:** C—Cl absorption: Check for a strong peak in the region of 12.5–16.6 μ (800–600 cm^{-1}).
(c) **Bromine:** C—Br absorption: The peak occurs at 16.6–20.0 μ, or beyond the region of most infrared spectrophotometers.
(d) **Iodine:** C—I absorption: The peak occurs at 20 μ, or beyond the region of most infrared spectrophotometers.

10.   Alkanes: C—H absorption: Check for a strong peak near 3.33 μ (3000 cm^{-1}) and medium peaks near 6.90 μ (1450 cm^{-1}) and 7.27 μ (1375 cm^{-1}). The spectrum will be very simple with no other functional group absorption peaks.

Even more detail to the important functional groups will be found in the following section.

35.14   SURVEY OF THE IMPORTANT FUNCTIONAL GROUPS

Alkanes

CH$_3$ C—H stretching: Methyl groups give strong peaks at 3.38 μ (2960 cm^{-1}) and 3.48 μ (2870 cm^{-1}).

C—H bending: Methyl groups give two medium peaks at 6.90 μ (1450 cm^{-1}) and 7.28 μ (1375 cm^{-1}). Interference with methylene peaks can occur.

CH$_2$ C—H stretching: Methylene groups give two strong peaks at 3.43 μ (2925 cm^{-1}) and 3.51 μ (2855 cm^{-1}).

C—H stretching: Methylene groups **in a ring** give a strong peak in the 3.23–3.34 μ (3100–2990 cm^{-1}) region.

C—H bending: Methylene groups give a medium peak at 6.83 μ (1465 cm^{-1}).

Alkenes

C=C stretching: One moderate or weak peak at 5.95–6.25 μ (1675–1600 cm^{-1}). Conjugation moves this peak to the right. Symmetrically substituted double bonds have no peak because of no change in the dipole moment. Highly substituted double bonds usually have very weak absorption.

Conjugated symmetrical dienes give one medium-to-weak peak at 6.25 μ (1600 cm^{-1}).

Conjugated unsymmetrical dienes give two medium-to-weak peaks at 6.06 μ (1650 cm^{-1}) and 6.25 μ (1600 cm^{-1}).

C—H stretching: One or more medium peaks in the region 3.23–3.33 μ (3100–3000 cm^{-1}).

C—H bending: Gives strong absorption in the region 10.0–15.4 μ (1000–650 cm^{-1}).

Alkynes

C≡C stretching: Single medium-to-weak peak at 4.43–4.76 μ (2260–2100 cm^{-1}).

Conjugation moves this peak to the right.

Disubstituted triple bonds either give weak or no absorption, due to little or no change in the dipole moment.

Symmetrical triple bonds give no absorption, due to no change in the dipole moment.

≡C—H stretching: Strong, sharp, narrow peak at 3.00–3.06 μ (3333–3267 cm^{-1}).

Aromatic rings

C—C ring stretching: Medium-to-variable, sharp peak or peaks around 6.25–6.33 μ (1600–1580 cm^{-1}).   Medium-to-variable, sharp peak or peaks around 6.67–6.90 μ (1500–1450 cm^{-1}).

C—H stretching: Medium-to-weak peaks in the region 3.23–3.33 μ (3100–3000 cm^{-1}).

C—H out-of-plane bending: Strong-to-medium peaks, very useful for determining location of ring substituents. Very reliable for alkyl substituents. Less reliable for polar substituents.

Benzene	14.9 μ (670 cm^{-1})
Monosubstituted benzene	12.99–13.70 μ (770–730 cm^{-1}) 14.08–14.99 μ (710–690 cm^{-1})
1,2-Disubstituted benzene	12.99–13.61 μ (770–735 cm^{-1})
1,3-Disubstituted benzene	11.11–11.63 μ (900–860 cm^{-1}) 12.35–13.33 μ (810–750 cm^{-1}) 13.74–14.71 μ (725–680 cm^{-1})
1,4-Disubstituted benzene and 1,2,3,4-tetrasubstituted benzene	11.63–12.50 μ (860–800 cm^{-1})
1,2,3-Trisubstituted benzene	12.50–12.99 μ (800–770 cm^{-1}) 13.89–14.60 μ (720–685 cm^{-1})
1,2,4-Trisubstituted benzene	11.11–11.63 μ (900–860 cm^{-1}) 11.63–12.50 μ (860–800 cm^{-1})
1,3,5-Trisubstituted benzene	11.11–11.63 μ (900–860 cm^{-1}) 11.56–12.35 μ (865–810 cm^{-1}) 13.70–14.81 μ (730–675 cm^{-1})

The weak peaks shown in Figure 35-8 are also useful for determining the location of ring substituents. They are very reliable for alkyl substituents, but less reliable for polar substituents.

Alcohols and phenols

O—H stretching: With hydrogen bonding (common with neat samples and concentrated solutions), there will be a broad, strong peak at 2.82–3.13 μ (3550–3200 cm^{-1}). It may overlap C—H stretching absorptions. Without hydrogen bonding (common with dilute solutions), there will be a sharp, strong peak at 2.74–2.79 μ (3650–3600 cm^{-1}).

C—O stretching: Produces a broad, strong-to-medium peak in the range 7.7–10.0 μ (1300–1000 cm^{-1}).

Ethers

C—O stretching: Alkyl ethers give a strong-to-medium, broad peak at 8.70–9.23 μ (1150–1085 cm^{-1}). A branched alkyl group produces multiple peaks. There will be **no** C=O or O—H peaks from an ester or alcohol.

Aryl alkyl ethers absorb at 7.84–8.33 μ (1275–1200 cm^{-1}) and at 9.30–9.80 μ (1075–1020 cm^{-1}). Again, there will be **no** C=O or O—H peaks from an ester or alcohol.

Amines

N—H stretching: Primary amines with hydrogen bonding (common with neat samples and concentrated solutions) give two medium peaks at 2.94–3.00 μ (3400–3330 cm^{-1}) and 3.00–3.08 μ

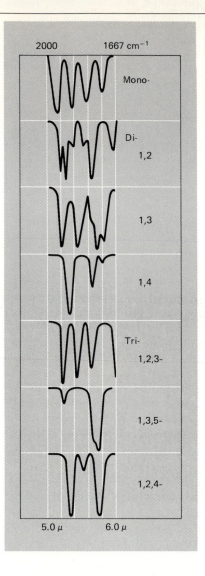

2000   1667 cm⁻¹

Mono-

Di-
1,2

1,3

1,4

Tri-
1,2,3-

1,3,5-

1,2,4-

5.0 μ   6.0 μ

FIGURE 35-8

The 5.0–6.0 μ (2000–1667 cm⁻¹) absorption patterns
for various substituted benzene compounds.

(3330–3250 cm⁻¹). Primary amines without hydrogen
bonding (common with dilute solutions) give two weak
peaks at 2.86 μ (3500 cm⁻¹) and 2.94 μ (3400 cm⁻¹). Primary aromatic amines have these peaks shifted slightly to
the right.

Secondary amines give one weak peak at 2.58–3.02 μ
(3350–3310 cm⁻¹), no matter how their spectrum is taken.

Tertiary amines have no N–H stretching.

C—N stretching: Gives medium-to-weak peaks in the range 7.4–10.0 μ
(1350–1000 cm⁻¹).

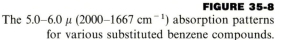

Carboxylic acids

C=O stretching: Produces a strong, broad peak at 5.81–5.86 μ (1720–1706 cm^{-1}).

Conjugation moves this peak to 5.85–5.95 μ (1710–1680 cm^{-1}).

Internal hydrogen bonding moves this peak as far to the right as 6.01 μ (1665 cm^{-1}).

O—H stretching: Hydrogen bonding usually produces a very broad, strong peak at 3.0–4.0 μ (3300–2500 cm^{-1}). The peak often interferes with C—H peaks.

C—O stretching: Produces a strong peak in the range 7.6–8.3 μ (1320–1210 cm^{-1}).

Esters

C=O stretching: Gives a strong peak at 5.71–5.76 μ (1750–1735 cm^{-1}).

Conjugation with the carbonyl carbon moves this peak to 5.78–5.83 μ (1730–1715 cm^{-1}).

Conjugation with the alcohol oxygen moves this peak to 5.62–5.65 μ (1780–1770 cm^{-1}).

Ring strain in lactones moves this peak to the left.

C—O stretching: Produces two or more peaks, one stronger than the others, in the range 7.0–10.0 μ (1430–1000 cm^{-1}).

Saturated esters have strong peaks at 8.26–8.60 μ (1210–1163 cm^{-1}).

Aromatic esters have strong peaks at 7.63–8.00 μ (1310–1250 cm^{-1}).

α,β unsaturated acid esters give peaks from 7.69–8.62 μ (1300–1160 cm^{-1}).

Amides

C=O stretching: Gives a strong peak at 5.90–6.06 μ (1700–1630 cm^{-1}).

N—H stretching: The location of the peaks depends upon how the spectrum was taken. Primary amides in KBr pellets give two medium peaks at about 2.99 μ (3350 cm^{-1}) and 3.15 μ (3180 cm^{-1}). Primary amides in nonpolar solvents give two medium peaks at about 2.84 μ (3520 cm^{-1}) and 2.94 μ (3400 cm^{-1}).

Secondary amides in KBr pellets give a peak at 3.00–3.27 μ (3330–3060 cm^{-1}). Secondary amides in nonpolar solvents give a peak at 2.86–2.94 μ (3500–3400 cm^{-1}).

Tertiary amides have no N—H stretching.

N—H bending: Primary amides have one or more weak peaks at 6.04–6.29 μ (1655–1590 cm^{-1}).

Secondary amides have one or more weak peaks at 6.37–6.62 μ (1570–1510 cm^{-1}).

Tertiary amides have no N—H bending.

C—N stretching: Gives a strong peak at around 7.14 μ (1400 cm^{-1}).

Anhydrides

C=O stretching: Gives two strong peaks at 5.46–5.56 μ (1830–1800 cm^{-1}) and 5.63–5.75 μ (1775–1740 cm^{-1}).

Ring strain in cyclic anhydrides moves these peaks to the left.

Conjugation moves these peaks to the right.

C—O stretching: Gives a strong peak at 7.7–11 μ (1300–900 cm^{-1}).

Aldehydes

C=O stretching: Aliphatic aldehydes have a strong peak at 5.75–5.82 μ (1740–1720 cm^{-1}).

Conjugation moves this peak to 5.85–5.94 μ (1710–1685 cm^{-1}).

C—H stretching: Gives two weak peaks near 3.50 μ (2850 cm^{-1}) and 3.65 μ (2750 cm^{-1}). C—H stretching in alkyl groups is more to the left.

Ketones

C=O stretching: Gives for unconjugated ketones a strong peak at 5.79–5.86 μ (1725–1705 cm^{-1}).

Conjugated ketones have this peak at 5.79–5.95 μ (1725–1680 cm^{-1}).

Cyclic ketones have this peak at 5.51–5.86 μ (1815–1705 cm^{-1}). Strained ring systems have this peak more to the left in this range.

Enolic, β-diketones give a strong, broad peak at 6.10–6.33 μ (1640–1580 cm^{-1}).

Nitro compounds

N=O stretching: Gives two strong peaks at 6.25–6.67 μ (1600–1500 cm^{-1}) and 7.2–7.7 μ (1390–1300 cm^{-1}). The first peak can interfere with an aromatic ring peak.

Nitriles

C≡N stretching: Gives one medium, sharp peak at 4.42–4.50 μ (2260–2220 cm^{-1}).

Conjugation moves this peak to the right.

Halogens

C—F stretching: Gives a strong peak in the region 7.1–10.0 μ (1400–1000 cm^{-1}).

C—Cl stretching: Gives a strong peak in the region 12.5–16.6 μ (800–600 cm^{-1}).

C—Br stretching: Gives a peak at 16.6–20.0 μ, or beyond the region of most infrared spectrometers.

C—I stretching: Gives a peak at 20 μ, or beyond the region of most infrared spectrometers.

These absorptions are given in convenient table form in Table 35-1.

TABLE 35-1.

Infrared Absorption-Structure Correlations

Vibration type	Wavelength (μ)	Wave number (cm^{-1})	Intensity
C—H stretching and bending			
Alkanes			
—CH$_3$ stretching	3.38	2960	Strong
	3.48	2870	Strong
—CH$_3$ bending	6.90	1450	Medium
	7.28	1375	Medium
—CH$_2$— stretching	3.43	2925	Strong
	3.51	2855	Strong
—CH$_2$— in ring stretching	3.23–3.34	3100–2990	Strong
—CH$_2$— bending	6.83	1465	Medium
Alkenes			
stretching	3.23–3.33	3100–3000	Medium
bending	10–15.4	1000–650	Strong
Alkynes			
stretching	3.00–3.06	3333–3267	Strong
Aromatic			
stretching	3.23–3.33	3100–3000	Medium to weak
bending			
Benzene	14.9	670	Strong
Monosubstituted benzene	12.99–13.70	770–730	Strong to medium
	14.08–14.99	710–690	Strong to medium
1,2-disubstituted benzene	12.99–13.61	770–735	Strong to medium
1,3-disubstituted benzene	11.11–11.63	900–860	Strong to medium
	12.35–13.33	810–750	Strong to medium
	13.74–14.71	725–680	Strong to medium

TABLE 35-1. (continued)

Vibration type	Wavelength (μ)	Wave number (cm^{-1})	Intensity
1,4-disubstituted benzene and 1,2,3,4-tetrasubstituted benzene	11.63–12.50	860–800	Strong to medium
1,2,3-trisubstituted benzene	12.50–12.99	800–770	Strong to medium
	13.89–14.60	720–685	Strong to medium
1,2,4-trisubstituted benzene	11.11–11.63	900–860	Strong to medium
	11.63–12.50	860–800	Strong to medium
1,3,5-trisubstituted benzene	11.11–11.63	900–860	Strong to medium
	11.56–12.35	865–810	Strong to medium
	13.70–14.81	730–675	Strong to medium
Aldehydes			
stretching	3.50	2850	Weak
	3.65	2750	Weak
O—H stretching			
Alcohols and phenols			
hydrogen bonded	2.82–3.13	3550–3200	Strong, broad
free	2.74–2.79	3650–3600	Strong, broad
Carboxylic acids	3.0–4.0	3300–2500	Strong, broad
N—H stretching and bending			
Amines, primary (stretching)			
hydrogen bonded	2.94–3.00	3400–3330	Medium
	3.00–3.08	3330–3250	Medium
free	2.86	3500	Medium
	2.94	3400	Medium
Amines, secondary (stretching)	2.58–3.02	3350–3310	Weak
Amides, primary (stretching)			
in KBr pellet	2.99	3350	Medium
	3.15	3180	Medium
in nonpolar solvent	2.84	3520	Medium
	2.94	3400	Medium
Amides, secondary (stretching)			
in KBr pellet	3.00–3.27	3330–3060	Medium
in nonpolar solvent	2.86–2.94	3500–3400	Medium
Amides, primary (bending)	6.04–6.29	1655–1590	Weak
Amides, secondary (bending)	6.37–6.62	1570–1510	Weak
C=C stretching			
Alkenes	5.95—6.25	1675—1600	Medium to weak
Conjugated dienes			
symmetrical	6.25	1600	Medium to weak
unsymmetrical	6.06	1650	Medium to weak
	6.25	1600	Medium to weak
Aromatic	6.25—6.33	1600—1580	Medium to variable
	6.67—6.90	1500—1450	Medium to variable
C≡C stretching			
Alkynes	4.43–4.76	2260–2100	Medium to weak

TABLE 35-1. (continued)

Vibration type	Wavelength (μ)	Wave number (cm^{-1})	Intensity
C=O stretching			
Carboxylic acids			
aliphatic	5.81–5.86	1720–1705	Strong
conjugated	5.85–5.95	1710–1680	Strong
internal hydrogen bonding	<6.01	>1665	Strong
Esters			
aliphatic	5.71–5.76	1750–1735	Strong
conjugated with the carbonyl carbon	5.78–5.83	1730–1715	Strong
conjugated with the alcohol oxygen	5.62–5.65	1780–1770	Strong
Amides	5.90–6.06	1700–1630	Strong
Anhydrides	5.46–5.56	1830–1800	Strong
	5.63–5.75	1775–1740	Strong
Aldehydes			
aliphatic	5.75–5.82	1740–1720	Strong
conjugated	5.85–5.94	1710–1685	Strong
Ketones			
aliphatic	5.79–5.86	1725–1705	Strong
conjugated	5.79–5.95	1725–1680	Strong
cyclic	5.51–5.86	1815–1705	Strong
enolic, β-diketones	6.10–6.33	1640–1580	Strong
C—O stretching			
Alcohols and phenols	7.7–10.0	1300–1000	Strong to medium
Ethers			
alkyl	8.70–9.23	1150–1085	Strong to medium
aryl, alkyl	7.84–8.33	1275–1200	Strong to medium
	9.30–9.80	1075–1020	Strong to medium
Carboxylic acids	7.6–8.3	1320–1210	Strong to medium
Esters			
saturated	8.26–8.60	1210–1165	Strong to medium
aromatic	7.63–8.00	1310–1250	Strong to medium
α,β-unsaturated acid esters	7.69–8.62	1300–1160	Strong to medium
C—N stretching			
Amines	7.4–10.0	1350–1000	Medium to weak
Amides	7.14	1400	Strong
C≡N stretching			
Nitriles	4.42–4.50	2260–2220	Medium
N=O stretching			
Nitro compounds	6.25–6.67	1600–1500	Strong
	7.2–7.7	1390–1300	Strong
C—X stretching			
Fluorine	7.1–10.0	1400–1000	Strong
Chlorine	12.5–16.6	800–600	Strong
Bromine	16.6–20.0	600–500	Strong
Iodine	20.0	500	Strong

35.15   SAMPLE INFRARED SPECTRUM INTERPRETATIONS

For ease in analysis, each of the following compounds has the types of vibration, the bonds in the molecule, and the appropriate peaks in the spectrum all numbered.

1. Methyl salicylate (oil of wintergreen) (Figure 35-9)

1. C=O stretching

2. O—H stretching

3. C—O stretching

4. C—O stretching

5. C—H stretching

6. C—C stretching

7. C—C stretching

8. C—H out-of-plane bending for 1,2-disubstituted benzene

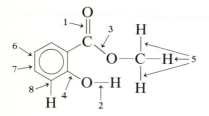

FIGURE 35-9

The infrared spectrum of methyl salicylate, neat. Spectrum courtesy of Sadtler Research Laboratories, division of Bio-Rad Laboratories, Inc., copyrighted 1969.

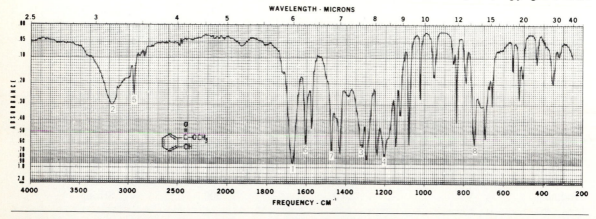

2. Phenacetin, the formerly used pain killer (Figure 35-10)

1. C=O stretching

2. C—N stretching

3. N—H stretching

4. N—H stretching

5. C—H stretching

6. C—H bending

7. C—H bending

8. C—H stretching

9. C—H out-of-plane bending for 1,4-disubstituted benzene

10. C—O stretching

11. C—O stretching

12. C—C stretching

13. C—C stretching

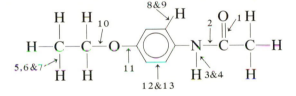

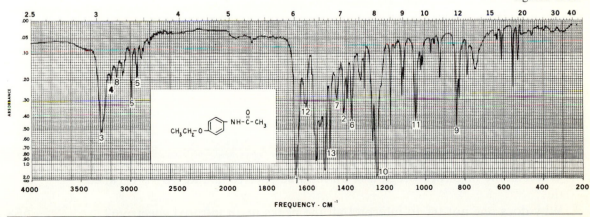

FIGURE 35-10

The infrared spectrum of phenacetin, the formerly used pain-killer, in a KBr pellet. Spectrum courtesy of Sadtler Research Laboratories, division of Bio-Rad Laboratories, Inc., copyrighted 1966.

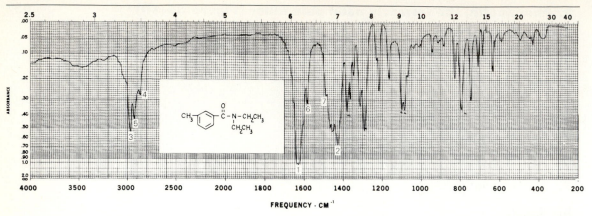

FIGURE 35-11

The infrared spectrum of *N,N*-diethyl-*m*-toluamide, the insect repellent in Off and 6-12, neat. Spectrum courtesy of Sadtler Research Laboratories, division of Bio-Rad Laboratories, Inc., copyrighted 1967.

3. *N,N*-Diethyl-*m*-toluamide, the insect repellent in Off and 6-12 (Figure 35-11)

1. C=O stretching

2. C—N stretching

3. C—H stretching

4. C—H stretching

5. C—H stretching

6. C—C stretching

7. C—C stretching

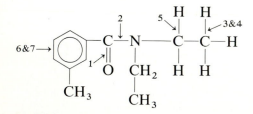

QUESTIONS

1. Where is the fingerprint region of an infrared spectrum, and why is it called this?

2. Why does one not try to interpret all the peaks in an infrared spectrum?

658

3. What is the frequency of vibration in sec^{-1} of the $C{=}O$ stretching peak of an aromatic ketone? What is the energy in kcal/mole of this stretching vibration?

4. Explain why a $C{=}O$ stretching peak appears to the left of a $C{-}O$ stretching peak.

5. Explain why 3,4-diethyl-3-hexene has no $C{=}C$ stretching peak.

6. Explain why a $C{=}O$ stretching peak has a stronger intensity than a $C{=}C$ stretching peak.

7. Interpret at least three of the uninterpreted infrared spectra given in this book or in your lecture textbook. Your instructor may choose which ones are to be done.

References

1. R. Conley, **Infrared Spectroscopy,** 2nd ed., Allyn and Bacon, Boston, 1972.

2. J. Dyer, **Applications of Absorption Spectroscopy of Organic Compounds,** Prentice-Hall, Englewood Cliffs, N.J., 1965.

3. K. Nakanishi and P. Solomon, **Infrared Absorption Spectroscopy,** 2nd ed., Holden-Day, San Francisco, 1977.

4. R. Silverstein, G. Bassler, and T. Morrill, **Spectrometric Identification of Organic Compounds,** 4th ed., Wiley, New York, 1981.

EXPERIMENT

36

NUCLEAR MAGNETIC RESONANCE SPECTROSCOPY: PREPARING THE SAMPLES AND INTERPRETING THE SPECTRA

36.1 INTRODUCTION

If our chemist in Experiment 35 were given a small sample of a pure organic compound to identify, he or she would also take its ^{1}H NMR spectrum and perhaps even its ^{13}C NMR spectrum. With them, he or she would know the number, kind, and relative locations of the hydrogens and carbons in the compound. From this, most if not all of the functional groups would be known, and also which specific alkyl groups or other groups were attached to the various functional groups. Valuable information could therefore be attained that is not possible with an infrared spectrum. In fact, even though NMR spectroscopy has been commonly available only since the early 1960s, it is frequently considered to be even more important than infrared spectroscopy for structure determination. The two methods, however, are usually used together and complement each other in the information obtained.

TABLE 36-1.

**Price for 100 Grams of Some Commonly
Used Deuterated Solvents**

$CDCl_3$	$25
D_2O	$60
Acetone-d_6	$140
DMSO-d_6	$140
Unisol-d	$140
CD_2Cl_2	$525

Other benefits of NMR spectroscopy are the very small quantity of sample needed, the speed with which the spectrum can be taken, and the fact that the sample can be recovered and used for other purposes, if necessary. In fact, the only negative feature of NMR spectroscopy is its expense. An "inexpensive" ^{1}H NMR spectrometer can easily cost over $60,000, and more expensive ^{1}H and ^{13}C models frequently cost well over $500,000. Thus, many colleges may not be able to afford one, and students may have to rely only on physical tests, chemical tests, and an infrared spectrometer for structure determination or verification of a laboratory compound.

Even after an NMR spectrometer is purchased, the expense does not stop. Depending on the quality, NMR sample tubes cost between $2 and $5 each. In addition, expensive deuterated solvents are frequently needed. The price of 100 grams of a number of commonly used deuterated solvents is given in Table 36-1. This price will, of course, vary with the amount purchased.

For those schools that cannot afford an actual NMR spectrometer, an NMR simulator is available to teach students how to operate and use an NMR spectrometer in the laboratory. Available from Norell, Inc., 314 Arbor Avenue, Landisville, N.J. 08326, it also makes sophisticated and accurate analysis of programmed compounds available to students.

36.2 THEORY

To understand the principle behind an NMR spectrum, one must first realize that atomic nuclei, like electrons, are able to spin clockwise or counterclockwise on an axis. In addition, if these nuclei contain an odd number of either protons or neutrons, this spinning creates a small magnetic field around each nucleus. Thus ^{1}H, ^{2}H, ^{11}B, ^{13}C, ^{14}N, ^{15}N, ^{17}O, ^{19}F, ^{31}P, and ^{33}S are all able to create a magnetic field, but ^{12}C, ^{16}O, and ^{32}S cannot.

If spinning, magnetic nuclei are placed in a strong, external magnetic field, they will align themselves to make the direction of their magnetic fields or their magnetic moments the same as the direction of the external field. However, if

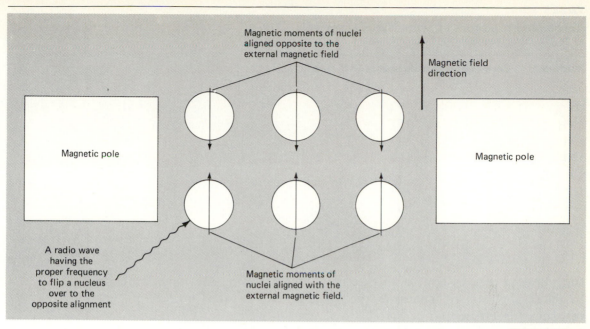

FIGURE 36-1
Spinning magnetic nuclei in an external magnetic field.

radio waves having the proper frequency are supplied, the aligned nuclei can absorb this energy and flip over to make their magnetic fields have a direction opposite to the external field. This is depicted in Figure 36-1.

In addition, an external magnetic field causes magnetic nuclei within it to precess like a child's toy top. This is depicted in Figure 36-2. When these nuclei flip over to the opposite alignment, the radio frequency causing it is exactly equal to the nuclei's angular precessional frequency. The term **resonance**, meaning

FIGURE 36-2
Precessional motion of nuclei in an external magnetic field.

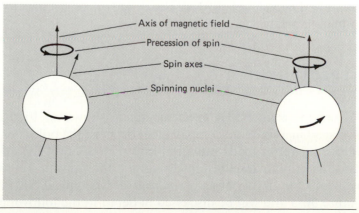

TABLE 36-2.

Resonance Frequencies of Some Nuclei Observed
in NMR Spectroscopy at 14,092 Gauss

Nucleus	Frequency in megahertz (10^6 sec^{-1})
^1H	60.0
^2H	9.2
^{13}C	15.1
^{14}N	4.3
^{15}N	6.1
^{17}O	8.1
^{19}F	56.5
^{31}P	24.2

the response of a system to an external stimulus having the same frequency, therefore applies here. Therefore, the whole process is called nuclear magnetic resonance spectroscopy.

The frequency needed for resonance will vary, but depends on both the strength of the external field and the magnetic moment of the nuclei. Table 36-2 gives the frequencies of some nuclei observed in NMR spectroscopy at a typical external field strength of 14,092 gauss. What is important about these frequencies is that they are significantly far away from each other so that one kind of nuclei does not interfere with another. Therefore, over the narrow range of a typical ^1H or ^{13}C NMR spectrum, we don't have to worry about the deuterium in the solvent or the ^{14}N in an amine absorbing in the same region being measured.

These frequencies also represent radiation having a much lower energy than for infrared, visible, or ultraviolet spectroscopy. For ^1H NMR spectroscopy, for example, only 5.7×10^{-6} kcal/mole is involved at 14,092 gauss, versus 2 to 11 kcal/mole for infrared spectroscopy, and as much as 150 kcal/mole for ultraviolet spectroscopy. The relation of the NMR region of the electromagnetic spectrum to others is illustrated in Figure 36-3.

36.3 NMR SPECTROMETER

A schematic diagram of an NMR spectrometer is shown in Figure 36-4. A sample tube is placed in a spinning compartment between the poles of a powerful magnet. A radio transmitter sends radiation into the region of the sample tube and, if any of this radiation is absorbed by the sample, the amount is electronically determined by a detector and sent to a graph recorder. A typical NMR spectrum is shown in Figure 36-5 and will be discussed in depth in the following sections.

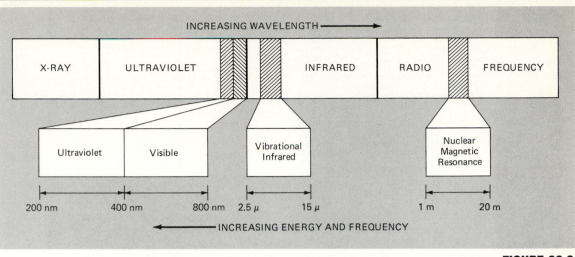

FIGURE 36-3
Portion of electromagnetic spectrum, showing the relation of
NMR radiation to other types.

Some NMR spectrometers operate by applying a constant magnetic field
while varying the radio frequency. Others utilize a constant radio frequency and
vary the magnetic field strength. The latter method is usually preferred, but
either one gives the same spectrum, since resonance frequencies again are
proportional to the magnetic field strength.

FIGURE 36-4
Schematic diagram of an NMR spectrometer.

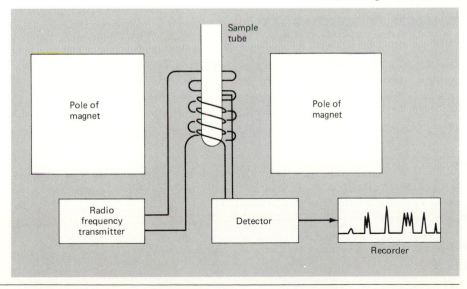

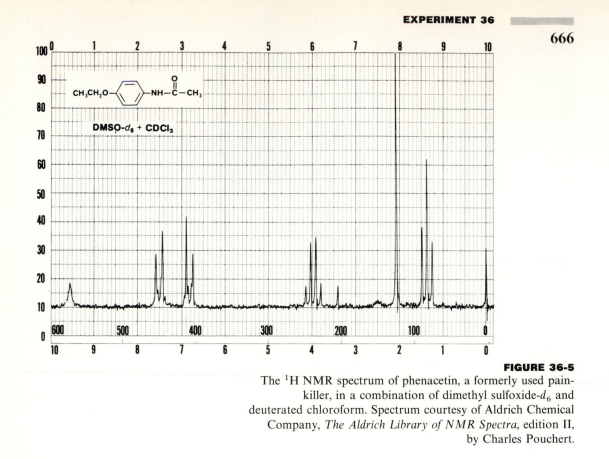

FIGURE 36-5

The ^{1}H NMR spectrum of phenacetin, a formerly used pain-
killer, in a combination of dimethyl sulfoxide-d_6 and
deuterated chloroform. Spectrum courtesy of Aldrich Chemical
Company, *The Aldrich Library of NMR Spectra*, edition II,
by Charles Pouchert.

36.4 PROTON NMR SPECTROSCOPY

^{1}H NMR spectra are the kind usually taken in organic chemistry. Contributing
to this are the facts that practically all organic compounds contain hydrogen
and the relative ease and lower cost compared with ^{13}C NMR spectra. The
following three types of basic information make ^{1}H NMR spectroscopy of
special value in structure determination. Each of them is discussed in more
detail later.

1. The area of a peak is directly proportional to the number of protons
causing the peak. This allows the number of each type of proton in a compound
to be directly determined.

2. The **chemical shift** or the position of a peak in the spectrum allows the
functional group attached to or attached near the proton to be directly
determined.

3. Spin-spin coupling to produce split peaks of known separations allows the number and the steric relationship of nearby protons to be directly determined.

36.5 PEAK AREAS

^{1}H NMR spectrometers have electronic **integrators** to compute the relative areas of all the peaks in the spectrum. By simply measuring the height of each integration line, one can then know the number of hydrogens responsible for each peak. This proves to be of tremendous value in structure determination.

For example, the three peaks in Figure 36-6 have heights of 6.6, 2.7, and 3.9 spaces in their accompanying integration lines. The number of hydrogens in each peak must therefore be proportional to these heights. Since benzyl acetate by analysis is known to have 10 hydrogens, we can compute the number of vertical spaces per hydrogen by dividing the total number of spaces by 10.

$$\frac{6.6 \text{ spaces} + 2.7 \text{ spaces} + 3.9 \text{ spaces}}{10 \text{ hydrogens}} = \frac{13.2 \text{ spaces}}{10 \text{ hydrogens}}$$

$$= \text{about 1.3 spaces per hydrogen}$$

FIGURE 36-6
The ^{1}H NMR spectrum of benzyl acetate in carbon tetrachloride. Spectrum courtesy of Aldrich Chemical Company, *The Aldrich Library of NMR Spectra*, edition II, by Charles Pouchert.

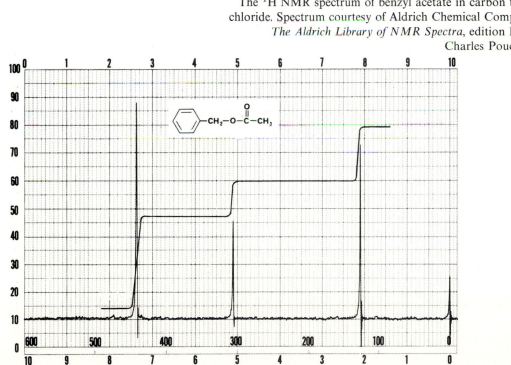

Dividing the number of vertical spaces in each integration line by the number of computed spaces per hydrogen then allows us to know the number of hydrogens in each peak.

1st peak $\qquad \dfrac{6.6 \text{ spaces}}{1.3 \text{ spaces/hydrogen}} = 5 \text{ hydrogens}$

2nd peak $\qquad \dfrac{2.7 \text{ spaces}}{1.3 \text{ spaces/hydrogen}} = 2 \text{ hydrogens}$

3rd peak $\qquad \dfrac{3.9 \text{ spaces}}{1.3 \text{ spaces/hydrogen}} = 3 \text{ hydrogens}$

Therefore, the first peak represents 5 hydrogens, the second one represents 2 hydrogens, and the third one represents 3 hydrogens. Frequently, errors of 5 to 10 percent are made in determining the heights of the integration lines, so the number of hydrogens computed must always be rounded off to the nearest whole number. Remember that fractional numbers of hydrogens are not possible.

36.6 CHEMICAL SHIFT

Fortunately, all protons do not absorb energy at the same frequency or the same external magnetic field strength. If they did, a ^{1}H NMR spectrum would be absolutely worthless in structure determination, since there would then be only one peak and it would be impossible to distinguish any of the different types of protons from each other.

To understand why a ^{1}H NMR spectrum usually consists of more than one peak, one must consider the electrons around the various protons. Under the influence of an external magnetic field, these electrons generate their own small magnetic fields, which partially oppose the external field. As a result, the actual magnetic field sensed or felt by the protons is slightly less than the external field. The electrons are said to partially **shield** the protons from the external field.

Of course, because of varying electron densities around different types of protons, the amount of this shielding will also vary. For example, if the protons are near a strong electron-withdrawing group, they will have a lower electron density and will be less shielded compared with no electron-withdrawing group being present. More shielded protons, of course, will absorb at higher frequencies or higher external field strengths than will less shielded protons, since it will take more energy to overcome the shielding.

The difference in where the various protons absorb, however, is generally no more than 10 to 12 parts per million in the magnetic field strength. The easiest way to accurately measure an amount this small is to determine the exact positions of all the peaks produced by the sample compound relative to a standard reference substance. The internationally accepted reference substance is tetramethylsilane (TMS), $(CH_3)_4Si$. It was chosen because its 12 protons all produce one peak at a higher field strength than are the peaks of practically all the protons found in organic compounds.

The **chemical shift** of a proton is a quantitative measure of the position of that proton's peak relative to TMS. It is given the symbol δ and may be calculated by the equation

$$\delta = \frac{\text{Observed distance or shift from TMS in hertz} \times 10^6}{\text{Operating frequency of the instrument in hertz}}$$

For example, if a certain peak was 275 Hz away from TMS in a 60.0-MHz instrument, its chemical shift would be

$$\delta = \frac{275 \text{ Hz} \times 10^6}{60.0 \times 10^6 \text{ Hz}} = 4.58$$

Chemical shift values are important in ^1H NMR spectroscopy because they help identify the various types of protons responsible for the peaks in the spectrum. A list of some common chemical shift values is given in Table 36-3. Remember that because of the way chemical shift values are calculated, they are **independent** of the external field strength or the frequency of the spectrometer. Thus, 60-MHz, 100-MHz, and 200-MHz instruments would all give the same chemical shift values for a particular sample compound. If it weren't this way, there would have to be a different chemical shift table for each type of spectrometer.

A typical ^1H NMR spectrum will have the reference peak for TMS at the far right ($\delta = 0$) with the sample compound peaks to the left of it. The chemical shift values for these peaks will all be indicated by the horizontal numbers on the spectrum, increasing from right to left. Sometimes there will also be a second set of horizontal numbers, increasing from left to right, and ending with 10 at the far right for TMS. This second set of numbers is another older way to express chemical shifts in τ (Greek letter tau) values. It is related to the δ system by the simple equation

$$\tau = 10 - \delta$$

Most chemical shift tables of ^1H NMR spectra, however, prefer to use only the δ system.

The words **upfield** and **downfield** are also frequently used with NMR spectra. Upfield refers to the higher field strengths toward the right-hand portion of the spectrum, while downfield refers to the lower field strengths toward the left-hand portion of the spectrum. Sometimes the relative positions of two peaks are given by these words. Thus if peak A is to the left of peak B, it will be said to be downfield of peak B.

TABLE 36-3.

Chemical Shifts of Various Protons

Compound or class	Type of proton	δ (ppm)
Primary aliphatic	$R-CH_3$	0.8–1.0
Secondary aliphatic	$R-CH_2-R$	1.2–1.5
Tertiary aliphatic	R_3C-H	1.4–1.7
Vinylic	$C=C-H$	4.6–5.9
Allylic	$C=C-CH_3$	1.6–1.9
Aliphatic acetylenic	$R-C\equiv C-H$	2.3–2.5
Aromatic acetylenic	$Ar-C\equiv C-H$	2.8–3.1
Aromatic	$Ar-H$	6–8.5
Primary benzylic	$Ar-CH_3$	2.2–2.5
Secondary benzylic	$Ar-CH_2-R$	2.5–2.9
Tertiary benzylic	$Ar-CHR_2$	2.8–3.2
Alcohols	$R-OH$	1–5.5
Alcohols	$HO-C\underline{H}$	3.4–4.0
Alcohols	$HO-CH_2-C\underline{H}$	1.2–1.6
Phenols	$Ar-OH$	4–12
Aliphatic amines	$R-NH_2$	0.6–2.5
Aliphatic amines	R'_2N-CH_3	2.2–2.6
Aliphatic amines	$R'_2N-C\underline{H}_2CH_3$	2.5–2.8
Aliphatic amines	$R'_2N-CH_2C\underline{H}_3$	1.0–1.3
Aromatic amines	$Ar-NH_2$	3–4.5
Aromatic amines	$Ar-NH-C\underline{H}_3$	2.8–3.1
Aromatic amines	$Ar-NH-C\underline{H}_2CH_3$	3.0–3.3
Aliphatic ethers	$R'O-CH_3$	3.2–3.5
Aliphatic ethers	$R'O-CH_2-R$	3.4–3.8
Aliphatic ethers	$R'O-CH_2C\underline{H}_3$	1.2–1.4
Aliphatic ethers	$R'O-CH_2CH_2C\underline{H}_3$	0.9–1.1
Aromatic ethers	$ArO-CH_3$	3.7–4.0
Aromatic ethers	$ArO-CH_2-R$	3.9–4.3
Aliphatic ketones	$R'-(C=O)-CH_3$	2.1–2.4
Aliphatic ketones	$R'-(C=O)-CH_2-R$	2.3–2.7
Aliphatic ketones	$R'-(C=O)-CH_2C\underline{H}_2-R$	1.1–1.4
Aromatic ketones	$Ar-(C=O)-CH_3$	2.4–2.6
Aromatic ketones	$Ar-(C=O)-CH_2-R$	2.5–2.8
Aliphatic aldehydes	$R-(C=O)-H$	9.4–9.9
Aliphatic aldehydes	$H-(C=O)-CH_2-R$	2.1–2.4
Aliphatic aldehydes	$H-(C=O)-CH_2C\underline{H}_2-R$	1.1–1.4
Aromatic aldehydes	$Ar-(C=O)-H$	9.7–10.3
Aliphatic esters	$R'O-(C=O)-CH_3$	1.9–2.2
Aliphatic esters	$R'O-(C=O)-C\underline{H}_2CH_3$	2.1–2.4

TABLE 36-3. (continued)

Compound or class	Type of proton	δ (ppm)
Aliphatic esters	R′O—(C=O)—CH$_2$C$\underline{H}_2$—R	1.2–1.4
Aliphatic esters	R′—(C=O)—O—CH$_3$	3.6–4.0
Aliphatic esters	R′—(C=O)—O—C$\underline{H}_2$CH$_3$	3.7–4.1
Aliphatic esters	R—O—(C=O)—H	8.0–8.2
Aromatic esters	ArO—(C=O)—CH$_3$	2.0–2.5
Aromatic esters	ArO—(C=O)—C$\underline{H}_2$CH$_3$	2.2–2.7
Aromatic esters	Ar—(C=O)—O—CH$_3$	4.0–4.2
Aromatic esters	Ar—(C=O)—O—C$\underline{H}_2$CH$_3$	4.2–4.5
Aliphatic carboxylic acids	R—(C=O)—OH	10.4–12.0
Aliphatic carboxylic acids	R—C$\underline{H}_2$—(C=O)—OH	2.2–2.4
Aliphatic carboxylic acids	R—C$\underline{H}_2$CH$_2$—(C=O)—OH	1.0–1.4
Aromatic carboxylic acids	Ar—(C=O)—OH	10.4–12.0
Aliphatic amides	R—(C=O)—NH$_2$	5.5–7.5
Aliphatic amides	R$_2$N—(C=O)—CH$_3$	1.8–2.2
Aliphatic amides	R—(C=O)—NH—C$\underline{H}_3$	2.8–3.0
Alkyl chlorides	R—CH$_2$—Cl	3.5–3.7
Alkyl chlorides	R—C$\underline{H}_2$CH$_2$—Cl	1.6–1.8
Methylene chloride	CH$_2$Cl$_2$	5.30
Chloroform	CHCl$_3$	7.27
Alkyl bromides	R—CH$_2$—Br	3.2–3.4
Alkyl bromides	R—C$\underline{H}_2$CH$_2$—Br	1.6–1.8
Alkyl bromides	R—C$\underline{H}_2$CH$_2$CH$_2$—Br	1.2–1.3
Alkyl iodides	R—CH$_2$—I	3.0–3.3
Alkyl iodides	R—C$\underline{H}_2$CH$_2$—I	1.7–1.9
Alkyl fluorides	R—CH$_2$—F	4.2–4.5
Alkyl fluorides	R—C$\underline{H}_2$CH$_2$—F	1.4–1.6
Enols	C=C—OH	15–17
Aliphatic thiols	R—S$\underline{H}$	1.0–2.0
Aromatic thiols	Ar—SH	3.0–4.0
Nitriles	R—CH$_2$—CN	2.0–2.3
Nitro compounds	R—CH$_2$—NO$_2$	4.2–4.6

From the values in Table 36-3, some observations may be made.

1. Primary protons absorb more upfield than secondary protons, which in turn absorb more upfield than tertiary protons. For example:

R—CH$_3$	R—CH$_2$—R	R$_3$C—H
δ 0.8–1.0	δ 1.2–1.5	δ 1.4–1.7

and

Ar—CH$_3$	Ar—CH$_2$—R	Ar—CHR$_2$
δ 2.2–2.5	δ 2.5–2.9	δ 2.8–3.2

2. Increasing the electronegativity of an element attached to carbon increases the chemical shift of protons on this carbon. For example:

Increasing electronegativity $\longrightarrow$

R—CH$_2$—I	R—CH$_2$—Br	R—CH$_2$—Cl	R—CH$_2$—F
δ 3.0–3.3	δ 3.2–3.4	δ 3.5–3.7	δ 4.2–4.5

3. Multiple electron-withdrawing substituents increase the chemical shift more than single substituents. For example:

CHCl$_3$	CH$_2$Cl$_2$	CH$_3$—Cl
δ 7.27	δ 5.30	δ 3.05

4. The deshielding effect of electron-withdrawing groups decreases with increasing distance. In fact, groups three carbons away have little measurable effect. For example:

R′O—CH$_3$	R′O—CH$_2$CH$_3$	R′O—CH$_2$CH$_2$CH$_3$
δ 3.2–3.5	δ 1.2–1.4	δ 0.9–1.1

5. As a substituent, aromatic rings produce a larger chemical shift than aliphatic groups. For example:

Ar—NH$_2$	R—NH$_2$
δ 3–4.5	δ 0.6–2.5
Ar—OH	R—OH
δ 4–12	δ 1–5.5
Ar—C≡C—H	R—C≡C—H
δ 2.8–3.1	δ 2.3–2.5
Ar—O—CH$_3$	R—O—CH$_3$
δ 3.7–4.0	δ 3.2–3.5
Ar—(C=O)—H	R—(C=O)—H
δ 9.7–10.3	δ 9.4–9.9
Ar—(C=O)—CH$_3$	R—(C=O)—CH$_3$
δ 2.4–2.6	δ 2.1–2.4

6. Aromatic, vinylic, acetylenic, and carbonyl protons have chemical shifts not easily explained by the electronegativity of the attached groups. In particular, aromatic and carbonyl protons are very far downfield from what would be expected, vinylic protons are quite far downfield, and acetylenic protons should be even further downfield than vinylic protons, but they are not. Anisotropic effects have been used to explain these phenomena.

In the case of aromatic rings, the circulating π electrons create another secondary magnetic field, which is illustrated in Figure 36-7. This secondary, anisotropic field has its lines of force by the aromatic protons going in the same direction as the external field. As a result, these lines of force reinforce the external field, the protons are more deshielded and have a chemical shift of $\delta\,6\text{--}8.5$, much higher than expected. This explanation also accounts for the results in observation 5.

Similar secondary, anisotropic fields are produced by the π electrons of alkenes and carbonyl compounds. These fields, shown in Figures 36-8 and 36-9, also have the lines of force by the vinylic and carbonyl protons going in the same direction as the external field. As a result, these protons are also deshielded and have chemical shifts much higher than expected. Vinylic protons absorb at $\delta\,4.6\text{--}5.9$, while carbonyl protons absorb at $\delta\,8.0\text{--}10.3$.

Since the π bond of an alkene deshields vinylic protons, we might expect the two π bonds of an alkyne to deshield acetylenic protons even more. The opposite is true, with acetylenic protons absorbing far upfield at $\delta\,2\text{--}3$. The secondary, anisotropic field produced by the two π bonds explains why. As

FIGURE 36-7
The secondary, anisotropic magnetic field of an aromatic ring.

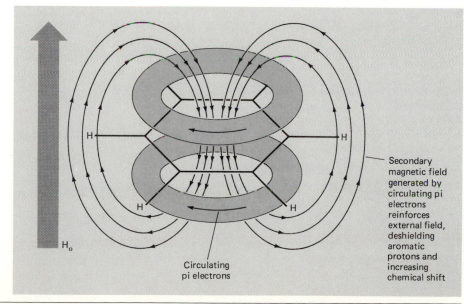

H_o

Circulating
pi electrons

Secondary
magnetic field
generated by
circulating pi
electrons
reinforces
external field,
deshielding
aromatic
protons and
increasing
chemical shift

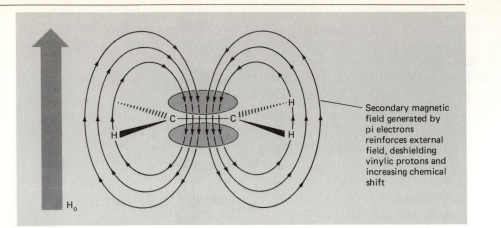

Secondary magnetic field generated by pi electrons reinforces external field, deshielding vinylic protons and increasing chemical shift

FIGURE 36-8
The secondary, anisotropic magnetic field of an alkene.

shown in Figure 36–10, this field has its lines of force by acetylenic protons going in the **opposite** direction of the external field, thus shielding rather than deshielding these protons.

36.7 SPIN-SPIN COUPLING

The spin-spin coupling of protons on adjacent carbons to produce the splitting of peaks is also very important in the structure determination of organic compounds. To understand it, we must realize that protons are not only affected

FIGURE 36-9
The secondary, anisotropic magnetic field of a carbonyl compound.

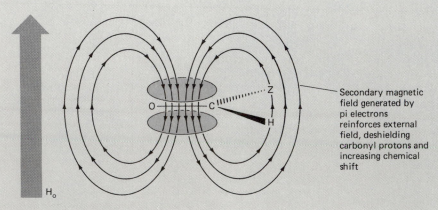

Secondary magnetic field generated by pi electrons reinforces external field, deshielding carbonyl protons and increasing chemical shift

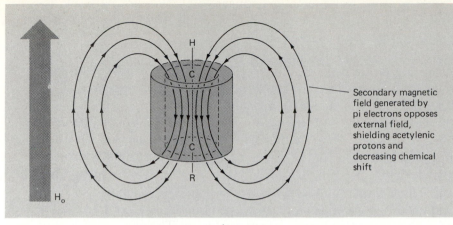

FIGURE 36-10
The secondary, anisotropic magnetic field of an alkyne.

by the external magnetic field but by the magnetic fields of all the protons on adjacent carbons.

As an example, consider the compound 1,1-dichloroethane. As shown in Figure 36-11, the three protons on carbon 2 can have the proton on carbon 1 either aligned with or against the external field. When the proton is aligned with

FIGURE 36-11
One possible type of spin-spin splitting in 1,1-dichloroethane. The proton on carbon 1 splits the NMR signal for the three protons on carbon 2 into a doublet.

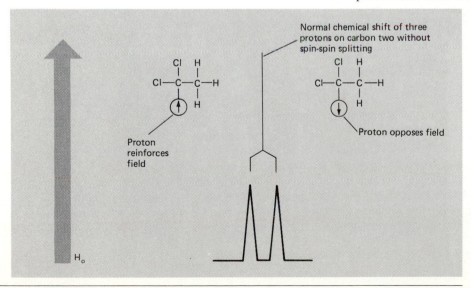

the field, the other protons feel a slightly stronger total field, are partially de-shielded, and absorb at a slightly lower field. But when the proton is aligned against the field, the other protons feel a slightly weaker total field, are partially shielded, and absorb at a slightly higher field. The end result is not a single peak, but the splitting of the peak into two equal size peaks or a doublet, each of which is equidistant from where the single peak would have appeared in the spectrum if there had been no spin-spin splitting.

Also, as shown in Figure 36-12, the proton on carbon 1 can have the three methyl protons on carbon 2 aligned either with or against the field. There are a total of eight possible combinations. In one of these, all three methyl protons are aligned with the field, and in another one all three methyl protons are aligned against the field. Three combinations have two methyl protons aligned with the field and one aligned against it. Finally, three combinations have one methyl proton aligned with the field and two aligned against it. The end result is thus not a single peak, but the splitting of the peak into four distinct peaks, or a quartet, having a ratio of intensities of 1:3:3:1. Furthermore, this quartet will

FIGURE 36-12

The other possible type of spin-spin splitting in 1,1-dichloro-ethane. The three protons on carbon 2 split the NMR signal for the proton on carbon 1 into a 1:3:3:1 quartet.

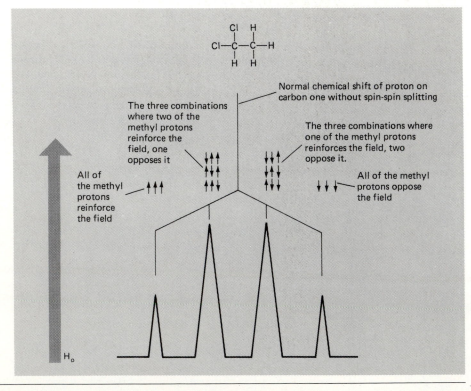

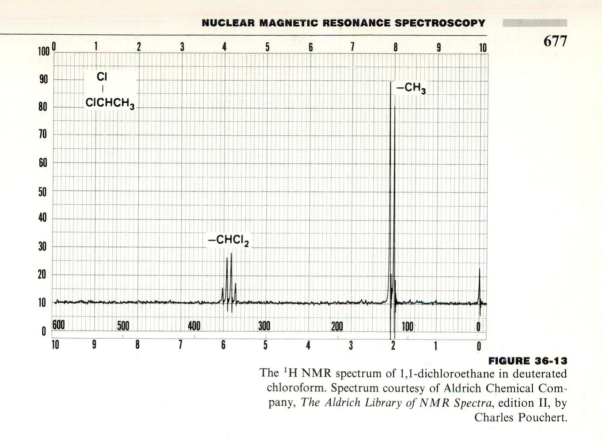

FIGURE 36-13
The ¹H NMR spectrum of 1,1-dichloroethane in deuterated chloroform. Spectrum courtesy of Aldrich Chemical Company, *The Aldrich Library of NMR Spectra*, edition II, by Charles Pouchert.

be centered at exactly where the single peak would have appeared in the spectrum if there had been no spin-spin splitting.

The actual, complete spectrum of 1,1-dichloroethane is shown in Figure 36-13. Observe that the two peaks forming the doublet are not exactly the same, nor do the four peaks in the quartet form a perfectly symmetrical 1:3:3:1 arrangement. This distortion is very common in ¹H NMR spectroscopy. In fact, perfectly symmetrical multiplets occur only when there is a very large distance between the different multiplets, compared to the distances between the peaks within the multiplets.

36.8   *N* + 1 RULE IN SPIN-SPIN SPLITTING

The analysis given for 1,1-dichloroethane can be extended to many other compounds. When this is done, one discovers that **when protons on adjacent carbon atoms have sufficiently differing chemical shifts, the number of peaks in a ¹H NMR signal is *N* + 1, where *N* is the total number of chemically equivalent protons on the adjacent carbon atoms**. Table 36-4 gives the number of peaks in a multiplet resulting from the *N* + 1

TABLE 36-4.

Splitting Patterns Resulting from *N* Equivalent Protons on the Adjacent Carbon Atoms

N	Appearance of multiplet	Relative intensities of peaks
0	Singlet	1
1	Doublet	1:1
2	Triplet	1:2:1
3	Quartet	1:3:3:1
4	Quintet	1:4:6:4:1
5	Sextet	1:5:10:10:5:1
6	Septet	1:6:15:20:15:6:1

rule, and the expected ideal intensities of each peak. We can see how this rule applies to some of the common splitting patterns observed in Figure 36-14.

36.9   COUPLING CONSTANTS

Also valuable in interpreting ^{1}H NMR spectra is the **coupling constant** or the distance between the individual peaks in a multiplet. Measured in hertz and given by the symbol *J*, it is a measure of the strength of the magnetic interaction between the protons on adjacent or conjugated atoms. When the interaction is stronger, for example, the coupling constant will be larger and the distance between the peaks within a multiplet will be greater. Because this interaction is independent of the strength of the external magnetic field, the coupling constant will be the same no matter what operating frequency is used for the spectrometer.

Some typical proton coupling constants are given in Table 36-5. From this, it should be evident that they are especially valuable in distinguishing the isomers of alkenes and aromatic compounds, and the size and stereochemistry of various ring systems. Also, in complicated spectra with many types of protons, coupling constants can be valuable in determining which sets of peaks belong to neighboring protons. These sets will have the same distances between the individual peaks.

36.10   AROMATIC COMPOUNDS

Aromatic proton peaks frequently are too complex to easily explain. However, the following simple generalizations are helpful. As shown in Figure 36-15(a), if there is an alkyl group on a monosubstituted ring, the five ring protons will give a broad singlet, due to their chemical shifts being nearly identical. An electron-withdrawing group will move the chemical shifts of all the protons down-

1. CH$_3$ -Singlet having a relative area of 3. Three protons are split by none.
|
Y

2. CH$_3$ -Triplet having a relative total area of 3. Three protons are split by two.
|
CH$_2$ -Quartet having a relative total area of 2. Two protons are split by three.
|
Y

3. CH$_3$ CH$_3$ -Doublet having a relative total area of 6. Six protons are split by one.

CH -Septet having a relative total area of 1. One proton is split by six.
|
Y

4. C(CH$_3$)$_3$ -Singlet having a relative area of 9. Nine protons are split by none.
|
Y

5.
|
X—C—H -Doublet having a relative total area of 1. One proton is split by one.
|
Y—C—H -Doublet having a relative total area of 1. One proton is split by one.
X ≠ Y

6. CH$_3$ -Doublet having a relative total area of 3. Three protons are split by one.
|
X—CH -Quartet having a relative total area of 1. One proton is split by three.
|
Y

7. X
|
CH$_2$ -Triplet having a relative total area of 2. Two protons are split by two.
|
CH$_2$ -Triplet having a relative total area of 2. Two protons are split by two.
|
Y
X ≠ Y

8. X
|
CH$_2$ -Doublet having a relative total area of 2. Two protons are split by one.
|
X—CH -Triplet having a relative total area of 1. One proton is split by two.
|
Y

FIGURE 36-14
Some common splitting patterns in ^{1}H NMR spectra.

field, but will influence the two closer ortho protons more. The end result, as shown in Figure 36-15(b), is often two differently sized and shaped peaks. If a ring is para-disubstituted with two identical groups, only one aromatic peak is produced, due to all the aromatic protons being equivalent. However, if the two groups differ somewhat in their electron-withdrawing ability, a spectrum like the one shown in Figure 36-15(c) is produced. If they differ markedly in their electron-withdrawing ability, the pattern resulting is like the one shown in Figure

TABLE 36-5.

Proton Spin-Spin Coupling Constants J in Hertz

Structural feature	J	Structural feature	J
H, C=C, H (cis)	6–14	O H H, −C−N−C−	5–9
H, C=C, H (trans)	11–18	H, −C−O−H	4–7
C=C, H H	0–5	H, H−C≡C−C−	2–3
(benzene) H H, ortho	6–10	H, H, −C−C≡C−C−	2–3
(benzene) H, meta	1–4	(cyclohexene) −H −H	8–11
(benzene) H, para	0–2	(cyclopentene) −H −H	5–7
H H, C=C−C=C	9–13	(cyclohexane) H H, e, e	0–5
C=C, H, CH	4–10	(cyclohexane) H, H, e, a	0–7
C=C, CH, H	0–3	(cyclohexane) H, H, a, a	5–14
C=C, H, C−H O	5–8	(Newman) H, H, anti	5–14
H O, −C−C−H	1–3		

TABLE 36-5. (continued)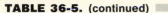

Structural feature	J	Structural feature	J
gauche	2–6	trans	5–8
cis	2–5	cis	3–6
trans	1–3	trans	0–4
cis	6–12	CH_3-CH_2-R	6–7
trans	4–8	CH_3, CH_3 CH—R	5–7
cis	8–10	CH—CH	2–9

36-15(d). Finally, aromatic rings with monosubstituted anisotropic groups such as carbon-oxygen and carbon-carbon double bonds frequently give two differently sized and shaped peaks as shown in Figure 36-15(e). All the ring protons are deshielded by the π bond of the anisotropic group, but the two ortho protons, being closer, are deshielded more and so appear downfield.

36.11 PROTONS ATTACHED TO OXYGEN AND NITROGEN

Because of proton exchange with the solvent, protons attached to oxygen and nitrogen frequently do not couple with protons on adjacent carbons to give spin-spin splitting. In addition, because of varying amounts of hydrogen bonding affecting the electron density around these protons, their peaks can have large chemical shift changes. Hydrogen bonding, proton exchange, and other factors also frequently make these peaks quite broad compared to other singlets, which can be a tremendous help in identifying these protons. Sometimes, protons attached to nitrogen atoms can even have the peak be so broad and flat that it almost cannot be recognized above the baseline.

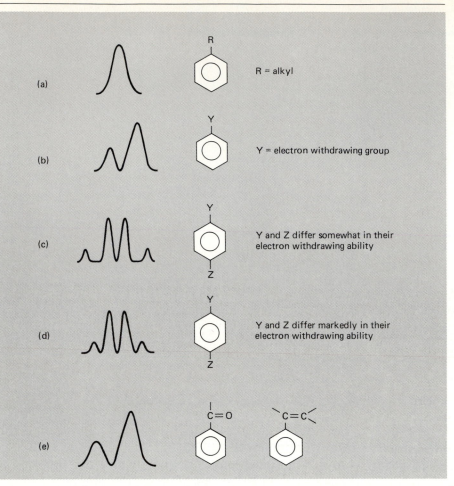

FIGURE 36-15
Some typical aromatic peaks.

36.12 OTHER CASES OF NO SPIN-SPIN SPLITTING

There are many other times when spin-spin splitting does not occur. For example, we do not observe it between the protons comprising a methyl group, since these protons all have the same chemical shift and are positionally equivalent. The protons are said to be **magnetically equivalent**. For the same reasons,

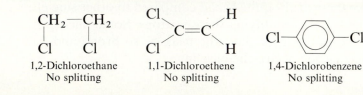

1,2-Dichloroethane
No splitting

1,1-Dichloroethene
No splitting

1,4-Dichlorobenzene
No splitting

we do not observe spin-spin splitting in compounds such as 1,2-dichloroethane, 1,1-dichloroethene, and 1,4-dichlorobenzene.

Spin-spin splitting also does not normally occur when the hydrogens are not on adjacent carbons. For example, there is no splitting between the six methyl protons and the two methylene protons in 1,2-dichloro-2-methylpropane. There is also no splitting between the aromatic protons and the benzylic protons of ethylbenzene. These protons are simply too far away from each other for there to be any significant magnetic influence.

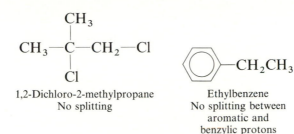

1,2-Dichloro-2-methylpropane
No splitting

Ethylbenzene
No splitting between aromatic and benzylic protons

36.13 CARBON NMR SPECTROSCOPY

It is not possible to obtain ^{12}C NMR spectra, since there are an even number of both protons and neutrons and, therefore, no magnetic spin. However, since 1970, with the development of new instrumental methods, ^{13}C NMR spectroscopy has become available to complement ^{1}H NMR spectroscopy.

Because ^{13}C comprises only 1.1 percent of all the carbon in a compound and has a low magnetic moment, a ^{13}C signal is only about $\frac{1}{6000}$ as intense as a ^{1}H signal. A signal this weak would normally be lost in with the noise, but using a ^{13}C spectrometer equipped with Fourier transform analysis allows the ^{13}C peaks to be easily determined.

Some typical ^{13}C chemical shifts are given in Table 36-6. Observe that these values are a tremendous help in identifying the various kinds of functional groups within a molecule. For example, the carbonyl carbon of an ester has a δ of about 170–175 ppm, whereas carbonyl carbons in other types of compounds usually have other chemical shift values. These chemical shifts are also usually about 15 to 20 times larger than comparable ^{1}H chemical shifts, since the carbons producing the peaks are one atom closer than the attached hydrogen to a shielding or deshielding group.

In contrast to ^{1}H NMR spectra, ^{13}C NMR spectra will usually consist only of many singlets rather than a number of multiplets. Spin-spin splitting of ^{13}C with protons does occur, but the spectra are frequently very complex and difficult to interpret. As a result, a special technique called **spin decoupling** is used to eliminate this splitting and give only singlets. Spin-spin splitting of ^{13}C atoms with each other can also occur, but is not likely, due to the very small chance of a ^{13}C atom being beside another ^{13}C atom.

TABLE 36-6.

Some Typical Chemical Shifts in ^{13}C NMR Spectra

Type of carbon	Chemical shift, δ (ppm)	Type of carbon	Chemical shift, δ (ppm)
R$\underline{\text{C}}$H$_2$CH$_3$	10–15	Ar$\underline{\text{C}}$	125–140
R$\underline{\text{C}}$H$_2$CH$_3$	15–25	Ar$\underline{\text{H}}$	125–150
R$_3$$\underline{\text{C}}$H	25–40		
$\underline{\text{C}}$H$_3$COR (C=O)	20	R$\underline{\text{C}}$NR$_2'$ (C=O)	150–170
$\underline{\text{C}}$H$_3$CR (C=O)	30	R$\underline{\text{C}}$NH$_2$ (C=O)	155–180
R$\underline{\text{C}}$H$_2$Br	25–35		
R$\underline{\text{C}}$H$_2$Cl	40–45	R$\underline{\text{C}}$OR' (C=O)	170–175
R$\underline{\text{C}}$H$_2$NH$_2$	35–50		
R$\underline{\text{C}}$H$_2$OH	50–65	R$\underline{\text{C}}$OH (C=O)	175–185
R$\underline{\text{C}}$H$_2$OR'	50–90		
R$\underline{\text{C}}$≡CH	65–70	R$\underline{\text{C}}$H (C=O)	190–200
R$\underline{\text{C}}$≡CH	75–85		
R$\underline{\text{C}}$H=CH$_2$	115–120	R$\underline{\text{C}}$CH$_3$ (C=O)	205–215
R$\underline{\text{C}}$H=CH$_2$	125–140		
R$\underline{\text{C}}$≡N	115–130		

36.14 PREPARING ^{1}H NMR SAMPLES

All ^{1}H NMR samples are run as liquids in 17.5 cm × 5 mm (7 inches × $\frac{3}{8}$ inch) sample tubes with tight-fitting plastic caps. As shown in Figure 36-16, the sample level must be between 3 and 4 cm high (1.2–1.6 inches). This ensures that the sample will be fully aligned between the magnetic poles when the sample tube is placed in the spectrometer. As a result, from 0.6–0.8 mL of liquid will be required. Care should be taken to avoid touching the sample tubes in this lower region to keep skin oils off the glass.

Nonviscous liquid compounds can be run "neat" or without a solvent. However, they are much more likely to give broadened peaks and poor resolution than liquids run in solution. Viscous liquids or solids must be run in solution. The sample concentration in solution generally is about 10–30 percent by weight or volume. Thus, about 0.2 mL of a liquid sample or 150 mg of a solid sample will be needed. Solid samples should be weighed to ensure a proper concentration.

It is very important to choose the proper solvent for any NMR solution. The three criteria for a solvent are as follows:

1. The sample must be very soluble in it.

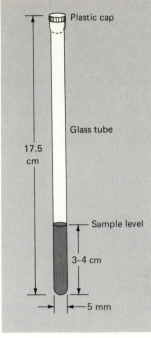

FIGURE 36-16
An NMR sample tube filled to the proper level.

2. It must have no protons which would give an NMR absorption peak.

3. It must not react with the sample. In particular, it should not allow proton exchange with the sample, if possible.

Carbon tetrachloride, CCl_4, satisfies the last two criteria and is a commonly used NMR solvent. But, unfortunately, many compounds are not very soluble in it, so more polar, deuterated solvents are used because deuterium does not absorb in the proton NMR region. Of the deuterated solvents, $CDCl_3$ is used most often, but dimethyl sulfoxide-d_6, CD_3SOCD_3, and hexadeuteroacetone, CD_3COCD_3, can be used if necessary. Deuterium oxide, D_2O, can be used for very polar compounds, but, as shown below, it frequently allows proton exchange with the sample. This causes the partial or complete disappearance of certain peaks with time and the appearance of a D—OH peak at $\delta\,4.5$.

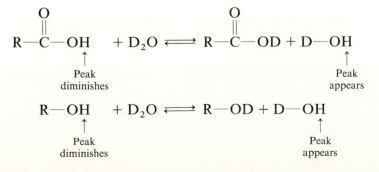

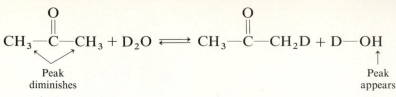

In addition, TMS is not soluble in D_2O. Therefore, sodium 2,2-dimethyl-2-silapentane-5-sulfonate (DSS) is used instead as the internal standard. DSS, being polar, is soluble in D_2O, and the protons on the three methyl groups attached to the silicon have almost the same chemical shift as the protons in TMS.

$$CH_3-\underset{\underset{CH_3}{|}}{\overset{\overset{CH_3}{|}}{Si}}-CH_2CH_2CH_2-SO_3^- \; Na^+$$

Sodium 2,2-dimethyl-2-silapentane-5-sulfonate (DSS)

If a viscous liquid or a solid has no specified NMR solvent, one normally tries CCl_4 first. The deuterated solvents are expensive, so they are used only when necessary. $CHCl_3$ is usually tried next, and if the sample is very soluble then $CDCl_3$ is used. In this way, the expensive waste of the deuterated solvent is avoided if it should prove to be unsatisfactory. If dealing with a highly polar compound, deuterated dimethyl sulfoxide or a combination of it and deuterated chloroform will often succeed. Or if the sample is not acid sensitive, trifluoroacetic acid can be used. The chemical shift for the single proton in trifluoroacetic acid is over 12, so it will not interfere with most other protons.

No matter what deuterated solvent is used, it will likely produce a very small peak from a few of its molecules being unavoidably protonated. Thus, samples run in $CDCl_3$ will have a very small peak at $\delta\,7.27$ from the $CHCl_3$ present, and samples run in DMSO-d_6 will have a very small peak at $\delta\,2.55$ from the CD_3SOCD_2H present.

In addition, if the sample is run in a solvent other than CCl_4 or $CDCl_3$, the chemical shift values may be changed by as much as 0.5–1.0 ppm, due to the sample's different solvent environment. Sometimes this can even be used to advantage to separate peaks which overlap when CCl_4 and $CDCl_3$ solutions are used.

NMR solvents other than D_2O may have the reference standard TMS already added. If not, or if a liquid sample is being run without a solvent, then about 5 to 10 μL (1 to 2 drops) of TMS must be added to the sample tube with a microsyringe or a pipet. Since TMS boils at 27°C, it must be kept in a refrigerator when not in use.

Some NMR solvents such as carbon tetrachloride, deuterated chloroform, and deuterated benzene are toxic and suspected carcinogens. Deuterated benzene, fortunately, is seldom needed, but carbon tetrachloride and deuterated chloroform are very commonly used. Therefore, they must be used with care to minimize the hazard. All sample solutions should be prepared in the hood, and the solvents themselves should be stored in the hood. Rubber or plastic gloves can also be used. Septum-capped bottles, which can be used only with a special hypodermic syringe, are recommended. Screw-capped bottles, if used, should have a pipet for each bottle stored in a test tube taped to the side of the bottle.

36.15   RECORDING ¹H NMR SPECTRA

At many schools having a ¹H NMR spectrometer, an experienced laboratory assistant will take the spectrum of student samples. If students are allowed to operate the spectrometer, the instructor will explain its use in detail and will likely have a laboratory assistant supervise students taking spectra. Remember that an NMR spectrometer is a complex and very expensive instrument.

Because of all the various makes and models of NMR spectrometers available, it would be impractical to describe all the different controls here. However, to be certain the spectrometer is working properly and is correctly adjusted, the spectrum of a standard sample of chloroform and TMS should be taken. By adjusting the sweep zero control, the $CHCl_3$ peak should appear at exactly 7.27 ppm and the TMS peak at exactly 0.00 ppm. The peaks should also be made as high and as narrow as possible by adjusting the resolution control, also called the Y-control or the homogeneity control.

36.16   INTERPRETING ¹H NMR SPECTRA

Learning to analyze ¹H NMR spectra requires studying a number of known examples and practicing with some unknowns. This section provides some helpful hints to make your spectral interpretations easier.

1.   Using the molecular formula, if given, and the number of integration spaces for each peak or multiplet, determine the number of protons represented by each peak or multiplet (see Section 36.4). If the molecular formula is not given, you can still determine the relative number of protons represented by each peak or multiplet. This will be a tremendous help to you in assigning a structure. Since the relative number must always be a whole number and will, for example, be a number such as 3 for a singlet, you can also determine the total number of protons, even if the molecular formula is not given.

2.   If the molecular formula is known, compare it with what would be expected for an open-chain, saturated molecule. Any difference suggests the possibility of rings, aromaticity, or double and triple bonds.

3.   You should learn to recognize methyl, ethyl, isopropyl, *tert*-butyl, and other groups by their known, characteristic peaks or multiplets (see Figure 36-14). Thus, a singlet of three protons would be a methyl group. A quartet of two protons and a triplet of three protons would be an ethyl group. A septet of one proton and a doublet of six protons would be an isopropyl group. And a singlet of nine protons would be a *tert*-butyl group, and so on. Sometimes the characteristic will not be perfect. For example, you may have difficulty counting all seven peaks in a septet for an isopropyl group.

4. Subtract any known portions of the molecule from the molecular formula, if given. This will greatly simplify your task of assigning the remaining portions.

5. Absorption around $\delta 7$ to $\delta 8$ suggests an aromatic ring. This will be supported by unsaturation indicated in the molecular formula, if given. From the splitting pattern and the coupling constants (see Figure 36-15 and Table 36-5), you may be able to know the relative locations of any ring substituents. Absorption further downfield than $\delta 7.2$ suggests the presence of an electron-withdrawing group.

6. Absorptions around $\delta 5$ to $\delta 6$ suggest vinyl protons. The coupling constants in Table 36-5 can help you determine the stereochemistry or other structural features of the alkene portion of the molecule.

7. A broad singlet of one proton at a chemical shift of $\delta 10$ or more suggests a carboxylic acid. A dicarboxylic acid would have a broad singlet of two protons. The molecular formula, if given, will support or negate these possibilities.

8. A broadened singlet from $\delta 1$ to $\delta 5.5$ of one proton might be due to the OH group of an alcohol. A broadened singlet from $\delta 4$ to $\delta 12$ might be due to the OH group of a phenol. Use the molecular formula, if given, or other known information (e.g., the presence or lack of aromatic absorption) to help you decide.

9. A broadened singlet from $\delta 0.6$ to $\delta 2.5$ of one or two protons might be due to the NH or NH_2 group of an aliphatic amine. If so, the molecular formula, if given, will indicate the presence of nitrogen. A broadened singlet from $\delta 3$ to $\delta 4.5$ of one or two protons might be due to the NH or NH_2 group of an aromatic amine. If so, the molecular formula, if given, will indicate the presence of nitrogen, and the spectrum will also have aromatic absorption.

10. Absorption around $\delta 9.4$ to $\delta 9.9$ of one proton suggests an aliphatic aldehyde. Absorption around $\delta 9.7$ to $\delta 10.3$ suggests an aromatic aldehyde. If so, the spectrum will also have aromatic absorption.

11. A singlet of one proton at $\delta 2.3$ to $\delta 2.5$ suggests a terminal aliphatic alkyne. A singlet of one proton at $\delta 2.8$ to $\delta 3.1$ suggests a terminal aromatic alkyne. If so, the spectrum will also have aromatic absorption.

12. Absorption around $\delta 1.9$ to $\delta 2.8$ suggests protons on the carbon adjacent to the carbonyl group of an aldehyde, ketone, ester, amide, or carboxylic acid. If so, the molecular formula, if given, and other absorption peaks will help you determine which type of carbonyl compound is present.

13. Absorption around $\delta\,2.8$ to $\delta\,4.5$ suggests protons on a carbon bearing a highly electronegative element such as oxygen, nitrogen, or a halogen. If so, the molecular formula, if given, will confirm the presence of the highly electronegative element..

14. Use the chemical shift values in Table 36-3 to confirm the presence of other specific structural groups. The coupling constants in Table 36-5 and the observations on pages 671–673 may also help you make a decision.

36.17 SAMPLE ^{1}H NMR PROBLEMS SOLVED

Compound A, Figure 36-17

Integration and the molecular formula $C_{12}H_{12}O$ tell us that the three multiplets from left to right represent seven, two, and three hydrogens, respectively. The triplet at $\delta\,1.4$ and the quartet at $\delta\,4.0$ suggest the presence of an ethyl group, which is supported by the number of hydrogens known to be present in these two multiplets. Furthermore, because of the downfield absorption at $\delta\,4.0$, the ethyl group must be attached to a highly electronegative element. In this case, from the molecular formula, this element can only be oxygen. Subtracting the

FIGURE 36-17

The ^{1}H NMR spectrum of compound A, $C_{12}H_{12}O$, in carbon tetrachloride. Spectrum courtesy of Sadtler Research Laboratories, division of Bio-Rad Laboratories, Inc., copyrighted 1973.

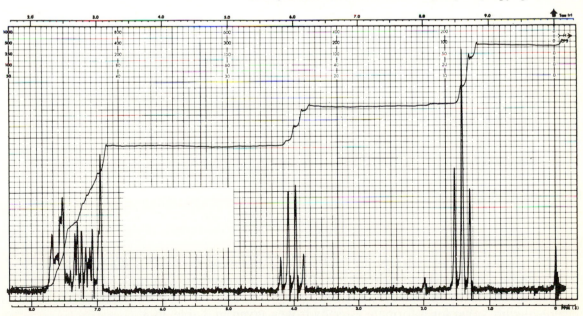

known portion from the molecular formula, we have

$$\begin{array}{r} C_{12}H_{12}O \\ -C_2\ H_5\ O \\ \hline C_{10}H_7 \end{array}$$

The seven hydrogens forming a complex multiplet in the aromatic region of the spectrum indicate the presence of more than one aromatic ring. With ten carbons remaining, the only possibility is to have two rings fused together to give us

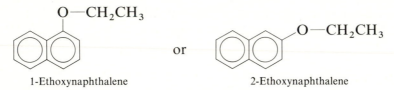

Unsubstituted naphthalene has a molecular formula of $C_{10}H_8$. Our seven aromatic hydrogens therefore indicate a monosubstituted naphthalene. Placing our known ethoxy group onto the naphthalene ring system, we have two possibilities:

O—CH$_2$CH$_3$

or

O—CH$_2$CH$_3$

1-Ethoxynaphthalene   2-Ethoxynaphthalene

Unfortunately, our limited knowledge of the complex aromatic splitting does not allow us to determine which structure is correct. The compound actually is 2-ethoxynaphthalene, which is the perfume fixative nerolin (see Experiment 11).

Compound B, Figure 36-18

Integration and the molecular formula $C_5H_{10}O_2$ tell us that the peaks or multiplets from left to right contain one, two, one, and six hydrogens, respectively. The apparent septet at $\delta 2.0$ and the doublet at $\delta 1.0$ suggest an isopropyl group, which is supported by the number of hydrogens known to be present in these two multiplets. Because both of these multiplets are quite far upfield, there is apparently no deshielding group or quite electronegative element nearby. Subtracting the known portion from the molecular formula, we have

$$\begin{array}{r} C_5H_{10}O_2 \\ -C_3H_7 \\ \hline C_2H_3O_2 \end{array}$$

The singlet of one hydrogen at $\delta 8.1$ is too far upfield to be an aldehyde hydrogen or a carboxylic acid hydrogen. It could, however, be a formate ester hydrogen, which is supported by the remaining atoms to be assigned and the chemical shift value for a formate ester hydrogen given in Table 36-3.

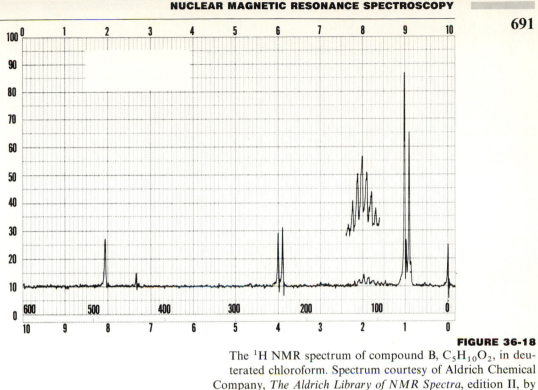

FIGURE 36-18

The ^{1}H NMR spectrum of compound B, $C_5H_{10}O_2$, in deuterated chloroform. Spectrum courtesy of Aldrich Chemical Company, *The Aldrich Library of NMR Spectra*, edition II, by Charles Pouchert.

Subtracting these atoms from those remaining, we have

$$\begin{array}{r} C_2H_3O_2 \\ - C\ H\ O_2 \\ \hline CH_2 \end{array}$$

The remaining methylene group can only go in between the noncarbonyl oxygen atom of the ester and the isopropyl group. In the spectrum, it is the doublet of two hydrogens at $\delta\,3.95$. This chemical shift is also what we would expect for hydrogens on a carbon atom adjacent to the oxygen atom of an ester. We have a doublet because these hydrogens are split by the one hydrogen on the isopropyl group. This hydrogen on the isopropyl group must therefore be split **not** into a septet but into a more complex multiplet, which the instrument was not able to fully resolve. The complete structure is isobutyl formate, which is the compound primarily responsible for the flavor of raspberries (see Experiment 9).

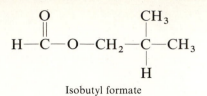

Isobutyl formate

Compound C, Figure 36-19

Integration and the molecular formula $C_9H_{12}N_2O_2$ tell us that the peaks or multiplets from left to right contain one, two, two, two, two, and three hydrogens, respectively. The triplet at δ 1.3 and the quartet at δ 3.9 suggest the presence of an ethyl group, which is supported by the numbers of hydrogens known to be present in these two multiplets. Furthermore, because of the downfield absorption at δ 3.9, the ethyl group must be attached to a highly electronegative element. From the molecular formula, this element could be either oxygen or nitrogen. However, according to Table 36-3, the chemical shift is too far downfield to be nitrogen. Therefore, we must have an ethoxy group, and subtracting this from the molecular formula, we have

$$\begin{array}{r} C_9H_{12}N_2O_2 \\ -\,C_2H_5O \\ \hline C_7H_7N_2O \end{array}$$

FIGURE 36-19
The ^{1}H NMR spectrum of compound C, $C_9H_{12}N_2O_2$, in polysol-d. Spectrum courtesy of Sadtler Research Laboratories, division of Bio-Rad Laboratories, Inc., copyrighted 1973.

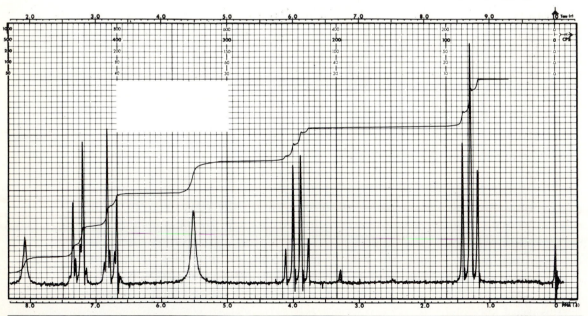

The multiplets of two hydrogens each at $\delta\,6.75$ and $\delta\,7.25$ appear to be aromatic absorptions. Furthermore, because of the symmetrical nature of the multiplets together, para disubstitution is suggested, with the ethoxy group being one of the two substituents. This gives us

$$-\langle\bigcirc\rangle-O-CH_2CH_3$$

Subtracting the carbons and hydrogens of the aromatic ring, we have

$$\begin{array}{r} C_7H_7N_2O \\ -C_6H_4 \\ \hline CH_3N_2O \end{array}$$

One of the three remaining hydrogens forms a broad singlet quite far down-field at $\delta\,8.1$, while the other two form a broad singlet further upfield from the aromatic hydrogens at $\delta\,5.5$. The broadness of these peaks suggests that these hydrogens are all connected to the nitrogen atoms, or two are connected to one nitrogen and the other is connected to oxygen.

$$-NH_2 \quad\text{and}\quad -\overset{|}{N}H \qquad or \qquad -NH_2 \quad\text{and}\quad -OH$$

However, with the latter combination, there is no possible satisfactory arrangement for all the remaining atoms together. Try it yourself and see. Therefore, this possibility must be rejected, and we know that all three hydrogens must be attached only to nitrogens. Subtracting three hydrogens and two nitrogens from all the remaining atoms, we have

$$\begin{array}{r} CH_3N_2O \\ -\ H_3N_2 \\ \hline CO \end{array}$$

The carbon and oxygen would appear to form a carbonyl group. Placing it with the other two groups, we have the possibilities

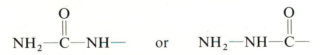

$$NH_2-\overset{\overset{\displaystyle O}{\|}}{C}-NH- \qquad or \qquad NH_2-NH-\overset{\overset{\displaystyle O}{\|}}{C}-$$

The arrangement on the right does not seem as likely. Amide hydrogens, as indicated in Table 36-3, typically absorb between $\delta\,5.5$ and $\delta\,7.5$. This could be considered consistent with the $\delta\,8.1$ observed for the one hydrogen, since it could be argued that the other nitrogen atom is further deshielding this hydrogen. However, it is unlikely from the chemical shift values given in Table 36-3 that an amide group could deshield the two hydrogens on the other nitrogen from $\delta\,0.6-2.5$ all the way up to $\delta\,5.5$.

Considering the other possibility, the two hydrogens absorbing at $\delta\,5.5$ do fall in the region of $\delta\,5.5$ to $\delta\,7.5$ expected for an amide. The single hydrogen absorbing at $\delta\,8.1$ is further downfield due to the aromatic ring. This gives us the following structure, which is the artificial sweetener dulcin (see Experiment 10).

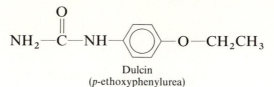

Dulcin
(*p*-ethoxyphenylurea)

Other ^{1}H NMR problems can be solved in a similar manner, but probably not as involved as the last portion of compound C.

QUESTIONS

1. Deduce the structure of each of the following compounds, based upon their ^{1}H NMR spectra and molecular formulas.

(a) C_3H_7ClO $\delta\,2.0$, 2H, quintet
     $\delta\,2.8$, 1H, singlet
     $\delta\,3.7$, 2H, triplet
     $\delta\,3.8$, 2H, triplet

(b) $C_7H_{12}O_3$ $\delta\,1.2$, 3H, triplet
     $\delta\,2.1$, 3H, singlet
     $\delta\,2.6$, 4H, multiplet
     $\delta\,4.0$, 2H, quartet

(c) C_6H_{14} $\delta\,0.8$, 12H, doublet
     $\delta\,1.4$, 2H, septet

(d) $C_3H_5Cl_3$ $\delta\,2.2$, 3H, singlet
     $\delta\,4.0$, 2H, singlet

(e) $C_{10}H_{13}Cl$ $\delta\,1.6$, 6H, singlet
     $\delta\,3.1$, 2H, singlet
     $\delta\,7.3$, 5H, singlet

(f) $C_4H_6Cl_2$ $\delta\,2.2$, 3H, singlet
     $\delta\,4.1$, 2H, doublet
     $\delta\,5.7$, 1H, triplet

(g) $C_9H_{10}O_2$ $\delta\,2.1$, 3H, singlet
     $\delta\,5.1$, 2H, singlet
     $\delta\,7.4$, 5H, singlet

(h) C_4H_9Br $\delta\,1.0$, 6H, doublet
     $\delta\,2.0$, 1H, multiplet
     $\delta\,3.3$, 2H, doublet

(i) C_7H_7ClO $\delta\,3.8$, 3H, singlet
     $\delta\,7.0$, 4H, symmetrical quartet

2. Predict the ^{1}H NMR spectra of the following compounds. Indicate for each proton signal the approximate chemical shift, the splitting pattern (doublet, triplet, etc.), and the number of hydrogens represented.

(a) CH_3—⟨◯⟩—O—$\overset{\overset{\displaystyle O}{\|}}{C}$—$CH_2CH_3$

(b) CH_3—$\overset{}{\underset{\overset{|}{H}}{N}}$—$\overset{\overset{\displaystyle O}{\|}}{C}$—$\overset{}{\underset{\overset{|}{CH_3}}{CH}}$—$CH_3$

(c) ⟨◯⟩—O—CH_2—$\overset{\overset{\displaystyle O}{\|}}{C}$—$CH_2CH_3$

(d) CH_3—$\overset{\overset{\displaystyle CH_3}{|}}{\underset{\overset{|}{CH_3}}{C}}$—$CH_2$—$Cl$

$$(e) \quad CH_3-\overset{\displaystyle CH}{\underset{\displaystyle CH_3}{|}}-\overset{\displaystyle O}{\overset{\|}{C}}-O-CH_3$$

$$(f) \quad H-\overset{\displaystyle O}{\overset{\|}{C}}-\bigcirc-\overset{\displaystyle CH_3}{\underset{\displaystyle CH_3}{\overset{|}{C}}}-\overset{\displaystyle O}{\overset{\|}{C}}-NH_2$$

3. Indicate the approximate ^{13}C chemical shift for each type of carbon for each of the compounds given in Problem 2.

References

1. R. Abraham, **The Analysis of High Resolution NMR Spectra,** Elsevier, New York, 1971.

2. A. Ault and G. Dudek, **An Introduction to Proton Nuclear Magnetic Resonance Spectroscopy,** Holden-Day, San Francisco, 1976.

3. R. Bible, **Interpretation of NMR Spectra, An Empirical Approach,** Plenum Press, New York, 1965.

4. F. Bovey, **Nuclear Magnetic Resonance Spectroscopy,** Academic Press, New York, 1969.

5. J. Dyer, **Applications of Absorption Spectroscopy of Organic Compounds,** Prentice-Hall, Englewood Cliffs, N.J., 1965.

6. L. Jackman and S. Sternhell, **Applications of Nuclear Magnetic Resonance in Organic Chemistry,** 2nd ed., Oxford University Press, New York, 1969.

7. L. Johnson and W. Jankowski, **Carbon-13 NMR Spectra,** Wiley-Interscience, New York, 1972.

8. G. Levy and G. Nelson, **Carbon-13 Nuclear Magnetic Resonance for Organic Chemists,** Wiley-Interscience, New York, 1972.

9. R. Silverstein, G. Bassler, and T. Morrill, **Spectrometric Identification of Organic Compounds,** 4th ed., Wiley, New York, 1981.

10. J. Stothers, **Carbon-13 NMR Spectroscopy,** Academic Press, New York, 1972.

EXPERIMENT

37

THE CHEMICAL LITERATURE

Literature is an investment of genius which pays dividends to all subsequent times.

John Burroughs, *Literary Fame*

37.1 INTRODUCTION

There are today over 9,300,000 known organic compounds, and almost 500,000 new ones are discovered or synthesized each year. As large as these numbers are, they are, however, almost nothing when one considers that the number of possible organic compounds approaches infinity. Organic chemistry has come a long way in the last 160 years, but it obviously still has a much longer way to go.

In organic chemistry, it is frequently necessary to know precisely what information is known about any of these 9,300,000 compounds. Otherwise, people would constantly be wasting their time and effort, unknowingly obtaining data or repeating investigations that have already been done. But with so many organic compounds, how can this information be obtained? The answer is the chemical literature.

The chemical literature can even be valuable to you as a student. Obviously, this textbook and all other textbooks cannot possibly present all the thousands of methods or techniques for synthesizing or isolating organic compounds. If this course includes a research problem or if you perform chemical research later in college or as part of your profession, you must know the best approach and the alternatives even before you go into the laboratory.

However, with over 9,300,000 known organic compounds and thousands of procedures, the chemical literature is obviously very vast and voluminous, consisting of many millions of pages in over 14,000 different journals. How can one possibly approach it in an efficient and systematic manner without being over-whelmed and just simply giving up? The following guide should prove to be very helpful.

37.2   THE STEPS IN A LITERATURE SEARCH

1.   Carefully name the compound, the reaction, or the synthetic method or technique. If there are several possibilities, write them all down. Be sure to include the IUPAC name if you are dealing with a compound.

2.   If dealing with a compound, consult one or more of the chemical hand-books (discussed later) to see if the compound is well known. If dealing with a reaction or a synthetic method or technique, consult one or more of the special books (discussed later) to see if the reaction or the synthetic method or technique is discussed in detail.

3.   If necessary, use the **Beilstein** reference number given in the chemical handbook to locate the compound in **Beilstein** (discussed later).

4.   Consult **Beilstein** or **Chemical Abstracts** (discussed later) to locate any compounds not found in any chemical handbook. Both sources will indicate briefly what is known about the compounds and will give the primary literature sources of this information. **Chemical Abstracts** and **Beilstein** may also be used to locate additional information on compounds found in the chemical handbooks.

5.   If necessary and if possible, consult the original literature to obtain the information in depth.

37.3   HANDBOOKS

As you learned in Experiment 2, rapid information on a compound's physical properties, such as its melting point, boiling point, solubility, and optical rotation, may be obtained by consulting a chemical handbook. The following represent the most common and useful sources.

1.   Handbook of Chemistry and Physics (Chemical Rubber Company, Boca Raton, Florida) is probably the handbook most frequently consulted. It contains information on the molecular weight and physical properties of over 15,000 organic compounds. It is also useful in giving the **Beilstein** reference. All compounds are listed alphabetically under the parent compound's name. Thus, ethyl acetate would be found alphabetically as the ethyl ester under its parent compound, acetic acid. If any difficulty is encountered with a name, a listing of the compounds according to their molecular formula is also given.

2. Lange's Handbook of Chemistry (McGraw-Hill, New York) contains information on over 7000 compounds. Many students prefer this handbook to the **Handbook of Chemistry and Physics** because the compounds are usually easier to find. It also is useful in giving the **Beilstein** reference.

3. The Merck Index (Merck Pharmaceutical Company, Rahway, New Jersey) gives the molecular weight, structural formula, physical and biological properties, and uses of more than 10,000 compounds. Unlike other handbooks, most of the compounds are either isolated from living organisms or they are synthesized compounds having a medicinal use or a physiological effect on humans. It is especially valuable in its extensive cross list of synonyms and trade names and in its literature references for the isolation and synthesis of many of the compounds. Another valuable feature is an extremely useful section called **Organic Name Reactions**, in which each reaction is described and references to original and review articles are given. The physical data of radioactive isotopes and their uses in medicine are also included.

4. Aldrich Catalog Handbook of Fine Chemicals (Aldrich Chemical Company, Milwaukee, Wisconsin, and available without charge from Aldrich) lists the prices, molecular weight, and physical properties of more than 12,000 compounds. This handbook is especially valuable in listing any special hazards of the compounds, how they may be disposed of, and the reference number to the **Registry of Toxic Effects of Chemical Substances**. Other valuable features include references to infrared and NMR spectra data, **Beilstein**, the **Merck Index**, and Fiesers' **Reagents for Organic Synthesis**.

5. Dictionary of Organic Compounds (Oxford University Press, New York) is a multivolume handbook that lists over 50,000 compounds alphabetically and gives the structural formula, IUPAC name, known physical data, and primary literature references. Especially valuable is the inclusion of physical data for derivatives of the parent compounds under the name of each parent compound.

6. Handbook of Tables for Identification of Organic Compounds (Chemical Rubber Company, Boca Raton, Florida) contains data on melting and boiling points of compounds and lists their derivatives according to functional groups.

37.4 REACTIONS, SYNTHESES, AND TECHNIQUES

Books that cover in depth what is known about certain organic reactions and synthetic methods and techniques can be very valuable and useful. Among those available are the following:

1. Organic Reactions (John Wiley and Sons, New York) is a multivolume series with over 100 general reactions discussed in detail with specific uses and limitations. Typical experimental procedures and extensive tables of examples are given with references. For ease in use, the subject index of the most recent volume can be checked for any desired reaction given in earlier volumes.

2. Organic Syntheses (John Wiley and Sons, New York) is a multivolume series containing the detailed procedures for synthesizing over 1000 compounds. Many of the syntheses are general methods that can be applied to the synthesis of numerous other compounds. Especially valuable are tables found at the end of each series of collective volumes. These tables classify the various methods according to the type of reaction, the type of compound prepared, the formula of the compound prepared, the use of specialized apparatus, and the preparation or purification of reagents and solvents.

3. Reagents for Organic Synthesis (Fieser and Fieser, John Wiley and Sons, New York) is a multivolume series dealing with the preparation and uses of a vast array of specific reagents for organic synthesis. Faulty or ill-advised conventional practices are stated, and the reagents are listed according to type. Author and subject indexes are included, along with the names and addresses of chemical and equipment suppliers.

4. Synthetic Methods in Organic Chemistry (Theilheimer, Interscience, New York) is a multivolume series indexed according to reaction type. Concise and comprehensive surveys of the current literature are given, along with references to the original literature, and information on yields and reaction conditions.

5. Technique of Organic Chemistry (Weissberger, Interscience, New York) is a multivolume series dealing with topics such as apparatus construction and performing complex reactions. It is the most extensive work dealing with organic laboratory techniques.

6. Synthetic Organic Chemistry (Wagner and Zook, John Wiley and Sons, New York) lists a number of preparative methods for different classes of organic compounds and gives the yields and references to the original literature.

7. Vogel's Textbook of Practical Organic Chemistry, Including Qualitative Organic Analysis (Longmans, New York) contains detailed laboratory procedures for synthesizing many organic compounds.

8. Newer Methods of Preparative Organic Chemistry (Interscience, New York) is a multivolume series with the same general format as **Organic Reactions**.

9. Advances in Organic Chemistry, Methods and Results (Interscience, New York) also is a multivolume series with the same general format as **Organic Reactions**.

10. Chemistry of Carbon Compounds (Rodd, Elsevier, New York) is a multivolume series that presents systematically the preparation and properties of numerous organic compounds. For ease in use, the last volume contains an extensive general index of all previous volumes.

11. Compendium of Organic Synthetic Methods (Harrison and Harrison, John Wiley and Sons, New York) gives a reaction and a literature reference for each entry. No data or laboratory procedures are provided.

12. Annual Reports in Organic Synthesis (McMurry and Miller, Academic Press, New York) is a multivolume series that cites only the new reactions published in the literature for a particular year. References are given, but laboratory procedures are not.

13. Organic Functional Group Preparations (Sandler and Karo, Academic Press, New York) presents actual laboratory procedures for synthesizing specific compounds.

14. Introduction to Modern Liquid Chromatography (Snyder and Kirkland, Interscience, New York).

15. Guides to Analysis by Gas Chromatography (Preston, Polyscience Corporation) is a multivolume series for various types of compounds.

37.5 JOURNAL ABSTRACTS

Chemical Abstracts

Chemical Abstracts is today the most important abstracting journal in chemistry. Every year it promptly summarizes over a half million papers, reports, and patents from more than 14,000 periodicals containing chemistry information. As you can probably imagine, it is an immense journal, and all its volumes occupy almost 400 feet of shelves in a library. It is also very expensive, and a yearly subscription price is now over $8000. Thus only colleges, companies, and research institutes can usually afford it.

Chemical Abstracts was first published in 1907. To make it easier to use, through 1956 a Collective Index of the previous 10 years' abstracts was provided. But since then, the Collective Index has been published every five years, and the latest one covers the period 1982 through 1986. For periods after 1986, noncollective indexes are published semiannually.

Compounds are indexed both by their molecular formulas and their chemical names. Since 1972, practically all trivial names have been abandoned. Thus acetone is now listed only under 2-propanone, and toluene is under benzene, methyl. If the proper chemical name is not known, one usually consults the Formula Index first, and the desired compound is located by name out of a list of perhaps 20 or more isomeric compounds. References to abstracts will generally be found in the Formula Index, but it usually pays to go immediately to the Chemical Substances Index, which may have additional entries and some indication of the specific information available in the reference. Thus the Chemical

Substances Index will help you to eliminate many references not having the specific information you may need on a particular compound.

Chemical Abstracts also has an Author Index, a General Subject Index, and a Numerical Patent Index, as well as descriptions and instructions for the use of all its indexes. The General Subject Index is very valuable for a broad category of chemical compounds such as "sweeteners."

Each index entry will refer to an abstract by using two numbers. The first one, given in boldface type, is the **Chemical Abstracts** volume number. The second indicates where the abstract can be found within this volume.

Until 1934, the second number always referred to the page number, with a smaller superscript number as a suffix indicating the fractional distance down the page on a scale of 1 to 9. Thus, $\mathbf{17}:492^4$ means that the entry would be found in Volume 17 about $\frac{4}{9}$ of the way down on page 492.

Then in 1934, **Chemical Abstracts** changed to two columns per page, so the second number became the column number instead of the page number. Also, in 1947, a letter replaced the superscript number, and the scale of distance down the column went from a to i.

Then in 1967, the system was changed again, and the second number refers specifically to the abstract number, with all abstracts being consecutively numbered. The letter that follows each abstract number has no meaning except as a computer check character.

Chemical Abstracts has the advantage of presenting the vast information of over 14,000 periodicals systematically within one source. It reviews the information and states the primary source of the information. Since no library could easily store or subscribe to all 14,000 periodicals, **Chemical Abstracts** obviously is a tremendous service. Also, since many of the periodicals are in foreign languages, including Japanese, Chinese, and Russian, the **Chemical Abstracts** review **in English** means that one does not have to be proficient in these languages. A special, but expensive, computer hookup, called **CAS Online**, also allows one to obtain information faster and much less laboriously than is possible by a manual search.

But on the negative side, **Chemical Abstracts** typically, out of necessity, provides only a brief review of each article. An abstract can easily be a scant one paragraph or even just a few sentences. Although tremendously helpful, it obviously does not equal the primary source itself, and one frequently wishes that the primary source was available or could be easily translated. Nevertheless, **Chemical Abstracts** typically does provide all vital information such as procedure, reaction conditions, amounts, yields, and the physical properties of all new compounds.

Beilstein's Handbuch der Organischen Chemie

Beilstein, until recently written only in German, has been issued in a number of series. The original edition (called the **Hauptwerk**, abbreviated H) rigorously and completely covers the organic chemistry literature up through 1909.

One must usually then use **Beilstein** for all organic chemistry literature before 1907 when **Chemical Abstracts** came into being. The first supplement of **Beilstein** (**Erstes Ergänzungswerk**, abbreviated E I) covers the literature rigorously and thoroughly from 1910 through 1919, the second supplement (**Zweites Ergänzungswerk**, abbreviated E II) rigorously and thoroughly covers the period of 1920 through 1929, and the third supplement (**Drittes Ergänzungswerk**, abbreviated E III) rigorously and thoroughly covers the period from 1930 through 1949. The fourth supplement (**Viertes Ergänzungswerk**, abbreviated E IV), covering the period from 1950 through 1959, is not yet complete. Thus, although **Beilstein** is even more thorough in its reporting than **Chemical Abstracts**, it is about 30 years behind. Furthermore, with the faster rate in which new compounds and processes are being reported, **Beilstein** may fall even further behind.

Beilstein is organized in a very sophisticated and complex manner. But if a compound is found in a chemical handbook, its **Beilstein** reference number or numbers will usually be given. For example, **Lange's Handbook of Chemistry** lists p-nitroacetanilide under "Nitroacetanilide (p)" and gives the **Beilstein** reference as XII-719. This means that p-nitroacetanilide can be found in the **Hauptwerk** on page 719 of volume 12. The **Aldrich Catalog Handbook of Fine Chemicals** lists this compound as "p-Nitroacetanilide" and says, "**Beil. 12**, 719," which also means it is found in the **Hauptwerk** on page 719 of volume 12. On the other hand, the **Handbook of Chemistry and Physics** uses the completely different entry "Acetic acid, amide, N(4-nitrophenyl)" and says "$B12^2$ 389," which means the compound is found in the second supplement of **Beilstein** on page 389 of volume 12.

Alternatively, one can consult the Formula Index (General Formelregister) of **Beilstein** to obtain the reference numbers. After determining, for example, that the molecular formula of p-nitroacetanilide is $C_8H_8N_2O_3$, one looks in the Formula Index to the second supplement (volume 29) and finds 102 entries with the formula. The 62nd is

"4-Nitro-acetanilide **12**, 719, I 351, II 389"

This means that information on p-nitroacetanilide is found in volume 12 of the **Hauptwerk** on page 719, in volume 12 of the **Erstes Ergänzungswerk** on page 351, and in volume 12 of the **Zweites Ergänzungswerk** on page 389.

Also important in **Beilstein** is the System Number, which is found on the top of each page. For example, for p-nitroacetanilide the given System Number is 1671. This means that this compound can be found in any incomplete supplements by looking under System Number 1671. There are no comprehensive formula indexes for incomplete supplements.

To thoroughly understand the information in **Beilstein**, one needs to understand German. But even without knowing much German, information such as primary literature references or a compound's physical properties can

be obtained without too much difficulty. In addition, German-English diction-
aries, especially those written for chemists, can help you translate key words.

Like **Chemical Abstracts, Beilstein** can also be searched now by using
a special computer hookup. Guidebooks for using **Beilstein** are also available.
Among them are the following:

1. How to Use Beilstein, Beilstein Institute, Frankfurt-am-Main, Berlin:
Springer-Verlag, 1982.

2. O. Weissbach, **The Beilstein Guide, A Manual for the Use of
Beilstein's Handbuch der Organischen Chemie**, New York: Springer-
Verlag, 1976.

Chemisches Zentralblatt

Chemisches Zentralblatt is the German equivalent of **Chemical Abstracts**. It goes back to 1856 and provides a more complete coverage of the
literature before 1939.

37.6 JOURNAL INDEXES

Science Citation Index

Started in 1961, the **Science Citation Index** covers every article from almost
3000 important journals and has the unique and valuable feature of listing every
more recent article that has cited a certain previous article as a footnote. Thus,
for example, if one is interested in all the applications of the Collins' reagent,
one can look up the original article by J. C. Collins, and all the later articles
referring to his reagent will be given.

Chemical Titles, Current Chemical Papers, and Current Contents

Chemical Titles, Current Chemical Papers, and **Current Contents**
publish a list of the titles, authors, and key words in the titles of papers recently
published or, sometimes, about to be published. This helps one keep more up to
date on current research, since there can frequently be a time gap between when
the paper was published and when the abstract appears.

37.7 THE PRIMARY LITERATURE

Although there are over 14,000 chemistry periodicals, it would be impractical
to include them all here. We list some of the more common ones. Boldface letters
indicate the standard form of abbreviation used by **Chemical Abstracts**.

1. Acta Chemica **Scand**inavica, English

2. **Angew**andte **Chem**ie, German, started in 1888

3. **Ann**alen der **Chem**ie, German, started in 1832

4. **Ann**ales de **Chim**ie, French, started in 1789

5. **Ann**ali di **Chim**ica, Italian

6. **Ark**iv für **Kemi**, English and German

7. **Aust**ralian **J**ournal of **Chem**istry, English

8. **Bull**etin of the **Chem**ical **Soc**iety of **Jap**an, English

9. **Bull**etin des **Soc**iete **Chim**iques **Belges**, French and English, started in 1887.

10. **Bull**etin de la **Soc**iete **Chim**ique de **Fr**ance, French, started in 1858

11. **Can**adian **J**ournal of **Chem**istry, English and French

12. **Chem**ische **Ber**ichte, German, started in 1868

13. **Chem**ical **Commun**ications, English

14. **Chem**istry and **Ind**ustry (London), English

15. **Chimia**, German, French, and English

16. **Collect**ion of **Czech**oslovak **Chem**ical **Commun**ications, English, German and Russian

17. **Comptes rendus Hebdomadaires, Series C**, French and English, started in 1835

18. **Dokl**ady **Akad**emii **Nauk SSSR**, Russian and English

19. **Experientia**, English, German and French

20. **Gazz**etta **Chim**ica **Ital**iana, Italian and English, started in 1871

21. **Helv**etica **Chim**ica **Acta**, German and French

22. **Israel J**ournal of **Chem**istry, English

23. Izvestiya **Akad**emii **Nauk SSSR, Otd**elenie **Khim**icheskikh **Nauk**, Russian and English

24. Journal of the **Amer**ican **Chem**ical **Soc**iety, English, started in 1879

25. Journal of **Bio**logical **Chem**istry, English

26. Journal of **Chem**ical **Ed**ucation, English

27. Journal of the **Chem**ical **Soc**iety, English, started in 1841

28. Journal of **Heterocycl**ic **Chem**istry, English

29. Journal of the **Indian Chem**ical **Soc**iety, English

30. Journal of **Med**icinal **Chem**istry, English

31. Journal of **Organometal**lic **Chem**istry, English, German, and French

32. Journal of **Org**anic **Chem**istry, English

33. Journal für **prakt**ische **Chem**ie, German and English, started in 1834

34. **Monatsh**efte für **Chem**ie, German, started in 1870

35. **Naturwissenschaften**, German and English

36. **Pure** and **Appl**ied **Chem**istry, English, German, and French

37. **Re**cueil des **tray**aux **chim**iques des **Paysbas**, English, German, and French, started in 1882

38. **Synthesis**, English and German

39. **Tetrahedron**, English, German, and French

40. **Tetrahedron Lett**ers, English, German, and French

41. **Z**eitschrift für **Naturforsch**ung, German

42. **Zh**urnal **Obshch**ei **Khim**ii, Russian and English, started in 1869

43. **Zh**urnal **Org**anicheskoi **Khim**ii, Russian and English

37.8 REVIEW AND CURRENT EVENTS JOURNALS

1. **Acc**ounts of **Chem**ical **Res**earch, English

2. **Adv**ances in **Carbohydrate Chem**istry, English

3. **Adv**ances in **Chromatography**, English

4. **Adv**ances in **Heterocyclic Chem**istry, English

5. **Adv**ances in **Org**anic **Chem**istry: Methods and Results, English

6. **Adv**ances in **Phys**ical **Org**anic **Chem**istry, English

7. **Ange**wandte **Chem**ie, German and English

8. **Ann**ual **Rep**orts on the **Prog**ress of **Chem**istry, English

9. **Ann**ual **Rev**iew of **Biochem**istry, English

10. **Chem**ical and **Eng**ineering **News**, English

11. **Chem**ical **Rev**iews, English

12. **Chem**ical **Soc**iety **Rev**iews, English

13. **Chem**istry in **Brit**ian, English

14. **Nature**, English, started in 1869

15. **Org**anic **Reactions**, English

16. **Science**, English

17. **Top**ics in **Stereo**chemistry, English

37.9   OTHER CHEMICAL LITERATURE SOURCES

1.   R. Bottle, ed., **The Use of Chemical Literature**, 2nd Edition, Hamden, Conn.: Archon Books, 1969.

2.   C. Burman, **How to Find Out in Chemistry**, 2nd Edition, New York: Oxford University Press, 1966.

3.   R. Gould, ed., **Advances in Chemistry Series**, No. 30, **Searching the Chemical Literature**, Washington: American Chemical Society, 1961.

4.   J. Hancock, "An Introduction to the Literature of Organic Chemistry," **Journal of Chemical Education, 45**, 193 (1968).

5.   R. Maizell, **How to Find Chemical Information: A Guide for Practicing Chemists, Teachers and Students**, New York: Interscience, 1979.

6.   M. Mellon, **Chemical Publications**, 4th Edition, New York: McGraw-Hill, 1965.

7.   H. Woodburn, **Using the Chemical Literature: A Practical Guide**, New York: Marcel Dekker, 1974.

37.10   COLLECTIONS OF SPECTRA

1.   A. Cornu and R. Massot, **Compilation of Mass Spectra Data**, 2nd Edition, London: Heyden and Son, 1975.

2.   **Handbook of Proton NMR Spectra and Data**, Asaki Research Center, Orlando, Florida: Academic Press, 1985.

3.   **High-Resolution NMR Spectra Catalog**, Palo Alto, California: Varian Associates, 1963.

4.   C. Pouchert, **The Aldrich Library of Infrared Spectra**, 3rd Edition, Milwaukee: Aldrich Chemical Company, 1981.

5.   C. Pouchert, **The Aldrich Library of NMR Spectra**, 2nd Edition, Milwaukee: Aldrich Chemical Company, 1983.

6. Sadtler Standard Spectra, Philadelphia: Sadtler Research Laboratories, Division of Bio-Rad Laboratories, a continuing collection of infrared and ^{1}H NMR spectra.

37.11 LITERATURE EXPERIMENT

Locate in the chemical literature a preparation for one of the following compounds, including all necessary experimental details, physical constants, and references. From the reference or references, write a complete laboratory synthesis procedure in enough detail so that another person could produce at least 5 grams of the compound without going to the given literature references.

1. Adrenaline
2. Azulene
3. Cubane
4. Cyclamate
5. Cyclooctatetraene
6. 7-Dehydrocholesterol
7. Dodecahedrane
8. 1-Ethyl-1,4-cyclohexadiene
9. Hexaphenylethane
10. Hexestrol

11. Ibuprofen
12. Prismane
13. Quinoline
14. Saccharin
15. Sulfamerazine
16. Tetracene
17. Thalidomide
18. Triphenylene
19. Tropinone
20. Tropolone

EXPERIMENT

38

ORGANIC QUALITATIVE ANALYSIS

38.1 INTRODUCTION

Imagine that you are an organic chemist and have just isolated a compound from an exotic tropical plant. The question now, of course, is exactly what is this compound? Is it one of the over 9,300,000 known organic compounds, or is it a new one with a completely undetermined structure? This question becomes especially important if, for example, the compound is suspected to be the active ingredient in an extract consumed by natives to effectively treat some previously incurable disease.

In earlier years, you would have relied very heavily on various reactions to determine the presence of certain functional groups or structural units in this compound. You might have even systematically degraded the compound to identifiable products, and then determined the structure by "piecing together" all the fragments. The entire process could be very long and tedious, with no guarantee of success.

But today, thanks to the development of highly effective spectroscopic methods, your task would be much easier. Why, you might even be able to determine the complete structure without any "wet chemistry," just by analyzing infrared, mass, ^{1}H NMR, and ^{13}C NMR spectra. If the structure was highly complex, x-ray crystallography could even be used.

In this course, you will not be asked to determine the structures of any new or extremely complex compounds. And you don't have to choose from well over 9,300,000 possibilities. Instead, your list of possible unknowns will be the much more limited number given in the accompanying tables at the end of the experiment. In addition, by using the guidelines in this experiment, the vast majority of these possibilities can be eliminated within one to two lab periods. Physical properties, spectroscopy, and chemical and solubility tests will all be used. Spectroscopy alone is frequently not enough, especially since the high cost prevents most schools from owning NMR and mass spectrometers. You will also learn much more chemistry by using this combined approach and will hopefully become a good chemical sleuth, attacking the problem of every unknown in a logical and efficient manner with a completely open mind.

38.2 PROCEDURE

Fortunately, the procedure for determining the structure of an unknown is quite straightforward and involves the following steps:

1. Determine the unknown's physical properties, such as its melting point or boiling point, color, odor, and so on.

2. Purify the unknown, if necessary, as indicated by its melting point, color, or its gas or thin-layer chromatography analysis.

3. Determine the unknown's infrared and, if possible, its ^{1}H NMR spectrum.

4. Classify the unknown by its solubility in various solvents.

5. Perform an elemental analysis.

6. Further classify the unknown, using relevant chemical tests.

7. Confirm the unknown's identity by taking the melting point of synthesized solid derivatives.

Each of these steps will now be covered in depth.

38.3 DETERMINE THE UNKNOWN'S PHYSICAL PROPERTIES

1. Determine the melting point or boiling point

Knowing the melting point or the boiling point of an unknown compound is critical. The compounds in the tables at the end are arranged by functional groups in order of increasing melting and boiling point. Thus, by knowing an unknown's melting point or boiling point, most of the possibilities can be eliminated. In addition, the melting point or boiling point can indicate the unknown's purity. If the unknown is known or is found to be impure, it must be purified by the procedure in Section 38.4 and the melting point or boiling point retaken.

Accuracy is extremely important. The thermometer used should be calibrated (page 75) and the reading taken carefully. Otherwise, you may have to expand your list of possibilities from the tables and may be even looking in the wrong portion of the tables! Remember that the observed melting point may be as much as 5°C below the literature value, so if your unknown is a solid, include as possibilities all compounds in the tables melting

5° higher. If done accurately, any observed melting point will usually not exceed the literature value.

An observed boiling point may conform to a literature value even less than a melting point, and will usually be lower than the literature value. In addition, if the boiling point is taken at less than one atmosphere of pressure, a correction must first be made using Figure 1-2 before comparing it to literature values.

Be aware that many liquids decompose or otherwise react at high temperatures. Therefore, do not initially commit all of your sample to a boiling point determination through regular distillation. Instead, utilize the capillary tube method of Siwoloboff, illustrated in Figure 38-1.

Place 2 to 5 drops of the unknown into a 5-cm length of 4-mm glass tubing, sealed at one end. Then place a melting-point capillary tube, sealed end up, into the larger tube. Fasten the larger sample tube to a thermometer with a rubber band, being sure the unknown is placed next to the thermometer bulb. Place the assembly into a melting-point bath and gradually heat. When a rapid stream of bubbles emerges from the capillary tube, the temperature is a few degrees **above** the boiling point of the unknown. Discontinue the heating and, as the system cools, a partial vacuum will cause liquid to be sucked up into the capillary tube. The temperature at which this occurs is the **approximate** boiling point of the unknown. Slowly resume the heating until bubbles again emerge, cease the heating and note at what temperature the bubbling again ceases. This is the boiling point of the unknown.

2. Note the color

A yellow solid may mean a nitro or nitroso compound, a quinone, an azo compound, an α-diketone, or a polyconjugated olefin or ketone. An orange to red

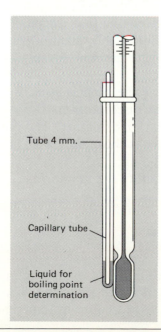

FIGURE 38-1

The apparatus for taking a boiling point by the Siwoloboff method.

Tube 4 mm.

Capillary tube

Liquid for boiling point determination

solid may mean a quinone, an azo compound or a polyconjugated olefin or ketone. A brown to dark purple liquid may indicate a phenol or amine which has been discolored by air oxidation products.

3. Carefully note the odor

Sometimes the odor of an unknown can aid its structural determination. For example, some amines have a characteristic fishlike odor. Many esters have a pleasant fruity or floral odor. Many carboxylic acids have a sharp and pungent odor. Mercaptans or thiols, sulfides and isonitriles also have unpleasant odors. Alcohols, ketones, alkanes, alkenes, and aromatic hydrocarbons all have characteristic odors.

Note an unknown's odor **only** by holding the sample away from you and fanning any fumes toward you. **Don't take a large sniff with your nose held directly over the sample. Furthermore, don't ever taste an unknown.** Every unknown should be treated as though it were toxic, because it probably is.

38.4   PURIFYING THE UNKNOWN, IF NECESSARY

Many unknowns given to you may only be from 95 to 99 percent pure. A wide melting point range or a colored liquid are indications of an impure sample. All impurities should be removed, since they can interfere with the unknown's chemical tests or spectra, giving you possibly misleading information.

All impure solids may be recrystallized from an appropriate solvent, after first trying a small amount of the solid in various possible solvents. The solid should be very soluble in the hot solvent, but only sparingly soluble in the cold solvent. Before committing your entire sample, make certain that recrystallizing a small amount has actually improved its purity, as indicated by an improvement in the melting point. Some unknowns may react with certain solvents or be otherwise changed by a recrystallization.

Impure liquids may be purified by distillation. If the boiling point by the Siwoloboff method is very high or if there is further discoloration, distillation at reduced pressure can help prevent decomposition or oxidation. The purity of colorless liquids can be checked by injection into a gas chromatograph having a column temperature 20 to 75° below the boiling point.

Column chromatography may also be used to purify the unknown.

38.5   DETERMINING THE INFRARED
AND, IF POSSIBLE, THE ^{1}H NMR SPECTRUM

Once you have a pure unknown with a known melting point or boiling point, you should obtain its infrared and, if possible, its ^{1}H NMR spectrum. The infrared spectrum can be taken and interpreted by the procedure in Experiment

35, and the ^{1}H NMR spectrum by the procedure in Experiment 36. It cannot be overstated how valuable these spectra can be in determining your unknown's structure. In fact, with them you can save yourself much valuable time and effort in the laboratory. Both types of spectra can indicate the presence or absence of certain functional groups in the molecule, and the ^{1}H NMR spectrum can tell you the number of each kind of hydrogen and other structural features.

38.6 CLASSIFYING THE UNKNOWN BY ITS SOLUBILITY IN VARIOUS SOLVENTS

Figure 38-2 classifies compounds on the basis of their solubility in water, ether, 5% NaOH, 5% NaHCO$_3$, 5% HCl, and concentrated H$_2$SO$_4$. Solubility tests should be performed on every unknown, since they are quick, reliable, and use only a small amount of sample. Except for the use of water and ether, these solubility tests determine the acid or base properties of organic compounds. Using the bases NaOH and NaHCO$_3$, you can tell how strongly acidic a compound is if it is an acid. Using HCl and H$_2$SO$_4$, you can tell how strongly basic a compound is if it is a base.

Do not perform any unnecessary tests. For example, if the unknown is water soluble, it will also be soluble in 5% HCl and 5% NaOH, since these solutions are mostly water. Therefore, performing solubility tests in these other reagents will tell you nothing, and will waste your time, your sample and the chemical reagents. As another example, if a compound is soluble in water, 5% HCl, 5% NaOH, or 5% NaHCO$_3$, it will also be soluble in concentrated H$_2$SO$_4$, so there is generally nothing to be gained from this test.

However, some extra solubility tests may provide valuable information. For example, if the unknown is soluble in water, it should also be tested with 5% NaHCO$_3$, especially if there is a sharp and pungent odor. Effervescence indicates a water-soluble carboxylic acid or sulfonic acid. Or if the unknown is insoluble in water, but soluble in 5% NaOH, its solubility in 5% HCl can pinpoint an amino acid.

Group I: Soluble in water, soluble in ether

Compounds containing up to about five carbons and also containing oxygen, nitrogen or sulfur are often soluble in both water and ether. However, increasing the number of carbons to five or six frequently results in insolubility or borderline solubility in water, due to the lower polar nature of the molecule. Furthermore, with five or six carbons, insolubility or borderline solubility in water is influenced by branching in the alkyl group. As the following examples illustrate, branched molecules are more water soluble than straight-chain molecules.

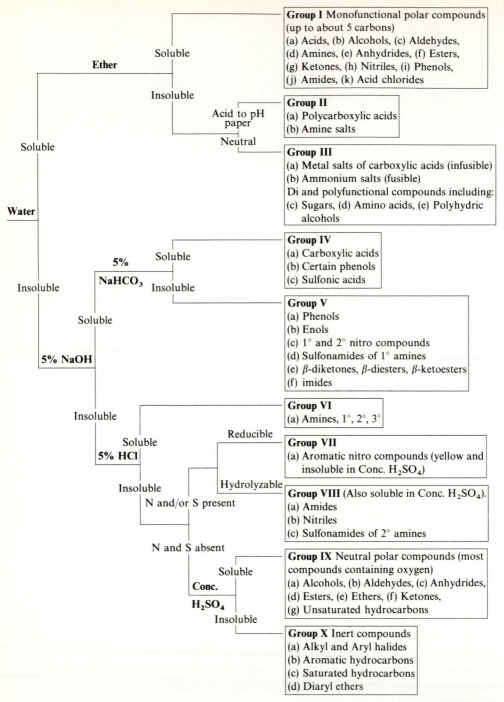

FIGURE 38-2

The classification of various compounds, based on their solubilities in water, ether, 5% NaOH, 5% NaHCO$_3$, 5% HCl, and concentrated H$_2$SO$_4$.

714

Soluble	Borderline	Insoluble
CH_3—$\underset{\underset{CH_3}{\mid}}{\overset{\overset{CH_3}{\mid}}{C}}$—$\overset{\overset{O}{\parallel}}{C}$—OH	CH_3—$\underset{\underset{CH_3}{\mid}}{CH}$—$CH_2$—$\overset{\overset{O}{\parallel}}{C}$—OH	$CH_3CH_2CH_2CH_2$—$\overset{\overset{O}{\parallel}}{C}$—OH
CH_3—$\underset{\underset{CH_3}{\mid}}{\overset{\overset{CH_3}{\mid}}{C}}$—$CH_2$—OH	CH_3—$\underset{\underset{OH}{\mid}}{CH}$—$\overset{\overset{CH_3}{\mid}}{CH}$—$CH_3$	$CH_3CH_2CH_2CH_2CH_2$—OH

Group I compounds should always have the pH of the solution tested with pH paper. Alkaline readings suggest the presence of an amine, while acidic readings suggest the presence of a carboxylic acid or a sulfonic acid. For acidic readings, also check the pH of the distilled water. Dissolved CO_2 can give a pH reading as low as 4 to 5.

Group II: Soluble in water, insoluble in ether, acid to pH paper

Group II compounds include polycarboxylic acids too polar to dissolve in ether, and amine salts.

Group III: Soluble in water, insoluble in ether, neutral to pH paper

Group III compounds include the metal and ammonium salts of carboxylic acids, and di- and polyfunctional compounds such as carbohydrates, amino acids, and polyhydric alcohols. All of these compounds are too polar to be soluble in ether.

Group IV: Insoluble in water, soluble in 5% NaOH and 5% NaHCO₃

Because $NaHCO_3$ is a weaker base than NaOH, compounds which dissolve in 5% $NaHCO_3$ must be appreciably acidic. This group therefore includes compounds like carboxylic acids and sulfonic acids, which react with $NaHCO_3$ to form a water-soluble sodium salt. Carbon dioxide gas is produced as a byproduct, and the observation of effervescence is evidence that a reaction occurred.

$$R—\overset{\overset{O}{\parallel}}{C}—OH + NaHCO_3 \longrightarrow R—\overset{\overset{O}{\parallel}}{C}—O^- Na^+ + H_2O + CO_2$$

$$R—SO_3H + NaHCO_3 \longrightarrow R—SO_3^- Na^+ + H_2O + CO_2$$

If the salts are formed from high-molecular-weight compounds, however, they are insoluble and precipitate. Examples are the salts of the long-chain carboxylic acids myristic (C_{14}), palmitic (C_{16}), and stearic (C_{18}) acids.

Phenols usually are not acidic enough to react with 5% $NaHCO_3$. However, if the phenol has one or more strong electron-withdrawing groups ortho and para to the hydroxyl group, the phenol now reacts. Resonance stabilization of the anion by the strong electron-withdrawing groups accounts for the difference. For example:

$$O_2N-\underset{NO_2}{\underset{|}{\bigcirc}}-OH + NaHCO_3 \longrightarrow O_2N-\underset{NO_2}{\underset{|}{\bigcirc}}-O^- Na^+ + CO_2$$
$$+ H_2O$$

Remember that some phenols such as phenol itself, 1,2-dihydroxybenzene, 1,3-dihydroxybenzene, and 1,4-dihydroxybenzene are polar enough to dissolve in water, and therefore belong in Group I.

Group V: Insoluble in water and 5% NaHCO₃, soluble in 5% NaOH

The compounds shown below, although typically not soluble in water or 5% $NaHCO_3$, are weakly acidic enough to react with and dissolve in 5% NaOH. In every case, the acidity is due to the resonance stabilization of the anion formed.

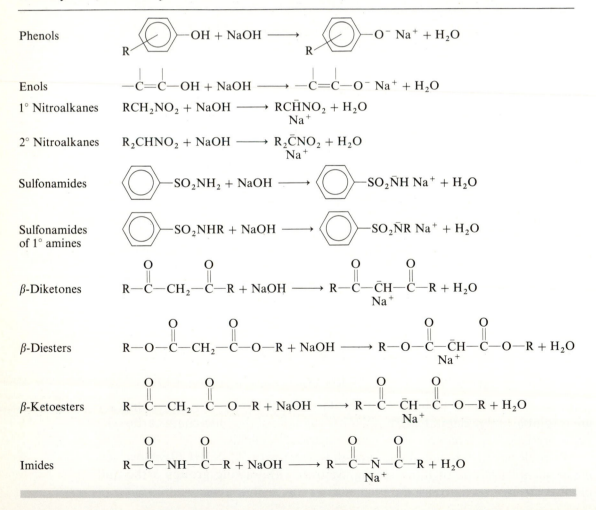

Phenols	
Enols	
1° Nitroalkanes	
2° Nitroalkanes	
Sulfonamides	
Sulfonamides of 1° amines	
β-Diketones	
β-Diesters	
β-Ketoesters	
Imides	

A special Group V case is the aminosulfonic acids. Although the dipolar ion or zwitterion is insoluble in water, the molecule is soluble in 5% NaOH, due to the removal of an acidic amino hydrogen. The dipolar ion, however, is insoluble in 5% HCl, due to the weak basicity of a sulfonate anion. For example:

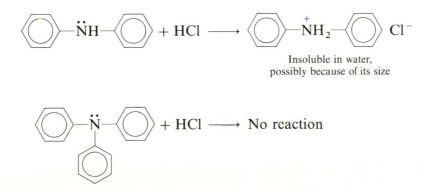

$$\bar{O}_3S - \langle \bigcirc \rangle - \overset{+}{N}H_3 + NaOH \longrightarrow \bar{O}_3S - \langle \bigcirc \rangle - NH_2 \; Na^+ + H_2O$$

Water insoluble   Water soluble

$$\bar{O}_3S - \langle \bigcirc \rangle - \overset{+}{N}H_3 + HCl \longrightarrow \text{No reaction}$$

Water insoluble

Group VI: Insoluble in water and 5% NaOH, soluble in 5% HCl

Practically all 1°, 2°, and 3° amines are soluble in 5% HCl, due to the formation of a water-soluble hydrochloride salt.

$$RNH_2 + HCl \longrightarrow RNH_3^+ \; Cl^-$$

$$R_2NH + HCl \longrightarrow R_2NH_2^+ \; Cl^-$$

$$R_3N + HCl \longrightarrow R_3NH^+ \; Cl^-$$

Diaryl and triaryl amines, however, do not dissolve, but for different reasons. Diaryl amines react with 5% HCl, but the amine hydrochloride formed is not water soluble, possibly because of its size. Triaryl amines, however, don't even react because the pair of electrons on nitrogen is so highly delocalized into the aromatic rings and therefore not available for sharing with the acid.

$$\langle \bigcirc \rangle - \ddot{N}H - \langle \bigcirc \rangle + HCl \longrightarrow \langle \bigcirc \rangle - \overset{+}{N}H_2 - \langle \bigcirc \rangle \; Cl^-$$

Insoluble in water,
possibly because of its size

$$\langle \bigcirc \rangle - \ddot{N} - \langle \bigcirc \rangle + HCl \longrightarrow \text{No reaction}$$

Some very high molecular weight amines may also react with dilute acid and yet be insoluble. They thus are like diaryl amines.

Group VII: Insoluble in water, 5% NaOH, 5% HCl and concentrated H₂SO₄.
Contains nitrogen and is reducible.

Group VII includes aromatic nitro compounds, which are reducible to amines and which are frequently yellow.

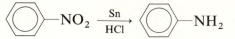

Group VIII: Insoluble in water, 5% NaOH and 5% HCl, soluble in concentrated H₂SO₄. Contains nitrogen and/or sulfur and is hydrolyzable.

Group VIII includes amides, nitriles, and sulfonamides of 2° amines. They dissolve in concentrated H_2SO_4 because they are weakly basic and are thus able to be protonated by H_2SO_4.

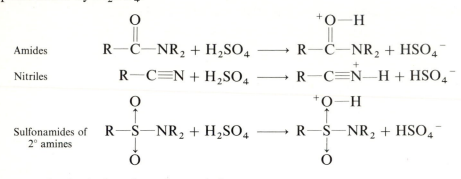

They may be hydrolyzed as shown below:

$$R\text{—}\overset{O}{\overset{\|}{C}}\text{—}NR_2 + H_2O \xrightarrow{H^+} R\text{—}\overset{O}{\overset{\|}{C}}\text{—}OH + H\text{—}\overset{+}{\underset{H}{N}}R_2$$

$$R\text{—}C\equiv N + H_2O \xrightarrow{H^+} R\text{—}\overset{O}{\overset{\|}{C}}\text{—}OH + {}^+NH_4$$

$$R\text{—}SO_2NR_2 + H_2O \xrightarrow{H^+} R\text{—}SO_3H + H\text{—}\overset{+}{\underset{H}{N}}R_2$$

Group IX: Insoluble in water, 5% NaOH and 5% HCl, soluble in concentrated H₂SO₄.

Group IX includes alcohols, aldehydes, ketones, esters, most ethers, alkenes, alkynes, and acid anhydrides. They dissolve in concentrated H_2SO_4 because they are weakly basic and thus able to be protonated by H_2SO_4.

Alcohols
$$R\!-\!OH + H_2SO_4 \longrightarrow R\!-\!\overset{+}{\underset{\underset{H}{|}}{O}}\!-\!H + HSO_4^-$$

Aldehydes
$$R\!-\!\overset{O}{\overset{\|}{C}}\!-\!H + H_2SO_4 \longrightarrow R\!-\!\overset{\overset{+}{O}\!-\!H}{\overset{\|}{C}}\!-\!H + HSO_4^-$$

Ketones
$$R\!-\!\overset{O}{\overset{\|}{C}}\!-\!R' + H_2SO_4 \longrightarrow R\!-\!\overset{\overset{+}{O}\!-\!H}{\overset{\|}{C}}\!-\!R' + HSO_4^-$$

Esters
$$R\!-\!\overset{O}{\overset{\|}{C}}\!-\!OR' + H_2SO_4 \longrightarrow R\!-\!\overset{\overset{+}{O}\!-\!H}{\overset{\|}{C}}\!-\!OR' + HSO_4^-$$

Most ethers
$$R\!-\!O\!-\!R' + H_2SO_4 \longrightarrow R\!-\!\overset{+}{\underset{\underset{H}{|}}{O}}\!-\!R' + HSO_4^-$$

Alkenes
$$\overset{}{>}\!C\!=\!C\!<\; + H_2SO_4 \longrightarrow \underset{\underset{H\quad OSO_3H}{|\quad\;\;\;|}}{-C\!-\!C-}$$

Alkynes
$$-C\!\equiv\!C- + H_2SO_4 \longrightarrow \underset{\underset{H\quad OSO_3H}{|\quad\;\;\;|}}{-C\!=\!C-}$$

Acid anhydrides
$$R\!-\!\overset{O}{\overset{\|}{C}}\!-\!O\!-\!\overset{O}{\overset{\|}{C}}\!-\!R + H_2SO_4 \longrightarrow R\!-\!\overset{O}{\overset{\|}{C}}\!-\!O\!-\!\overset{\overset{+}{O}\!-\!H}{\overset{\|}{C}}\!-\!R + HSO_4^-$$

Group X: Insoluble in water, 5% NaOH, 5% HCl and concentrated H₂SO₄

Group X includes alkyl and aryl halides, aromatic hydrocarbons, alkanes, and diaryl ethers. Diaryl ethers are included because the electron pairs on oxygen are delocalized into the aromatic rings and, therefore, evidently not available for sharing, even with concentrated H_2SO_4. Some aromatic hydrocarbons, however, become sulfonated in concentrated H_2SO_4 and then dissolve.

PROCEDURE

Perform each solubility step by adding 1 drop of the unknown liquid or a few crystals of the unknown solid into about 1 mL of the solvent in a small test tube. Tap the test tube gently with your finger to mix. The disappearance of the liquid or solid indicates that it is dissolving. If the liquid or solid dissolves completely, add some more of it in small amounts with gentle tapping to determine the extent of the solubility.

If the sample doesn't dissolve or only partially dissolves, heat the mixture **gently** on a steam bath to see if this helps. Avoid strong heating since it may cause a reaction to occur. Extra solvent may also be added. If a large amount of solvent is needed, the sample should be considered insoluble.

Sometimes mixing lines may appear when a liquid dissolves, due to changes in the refractive index. Also, lighter organic liquids may float on top of concentrated sulfuric acid rather than dissolve immediately. In this case, stir the mixture with a small glass rod. If a color change occurs with sulfuric acid, rather than dissolution, this should be considered a positive solubility test. In addition, if the unknown has color and dissolves in a certain solvent, the solution likely will become this color.

It may take several minutes to dissolve solids, especially if large crystals are present. If the crystals are large, it is helpful to pulverize them first with a mortar and pestle.

If necessary, use known samples as a basis of comparison. For example, benzoic acid will dissolve in both 5% NaOH and 5% $NaHCO_3$, but 2-naphthol will dissolve in 5% NaOH but **not** in 5% $NaHCO_3$. Acetanilide will not dissolve in 5% NaOH and 5% HCl, but it will dissolve in concentrated H_2SO_4.

Using Figure 38-2 as a guide, determine what group your unknown is in by performing the following steps:

Step 1. Test the solubility of your unknown in water. If it is soluble, go to step 2. If it is insoluble, go to step 4.

Step 2. Test the aqueous solution of your unknown with pH paper to determine if it is acidic, neutral, or alkaline. Go to step 3.

Step 3. Test the solubility of your unknown in ether. If it is soluble, it belongs in Group I. If it is insoluble, it belongs in Group II or III. Use your pH paper result to make a decision about Group II or III.

Step 4. Test the solubility of your unknown in 5% NaOH. If it is soluble, go to step 5. If it is insoluble, go to step 6.

Step 5. Test the solubility of your unknown in 5% $NaHCO_3$. If it is soluble and the solution emits a gas, the unknown belongs in Group IV. If it is insoluble, it belongs in Group V.

Step 6. Test the solubility of your unknown in 5% HCl. If it is soluble, it belongs in Group VI. If it is insoluble, go to step 7.

Step 7. Test the solubility of your unknown in concentrated H_2SO_4. **(BE CAREFUL: CONCENTRATED H_2SO_4 CAUSES SEVERE BURNS.)** If the unknown is insoluble and is later shown to contain nitrogen, it belongs in Group VII. If it is soluble and is later shown to contain nitrogen and/or sulfur, it belongs in Group VIII. If it is soluble and is later shown to contain no nitrogen or sulfur, it belongs in Group IX. If it is insoluble and is later shown to contain no nitrogen or sulfur, it belongs in Group X.

The previous solubility tests are, by themselves, unable to distinguish between compounds in Group VII and Group X or between compounds in Group VIII and Group IX. In addition, even if the group number is known, there are still many possible types of compounds within that group in which the unknown could be. Performing an elemental analysis for nitrogen, sulfur, and the halogens frequently helps one narrow the possibilities. The presence of more than one functional group can also be indicated by this analysis. There are two principal analysis tests: the Beilstein test for halogens, and the sodium fusion tests for nitrogen, sulfur, and the halogens.

A. Beilstein test

The Beilstein test is the easiest, simplest, and fastest method for detecting the presence of a halogen. Chlorine-, bromine-, and iodine-containing compounds all give a green flame when burned with copper wire. A volatile copper halide is responsible for the color. Fluorine cannot be determined by the test because copper fluoride is not volatile at the flame temperature. Furthermore, the test cannot differentiate between chlorine, bromine, and iodine. Nevertheless, knowing that one of these elements is present can be very valuable in eliminating many possible compounds.

The Beilstein test is very sensitive and gives a positive test even if trace impurities containing a halogen are present. Therefore, one should use caution in interpreting the results of the test if only a weak green color is obtained. In addition, the copper wire should first be heated in the burner flame until all traces of green color are gone.

PROCEDURE

Step 1. Form a flat coil at the end of the copper wire.

Step 2. Heat the copper coil in a microburner flame until the wire produces no green color or other color change in the flame.

Step 3. Allow the copper coil to partially cool, then dip it into the solid or liquid unknown.

Step 4. Place the copper coil coated with the sample into the flame. A flash of green is a positive test for chlorine, bromine, or iodine. Record your observation and your conclusion in your notebook.

Step 5. Repeat the above test with both bromobenzene and toluene, and compare the results with that of the unknown.

B. Sodium fusion tests

Nitrogen, sulfur, and the halogens usually are covalently bonded in organic compounds and not easily analyzed in this form. A **sodium fusion** or the reaction of the organic compounds with molten sodium, however, converts these elements to more conveniently analyzed sodium salts.

$$[N,S,X] \xrightarrow[\Delta]{Na} NaCN, Na_2S, NaX$$

The reaction mixture is then dissolved in distilled water, and the cyanide, sulfide, and halide ions are detected by standard qualitative inorganic analysis. Perform the tests with known compounds first. In this way, you can be certain of exactly what a positive test looks like, and any possibly defective test reagents can be identified and replaced.

PROCEDURE

CAUTION: Wear safety glasses! Molten sodium in the eyes can cause blindness. Also be sure to always handle sodium metal with a knife, forceps, or gloves. Don't touch it with your fingers. Keep sodium metal away from water and destroy any waste with 1-butanol or anhydrous ethanol.

Step 1. Vertically clamp a 10×75 mm test tube to a ring stand.

Step 2. Cut a small piece of clean sodium about the size of a small pea (3 mm on a side) and place it in the test tube.

Step 3. Place 20 mg of a solid sample on a spatula, or 4 drops of a liquid sample in a disposable eyedropper. Keep it ready to add to the sodium metal. **Note:** For most compounds, go to steps 4 and 5. But if you have a volatile liquid, it may vaporize before it reaches the molten sodium. Instead, add the sample directly to the very center of the unheated test tube. If a reaction occurs, let it subside, then heat the tube until it is red hot. Go to step 6.

Step 4. Heat the bottom of the test tube with a microburner until the sodium melts and its vapor rises about one-third the distance up the tube. The bottom of the tube will likely have a red glow.

Step 5. Remove the burner and **immediately** drop the sample directly into the very center of the tube. The sample must hit the sodium directly and not adhere to the side of the tube. A flash or a small explosion will likely occur. If it doesn't, heat the tube strongly until it is red hot to ensure a complete reaction. Charring of the sample will likely occur.

Step 6. Let the test tube cool to room temperature.

Step 7. Carefully add 10 drops of methanol, 1 drop at a time, to the reaction mixture. The methanol will react with any excess sodium metal.

Step 8. Stir the contents of the test tube with a glass rod to ensure that all the sodium metal has reacted.

Step 9. Empty the smaller tube into a 13 × 100 mm test tube.

Step 10. Holding the smaller tube away from you or a neighbor, carefully add 2 mL of **distilled** water. Gently boil and stir the mixture, then pour the water into the larger test tube. Repeat with another 2 mL of water.

Step 11. Carefully filter the contents of the larger tube into a funnel containing a small piece of fluted filter paper. Collect the alkaline filtrate in a second 13 × 100 mm test tube and save for the elemental analysis tests.

Analysis for nitrogen. Sodium cyanide reacts with ferrous ammonium sulfate to yield sodium ferrocyanide, which reacts with ferric salts to give a deep blue precipitate or solution of sodium ferricferrocyanide (Prussian blue). The formation of a greenish-blue color suggests the presence of nitrogen but an incomplete sodium fusion.

$$6 \, \text{NaCN} + \text{Fe(NH}_4)_2(\text{SO}_4)_2 \longrightarrow \text{Na}_4\text{Fe(CN)}_6 + \text{Na}_2\text{SO}_4 + (\text{NH}_4)_2\text{SO}_4$$

Sodium   Ferrous   Sodium
cyanide   ammonium   ferrocyanide
             sulfate

$$\text{Na}_4\text{Fe(CN)}_6 + \text{Fe}^{+3} \longrightarrow \text{NaFe}[\text{Fe(CN)}_6] + 3 \, \text{Na}^+$$

Sodium
ferricferrocyanide
(Prussian blue)

The ferric salts in the test are produced by ferrous ion being air oxidized in a basic solution. Ferric hydroxide and ferrous hydroxide also precipitate, but dissolve with added acid.

Potassium fluoride increases the sensitivity of the Prussian blue test. It forms a stable complex with ferric ion and thus prevents the ferric ion concentration from getting so high that insoluble ferric ferrocyanide is formed rather than Prussian blue. Potassium fluoride also minimizes the formation of ferricyanide ion.

Remember that the presence of sulfur in an organic compound often interferes with the test for nitrogen. But the test can fail even if no sulfur is present. In particular, nitro groups frequently give poor results. Always check for the presence of nitro groups as a separate test (Section 38.21A).

PROCEDURE

Step 1. Place 10 drops of the alkaline filtrate from the sodium fusion in a small test tube. Add 10% NaOH dropwise until pH paper indicates a pH of 12–14.

Step 2. Add 2 drops of 30% KF solution and 2 drops of saturated $Fe(NH_4)_2(SO_4)_2$ to the solution.

Step 3. Gently boil the solution for 1 minute.

Step 4. Acidify the hot solution by adding 30% H_2SO_4 dropwise until the ferric hydroxide and ferrous hydroxide dissolve with gentle tapping. Do not add excess acid. A dark blue solution or precipitate (not green) indicates the presence of nitrogen. Record your observation and your conclusion in your notebook.

Step 5. If the color is faint, filter the solution and see if there is any blue color on the paper.

Analysis for sulfur. Sulfur may be detected by two tests:

1. Sodium sulfide produces a purple color when placed with sodium nitro-prusside, $Na_2(NO)Fe(CN)_6 \cdot 2H_2O$.

2. Sodium sulfide produces a black precipitate of lead sulfide when placed with lead acetate.

$$Na_2S + Pb(C_2H_3O_2)_2 \longrightarrow \underset{\text{Black}}{PbS} + 2\,NaC_2H_3O_2$$

PROCEDURE

CAUTION: Lead acetate is an OSHA Class I carcinogen. Handle it with care. Your instructor may prefer that you test for sulfur only by test 1.

Test 1. Dilute 1 drop of the alkaline filtrate from the sodium fusion with 1 mL of distilled water, and add 1 drop of sodium nitroprusside. A purple color indicates the presence of sulfur. Record your observation and your conclusion in your notebook.

Test 2. Acidify about 10 drops of the alkaline filtrate from the sodium fusion by adding glacial acetic acid dropwise until blue litmus paper turns red. Then add a few drops of 1% lead acetate solution. A black precipitate indicates the presence of sulfur. Record your observation and your conclusion in your notebook.

Analysis for halogens. Sodium halides, except for sodium fluoride, react with silver nitrate to give a silver halide precipitate. The presence of a halogen can therefore be detected by this method.

$$NaX + AgNO_3 \longrightarrow AgX\downarrow + NaNO_3$$

Silver chloride is white, silver bromide is off-white, and silver iodide is yellow.

Sodium cyanide and sodium sulfide interfere with the test, however. Therefore, the alkaline solution is first acidified with diluted HNO_3, and then boiled to eliminate hydrogen cyanide and hydrogen sulfide gas.

$$NaCN + HNO_3 \longrightarrow HCN\uparrow + NaNO_3$$

$$Na_2S + 2\,HNO_3 \longrightarrow H_2S\uparrow + 2\,NaNO_3$$

After the addition of silver nitrate, concentrated NH_4OH is added if there is any precipitate. Silver chloride readily dissolves by forming a water-soluble complex ion. Silver bromide, however, only partially forms this complex, while silver iodide doesn't form it at all. Therefore, the three halogens may again be differentiated.

$$AgCl + 2\,NH_4OH \longrightarrow Ag(NH_3)_2{}^+\,Cl^- + 2\,H_2O$$

Since chlorine is also more reactive than bromine or iodine, it will react with sodium bromide or sodium iodide to give free bromine or iodine. Bromine has a distinctive orange to brown color in methylene chloride, iodine has a violet color, while chlorine is either colorless or a light yellow. The three halogens, therefore, can again be differentiated.

$$Cl_2 + 2\,NaBr \longrightarrow Br_2 + 2\,NaCl$$

$$Cl_2 + 2\,NaI \longrightarrow I_2 + 2\,NaCl$$

PROCEDURE

CAUTION: HCN and H$_2$S are both very toxic. Boil the solution in step 1 and step 5 in the hood.

In addition, silver ion severely stains the skin and is a heavy-metal toxin. Use it carefully and dispose of it in a designated waste jar.

Step 1. Place 20 drops of the alkaline filtrate from the sodium fusion in a test tube. Add 6 M HNO_3 dropwise with mixing until blue litmus paper turns red. If a positive test for nitrogen or sulfur has been obtained, gently boil the solution for 2 to 3 minutes **in the hood**.

Step 2. Let the solution cool, then add 4 to 5 drops of 5% $AgNO_3$. Note and record in your notebook the color of any precipitate formed. If only a slight turbidity is formed, this should be considered a negative test. Stop here.

Step 3. If a precipitate has been formed, add concentrated NH_4OH with mixing until the solution tests highly basic to pH paper. If you have a white precipitate which dissolves readily, silver chloride is indicated. If you have an off-white precipitate which dissolves with difficulty, silver bromide is indicated.

If you have a yellow precipitate which doesn't dissolve, silver iodide is indicated. Record and interpret your results in your notebook.

Step 4. If you need to further establish the presence of a certain halide ion, add 10% H_2SO_4 dropwise with mixing to 20 drops of the alkaline filtrate from the sodium fusion until blue litmus paper turns red.

Step 5. Gently boil the solution for 2 to 3 minutes **in the hood**.

Step 6. Let the solution cool to room temperature; then add 10 drops of methylene chloride and 5 drops of chlorine water or household chlorine bleach or 2 to 4 mg of calcium hypochlorite.

Step 7. Check to see if the solution is still acidic and, if necessary, add 10% H_2SO_4 dropwise.

Step 8. Stopper the tube, shake it vigorously for 1 minute, and allow the two layers to separate. An orange to brown color in the lower methylene chloride layer indicates the presence of bromine, a violet color indicates the presence of iodine, while a colorless to light yellow color indicates the presence of chlorine from the added reagent. Record your observation and your conclusion in your notebook.

38.8 FURTHER CLASSIFYING THE UNKNOWN USING RELEVANT CHEMICAL TESTS

With the results of your unknown's infrared spectrum, solubility tests, and elemental analysis tests, it is now time to further classify it with simple chemical tests. The possible types of unknowns and their chemical tests are listed in Table 38-1 and on the back inside cover of your book.

Remember to do only those tests considered necessary. For example, if the unknown's infrared spectrum shows no carbonyl group, it would be a waste of your sample, time, and effort to do any of the tests for an aldehyde or ketone. In addition, you may needlessly become frustrated and confused. Therefore, try to use a systematic and efficient approach, utilizing logic and common sense to pinpoint the type of compound you have. However, be aware that some unknowns may have more than one functional group. So if your compound, for example, tests positive for an aldehyde, don't exclude the possibility that it could also contain a carbon-carbon double bond. Try to keep an open mind.

In addition, always perform an unfamiliar chemical test on several reference compounds before performing it on your unknown. Otherwise, you may not be able to know precisely what a positive test and a negative test looks like, and you could possibly make a wrong decision.

TABLE 38-1.
Chemical Classification Tests

Type of compound	Chemical test	Section	Page
Aldehydes	2,4-Dinitrophenylhydrazine	38.9A	728
	Ferric chloride	38.9B	729
	Tollens	38.9C	731
	Chromic acid	38.9D	732
	Schiff	38.9E	733
Ketones	2,4-Dinitrophenylhydrazine	38.9A	728
	Ferric chloride	38.9B	729
	Iodoform	38.9F	733
Alcohols	Acetyl chloride	38.10A	736
	Lucas	38.10B	736
	Chromic acid	38.10C	737
	Ceric nitrate	38.10D	739
	Iodoform	38.10E	739
Phenols	Ferric chloride	38.11A	740
	Bromine water	38.11B	740
	Ceric nitrate	38.11C	741
	Sodium hydroxide	38.11D	742
Carboxylic acids	pH of aqueous solution	38.12A	742
	Sodium bicarbonate	38.12B	743
	Silver nitrate	38.12C	744
	Neutralization equivalent	38.12D	744
Amines	pH of aqueous solution	38.13A	745
	Hinsberg	38.13B	746
	Nitrous acid	38.13C	748
	Acetyl chloride	38.13D	750
Amides	Ammonia	38.14A	751
	Ferric hydroxamate	38.14B	752
Esters	Ferric hydroxamate	38.15A	753
	Alkaline hydrolysis	38.15B	754
Ethers	Ferrox	38.16A	755
Nitriles	Ammonia	38.17A	756
	Ferric hydroxamate	38.17B	756
Alkenes and alkynes	Bromine in CH_2Cl_2	38.18A	757
	Dilute $KMnO_4$ (Baeyer test)	38.18B	759
Alkyl halides	Sodium fusion	38.19A	760
	Silver nitrate	38.19B	760
	Sodium iodide	38.19C	761
Aryl halides	Sodium fusion	38.20A	763
	Ignition	38.20B	763
	Aluminum chloride, chloroform	38.20C	763
Nitro groups	Iron(II) hydroxide	38.21A	764
Aromatic hydrocarbons	Ignition	38.22A	765
	Aluminum chloride, chloroform	38.22B	765
Alkanes	None	38.23	765

38.9 ALDEHYDES AND KETONES
(see Tables 38-6 and 38-7, pages 799 and 801)

■ **SOLUBILITY GROUPS I, V, AND IX**
■ **CLASSIFICATION TESTS SUMMARY**

38.9A.   2,4-Dinitrophenylhydrazine: both aldehydes and ketones

38.9B.   Ferric chloride: aldehydes and ketones with a high enol content

38.9C.   Tollens': most aldehydes and a few ketones

38.9D.   Chromic acid: aldehydes only

38.9E.   Schiff's: aldehydes only

38.9F.   Iodoform: methyl ketones and acetaldehyde

38.9A. 2,4-Dinitrophenylhydrazine test

Most aldehydes and ketones react with 2,4-dinitrophenylhydrazine in an acid medium to give a yellow to red precipitate of 2,4-dinitrophenylhydrazone.

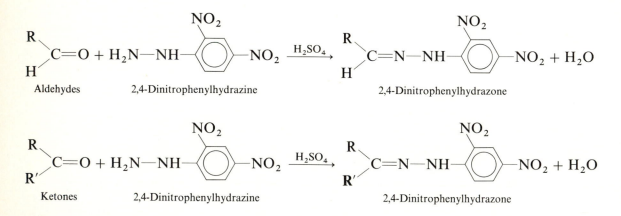

The color of the precipitate frequently indicates the amount of conjugation in the original aldehyde or ketone. Unconjugated aldehydes and ketones give yellow precipitates, conjugated ones give orange to red precipitates, while highly conjugated molecules give red precipitates. But since the 2,4-dinitrophenylhydrazine reagent itself is orange-red, the color of the precipitate must be judged cautiously unless the precipitate is freed of any starting material. Also, compounds that are strongly basic or acidic can sometimes precipitate the unreacted reagent, either giving a false positive test or the wrong color of the product.

A false positive test can also be given by some allylic and benzylic alcohols, which are oxidized by 2,4-dinitrophenylhydrazine to aldehydes and ketones which then react. To rule this out, one should be sure that the unknown gives carbonyl absorption in its infrared spectrum.

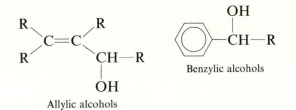

Allylic alcohols

Benzylic alcohols

A false positive test can also be given by some alcohols containing aldehyde and ketone impurities. In this case, the amount of precipitate formed will usually be small and can be ignored. Weak carbonyl absorption in the infrared spectrum will also establish the presence of these impurities.

Esters, amides, carboxylic acids, acid halides, and acid anhydrides do not yield 2,4-dinitrophenylhydrazine derivatives. Therefore, a positive test can usually rule out these compounds.

PROCEDURE

CAUTION: Many phenylhydrazine derivatives are carcinogenic. Handle the reagent and the product with care, avoiding direct contact.

Reference compounds. Benzaldehyde, cyclohexanone, and benzophenone

Step 1. Place one drop of a liquid sample in a test tube, or dissolve about 10 mg of a solid sample in the minimum amount of 95% ethanol or diglyme in the tube.

Step 2. Add 20 drops of 2,4-dinitrophenylhydrazine reagent, stopper the tube, and shake vigorously. Most aldehydes and ketones will give a yellow to red precipitate immediately.

Step 3. If no precipitate forms immediately, allow the solution to stand for 15 minutes. If no precipitate has formed by then, heat the solution **gently** and wait another 15 minutes. Record your observations and interpret the results.

38.9B. Ferric chloride test

Aldehydes and ketones having an α-hydrogen can undergo tautomerism to form an enol.

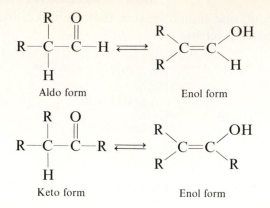

If the equilibrium lies substantially to the right, due to stabilization of the enol, then a violet, red, or tan-colored complex is formed with ferric chloride. For example, 2,4-pentanedione gives a positive test, since the enol form is stabilized by intramolecular hydrogen bonding.

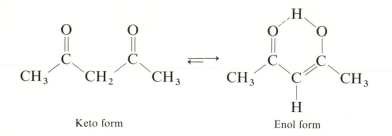

Most phenols also give a positive test. Therefore, the test is only worthwhile if the compound is already known to be an aldehyde or ketone.

PROCEDURE

Reference compounds.   2-pentanone and 2,4-pentanedione

Step 1.   Dissolve 2 drops of a liquid sample or about 20 mg of a solid sample in water or 95% ethanol.

Step 2.   Add 3 to 5 drops of a 3% aqueous ferric chloride solution. Record your observation and interpret the result. Note that the color may be only transient.

Step 3.   If no positive result is attained, repeat the test by adding 2 drops of a liquid sample or about 20 mg of a solid sample to 1 mL of methylene chloride. Add 1 drop of pyridine and 3 to 5 drops of a 1% ferric chloride in methylene chloride solution. Record your observation and interpret the result. This test is somewhat more sensitive than the first one and may now give a positive result.

38.9C. Tollens' test

Tollens' test distinguishes between most aldehydes and ketones. As shown below, aldehydes are easily oxidized by silver ion in alkaline solution to give the salt of a carboxylic acid plus a silver metal precipitate. Ketones, however, usually don't react. To prevent the silver ion from forming insoluble silver hydroxide or silver oxide in the alkaline solution, it is complexed with ammonia.

$$\underset{\text{Aldehyde}}{R-\overset{\overset{\displaystyle O}{\|}}{C}-H} + \underset{\text{Tollens' reagent}}{2\ Ag(NH_3)_2{}^+\ OH^-} \longrightarrow \underset{\substack{\text{Silver}\\\text{mirror}\\\text{precipitate}}}{2\ Ag} + \underset{\substack{\text{Ammonium}\\\text{salt of a}\\\text{carboxylic acid}}}{RCO_2{}^-\ NH_4{}^+} + 3\ NH_3 + H_2O$$

$$\underset{\text{Ketone}}{R-\overset{\overset{\displaystyle O}{\|}}{C}-R'} + \underset{\text{Tollens' reagent}}{2\ Ag(NH_3)_2{}^+\ OH^-} \longrightarrow \text{Usually no reaction}$$

Most aldehydes will give a positive test. But to save time, effort, and your sample, do not perform the Tollens test on your unknown unless you already know that it is an aldehyde or ketone. The test otherwise has no value.

The Tollens reagent is prepared by the following reactions:

$$AgNO_3 \xrightarrow{NaOH} AgOH \quad \text{or} \quad Ag_2O \xrightarrow{NH_4OH} Ag(NH_3)_2{}^+\ OH^-$$

PROCEDURE

CAUTION: Silver ion severely stains the skin and is a heavy-metal toxin. Use it carefully, disposing of it and all residues in a designated waste jar containing dilute nitric acid.

In addition, the Tollens reagent must be prepared immediately before use. It should never be stored, since it may become highly explosive. It may be destroyed with dilute nitric acid.

Reference compounds. Acetone and benzaldehyde

Step 1. In a medium-size test tube, add 20 drops of 10% NaOH to 20 drops of aqueous, freshly prepared, 10% $AgNO_3$.

Step 2. To the resulting brown precipitate, add 6 M NH_4OH dropwise, with stirring, until the precipitate has dissolved.

Step 3. Dissolve 1 drop of a liquid sample or about 15 mg of a solid sample in the minimum amount of water, 95% ethanol, or diglyme.

Step 4. Add the sample solution from step 3 dropwise to the Tollens reagent from step 2. Observe whether a silver precipitate or mirror is formed.

Step 5. If no precipitate or mirror forms, **gently** heat the test tube in a steam bath or a warm-water bath. Observe whether a mirror now forms. Record your observation and interpret the result.

38.9D. Chromic acid test

Another way to distinguish aldehydes from ketones involves the oxidation of the aldehyde by chromic acid. In the reaction, the color of the mixture changes from orange to green, as chromic acid is converted into chromium(III) sulfate. Ketones do not undergo the reaction. Therefore, this test complements Tollens' test, but it can also distinguish between aliphatic and aromatic aldehydes, based upon how long it takes for the color change to occur. The chromic acid needed for the test is prepared by the reaction of chromium trioxide with water in sulfuric acid.

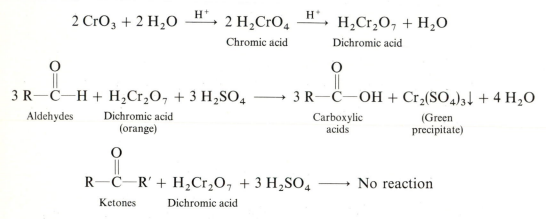

$$2\,CrO_3 + 2\,H_2O \xrightarrow{\;H^+\;} 2\,H_2CrO_4 \xrightarrow{\;H^+\;} H_2Cr_2O_7 + H_2O$$

Chromic acid   Dichromic acid

$$3\,R\overset{\overset{\displaystyle O}{\|}}{-}C-H + H_2Cr_2O_7 + 3\,H_2SO_4 \longrightarrow 3\,R\overset{\overset{\displaystyle O}{\|}}{-}C-OH + Cr_2(SO_4)_3{\downarrow} + 4\,H_2O$$

Aldehydes   Dichromic acid   Carboxylic   (Green
           (orange)   acids   precipitate)

$$R\overset{\overset{\displaystyle O}{\|}}{-}C-R' + H_2Cr_2O_7 + 3\,H_2SO_4 \longrightarrow \text{No reaction}$$

Ketones   Dichromic acid

Primary and secondary alcohols also give a positive test, but in less time. Therefore, be certain you actually have an aldehyde or ketone before performing this test, to avoid possible confusion. Aldehydes and ketones give strong carbonyl absorption in their infrared spectra, and they give a positive 2,4-dinitrophenylhydrazine test. However, alcohols do not.

PROCEDURE

CAUTION: Many chromium(VI) compounds are carcinogenic. Handle the reagent with care, avoiding direct contact. All chromium compounds should also be discarded in a designated waste container, not in a sink or wastebasket.

Reference compounds. Benzaldehyde and cyclohexanone

Step 1. Check the reagent-grade acetone for purity by adding 5 drops of chromic acid to 10 drops of acetone. If no green precipitate forms within 3 to 5 minutes, the acetone is pure. If a green precipitate forms, obtain fresh reagent-grade acetone and test again, or distill some acetone from potassium permanganate.

Step 2. Dissolve 2 drops of a liquid sample or about 15 mg of a solid sample in 20 drops of the pure acetone.

Step 3. Add 5 drops of chromic acid, 1 drop at a time, to the acetone solution with agitation.

Step 4. Observe how long it takes for any green precipitate to form. With aliphatic aldehydes, the precipitate usually appears within 30 seconds, but with aromatic aldehydes it usually takes from 30 seconds to over 2 minutes. Record and interpret your result.

38.9E. Schiff's test

Schiff's test also distinguishes between aldehydes and ketones, and therefore complements Tollens' test and the chromic acid test. Schiff's reagent reacts with aldehydes to give a purplish-red color. Ketones do not give this color.

Again, this test should not be performed unless you know that your unknown is an aldehyde or ketone. It is also not necessary if good results have been obtained with the Tollens and chromic acid tests.

PROCEDURE

Reference compounds. Benzaldehyde and acetone

Step 1. Add 2 drops of a liquid sample to 20 drops of Schiff's *p*-rosaniline hydrochloride (fuchsia) reagent. If the sample is a solid, first dissolve about 15 mg of it in the minimum amount of water or 95% ethanol. Then add it to 20 drops of Schiff's reagent.

Step 2. Observe the color after 10 minutes. Record and interpret your result.

38.9F. Iodoform test

The iodoform test is used to detect methyl ketones, which react with iodine under alkaline conditions to give the yellow precipitate iodoform.

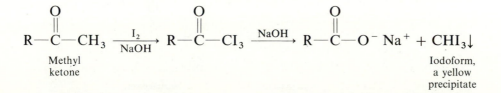

$$\underset{\substack{\text{Methyl}\\\text{ketone}}}{R-\overset{\overset{\textstyle O}{\|}}{C}-CH_3} \xrightarrow[\text{NaOH}]{I_2} R-\overset{\overset{\textstyle O}{\|}}{C}-CI_3 \xrightarrow{\text{NaOH}} R-\overset{\overset{\textstyle O}{\|}}{C}-O^- Na^+ + \underset{\substack{\text{Iodoform,}\\\text{a yellow}\\\text{precipitate}}}{CHI_3\downarrow}$$

Acetaldehyde, ethanol, and secondary alcohols with the hydroxyl group at position 2 also give a positive test.

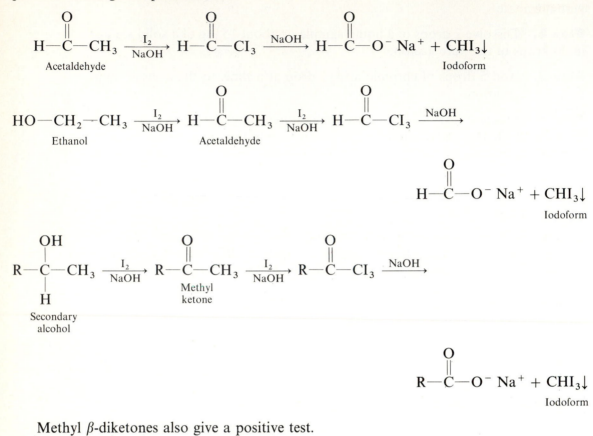

Methyl β-diketones also give a positive test.

R—C(=O)—CH₂—C(=O)—CH₃ —I₂/NaOH→ R—C(=O)—CH₂—C(=O)—CI₃ —NaOH→
Methyl β-diketone

R—C(=O)—CH₂—C(=O)—O⁻ Na⁺ + CHI₃↓
Iodoform

However, the following compounds do not.

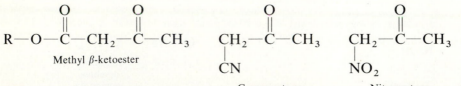

Methyl β-ketoester   Cyanoacetone   Nitroacetone

PROCEDURE

Reference compounds.   2-butanone, 2-pentanol, 3-pentanone

Step 1.   In a 20×150 mm test tube, dissolve 5 drops of a liquid sample or about 100 mg of a solid sample in either 5 mL of water or 2 mL of *bis*(2-ethoxyethyl) ether and 5 mL of water.

Step 2.   Add 20 drops of 10% NaOH.

Step 3.   Add the iodine-potassium iodide solution dropwise, with mixing, until the brown color of iodine persists.

Step 4.   If less than 2 mL of the iodine-potassium iodide solution was added, heat the test tube in a beaker of water at 60°C. If you have added more than 2 mL, go to step 6.

Step 5.   If the heating in step 4 causes the brown color of iodine to disappear, add more iodine-potassium iodide solution dropwise, with mixing, until the reaction mixture remains pale brown for 2 minutes at 60°C.

Step 6.   Add 10% NaOH solution dropwise, with mixing, until the brown color of iodine disappears. If iodoform is present, only its intense yellow color will remain.

Step 7.   If a yellow color is present, fill the test tube with distilled water, stopper, shake vigorously, and allow the tube to stand for 15 minutes.

Step 8.   Collect the yellow precipitate by suction filtration, dry it, and determine its melting point to ensure that it is actually iodoform. The melting point of iodoform is 119–121°. Record your observations and interpret the results.

38.10   ALCOHOLS (Table 38-8, page 804)

SOLUBILITY GROUPS   I, III, AND IX
CLASSIFICATION TESTS SUMMARY

38.10A.   Acetyl chloride

38.10B.   Lucas: distinguishes between 1°, 2°, and 3° alcohols

38.10C.   Chromic acid: distinguishes 1° and 2° alcohols from 3° alcohols

38.10D.   Ceric nitrate: distinguishes between alcohols of less than 10 carbons and alcohols of 10 or more carbons

38.10E.   Iodoform: distinguishes ethanol and secondary alcohols with the hydroxyl group at position 2 from all other alcohols

38.10A. Acetyl chloride test

All alcohols react with acetyl chloride to form esters.

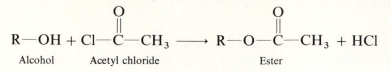

$$R{-}OH + Cl{-}\overset{\overset{\displaystyle O}{\|}}{C}{-}CH_3 \longrightarrow R{-}O{-}\overset{\overset{\displaystyle O}{\|}}{C}{-}CH_3 + HCl$$

Alcohol   Acetyl chloride   Ester

The reaction is exothermic, and the heat evolved is usually easily detected. Many times, the ester produced will also have a sweet, fruity odor. Some esters also form cloudy solutions when water is added.

Phenols and primary and secondary amines, however, also react with acetyl chloride. Therefore, one cannot be certain from the test that the unknown is an alcohol until other tests confirm an alcohol or reject a phenol or amine.

PROCEDURE

Reference compounds.   1-butanol and 2-butanol

Step 1.   Cautiously add about 5 drops of acetyl chloride to about 10 drops of a liquid sample in a small test tube. If the sample is a solid, dissolve about 50 mg of it in a small amount of **anhydrous** ether before adding the acetyl chloride. The evolution of heat and hydrogen chloride gas possibly indicates the presence of an alcohol.

Step 2.   If a reaction occurred in step 1, **cautiously** add water to the reaction mixture. Note the odor and the amount of cloudiness, and try to interpret the results. **Do not add water if no reaction occurred in step 1, since acetyl chloride reacts very vigorously with water.**

38.10B. Lucas test

The Lucas test distinguishes between 1°, 2°, and 3° alcohols having less than eight carbon atoms. The test is not used on larger-size alcohols, due to their insolubility in the Lucas reagent. The test should also not be performed if the unknown gave a negative acetyl chloride test in Section 38.10A.

The Lucas test is based upon how rapidly a less-than-eight-carbon alcohol will form an insoluble alkyl chloride when placed wth $ZnCl_2$ and concentrated HCl.

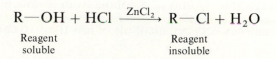

$$R{-}OH + HCl \xrightarrow{\;ZnCl_2\;} R{-}Cl + H_2O$$

Reagent   Reagent
soluble   insoluble

Since the reaction involves a carbocation intermediate, alcohols that easily form this intermediate will react the fastest.

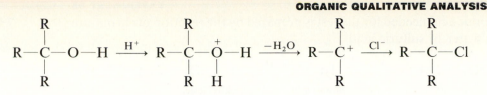

The order of stability of carbocations is

$$\text{Benzylic} > \text{allylic} > 3° > 2° \gg 1°$$

Therefore, the rate of reaction of reagent-soluble alcohols in the test is the same.

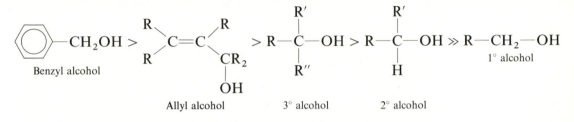

Benzyl, allyl, and tertiary alcohols react immediately with the Lucas reagent at room temperature. Secondary alcohols react within 2 to 5 minutes, while primary alcohols do not react at all but merely dissolve in the Lucas reagent. Some secondary alcohols may have to be heated slightly to react.

PROCEDURE

Reference compounds.   Benzyl alcohol, 2-methyl-2-propanol, 2-butanol, and 1-butanol

Step 1.   Add 4 to 5 drops of the sample to 2 to 3 mL of the Lucas reagent in a test tube.

Step 2.   Observe how soon any cloudiness appears. A layer of alkyl chloride will also appear if a reaction occurs.

Step 3.   If no reaction has occurred after 5 minutes, warm the mixture gently for 2 to 3 minutes. Wait another 5 minutes. Cloudiness now indicates a secondary alcohol. A clear solution indicates a primary alcohol. Record your observation and interpret the result.

38.10C. Chromic acid test

Primary and secondary alcohols can be distinguished from tertiary alcohols by the chromic acid test. Primary and secondary alcohols are oxidized by chromic acid to yield the organic product plus a green precipitate of chromium(III) sulfate. Tertiary alcohols fail to react because they cannot be oxidized by chromic acid. The chromic acid needed for the test is prepared by the reaction of chromium trioxide with water in sulfuric acid.

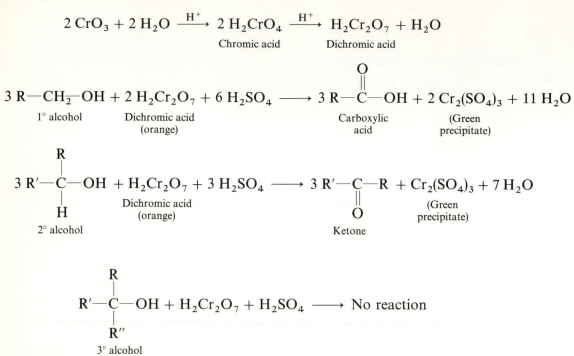

$$2\ CrO_3 + 2\ H_2O \xrightarrow{H^+} 2\ H_2CrO_4 \xrightarrow{H^+} H_2Cr_2O_7 + H_2O$$

Chromic acid   Dichromic acid

$$3\ R\!-\!CH_2\!-\!OH + 2\ H_2Cr_2O_7 + 6\ H_2SO_4 \longrightarrow 3\ R\!-\!\overset{\displaystyle O}{\overset{\|}{C}}\!-\!OH + 2\ Cr_2(SO_4)_3 + 11\ H_2O$$

1° alcohol   Dichromic acid (orange)   Carboxylic acid   (Green precipitate)

$$3\ R'\!-\!\underset{\displaystyle H}{\overset{\displaystyle R}{\underset{|}{\overset{|}{C}}}}\!-\!OH + H_2Cr_2O_7 + 3\ H_2SO_4 \longrightarrow 3\ R'\!-\!\underset{\displaystyle O}{\overset{\|}{C}}\!-\!R + Cr_2(SO_4)_3 + 7\ H_2O$$

2° alcohol   Dichromic acid (orange)   Ketone   (Green precipitate)

$$R'\!-\!\underset{\displaystyle R''}{\overset{\displaystyle R}{\underset{|}{\overset{|}{C}}}}\!-\!OH + H_2Cr_2O_7 + H_2SO_4 \longrightarrow \text{No reaction}$$

3° alcohol

If the Lucas test (Section 38.10B) has already classified the alcohol, then the chromic acid test is not needed. But one advantage of the chromic acid test over the Lucas test is that it can be used for all alcohols, regardless of their molecular weight and solubility. In addition, the chromic acid test can distinguish benzylic and allylic alcohols from tertiary alcohols, but the Lucas test cannot.

Remember that aldehydes will also give a positive chromic acid test, so don't perform this test unless aldehydes have already been ruled out. Aldehydes, however, typically take from 30 seconds to over 2 minutes to react, while primary and secondary alcohols will react in 2 to 5 seconds.

PROCEDURE

CAUTION: Many chromium(VI) compounds are carcinogenic. Handle the reagent with care, avoiding direct contact. All chromium compounds should also be discarded in a designated waste container, not in a sink or wastebasket.

Reference compounds.   1-butanol, 2-methyl-2-propanol

Step 1.   Check the reagent-grade acetone for purity by adding 5 drops of chromic acid to 10 drops of acetone. If no green precipitate forms within

10 seconds, the acetone is pure enough to use. If a green precipitate forms, obtain fresh reagent-grade acetone and test again, or distill some acetone from potassium permanganate.

Step 2. Dissolve 2 drops of a liquid sample or about 15 mg of a solid sample in 20 drops of the pure acetone.

Step 3. Add 5 drops of chromic acid, 1 drop at a time, to the acetone solution. A green precipitate, appearing within 2 to 5 seconds, indicates a primary or secondary alcohol. Record and interpret your result.

38.10D. Ceric nitrate test

The ceric nitrate test distinguishes alcohols having 10 carbons or less from alcohols having more than 10 carbons. Alcohols with 10 carbons or less will give a red color in the presence of ceric ammonium nitrate.

$$(NH_4)_2Ce(NO_3)_6 + ROH \longrightarrow (NH_4)_2Ce(OR)(NO_3)_5 + HNO_3$$

Yellow   Red

Note that the test is not necessary if the boiling point of the alcohol already indicates 10 carbons or less. Phenols can also give a positive test, with a brown to greenish-brown precipitate in aqueous solution, and a red to brown solution in diglyme. In addition, aromatic amines may be oxidized by the reagent and give a positive color test.

PROCEDURE

Reference compounds. 1-butanol, 1-octanol, 1-dodecanol

Step 1. In a test tube, dissolve 5 drops of a liquid sample or about 30 mg of a solid sample in 1 to 2 mL of water or diglyme.

Step 2. Add 10 drops of ceric ammonium nitrate reagent. Record and interpret the result.

38.10E. Iodoform test

The iodoform test is useful for distinguishing ethanol and secondary alcohols with the hydroxyl group in position 2 from all other alcohols. Refer to Section 38.9F for a complete discussion and the test procedure.

$$\underset{H}{\overset{OH}{\underset{|}{\overset{|}{R-C-CH_3}}}}$$ Gives a positive iodoform test

38.11 PHENOLS (Table 38-9, page 806)

- **SOLUBILITY GROUPS I, IV, AND V**
- **CLASSIFICATION TESTS SUMMARY**

38.11A. Ferric chloride

38.11B. Bromine water

38.11C. Ceric nitrate

38.11D. Sodium hydroxide

38.11A. Ferric chloride test

Many phenols react with ferric chloride to give red, blue, violet, or green complexes. The colors, however, may only be transient, and some phenols don't form any color at all. Therefore, a negative test should not be considered significant without other evidence. Aldehydes and ketones which exist significantly in the enol form also give a positive test, so they should be ruled out before performing this test.

PROCEDURE

Reference compounds. Phenol, β-naphthol

Step 1. Dissolve 2 drops of a liquid sample or about 20 mg of a solid sample in water or 95% ethanol.

Step 2. Add 3 to 5 drops of a 3% aqueous ferric chloride solution. Record your observation and interpret the result.

Step 3. If no positive result is attained, repeat the test by adding 2 drops of a liquid sample or about 20 mg of a solid sample to 1 mL of methylene chloride. Add 1 drop of pyridine and 3 to 5 drops of a 1% ferric chloride in methylene chloride solution. Record your observation and interpret the result. This test is somewhat more sensitive than the first one and may now give a positive result.

38.11B. Bromine water test

Phenols are highly activated toward electrophilic aromatic substitution, and they react with bromine in water to give a substitution product. Usually, all the open positions, ortho and para to the hydroxyl group, are substituted. For example:

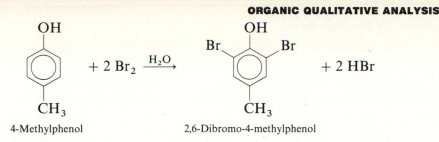

4-Methylphenol   2,6-Dibromo-4-methylphenol

A positive test is both the loss of color by the aqueous bromine solution and the precipitation of the substitution product. However, other highly activated aromatic compounds, such as anilines or ethers, also react in a similar manner.

PROCEDURE

CAUTION: Bromine causes severe burns. Be careful not to get it on your skin. Bromine burns may be treated by flushing the affected area with plenty of water, and then applying a wet dressing of 10% sodium thiosulfate.

Reference compound. Phenol

Step 1. Dissolve 5 drops of a liquid sample or about 30 mg of a solid sample in water or an ethanol-water mixture.

Step 2. Add a saturated solution of bromine in water, drop by drop, with mixing until the bromine color is no longer discharged. Record your observations and interpret the result.

38.11C. Ceric nitrate test

Phenols having no more than 10 carbons, like alcohols, can react with ceric ammonium nitrate. A brown to greenish-brown precipitate is formed in aqueous solutions, while a red to brown solution is formed in diglyme.

$$(NH_4)_2Ce(NO_3)_6 + ArOH \longrightarrow (NH_4)_2Ce(OAr)(NO_3)_5 + HNO_3$$

Yellow   Typically brown to red

PROCEDURE

Reference compound. Phenol

Step 1. In a test tube, dissolve 5 drops of a liquid sample or about 30 mg of a solid sample in 1 to 2 mL of water or diglyme.

Step 2. Add 10 drops of ceric ammonium nitrate reagent. Record and interpret the result.

38.11D. Sodium hydroxide test

Phenols having much conjugation in the phenolate anion will often become colored when dissolved in 10% NaOH. Others may not have any color originally in alkaline solution, but will become oxidized to brown-colored products. Sometimes the phenolate anions will also precipitate in 10% NaOH, but will dissolve in water.

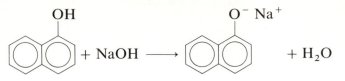

PROCEDURE

Reference compounds. Phenol, α-naphthol, β-naphthol

Step 1. Dissolve a small amount of the sample in 20 drops of 10% NaOH. Note any change in color.

Step 2. If there is no change in color, allow the solution to stand for 30 minutes to see if any change occurs.

Step 3. If a precipitate forms, dilute the solution with water and stir. Interpret and record the results.

38.12 CARBOXYLIC ACIDS (Table 38-10, page 808)

■ **SOLUBILITY GROUPS   I, II, III, AND IV**
■ **CLASSIFICATION TESTS SUMMARY**

38.12A.   pH of aqueous solution

38.12B.   Sodium bicarbonate

38.12C.   Silver nitrate

38.12D.   Neutralization equivalent

38.12A. pH of aqueous solution

All carboxylic acids react with water to form a carboxylate anion and the hydronium ion. Checking the pH with pH paper then confirms the acidity of the solution. If the carboxylic acid is insoluble in water, ethanol or methanol can be added to dissolve it.

$$R-\overset{\overset{\displaystyle O}{\|}}{C}-OH + H_2O \rightleftharpoons R-\overset{\overset{\displaystyle O}{\|}}{C}-O^- + H_3O^+$$

This test may already have been done in the determination of the unknown's solubility group.

PROCEDURE

Reference compounds. Acetic acid and benzoic acid

Water-soluble acids

Step 1. Since any dissolved carbon dioxide makes distilled water acidic, check the pH of the water with pH paper before adding the sample.

Step 2. Dissolve 2 drops of a liquid sample or about 15 mg of a solid sample in 1 mL of water.

Step 3. Determine the new pH of the solution with pH paper. Record and interpret your result.

Water-insoluble acids

Step 1. Since any dissolved carbon dioxide makes distilled water acidic, check the pH of an alcohol-water mixture before adding the sample.

Step 2. Dissolve 2 drops of a liquid sample or about 15 mg of a solid sample in methanol or 95% ethanol. Add water dropwise, with stirring, until the solution just turns cloudy, then add methanol or 95% ethanol dropwise, with stirring, until the solution is again clear.

Step 3. Determine the new pH of the solution with pH paper. Record and interpret the result.

38.12B. Sodium bicarbonate test

Carboxylic acids when placed in 5% $NaHCO_3$ cause bubbles of carbon dioxide gas to be released.

$$\underset{\displaystyle R-\overset{\displaystyle \overset{O}{\|}}{C}-OH}{} + NaHCO_3 \longrightarrow R-\overset{\overset{\displaystyle O}{\|}}{C}-O^-\ Na^+ + CO_2\uparrow + H_2O$$

This test may already have been performed in the determination of the unknown's solubility group.

PROCEDURE

Reference compounds. Acetic acid and benzoic acid

Step 1. Place a small amount of the sample in 5% $NaHCO_3$ solution.

Step 2. If the sample is an acid, it should dissolve, causing bubbles of carbon dioxide gas to be released. Record and interpret your result.

38.12C. Silver nitrate test

Carboxylic acids react with silver nitrate to give a precipitate of the silver salt of the carboxylic acid.

$$\underset{\substack{\displaystyle \| \\ R-C-OH}}{O} + AgNO_3 \longrightarrow \underset{\substack{\displaystyle \| \\ R-C-O^-}}{O} Ag^+\downarrow + HNO_3$$

Alkyl halides, except fluorides, also react with silver nitrate to give a precipitate of the silver halide.

$$R-X + AgNO_3 \longrightarrow R^+\ NO_3{}^- + AgX\downarrow$$

The two possibilities, however, can be distinguished because the silver salts of carboxylic acids are soluble in nitric acid, whereas silver halides are not.

PROCEDURE

Reference compounds. Acetic acid and benzoic acid

Step 1. Dissolve 2 drops of a liquid sample or about 20 mg of a solid sample in 95% ethanol. Add to 2 mL of a 2% solution of $AgNO_3$ in 95% ethanol.

Step 2. If a precipitate forms, add 5% nitric acid and observe if the precipitate dissolves. Record and interpret your results.

38.12D. Neutralization equivalent

Although not a true classification test, the neutralization equivalent of a carboxylic acid can be very valuable in identifying the carboxylic acid. The molecular weight of the acid will either be the neutralization equivalent or some whole number multiple of it, depending upon how many carboxyl groups are present in the molecule.

PROCEDURE

Reference compound. Benzoic acid

Step 1. Accurately weigh to three significant figures about 0.2 gram of the acid.

Step 2. Place the acid in a 125-mL Erlenmeyer flask, and dissolve it in about 50 mL of water or a mixture of water and ethanol. All the acid doesn't have to dissolve, since it will dissolve during the titration.

Step 3. Add a few drops of phenolphthalein indicator to the solution.

Step 4. Using a standard NaOH solution of about 0.1 M, titrate the carboxylic acid to a pink endpoint.

Step 5. Calculate the neutralization equivalent, using the equation

$$\text{Neutralization equivalent} = \frac{\text{Milligrams acid}}{\text{Molarity of NaOH} \times \text{mL of NaOH added}}$$

38.13   AMINES (Tables 38-11, 38-12, and 38-13, pages 811–814)

▪ SOLUBILITY GROUPS   I, II, III, AND VI
▪ CLASSIFICATION TESTS SUMMARY

38.13A.   pH of an aqueous solution

38.13B.   Hinsberg: distinguishes between 1°, 2°, and 3° amines

38.13C.   Nitrous acid: distinguishes 1° amines from 2° and 3° amines, and 1° aromatic amines from 1° aliphatic amines

38.13D.   Acetyl chloride

38.13A. pH of aqueous solution

Most amines will react with water to form the protonated amine and hydroxide ion. Checking the pH with pH paper then confirms the basicity of the solution. If the amine is insoluble in water, ethanol or methanol may be added to dissolve it.

$$\underset{\substack{\\ 1°, 2°, \text{ or } 3°}}{\overset{R'}{\underset{R''}{R-N:}}} + H_2O \rightleftharpoons \overset{R'}{\underset{R''}{R-\overset{+}{N}-H}} + OH^-$$

This test may already have been done in the determination of the unknown's solubility group.

PROCEDURE

Reference compounds.   Triethylamine and aniline

Water-soluble amines

Step 1.   Since any dissolved carbon dioxide makes distilled water acidic, check the pH of the water with pH paper before adding the sample.

Step 2.   Dissolve 2 drops of a liquid sample or about 15 mg of a solid sample in 1 mL of water.

Step 3.   Determine the new pH of the solution with pH paper. Record and interpret your result.

Water-insoluble amines

Step 1.   Since any dissolved carbon dioxide makes distilled water acidic, check the pH of a water-alcohol mixture with pH paper before adding the sample.

Step 2.   Dissolve 2 drops of a liquid sample or about 15 mg of a solid sample in methanol or 95% ethanol. Add water dropwise, with mixing, until the solution just turns cloudy. Then add methanol or 95% ethanol dropwise, with mixing, until the solution is again clear.

Step 3.   Determine the new pH of the solution with pH paper. Record and interpret the result.

38.13B. Hinsberg test

The Hinsberg test distinguishes between 1°, 2°, and 3° amines. It is based upon the reactivity of the amine with benzenesulfonyl chloride and upon the solubility or insolubility of the product in both sodium hydroxide solution and hydrochloric acid.

   Primary amines react with benzenesulfonyl chloride to yield the benzenesulfonamide of a primary amine. This sulfonamide is soluble in base, due to the removal of the slightly acidic hydrogen on nitrogen, but it is insoluble in acid.

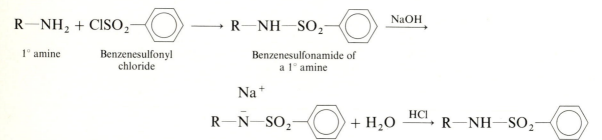

   Secondary amines react with benzenesulfonyl chloride to yield the benzenesulfonamide of a secondary amine. This sulfonamide is insoluble in base, since it has no slightly acidic hydrogen on nitrogen, and it is also insoluble in acid.

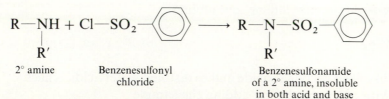

Tertiary amines give no apparent reaction with benzenesulfonyl chloride. But in reality, the following reactions occur:

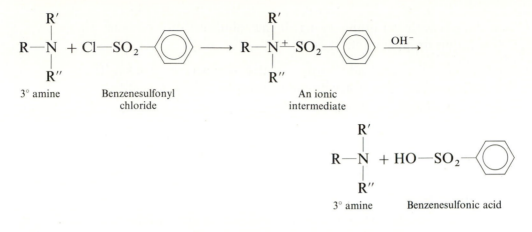

3° amine / Benzenesulfonyl chloride / An ionic intermediate

3° amine / Benzenesulfonic acid

The reaction mixture also forms no precipitate with hydrochloric acid, the exact opposite of what occurs with primary and secondary amines.

In summary, the following observations distinguish the three classes of amines in the test.

1° Amine + benzenesulfonyl chloride $\xrightarrow{\text{NaOH}}$ solution $\xrightarrow{\text{HCl}}$ precipitate

2° Amine + benzenesulfonyl chloride $\xrightarrow{\text{NaOH}}$ precipitate $\xrightarrow{\text{HCl}}$ precipitate

3° Amine + benzenesulfonyl chloride $\xrightarrow{\text{NaOH}}$ no apparent reaction $\xrightarrow{\text{HCl}}$ clear solution

Although the Hinsberg test works well with reagent-grade amines, practical grades contain impurities which interfere with the results. Therefore, trace amounts of any precipitate should be ignored. In addition, the reaction time must be short, and any heating must be gentle. Otherwise, additional reactions may occur which can interfere with the results.

PROCEDURE

CAUTION: Benzenesulfonyl chloride is a lachrymator. Perform the Hinsberg test only in the hood.

Reference compounds. Propyl amine or aniline, diethylamine or *N*-methylaniline, triethylamine

Step 1. Place 3 mL of 10% NaOH in a large test tube.

Step 2. Add 3 drops of a liquid sample or about 100 mg of a solid sample, along with 7 drops of benzenesulfonyl chloride.

Step 3. Stopper the test tube and shake it for 5 minutes. If no reaction appears to have occurred, the amine is probably tertiary. The addition of HCl will produce a clear solution.

Step 4. Check the pH with red litmus paper. If the solution is not alkaline, add 10% NaOH solution dropwise until it is.

Step 5. If a precipitate is present, add 5 mL of water, stopper, and shake. If the precipitate is still present, a secondary amine is indicated. If the precipitate dissolves, a primary amine may be present.

Step 6. Add 3 M HCl dropwise until the mixture is acidic to blue litmus paper. If a precipitate forms, a primary amine is indicated. Record and interpret the results.

38.13C. Nitrous acid test

The nitrous acid test distinguishes primary amines from secondary and tertiary amines. Unlike the Hinsberg test, it distinguishes primary aromatic amines from primary aliphatic amines. Therefore, if the Hinsberg test indicates that your amine is primary, you should also do the nitrous acid test.

Nitrous acid is formed by the reaction of sodium nitrite with an acid such as hydrochloric acid.

$$NaNO_2 + HCl \longrightarrow HNO_2 + NaCl$$
Nitrous
acid

It reacts with primary aliphatic amines to yield an unstable diazonium ion, which decomposes to give nitrogen gas and a frequently complex mixture of alcohols, alkenes, and alkyl chlorides.

$$R-NH_2 \xrightarrow[\text{HCl}]{\text{NaNO}_2} [R-\overset{+}{N}\equiv N\ Cl^-] \longrightarrow N_2\uparrow + R^+\ Cl^- \xrightarrow{H_2O} \text{Alcohols, alkenes,}$$

1° aliphatic   Unstable   and alkyl chlorides
diazonium ion

Primary aromatic amines, however, give a much more stable diazonium ion, which can react with a highly activated aromatic ring to give an azo dye. For example, the diazonium ion reacts with β-naphthol to yield an orange to red dye. Mild heating also causes this diazonium ion to decompose, yielding nitrogen gas.

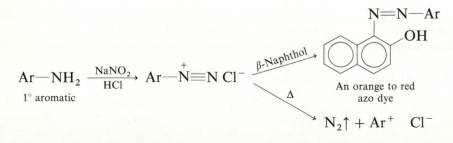

Secondary aliphatic and aromatic amines react with nitrous acid to yield *N*-nitrosoamines, which are yellow to orange liquids and solids. No nitrogen gas is evolved.

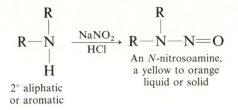

2° aliphatic or aromatic

An *N*-nitrosoamine, a yellow to orange liquid or solid

Tertiary aliphatic and aromatic amines also react with nitrous acid to yield yellow to orange nitrosoamines. With aliphatic amines, the nitroso group replaces one of the alkyl groups, which ends up as an aldehyde or ketone. With aromatic amines, substitution occurs at the para position of the ring, if it is open. Otherwise, substitution occurs at the more sterically hindered ortho position.

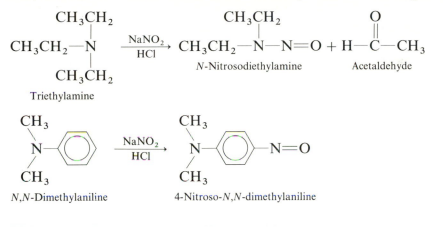

PROCEDURE

CAUTION: Nitrosoamines, formed from secondary and possibly tertiary amines, are carcinogenic. Avoid direct contact with these products and discard them in a designated waste container, not in a sink or wastebasket.

Reference compounds.   Butylamine, aniline, *N*-methylaniline

Step 1.   In a test tube, dissolve 3 drops of a liquid sample or about 50 mg of a solid sample in 2 mL of 2 *M* HCl.

Step 2.   Cool the solution to 0–5°C in an ice bath.

Step 3.   Add 5 drops of a cold 20% aqueous solution of sodium nitrite, dropwise, with mixing.

Step 4. Look for bubbles of **colorless** nitrogen gas. The substantial evolution of a colorless gas at 0–5°C indicates a primary aliphatic amine. The gas must not be brown nitrogen oxide.

Step 5. If no substantial evolution of colorless gas occurs, look for the formation of a yellow to orange liquid or solid. This indicates either a secondary amine or a tertiary aromatic amine. Tertiary aliphatic amines usually do not react at this temperature.

Step 6. If there is neither the substantial formation of a colorless gas nor the formation of a yellow to orange liquid or solid, warm **half** of the solution gently to about room temperature. The evolution of colorless nitrogen gas at this higher temperature indicates a primary aromatic amine. To the remaining half, add drop by drop an ice-cold solution of about 50 mg of β-naphthol in 2 mL of 2 M NaOH. The formation of an orange to red azo dye precipitate further indicates a primary aromatic amine. Record your observations and interpret the results.

38.13D. Acetyl chloride test

Primary and secondary amines react with acetyl chloride to form acetamide derivatives. The reaction is exothermic, and the heat evolved is usually easily detected. The acetamide formed also frequently precipitates when the test mixture is diluted with water.

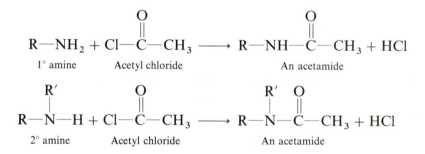

Phenols and alcohols, however, also react with acetyl chloride. Therefore, this test alone cannot establish the presence of an amine and frequently may not even be necessary.

PROCEDURE

Reference compounds. Aniline, diethylamine

Step 1. Cautiously add about 5 drops of acetyl chloride to about 10 drops of a liquid sample in a small test tube. If the sample is a solid, dissolve about 50 mg of it in a small amount of **anhydrous** ether before adding the acetyl chloride. The evolution of heat and hydrogen chloride gas possibly indicates the presence of an amine.

Step 2. If a reaction occurred in Step 1, **cautiously** add water to the reaction mixture. Observe whether a solid forms and interpret the results. **Do not add water if no reaction occurred in step 1, since acetyl chloride reacts very vigorously with water.**

38.14 AMIDES (Table 38-14, page 815)

▪ SOLUBILITY GROUPS I AND VIII
▪ CLASSIFICATION TESTS SUMMARY

38.14A. Ammonia: distinguishes 1° amides from 2° and 3° amides

38.14B. Ferric hydroxamate

38.14A. Ammonia test

Primary amides react with aqueous sodium hydroxide solution to yield ammonia, which can be detected by the blue color formed with copper ion. Secondary and tertiary amides give a negative test, since they do not yield ammonia when reacted with aqueous sodium hydroxide.

$$\underset{\text{1° amide}}{R-\overset{\overset{\displaystyle O}{\|}}{C}-NH_2} + NaOH \xrightarrow{\text{H}_2\text{O}} R-\overset{\overset{\displaystyle O}{\|}}{C}-O^- \; Na^+ + NH_3$$

Nitriles and ammonium salts also give a positive test. Nitriles, however, will not give strong carbonyl absorption in the infrared spectrum, while ammonium salts are insoluble in ether.

PROCEDURE

Reference compounds. Acetamide and benzamide

Step 1. Place 3 drops of a liquid sample or about 50 mg of a solid sample in a small test tube.

Step 2. Add 2 mL of 20% NaOH solution.

Step 3. Fasten a small circle of filter paper tightly over the top of the test tube and wet it with several drops of copper sulfate solution.

Step 4. Boil the contents in the test tube for about 1 minute and see if the paper turns a darker blue. Record and interpret the result.

38.14B. Ferric hydroxamate test

Amides react with hydroxylamine to give a hydroxamic acid, which in turn reacts with ferric chloride to yield a ferric hydroxamate. A red to violet or a blue to violet color indicates a positive test.

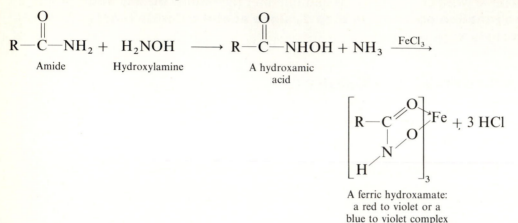

A ferric hydroxamate:
a red to violet or a
blue to violet complex

Esters, acid halides, acid anhydrides, and nitriles give the same result, so a positive test is not conclusive proof of an amide. In addition, phenols, enols, and other compounds react with ferric chloride to give similarly colored products. The unknown must, therefore, first be tested with ferric chloride without any hydroxylamine present.

PROCEDURE

Reference compounds.   Acetamide and benzamide

Step 1.   Dissolve 2 to 3 drops of a liquid sample or about 30 mg of a solid sample in 1 mL of 95% ethanol. Add 1 mL of 1 M HCl and 2 drops of 5% ferric chloride solution. If a blue, violet, or red color is produced, stop here. The ferric hydroxamate test will give meaningless results. If a yellow color appears, it is okay to proceed.

Step 2.   In another test tube, dissolve 2 or 3 drops of a liquid sample or about 30 mg of a solid sample in the minimum amount of propylene glycol.

Step 3.   Add 2 mL of 1 M hydroxylamine hydrochloride solution in propylene glycol, 1 mL of 1 M potassium hydroxide in propylene glycol, and then boil the mixture for 2 minutes.

Step 4.   Cool the solution to room temperature, then add 0.5–1.0 mL of 5% aqueous ferric chloride. A red to violet or a blue to violet color is a positive test. A yellow color is a negative test. Brown colors or precipitates are indeterminate. Record and interpret your result.

38.15   ESTERS (Table 38-15, page 816)

■ SOLUBILITY GROUPS   I, V, AND IX
■ CLASSIFICATION TESTS SUMMARY

38.15A.   Ferric hydroxamate

38.15B.   Alkaline hydrolysis

38.15A. Ferric hydroxamate test

Esters react with hydroxylamine to give a hydroxamic acid, which in turn reacts with ferric chloride to yield a ferric hydroxamate. A red to violet or a blue to violet color indicates a positive test.

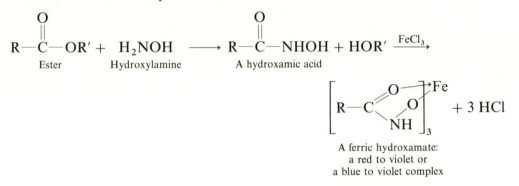

A ferric hydroxamate:
a red to violet or
a blue to violet complex

Amides, acid halides, acid anhydrides, and nitriles give the same result, so a positive test is not conclusive proof of an ester. In addition, phenols, enols, and other compounds react with ferric chloride to give similarly colored products. The unknown must, therefore, first be tested with ferric chloride without any hydroxylamine present.

PROCEDURE

Reference compounds.   Ethyl acetate and ethyl benzoate

Step 1.   Dissolve 2 or 3 drops of a liquid sample or about 30 mg of a solid sample in 1 mL of 95% ethanol. Add 1 mL of 1 M HCl and 2 drops of 5% ferric chloride solution. If a blue, violet, or red color is produced, stop here. The ferric hydroxamate test will give meaningless results. If a yellow color appears, it is okay to proceed.

Step 2.   In another test tube, dissolve 2 or 3 drops of a liquid sample or about 30 mg of a solid sample in 1 mL of 0.5 M hydroxylamine hydrochloride in ethanol.

Step 3.   Add 4 drops of 20% NaOH solution and heat the solution to boiling.

Step 4. Cool slightly and add 2 mL of 1 M HCl. If the solution becomes cloudy, add 95% ethanol until it is clear again.

Step 5. Add 5% ferric chloride solution dropwise, with mixing. If the color fades, continue adding the reagent until a color persists. Record and interpret the result.

38.15B. Alkaline hydrolysis

Esters are hydrolyzed in aqueous sodium hydroxide to yield an alcohol and the sodium salt of a carboxylic acid.

$$R-O-\overset{\overset{\displaystyle O}{\|}}{C}-R' \xrightarrow{\text{NaOH}} R-OH + Na^+ \ \ ^-O-\overset{\overset{\displaystyle O}{\|}}{C}-R' \xrightarrow{\text{HCl}} H-O-\overset{\overset{\displaystyle O}{\|}}{C}-R'$$

Ester   Alcohol   Sodium salt of a   Carboxylic
                                        carboxylic acid   acid

Insoluble esters will usually dissolve when they react, because both the alcohol portion (if no more than five carbons) and the sodium salt of the carboxylic acid are water soluble. Both products can be isolated, and derivatives can then establish what the ester is. In addition, the melting point of a solid carboxylic acid can be valuable.

Amides and nitriles also yield a carboxylic acid when hydrolyzed. But other tests, such as the sodium fusion test for nitrogen, can be used to distinguish these possibilities.

PROCEDURE

Reference compound. Ethyl benzoate

Step 1. Place 10 mL of 25% NaOH in a 25-mL round-bottom flask. Add 20 drops of a liquid sample or 1 gram of a solid sample. Attach a reflux condenser, and reflux the mixture for 30 minutes.

Step 2. Stop the heating and observe the mixture. If the oily sample layer, the sample crystals, or the odor of an ester have disappeared, go to step 3. But if the sample is still present, continue refluxing for 1 to 2 hours. Samples which don't react in this extended time are either unreactive esters or are **not** esters. If this is the case, stop here.

Step 3. If the sample has completely hydrolyzed, cool the reaction mixture in an ice bath and extract it three times with 5-mL portions of ether. Combine the ether extracts and save the remaining reaction mixture for step 6.

Step 4. Dry the combined ether extracts with anhydrous magnesium sulfate, filter or decant, then separate the ether from any alcohol present by simple distillation, using a warm-water bath.

Step 5. Identify the alcohol by preparing a derivative (see Section 38.27).

Step 6. To the remaining reaction mixture from step 3, add 3 M HCl until the mixture is acidic to blue litmus paper. If the acid separates as a solid, collect it by suction filtration on a Hirsch funnel. Determine its melting point and go to step 8. If it does not separate as a solid, go to step 7.

Step 7. Extract the solution three times with 5-mL portions of ether. Combine the ether extracts, dry them with anhydrous magnesium sulfate, filter or decant the ether, then evaporate it in the hood to recover the acid.

Step 8. Identify the acid by preparing a derivative (see Section 38.28).

38.16   ETHERS (Table 38-16, page 818)

■ SOLUBILITY GROUPS   IX AND X
■ CLASSIFICATION TESTS SUMMARY

38.16A.   Ferrox

38.16A. Ferrox test

The ferrox test distinguishes compounds containing oxygen (such as ethers) from hydrocarbons and halocarbons. Iron(III) ammonium sulfate and potassium thiocyanate together react to form iron(III) hexathiocyanatoferrate(III). This compound dissolves in the presence of oxygen-containing compounds to give a red to reddish-purple color. Hydrocarbons, halocarbons, and diaryl ethers do not dissolve this compound or give this color.

$$2 \, Fe(NH_4)(SO_4)_2 + 6 \, KSCN \longrightarrow Fe[Fe(SCN)_6] + 3 \, K_2SO_4 + (NH_4)_2SO_4$$

Iron(III) ammonium sulfate	Potassium thiocyanate	Iron(III) hexathiocyanato- ferrate(III)

PROCEDURE

Reference compounds.   Ethyl ether and methoxybenzene

Step 1. In a dry test tube, carefully grind together with a stirring rod a crystal of ferric ammonium sulfate and a crystal of potassium thiocyanate. Iron(III) hexathiocyanatoferrate(III) will adhere to the stirring rod as a colored mass.

Step 2. In another test tube, dissolve 3 drops of a liquid sample or about 30 mg of a solid sample in toluene.

Step 3. Stir the solution with the stirring rod holding the iron(III) hexathiocyanatoferrate(III). Record and interpret the result.

38.17 NITRILES (Table 38-17, page 819)

- ■ SOLUBILITY GROUPS I AND VIII
- ■ CLASSIFICATION TESTS SUMMARY

38.17A. Ammonia

38.17B. Ferric hydroxamate

38.17A. Ammonia test

Nitriles react with aqueous sodium hydroxide solution to yield ammonia, which can be detected by the blue color formed with copper ion.

$$R—C\equiv N + NaOH + H_2O \longrightarrow R—\overset{\overset{\displaystyle O}{\|}}{C}—O^- \ Na^+ + NH_3$$

Amides and ammonium salts also give a positive test. Amides, however, will give strong carbonyl absorption in the infrared spectrum, while ammonium salts are insoluble in ether.

PROCEDURE

Reference compound.   Benzonitrile

Step 1.   Place 3 drops of a liquid sample or about 50 mg of a solid sample in a small test tube.

Step 2.   Add 2 mL of 20% NaOH solution.

Step 3.   Fasten a small circle of filter paper tightly over the top of the test tube and wet it with several drops of copper sulfate solution.

Step 4.   Boil the contents in the test tube for about 1 minute and see if the paper has turned a darker blue. Record and interpret the result.

38.17B. Ferric hydroxamate test

Nitriles react with hydroxylamine to give a hydroxamic acid, which in turn reacts with ferric chloride to yield a ferric hydroxamate. A red to violet or a blue to violet color indicates a positive test.

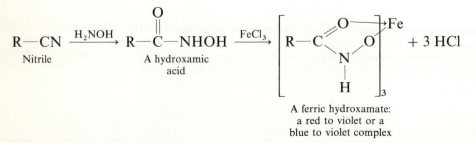

Esters, amides, acid halides, and acid anhydrides give the same result, so a positive test is not conclusive proof of a nitrile. In addition, phenols, enols, and other compounds react with ferric chloride to give similarly colored products. The unknown must, therefore, first be tested with ferric chloride without any hydroxylamine present.

PROCEDURE

Reference compound. Benzonitrile

Follow the same procedure given for amides (Section 38.14B). Record and interpret your result.

38.18 ALKENES AND ALKYNES (Table 38-18, page 819)

SOLUBILITY GROUP IX
CLASSIFICATION TESTS SUMMARY

38.18A. Bromine in methylene chloride

38.18B. Dilute potassium permanganate (the Baeyer test)

38.18A. Bromine in methylene chloride

Alkenes react with bromine in methylene chloride to yield a colorless solution of the vicinal dibromide.

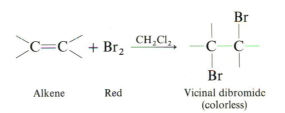

Alkyne   Red   Vicinal dibromide
                                   (colorless)

Alkynes also react with bromine, so the test qualitatively is not able to distinguish between alkenes and alkynes. Other functional groups may also be present, and no reaction may occur if the alkene or alkyne is part of a conjugated system or has strong electron-withdrawing groups present. For example:

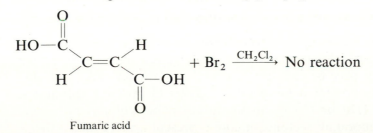

Fumaric acid

Most aromatic compounds do not react with bromine without a catalyst. But if the aromatic ring has a strongly activating ortho, para director present (e.g., —OH, —OR, —NR$_2$), a reaction occurs. Furthermore, the evolution of hydrogen bromide in this reaction frequently allows one to distinguish it from the alkene or alkyne reaction.

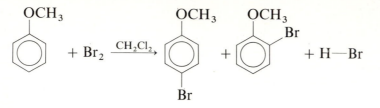

Acetone and aldehydes and ketones with a high enol content also will react with bromine. Again, the evolution of hydrogen bromide in this reaction frequently allows one to distinguish it from the alkene or alkyne reaction.

$$CH_3-\overset{\overset{\displaystyle O}{\|}}{C}-CH_3 + Br_2 \xrightarrow{CH_2Cl_2} CH_3-\overset{\overset{\displaystyle O}{\|}}{C}-\underset{\underset{\displaystyle Br}{|}}{C}H_2 + H-Br$$

PROCEDURE

CAUTION: Bromine causes severe burns. Be careful not to get it on your skin. Bromine burns may be treated by flushing the affected area with plenty of water, and applying a wet dressing of 10% sodium thiosulfate.

Bromine also reacts slowly with methylene chloride, causing a noticeable loss of color and the formation of hydrogen bromide. This hydrogen bromide may make it impossible to distinguish between an addition and a substitution reaction. To prevent this, use a reagent that has been freshly prepared and stored in brown glass bottles.

Reference compounds. Cyclohexene, cyclohexane, acetone, toluene, and phenol

Step 1. Dissolve 2 drops of a liquid sample or about 50 mg of a solid sample in about 2 mL of methylene chloride in a test tube.

Step 2. Add a 2% bromine in methylene chloride solution dropwise, with mixing, until the reddish color remains. Also blow gently over the test tube opening or place a wet piece of blue litmus paper over the test tube opening. An alkene or alkyne is present if more than 4 drops of bromine solution is needed and **no** fog of HBr **or** change in the litmus paper color is produced. A reactive aromatic compound, acetone, or enol is present if more than 4 drops of bromine solution is needed **and** a fog of HBr **or** a change in the litmus paper color is produced. No reaction is indicated if 4 drops or less of bromine solution is needed. Record your observations and interpret the result.

38.18B. Dilute potassium permanganate (the Baeyer test)

Alkenes react with dilute potassium permanganate to yield a glycol and a brown precipitate of manganese dioxide.

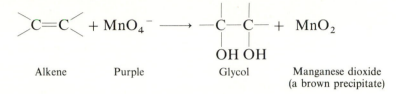

| Alkene | Purple | Glycol | Manganese dioxide (a brown precipitate) |

Alkynes also react with dilute potassium permanganate, so the test qualitatively is not able to distinguish between alkenes and alkynes. In addition, aldehydes, primary and secondary alcohols, phenols, and aromatic amines give a positive result. Therefore, the test results must be interpreted with caution if any of these compounds are present.

PROCEDURE

Reference compounds. Cyclohexene, toluene, benzaldehyde, and methanol

Step 1. Dissolve 2 drops of a liquid sample or about 30 mg of a solid sample in 1 to 2 mL of water or acetone.

Step 2. Add 4 to 5 drops of a 1% aqueous, $KMnO_4$ solution with mixing.

Step 3. Let the solution stand for 1 to 2 minutes. The disappearance of the purple color and the formation of a brown precipitate indicates a positive test. Record and interpret the result. (**CAUTION:** Potassium permanganate solutions undergo some decomposition to manganese dioxide on standing. Interpret any small amounts of brown precipitate with caution.)

38.19   ALKYL HALIDES (Table 38-19, page 820)

■ **SOLUBILITY GROUP   X**
■ **CLASSIFICATION TESTS SUMMARY**

38.19A. Sodium fusion

38.19B. Silver nitrate

38.19C. Sodium iodide

38.19A. Sodium fusion test

All alkyl halides (except fluorides) will give a positive test for a specific halogen in the sodium fusion test (see Section 38.7B).

38.19B. Silver nitrate test

Alkyl halides (except fluorides) react with silver nitrate to give a silver halide precipitate.

$$R-X + AgNO_3 \longrightarrow R^+ NO_3^- + AgX\downarrow$$

The rate of this S_N1 reaction indicates the type of organic halide present. Table 38-2 gives the results to be expected. Observe that unactivated aryl halides, vinyl halides, chloroform, and carbon tetrachloride all do not react, even at higher temperatures, due to their carbocations being so unstable.

Amine salts, carboxylic acids, and acid halides also react with silver nitrate. These compounds, however, have different solubility properties, amine salts contain nitrogen, and both carboxylic acids and acid halides show strong carbonyl absorption in the infrared spectrum. In addition, the precipitate formed from a carboxylic acid is soluble in added nitric acid, whereas silver halides are not.

$$\underset{\substack{\text{Carboxylic} \\ \text{acid}}}{R-\overset{\overset{\displaystyle O}{\|}}{C}-OH} + AgNO_3 \longrightarrow \underset{\substack{\text{Precipitate soluble in} \\ \text{added nitric acid}}}{R-\overset{\overset{\displaystyle O}{\|}}{C}-O^- Ag^+} + HNO_3$$

TABLE 38-2.
Alcoholic Silver Nitrate Test Results

1. Give an immediate precipitate at room temperature
   Tertiary alkyl halides
   Allylic halides
   Benzylic halides
   Alkyl iodides
   α-Haloethers
   1,2-Dibromides

2. Do not react at room temperature, but give a precipitate at higher temperatures
   Primary and secondary alkyl chlorides and bromides
   Geminal dibromides
   Activated aryl halides

3. Do not give a precipitate, even at higher temperatures
   Unactivated aryl halides
   Vinyl halides
   Chloroform and carbon tetrachloride

PROCEDURE

Reference compounds.   *t*-Butyl chloride, butyl chloride, bromobenzene

Step 1.   In a small test tube, add 1 drop of a liquid sample or about 5 drops of an ethanol solution of a solid sample to 2 mL of 2% ethanolic $AgNO_3$ solution.

Step 2.   Observe how long it takes for a precipitate to form. If no reaction has occurred after 2 minutes, heat the solution to boiling and observe if a precipitate forms.

Step 3.   If a precipitate forms, add 5 drops of 5% HNO_3 and stir. The test is positive if the precipitate does not dissolve. Record and interpret your results.

38.19C. Sodium iodide test

Alkyl chlorides and bromides react with sodium iodide in acetone to yield an alkyl iodide and either sodium chloride or sodium bromide.

$$R-Cl + NaI \xrightarrow{\text{acetone}} R-I + NaCl\downarrow$$

$$R-Br + NaI \xrightarrow{\text{acetone}} R-I + NaBr\downarrow$$

The reaction occurs by an S_N2 mechanism and depends on the fact that both sodium chloride and sodium bromide are not very soluble in acetone, whereas sodium iodide is. The formation of a precipitate therefore indicates a positive test. With 1,2-dichloro and 1,2-dibromo compounds, a positive test is also the formation of a reddish-brown color, due to the elemental iodine produced in the reaction.

$$\overset{|}{\underset{Cl}{C}}-\overset{|}{\underset{Cl}{C}}- + 2\,NaI \xrightarrow{\text{acetone}} -\overset{|}{C}=\overset{|}{C}- + 2\,NaCl\downarrow + I_2$$

$$\overset{|}{\underset{Br}{C}}-\overset{|}{\underset{Br}{C}}- + 2\,NaI \xrightarrow{\text{acetone}} -\overset{|}{C}=\overset{|}{C}- + 2\,NaBr\downarrow + I_2$$

The sodium iodide test is often used to complement the silver nitrate test (see Section 38.19B). When used together, these tests allow one to determine quite accurately the type of organic halide. Of course, if the compound is known to be an alkyl iodide, the sodium iodide test has no value.

TABLE 38-3.

Sodium Iodide in Acetone Test Results

1. Form a precipitate within 3 minutes at 25°C
Primary alkyl bromides
Allylic chlorides and bromides
Benzylic chlorides and bromides
α-Halo ketones, esters, amides, and nitriles
Acid chlorides and bromides

2. Form a precipitate within 6 minutes at 50°C
Primary and secondary alkyl chlorides
Secondary and tertiary alkyl bromides
Cyclopentyl chloride
Benzal chloride, benzotrichloride, bromoform, 1,1,2,2-tetrabromoethane

3. Unreactive under the specified conditions
Tertiary alkyl chlorides
Cyclohexyl chloride and bromide
Vinyl halides
Aryl halides
Chloroform, carbon tetrachloride, trichloroacetic acid

Table 38-3 gives the results to be expected in the sodium iodide test. Observe that vinyl halides, aryl halides, chloroform, and carbon tetrachloride are again unreactive, as they were in the silver nitrate test (see Section 38.19B).

PROCEDURE

Reference compounds. *t*-Butyl chloride, butyl chloride, butyl bromide

Step 1. Add 2 drops of the chloro or bromo compound to 1 mL of the sodium iodide–acetone test solution. If the compound is a solid, dissolve about 50 mg of it in the minimum amount of acetone and add this solution to the reagent.

Step 2. Mix the contents; then allow the tube to stand for 3 minutes at room temperature.

Step 3. Note whether a white precipitate of sodium chloride or sodium bromide has formed, or whether the solution has become reddish-brown from the formation of iodine.

Step 4. If no change has occurred, heat the solution in a beaker of water at 50°C for 6 minutes. Record and interpret the results.

- ## SOLUBILITY GROUP   X
- ## CLASSIFICATION TESTS SUMMARY

38.20A.   Sodium fusion, silver nitrate, and sodium iodide

38.20B.   Ignition

38.20C.   Reaction with aluminum chloride and chloroform

38.20A. Sodium fusion, silver nitrate, and sodium iodide tests

All aryl halides (except fluorides) will give a positive test for a specific halogen in the sodium fusion test (see Section 38.7B). But unless the aryl halide is activated, it will **not** give a positive silver nitrate test (see Section 38.19B). All aryl halides give a **negative** sodium iodide test (see Section 38.19C).

38.20B. Ignition test

Being highly unsaturated, aryl halides will burn with a yellow sooty or smoky flame when ignited. Saturated compounds, on the other hand, burn with a clean yellow to blue flame.

PROCEDURE

Reference compounds.   Bromobenzene and chlorobenzene

Step 1.   Place a few drops or milligrams of the sample in a crucible cover.

Step 2.   While holding the cover with tongs, bring it near a Bunsen burner flame and allow the sample to ignite. Record and interpret the results.

38.20C. Reaction with aluminum chloride and chloroform

Aryl halides usually give characteristic colors when placed with a mixture of aluminum chloride and chloroform. These colors result from a Friedel-Crafts reaction between the aryl halide and the chloroform. But if the aryl halide has nitro or other highly deactivating groups present, no reaction occurs. The color produced from various types of aromatic rings is given in Table 38-4.

TABLE 38-4.

Colors Produced by Aryl Halides Reacting with Aluminum Chloride and Chloroform

Aromatic ring system	Color
Benzene	Orange to red
Naphthalene	Blue to purple
Anthracene	Green
Biphenyl or phenanthrene	Purple

PROCEDURE

Reference compounds. Bromobenzene or chlorobenzene, 4-nitrobromo-benzene or 4-nitrochlorobenzene

Step 1. Heat with a burner about 0.1 gram of fresh, anhydrous $AlCl_3$ in a test tube to sublime the $AlCl_3$ onto the side of the tube. Let the tube air cool to room temperature.

Step 2. Dissolve 2 drops of a liquid sample or about 25 mg of a solid sample in the minimum amount of chloroform.

Step 3. Add the chloroform solution so that it touches the sublimed $AlCl_3$ on the side of the test tube. Record the color formed and interpret the result.

38.21   NITRO COMPOUNDS (Table 38-21, page 823)

- **SOLUBILITY GROUPS   V AND VII (unless other functional groups are present)**
- **CLASSIFICATION TEST SUMMARY**

38.21A.   Ferrous hydroxide

38.21A. Ferrous hydroxide test

Essentially all nitro compounds oxidize ferrous ion in alkaline solution to produce ferric hydroxide, which precipitates as a reddish-brown solid. At the same time, the nitro group is reduced to an amine.

$$R{-}NO_2 + 6\,Fe^{+2} + 12\,OH^- + 4\,H_2O \longrightarrow R{-}NH_2 + 6\,Fe(OH)_3\downarrow$$

Reddish-brown
precipitate

PROCEDURE

Reference compound.   4-Nitrotoluene

Step 1. Place 1.5 mL of freshly prepared 5% aqueous ferrous ammonium sulfate in a small test tube. Then add 1 drop of a liquid sample or about 10 mg of a solid sample.

Step 2. Mix the contents, then add 1 drop of 3 M H_2SO_4 and 20 drops of 2 M KOH in methanol.

Step 3. Stopper the tube and shake it. A positive test is the formation of a reddish-brown precipitate within 1 minute. Record and interpret your result.

38.22 AROMATIC HYDROCARBONS (Table 38-22, page 824)

■ **SOLUBILITY GROUP X**
■ **CLASSIFICATION TESTS SUMMARY**

38.22A. Ignition

38.22B. Aluminum chloride and chloroform

38.22A. Ignition test

Aromatic hydrocarbons, being highly unsaturated, will burn with a yellow sooty or smoky flame when ignited. See Section 38.20B for further details and the test procedure, using toluene as a reference compound. Other types of aromatic compounds will also give the same result.

38.22B. Aluminum chloride and chloroform test

Aromatic hydrocarbons usually give characteristic colors when placed with a mixture of aluminum chloride and chloroform. These colors result from a Friedel-Crafts reaction between the aromatic hydrocarbon and the chloroform. See Section 38.20C for further details and the test procedure, using toluene and naphthalene as reference compounds. Other types of aromatic compounds will also give the same result.

38.23 ALKANES (Table 38-23, page 825)

■ **SOLUBILITY GROUP X**
■ **CHEMICAL CLASSIFICATION TESTS**

Alkanes are generally classified by their lack of giving any positive chemical tests.

38.24 CONFIRMING THE UNKNOWN'S IDENTITY BY SYNTHESIZING SOLID DERIVATIVES

By following the stated procedure and utilizing the tables at the end of the experiment, you should have by now either pinpointed the identity of your unknown or greatly narrowed the number of possibilities. It is now time to complete the identification by, if possible, synthesizing one or two solid derivatives. For the unknown to be a certain possibility, the melting points of its derivatives must agree with the listed literature values for this possibility.

Remember that obtaining an accurate melting point of a derivative is very important. Otherwise, you could easily reach the wrong conclusion regarding the identity of an unknown. Since the derivative prepared may be very impure, it must first be recrystallized and dried before the melting point is taken. Several recrystallizations may even be necessary. Use the melting-point range and possibly the color as criteria of the purity.

It is also very important when preparing a derivative to choose one which will give you a definite answer. For example, suppose your unknown is a liquid with a boiling point of 183–184°C at 745 torr. Chemical and solubility tests also indicate it to be a primary amine. Looking at Table 38-11, you find the following possibilities:

Compound	BP	MP	Benzamide	Picrate	Acetamide	Benzene-sulfonamide
Benzylamine	184	—	105	194	60	88
Aniline	184	—	163	198	114	112
4-Fluoroaniline	186	—	185	—	152	—
1-Phenylethylamine	187	—	120	—	57	—

Choosing to make the picrate derivative would not be wise. Two of the possibilities have no picrate derivative listed, and the other two form picrate derivatives having nearly the same melting point. Therefore, how can you make a definite decision?

You would also not want to make the acetamide derivative. Two of these possibilities form acetamide derivatives having nearly the same melting point, so again no definite decision may be possible.

But observe how the benzamide derivatives of all four possibilities have melting points widely separated, so a definite decision can be made. A little thinking and planning can again save you much needless frustration and wasted time and effort in the laboratory.

Sometimes you will find several values listed for the same derivative. This results from more than one literature value being reported. In these cases, either value may be considered correct, and you should use whichever one corresponds better to the melting point of your derivative.

For your convenience, a list of the possible classes of compounds, the table and page where they and their derivatives can be found, and the section and page for preparing the derivatives are given in Table 38-5 and on the back inside cover of the book.

38.25 REPORTING THE RESULTS

Properly reporting the results of your qualitative analysis is very important. A few examples here will illustrate the basic procedure.

Example 1

Physical examination. The compound is a white solid with no detectable odor.

TABLE 38-5. ▨

Derivatives of Various Classes of Compounds

Class of compound	Table	Page	Derivatives	Section	Page
Aldehydes	38-6	799	Semicarbazone	38.26A	772
			2,4-Dinitrophenyl-hydrazone	38.26B	773
			4-Nitrophenylhydrazone	38.26C	774
			Phenylhydrazone	38.26D	774
Ketones	38-7	801	Semicarbazone	38.26A	772
			2,4-Dinitrophenyl-hydrazone	38.26B	773
			4-Nitrophenylhydrazone	38.26C	774
			Phenylhydrazone	38.26D	774
Alcohols	38-8	804	3,5-Dinitrobenzoate	38.27A	776
			1-Naphthylurethane	38.27B	777
			Phenylurethane	38.27C	778
Phenols	38-9	806	3,5-Dinitrobenzoate	38.27A	776
			1-Naphthylurethane	38.27B	777
			Ring bromination	38.27D	779
Carboxylic acids	38-10	808	p-Toluidide	38.28A	780
			Anilide	38.28B	782
			Amide	38.28C	782
Primary amines	38-11	811	Benzamide	38.29A	783
			Acetamide	38.29B	785
			Benzenesulfonamide	38.29C	786
			Picrate	38.29D	787
Secondary amines	38-12	813	Benzamide	38.29A	783
			Acetamide	38.29B	785
			Benzenesulfonamide	38.29C	786
			Picrate	38.29D	787
Tertiary amines	38-13	814	Picrate	38.29D	787
			Methiodide	38.29E	788
Amides	38-14	815	Hydrolysis products and their derivatives	38.30	789
Esters	38-15	816	Hydrolysis products and their derivatives	38.31	790
Ethers	38-16	818	Picrate	38.32A	790
			Bromination product	38.32B	791
			Nitration product	38.32C	791
Nitriles	38-17	819	Carboxylic acids and their derivatives	38.33	792
Alkenes	38-18	819	None	—	—
Alkyl halides	38-19	820	Anilide	38.34A	793
			1-Naphthalide	38.34B	794
Aryl halides	38-20	822	Nitration product	38.35A	795
			Oxidation product	38.35B	795
Nitro compounds	38-21	823	Reduced amine	38.36A	796
			Acyl derivative of reduced amine	38.36B	797
			Nitration product	38.36C	797
Aromatic hydrocarbons	38-22	824	Nitration product	38.37A	798
			Picrate	38.37B	798
Alkanes	38-23	825	None	—	—

Melting point. 52–53°, corrected 53–54°.

Infrared spectrum. Gives strong carbonyl absorption at 5.80 μ (1720 cm^{-1}). The compound appears from the spectrum to be a ketone for the following reasons:

1. No broad, strong peak in the range 3.0–4.0 μ. **Conclusion:** no carboxylic acid.

2. Peak at 3.50 μ (2850 cm^{-1}), but no peak at 3.65 μ (2750 cm^{-1}). **Conclusion:** probably not an aldehyde.

3. No narrow or broad peak at 2.85 μ (3500 cm^{-1}). **Conclusion:** no primary or secondary amide. Carbonyl absorption is also not within the range of 5.90–6.06 μ (1700–1630 cm^{-1}) for an amide.

4. There are not two medium peaks between 7.7–10.0 μ (1300–1000 cm^{-1}). **Conclusion:** probably not an ester.

Solubility tests

Water	does not dissolve
5% NaOH	does not dissolve
5% HCl	does not dissolve
Conc. H$_2$SO$_4$	dissolves

Conclusion: Compound is in Group VIII or Group IX. Insolubility in 5% NaOH confirms infrared determination of no carboxylic acid.

Na fusion results

Nitrogen	negative: no Prussian blue color. This confirms no amide present.
Sulfur	negative: no purple color with sodium nitroprusside.
Halogens	negative: no precipitate with silver nitrate solution.

Classification tests

1. 2,4-Dinitrophenylhydrazine positive: red precipitate obtained. This confirms either an aldehyde or ketone.

2. Tollens no silver mirror: therefore no aldehyde. Compound appears to be ketone, confirming infrared spectrum.

Possibilities from Table 38-7

Compound	MP	Semi-carbazone	Phenyl-hydrazone	2,4-Dinitro-phenyl-hydrazone	4-Nitro-phenyl-hydrazone
Benzophenone	48	167	137	238(229)	154(144)
4-Bromoacetophenone	51	208	126	230(237)	—
2-Acetylnaphthalene	54	235	176	262	—
Phenyl p-tolyl ketone	54	122	109	200	—
Benzalacetophenone	58(62)	168(179)	120	245	—

Of these, 4-bromoacetophenone seems unlikely because of negative halogen test.

Prepared derivatives

1. Semicarbazone melts at 231–233°.

2. Phenylhydrazone melts at 174–175°.

Therefore, the only possibility from Table 38-7 is 2-acetylnaphthalene.

Example 2

Physical examination. The compound is a colorless liquid with an undistinguishable odor.

Boiling point. 115–117° at 745 torr, corrected 116–118°.

Infrared spectrum. No strong peak at around 5.8 μ. Compound, therefore, has no carbonyl group. It cannot be an aldehyde, ketone, carboxylic acid, amide, ester, or anhydride. Compound has strong, broad peak at 2.9 μ (3500 cm^{-1}). Could be an alcohol, phenol, or amine. The broad, strong peak at 8.0 μ (1250 cm^{-1}) supports an alcohol, phenol, or amine. However, the strength of both peaks makes an amine less likely.

Solubility tests

Water	dissolves
Ether	dissolves
Conclusion:	Compound is in Group I. pH of aqueous solution of compound is 6.8, no change. Therefore, compound is probably not an amine. It also lacks the "fishy" odor of water-soluble amines.

Na fusion results

Nitrogen   negative: no Prussian blue color. This supports no amine.
Sulfur   negative: no purple color with sodium nitroprusside.
Halogens   negative: no precipitate with silver nitrate solution.

Classification tests

1. Lucas   negative: no insolubility after 5 minutes. This rules out 2° and 3° alcohols.

2. Chromic acid   positive: green precipitate. Compound seems to be 1° alcohol.

3. Bromine water   negative: no appreciable loss of color. This rules out phenols and double bond.

Possibilities from Table 38-8

Compound	BP	3,5-Dinitrobenzoate	1-Naphthylurethane	Phenylurethane
2-Propyn-1-ol	114	—	—	63
1-Butanol	116	64	71	61
3-Pentanol	116	99	95	48
2-Pentanol	119	61	74	—
1-Methoxy-2-propanol	119	85	—	—
3-Methyl-3-pentanol	123	96(62)	83	43
2-Methoxyethanol	124	—	112	—

Of these possibilities, 2-propyn-1-ol is eliminated because of the negative bromine water test. 3-Pentanol, 2-pentanol, 1-methoxy-2-propanol, and 3-methyl-3-pentanol are all eliminated because of the negative Lucas test. Because of the boiling point, 2-methoxyethanol seems unlikely.

Prepared derivatives

1.   1-Naphthylurethane melts at 69–70°.

2.   3,5-Dinitrobenzoate melts at 63–64°.

Therefore, the only possibility from Table 38-8 is 1-butanol.

Example 3

Physical examination.   The compound is a white solid with no detectable odor.

Melting point.   66–67°, corrected 68–69°.

Infrared spectrum.   No strong peak at around 5.8 μ. Compound, therefore, has no carbonyl group. It cannot be an aldehyde, ketone, carboxylic acid, amide, ester, or anhydride.

Compound has a medium peak at 2.9 μ (3450 cm^{-1}). Probably too weak to be an alcohol or phenol. Could be an amine.

Solubility tests

Water	does not dissolve
5% NaOH	does not dissolve
5% HCl	dissolves

Conclusion:	Compound is in Group VI and is an amine.

Na fusion results

Nitrogen	positive: Prussian blue color. Further evidence of amine.
Sulfur	negative: no purple color with sodium nitroprusside.
Halogens	positive for chlorine: white precipitate with silver nitrate solution. Precipitate dissolves readily when NH$_4$OH added, so it can't be silver bromide.

Classification tests

1.   Hinsberg   positive for 1° amine: benzenesulfonamide is soluble in NaOH, but insoluble in HCl.

2.   Nitrous acid   test not done, since all possible 1° amines listed are aromatic.

Possibilities from Table 38-11

Compound	MP	Benzamide	Picrate	Acetamide	Benzenesulfonamide
4-Bromoaniline	64	204	180	168	134
2,4,5-Trimethylaniline	64	167	—	162	—
3-Phenylenediamine	66	125(mono) 240(di)	184	87(mono) 191(di)	—
4-Iodoaniline	67	222	—	184	—
4-Chloroaniline	70	192	178	179	122
2-Nitroaniline	72	110(98)	73	92	104

Of these, the only one containing chlorine is 4-chloroaniline.

1. Benzamide melts at 190–191°.

2. Acetamide melts at 178–179°.

Therefore, the only possibility from Table 38-11 is 4-chloroaniline.

38.26 ALDEHYDES AND KETONES
(Tables 38-6 and 38-7, pages 799 and 801)

38.26A. Semicarbazone derivative

Most aldehydes and ketones will react with semicarbazide hydrochloride to give a solid semicarbazone derivative.

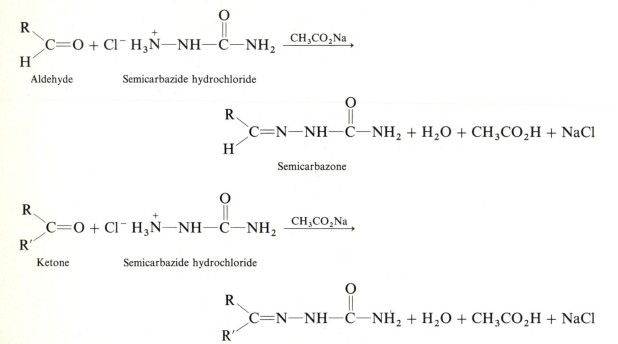

PROCEDURE

Step 1. Dissolve 0.2 gram of semicarbazide hydrochloride and 0.3 gram of anhydrous sodium acetate in 2 mL of water in a small test tube.

Step 2. In another small test tube, dissolve 10 drops of a liquid sample or about 200 mg of a solid sample in the minimum amount of 95% ethanol. Add this solution to the semicarbazide hydrochloride solution.

Step 3. Heat the solution on a steam bath or in a hot-water bath for 10 minutes. Then cool it in an ice bath.

Step 4. Collect the precipitate by suction filtration on a Hirsch funnel.

Step 5. Recrystallize the product from ethanol, dry it, and determine its melting point. Compare the value with those listed in Table 38-6 or 38-7 for the possible aldehydes or ketones.

38.26B. 2,4-Dinitrophenylhydrazone derivative

Most aldehydes and ketones will react with 2,4-dinitrophenylhydrazine in an acidic solution to yield a yellow to red solid of the 2,4-dinitrophenylhydrazone derivative.

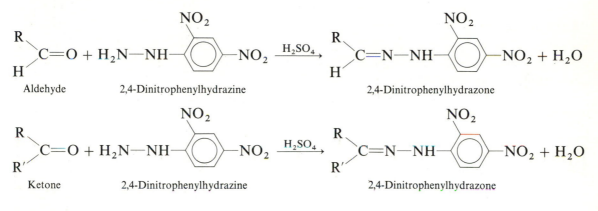

PROCEDURE

CAUTION: Many phenylhydrazine derivatives are carcinogenic. Handle with care, avoiding direct contact.

Step 1. Add an estimated 0.001 mole of the unknown compound to 10 mL of 2,4-dinitrophenylhydrazine reagent and mix. If the unknown is a solid, dissolve it first in the minimum amount of 95% ethanol.

Step 2. Allow the solution to stand for 15 minutes if an immediate precipitate is not formed. If necessary, heat the solution gently for 1 minute and allow it to stand for another 15 minutes.

Step 3. Collect the product by suction filtration on a Hirsch funnel.

Step 4. Recrystallize the product from ethanol, ethanol-water, or ethyl acetate.

Step 5. Dry the product and determine its melting point. Compare the value with those listed in Table 38-6 or 38-7 for the possible aldehydes or ketones.

38.26C. 4-Nitrophenylhydrazone derivative

Many aldehydes and ketones will react with 4-nitrophenylhydrazine to yield a solid 4-nitrophenylhydrazone derivative.

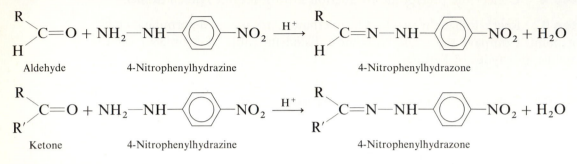

Aldehyde   4-Nitrophenylhydrazine   4-Nitrophenylhydrazone

Ketone   4-Nitrophenylhydrazine   4-Nitrophenylhydrazone

PROCEDURE

CAUTION: Many phenylhydrazine derivatives are carcinogenic. Handle with care, avoiding direct contact.

    Also be careful when using glacial acetic acid, since it can cause severe burns.

Step 1. Heat a mixture of 0.3 gram of 4-nitrophenylhydrazine, 0.3 gram or 0.4 mL of the aldehyde or ketone, 7 to 10 mL of 95% ethanol, and 1 drop of glacial acetic acid to the boiling point.

Step 2. Keep the mixture hot for a few minutes, adding more 95% ethanol if necessary to obtain a clear solution.

Step 3. Cool the solution and collect the precipitated 4-nitrophenylhydrazone by suction filtration on a Hirsch funnel. If the product does not precipitate with cooling, reheat the mixture to the boiling point, add hot water dropwise until the mixture becomes cloudy, then add hot 95% ethanol to obtain a clear solution. Cool the solution again and collect the product by suction filtration.

Step 4. To recrystallize the product, dissolve it in the minimum amount of hot 95% ethanol, add hot water dropwise until the mixture becomes cloudy, then add hot 95% ethanol dropwise to obtain a clear solution. Cool the solution and collect the product by suction filtration.

Step 5. Dry the product and determine its melting point. Compare the value with those listed in Table 38-6 or 38-7 for the possible aldehydes or ketones.

38.26D. Phenylhydrazone derivative

Many aldehydes and ketones will react with phenylhydrazine to yield a solid phenylhydrazone derivative.

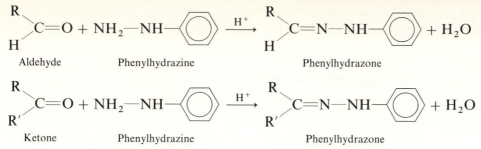

PROCEDURE

CAUTION: Phenylhydrazine is very toxic, and many phenylhydrazine derivatives are carcinogenic. Handle with care, avoiding direct contact.

Also be careful when using glacial acetic acid, since it can cause severe burns.

Water-insoluble aldehydes and ketones

Step 1. Dissolve 0.3 mL or about 0.25 gram of the aldehyde or ketone in about 1 mL of 95% ethanol in a test tube. Add water dropwise until the mixture becomes cloudy; then add 95% ethanol dropwise until the mixture is again clear.

Step 2. Add 5 drops of phenylhydrazine and 2 drops of glacial acetic acid. Heat the mixture in a hot-water bath for several minutes, followed by cooling.

Step 3. Collect any precipitated phenylhydrazone by suction filtration on a Hirsch funnel.

Step 4. To recrystallize any precipitate obtained, dissolve it in the minimum amount of hot 95% ethanol, add hot water dropwise until the mixture becomes cloudy, then add hot 95% ethanol dropwise to obtain a clear solution. Cool the solution and collect the product by suction filtration.

Step 5. Dry the product and determine its melting point. Compare the value with those listed in Table 38-6 or 38-7 for the possible aldehydes or ketones.

Water-soluble aldehydes and ketones

Step 1. Add 0.3 mL of phenylhydrazine to 3 mL of water in a test tube. Add glacial acetic acid dropwise until the phenylhydrazine just dissolves, then add 0.3 mL or about 0.25 gram of your aldehyde or ketone.

Step 2. Heat the mixture in a hot-water bath for 1 minute; then add 1 mL of hot water and cool.

Step 3. If a precipitate forms, follow steps 3 through 5 given for water-insoluble aldehydes and ketones.

38.27 ALCOHOLS AND PHENOLS
(Tables 38-8 and 38-9, pages 804 and 806)

38.27A. 3,5-Dinitrobenzoate derivative

Many alcohols and phenols react with 3,5-dinitrobenzoyl chloride to form a solid 3,5-dinitrobenzoate derivative.

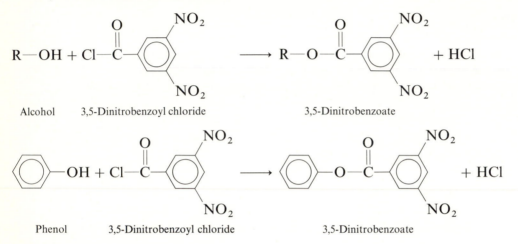

Alcohol	3,5-Dinitrobenzoyl chloride	3,5-Dinitrobenzoate

Phenol	3,5-Dinitrobenzoyl chloride	3,5-Dinitrobenzoate

PROCEDURE

CAUTION: 3,5-Dinitrobenzoyl chloride readily hydrolyzes. First check the purity of the reagent by determining its melting point. It must be within 3°C of its literature value of 74°C. The melting point will be high when the carboxylic acid is present.

Pyridine, if used, is very toxic and has an extremely disagreeable odor. Avoid direct contact with it and use it only in the hood.

Step 1. Gently boil for 5 minutes in the hood a mixture of 0.3 gram of 3,5-dinitrobenzoyl chloride and 0.3 mL of the liquid unknown. If the unknown is a solid, reflux in the hood about 300 mg of it and 0.3 gram of 3,5-dinitrobenzoyl chloride in 2 mL of dry pyridine for 15 minutes.

Step 2. Cool the reaction mixture. If the unknown is a liquid, add 2 mL of 5% Na_2CO_3 solution and 1 mL of water. If the unknown is a solid, add 3 mL of 5% Na_2CO_3 solution and 3 mL of water.

Step 3. Stir the mixture for several minutes, carefully grinding up any solid present.

Step 4. Collect the product by suction filtration on a Hirsch funnel, and wash it with a small amount of cold water.

Step 5. Place the crystals in a 25-mL Erlenmeyer flask and dissolve them in the minimum amount of hot 95% ethanol. Add hot water dropwise until the solution begins to turn cloudy, then add hot 95% ethanol dropwise until the mixture is again clear. Allow the hot solution to slowly cool.

Step 6. Collect the recrystallized product by suction filtration, wash it with a small amount of cold 50% aqueous ethanol solution, and dry.

Step 7. Determine the melting point and compare the value with those listed in Table 38-8 or 38-9 for the possible alcohols or phenols.

38.27B. 1-Naphthylurethane derivative

Many alcohols and phenols react with 1-naphthylisocyanate by nucleophilic addition to form a solid 1-naphthylurethane derivative.

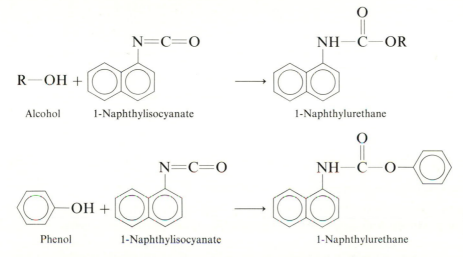

The alcohol or phenol used must be completely free of water. Water reacts with 1-naphthylisocyanate to form 1,3-di(1-naphthyl)urea, which will crystallize with the product and contaminate it.

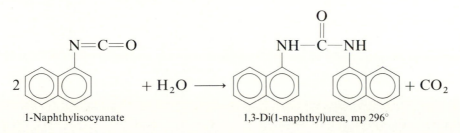

In addition, the 1-naphthylurethanes of tertiary alcohols and other sterically hindered alcohols and phenols are difficult to prepare by this method.

PROCEDURE

CAUTION: 1-Naphthylisocyanate is very toxic. Perform this reaction only in the hood, avoiding direct contact with the reagent.

Pyridine, if used, is also very toxic and has an extremely disagreeable odor. Avoid direct contact with it and use it only in the hood.

Step 1. Place 0.5 gram of the **dry** alcohol or phenol in a **dry** test tube, and cautiously add 0.8 mL of 1-naphthylisocyanate. If the unknown is a phenol, add several drops of pyridine to catalyze the reaction.

Step 2. If the reaction is not spontaneous, heat the mixture on a steam bath for 10 to 15 minutes.

Step 3. Cool the test tube in an ice bath, and induce crystallization by scratching the tube with a stirring rod.

Step 4. Collect the product by suction filtration on a Hirsch funnel.

Step 5. Dissolve the product in about 5 to 6 mL of hot hexane, and gravity filter the mixture, using a preheated funnel. This will remove any insoluble 1,3-di(1-naphthyl)urea present.

Step 6. Cool the filtrate, collect the resulting precipitate by suction filtration, and dry it.

Step 7. Determine the melting point and compare the value with those listed in Table 38-8 or 38-9 for the possible alcohols or phenols.

38.27C. Phenylurethane derivative of alcohols

Many alcohols and phenols react with phenylisocyanate by nucleophilic addition to form a solid phenylurethane derivative. The reaction, however, occurs more slowly than the corresponding 1-naphthylisocyanate one (see Section 38.27B). In addition, tables of phenylurethane derivatives of phenols generally are not given in qualitative analysis books, since the 1-naphthylurethane usually is the more preferred derivative of phenols.

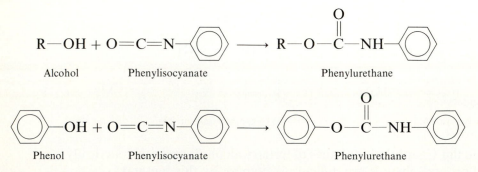

The alcohol used must be completely free of water. Water reacts with phenyl-isocyanate to form 1,3-diphenylurea, which will crystallize with the product and contaminate it.

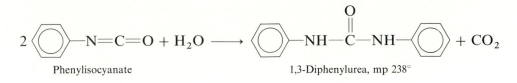

In addition, phenylurethanes of tertiary alcohols and other sterically hindered alcohols and phenols are difficult to prepare by this method.

PROCEDURE

CAUTION: Phenylisocyanate is very toxic. Perform this reaction only in the hood, avoiding direct contact with the reagent.

Step 1. Place 0.5 gram of the **dry** alcohol in a **dry** test tube and cautiously add 0.6 mL of phenylisocyanate.

Step 2. If the reaction is not spontaneous, heat the mixture on a steam bath for 10 to 15 minutes.

Step 3. Cool the test tube in an ice bath, and induce crystallization by scratching the test tube with a stirring rod.

Step 4. Collect the product by suction filtration on a Hirsch funnel.

Step 5. Dissolve the product in about 5 to 6 mL of hot hexane, and gravity filter the mixture, using a preheated funnel. This will remove any insoluble 1,3-diphenylurea present.

Step 6. Cool the filtrate, collect the resulting precipitate by suction filtration, and dry it.

Step 7. Determine the melting point and compare the value with those listed in Table 38-8 for the possible alcohols.

38.27D. Ring bromination of phenols

Being highly activated toward electrophilic aromatic substitution, phenols will readily undergo ring bromination. If adequate bromine is present, all the unsubstituted positions ortho and para to the hydroxyl group will be attacked.

For example:

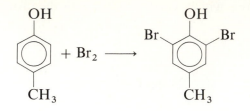

PROCEDURE

CAUTION: Bromine causes severe burns. Be careful not to get it on your skin. Bromine burns may be treated by flushing the affected area with plenty of water and applying a wet dressing of 10% sodium thiosulfate.

Step 1. If a stock brominating solution is not available, prepare one by dissolving 0.8 gram of KBr in 5 mL of water, and then add 0.5 gram of bromine.

Step 2. In a separate test tube, dissolve 0.1 gram of a solid sample or 4 drops of a liquid sample in 1 mL of methanol. Then add 1 mL of water.

Step 3. Add the bromine solution from step 1 dropwise with stirring to the sample solution prepared in step 2. Continue the addition until the yellow color of bromine persists, even with vigorous shaking of the **stoppered** tube.

Step 4. Add 3 to 5 mL of water and shake the stoppered tube vigorously. Collect the precipitated product by suction filtration on a Hirsch funnel.

Step 5. To recrystallize the product, dissolve it in the minimum amount of hot 95% ethanol, add hot water dropwise until the solution becomes cloudy, then add hot 95% ethanol until the mixture is again clear. Cool the solution and collect the precipitated product by suction filtration.

Step 6. Dry the product and determine its melting point. Compare the value with those listed in Table 38-9 for the possible phenols.

38.28 Carboxylic Acids (Table 38-10, page 808)

38.28A. *p*-Toluidide derivative

p-Toluidides are useful, solid derivatives of most carboxylic acids. They are best prepared by converting the carboxylic acid to the much more reactive acid chloride, followed by the reaction with *p*-toluidine. Thionyl chloride is the preferred reagent in the first step, since all the by-products are gaseous and, therefore, easily removed.

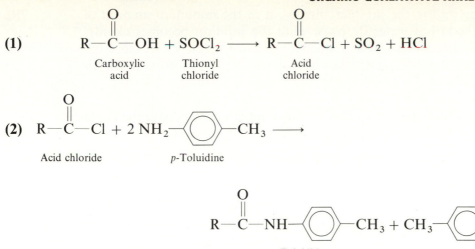

(1) Carboxylic acid + Thionyl chloride → Acid chloride + SO_2 + HCl

(2) Acid chloride + 2 p-Toluidine →

p-Toluidide

PROCEDURE

CAUTION: Thionyl chloride is a toxic, highly volatile, reactive chemical, which causes serious burns if spilled on the skin. Handle it with care, avoiding direct contact.

The reagent should also be used only in the hood to avoid exposure to its fumes and to the SO_2 and HCl gas produced in the reaction. Avoid leaving the reagent standing in an uncovered container, since it will also react with the moisture in the atmosphere, producing SO_2 and HCl gas.

p-Toluidine is also toxic. Handle it with care, avoiding direct contact.

Step 1. Place 1.0 gram of the carboxylic acid and 4 mL of thionyl chloride into a **dry** 25-mL round-bottom flask. Attach a reflux condenser and heat the mixture for 30 minutes, using a steam bath.

Step 2. Cool the solution and separate it into two equal portions. One portion can be used to prepare the p-toluidide derivative, and the other can be used for either the anilide or amide derivative.

Step 3. Dissolve 1 gram of p-toluidine in 20 mL of toluene. Add one portion of the acid chloride mixture and warm for 10 minutes on a steam bath.

Step 4. Extract the reaction mixture from step 3 first with 5 mL of water, then with 5 mL of 5% HCl, then with 5 mL of 5% NaOH, and finally with 5 mL of water.

Step 5. Dry the toluene solution with anhydrous magnesium sulfate. Gravity filter the liquid into a small beaker, then evaporate the filtrate to dryness on a hot plate in the hood.

Step 6. To recrystallize the product, dissolve it in the minimum amount of hot 95% ethanol, add hot water dropwise until the solution becomes cloudy, then add hot 95% ethanol dropwise until the mixture is again clear. Cool the solution and collect the precipitated product by suction filtration on a Hirsch funnel.

Step 7. Dry the product and determine its melting point. Compare the value with those listed in Table 38-10 for the possible carboxylic acids.

38.28B. Anilide derivative

Anilides are also useful, solid derivatives of most carboxylic acids. They are best prepared by converting the carboxylic acid to the much more reactive acid chloride, followed by the reaction with aniline. Thionyl chloride is the preferred reagent in the first step, since all the by-products are gaseous and, therefore, easily removed.

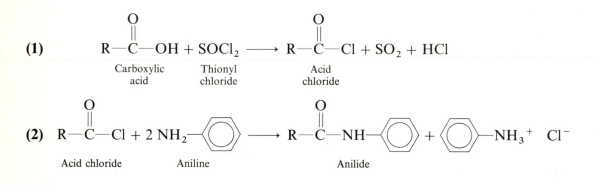

PROCEDURE

The procedure for making anilides is identical to the one for making *p*-toluidides (Section 38.28A), except that aniline is used instead of *p*-toluidine. **Be sure to follow all the safety precautions indicated beforehand, and remember that aniline is toxic. Handle it with care, avoiding direct contact.**

38.28C. Amide derivative

Amides are also useful, solid derivatives of most carboxylic acids. They are best prepared by converting the carboxylic acid to the much more reactive acid chloride, followed by the reaction with ammonium hydroxide. Thionyl chloride is the preferred reagent in the first step, since all the by-products are gaseous and, therefore, easily removed. Amide derivatives of water-soluble carboxylic acids are also water soluble and, as a result, more difficult to isolate.

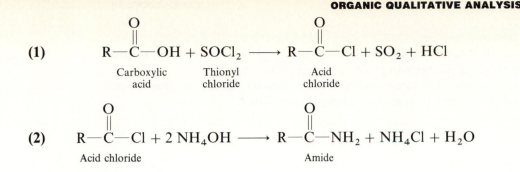

(1) Carboxylic acid + Thionyl chloride $\longrightarrow$ Acid chloride + SO_2 + HCl

$$R-\overset{O}{\underset{\|}{C}}-OH + SOCl_2 \longrightarrow R-\overset{O}{\underset{\|}{C}}-Cl + SO_2 + HCl$$

(2)

$$R-\overset{O}{\underset{\|}{C}}-Cl + 2\,NH_4OH \longrightarrow R-\overset{O}{\underset{\|}{C}}-NH_2 + NH_4Cl + H_2O$$

Acid chloride   Amide

PROCEDURE

Step 1. Prepare the acid chloride by the procedure in steps 1 and 2 of Section 38.28A. Be sure to follow all the safety precautions indicated beforehand.

Step 2. Cool 10 mL of concentrated NH_4OH in a 50-mL beaker in an ice bath in the hood. Then add one portion of the prepared acid chloride mixture with stirring.

Step 3. When the reaction is complete, collect the precipitated product by suction filtration on a Hirsch funnel.

Step 4. To recrystallize the product, dissolve it in the minimum amount of hot 95% ethanol, add hot water dropwise until the solution becomes cloudy, then add hot 95% ethanol dropwise until the mixture is again clear. Cool the solution and collect the precipitated product by suction filtration.

Step 5. Dry the product and determine its melting point. Compare the value with those listed in Table 38-10 for the possible carboxylic acids.

38.29 AMINES (Tables 38-11, 38-12, and 38-13, pages 811–814)

38.29A. Benzamide derivative

Most primary and secondary amines react with benzoyl chloride to form solid benzamide derivatives. However, tertiary amines do not.

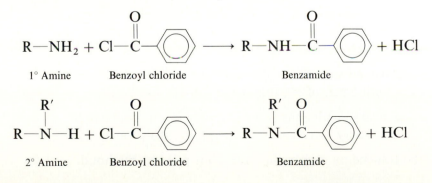

PROCEDURE

> **CAUTION: Benzoyl chloride is toxic, a lachrymator, and a skin irritant. Avoid direct contact with it, performing the reaction in the hood.**
>
> **Pyridine, if used, is also toxic and has an extremely disagreeable odor. Avoid direct contact with it and use it only in the hood.**

Water-soluble amines

Step 1. Place 4 to 5 drops of a liquid sample or about 100 mg of a solid sample in a test tube. Add 20 drops of 10% NaOH and 10 drops of fresh benzoyl chloride.

Step 2. Stopper the test tube and shake it for 10 minutes.

Step 3. Add 6 M HCl dropwise until pH paper indicates a pH of 7 to 8. Cool the mixture in an ice bath.

Step 4. Collect the crystals by suction filtration on a Hirsch funnel, and wash them with a small amount of cold water.

Step 5. To recrystallize the product, dissolve it in the minimum amount of hot 95% ethanol, add hot water dropwise until the solution becomes cloudy, then add hot 95% ethanol dropwise until the mixture is again clear. Cool the solution and collect the precipitated product by suction filtration.

Step 6. Dry the product and determine its melting point. Compare the value with those listed in Table 38-11 or 38-12 for the possible amines.

Water-insoluble amines

Step 1. Place 5 mL of pyridine, 0.5 mL of fresh benzoyl chloride, and 10 mL of toluene in a 50-mL round-bottom flask. Add 10 drops of a liquid sample or about 0.5 gram of a solid sample, attach a reflux condenser, and heat the mixture under reflux for about 30 minutes.

Step 2. Pour the reaction mixture into about 50 mL of water in a separatory funnel. Shake the two to hydrolyze any excess benzoyl chloride and dissolve much of the pyridine.

Step 3. Separate the two layers. Wash the upper toluene layer first with 5 mL of water and then with 5 mL of 5% sodium carbonate solution.

Step 4. Dry the toluene layer with anhydrous magnesium sulfate. Then separate the toluene by either gravity filtration or decantation into a small beaker.

Step 5. Remove the toluene by evaporation on a hot plate in the hood.

Step 6. Recrystallize the benzamide from 95% ethanol, collect it by suction filtration on a Hirsch funnel, and dry it.

Step 7. Determine the melting point and compare the value with those listed in Table 38-11 or 38-12 for the possible amines.

38.29B. Acetamide derivative

Most primary and secondary amines react with acetic anhydride to form solid acetamide derivatives. However, tertiary amines do not.

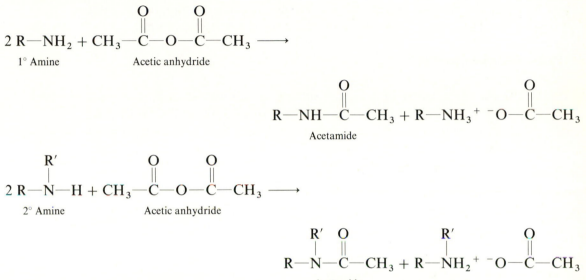

PROCEDURE

CAUTION: Acetic anhydride is a toxic and irritating substance. Avoid breathing it and getting it on your skin.

Pyridine, if used, is also toxic and has an extremely disagreeable odor. Avoid direct contact with it and use it only in the hood.

Water-soluble amines

Step 1. Place 10 drops of acetic anhydride in a test tube. Then add 4 to 5 drops of a liquid sample or about 100 mg of a solid sample.

Step 2. Heat the mixture for about 5 minutes, cool, and add 5 mL of water. Stir vigorously to precipitate the product and hydrolyze any excess acetic anhydride.

Step 3. If no product crystallizes, cool further in an ice bath, and scratch the sides of the test tube with a stirring rod.

Step 4. Collect the crystals by suction filtration on a Hirsch funnel, and wash them with several small portions of cold 5% HCl.

Step 5. To recrystallize the product, dissolve it in the minimum amount of hot 95% ethanol, add hot water dropwise until the solution becomes cloudy, then add hot 95% ethanol dropwise until the mixture is again clear. Cool the solution and collect the crystals by suction filtration.

Step 6. Dry the product and determine its melting point. Compare the value with those listed in either Table 38-11 or 38-12 for the possible amines.

Water-insoluble amines

Step 1. Dissolve 5 drops of a liquid sample or about 100 mg of a solid sample in 2 mL of pyridine in a 25-mL round-bottom flask in the hood. Add 10 drops of acetic anhydride, attach a reflux condenser, and reflux the mixture for 1 hour.

Step 2. After refluxing, cool the reaction mixture and extract it with 10 mL of 5% H_2SO_4 to remove the pyridine.

Step 3. Place the remaining organic material in a test tube. If necessary, induce crystallization by scratching the inside of the test tube with a glass rod.

Step 4. To recrystallize the product, dissolve it in the minimum amount of hot 95% ethanol, add hot water dropwise until the solution becomes cloudy, then add hot 95% ethanol dropwise until the mixture is again clear. Cool the solution and collect the crystals by suction filtration on a Hirsch funnel.

Step 5. Dry the product and determine its melting point. Compare the value with those listed in either Table 38-11 or 38-12 for the possible amines.

38.29C. Benzenesulfonamide derivative

Most primary and secondary amines react with benzenesulfonyl chloride to form solid benzenesulfonamide derivatives. However, tertiary amines do not.

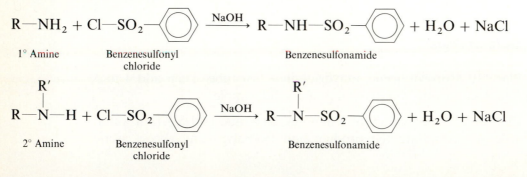

PROCEDURE

CAUTION: Benzenesulfonyl chloride is a lachrymator and is toxic. Avoid direct contact with it and use it only in the hood.

Step 1. Place 0.2 gram of the amine in a test tube. Then add 0.5 mL of benzenesulfonyl chloride and 5 mL of 10% NaOH solution.

Step 2. Stopper the tube and shake it for 3 to 5 minutes.

Step 3. Cautiously acidify the reaction mixture with 6 M HCl until pH paper indicates a pH of 5.

Step 4. Collect the precipitate by suction filtration on a Hirsch funnel and wash it with several small portions of cold water.

Step 5. To recrystallize the product, dissolve it in the minimum amount of hot 95% ethanol, add hot water dropwise until the solution becomes cloudy, then add hot 95% ethanol dropwise until the mixture is again clear. Cool the solution and collect the product by suction filtration.

Step 6. Dry the product and determine its melting point. Compare the value with those listed in Table 38-11 or 38-12 for the possible amines.

38.29D. Picrate derivative

Many primary, secondary, and tertiary amines react with picric acid to form a picrate salt having a measurable, sharp melting point. The reaction with tertiary amines is shown below.

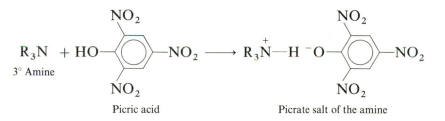

Picric acid   Picrate salt of the amine

PROCEDURE

CAUTION: Picric acid when dry is a potential explosive. Be sure it always remains moist or dissolved in solution.

Step 1. In a 25-mL Erlenmeyer flask, dissolve 10 drops of a liquid sample or about 0.2 gram of a solid sample in about 5 mL of 95% ethanol. Add 5 mL of a saturated solution of picric acid in 95% ethanol.

Step 2.   Heat the solution to boiling; then allow it to slowly cool.

Step 3.   Collect the product by suction filtration on a Hirsch funnel, and wash it with a small amount of cold 95% ethanol. Recrystallization is usually not necessary.

Step 4.   Dry the product and determine its melting point. Compare the value with those listed in Table 38-11, 38-12, or 38-13 for the possible amines.

38.29E. Methiodide derivative of tertiary amines

Tertiary amines frequently react with methyl iodide to form a methiodide derivative having a measurable, sharp melting point.

$$R_3N \ + \ CH_3{-}I \ \longrightarrow \ R_3\overset{+}{N}{-}CH_3 \ I^-$$

| 3° Amine | Methyl iodide | Methiodide |

PROCEDURE

CAUTION: Methyl iodide is toxic and a suspected carcinogen. Avoid breathing it and getting it on your skin.

Step 1.   In a large test tube, mix 10 drops of a liquid sample or about 0.3 gram of a solid sample with 10 drops of methyl iodide.

Step 2.   Allow the reaction mixture to stand for 5 minutes. Then vertically clamp the test tube in the hood, and place a smaller test tube containing an ice-water mixture about halfway into the larger one. Clamp the smaller test tube to hold it in place.

Step 3.   Reflux the reaction mixture on a steam bath for about 5 minutes, then allow it to cool.

Step 4.   If crystallization has not occurred, cool further in an ice bath and scratch the sides of the test tube with a stirring rod.

Step 5.   Collect the crystals by suction filtration on a Hirsch funnel, and recrystallize them from 95% ethanol.

Step 6.   Collect and dry the purified product, and determine its melting point. Compare the value with those listed in Table 38-13 for the possible amines.

There are no convenient derivatives of amides. Amides may be hydrolyzed to yield the carboxylic acid and the amine or ammonia. Derivatives of one or both hydrolysis products can then be made to identify the original amide.

$$R-\overset{\overset{\displaystyle O}{\|}}{C}-NHR' + H_2O \xrightarrow{H_2SO_4} R-\overset{\overset{\displaystyle O}{\|}}{C}-OH + R'-NH_2$$

PROCEDURE

Step 1.   Add 1 gram or 1 mL of the amide to 10 mL of 6 M sulfuric acid in a 25-mL round-bottom flask. Attach a reflux condenser and reflux the mixture for 1 hour.

Step 2.   Cool the reaction mixture in ice, then pour into 20 mL of cold water. If a solid carboxylic acid separates, collect it by suction filtration on a Hirsch funnel, wash it with cold water, and dry. Take its melting point to help determine its structure and, if necessary, prepare a derivative or derivatives (see Section 38.28). Go to step 6.

Step 3.   If no precipitate formed in step 2, extract the acidic, aqueous solution three times with 10-mL portions of ether. Combine the ether extracts, saving the remaining aqueous solution for step 6.

Step 4.   Dry the combined ether extracts with anhydrous magnesium sulfate, filter or decant the ether, then separate the ether from the carboxylic acid by simple distillation.

Step 5.   Identify the carboxylic acid by preparing a derivative or derivatives (see Section 38.28).

Step 6.   Add 10% NaOH solution to the aqueous solution from step 2 or step 3 and **carefully** note the odor. If an inorganic precipitate forms, add more water to dissolve it.

Step 7.   If an insoluble amine is present or if an intermediate molecular weight amine is suspected from the odor, extract the aqueous solution three times with 10-mL portions of ether.

Step 8.   Dry the combined ether extracts with anhydrous magnesium sulfate, filter or decant the ether, then separate the ether from the amine by simple distillation. Note that an aliphatic amine of low molecular weight will probably be lost during these steps, due to its solubility in water and its volatility during the removal of the ether.

Step 9. Identify the amine, if possible, by preparing a derivative or derivatives (see Section 38.29).

38.31 ESTERS (Table 38-15, page 816)

There are no convenient derivatives of esters. Esters may be hydrolyzed to yield the carboxylic acid and alcohol. These may then be separated and derivatives prepared to identify the original ester (see Section 38.15B).

$$R-\overset{\overset{\displaystyle O}{\|}}{C}-OR' + H_2O \xrightarrow{OH^-} R-\overset{\overset{\displaystyle O}{\|}}{C}-OH + R'-OH$$

38.32 ETHERS (Table 38-16, page 818)

There are no convenient derivatives of aliphatic ethers. Any aliphatic ethers having a similar boiling point may have their identity confirmed by a ^{1}H NMR spectrum, if a spectrometer is available.

Aromatic ethers may be identified by preparing the following derivatives.

38.32A. Picrate derivative of aromatic ethers

Many aromatic ethers are sufficiently electron-rich in the aromatic ring to react with picric acid to form a picric acid complex having a measurable, sharp melting point.

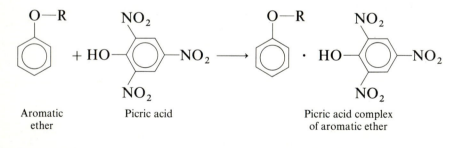

Aromatic ether   Picric acid   Picric acid complex of aromatic ether

PROCEDURE

Follow the procedure given in Section 38.29D, using the same amount of the aromatic ether instead of the tertiary amine. **Be sure to follow the safety precaution indicated beforehand.** Compare the melting point obtained with those listed in Table 38-16 for the possible aromatic ethers. The melting point should be taken as soon as possible, since some picrate complexes decompose on standing.

38.32B. Bromination derivative of aromatic ethers

Aromatic ethers, being activated toward electrophilic aromatic substitution, readily undergo ring bromination. If adequate bromine is present, sometimes all unsubstituted positions ortho and para to the ether group will be attacked, unless there is steric hindrance. For example:

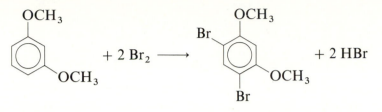

PROCEDURE

CAUTION: Bromine causes severe burns. Be careful not to get it on your skin. Bromine burns may be treated by flushing the affected area with plenty of water and applying a wet dressing of 10% sodium thiosulfate.

Glacial acetic acid can also cause burns. Be careful not to get it on your skin.

Step 1. Check to see if the melting point of a bromo derivative is given for the possible ethers.

Step 2. If so, dissolve 10 drops of a liquid sample or about 200 mg of a solid sample in 2 mL of glacial acetic acid.

Step 3. Add a solution of 10% bromine in glacial acetic acid in 10-drop portions until the yellow color persists for at least 1 minute after mixing.

Step 4. Add 5 mL of water, mix well, and collect the precipitate by suction filtration on a Hirsch funnel. Wash the solid with several small portions of cold water.

Step 5. To recrystallize the product, dissolve it in the minimum amount of hot 95% ethanol, add hot water dropwise until the solution becomes cloudy, then add hot 95% ethanol until the mixture is again clear. Allow the hot solution to slowly cool, and collect the product by suction filtration.

Step 6. Dry the product and determine its melting point. Compare the value with those listed in Table 38-16 for the possible ethers.

38.32C. Nitration derivative of aromatic ethers

Aromatic ethers readily undergo nitration at the ortho and para positions. But since the added nitro group greatly deactivates the ring, generally no more than one to two groups will enter the ring at ice-bath temperatures. For example:

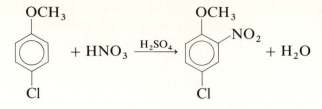

PROCEDURE

CAUTION: Remember that both concentrated nitric and sulfuric acids cause severe burns. Handle them with extreme care. Nitric acid will also turn the skin yellow.

Step 1. Check to see if the melting point of a nitro derivative is given for the possible ethers.

Step 2. If so, place 10 drops of a liquid sample or about 200 mg of a solid sample into a test tube. Add 20 drops of concentrated sulfuric acid and cool the mixture in an ice bath.

Step 3. Over a period of 10 minutes, add 5 drops of concentrated nitric acid to the cold reaction mixture, swirling the mixture in the ice bath after each addition.

Step 4. Allow the mixture to stand for another 10 minutes in the ice bath; then **carefully** add 5 mL of a mixture of ice and water.

Step 5. Collect the precipitate by suction filtration on a Hirsch funnel and wash with several small portions of cold water.

Step 6. To recrystallize the product, dissolve it in the minimum amount of hot 95% ethanol, add hot water dropwise until the solution becomes cloudy, then add hot 95% ethanol dropwise until the mixture is again clear. Allow the hot solution to slowly cool, and collect the product by suction filtration.

Step 7. Dry the product and determine its melting point. Compare the value with those given in Table 38-16 for the possible ethers.

38.33 NITRILES (Table 38-17, page 819)

There are no convenient derivatives of nitriles. Nitriles may be hydrolyzed to yield the carboxylic acid. Derivatives of the carboxylic acid can then be made to identify it and the original nitrile.

$$R-C\equiv N + 2\,H_2O \xrightarrow{H_2SO_4} R-\overset{\displaystyle O}{\overset{\|}{C}}-OH + NH_3$$

Nitrile   Carboxylic acid

PROCEDURE

Follow the first five steps of the procedure given in Section 38.30 for amides. Substitute 1 gram or 1 mL of the nitrile for the amide.

38.34 ALKYL HALIDES (Table 38-19, page 820)

38.34A. Anilide derivative

Many alkyl halides will react with magnesium metal in anhydrous ether to give the Grignard reagent. Reaction of the Grignard reagent with phenylisocyanate by nucleophilic addition, followed by a reaction workup with aqueous acid then yields the anilide derivative.

(1)
$$R-X + Mg \xrightarrow[\text{ether}]{\text{dry}} R-MgX$$

Alkyl   Grignard
halide   reagent

(2)
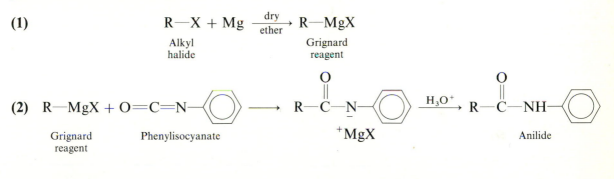

Grignard   Phenylisocyanate ^+MgX Anilide
reagent

PROCEDURE

CAUTION: Phenylisocyanate is very toxic. Perform the second reaction only in the hood, avoiding direct contact with the reagent.

Step 1. Place 1 mL of the alkyl halide into a clean, **dry** test tube. Add 0.3 gram of magnesium turnings, 5 mL of anhydrous ether, and an iodine crystal. Gently warm the reaction to get it started, but be prepared to immerse the test tube in cold water if the reaction begins to get out of control. Add extra ether, if necessary, to replace what may have evaporated.

Step 2. In the hood, prepare a solution of 0.5 mL of phenylisocyanate in 10 mL of absolute ether. When the reaction in step 1 is over, pour the reaction mixture into the phenylisocyanate solution. Swirl the mixture and allow it to stand for 10 minutes.

Step 3. Add to the reaction mixture 20 mL of cold 1 M HCl and stir vigorously. When the reaction is completed, separate the ether layer and dry it over anhydrous magnesium sulfate. Filter or decant the ether, and separate it from the amide by simple distillation.

Step 4. Dissolve the product in about 5 to 6 mL of hot hexane, and gravity filter the mixture, using a preheated funnel. This will remove any insoluble 1,3-diphenylurea, which may have been formed by the reaction of phenylisocyanate with water (see Section 38.27C).

Step 5. Cool the filtrate and collect the resulting precipitate by suction filtration on a Hirsch funnel.

Step 6. Dry the product and determine its melting point. Compare the value with those listed in Table 38-19 for the possible alkyl halides.

38.34B. 1-Naphthalide derivative

Many alkyl halides will react with magnesium metal in anhydrous ether to give the Grignard reagent. Reaction of the Grignard reagent with 1-naphthylisocyanate by nucleophilic addition, followed by a reaction workup with aqueous acid then yields the 1-naphthalide derivative.

(1) R—X + Mg $\xrightarrow[\text{ether}]{\text{dry}}$ R—MgX

      Alkyl   Grignard
      halide   reagent

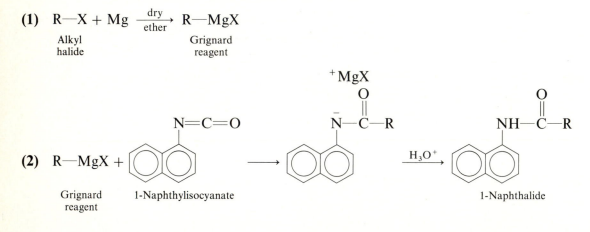

(2) R—MgX +

   Grignard   1-Naphthylisocyanate   1-Naphthalide
   reagent

PROCEDURE

Follow the procedure given in Section 38.34A, substituting 1-naphthylisocyanate for phenylisocyanate. **Remember that 1-naphthylisocyanate is very toxic. Perform the second reaction only in the hood, avoiding direct contact with the reagent.**

38.35 ARYL HALIDES (Table 38-20, page 822)

38.35A. Nitration derivative

Aryl halides undergo nitration at the ortho and para positions. Since the added nitro group greatly deactivates the ring, generally no more than one to two groups will enter the ring. For example:

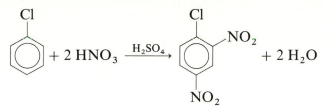

PROCEDURE

CAUTION: Remember that both fuming nitric acid and concentrated sulfuric acid cause severe burns. Handle them with extreme care. Nitric acid will also turn the skin yellow.

Step 1. Add 2 mL of concentrated sulfuric acid to 0.5 gram of the aryl halide in a test tube and mix.

Step 2. Add 2 mL of fuming nitric acid dropwise with mixing, cooling the reaction mixture in a beaker of water, if necessary.

Step 3. After all the fuming nitric acid has been added, warm the reaction mixture in a steam bath for 10 minutes.

Step 4. Cool the reaction mixture; then **carefully** pour it into 15 mL of a mixture of ice and water. Collect the resulting precipitate by suction filtration on a Hirsch funnel, and wash it with several small portions of cold water.

Step 5. Recrystallize the product from 95% ethanol, collect it by suction filtration, and dry.

Step 6. Determine the melting point and compare it with those listed in Table 38-20 for the possible aryl halides.

38.35B. Oxidation derivative

Aryl halides containing alkyl side chains may have the alkyl group oxidized to an aromatic carboxylic acid by alkaline potassium permanganate. If more than one carbon is present in the side chain, all the extra carbons are oxidized to carbon dioxide. For example:

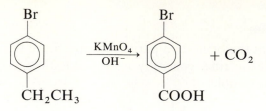

PROCEDURE

Step 1. In a 50-mL round-bottom flask, place 2.0 grams of $KMnO_4$, 25 mL of water, 5 mL of 10% NaOH, and 0.5 gram or 0.5 mL of the aryl halide.

Step 2. Attach a reflux condenser and reflux the mixture over a hot plate for 1 hour or until the purple color is gone.

Step 3. Cool the reaction mixture and **carefully** add 3 M sulfuric acid until the mixture is acid to litmus.

Step 4. Remove the brown manganese dioxide precipitate by adding a small amount of sodium bisulfite solution with stirring. Reheat the reaction mixture, if necessary.

Step 5. Cool the solution, collect the carboxylic acid by suction filtration on a Hirsch funnel, and wash the crystals with several small portions of cold water.

Step 6. To recrystallize the product, dissolve it in the minimum amount of hot 95% ethanol, add hot water dropwise until the solution becomes cloudy, then add hot 95% ethanol dropwise until the mixture is again clear. Cool the solution and collect the resulting precipitate by suction filtration.

Step 7. Dry the product and determine its melting point. Compare the value obtained with those listed in Table 38-20 for the possible aryl halides.

38.36   NITRO COMPOUNDS (Table 38-21, page 823)

38.36A. Reduced amine derivative

Nitro compounds will yield an amine derivative by chemical reduction with tin and hydrochloric acid. For example:

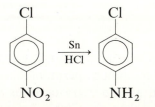

PROCEDURE

Step 1. In a 25-mL round-bottom flask, place 0.5 gram or 0.5 mL of the sample and 1 gram of tin. Then add 10 mL of 10% hydrochloric acid in portions with stirring.

Step 2. Attach a reflux condenser and reflux the mixture over a hot plate for 30 minutes. For very insoluble nitro compounds, 3 mL of 95% ethanol may be added to hasten the reaction.

Step 3. Add 5 mL of water to the reaction mixture, cool thoroughly, and then add 40% sodium hydroxide solution with stirring until the tin hydroxide dissolves.

Step 4. Extract the reaction mixture three times with 10-mL portions of ether, dry the combined ether extracts over anhydrous magnesium sulfate, decant or filter the ether, then evaporate the ether to leave the liquid or solid amine.

Step 5. Determine the melting point or boiling point of the amine, and compare the value with those listed in Table 38-21 for the possible nitro compounds.

38.36B. Acyl derivative of the reduced amine

The amine derivative produced above may be converted to an acyl derivative by following the procedure given in Sections 38.29A, B, or C. **Be sure to follow the safety precautions indicated.** Determine the melting point of the acyl derivative, and compare it with the values listed in Table 38-21 for the possible nitro compounds.

38.36C. Nitro derivative

Some aromatic nitro compounds can be further nitrated in the aromatic ring to yield a derivative. For example:

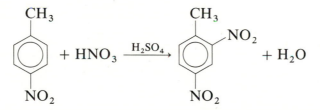

PROCEDURE

First check to see if a nitro derivative is given for the possible nitro compounds in Table 38-21. If so, follow the procedure given in Section 38.35A. **Be sure to note the safety precaution indicated beforehand.**

38.37 AROMATIC HYDROCARBONS (Table 38-22, page 824)

38.37A. Nitration derivative

Aromatic hydrocarbons will react with a mixture of concentrated nitric acid and sulfuric acid to yield nitro derivatives. For example:

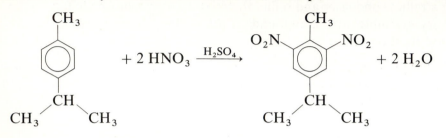

PROCEDURE

The best nitration procedure depends upon how many nitro groups are entering the ring. If the nitro derivative in Table 38-22 indicates only one nitro group per ring, you should follow the procedure in Section 38.32C. But if the nitro derivative has two or more nitro groups per ring, you should instead follow the more vigorous reaction procedure in Section 38.35A. **In either case, be very careful when handling concentrated nitric acid and sulfuric acid. They cause severe burns.**

38.37B. Picrate derivative

Many aromatic hydrocarbons are sufficiently electron-rich in the aromatic ring to react with picric acid to form a picric acid complex having a measurable, sharp melting point.

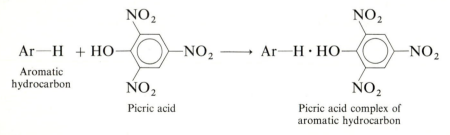

PROCEDURE

Follow the procedure given in Section 38.29D, using the same amount of the aromatic hydrocarbon instead of the tertiary amine. **Be sure to follow the safety precaution indicated beforehand.** Compare the melting point obtained with those listed in Table 38-22 for the possible aromatic hydrocarbons. The melting point should be taken as soon as possible, since some picrate complexes decompose on standing.

TABLE 38-6.

Aldehydes

Compound	BP	MP	Semi-carbazone	Phenyl-hydrazone	2,4-Dinitro-phenyl-hydrazone	4-Nitro-phenyl-hydrazone
Ethanal	21	—	162	63(99)	168	129
Propanal	48	—	89(154)	—	148(156)	124
Glyoxal	50	—	270	180	328	311
Propenal	52	—	171	50	165	150
2-Methylpropanal	64	—	125(119)	—	187	132
Butanal	75	—	95(106)	—	123	92
Trimethylacetaldehyde	75	—	190	—	209	—
3-Methylbutanal	92	—	107	—	123	110
2-Methylbutanal	93	—	103	—	120	—
Chloral	98	—	90	—	131	131
Pentanal	102	—	—	—	106(98)	—
2-Butenal	104	—	199	56	190	184
2-Ethylbutanal	117	—	99	—	95(130)	—
Hexanal	130	—	106	—	104	—
Heptanal	156	—	109	—	108	73
Cyclohexanecarbox-aldehyde	161	—	173	—	172	—
2-Furaldehyde (furfural)	162	—	202	97	212(230)	54
2-Ethylhexanal	163	—	254	—	114(120)	—
3-Cyclohexenecarbox-aldehyde	165	—	155	—	—	—
Octanal	171	—	101	—	106	80
Benzaldehyde	179	—	222	158	237	261(234)
Phenylethanal	195	33	153	58	121(110)	—
2-Hydroxybenzaldehyde	197	—	231	142	248	227
2-Methylbenzaldehyde	200	—	212	101	195	222

TABLE 38-6. (continued)

Aldehydes

Compound	BP	MP	Semi-carbazone	Phenyl-hydrazone	2,4-Dinitro-phenyl-hydrazone	4-Nitro-phenyl-hydrazone
4-Methylbenzaldehyde	204	—	234(215)	114	234	200
3,7-Dimethyl-6-octenal (citronellal)	207	—	82(92)	—	77	—
2-Chlorobenzaldehyde	213	11	225(146)	86	214	249
3-Phenylpropanal	224	—	127	—	149	122
3-Bromobenzaldehyde	235	—	205	141	256	220
2-Ethoxybenzaldehyde	247	—	219	—	—	—
4-Methoxybenzaldehyde	248	2	210	120	253	160
trans-Cinnamaldehyde	250(dec)	—	215	168	255(dec)	195
1-Naphthaldehyde	—	34	221	82	254	237
3,4-Methylenedioxybenz-aldehyde (piperonal)	263	37	230	100	266(dec)	199
2-Methoxybenzaldehyde	245	38	215(dec)	94	254	204
3,4-Dimethoxybenzalde-hyde	—	44	177	121	261	—
4-Chlorobenzaldehyde	214	48	230	126	254(270)	218
4-Bromobenzaldehyde	—	57	228	113	128(257)	207
3-Nitrobenzaldehyde	—	58	246	124	293(dec)	247
4-Dimethylaminobenzal-dehyde	—	74	222	148	325	182
Vanillin (4-hydroxy-3-methoxybenzaldehyde)	285(dec)	82	230	105	271(dec)	227
3-Hydroxybenzaldehyde	—	104	198	131	257(dec)	221
4-Nitrobenzaldehyde	—	106	221(211)	159	320(dec)	249
4-Hydroxybenzaldehyde	—	116	224	177	280(dec)	266
d,l-Glyceraldehyde	—	142	160(dec)	—	167	—

TABLE 38-7.

Ketones

Compound	BP	MP	Semi-carbazone	Phenyl-hydrazone	2,4-Dinitrophenyl-hydrazone	4-Nitrophenyl-hydrazone
2-Propanone	56	—	187	42	126	152
2-Butanone	80	—	146(136)	—	117	129
2,3-Butanedione	88	—	235(mono) 278(di)	245(di)	314(di)	230(mono)
3-Methyl-2-butanone	94	—	112	—	120	108
2-Pentanone	101	—	112	—	143	117
3-Pentanone	102	—	138	—	156	141
Pinacolone	106	—	157	—	125	139
4-Methyl-2-pentanone	117	—	132	—	95	—
Chloroacetone	119	—	164(dec)	—	125	—
2,4-Dimethyl-3-pentanone	124	—	160	—	95(88)	—
2-Hexanone	128	—	125	—	106	88
5-Hexene-2-one	129	—	102	—	108	—
4-Methyl-3-penten-2-one	130	—	164	142	205	134
Cyclopentanone	131	—	210	50	146	154
2,3-Pentanedione	134	—	122(mono) 209(di)	—	209	—
2,4-Pentanedione	139	—	—	—	122(mono) 209(di)	
4-Heptanone	144	—	132	—	75	—
5-Methyl-2-Hexanone	145	—	147	—	95	—
3-Hydroxy-2-butanone	145	—	185(202)	—	318	—
2-Heptanone	147	—	123	207	89	—
3-Heptanone	149	—	101(152)	—	—	—
Cyclohexanone	156	—	166	77	162	146
6-Methyl-3-heptanone	160	—	132	—	—	—
5-Methyl-3-heptanone	160	—	102	—	—	—
2-Methyl-cyclohexanone	163	—	195	—	137	132
4-Hydroxy-4-methyl-2-pentanone	166	—	—	—	203	—
3-Octanone	167	—	112	—	67	—
2,6-Dimethyl-4-heptanone	168	—	122	—	66(92)	—

TABLE 38-7. (continued)

Ketones

Compound	BP	MP	Semi-carbazone	Phenyl-hydrazone	2,4-Dinitrophenyl-hydrazone	4-Nitrophenyl-hydrazone
3-Methyl-cyclohexanone	169	—	179(191)	94	155	119
Methyl acetoacetate	170	—	152	—	—	—
4-Methyl-cyclohexanone	171	—	199	109	130	128
2-Octanone	173	—	122	—	58	92
Methyl cyclohexyl ketone	180	—	177	—	140	154
Cycloheptanone	181	—	163	—	148	137
Ethyl acetoacetate	181	—	133	—	93	—
5-Nonanone	187	—	90	—	41	—
2,5-Hexanedione	191	—	185(mono) 224(di)	120(di)	257(di)	210(di)
Phorone	198	27	221(186)	—	118	—
Acetophenone	202	20	198	105	238(250)	184
Ethyl levulinate	206	—	148	104	102	—
l-Menthone	209	—	189	53	146	—
3,5,5-Trimethyl-cyclohexene-3-one	215	—	200	68	—	—
Phenyl-2-propanone	216	27	198	87	156	—
Ethyl phenyl ketone	218	21	182(174)	—	191	—
Isobutyl phenyl ketone	222	—	181	73	163	—
1-Phenyl-2-butanone	226	—	135(146)	—	—	—
4-Methyl-acetophenone	226	28	205	94	258	198
l-Carvone	230		162	106	191	174
2-Undecanone	231	12	122	—	63	90
4-Chloro-acetophenone	232	12	204(160)	114	236	239
Butyl phenyl ketone	232	—	188	—	—	—
4-Phenyl-2-butanone	235	—	142	—	127	—

TABLE 38-7. (continued)
Ketones

Compound	BP	MP	Semi-carbazone	Phenyl-hydrazone	2,4-Dinitrophenyl-hydrazone	4-Nitrophenyl-hydrazone
Dibenzyl ketone	330	34	146(126)	120	100	—
4-Chloro-propiophenone	—	36	176	—	223	—
4-Phenyl-3-buten-2-one	—	37	187	—	227	—
4-Methoxy-acetophenone	258	38	198	142	228	195
Ethyl benzoylacetate	265	—	125	—	—	—
1-Acetyl-naphthalene	302	—	289(233)	146	—	—
2-Benzoyl-pyridine	—	42	—	—	199	—
Benzophenone	305	48	167	137	238(229)	154(144)
4-Bromo-acetophenone	225	51	208	126	230(237)	—
2-Acetyl-naphthalene	—	54	235	176	262	—
Phenyl-*p*-tolyl ketone	—	54	122	109	200	—
Benzal-acetophenone	348	58(62)	168(179)	120	245	—
Desoxybenzoin	320	60	148	116	204	163
3-Nitro-acetophenone	202	80	257	128	228	—
4-Nitro-acetophenone	—	80	—	132	257	—
9-Fluorenone	345	83	234	151	283	269
Benzil	—	95	174(mono) 244(di)	225(di)	189(di)	—
Dibenzalacetone	—	112	190	152	180	173
4-Hydroxy-benzophenone	—	134	194	144	242	—
Benzoin	344	136	206(dec)	106	245	—
4-Hydroxy-propiophenone	—	148	—	—	229(240)	—
d,l-Camphor	205	179	237(248)	233	177(164)	217
Ninhydrin	—	243	—	207(di)	—	—

TABLE 38-8.

Alcohols

Compound	BP	MP	3,5-Dinitrobenzoate	1-Naphthylurethane	Phenylurethane
Methanol	65	—	107	124	47
Ethanol	78	—	93	79	52
2-Propanol	83	—	122	106	76(88)
2-Methyl-2-propanol	83	25	142	101	136
2-Propen-1-ol	97	—	49	108	70
1-Propanol	97	—	74	80	57
2-Butanol	99	—	75	97	65
2-Methyl-2-butanol	102	—	116	72	42
2-Methyl-3-butyn-2-ol	104	—	112	—	—
2-Methyl-1-propanol	108	—	86	104	86
2-Propyn-1-ol	114	—	—	—	63
1-Butanol	116	—	64	71	61
3-Pentanol	116	—	99	95	48
2-Pentanol	119	—	61	74	—
1-Methoxy-2-propanol	119	—	85	—	—
3-Methyl-3-pentanol	123	—	96(62)	83	43
2-Methoxy-ethanol	124	—	—	112	—
1-Chloro-2-propanol	127	—	77	—	—
2-Methyl-1-butanol	129	—	70	82	31
3-Methyl-1-butanol	131	—	61	68	55
2-Chloroethanol	131	—	95	101	51
4-Methyl-2-pentanol	132	—	65	88	143
2-Ethoxyethanol	135	—	75	67	—
3-Hexanol	135	—	97(77)	—	—
1-Pentanol	138	—	46	68	46
2-Hexanol	139	—	38	60	—
Cyclopentanol	140	—	115	118	132
2,4-Dimethyl-3-pentanol	140	—	—	95	95
2-Ethyl-1-butanol	146	—	51	—	—
2-Methyl-1-pentanol	148	—	51	75	—
2,2,2-Trichloro-ethanol	151	19	142	120	87
1-Hexanol	157	—	58	60	42

TABLE 38-8. (continued)

Alcohols

Compound	BP	MP	3,5-Dinitrobenzoate	1-Naphthylurethane	Phenylurethane
2-Heptanol	159	—	49	54	—
Cyclohexanol	161	25	112	129	82
4-Hydroxy-4-methyl-2-pentanone	166	—	55	—	—
2-Furyl-methanol	171	—	80	130	45
1-Heptanol	176	—	47	62	60(68)
Tetrahydro-furfuryl alcohol	178	—	83	—	61
2-Octanol	179	—	32	63	114
2,3-Butanediol	184	—	—	—	201(di)
2-Ethyl-1-hexanol	186	—		61	34
1,2-Propanediol	187	—	—	153(di)	—
1-Octanol	194	—	61	67	74
3,7-Dimethyl-1,6-octadiene-3-ol (linalool)	196	—	—	53	66
1,2-Ethanediol	197	—	169(di)	176(di)	157(di)
1-Phenylethanol	202	20	93	106	92
1,3-Butanediol	204	—	—	184	122
Benzyl alcohol	205	—	112	134	77
1,3-Propanediol	214	—	178(di)	164(di)	137(di)
1-Nonanol	215		52	65	62
2-Phenylethanol	219	—	108	119	78
1,4-Butanediol	230	—	—	199(di)	183(di)
1-Decanol	231	—	57	73	59
3-Phenyl-1-propanol	236	—	92	—	45
Cinnamyl alcohol	257	33	121	114	90
1-Dodecanol	259	24	60	80	74
4-Methoxy-benzyl alcohol	259	24	—	—	92
Glycerol	290(dec)	—	76(tri)	191(tri)	180(tri)
1-Tetradecanol	—	39	67	82	74
(−)-Menthol	—	43	153	119	111
1-Hexadecanol	—	49	66	82	73
2,2-Dimethyl-1-propanol	113	53	—	100	144
1-Octadecanol	—	59	77(66)	—	79
Diphenyl-methanol	288	69	141	135	139
Benzoin	—	133	—	140	165
(−)-Cholesterol	—	148	195	176	168
(−)-Borneol	—	208	154	132	138

TABLE 38-9.

Phenols

Compound	BP	MP	1-Naphthyl-urethane	Bromo Derivatives				3,5-Dinitro-benzoate
				Mono	Di	Tri	Tetra	
2-Chlorophenol	176	7	120	48	76	—	—	143
2-Bromophenol	195	—	129	—	—	95	—	—
2-Hydroxybenzaldehyde	196	—	—	—	—	—	—	—
Methyl salicylate	223	—	—	—	—	—	—	—
2-tert-Butylphenol	224	—	—	—	—	—	—	—
Ethyl salicylate	234	—	—	—	—	—	—	—
3-Methylphenol	203	12	128	—	—	84	—	165
2-Methylphenol	191	32	142	—	56	—	—	133
2-Methoxyphenol	204	32	118	—	—	116	—	141
4-Methylphenol	202	36	146	—	49	—	198(108)	188
Phenol	181	42	133	—	—	95	—	145
Phenylsalicylate	—	42	—	—	—	—	—	—
4-Chlorophenol	217	43	166	33	90	—	—	186
2,4-Dichlorophenol	210	45	—	68	—	—	—	—
2-Nitrophenol	216	45	113	—	117	—	—	155(142)
4-Ethylphenol	219	45	128	—	—	—	—	132
2,6-Dimethylphenol	—	45	176	79	—	—	—	158
2-Isopropyl-5-methyl-phenol (thymol)	234	51	160	55	—	—	—	103
3,4-Dimethylphenol	228	64	141	—	—	171	—	181
4-Bromophenol	238	64	169	—	—	95	—	191
2,4,6-Trichlorophenol	—	67	188	—	—	—	—	—
3,5-Dimethylphenol	220	68	109	—	—	166	—	195
3,5-Dichlorophenol	—	68	—	—	—	189	—	—
2,4,6-Trimethylphenol	—	69	—	—	158	—	—	—
2,5-Dimethylphenol	212	75	173	—	—	178	—	137
4-Hydroxy-3-methoxy-benzaldehyde (vanillin)	—	80	—	—	—	—	—	—
2,3,5-Trimethylphenol	—	94	174	—	—	—	—	—

TABLE 38-9. (continued)

Phenols

Compound	BP	MP	1-Naphthyl-urethane	Bromo Derivatives				3,5-Dinitro-benzoate
				Mono	Di	Tri	Tetra	
1-Naphthol	278	96	152	—	105	—	—	217
3-Nitrophenol	—	97	167	—	91	—	—	159
4-*tert*-Butylphenol	—	99	110	50	67	—	—	—
3-Hydroxybenzaldehyde	—	104	—	—	—	—	—	—
2-Hydroxyphenol	245	104	175	—	—	—	192	152(di)
3-Hydroxyphenol	281	109	275	—	—	112	—	201(di)
2-Chloro-4-nitrophenol	—	111	—	—	—	—	—	—
4-Nitrophenol	—	114	150	—	142	—	—	186
4-Hydroxybenzaldehyde	—	115	—	—	181	—	—	—
Ethyl 4-hydroxybenzoate	—	116	—	—	—	—	—	—
1,3,5-Trihydroxybenzene	—	117(dihyd)	—	—	—	151	—	162(tri)
3-Aminophenol	—	122	—	—	—	—	—	179
2-Naphthol	286	123	157	84	—	—	—	210
1,2,3-Trihydroxyphenol	309	133	173 230(tri)	—	158	—	—	205(tri)
4-Hydroxybenzophenone	—	134	—	—	—	—	—	—
Salicylanilide	—	136	—	—	—	—	—	—
Salicylamide	—	141	—	—	—	—	—	—
2-Hydroxybenzoic acid	—	159	—	—	—	—	—	—
4-Phenylphenol	305	164	—	—	—	—	—	—
4-Acetamidophenol	—	169	—	—	—	—	—	—
1,4-Dihydroxybenzene (hydroquinone)	—	171	247	—	186	—	—	317(di)
2-Aminophenol	—	174	—	—	—	—	—	—
4-Aminophenol	—	184	—	—	—	—	—	178
4-Hydroxybenzoic acid	—	215	—	—	—	—	—	—
1,3,5-Trihydroxybenzene	—	219(anhy)	—	—	—	151	—	162(tri)
Phenolphthalein	—	265	—	—	—	—	—	—

TABLE 38-10.

Carboxylic Acids

Compound	BP	MP	*p*-Toluidide	Anilide	Amide
Formic acid	101	8	53	47	43
Acetic acid	118	17	148	114	82
Propenoic acid	139	13	141	104	85
Propanoic acid	141	—	124	103	81
2-Methylpropanoic acid	154	—	104	105	128
Butanoic acid	162	—	72	95	115
2-Methylpropenoic acid	163	16	—	87	102
Trimethylacetic acid	164	35	120	127	178
Pyruvic acid	165(dec)	14	109	104	124
3-Methylbutanoic acid	176	—	109	109	135
Pentanoic acid	186	—	70	63	105
2-Methylpentanoic acid	186	—	80	95	79
2-Chloropropanoic acid	186	—	124	92	80
Dichloroacetic acid	194	6	153	118	98
Hexanoic acid	205	—	75	95	101
2-Bromopropanoic acid	205	24	125	99	123
Heptanoic acid	224	—	80	71	96
Octanoic acid	237	16	70	57	107
Nonanoic acid	254	12	84	57	99
Cyclohexylcarboxylic acid	—	30	—	146	186
Decanoic acid	268	32	78	70	108
4-Oxopentanoic acid	246	33	108	102	108(dec)
3-Chloropropanoic acid	204	41	—	—	101
Dodecanoic acid	299	43	87	78	100
3-Phenylpropanoic acid	279	48	135	92	105(82)
Bromoacetic acid	208	50	—	131	91
Tetradecanoic acid	—	54	93	84	103
Trichloroacetic acid	198	57	113	97	141
Hexadecanoic acid	—	62	98	90	106
Chloroacetic acid	189	63	162	137	121
3,3-Dimethylacrylic acid	—	69	—	126	107
Octadecanoic acid	—	69	102	95	109
trans-2-Butenoic acid	—	72	132	118	158
Phenylacetic acid	—	77	136	118	156
Glycolic acid	—	80	143	97	120
Pentanedioic acid	—	97	—	224(di)	175(di)
Phenoxyacetic acid	—	99	—	99	101
2-Methoxybenzoic acid	200	101	—	131	129

TABLE 38-10. (continued)

Carboxylic Acids

Compound	BP	MP	*p*-Toluidide	Anilide	Amide
Oxalic acid dihydrate	—	102	169(mono) 268(di)	148(mono) 257(di)	219(mono) 400(dec)(di)
2-Methylbenzoic acid	—	104	144	125	142
Nonanedioic acid	—	104	201(di)	107(mono) 186(di)	93(mono) 175(di)
3-Methylbenzoic acid	263	110	118	126	94
d,l-Phenylhydroxyacetic acid	—	120	172	151	133
Benzoic acid	249	122	158	163	130
2-Benzoylbenzoic acid	—	127	—	195	165
2-Furoic acid	—	129	107	123	143
Maleic acid	—	133	142(di)	198(mono) 187(di)	172(mono) 260(di)
Decanedioic acid	—	133	201(di)	122(mono) 200(di)	170(mono) 210(di)
trans-Cinnamic acid	300	133	168	153	147
d,l-Malic acid	—	133	178(mono) 207(di)	155(mono) 198(di)	163(di)
Acetylsalicylic acid	—	136	136	136	138
Malonic acid	—	137	86(mono) 253(di)	132(mono) 230(di)	170(di)
meso-Tartaric acid	—	140	—	193(mono)	190(di)
2-Chlorobenzoic acid	—	140	131	118	139
3-Nitrobenzoic acid	—	140	162	155	143(174)
2-Aminobenzoic acid	—	146	151	131	109
Diphenylacetic acid	—	148	172	180	167
2-Bromobenzoic acid	—	150	—	141	155
Benzilic acid	—	150	190	175	154
Hexanedioic acid	—	152	239	151(mono) 241(di)	125(mono) 220(di)
Citric acid	—	153	189(tri)	199(tri)	210(tri)
4-Chlorophenoxyacetic acid	—	158	—	125	133
3-Chlorobenzoic acid	—	158	—	122	134
2-Hydroxybenzoic acid	—	159	156	136	142
2-Iodobenzoic acid	—	162	—	141	110
2,4-Dichlorobenzoic acid	—	162	—	—	194
1-Naphthoic acid	—	162	—	161	205
5-Bromo-2-hydroxybenzoic acid	—	165	—	222	232
Methylenesuccinic acid	—	166(dec)	—	152(mono)	191(di)
d-Tartaric acid	—	173	—	180(mono) 264(di)	171(mono) 196(di)
4-Chloro-3-nitrobenzoic acid	—	180	—	131	156

TABLE 38-10. (continued)

Carboxylic Acids

Compound	BP	MP	*p*-Toluidide	Anilide	Amide
4-Methylbenzoic acid	—	180	160	145	160
3,4-Dimethoxybenzoic acid	—	182	—	154	164
4-Methoxybenzoic acid	280	184	186	169	167
Butanedioic acid	235(dec)	188	180(mono) 255(di)	143(mono) 230(di)	157(mono) 260(di)
4-Aminobenzoic acid	—	189	—	—	—
N-Benzoylglycine	—	190	208	—	183
β-Alanine	—	200(dec)	—	—	—
3-Hydroxybenzoic acid	—	201	163	157	170
3,5-Dinitrobenzoic acid	—	202	—	234	183
Phthalic acid	—	210(dec)	150(mono) 201(di)	169(mono) 253(di)	149(mono) 220(di)
4-Hydroxybenzoic acid	—	214	204	197	162
2,4-Dihydroxybenzoic acid	—	217	—	126	228
Glycine	—	230(dec)	—	—	—
Pyridine-3-carboxylic acid	—	238	150	132	128
4-Nitrobenzoic acid	—	240	204	211	201
4-Chlorobenzoic acid	—	242	—	194	179(170)
Fumaric acid	—	293	—	233(mono) 314(di)	270(mono) 266(di)
Terephthalic acid	—	300	—	334(di)	>225(di)

TABLE 38-11. ▰▰
Primary Amines

Compound	BP	MP	Benz-amide	Picrate	Acet-amide	Benzene-sulfonamide
Isopropylamine	34	—	71(100)	165	—	26
t-Butylamine	46	—	134	198	101	—
Propylamine	48	—	84	135	—	36
Allylamine	56	—	—	140	—	39
sec-Butylamine	63	—	76	139	—	70
Isobutylamine	69	—	57	150	—	53
2-Amino-2-methylbutane	77	—	—	183	—	—
Butylamine	78	—	42	151	—	—
Isoamylamine	95	—	—	138	—	—
Amylamine	104	—	—	139	—	—
Ethylenediamine	118	—	244(di)	233	172(di)	168(di)
1-Aminohexane	129	—	40	126	—	96
Cyclohexylamine	135	—	149	—	104	89
2-Aminoheptane	143	—	—	—	—	—
Furfurylamine	145	—	—	150	—	—
1-Aminoheptane	155	—	—	121	—	—
1-Amino-2-propanol	163	—	—	142	—	—
2-Fluoroaniline	176	—	113	—	80	—
Benzylamine	184	—	105	194	60	88
Aniline	184	—	163	198	114	112
4-Fluoroaniline	186	—	185	—	152	—
1-Phenylethylamine	187	—	120	—	57	—
2-Phenylethylamine	198	—	116	174	114	69
2-Methylaniline	200	—	144	213	110	124
3-Methylaniline	203	—	125	200	65	95
2-Chloroaniline	208	—	99	134	87	129
2-Ethylaniline	210	—	147	194	111	—
4-Ethylaniline	216	—	151	—	94	—
2,6-Dimethylaniline	216	11	168	180	177	—
2,4-Dimethylaniline	218	—	192	209	133	130
2,5-Dimethylaniline	218	—	140	171	139	—
2-Methoxyaniline	225	6	60	200	85	89
3-Chloroaniline	230	—	120	177	74	121
2-Ethoxyaniline	231	—	104	—	79	102
3-Chloro-4-methylaniline	242	—	122	—	105	—
4-Ethoxyaniline	250	2	173	69	137	143
4-Chloro-2-methylaniline	241	29	142	—	140	—

TABLE 38-11. (continued)

Primary Amines

Compound	BP	MP	Benz-amide	Picrate	Acet-amide	Benzene-sulfonamide
4-Methylaniline	200	43	158	182	147	120
2-Aminobiphenyl	—	49	102(86)	—	119	—
2,5-Dichloroaniline	251	50	120	86	132	—
1-Naphthylamine	—	50	160	163(181)	159	167
4-Aminobiphenyl	—	53	230	—	171	
4-Methoxyaniline	240	58	154	170	130	95
2-Aminopyridine	—	60	165(di)	216(233)	71	165(di)
4-Bromoaniline	245(dec)	64	204	180	168	134
2,4,5-Trimethylaniline	—	64	167	—	162	
3-Phenylenediamine	—	66	125(mono) 240(di)	184	87(mono) 191(di)	—
4-Iodoaniline	—	67	222	—	184	—
4-Chloroaniline	232	70	192	178	179	122
2-Nitroaniline	—	72	110(98)	73	92	104
Ethyl p-aminobenzoate	—	89	148	—	110	—
2,4-Diaminotoluene	—	99	224(di)	—	224(di)	178
2-Phenylenediamine	258	102	301(di)	208	185(di)	185(di)
2-Methyl-5-nitroaniline	—	106	186	—	151	172
2-Chloro-4-nitroaniline	—	108	161	—	139	—
3-Nitroaniline		114	157	143	155(76)	136
4-Methyl-2-nitroaniline	—	116	148	—	99	102
4-Chloro-2-nitroaniline	—	118	—	—	104	—
2,4,6-Tribromoaniline	—	120	200	—	232	—
3-Aminophenol	—	122	174(198) (153)	—	148(mono) 101(di)	—
2-Methyl-4-nitroaniline		132	—	—	202	158
2-Methoxy-4-nitroaniline	—	140	149	—	153	181
4-Phenylenediamine	267	140	128(mono) 300(di)	—	162(mono) 304(di)	247(di)
2-Aminobenzoic acid	—	147	182	104	—	214
4-Nitroaniline	—	148	199	100	215	139
4-Aminoacetanilide	—	164	—	—	304	—
3-Aminobenzoic acid	—	173	248	—	—	—
2-Aminophenol	—	174	182	—	201	141
2,4-Dinitroaniline	—	180	202(220)	—	120	—
4-Aminophenol	—	184	234(di)	—	168(mono) 150(di)	125
4-Aminobenzoic acid	—	189	278	—	—	—
β-Alanine	—	200(dec)	120	—	—	—
Glycine	—	230(dec)	188	202	—	—

TABLE 38-12.

Secondary Amines

Compound	BP	MP	Benzamide	Picrate	Acetamide	Benzenesulfonamide
Diethylamine	56	—	42	155	—	42
Diisopropylamine	84	—	—	140	—	—
Pyrrolidine	88	—	—	112(163)	—	51
Piperidine	106	—	48	152	—	93
Dipropylamine	110	—	—	75	—	51
Morpholine	129	—	75	146	—	118
Diisobutylamine	139	—	—	121	86	55
N-Methylcyclohexylamine	148	—	85	170	—	—
Dibutylamine	159	—	—	59	—	—
Indole-3-acetic acid	168	—	—	—	—	—
Benzylmethylamine	184	—	—	117	—	—
N-Methylaniline	196	—	63	145	102	79
Diamylamine	205	—	—	—	—	—
N-Ethylaniline	205	—	60	138(132)	54	—
N-Ethyl-*m*-toluidine	221	—	72	—	—	—
Dicyclohexylamine	256	—	153(57)	173	103	—
Dibenzylamine	300	—	112	—	—	68
N-Benzylaniline	298	37	107	48	58	119
Indole	254	52	68	—	157	254
Diphenylamine	302	52	180	182	101	124
N-Phenyl-1-naphthylamine	335	62	152	—	115	—
N-Phenyl-2-naphthylamine	—	108	148(136)	—	93	—
4-Nitro-*N*-methylaniline	—	152	112	—	153	121

TABLE 38-13.

Tertiary Amines

Compound	BP	MP	Picrate	Methiodide
Triethylamine	89	—	173	280
Pyridine	115	—	167	117
2-Methylpyridine	129	—	169	230
2,6-Lutidine	143	—	168(161)	233
4-Methylpyridine	143	—	167	152
3-Methylpyridine	144	—	150	92(36)
Tripropylamine	157	—	116	207
2,4-Lutidine	159	—	180	113
2,4,6-Collidine	172	—	156	—
2-Methyl-4-ethylpyridine	173	—	123	—
N,N-Dimethylbenzylamine	183	—	93	179
N,N-Dimethylaniline	193	—	163	228(dec)
Tributylamine	216	—	105	186
N,N-Diethylaniline	217	—	142	102
Quinoline	237	—	203	133(72)
Isoquinoline	243	—	222	159
2-Benzoylpyridine	—	42	130	—
Tribenzylamine	—	91	190	184

TABLE 38-14.

Amides

Compound	BP	MP	Compound	BP	MP
N,N-Dimethylformamide	153	—	o-Toluanilide	—	125
N,N-Dimethylacetamide	165	—	m-Toluanilide	—	126
N,N-Diethylformamide	176	—	p-Methoxyacetanilide	—	127
Formamide	195(dec)	—	Benzamide	—	128
N-Methylacetamide	205	27	Isobutyramide	—	128
Acetylpiperidine	226	—	Nicotinamide	—	132
N-Methylformanilide	244	—	m-Chlorobenzamide	—	134
Formanilide	—	48	Salicylamide	—	140
Benzoylpiperidine	—	48	o-Toluamide	—	140
N-n-Propylacetanilide	—	50	o-Bromobenzanilide	—	141
m-Acetotoluide	—	66	Furamide	—	142
ε-Caprolactam	—	70	o-Benzotoluidide	—	142
Propionamide	—	81	m-Nitrobenzamide	—	143
Acetamide	—	82	p-Toluanilide	—	145
Acrylamide	—	84	Phenylurea	—	147
2-Chloroacetanilide	—	88	Cinnamanilide	—	151
2-Nitroacetanilide	—	92	p-Acetamidotoluene	—	153
Heptamide	—	94	m-Nitrobenzanilide	—	154
Butyranilide	—	95	o-Bromobenzamide	—	155
Methylurea	—	101	m-Nitroacetanilide	—	155
Caproamide	—	101	p-Benzotoluidide	—	158
N-Methylacetanilide	—	102	N-α-Naphthylacetamide	—	159
Propionanilide	—	103	p-Toluamide	—	160
Isovaleranilide	—	109	Benzanilide	—	163
Methacrylamide	—	111	p-Bromoacetanilide	—	167
2-Acetamidotoluene	—	112	p-Acetamidophenol	—	169
Acetanilide	—	114	p-Chloroacetanilide	—	179
Butyramide	—	115	p-Chlorobenzanilide	—	194
2-Chloroacetamide	—	117	p-Nitrobenzamide	—	201

TABLE 38-15.
Esters

Compound	BP	MP	Compound	BP	MP
Methyl formate	34	—	Methyl crotonate (trans)	119	—
Ethyl formate	54	—	Ethyl butyrate	120	—
Methyl acetate	57	—	Propyl propionate	122	—
Ethyl trifluoroacetate	61	—	Butyl acetate	127	—
Vinyl acetate	72	—	Methyl pentanoate	128	—
Ethyl acetate	77	—	Methyl chloroacetate	130	—
Methyl propionate	77	—	Ethyl 3-methylbutanoate (ethyl isovalerate)	132	
Methyl acrylate	80	—			
Propyl formate	81	—	Isobutyl propionate	137	—
Isopropyl formate	90	—	Isoamyl acetate	142	—
Ethyl chloroformate	93	—	Ethyl chloroacetate	143	—
Methyl isobutyrate	93	—	Propyl butyrate	143	—
Isopropenyl acetate	94	—	Butyl propionate	146	—
sec-Butyl formate	97	—	Amyl acetate	149	—
t-Butyl acetate	98	—	Ethyl lactate	154	—
Isobutyl formate	98	—	Ethyl pyruvate	155	—
Ethyl acrylate	99	—	Butyl butyrate	165	—
Ethyl propionate	99	—	Ethyl trichloroacetate	168	—
Methyl methacrylate	100	—	Ethyl hexanoate	168	—
Methyl trimethylacetate (methyl pivalate)	101	—	Methyl acetoacetate	170	—
			Cyclohexyl acetate	175	—
Propyl acetate	102	—	2-Furylmethyl acetate	176	—
Methyl butyrate	102	—	Dimethyl malonate	180	—
Allyl acetate	103	—	Ethyl acetoacetate	181	—
Butyl formate	107	—	Methyl 2-furoate	181	—
Ethyl isobutyrate	110	—	Diethyl oxalate	185	—
sec-Butyl acetate	111	—	Heptyl acetate	192	—
Isobutyl acetate	117	—	Dimethyl succinate	196	—
Methyl 3-methylbutanoate (methyl isovalerate)	117	—	Phenyl acetate	197	—
			Methyl benzoate	199	—

TABLE 38-15. (continued)

Esters

Compound	BP	MP	Compound	BP	MP
γ-Butyrolactone	204	—	Ethyl cinnamate (trans)	271	—
Ethyl levulinate	206	—	Dimethyl phthalate	284	—
γ-Valerolactone	207	—	Diethyl phthalate	298	—
Ethyl octanoate	207	—	Dibutyl phthalate	340	—
o-Cresyl acetate	208	—	Benzyl benzoate	323	21
Ethyl cyanoacetate	210	—	Methyl p-toluate	—	33
Ethyl benzoate	212	—	Ethyl 2-furoate	—	33
m-Cresyl acetate	212	—	Methyl cinnamate	—	36
p-Cresyl acetate	213	—	Benzyl cinnamate	—	39
methyl o-toluate	215	—	Methyl nicotinate	209	42
Benzyl acetate	216	—	Phenyl salicylate	—	42
Diethyl succinate	217	—	Methyl p-chlorobenzoate	—	44
Methyl phenylacetate	218	—	Ethyl m-nitrobenzoate	—	47
Isopropyl benzoate	218	—	1-Naphthyl acetate	—	48
Diethyl fumarate	219	—	Methyl mandelate	—	52
Methyl salicylate	222	—	Ethyl p-nitrobenzoate	—	56
(-)-Bornyl acetate	223	29	Phenyl benzoate	314	70
Diethyl maleate	225	—	2-Naphthyl acetate	—	70
Ethyl phenylacetate	229	—	p-Cresyl benzoate	—	71
Propyl benzoate	230	—	Diphenyl phthalate	—	75
Methyl m-chlorobenzoate	231	—	Methyl m-nitrobenzoate	—	78
Ethyl salicylate	234	—	Methyl p-bromobenzoate	—	81
Butyl levulinate	237	—	Ethyl p-aminobenzoate	—	90
Ethyl decanoate	244	—	Methyl p-nitrobenzoate	—	94
Butyl benzoate	249	—	Propyl p-hydroxybenzoate	—	98
Glyceryl triacetate	258	—	Cholesteryl acetate	—	116 (94)
Ethyl benzoylacetate	265	—	Ethyl p-hydroxybenzoate	—	118
Dimethyl suberate	268	—	Methyl p-hydroxybenzoate	—	128
Ethyl dodecanoate	269	—	Acetylsalicylic acid	—	136

TABLE 38-16.

Ethers

Compound	BP	MP	Picrate	Other Derivatives	
Ethyl vinyl ether	33	—	—	—	
Ethyl ether	37	—	—	—	
Tetrahydrofuran	66	—	—	—	
Isopropyl ether	68	—	—	—	
1,2-Dimethoxyethane	85	—	—	—	
Propyl ether	90	—	—	—	
Ethyl butyl ether	92	—	—	—	
Butyl vinyl ether	94	—	—	—	
Isobutyl ether	122	—	—	—	
Butyl ether	143	—	—	—	
Methoxybenzene	154	—	81	Dinitro	87
Ethoxybenzene	172	—	92	Nitro	58
4-Methyl-1-methoxybenzene	174	—	89	—	
3-Methyl-1-methoxybenzene	176	—	114	Trinitro	91
2-Chloro-1-methoxybenzene	196	—	—	Nitro	95
4-Chloro-1-methoxybenzene	202	—	—	Nitro	98
1,2-Dimethoxybenzene	207	—	57	Dibromo	92
Butyl phenyl ether	210	—	112	—	
4-Bromo-1-methoxybenzene	215	—	—	Nitro	88
2-Bromo-1-methoxybenzene	216	—	—	Nitro	106
4-Propenyl-1-methoxybenzene (anethole)	237	23	70	Tribromo	108
2-Nitro-1-methoxybenzene	273	—	—	—	
Benzyl ether	298	—	78	Dibromo	107
Phenyl ether	259	28	110	Dinitro	135
				Dibromo	54
4-Nitro-1-methoxybenzene	—	52	—	—	
1,4-Dimethoxybenzene	213	60	119	Dibromo	142
2-Methoxynaphthalene	—	75	108	Bromo	86

TABLE 38-17.

Nitriles

Compound	BP	MP	Compound	BP	MP
Acetonitrile	81	—	Benzyl cyanide	234	—
Trichloroacetonitrile	84	—	Adiponitrile	295	—
Propionitrile	97	—	4-Chlorobenzyl cyanide	—	30
Isobutyronitrile	108	—	Malononitrile	—	33
Butyronitrile	116	—	Stearonitrile	—	39
Crotononitrile (trans)	119	—	3-Chlorobenzonitrile	—	41
Chloroacetonitrile	127	—	2-Chlorobenzonitrile	232	47
Isovaleronitrile	130	—	Succinonitrile	—	57
Furonitrile	146	—	4-Methoxybenzonitrile	—	62
3-Chloropropionitrile	175	—	Diphenylacetonitrile	—	72
Benzonitrile	191	—	4-Chlorobenzonitrile	—	92
2-Tolunitrile	205	—	2-Nitrobenzonitrile	—	110
3-Tolunitrile	212	—	3-Nitrobenzonitrile	—	118
4-Tolunitrile	217	27	4-Nitrobenzonitrile	—	147

TABLE 38-18.

Alkenes

Compound	BP	MP	Compound	BP	MP
Isoprene	34	—	Styrene	146	—
Cyclopentene	44	—	1,5-Cyclooctadiene	150	—
1-Hexene	64	—	*d,l*-α-Pinene	152	—
Cyclohexene	83	—	*l*-Camphene	160	51
1-Heptene	94	—	*l*-β-Pinene	166	—
Cycloheptene	114	—	α-Methylstyrene	167	—
5-Methyl-2-norbornene	116	—	*d*-Limonene	176	—
1-Octene	123	—	1-Decene	181	—
4-Vinyl-1-cyclohexene	127	—	Indene	181	—
2,5-Dimethyl-2,4-hexadiene	133	—	1-Tetradecene	251	—
1,3-Cyclooctadiene	143	—	1-Hexadecene	274	—
Cyclooctene	146	—	*trans*-Stilbene	—	124

TABLE 38-19.
Alkyl Halides

Compound	BP	Anilide	1-Naphthalide
2-Chloropropane	36	103	—
Bromoethane	38	104	126
Dichloromethane	41	—	—
Iodomethane	43	114	160
Allyl chloride	45	114	—
1-Chloropropane	46	92	121
2-Chloro-2-methylpropane	51	128	147
1,1-Dichloroethane	57	—	—
2-Bromopropane	59	103	—
Trichloromethane	62	—	—
Bromochloromethane	68	—	—
2-Chlorobutane	69	108	129
1-Chloro-2-methylpropane	69	109	125
1-Bromopropane	71	92	121
Iodoethane	72	104	126
Allyl bromide	72	114	—
2-Bromo-2-methylpropane	73	128	147
1,1,1-Trichloroethane	75	—	—
Tetrachloromethane	76	—	—
1-Chlorobutane	77	63	112
1,2-Dichloroethane	83	—	—
2-Chloro-2-methylbutane	86	92	138
2-Iodopropane	89	103	—
2-Bromobutane	90	108	129
1-Bromo-2-methylpropane	91	109	125
2,3-Dichloro-1-propene	94	—	—
1,2-Dichloropropane	95	—	—
Dibromomethane	97	—	—
2-Iodo-2-methylpropane	98	128	147
1-Bromobutane	100	63	112
1-Chloro-3-methylbutane	100	108	111
1-Iodopropane	102	92	121
Bromotrichloromethane	105	—	—
1-Chloropentane	107	96	112
2-Bromo-2-methylbutane	108	92	138
1,1,2-Trichloroethane	112	—	—
Chlorocyclopentane	114	—	—
1-Bromo-3-methylbutane	118	108	111
2-Bromopentane	118	93	104

TABLE 38-19. (continued)

Alkyl Halides

Compound	BP	Anilide	1-Naphthalide
3-Bromopentane	119	121	—
2-Iodobutane	119	108	129
1-Iodo-2-methylpropane	120	109	125
3,4-Dichloro-1-butene	123	—	—
1,3-Dichloropropane	124	—	—
1,3-Dichloro-2-butene	125	—	—
2-Iodo-2-methylbutane	128	92	138
1-Bromopentane	129	96	112
1-Iodobutane	130	63	112
1,2-Dibromoethane	131	—	—
1-Chlorohexane	134	69	106
Bromocyclopentane	137	—	—
1,2-Dibromopropane	141	—	—
Chlorocyclohexane	142	146	188
1-Bromo-3-chloropropane	144	—	—
1-Iodo-3-methylbutane	148	108	111
Tribromomethane	148	—	—
1,1,2,2-Tetrachloroethane	148	—	—
1-Iodopentane	156	96	112
1,2,3-Trichloropropane	156	—	—
1-Bromohexane	157	69	106
1-Chloroheptane	160	57	95
1,4-Dichlorobutane	162	—	—
Bromocyclohexane	165	146	188
1,3-Dibromopropane	167	—	—
1-Bromoheptane	174	57	95
Benzyl chloride	179	117	166
1-Iodohexane	180	69	106
1-Chlorooctane	184	57	91
1-Chloro-1-phenylethane	195	133	—
1-Chloro-2-phenylethane	197	97	—
Benzyl bromide	198	117	166
1-Iodoheptane	204	57	95
1-Bromooctane	204	57	91
1-Bromo-1-phenylethane	205	133	—
Bornyl chloride	207	—	—
1-Bromo-2-phenylethane	218	97	—
Benzotrichloride	221	—	—
1-Bromodecane	238	—	—
Triiodomethane	119(MP)	—	—

TABLE 38-20.
Aryl Halides

Compound	BP	MP	Nitration product		Oxidation product
			Position	MP	
Chlorobenzene	132	—	2,4	52	—
Bromobenzene	156	—	2,4	75	—
2-Chlorotoluene	159	—	3,5	63	2-Chlorobenzoic acid, 141
4-Chlorotoluene	162	—	2	38	4-Chlorobenzoic acid, 240
3-Chlorotoluene	162	—	4,6	91	3-Chlorobenzoic acid, 158
1,3-Dichlorobenzene	173	—	4,6	103	—
1,2-Dichlorobenzene	178	—	4,5	110	—
2-Bromotoluene	182	—	3,5	82	2-Bromobenzoic acid, 147
3-Bromotoluene	183	—	4,6	103	3-Bromobenzoic acid, 155
4-Bromotoluene	184	28	2	47	4-Bromobenzoic acid, 251
Iodobenzene	188	—	4	171	—
2,4-Dichlorotoluene	197	—	3,5	104	2,4-Dichlorobenzoic acid, 164
3,4-Dichlorotoluene	201	—	6	63	3,4-Dichlorobenzoic acid, 206
1,2,4-Trichlorobenzene	214	—	5	56	—
1,3-Dibromobenzene	219	—	4	61	—
1,2-Dibromobenzene	224	—	4,5	114	—
1-Chloronaphthalene	259	—	4,5	180	—
1-Bromonaphthalene	280	—	4	85	—
1,2,3,4-Tetrachlorobenzene	—	44	5	64	—
			5,6	151	
1,2,3-Trichlorobenzene	—	52	4	56	—
1,4-Dichlorobenzene	—	53	2	54	—
2-Chloronaphthalene	—	56	1,8	175	—
1,3,5-Trichlorobenzene	—	63	2	71	—
1,4-Bromochlorobenzene	—	67	2	72	—
1,4-Dibromobenzene	—	89	2,5	84	—
1,2,4,5-Tetrachlorobenzene	—	139	3	99	—
			3,6	227	
Hexachlorobenzene	—	229	—	—	—

TABLE 38-21.

Nitro Compounds

Compound	BP	MP	BP	MP	Benzene-sulfon-amide	Acet-amide	Benz-amide	Other derivatives
			Reduced amine		Acyl derivatives of reduced amine			
Nitromethane	101	—	—	—	30	28	80	
Nitroethane	114	—	17	—	58	—	71	
2-Nitropropane	120	—	34	—	26	—	—	Amine hydrochloride, 140
1-Nitropropane	130	—	49	—	36	—	84	Amine hydrochloride, 157
Nitrobenzene	210	—	184	—	112	114	160	*m*-Dinitrobenzene, 90
2-Nitrotoluene	225	—	200	—	124	110	147	2,4-Dinitrotoluene, 70
2-Nitro-*m*-xylene	225	—	215	—		177	168	2-Nitroisophthalic acid, 300
3-Nitrotoluene	231	16	203	—	95	65	125	*m*-Nitrobenzoic acid, 140
3-Nitro-*o*-xylene	245	—	221	—	—	135	189	
4-Ethylnitrobenzene	246	—	216	—	—	94	151	
2-Chloronitrobenzene	246	32	209	—	129	—	99	2,4-Dinitrochloro-benzene, 52
2-Nitroanisole	265	—	224	—	89	85	60(84)	2,4,6-Trinitroanisole, 68
2-Chloro-6-nitrotoluene	—	35	245	—	—	157(136)	173	
4-Chloro-2-nitrotoluene	—	37	—	21	—	139(131)	—	
3,4-Dichloronitrobenzene	—	41	—	72	—	121	—	
2-Bromonitrobenzene	261	44	229	32	—	99	116	2,4-Dinitrobromo-benzene, 72
2,4-Dinitrochlorobenzene	—	52	—	91	—	242(di)	178(di)	2,4-Dinitrophenol, 114
4-Nitrotoluene	—	53	—	45	120	147	158	2,4-Dinitrotoluene, 70
4-Nitroanisole	258	54	243	57	95	127	154	2,4-Dinitroanisole, 89
1-Nitronaphthalene	—	61	—	50	167	159	160	
2-Nitronaphthalene	—	78	306	113	136(102)	132	162	
1-Chloro-4-nitrobenzene	—	83	—	72	121	179	192	4-Nitrophenol, 114
1,3-Dinitrobenzene	302	90	—	63	194(di)	87(mono) 191(di)	125(mono) 240(di)	3-Nitroaniline, 114
4-Nitrobiphenyl	—	114	302	53	—	171	230	
1-Bromo-4-nitrobenzene	259	126	—	66	134	168	204	4-Nitrophenol, 114

TABLE 38-22.

Aromatic Hydrocarbons

Compound	BP	MP	Nitration product		Picrate
			Position	MP	
Benzene	80	—	1,3	89	84
Toluene	111	—	2,4	70	88
Ethylbenzene	136	—	2,4,6	37	96
p-Xylene	138	—	2,3,5	139	90
m-Xylene	139	—	2,4,6	183	91
o-Xylene	144	—	4,5	118(71)	88
Isopropylbenzene	153	—	2,4,6	109	—
Propylbenzene	159	—	—	—	103
1,3,5-Trimethylbenzene	164	—	2,4	86	97
			2,4,6	235	
1,2,4-Trimethylbenzene	168	—	3,5,6	185	97
tert-Butylbenzene	169	—	2,4	62	
sec-Butylbenzene	173	—	—	—	—
Indane	176	—	—	—	—
4-Isopropyltoluene	177	—	2,6	54	—
m-Diethylbenzene	181	—	2,4,6	62	
4-tert-Butyltoluene	190	—	—.	—	
1,2,3,5-Tetramethylbenzene	198	—	4,6	181(157)	—
p-Diisopropylbenzene	203	—	—	—	—
1,2,3,4-Tetramethylbenzene	204	—	5,6	176	92
1,2,3,4-Tetrahydronaphthalene	207	—	5,7	95	—
1,3,5-Triethylbenzene	218	—	2,4,6	108	
Cyclohexylbenzene	237	—	4	58	—
1-Methylnaphthalene	243	—	4	71	142
Diphenylmethane	264	25	2,4,2′,4′	172	—
2-Methylnaphthalene	241	36	1	81	116
Pentamethylbenzene	—	51	6	154	131
Dibenzyl	—	53	4,4′	180	—
Biphenyl	—	70	4,4′	237(229)	—
p-Di-tert-butylbenzene	—	76	—	—	—
1,2,4,5-Tetramethylbenzene	—	81	3,6	205	—
Naphthalene	—	81	1	61(57)	149
Triphenylmethane	—	92	4,4′,4″	206	—
Acenaphthene	—	95	5	101	161
Phenanthrene	—	100	—	—	144(133)
Fluorene	—	114	2	156	87(77)
			2,7	199	
Hexamethylbenzene	—	164	—	—	170
Anthracene	—	216	—	—	138

TABLE 38-23.

Alkanes

Compound	BP	Sp gr	n_D^{20}
Pentane	36	0.631	1.3570
Cyclopentane	49	0.750	1.4093
2-Methylpentane	60	0.6532	1.3715
3-Methylpentane	63	0.6645	1.3765
Hexane	69	0.660	1.3754
Cyclohexane	80	0.790	1.4263
Heptane	98	0.684	1.385
Methylcyclohexane	101	0.769	1.4235
Octane	125	0.703	1.3890
Ethylcyclohexane	131	0.788	1.4330
Cyclooctane	150	0.835	1.4586
Decane	173	0.730	1.415

QUESTIONS

1. Select a quick and easy chemical test which would allow you to distinguish between each of the following pairs of compounds. Also state what you would observe in performing the test. Sometimes there will be more than one possible answer.
   (a) Cyclohexane and cyclohexene
   (b) Toluene and 4-methylaniline
   (c) Benzaldehyde and acetophenone
   (d) Acetone and 3-pentanone
   (e) Toluene and benzyl chloride
   (f) t-Butyl alcohol and butyl alcohol
   (g) Aniline and N-methylaniline
   (h) Nitrobenzene and aniline
   (i) 4-Methylcyclohexanol and p-cresol
   (j) Benzoic acid and benzamide
   (k) Ethoxybenzene and ethylbenzene
   (l) Pentanol and 3-pentanone
   (m) Benzyl chloride and chlorobenzene
   (n) Ethyl benzoate and acetophenone
2. Write the structure of a compound which will do the following. Explain your answer. Many times, there will be more than one possible answer.
   (a) An alcohol which gives a negative iodoform test and a negative Lucas test.
   (b) An alcohol which gives a positive iodoform test and a negative Lucas test.

(c) An amine which fails to react with acetyl chloride, but yields a derivative with nitrous acid.

(d) A water-soluble compound which decolorizes an alkaline solution of potassium permanganate, but is not affected by bromine in methylene chloride.

(e) A compound containing only carbon, hydrogen, and oxygen, which yields two products when refluxed with sodium hydroxide solution.

(f) A compound which dissolves in dilute sodium hydroxide and decolorizes a cold, dilute solution of potassium permanganate.

(g) A water-soluble compound which is inert to Tollens' reagent, but reacts with 2,4-dinitrophenylhydrazine.

(h) A nitrogen-containing compound, soluble only in ether and in cold, concentrated sulfuric acid. The compound gives a positive ferric hydroxamate test and produces ammonia when reacted with aqueous sodium hydroxide.

(i) A halogen-containing compound, soluble in cold, concentrated sulfuric acid, which fails to react with alcoholic silver nitrate solution.

(j) A water-soluble compound, containing no halogen or nitrogen, which forms a precipitate with alcoholic silver nitrate solution.

3. State in which solubility group you would expect to find the following compounds. Explain your answer.

(a) Propyl alcohol   (h) Cyclohexanone
(b) *N,N*-Dimethylaniline   (i) Benzanilide
(c) Glycerol   (j) Ethylamine
(d) 1-Naphthol   (k) Diphenyl ether
(e) 2-Bromobutane   (l) Aniline hydrochloride
(f) Toluene   (m) 3-Heptyne
(g) Propanoic acid   (n) 2-Nitrohexane

4. Specifically identify each of the following compounds, utilizing the tables given in this experiment. Thoroughly explain your answer, writing equations for all positive tests and the formation of any derivatives.

(a) A brown liquid having a boiling point of 191–193° was insoluble in water but soluble in dilute acid. A sodium fusion gave a positive test for nitrogen but negative tests for sulfur and the halogens. The compound did not react with acetyl chloride or benzenesulfonyl chloride, but did react with nitrous acid. A picrate derivative had a melting point of 162–163°.

(b) A compound having a melting point of 140–142° was insoluble in water, dilute acid, and dilute base. It was soluble in cold, concentrated sulfuric acid and contained nitrogen but no sulfur or halogen. When the compound was refluxed for 1 hour in sodium hydroxide solution, it reacted to yield an insoluble oil. This oil was soluble in dilute hydrochloric acid and reacted with benzoyl chloride to form a solid having a melting point of 142–144°. Acidifying the alkaline solution from which the oil was obtained yielded a solid melting at 119–121°. This solid when reacted with thionyl chloride and aniline yielded a solid melting at 160–162°.

(c) A colorless liquid having a boiling point of 94–97° was soluble in water and ether. A sodium fusion gave negative tests for nitrogen, sulfur, and the halogens. The compound reacted with acetyl chloride, decolorized bromine in methylene chloride, and reduced a dilute potassium permanganate solution. The compound did not react with 2,4-dinitrophenylhydrazine, but reacted with 1-naphthylisocyanate to form a compound melting at 106–107°.

(d) A compound having a boiling point of 159–161° was soluble in ether, but insoluble in water, dilute acid, dilute base, and cold, concentrated sulfuric acid. A sodium fusion gave a positive test for chlorine but negative tests for sulfur and nitrogen. The compound gave no precipitate with alcoholic silver nitrate solution. Refluxing the compound in alkaline potassium permanganate solution made it slowly dissolve. Acidification produced a solid having a melting point of 156–157°.

(e) A colorless liquid having a boiling point of 212–214° was soluble only in ether and in cold, concentrated sulfuric acid. A sodium fusion showed no nitrogen, halogen, or sulfur. The compound did not react with 2,4-dinitrophenylhydrazine, acetyl chloride, or bromine in methylene chloride. When refluxed with sodium hydroxide solution, the compound slowly dissolved. Extraction with ether gave a compound boiling at 203–205°, which reacted with 3,5-dinitrobenzoyl chloride to yield a compound melting at 110–111°. Acidification of the sodium hydroxide solution from the refluxing produced a distinct odor of vinegar.

(f) A compound having a boiling point of 167–171° was soluble only in ether and in cold, concentrated sulfuric acid. A sodium fusion gave negative tests for nitrogen, sulfur, and the halogens. The compound did not react with acetyl chloride, Tollens' reagent, or dilute potassium permanganate, but it did yield a yellow solid when treated with sodium hypoiodite. When placed with 2,4-dinitrophenylhydrazine, the compound yielded a solid melting at 56–57°.

References

1. N. Cheronis and J. Entrikin, **Identification of Organic Compounds,** Wiley-Interscience, New York, 1963.

2. D. Pasto and C. Johnson, **Organic Structure Determination,** Prentice-Hall, Englewood Cliffs, N.J., 1969.

3. Z. Rappoport, **Handbook of Tables for Organic Compound Identification,** Chemical Rubber Company, Cleveland, 1967.

4. R. Shriner, R. Fuson, D. Curtin, and T. Morrill, **The Systematic Identification of Organic Compounds,** 6th ed., Wiley, New York, 1980.

A

SUBLIMATION

A.1 INTRODUCTION

The odors of many solids, such as camphor, naphthalene (moth crystals), and 1,4-dichlorobenzene (moth crystals) are probably quite familiar to you. You may have, however, never considered exactly why these solids have an odor. Since any odor results from gas molecules being inhaled, somehow the camphor, naphthalene, and 1,4-dichlorobenzene must have vaporized directly from the solid phase, completely avoiding the liquid phase. This direct vaporization from the solid phase is called **sublimation**. Many other solids also undergo sublimation, among them dry ice and iodine. Solid carbon dioxide is called dry ice because at normal pressures it vaporizes without ever melting or getting wet.

A.2 THEORY OF SUBLIMATION

Figure A-1 helps us understand why a solid such as camphor, naphthalene, or 1,4-dichlorobenzene should sublime. Along line AB, the sublimation curve, the solid and vapor are in equilibrium, with the solid phase to the left of the line and the vapor phase to the right of the line. Observe that as long as the temperature is kept below melting point B, the solid will sublime and will have a vapor pressure corresponding to its temperature. For example, at temperature T_1, the corresponding vapor pressure will be VP_1. At the higher temperature T_2, the corresponding higher vapor pressure will be VP_2. Above temperature B, the liquid will exist in equilibrium with the vapor, and at temperature C it will boil if the atmospheric pressure is 760 torr.

A solid will only sublime appreciably if it has a significant vapor pressure before its melting point is reached. This usually occurs with nonpolar substances having highly symmetrical structures. Nonpolar substances generally have higher vapor pressures because the lower dipole moment lowers the attractive forces between molecules, making it easier for individual surface molecules

829

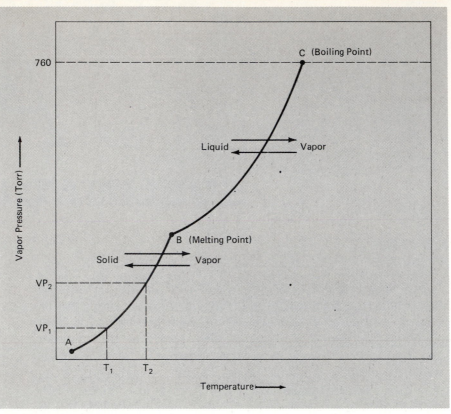

FIGURE A-1
Vapor pressure curves for a sublimable solid below and above
its melting point.

to escape from their neighbors. On the other hand, highly symmetrical substances generally have higher melting points because they can better fit together in the crystal lattice. Before the higher melting-point temperature is reached, many more surface molecules may thus be able to attain enough energy to vaporize.

Table A-1 gives the vapor pressures of some solids at their melting points. Notice that carbon dioxide, perfluorocyclohexane, and hexachloroethane all have vapor pressures greater than 1 atmosphere (atm). This means that none of these three substances will exhibit normal melting behavior at atmospheric pressure. For example, if you tried to take the melting point of hexachloroethane at atmospheric pressure, you would see vapor being released from the end of the capillary tube. But in a sealed capillary tube, hexachloroethane will melt at 186°C. Not even a sealed capillary tube, however, would be satisfactory for dry ice, since a minimum internal pressure of 5.1 atm would be needed for melting to occur, and it is unlikely that a sealed capillary tube could withstand this pressure.

TABLE A-1.

Vapor Pressures of Solids at Their Melting Points

Compound	Vapor pressure of solid at melting point (torr)	Melting point (°C)
Carbon dioxide	3876 (5.10 atm)	−57
Perfluorocyclohexane	950 (1.25 atm)	59
Hexachloroethane	780 (1.03 atm)	186
Camphor	370	179
Iodine	90	114
Anthracene	41	218
Phthalic anhydride	9	131
1,4-Dibromobenzene	9	87
1,4-Dichlorobenzene	8.5	53
Naphthalene	7	80
Benzoic acid	6	122

All the other substances in the table have vapor pressures of less than 1 atm at their melting points, so they exhibit normal melting behavior. All of them at room temperature also have vapor pressures of less than 1 torr (0.0013 atm), so they would be expected to only sublime very slowly in an unheated open container exposed to the atmosphere. To speed up the process, you would have to heat the sample in a reduced pressure apparatus. For example, if you wanted to sublime in a reasonably short time a sample of anthracene, it would be necessary to heat the anthracene to a temperature **less than** 218°C in an apparatus with a pressure of **less than** 41 torr. If the pressure in the apparatus were much less than 41 torr, then the temperature could also be much less than 218°C. For example, at 1.0 torr the necessary temperature would be 145°C. All this information can be obtained from a chemistry handbook.

A.3   USES OF SUBLIMATION

Sublimation is primarily used in organic chemistry to purify small amounts of solids. For example, if anthracene, camphor, naphthalene, or other vaporizable solids are contaminated with essentially nonvolatile impurities, then sublimation may be used to separate the vaporizable solid from these impurities. However, if the impurities have about the same vapor pressure as the desired compound, they will also sublime and little separation will occur. In this case, some other method, such as recrystallization or chromatography, would have to be used.

Sublimation is also sometimes used to locate colorless organic compounds which have been separated by thin-layer or paper chromatography. The chromatogram is placed in a chamber with several iodine crystals and is mildly heated. The iodine vapor produced reacts with the colorless compounds to form visible colored complexes. Since this reaction is usually reversible, the location of each complex should be marked before the color disappears.

A.4 APPARATUS AND TECHNIQUE

To purify solids by sublimation, you can use an apparatus such as those shown in Figure A-2 or Figure A-3. To prepare the apparatus in Figure A-2, place the solid to be sublimed in a filter flask. A 15 × 125 mm or larger test tube is then inserted into a cork or rubber stopper having a drilled hole of the size necessary for a tight seal. If necessary, glycerine may be used to lubricate the hole. The cork and test tube are then placed inside the filter flask, with the bottom of the test tube about 2 cm above the bottom of the filter flask. As an alternative, a neoprene adapter may be used to give a tight seal between the test tube and filter flask. Ice water is placed inside the test tube, and the side arm of the filter flask is connected to a safety trap and a water aspirator or vacuum pump.

To prepare the apparatus in Figure A-3, place the solid to be sublimed into a 20 × 155 mm side-arm test tube. A 15 × 125 mm test tube is then inserted into a No. 2 neoprene adapter, and the adapter is fitted into the side-arm test tube. The bottom of the inner tube should be about 2 cm from the bottom of the outer one. Ice water is next placed inside the inner test tube, and the side arm of the outer test tube is connected to a safety trap and a water aspirator or vacuum pump.

FIGURE A-2
Sublimation apparatus.

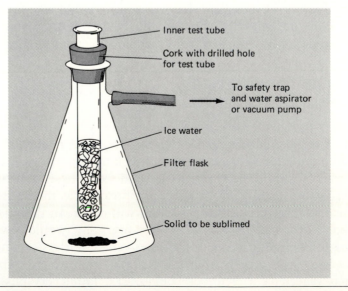

Inner test tube

Cork with drilled hole for test tube

To safety trap and water aspirator or vacuum pump

Ice water

Filter flask

Solid to be sublimed

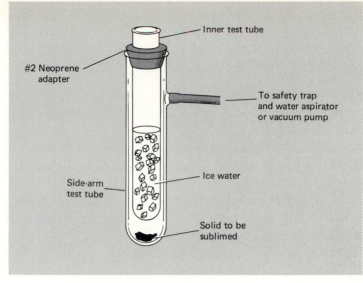

FIGURE A-3
Sublimation apparatus.

After the water aspirator or the vacuum pump has been turned on and the pressure inside the system has been reduced, the sample is **carefully** heated with a small flame, moving the burner back and forth. If the sample begins to melt, the burner is removed for a few seconds before continuing. As alternative methods, a sand bath, oil bath, hot-water bath, or steam bath may be used to heat the sample. These other methods may take longer, but they better control the temperature and can prevent charring of the sample.

During the heating, the sample will be seen vaporizing from the bottom of the filter flask or the side-arm test tube and being deposited onto the cold test tube. The reason for the ice water or other cooling material inside the inner test tube should be obvious. If the sublimed vapor were not cooled, it would not be conveniently deposited on this test tube, but would be removed from the apparatus and perhaps deposited unwantedly all over the inside of the vacuum hoses and the safety trap. The value of the procedure would then be lost.

The distance between the inner test tube and the filter flask or side-arm test tube is very important. If the distance is too small, the impurities in the original sample can spatter up and contaminate the purified sublimed material. But if the distance is too large, the vapor will have to travel further, a higher temperature may be needed, and decomposition of the sample may occur.

When the sublimation is complete, remove the heat and let the apparatus cool. **Slowly** release the vacuum by opening the stopcock or the clamp on the hose coming up from the safety trap. Otherwise the surge of air might dislodge the collected crystals from the cold test tube back into the residue. Then **carefully** remove the inner tube of the apparatus, trying not to have any of the sublimed crystals fall off the tube into the residue. Pour off the remaining ice and water from the inner tube, and scrape the sublimed product onto weighing paper with a spatula.

QUESTIONS

1. Why is the odor of a spice such as pepper or cinnamon **not** considered to be an example of sublimation?
2. Under what conditions could one have liquid carbon dioxide?
3. Why could one **not** effectively separate 1,4-dichlorobenzene from 1,4-dibromobenzene by sublimation?
4. A substance has a vapor pressure of 75 torr at its melting point of 140°C. Describe the physical conditions you would use to experimentally sublime this substance in a reasonably short time.
5. Look at Table A-1. Why would you use a vacuum pump and not a water aspirator to effectively sublime a sample of benzoic acid in a reasonably short time?

B

USE OF THE REFRACTOMETER

B.1 DISCUSSION

The bending of light as it travels from one medium to another is familiar to all of you. As a result of it, objects in water appear to be closer to the surface than they actually are. If these objects are also partially above water, they may either look bent or broken at the point where they enter the water. The lenses in eyeglasses bend light to allow us to see better. The brilliance of cut and polished diamonds is partially the result of the bending of light as it travels from air into the stone and back out into the air.

This bending of the light as it travels from one medium to another is called **refraction**. The light bends because its velocity in air is different from its velocity in a liquid or solid. The ratio of the velocity in air to the velocity in a liquid or solid is defined as the **refractive index**, n, of the liquid or solid.

$$n = \frac{\text{Velocity}_{\text{air}}}{\text{Velocity}_{\text{liquid or solid}}}$$

Since the velocity of light in air is always greater than its velocity in a liquid or solid, the refractive index as a result will always be greater than 1.

The refractive index is also equal to the sine of the angle of incidence, i, divided by the sine of the angle of refraction, r.

$$n = \frac{\sin i}{\sin r}$$

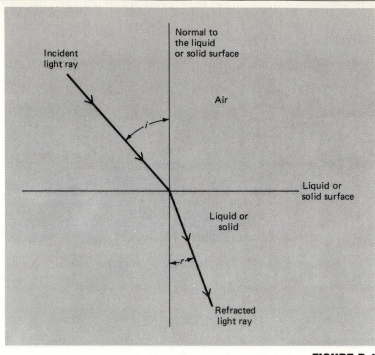

FIGURE B-1
The angle of incidence i and the angle of refraction r for the refraction of light from air into a liquid or solid.

These angles are illustrated in Figure B-1. For example, if i is 43.1° and r is 28.4°, the refractive index is

$$n = \frac{\sin i}{\sin r} = \frac{\sin 43.1°}{\sin 28.4°} = \frac{0.6833}{0.4756} = 1.437$$

For any given liquid or solid, the refractive index depends on both the **wavelength of light used** and the **temperature**. As the "rainbow" of colors produced by a prism illustrates, light of differing wavelengths is refracted to different extents in the same medium, giving different refractive indexes in that medium. Temperature affects the refractive index by altering the density of the medium and thus the speed of light traveling through that medium. Usually refractive indexes are measured at 20°C and 589 nanometers wavelength, the D line of sodium light (1 nanometer = 10^{-9} meter). Using these conditions, the refractive index is reported in the following way:

$$n_{\text{D}}^{20} = 1.4372$$

Observe that the superscript indicates the temperature for the measurement, while the subscript indicates that the wavelength used was the sodium D line.

If another temperature or wavelength were used, the 20 or the D would be replaced by the appropriate value, with the wavelength generally being expressed in nanometers.

B.2 IMPORTANCE OF THE REFRACTIVE INDEX IN ORGANIC CHEMISTRY

The refractive index is a useful physical property of organic liquids. It can therefore be valuable, along with the boiling point and various spectra, for the identification of organic liquids. For example, the reported n_D^{20} for ethylbenzene is 1.4959. Therefore, if a pure unknown liquid is thought to be ethylbenzene, then it must have a refractive index very close to this value under these conditions, along with a boiling point of 136.2°C at 760 torr and the same spectra as ethylbenzene.

Note that the refractive index of ethylbenzene is reported to four decimal places. As a result, the refractive index is a very accurate physical constant and can be used as a measure of the purity of liquid organic compounds. The higher the purity, the closer the refractive index of the liquid must be to the reported value. Usually even very small amounts of impurities will significantly change the reading. Therefore, unless a liquid has been extensively purified, it will not be possible to reproduce the last two digits of the reported value.

B.3 MEASURING THE REFRACTIVE INDEX

Usually the refractive index does not have to be calculated as was done in the one example in the discussion section. Instead, with an instrument called a **refractometer**, it may be measured directly. Many models and styles of refractometers are available, but the type most common to organic chemistry laboratories is the Abbé refractometer shown in Figure B-2. This refractometer has the following advantages:

1. White light is used for illumination, but the instrument compensates for this with a prism system to make the refractive index actually obtained the same as that for the sodium D line of 589 nanometers.

2. Water from a constant-temperature bath can be circulated through the sample region to maintain a temperature of 20°C.

3. Only a few drops of sample are required.

Although the inside of an Abbé refractometer is quite complex, a simplified diagram of its internal structure is shown in Figure B-3. Light goes from the lamp through the sample, which is held in place between two hinged prisms. The

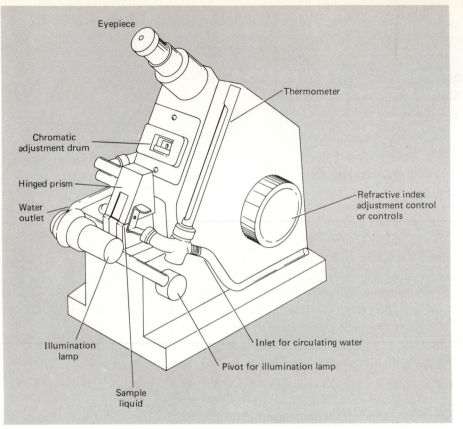

Eyepiece

Thermometer

Chromatic
adjustment drum

Hinged prism

Water
outlet

Refractive index
adjustment control
or controls

Illumination
lamp

Inlet for circulating water

Pivot for illumination lamp

Sample
liquid

FIGURE B-2
The Abbé refractometer (Bausch and Lomb Abbé 3L).

light, as can be seen, is bent to some degree by the sample and these prisms, and goes to an adjustable mirror. Only when the mirror is at the proper angle, can the light from it be properly reflected into the system of Amici prisms (to compensate for using white light) and the lens with crosshairs. The light then travels further to the eyepiece where it is viewed by the person measuring the refractive index. Note from the diagram how the angle of the mirror is controlled by coarse- and fine-adjustment wheels, and how the proper angle of the mirror determines what the refractive index is on the scale.

B.4   PROCEDURE FOR USING THE ABBÉ REFRACTOMETER

Step 1.   If the refractometer is to be used at other than room temperature, turn on the water from the constant-temperature bath or other source to cool the system.

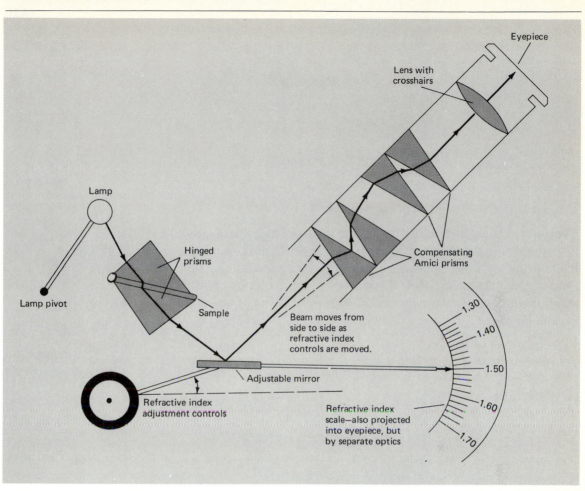

Labels in figure:
Eyepiece
Lens with crosshairs
Lamp
Hinged prisms
Compensating Amici prisms
Lamp pivot
Sample
Beam moves from side to side as refractive index controls are moved.
Adjustable mirror
Refractive index adjustment controls
Refractive index scale—also projected into eyepiece, but by separate optics
1.30
1.40
1.50
1.60
1.70

FIGURE B-3
Simplified diagram of an Abbé refractometer.

Step 2. Introduce the liquid sample between the two hinged prisms. If the liquid is not very viscous, place several drops into the groove on the side of the prisms, using an eyedropper. If the liquid is viscous, open the hinged prisms by lifting the upper one. Then add a few drops of the liquid to the lower prism with a wooden applicator or an eyedropper. **CAUTION: Do not touch the prisms, especially with an eyedropper. They are easily scratched and will then be ruined.** Close the prisms. This should spread the liquid evenly to form a thin film.

Step 3. Turn on the illumination lamp and look into the eyepiece. Adjust the lamp, moving it up or down on its pivot to give the most illumination inside the eyepiece. If necessary, adjust the eyepiece until the crosshairs seen inside are in focus (see Figure B-4).

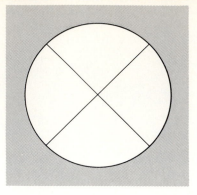

The focused crosshairs seen inside the eyepiece.

Step 4. Rotate the refractive index adjustment control until you observe a horizontal line dividing the field into a light and dark section (see Figure B-5). If the horizontal line appears as a diffuse colored band (see Figure B-6), the refractometer is showing color dispersion. Correct this by rotating the chromatic adjustment drum (shown in Figure B-2) until a sharp, uncolored dividing line is obtained.

FIGURE B-5
Sharp, uncolored, horizontal line separating the field into light and dark sections.

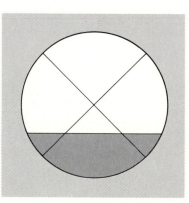

FIGURE B-6
Refractometer showing color dispersion.

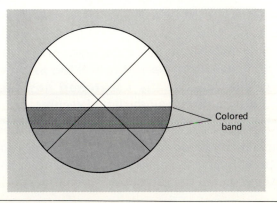

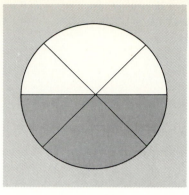

FIGURE B-7
Position of sharp, uncolored, horizontal line at the correct refractive index. The line is exactly at the intersection of the crosshairs.

Step 5. Continue to rotate the refractive index adjustment control or controls until you observe the sharp, horizontal line exactly at the intersection of the crosshairs (see Figure B-7). At this point, the reflecting mirror inside has been adjusted to the proper position inside for the correct refractive index.

Step 6. Press a small button on the left side of the instrument. The refractive index scale will become visible in the eyepiece. It can then be read and its value recorded. Note that in some models of refractometers, the refractive index scale is visible at all times through a separate eyepiece.

Step 7. When the measurement is completed, open the hinged prisms by lifting the upper one. Thoroughly clean both prisms by wiping them **gently** with lens paper, moistened with ethanol, petroleum ether, or acetone. **CAUTION: Do not use paper towels or other paper products, since they can scratch and ruin the prisms.**

Step 8. After the cleaning solvent has evaporated, close the prisms so dust won't collect on them. Turn off the instrument when it is no longer in use.

B.5 TEMPERATURE CORRECTIONS

The refractive index **decreases** with **increasing** temperature. Therefore, if the measurement is not taken in a room at 20°C, or if the cooling water circulating through the instrument is not at 20°C, a temperature correction must be made. For most organic liquids, the refractive index changes between 0.00035 to 0.00055 unit/°C. If the exact correction factor for a particular liquid is not known, 0.00045 unit/°C is generally chosen. Using this value, the following formula allows us to compute the refractive index at 20°C when taken at another temperature, t.

$$n_{\mathrm{D}}^{20} = n_{\mathrm{D}}^{t} + (0.00045)(t - 20°C)$$

For example, if the refractive index of ethylbenzene were taken at 24.8°, you would observe a value of 1.4937. Using the temperature correction formula, we obtain

$$n_D^{20} = 1.4937 + (0.00045)(24.8°C - 20.0°C)$$
$$= 1.4937 + (0.00045)(4.8°C)$$
$$= 1.4959$$

APPENDIX

C

APPROXIMATE TIME AND THE NECESSARY CHEMICALS AND SUPPLIES FOR EACH EXPERIMENT

The following quantities of chemicals are for a class of 20 students. In each case, an additional amount has been included to allow for spillage, waste, and the inadvertent need to repeat an experiment. This ensures that additional material will usually not have to be obtained from the stockroom.

1. DISTILLATION—FRACTIONAL DISTILLATION

Estimated Time   3 to 4 hours

50% Methanol and 50% water mixture (volume-volume)	1500 mL
Acetone	750 mL
90% 2-Propanol and 10% water mixture (volume-volume)	1500 mL
or 76% 1-propanol and 24% water mixture (volume-volume)	1500 mL

or 750 mL of each of the last two mixtures
Crushed ice
Boiling stones
Distilled water
Glass wool and glass helices, glass beads, or some other packing
    material

843

2. MELTING POINTS

Estimated Time   4 hours

A high-boiling liquid such as silicone oil, paraffin oil, etc.	3000 mL
p-Dichlorobenzene	10 grams
Naphthalene	10 grams
Acetanilide	10 grams
Salicylic acid	10 grams
Oxalic acid	10 grams
p-Nitrobenzoic acid	10 grams
Urea	10 grams
Benzamide	10 grams
75% Urea, 25% benzamide mixture (mole-mole percentage). Mix 6.0 grams of urea and 4.0 grams of benzamide.	10 grams
50% Urea, 50% benzamide mixture (mole-mole percentage). Mix 3.3 grams of urea and 6.7 grams of benzamide.	10 grams
25% Urea, 75% benzamide mixture (mole-mole percentage). Mix 1.4 grams of urea and 8.6 grams of benzamide.	10 grams
Maleic anhydride	1 gram
Stearic acid	1 gram
Acetamide	1 gram
1-Naphthol	1 gram
o-Toluic acid	1 gram
Benzoic acid	1 gram
Cinnamic acid	1 gram
Salicylamide	1 gram
Adipic acid	1 gram
Benzanilide	1 gram
d-Tartaric acid	1 gram
4-Methylbenzoic acid	1 gram
Succinic acid	1 gram
3,5-Dinitrobenzoic acid	1 gram
Anthracene	1 gram
Capillary tubes	

3. RECRYSTALLIZATION

Estimated Time   3 to 4 hours

Benzoic acid	10 grams
Sodium benzoate	10 grams
Resorcinol	10 grams
Anthracene	10 grams

Stearic acid	10 grams
95% Ethanol	300 mL
Ligroin	300 mL
Impure benzoic acid (a mixture of 100 grams benzoic acid, 10 grams sodium chloride, and perhaps a variable amount of sawdust and 0.2 gram of a dye such as Congo red)	
Dark brown sugar	300 grams
Decolorizing carbon	40 grams
Ice	
Capillary tubes	

4. EXTRACTION OF ASPIRIN, β-NAPHTHOL, AND NAPHTHALENE

Estimated Time 4 to 5 hours, plus optional spectra time

A prepared mixture of about 30% aspirin, 30% β-naphthol, and 40% naphthalene by weight	120 grams
Ethyl ether	2000 mL
5% Sodium bicarbonate solution	1000 mL
3 M hydrochloric acid	2000 mL
5% Sodium hydroxide solution	1000 mL
Saturated sodium chloride solution	500 mL
Anhydrous calcium chloride	100 grams
Petroleum ether	500 mL
95% ethanol	700 mL
Hydrion paper (pH 1–2)	
Cotton	
Ice	

5. FERMENTATION OF SUCROSE

Estimated Time $\frac{1}{2}$ hour the first period and 2 to 3 hours the second period

Sucrose	1000 grams
Pasteur's nutrient (Dissolve 2.0 grams of potassium phosphate, 0.20 gram of calcium phosphate, 0.20 gram of magnesium sulfate, and 10.0 grams of ammonium tartrate in enough water to make 1000 mL of solution.)	1000 mL
Fresh yeast	300 grams
Mineral oil	1000 mL
Lime water	3000 mL
Filter aid	400 grams
Anhydrous potassium carbonate	800 grams
Boiling chips	

6A. PAPER CHROMATOGRAPHY OF ARTIFICIAL FOOD COLORING AND KOOL-AID POWDERED DRINK MIX

Estimated Time $2\frac{1}{2}$ to 3 hours

Black cherry Kool-Aid	10 grams
Cherry Kool-Aid	10 grams
Grape Kool-Aid	10 grams
Lime Kool-Aid	10 grams
Punch Kool-Aid	10 grams
Orange Kool-Aid	10 grams
Raspberry Kool-Aid	10 grams
Strawberry Kool-Aid	10 grams
Green commercial food coloring	3 vials
Red commercial food coloring	3 vials
Blue commercial food coloring	3 vials
Yellow commercial food coloring	3 vials
2% Blue dye No. 1 solution (if available)	25 mL
2% Blue dye No. 2 solution (if available)	25 mL
2% Red dye No. 3 solution (if available)	25 mL
2% Red dye No. 40 solution (if available)	25 mL
2% Yellow dye No. 5 solution (if available)	25 mL
2% Yellow dye No. 6 solution (if available)	25 mL
Solvent No. 1: Equal volumes of 2 M ammonium hydroxide, 95% ethanol and 1-butanol	1500 mL
Solvent No. 2: 60% isopropanol and 40% water (by volume)	1500 mL
Solvent No. 3: 95% ethanol	1500 mL
Capillary tubes for preparing micropipets	
Whatman No. 1 paper, cut to at least 16 × 28 cm in size	150 pieces
Solvent chambers and covers	
Millimeter rulers	

6B. THIN-LAYER CHROMATOGRAPHY OF PEN INKS

Estimated Time   2 to $2\frac{1}{2}$ hours

Various narrow felt-tip pens of different colors and manufacturers	3 pens each
Commercial, plastic-backed, thin-layer sheets with either a cellulose or an activated silica gel coating	25
A solution of 70% ethyl acetate, 20% formic acid (90%) and 10% water (by volume)	1500 mL
Filter paper liners (cut to size)	20
Rectangular solvent chambers and covers	
Millimeter rulers	

6C. COLUMN CHROMATOGRAPHY OF TOMATO PASTE

Estimated Time $1\frac{1}{2}$ to 2 hours the first period, 3 hours the second period, plus optional spectra time

Tomato paste	300 grams
95% Ethanol	500 mL
Methylene chloride	600 mL
Saturated sodium chloride solution	300 mL
Anhydrous magnesium sulfate	100 grams
Alumina, acid-washed (Fisher A-540)	400 grams
Clean, white sand	150 grams
Petroleum ether	4000 mL
Toluene	4000 mL
Glass wool	

6D. COLUMN AND THIN-LAYER CHROMATOGRAPHY OF PAPRIKA PIGMENTS

Estimated Time 2 hours the first period, 3 hours the second period, plus optional spectra time

Ground paprika	45 grams
Commercial, plastic-backed, silica gel G sheets, cut to 2 × 10 cm	25 sheets
Methylene chloride	5000 mL
Silica gel, 60–200 mesh	350 grams
Clean, white sand	150 grams
Solvent chambers and covers	
Paper liner wicks (cut to size)	
Boiling chips	
Capillary tubes for preparing micropipets	
Millimeter rulers	

6E. GAS CHROMATOGRAPHIC ANALYSIS OF GASOLINES

Estimated Time 3 to 4 hours

Simulated gasoline mixture containing by volume:		10 mL or the smallest amount over 10 mL that can be accurately prepared
Pentane	12.0 to 14.0%	
2-Methylpentane	3.0%	
3-Methylpentane	7.0%	
Hexane	14.0%	
Benzene (optional)	0.0 to 2.0%	
Cyclohexane	12.0%	
2-Methylhexane	3.0%	

3-Methylhexane	6.0%
Heptane	2.0%
2,2,4-Trimethylpentane	8.0%
Toluene	20.0%
p-Xylene	2.0%
m-Xylene	3.0%
o-Xylene	6.0%

Several brands of "regular" gasolines   10 mL each

Several brands of "premium" gasolines   10 mL each

Several brands of Gasohol (if available)   10 mL each

Gas chromatographs with 12- to 15-foot Chromosorb
    W columns, coated with silicone oil (SE-30) as the
    adsorbent

Hypodermic syringes (10 microliters or smaller)

7A. SYNTHESIS OF PLEXIGLAS

Estimated Time $\frac{3}{4}$ hour; other experiments may be done during the 20-minute heating period

Methyl methacrylate	150 mL
Benzoyl peroxide	5 grams
Glazed weighing paper	
3 × 5 cards	

7B. SYNTHESIS OF POLYSTYRENE

Estimated Time $1\frac{1}{2}$ hours; other experiments may be done during the 1-hour heating period

Styrene	150 mL
Benzoyl peroxide	10 grams
Glazed weighing paper	
3 × 5 cards	

7C. SYNTHESIS OF THIOKOL RUBBER

Estimated Time 1 to $1\frac{1}{2}$ hours

10% Sodium hydroxide solution	500 mL
Powdered sulfur	75 grams
1,2-Dichloroethane	200 mL
Liquid dishwashing detergent	10 mL

7D. SYNTHESIS OF NYLON

Estimated Time $\frac{1}{2}$ hour

5% Aqueous solution of 1,6-hexanediamine	300 mL
20% Sodium hydroxide solution	20 mL
9% Solution of sebacoyl chloride in cyclohexane	300 mL
Methyl red solution	20 mL
Methyl orange solution	20 mL
Bromocresol green solution	20 mL
6-inch lengths of thick wire	20 pieces

7E. SYNTHESIS OF POLYESTERS

Estimated Time $1\frac{1}{2}$ hours

Powdered phthalic anhydride	150 grams
Ethylene glycol	30 mL
Glycerol	60 mL
20% Sodium hydroxide solution	300 mL
Acetone	200 mL
Chloroform	200 mL
Toluene	200 mL
Ethyl ether	200 mL
95% Ethanol	200 mL

7F. SYNTHESIS OF POLYURETHANE FOAMS

Estimated Time 3 to 4 hours; other experiments may be done during the 2- to 3-hour setting time

Toluene-2,4-diisocyanate (keep tightly closed in hood)	200 mL
Castor oil	100 mL
Glycerol	50 mL
Triethylene glycol	80 mL
Triethylamine	10 mL
Silicone oil	10 mL
Ethylene glycol (for interested students)	20 mL
Polyethylene glycol (for interested students)	20 mL
Propylene glycol (for interested students)	20 mL
Polypropylene glycol (for interested students)	20 mL
Small **waxed** paper cups	100

8A. STEAM DISTILLATION OF ESSENTIAL OILS FROM SPICES

Estimated Time   3 to 4 hours, plus optional spectra time

Cloves	80 grams
Allspice	80 grams
Cinnamon	80 grams
Caraway	80 grams
Cumin	80 grams
Anise	80 grams
Nutmeg	50 grams
Pepper	50 grams
Ginger	50 grams
Sage	50 grams
Thyme	50 grams
Oregano	50 grams
Methylene chloride	1200 mL
Anhydrous magnesium sulfate	100 grams
1 M sodium hydroxide solution	50 mL
Benzoyl chloride	10 mL
Methanol	200 mL
Semicarbazide hydrochloride	10 grams
Anhydrous sodium acetate	10 grams
Absolute ethanol	50 mL
2,4-Dinitrophenylhydrazine solution (Dissolve 0.60 gram of 2,4-dinitrophenylhydrazine in 3.0 mL of concentrated sulfuric acid. Mix together in a beaker 4 mL water and 14 mL 95% ethanol. Slowly add the acidic 2,4-dinitrophenylhydrazine solution with vigorous stirring to the aqueous ethanol solution. After thorough mixing, gravity filter the solution through a fluted filter.)	10 mL

Boiling stones
Ice
Capillary tubes

8B. STEAM DISTILLATION OF LIMONENE FROM GRAPEFRUIT AND ORANGE PEELS

Estimated Time $2\frac{1}{2}$ to 3 hours, plus optional spectra time

Peels from 35 grapefruits or 70 oranges, or a combination of both, such as 20 grapefruits and 40 oranges

Methylene chloride	1000 mL
Anhydrous magnesium sulfate	100 grams

8C. SYNTHESIS OF CAMPHOR FROM CAMPHENE

Estimated Time   6 hours, plus optional spectra time

Camphene	125 grams
Glacial acetic acid	600 mL
6 M sulfuric acid	40 mL
10% Sodium carbonate solution	1200 mL
Potassium hydroxide pellets	70 grams
95% Ethanol	400 mL
Sodium bisulfite	150 grams
Household chlorine bleach	1500 mL
Saturated sodium bisulfite solution	250 mL
Saturated sodium carbonate solution	1200 mL
Starch-iodide test paper	
Capillary tubes	
Ice	
Glass wool	

8D. SODIUM BOROHYDRIDE REDUCTION OF CAMPHOR TO BORNEOL AND ISOBORNEOL

Estimated Time   3 to 4 hours

Camphor	90 grams
Absolute ethanol	1200 mL
Sodium borohydride	50 grams
Reagent-grade acetone **or** reagent-grade methylene chloride	50 mL
Ice	
Gas chromatographs with 5-foot columns of 5% Carbowax 4000 on acid-washed firebrick	
Hypodermic syringes (10 microliters or smaller)	

8E. HOUSEHOLD BLEACH OXIDATION OF MENTHOL TO MENTHONE

Estimated Time   4 to 5 hours, plus optional spectra time

Menthol	300 mL
Glacial acetic acid	300 mL

Sodium bisulfite 150 grams
Household chlorine bleach 2500 mL
Saturated sodium bisulfite solution 300 mL
Saturated sodium carbonate solution 1200 mL
Methylene chloride 1600 mL
Anhydrous magnesium sulfate 120 grams
Starch-iodide test paper
Ice
Glass wool

8F. CHROMIUM(VI) OXIDE OXIDATION OF CINNAMYL ALCOHOL TO CINNAMALDEHYDE

Estimated Time   4 to 5 hours, plus optional spectra time

Pyridine 150 mL
Methylene chloride 2500 mL
Chromium trioxide 90 grams
Drierite 200 grams
Cinnamyl alcohol 50 grams
Ethyl ether 2500 mL
5% Sodium hydroxide solution 3000 mL
5% Hydrochloric acid solution 3000 mL
5% Sodium bicarbonate solution 1500 mL
Saturated sodium chloride solution 1500 mL
Anhydrous magnesium sulfate 200 grams
Semicarbazide hydrochloride 10 grams
Anhydrous sodium acetate 10 grams
Absolute ethanol 50 mL
Methanol 200 mL
Cotton
Ice

8G. GAS CHROMATOGRAPHIC ANALYSIS OF TURPENTINE

Estimated Time   3 hours

Standard solution of terpenes containing by volume 50.0% α-pinene,   10 mL
    30.0% β-pinene, 15.0% Δ^3-carene, and 5.0% limonene
Turpentine   10 mL
Gas chromatographs with 3-meter Chromosorb W columns, coated
    with 10% SE-30 silicone rubber
Hypodermic syringes (10 microliters or smaller)

9. ARTIFICIAL FLAVORINGS—THE SYNTHESIS OF ESTERS

Estimated Time 5 to 7 hours, plus the optional spectra time; other experiments may be done during the 2- to 3-hour reflux period

Glacial acetic acid	300 mL
Butyric acid	200 mL
Formic acid	100 mL
Salicylic acid	100 grams
Isopentanol	100 mL
Benzyl alcohol	75 mL
Propanol	100 mL
Absolute ethanol containing **no** other alcohols	150 mL
Isobutanol	100 mL
Methanol	400 mL
Concentrated sulfuric acid	125 mL
15% Sodium chloride solution	2500 mL
Methylene chloride	500 mL
10% Sodium bicarbonate solution	2000 mL
Anhydrous magnesium sulfate	100 grams
Boiling chips	
Ice	

10. ARTIFICIAL SWEETENERS—THE SYNTHESIS OF DULCIN

Estimated Time $2\frac{1}{2}$ hours the first lab period, $\frac{1}{2}$ hour the second lab period, plus optional spectra time; one hour of the first period may be used for other experiments

p-Ethoxyanilinc (p-phenetidine), recently distilled	50 mL
Glacial acetic acid	60 mL
Potassium cyanate	50 grams
Decolorizing carbon	20 grams
Capillary tubes	
Ice	

11. PERFUMES—THE SYNTHESIS OF NEROLIN

Estimated Time 6 to 7 hours, plus optional spectra time; two hours of the first period may be used for other experiments

Methanol	1200 mL
β-Naphthol	100 grams

Potassium hydroxide pellets	50 grams
Ethyl iodide	60 mL
Decolorizing carbon	20 grams
Salt for an ice-salt bath	1000 grams
Capillary tubes	
Boiling chips	
Ice	

12. LOCAL ANESTHETICS—THE SYNTHESIS OF BENZOCAINE

Estimated Time   5 to 6 hours, plus optional spectra time; two hours of the first period may be used for other experiments

p-Aminobenzoic acid	90 grams
Absolute ethanol containing **no** other alcohols	1000 mL
Concentrated sulfuric acid	75 mL
20% Sodium carbonate solution	1200 mL
Ethyl ether	1500 mL
Anhydrous magnesium sulfate	150 grams
95% Ethanol	500 mL
pH paper (pH 9 and up)	
Capillary tubes	
Boiling chips	
Ice	

13A. SYNTHESIS OF ASPIRIN

Estimated Time   4 hours, plus optional spectra time

Salicylic acid	60 grams
Acetic anhydride	150 mL
Concentrated sulfuric acid (in dropper bottles)	10 mL
Saturated sodium bicarbonate solution	800 mL
3 *M* hydrochloric acid	1000 mL
Ethyl ether	700 mL
Petroleum ether	700 mL
Aqueous iodine solution (Dissolve 0.25 gram of iodine and 0.5 gram of potassium iodide in 100 mL of water; place in dropper bottles.)	10 mL
Commercial aspirin	40 tablets
Blue litmus paper	
Capillary tubes	
Ice	

13B. SYNTHESIS OF PHENACETIN

Estimated Time 2 hours the first period, $\frac{1}{2}$ hour the second period, plus optional spectra time

Sodium acetate trihydrate	70 grams
p-Ethoxyaniline (p-phenetidine), recently distilled	60 mL
3 M hydrochloric acid	300 mL
Decolorizing carbon	20 grams
Acetic anhydride	80 mL
Capillary tubes	
Ice	

13C. THIN-LAYER CHROMATOGRAPHIC ANALYSIS OF OVER-THE-COUNTER ANALGESICS

Estimated Time $2\frac{1}{2}$ to 3 hours

Commercial, 10 cm × 10 cm, plastic-backed, silica gel G plates with a fluorescent indicator	25 plates
Anacin	30 tablets
Excedrin	30 tablets
Advil	30 tablets
Tylenol	30 tablets
5% Solution of pure caffeine in 50:50 methylene chloride-methanol (volume-volume)	20 mL
5% Solution of pure aspirin in 50:50 methylene chloride-methanol (volume-volume)	20 mL
5% Solution of pure acetaminophen in 50:50 methylene chloride-methanol (volume-volume)	20 mL
5% Solution of pure ibuprofen in 50:50 methylene chloride-methanol (volume-volume)	20 mL
Various unknown solutions containing any combination of caffeine, aspirin, acetaminophen and ibuprofen in 50:50 methylene chloride-methanol (volume-volume). Ten possibilities are	5 mL each

Caffeine and aspirin
Caffeine and acetaminophen
Caffeine and ibuprofen
Aspirin and acetaminophen
Aspirin and ibuprofen
Acetaminophen and ibuprofen
Caffeine, aspirin, and acetaminophen

Caffeine, aspirin, and ibuprofen
Caffeine, acetaminophen, and ibuprofen
Aspirin, acetaminophen, and ibuprofen

50:50 Mixture of methylene chloride and methanol (volume-volume)	700 mL
Ethyl acetate	1500 mL
Capillary tubes	
Rectangular solvent chambers and covers	
Paper liner wicks	
Millimeter rulers	
Ultraviolet lamps in a darkened area	

14. SULFA DRUGS—THE SYNTHESIS OF SULFANILAMIDE

Estimated Time   4 to 5 hours, plus optional spectra time

Acetanilide	120 grams
Chlorosulfonic acid	300 mL
Concentrated ammonium hydroxide	400 mL
3 M hydrochloric acid	450 mL
Decolorizing carbon	20 grams
Saturated sodium bicarbonate solution	2000 mL
Red litmus paper	
Ice	
Capillary tubes	

15. BARBITURATES—THE SYNTHESIS OF BARBITURIC ACID

Estimated Time   5 to 6 hours, plus 4 hours drying time in an oven, plus optional spectra time. The 4-hour drying time should be done between lab periods so that various parts of 3 lab periods will not be necessary. During the 3 hours of refluxing, other experiments may be done.

Absolute ethanol	1500 mL
Sodium metal	40 grams
Ligroin or xylene	600 mL
Diethyl malonate	150 mL
Urea (dried beforehand by heating for 1 hour in an oven at 105 to 110°C)	60 grams
Anhydrous calcium chloride (for drying tubes)	300 grams
Concentrated hydrochloric acid	120 mL
Ice	
Capillary tubes	

16A. ISOLATION OF CAFFEINE FROM TEA

Estimated Time   3 to 4 hours, plus optional spectra time

Tea bags	400 grams
Powdered sodium carbonate	300 grams
Methylene chloride	2000 mL
Anhydrous magnesium sulfate	100 grams
95% Ethanol	300 mL
Capillary tubes	
Cotton	
Ice	

16B. ISOLATION OF CAFFEINE FROM COFFEE

Estimated Time   3 hours, plus 1 hour if sublimation done, plus optional spectra time

Instant coffee	300 grams
Powdered sodium carbonate	300 grams
Methylene chloride	2000 mL
Anhydrous magnesium sulfate	100 grams
95% Ethanol	300 mL
Capillary tubes	
Cotton	
Ice	

16C. ISOLATION OF CAFFEINE FROM NODOZ

Estimated Time $2\frac{1}{2}$ to 3 hours, plus optional spectra time

NoDoz	130 tablets
Methylene chloride	2000 mL
Anhydrous magnesium sulfate	100 grams
95% Ethanol	300 mL
Capillary tubes	
Cotton	
Ice	

17. ISOLATION OF NICOTINE FROM TOBACCO

Estimated Time   3 to 4 hours the first period, plus $\frac{1}{2}$ hour the second period

Cigar tobacco (dried)	300 grams
50% Sodium hydroxide solution	450 mL

Ethyl ether	2500 mL
Methanol	300 mL
Saturated solution of picric acid in methanol	300 mL
50% Ethanol and water solution (volume-volume)	500 mL
Ice	
Glass wool	
Wooden applicator sticks	
Capillary tubes	

18A. SYNTHESIS OF INSECT REPELLENTS "OFF" AND "6-12"

Estimated Time   3 hours the first period, $1\frac{1}{2}$ to 2 hours the second period, plus optional spectra time

m-Toluic acid (3-methylbenzoic acid)	120 grams
Thionyl chloride	150 mL
Anhydrous calcium chloride (for drying tubes)	300 grams
Diethylamine	300 mL
Anhydrous ethyl ether	1200 mL
5% Sodium hydroxide solution	1200 mL
10% Hydrochloric acid	600 mL
Anhydrous magnesium sulfate	100 grams
Alumina	700 grams
Petroleum ether or ligroin	2500 mL
Cotton	
Boiling stones	

18B. PHEROMONES FROM THE GRIGNARD SYNTHESIS— 4-METHYL-3-HEPTANOL AND 4-METHYL-3-HEPTANONE

Estimated Time   7 to 8 hours, plus optional spectra time

Magnesium turnings (oven dried)	150 grams
Anhydrous ethyl ether	3000 mL
2-Bromopentane	550 mL
Iodine crystals	5 grams
Anhydrous calcium chloride (for drying tubes)	500 grams
Propanal (recently distilled)	300 mL
10% Hydrochloric acid	2000 mL
10% Sodium hydroxide solution	2000 mL
Saturated salt solution	700 mL
Anhydrous magnesium sulfate	200 grams
Glacial acetic acid	300 mL
Sodium bisulfite	150 grams

Household chlorine bleach	2000 mL
Saturated sodium bisulfite solution	300 mL
Saturated sodium carbonate solution	1200 mL
Methylene chloride	1800 mL
Glass wool	
Starch-iodine test paper	
Ice	

18C. SYNTHESIS OF A DIELS-ALDER INSECTICIDE ANALOG—*CIS*-NORBORNENE-5,6-*ENDO*-DICARBOXYLIC ANHYDRIDE

Estimated Time   3 hours, plus optional spectra time

Dicyclopentadiene	300 mL
Anhydrous calcium chloride	25 grams
Maleic anhydride	75 grams
Ethyl acetate	300 mL
Ligroin, bp 60 to 90°C	300 mL
Ice	
Capillary tubes	

19A. SYNTHESIS OF LUMINOL (3-AMINOPHTHALHYDRAZIDE)—THE LIGHT OF THE FIREFLY

Estimated Time $1\frac{1}{2}$ to 2 hours, plus optional spectra time; another short experiment may be scheduled during the same period

3-Nitrophthalic acid	40 grams
10% Aqueous hydrazine solution	60 mL
Triethylene glycol	120 mL
10% Sodium hydroxide solution	300 mL
Sodium hydrosulfite dihydrate (sodium dithionite)	120 grams
Glacial acetic acid	100 mL
Fluorescein	1 gram
Rhodamine B	1 gram
Eosin	1 gram
Phenolphthalein	1 gram
Boiling stones	
Ice	
Either potassium hydroxide pellets	250 grams
and dimethyl sulfoxide	700 mL
Or 10% sodium hydroxide solution,	1000 mL
3% aqueous potassium ferricyanide solution,	2000 mL
and 3% hydrogen peroxide solution	2000 mL

19B. SYNTHESIS OF 2-(2,4-DINITROBENZYL)PYRIDINE—A PHOTOCHROMIC COMPOUND

Estimated Time   4 to 5 hours, plus optional spectra time

Concentrated sulfuric acid	300 mL
2-Benzylpyridine	90 mL
Red fuming nitric acid	50 mL
20% Sodium hydroxide solution	2500 mL
Ethyl ether	3000 mL
Saturated salt solution	700 mL
Anhydrous magnesium sulfate	150 grams
95% Ethanol	600 mL
Ice	
Capillary tubes	

19C. SYNTHESIS OF DIXANTHYLENE—A THERMOCHROMIC COMPOUND FROM A PHOTOCHEMICAL REACTION

Estimated Time $\frac{1}{2}$ hour the first period, 5 to 14 days exposure time in bright sunlight, 3 to 4 hours the second period, plus optional spectra time

Xanthene	25 grams
Xanthone	25 grams
Toluene	600 mL
Acetic anhydride	450 mL
p-Toluenesulfonic acid monohydrate	200 grams
Glacial acetic acid	200 mL
Mesitylene	400 mL
Capillary tubes	

19D. PHOTOREDUCTION OF BENZOPHENONE TO BENZOPINACOL

Estimated Time   1 hour the first period, 5 to 10 days exposure time in bright sunlight, 3 hours the second period, plus optional spectra time

Benzophenone	250 grams
Naphthalene	10 grams
Isopropyl alcohol	3000 mL
Glacial acetic acid (in dropper bottles)	10 mL
Sodium metal	10 grams
1 M hydrochloric acid	200 mL
Glacial acetic acid	200 mL
Iodine crystals	5 grams

Ethanol   400 mL
Ice
Capillary tubes

19E. PHOTOCHEMICAL DIMERIZATION OF *TRANS*-CINNAMIC ACID

Estimated Time $\frac{1}{2}$ hour the first period, 2 weeks exposure time in bright
sunlight, 2 hours the second period, plus optional spectra time; another short
experiment may also be scheduled the second period

trans-Cinnamic acid	50 grams
Tetrahydrofuran	100 mL
Toluene	600 mL
95% Ethanol	500 mL
Decolorizing carbon	10 grams
Ice	
Capillary tubes	

20. AZO DYES—SYNTHESIS OF AMERICAN FLAG RED

Estimated Time 2 to $2\frac{1}{2}$ hours, plus optional spectra time

p-Nitroaniline	25 grams
3 *M* sulfuric acid	300 mL
Sodium nitrite	15 grams
β-Naphthol	25 grams
10% Sodium hydroxide solution	400 mL
Toluene	800 mL
Ice	

21. DYEING OF FABRICS

Estimated Time 4 to 5 hours

Multifiber Fabric 10A[1], cut into pieces approximately 2 × 4 inches	250 pieces
Laundered white cotton fabric, approximately 1 × 2 inches	250 pieces
0.5% Picric acid solution	900 mL
Concentrated sulfuric acid (in dropper bottle)	20 mL
Indigo	5 grams
Sodium hydrosulfite (sodium dithionite)	5 grams
Sodium hydroxide pellets	10 grams
Methylene blue	5 grams

---

[1] Available from Testfabrics, Inc., P.O. Box 118, 200 Blackford Ave., Middlesex, New Jersey 08846

p-Nitroaniline	10 grams
2% Hydrochloric acid	300 mL
Sodium nitrite	5 grams
β-Naphthol	5 grams
10% Sodium hydroxide solution	300 mL
Congo red	5 grams
10% Sodium carbonate solution	25 mL
10% Sodium sulfate solution	25 mL
Malachite green	5 grams
Eosine	5 grams
Alizarin	5 grams
Sodium carbonate	5 grams
Strong laundry detergent	500 grams
Undiluted household chlorine bleach	4000 mL
5% Acetic acid solution	4000 mL
Ice	
Paper toweling	
Scissors	

22A. FRIEDEL-CRAFTS SYNTHESIS OF 1,4-DI-*t*-BUTYL-2,5-DIMETHOXYBENZENE

Estimated Time   3 hours, plus optional spectra time

1,4-Dimethoxybenzene	120 grams
t-Butyl alcohol	200 mL
Glacial acetic acid	400 mL
Concentrated sulfuric acid	600 mL
Methanol	1200 mL
Methylene chloride	800 mL
Anhydrous magnesium sulfate	100 grams
Cotton	
Ice	
Glass fiber filter paper	
Capillary tubes	

22B. FRIEDEL-CRAFTS SYNTHESIS OF RESACETOPHENONE (2,4-DIHYDROXYACETOPHENONE)

Estimated Time   3 hours the first period, $\frac{1}{2}$ hour the second period, plus optional spectra time

Glacial acetic acid	180 mL
Zinc chloride, finely powdered	180 grams

Resorcinol	120 grams
6 M hydrochloric acid	900 mL
1 M hydrochloric acid	2000 mL
Decolorizing carbon	20 grams
Ice	
Capillary tubes	

23A. ALDOL SYNTHESIS OF DIBENZALACETONE FROM BENZALDEHYDE AND ACETONE

Estimated Time   2 to $2\frac{1}{2}$ hours, plus optional spectra time

Benzaldehyde	90 mL
Acetone, reagent grade	35 mL
95% Ethanol	1000 mL
10% Sodium hydroxide solution	900 mL
Glacial acetic acid	25 mL
95% Ethanol or 2-propanol	1000 mL
Ice	
Capillary tubes	

23B. ALDOL SYNTHESIS OF 2-METHYL-2-PENTENAL FROM PROPIONALDEHYDE

Estimated Time $2\frac{1}{2}$ to 3 hours, plus optional spectra time

10% Sodium hydroxide solution	150 mL
Propionaldehyde, recently distilled	600 mL
Anhydrous calcium chloride	200 grams

24. CANNIZZARO SYNTHESIS OF CUMIC ALCOHOL AND CUMIC ACID FROM CUMALDEHYDE

Estimated Time $\frac{1}{2}$ hour the first period, 3 hours the second period, plus optional spectra time

Potassium hydroxide pellets, 85%	120 grams
95% Ethanol	1500 mL
Cumaldehyde	150 mL
Ethyl ether	1500 mL
20% Sodium bisulfite	300 mL
Anhydrous magnesium sulfate	100 grams
3 M hydrochloric acid	800 mL
Decolorizing carbon	10 grams

pH paper (pH of 1 to 2)
Ice
Capillary tubes

25. WITTIG SYNTHESIS OF
E,E-1,4-DIPHENYL-1, 3-BUTADIENE

Estimated Time   3 to 4 hours, plus optional spectra time

Benzyl chloride	110 mL
Triethyl phosphite	170 mL
Sodium methoxide	60 grams
N,N-Dimethylformamide	900 mL
Cinnamaldehyde	120 mL
Methanol	1500 mL
Absolute ethanol or methylcyclohexane	1000 mL
Boiling stones	
Ice	
Capillary tubes	

26. BENZYNE SYNTHESIS OF TRIPTYCENE

Estimated Time   4 hours, plus optional spectra time

Anthracene	60 grams
Isoamyl nitrite	120 mL
1,2-Dimethoxyethane	1200 mL
Anthranilic acid	160 grams
95% Ethanol	600 mL
Sodium hydroxide pellets	180 grams
80% Methanol, 20% water solution (by volume)	2000 mL
Maleic anhydride	30 grams
Triethylene glycol dimethyl ether (triglyme)	600 mL
Methylcyclohexane	600 mL
Boiling chips	
Ice	
Capillary tubes	

27. SYNTHESIS OF FERROCENE

Estimated Time   4 to 5 hours, plus optional spectra time

Dicyclopentadiene	600 mL
Anhydrous calcium chloride	30 grams

Potassium hydroxide flakes or pellets	750 grams
1,2-Dimethoxyethane	2000 mL
Iron(II) chloride tetrahydrate	200 grams
Dimethyl sulfoxide	900 mL
6 M hydrochloric acid	2500 mL
Tin(II) chloride	30 grams
Hexane or ligroin	1000 mL
Nitrogen gas	
Boiling chips	
Ice	
Capillary tubes	

28A. CHEMICAL AND PHYSICAL PROPERTIES OF FATS AND OILS

Estimated Time $1\frac{1}{2}$ to 2 hours; another experiment may be scheduled during the same period

Cottonseed oil	50 mL
Olive oil	50 mL
Corn oil	50 mL
Linseed oil	50 mL
Margarine	30 grams
Butter	30 grams
Lard	30 grams
Tallow	30 grams
95% Ethanol	40 mL
Ethyl ether	40 mL
Acetone	40 mL
Toluene	40 mL
Methylene chloride	1000 mL
Glycerol	10 mL
Potassium bisulfate	10 grams
0.1 M solution of bromine in methylene chloride	700 mL
10% Sodium thiosulfate solution (for safety)	500 mL

28B. GAS CHROMATOGRAPHIC ANALYSIS OF FATS AND OILS

Estimated Time   3 to 4 hours

Wesson oil	10 mL
Crisco oil	10 mL
Mazola oil	10 mL
Puritan oil	10 mL

Cottonseed oil	10 mL
Linseed oil	10 mL
Shortening	10 grams
Lard	10 grams
0.5 *M* solution of potassium hydroxide in methanol	150 mL
12% Boron trifluoride in methanol solution	300 mL
Petroleum ether	1000 mL
Saturated salt solution	1800 mL
Anhydrous magnesium sulfate	100 grams
Methyl myristate in concentrated methylene chloride solution	10 mL
Methyl palmitate in concentrated methylene chloride solution	10 mL
Methyl palmitoleate in concentrated methylene chloride solution	10 mL
Methyl stearate in concentrated methylene chloride solution	10 mL
Methyl oleate in concentrated methylene chloride solution	10 mL
Methyl linoleate in concentrated methylene chloride solution	10 mL
Methyl linolenate in concentrated methylene chloride solution	10 mL

Gas chromatographs with 9-foot Chromosorb W columns coated
    with 15% diethyleneglycol succinate
Hypodermic syringes (10 microliters or smaller)

28C. ISOLATION OF TRIMYRISTIN AND MYRISTIC ACID FROM NUTMEG

Estimated Time   3 hours the first period, $\frac{1}{2}$ hour the second period, plus optional spectra time; portions of other experiments may be done during the distilling and refluxing time

Powdered nutmeg	240 grams
Ethyl ether	2000 mL
Acetone	750 mL
6 *M* sodium hydroxide	400 mL
95% Ethanol	400 mL
Concentrated hydrochloric acid	400 mL
Ice	
Boiling chips	
Capillary tubes	

28D. ISOLATION OF OLEIC ACID FROM OLIVE OIL

Estimated Time   4 hours, plus optional spectra time

Olive oil	300 mL
Potassium hydroxide pellets	70 grams

Triethylene glycol	600 mL
Concentrated hydrochloric acid	300 mL
Ethyl ether	3500 mL
Saturated salt solution	1500 mL
Anhydrous magnesium sulfate	200 grams
Acetone	2000 mL
Urea	300 grams
Methanol	1800 mL
Ice	
Either sodium chloride for an ice-salt bath	1500 grams
Or practical-grade ethanol or isopropanol for a dry ice bath	4000 mL
Dry ice	2500 grams

28E. SAPONIFICATION NUMBER OF FATS AND OILS

Estimated Time $1\frac{1}{2}$ hours, assuming each student saponifies one fat or oil and shares his or her results with the other students; the amounts of the reagents are also based on this assumption

Coconut oil	10 mL
Butter	10 grams
Corn oil	10 mL
Olive oil	10 mL
Safflower oil	10 mL
95% Ethanol	600 mL
0.5 M potassium hydroxide solution	1800 mL
Standardized 0.500 M hydrochloric acid	1500 mL
Phenolphthalein indicator	20 mL
Boiling stones	
Burets	40

28F. IODINE NUMBER OF FATS AND OILS

Estimated Time 3 hours; portions of other experiments may be done during the 1-hour waiting period

Butter	50 grams
Olive oil	50 mL
Safflower oil	50 mL
Methylene chloride	4000 mL
0.15 M iodine in glacial acetic acid solution	2500 mL
5% Potassium iodide solution	1200 mL
Standardized 0.250 M sodium thiosulfate solution	2500 mL

1% Starch solution	150 mL
50-mL Volumetric flasks	60
250-mL Glass stoppered bottles	80
10-mL Pipets	20
Burets	20

29A. SYNTHESIS OF SOAP

Estimated Time   1 hour, plus drying time

95% Ethanol	300 mL
Sodium hydroxide pellets	120 grams
Sodium chloride (not table salt)	2400 grams
Crisco oil	25 mL
Wesson oil	25 mL
Shortening	25 grams
Lard	25 grams
Olive oil	25 mL
Boiling chips	
Ice	

29B. SYNTHESIS OF A DETERGENT

Estimated Time   1 hour, plus drying time

Glacial acetic acid	150 mL
Chlorosulfonic acid (keep in hood with gloves)	40 mL
Lauryl alcohol (dodecyl alcohol)	175 grams
Sodium chloride (not table salt)	200 grams
Saturated sodium carbonate solution	1000 mL
Ice	
Red litmus paper	

29C. TESTS ON SOAPS AND DETERGENTS

Estimated Time   1 hour

5% Calcium chloride solution (in dropper bottles)	25 mL
5% Magnesium chloride solution (in dropper bottles)	25 mL
Trisodium phosphate or sodium carbonate	20 grams
Commercial detergent (if Experiment 29B was not done)	15 grams
Vegetable oil (in dropper bottles)	30 mL
1.0 M sulfuric acid solution	150 mL
Blue litmus paper	

30A. ISOLATION OF CHOLESTEROL FROM GALLSTONES

Estimated Time 4 to 5 hours, plus optional spectra time

Gallstones (from a hospital surgery or pathology department)	60 grams
Anhydrous ethyl ether	1500 mL
Methanol	1200 mL
Decolorizing carbon	20 grams
Brominating solution of 17.0 grams of bromine and 1.3 grams of anhydrous sodium acetate in 140 mL of glacial acetic acid	150 mL
Glacial acetic acid	350 mL
Zinc dust (fresh)	10 grams
10% Sodium hydroxide solution	1500 mL
Saturated salt solution	700 mL
Anhydrous magnesium sulfate	100 grams
10% Sodium thiosulfate solution (for safety)	500 mL
Ice	
Capillary tubes	
Blue litmus paper	

30B. ISOLATION OF CHOLESTEROL FROM EGG YOLKS

Estimated Time 5 to 6 hours, plus optional spectra time

Eggs (medium size)	4 dozen
Methanol	2000 mL
Ethyl ether	5000 mL
Acetone	4000 mL
15% Potassium hydroxide solution in ethanol	700 mL
Saturated salt solution	900 mL
Anhydrous magnesium sulfate	150 grams
Decolorizing carbon	20 grams
Concentrated hydrochloric acid	200 mL
Blue litmus paper	
Cotton	
Capillary tubes	

30C. CONCENTRATION OF CHOLESTEROL IN BLOOD

Estimated Time 3 to 4 hours

Human blood (outdated from a blood bank)	60 mL
Alcoholic potassium hydroxide solution (freshly made by adding 1.0 mL of 33% aqueous potassium hydroxide solution to 16 mL of absolute ethanol)	200 mL

Hexane	600 mL
Concentrated sulfuric acid	50 mL
Standard solution containing 200 mg of cholesterol per liter of absolute ethanol	150 mL
Acetic anhydride	800 mL
Glacial acetic acid	400 mL
25-mL Volumetric flasks	40
Ice	
Centrifuges	
Visible spectrophotometers	

31. ANALYSIS OF VITAMIN C IN TABLETS AND FRUIT JUICE

Estimated Time   3 to 4 hours

Vitamin C tablets, 100 mg or 250 mg	30 250 mg or 60 100 mg
1% Oxalic acid solution	14,000 mL
Standardized DCIP solution (2,6-dichloroindophenol). To have 1.00 mL of DCIP solution be equivalent to 0.300 mg of Vitamin C, dissolve 0.457 gram of DCIP in 980 mL of water. Then add a phosphate buffer, made by dissolving 0.066 gram of K_2HPO_4 and 0.081 gram of KH_2PO_4 in 20 mL of water. Stir or shake thoroughly. The solution keeps for several days.	2400 mL
Either citrus juice, such as orange or grapefruit juice (the juice may be fresh, bottled, or canned, or it may be prepared from frozen concentrate)	5000 mL
Or Hi-C apple drink	4500 mL
250-mL Volumetric flasks	25
100-mL Volumetric flasks	25
Burets	50
1.0-, 2.0-, and 5.0-mL Pipets (for greater accuracy)	
Waxed paper	

32A. ISOLATION OF CASEIN FROM MILK

Estimated Time   1 hour, plus drying time

Powdered, nonfat dry milk	300 grams
10% Acetic acid solution	1000 mL
Powdered calcium carbonate	45 grams
Ethyl ether	600 mL

32B. ISOLATION OF LACTOSE FROM MILK

Estimated Time 1 hour the first period, plus $\frac{1}{2}$ hour the second period, plus drying time

Original liquid from Experiment 32A	
95% Ethanol	2000 mL
Decolorizing carbon	20 grams
Filter aid (Celite)	70 grams
Ice	
Capillary tubes	

33A. CHARACTERIZATION REACTIONS OF CARBOHYDRATES

Estimated Time 6 hours

1% Glucose solution	300 mL
1% Fructose solution	300 mL
1% Xylose solution	300 mL
1% Arabinose solution	300 mL
1% Galactose solution	300 mL
1% Sucrose solution	450 mL
1% Lactose solution	450 mL
1% Maltose solution	450 mL
1% Starch solution	450 mL
1% Glycogen solution	450 mL
Concentrated sulfuric acid	2000 mL
Tes-Tape	400 cm
Concentrated hydrochloric acid (in dropper bottles)	100 mL
10% Sodium hydroxide solution	300 mL
Sucrose	1500 grams
1% Brown sugar solution	400 mL
2% Honey solution	400 mL
2% Molasses solution	400 mL
2% Pure maple syrup solution	400 mL
2% White corn syrup solution	400 mL
5% Canned fruit syrup solution	400 mL
Unsweetened apple juice	400 mL
Unsweetened white grape juice	400 mL
Molisch reagent (Dissolve 5 grams of α-naphthol in 100 mL of 95% ethanol; place in dropper bottles.)	100 mL
Bial's reagent (Dissolve 3 grams of orcinol in 1000 mL of concentrated hydrochloric acid, and add 3 mL of 10% aqueous ferric chloride solution.)	1000 mL

Seliwanoff's reagent (Dissolve 1.0 gram of resorcinol in 650 mL of   2000 mL
    concentrated hydrochloric acid and dilute to 2000 mL with
    water.)

Benedict's reagent (Dissolve 520 grams of hydrated sodium citrate   3000 mL
    and 300 grams of anhydrous sodium carbonate in 2700 mL
    of water; add to it 300 mL of a solution of 52 grams of hy-
    drated cupric sulfate in water.)

Barfoed's reagent (Dissolve 200 grams of cupric acetate in 3000 mL   3000 mL
    of water; filter, if necessary, and add 27 mL of glacial acetic
    acid.)

Iodine solution (Dissolve 2 grams of potassium iodide in 100 mL   100 mL
    of water, add 1 gram of iodine and shake, and place in
    dropper bottles.)

Red and blue litmus paper

33B. THIN-LAYER CHROMATOGRAPHY OF SUGARS AND FOOD PRODUCTS

Estimated Time   3 hours

A solution of 10 volumes 1-butanol, 6 volumes glacial acetic acid, 3 volumes ethyl ether, and 1 volume water	1500 mL
Commercial, plastic-backed, silica gel G plates	25 plates
5% Glucose solution	10 mL
5% Fructose solution	10 mL
5% Xylose solution	10 mL
5% Arabinose solution	10 mL
5% Galactose solution	10 mL
5% Sucrose solution	10 mL
5% Maltose solution	10 mL
5% Brown sugar solution	10 mL
5% Honey solution	10 mL
5% Molasses solution	10 mL
5% Maple syrup solution	10 mL
5% Corn syrup solution	10 mL
25% Canned fruit syrup solution	10 mL
Unsweetened apple juice	10 mL
Unsweetened white grape juice	10 mL
Prepared solution of 2.8 grams aniline and 5.0 grams of phthalic acid in 300 mL of 95% ethanol, **or** 3.7 grams of p-anisidine and 5.0 grams of phthalic acid in 300 mL of 95% ethanol. Place solution in commercial spray bottles.	300 mL

Capillary tubes (for micropipets)
Rectangular solvent chambers and covers
Wick paper
Millimeter rulers
Plastic gloves

33C. OPTICAL ROTATION OF CARBOHYDRATES

Estimated Time   3 hours; with a limited number of polarimeters, this experiment can be done by the students working in teams, or over a number of lab periods, along with other experiments

10.0% Glucose solution	1000 mL
10.0% Fructose solution	1000 mL
10.0% Sucrose solution	1000 mL
10.0% Lactose solution	1000 mL
5.00% Glucose solution	1000 mL
5.00% Fructose solution	1000 mL
5.00% Sucrose solution	1000 mL
5.00% Lactose solution	1000 mL
2.00% Glucose solution	1000 mL
2.00% Fructose solution	1000 mL
2.00% Sucrose solution	1000 mL
2.00% Lactose solution	1000 mL
Polarimeters	

33D. KINETICS OF THE ENZYMATIC AND ACID-CATALYZED HYDROLYSIS OF SUCROSE

Estimated Time $2\frac{1}{2}$ to 3 hours; with a limited number of polarimeters, this experiment can be done by the students working in teams, or some students may do it one period while other students are performing other experiments

Sucrose	1200 grams
6.0 M hydrochloric acid	300 mL
Aqueous invertase solution containing 5 mg invertase per mL	10 mL
Timers	20
50-mL Volumetric flasks	20
Mohr pipets for measuring 0.20 mL of liquid	
Polarimeters	
Graph paper	

33E. KINETICS OF THE MUTAROTATION OF D-GLUCOSE

Estimated Time $\frac{1}{2}$ hour the first period, 3 hours the second period, and $\frac{1}{2}$ hour the third period; with a limited number of polarimeters, this experiment may be done by the students working in teams, or some students may do the polarimeter portion one period while other students are performing other experiments

Anhydrous glucose (dextrose)	300 grams
Glacial acetic acid	600 mL
95% Ethanol	400 mL
Absolute ethanol	400 mL
50-mL Volumetric flasks	20
Timers	20
Polarimeters	
Ice	
Graph paper	

34A. PAPER CHROMATOGRAPHY OF AMINO ACIDS

Estimated Time $1\frac{1}{2}$ hours the first period for hydrolyzing the proteins, plus 3 to 4 hours the second period; other experiments may be done while the proteins are being hydrolyzed and the chromatogram is being developed

Men's hair (from a barber shop)	5 grams
Undyed wool	5 grams
Gelatin	5 grams
Casein	5 grams
6 M hydrochloric acid	600 mL
Decolorizing carbon	20 grams
Whatman No. 1 paper, 15 to 20 cm high and 20 to 25 cm wide (handle **only** with gloves and also place a letter T in pencil to identify the top edge)	30 pieces
Solvent 1: 70% isopropyl alcohol, 10% formic acid, and 20% water (by volume)	1000 mL
Solvent 2: 70% ethyl acetate, 20% formic acid, and 10% water (by volume)	1000 mL
Solvent 3: 80% phenol and 20% water (by volume and kept below a protective layer of ligroin until needed)	1000 mL
Ninhydrin spray	4 cans
0.1 M leucine (mol wt 131, acidified with 20 drops of 6 M hydrochloric acid)	20 mL
0.1 M lysine (mol wt 146, acidified with 20 drops of 6 M hydrochloric acid)	20 mL

0.1 *M* threonine (mol wt 119, acidified with 20 drops of 6 *M* hydrochloric acid) — 20 mL

0.1 *M* tryptophan (mol wt 204, acidified with 20 drops of 6 *M* hydrochloric acid) — 20 mL

0.1 *M* valine (mol wt 117, acidified with 20 drops of 6 *M* hydrochloric acid) — 20 mL

0.1 *M* alanine (mol wt 89, acidified with 20 drops of 6 *M* hydrochloric acid) — 20 mL

0.1 *M* aspartic acid (mol wt 133, acidified with 20 drops of 6 *M* hydrochloric acid) — 20 mL

0.1 *M* glutamic acid (mol wt 147, acidified with 20 drops of 6 *M* hydrochloric acid) — 20 mL

0.1 *M* glycine (mol wt 75, acidified with 20 drops of 6 *M* hydrochloric acid) — 20 mL

0.1 *M* proline (mol wt 115, acidified with 20 drops of 6 *M* hydrochloric acid) — 20 mL

0.1 *M* serine (mol wt 105, acidified with 20 drops of 6 *M* hydrochloric acid) — 20 mL

0.1 *M* tyrosine (mol wt 181, acidified with 20 drops of 6 *M* hydrochloric acid) — 20 mL

Solvent chambers and covers (circular)

Paper wicks

Plastic gloves

Boiling chips

Millimeter rulers

Capillary tubes (for micropipets)

34B. CHEMICAL TESTS OF PROTEINS AND AMINO ACIDS

Estimated Time 2 to 3 hours

Millon's reagent (Dissolve 20 grams of mercury in 30 mL of concentrated nitric acid by gently heating the mixture in the hood; dilute the resulting solution to 100 mL with water and place in dropper bottles.) — 100 mL

Hopkins-Cole reagent (Place 10 grams of powdered magnesium in a 500-mL Erlenmeyer and cover with water; slowly add to the flask 250 mL of a cold, saturated oxalic acid solution with cooling and stirring; filter and to the filtrate add 25 mL of glacial acetic acid; dilute the resulting solution to 500 mL with water.) — 500 mL

0.5% Copper sulfate solution (in dropper bottles) — 100 mL

0.1% Ninhydrin solution — 150 mL

10% Sodium hydroxide solution	2000 mL
Concentrated nitric acid (in dropper bottles)	100 mL
Concentrated sulfuric acid	500 mL
5% Lead acetate solution (in dropper bottles)	100 mL
5% Lead nitrate solution (in dropper bottles)	100 mL
5% Mercuric chloride solution (in dropper bottles)	100 mL
95% Ethanol	300 mL
10% Tannic acid solution	200 mL
Saturated picric acid solution	200 mL
2% Gelatin solution	2000 mL
2% Casein solution	2000 mL
2% Egg albumin solution	2000 mL
0.1 M tyrosine solution (mol wt 181)	200 mL
0.1 M tryptophan solution (mol wt 204)	200 mL
0.1 M glycine solution (mol wt 75)	200 mL
0.1 M cysteine solution (mol wt 240)	200 mL
Blue litmus paper	

34C. ENZYMATIC RESOLUTION OF D,L-METHIONINE

Estimated Time 3 to 4 hours the first period, 2 to 7 days for enzymatic reaction, $\frac{1}{2}$ hour the second period, and 3 hours the third period

D,L-Methionine	60 grams
2.0 M sodium hydroxide solution	200 mL
Acetic anhydride	150 mL
1.0 M sulfuric acid	200 mL
Anhydrous magnesium sulfate	100 grams
Ethyl acetate	700 mL
1% Ninhydrin solution in ethanol (in dropper bottles)	20 mL
Acetone	300 mL
Anhydrous ethyl ether	200 mL
5% Sodium hydroxide solution	300 mL
Glacial acetic acid	200 mL
Acylase enzyme powder	0.5 grams
Absolute ethanol	1600 mL
1.0 M hydrochloric acid	500 mL
Cotton	
Ice	
Capillary tubes	
Hydrion paper (pH 7 to 8)	
Hydrion paper (pH 4 and under)	

Centrifuges
Centrifuge tubes
Polarimeters
25-mL Volumetric flasks

35. INFRARED SPECTROSCOPY

The estimated time and the amount of materials needed depend upon how many spectra are taken by the students.

36. NUCLEAR MAGNETIC RESONANCE SPECTROSCOPY

The estimated time and the amount of materials needed depend upon how many spectra are taken.

37. THE CHEMICAL LITERATURE

No laboratory time or materials needed. The synthetic procedure can be done by the students outside of class.

38. ORGANIC QUALITATIVE ANALYSIS

Estimated Time   Depends on how many unknowns the students are assigned; minimum time is 6 to 8 hours

The following quantities of reagents, solvents, and other items will also vary, depending on how many and what kinds of unknowns the students are assigned. The following quantities are for one unknown. Two or more unknowns generally will require less than double or triple the stated amounts.

Sodium metal	20 grams
30% Potassium fluoride solution (in dropper bottles)	20 mL
Saturated ferrous ammonium sulfate solution (in dropper bottles)	20 mL
10% Sodium nitroprusside solution (in dropper bottles)	20 mL
1% Lead acetate solution (optional, in dropper bottles)	20 mL
5% Silver nitrate solution (in dropper bottles)	20 mL
Chlorine water or household chlorine bleach (in dropper bottles)	20 mL
Or calcium hypochlorite	0.5 grams
2,4-Dinitrophenylhydrazine reagent (Dissolve 6.0 grams of 2,4-dinitrophenylhydrazine in 30 mL of concentrated sulfuric acid; in a flask, mix 40 mL of water with 140 mL of 95%	200 mL

ethanol; slowly add the 2,4-dinitrophenylhydrazine solution to the aqueous ethanol solution with stirring; after thorough mixing, gravity filter the solution; test it on a known compound.)

3% Aqueous ferric chloride solution (in dropper bottles)	30 mL
1% Ferric chloride in methylene chloride solution (in dropper bottles)	30 mL
10% Silver nitrate solution (freshly prepared)	50 mL
Chromic acid solution (Dissolve 10 grams of chromium trioxide in 10 mL of concentrated sulfuric acid; carefully add to 30 mL of water.)	40 mL
Schiff's fuchsin reagent (Dissolve 0.05 gram of fuchsin and 0.9 gram of sodium bisulfite in 50 mL of water, and add 1 mL of concentrated hydrochloric acid; keep in well-stoppered bottles, protected from light.)	30 mL
Iodoform reagent (Dissolve 20 grams of potassium iodide and 10 grams of iodine in 100 mL of water.)	100 mL
Acetyl chloride	30 mL
Lucas reagent (Cool 100 mL of concentrated hydrochloric acid in a flask in an ice bath; then add, with stirring and continued cooling, 160 grams of anhydrous zinc chloride to the acid.)	120 mL
5% Ceric ammonium nitrate solution (in dropper bottles)	30 mL
Saturated bromine in water solution (Add bromine to water until it forms a red pool in the bottom of the bottle.)	40 mL
2% Silver nitrate in 95% ethanol solution	50 mL
Benzenesulfonyl chloride	30 mL
20% Sodium nitrite solution (in dropper bottles)	20 mL
Copper sulfate solution	50 mL
5% Ferric chloride solution (in dropper bottles)	50 mL
0.5 M hydroxylamine hydrochloride in ethanol solution (in dropper bottles)	20 mL
1 M hydroxylamine hydrochloride in propylene glycol solution (in dropper bottles)	50 mL
Ferric ammonium sulfate	5 grams
Potassium thiocyanate	5 grams
2% Bromine in methylene chloride solution	50 mL
1% Potassium permanganate solution (in dropper bottles)	20 mL
15% Sodium iodide in acetone solution	50 mL
Anhydrous aluminum chloride	10 grams
Chloroform	40 mL
5% Ferrous sulfate solution (fresh, in dropper bottles)	30 mL
Semicarbazide hydrochloride	10 grams
Anhydrous sodium acetate	10 grams

4-Nitrophenylhydrazine	10 grams
Phenylhydrazine	10 mL
3,5-Dinitrobenzoyl chloride	10 grams
1-Naphthylisocyanate	15 mL
Phenylisocyanate	15 mL
Potassium permanganate	10 grams
Aqueous bromine solution (for phenol derivatives; dissolve 8 grams of potassium bromide in 50 mL of water, then add 5 grams of bromine)	50 mL
Thionyl chloride	30 mL
p-Toluidine	10 mL
Aniline	40 mL
Benzoyl chloride (fresh)	10 mL
Acetic anhydride	10 mL
Saturated picric acid in 95% ethanol solution	80 mL
Methyl iodide	10 mL
Tin metal	5 grams
10% Bromine in glacial acetic acid	50 mL
Magnesium turnings	5 grams
Iodine crystals	1 gram
Ethyl ether	1000 mL
Absolute ethyl ether	150 mL
95% Ethanol	2000 mL
Methanol	100 mL
Hexane	100 mL
Glacial acetic acid	200 mL
Methylene chloride	100 mL
Pyridine	100 mL
Acetone (reagent grade)	200 mL
Toluene	400 mL
Ethyl acetate	100 mL
Propylene glycol	50 mL
Diglyme	50 mL
Bis(2-ethoxyethyl) ether	50 mL
10% Sodium bisulfite solution	30 mL
5% Sodium carbonate solution	100 mL
10% Sodium thiosulfate solution (for bromine burns)	500 mL
Anhydrous magnesium sulfate	30 grams
5% Sodium bicarbonate solution	200 mL
Concentrated sulfuric acid	100 mL
6 *M* sulfuric acid	50 mL
30% Sulfuric acid	30 mL
10% Sulfuric acid	30 mL
5% Sulfuric acid	50 mL

6 *M* hydrochloric acid	100 mL
3 *M* hydrochloric acid	200 mL
2 *M* hydrochloric acid	100 mL
1 *M* hydrochloric acid	300 mL
10% Hydrochloric acid	50 mL
5% Hydrochloric acid	300 mL
Concentrated nitric acid	10 mL
Fuming nitric acid	10 mL
5% Nitric acid	50 mL
6 *M* nitric acid	30 mL
40% Sodium hydroxide solution	30 mL
25% Sodium hydroxide solution	200 mL
20% Sodium hydroxide solution	150 mL
10% Sodium hydroxide solution	800 mL
5% Sodium hydroxide solution	300 mL
2 *M* sodium hydroxide solution	100 mL
Standardized 0.1 *M* sodium hydroxide solution	200 mL
2 *M* potassium hydroxide in methanol solution	20 mL
1 *M* potassium hydroxide in propylene glycol solution	30 mL
Concentrated ammonium hydroxide	100 mL
6 *M* ammonium hydroxide	30 mL
Phenolphthalein indicator (in dropper bottles)	10 mL
Copper wire (thick, about 6 inches long)	20 lengths
Glass tubing (4 mm in diameter, cut into 5-cm lengths)	15 pieces
Acetamide (as reference compound)	5 grams
Acetanilide (as reference compound)	10 grams
Anisole (as reference compound)	10 mL
Benzaldehyde (as reference compound)	40 mL
Benzamide (as reference compound)	5 grams
Benzoic acid (as reference compound)	10 grams
Benzonitrile (as reference compound)	5 mL
Benzophenone (as reference compound)	10 grams
Benzyl alcohol (as reference compound)	10 mL
Bromobenzene (as reference compound)	30 mL
1-Butanol (as reference compound)	30 mL
2-Butanol (as reference compound)	30 mL
2-Butanone (as reference compound)	10 mL
t-Butyl alcohol (as reference compound)	20 mL
Butylamine (as reference compound)	20 mL
Butyl bromide (as reference compound)	20 mL
Butyl chloride (as reference compound)	30 mL
t-Butyl chloride (as reference compound)	30 mL
Cyclohexane (as reference compound)	20 mL
Cyclohexanone (as reference compound)	20 mL

Cyclohexene (as reference compound)	20 mL
p-Dichlorobenzene (as reference compound)	5 grams
Diethylamine (as reference compound)	20 mL
1-Dodecanol (as reference compound)	10 grams
Ethyl benzoate (as reference compound)	20 mL
N-Methylaniline (as reference compound)	20 mL
Naphthalene (as reference compound)	5 grams
1-Naphthol (as reference compound)	5 grams
2-Naphthol (as reference compound)	15 grams
4-Nitrobromobenzene or 4-nitrochlorobenzene (as reference compound)	5 mL
4-Nitrotoluene (as reference compound)	5 mL
1-Octanol (as reference compound)	10 mL
2,4-Pentanedione (as reference compound)	10 mL
2-Pentanol (as reference compound)	10 mL
2-Pentanone (as reference compound)	10 mL
3-Pentanone (as reference compound)	10 mL
Phenol (as reference compound)	20 mL
Propylamine (as reference compound)	20 mL
Sulfanilic acid (as sodium fusion reference)	5 grams
Triethylamine (as reference compound)	30 mL

pH paper of various pH values
Blue and red litmus paper
Capillary tubes
Ice

REACTION OR EXPERIMENT TYPE INDEX

Acetylation

Addition (to a Double Bond)

Aldol Condensation

Amide Formation

883

Analysis

Aromatic Substitution

Azo Coupling

Benzyne Intermediate

Bromination

Cannizzaro Reaction

Carbanions

Nucleophilic Aliphatic Substitution

Oxidation

Photochemistry

Polymerization

Qualitative Organic Analysis

Rearrangements

Reduction

APPENDIX

E

EXPERIMENTAL TECHNIQUES INDEX

Please note that experimental techniques such as simple distillation, evaporation of a solvent, melting-point determination, refluxing, extraction, drying a solution, and recrystallization are excluded from this index because they occur so frequently throughout the book.

Boiling-Point Determination (Siwoloboff Method)

Chromatography

Column Chromatography

891

SPECTRA INDEX

897

INDEX

INDEX

903